AIR POLLUTION RESEARCH REPORT 2

PHYSICO-CHEMICAL BEHAVIOUR OF ATMOSPHERIC POLLUTANTS

Commission of the European Communities

PHYSICO-CHEMICAL BEHAVIOUR OF ATMOSPHERIC POLLUTANTS

Proceedings of the Fourth European Symposium
held in Stresa, Italy,
23-25 September 1986

Organized within the framework
of the Concerted Action COST 611

Edited by

G. ANGELETTI and G. RESTELLI

Commission of the European Communities,
Directorate-General Science, Research and Development

D. REIDEL PUBLISHING COMPANY

A MEMBER OF THE KLUWER ACADEMIC PUBLISHERS GROUP

DORDRECHT / BOSTON / LANCASTER / TOKYO

Library of Congress Cataloging in Publication Data

CIP

"Fourth European Symposium on Physico-Chemical Behaviour of Atmospheric Pollutants" held in Stresa (Italy) from 23 to 25 September 1986."

"Organized by the Commission of the European Communities, Joint Research Centre Ispra, Directorate-General for Research, Science and Development, Brussels"—Verso of t.p.

Includes index.

1. Air—Pollution—Congresses. 2. Pollutants—Congresses. 3. Environmental chemistry—Congresses. 4. Acid deposition—Congresses. 5. Photochemical oxidants—Congresses. I. Angeletti, G., 1943– . II. Restelli, G. III. European Symposium on Physico-Chemical Behaviour of Atmospheric Pollutants (4th: 1986: Stresa, Italy) IV. Title: Atmospheric pollutants. V. Commission of the European Communities. Joint Research Centre. Ispra Establishment. VI. Commission of the European Communities. Directorate-General for Science, Research and Development.

TD881.P485 1987 628.5'3 87–4628

ISBN-13:978-94-010-8210-5 e-ISBN-13:978-94-009-3841-0

DOI: 10.1007/978-94-009-3841-0

The Symposium was organized by the
Commission of the European Communities
Joint Research Centre Ispra
Directorate-General for Research, Science and Development, Brussels

Publication arrangements: P. P. Rotondó
Commission of the European Communities,
Directorate-General Telecommunications, Information Industries and Innovation, Luxembourg

EUR 10832

Softcover reprint of the hardcover 1st edition 1987

Published by D. Reidel Publishing Company
P.O. Box 17, 3300 AA Dordrecht, Holland

Sold and distributed in the U.S.A. and Canada
by Kluwer Academic Publishers,
101 Philip Drive, Assinippi Park, Norwell, MA 02061, U.S.A.

In all other countries, sold and distributed
by Kluwer Academic Publishers Group,
P.O. Box 322, 3300 AH Dordrecht, Holland

PREFACE

This volume contains oral papers and posters presented at the "Fourth European Symposium on Physico-Chemical Behaviour of Atmospheric Pollutants" held in Stresa (Italy) from 23 to 25 September 1986.

The Symposium was organised within the framework of the Concerted Action COST 611* which is included in the R&D Programme on Environmental Research of the Commission of the European Communities. As in the past, the aim of the Symposium was to review current studies and technical progress achieved in the various sectors of the Concerted Action, in this case since the third Symposium, held in Varese (Italy) in April 1984.

The presentation of the papers in the Proceedings is organised according to the five sectors, into which the scientific content of the Concerted Action is divided: analysis of pollutants; chemical and photochemical reactions, mechanisms and rates; aerosol characterisation and particle formation; pollutants cycles; transport and modelling. From these papers it is clear that the aspects related to the physico-chemistry of acid deposition and formation of photochemical oxidants in the troposphere have received particular attention.

We believe that these Proceedings give a useful overview of current activities in this field in Europe. We also hope that they constitute a valuable contribution to a better understanding of the complex physico-chemical processes which occur in the planet's atmosphere, understanding of which is essential for effective management of the environment.

Brussels/Ispra, October 1986

G. ANGELETTI G. RESTELLI

* COST 611 : Scientific and Technical Cooperation among European Community Member Countries and the Non-Member Countries Finland, Norway, Sweden and Switzerland in the field of "Physico-Chemical Behaviour of Atmospheric Pollutants"

CONTENTS

WORKING PARTY 3: AEROSOL CHARACTERIZATION AND PARTICLE FORMATION
Chairman: J.A. SCOTT (University College Dublin, IRL)

WORKING PARTY 4: POLLUTANTS CYCLES
Chairman: S. BEILKE (Umweltbundesamt, Frankfurt, FRG)

WORKING PARTY 5: TRANSPORT AND MODELLING
Chairman: P. BESSEMOULIN (National Meteorological RC, FR)

WORKING PARTY 1
ANALYSIS OF POLLUTANTS

Chairman: A. LIBERTI

Detection of gaseous nitric acid by OH (A—X) - fluorescence generated by ArF-laser irradiation

Preparation of standard atmospheres of nitrogen acid compounds: the NO_2 permeation tube

Automated denuder systems

A battery-powered light-weight ozone analyzer for use in the troposphere and stratosphere

H_2O_2 in solid precipitation

Detection of nitrated and oxygenated polyaromatic hydrocarbons (PAH) in suspended particulate matter sampled in urban areas and their relation with anthropogenic emission and photochemical smog formation

Field intercomparison of sampling and analytical methods for SO_2 and SO_4 in ambient air

Field measurements on the use of denuders for the evaluation of acid deposition

Use of air - SO_2 mixtures for intercomparison of SO_2 monitors in the EEC

Determination of polychlorinated dibenzo-P-dioxins and dibenzofurans in outdoor air

Measurement of atmospheric nitrous and nitric acid by means of annular denuders

Evaluation of analytical methods for the determination of degradation products of chlorinated hydrocarbons

Sampling of aldehydes and carbonyl compounds in air and their determination by liquid chromatographic techniques

The use of a filter pack for the monitoring organic and inorganic acid species in air

IR-Tunable diode laser techniques for trace gas detection at PPBV level

Measurement of airborne hydrocarbons by headspace-chromatography

A study of atmospheric deposition chemistry in peninsular and insular Italy

Chemistry of atmospheric depositions at L. TOGGIA (2160 m.a.s.l.), with emphasis on the sampling procedure

Analysis of carbonyl compounds in the atmospheric liquid phase

SUMMARY BY THE CHAIRMAN

A. LIBERTI
CNR - Roma, I

Most activity related to analysis of pollutants has been devoted to the development and improvement of analytical procedures for the determination of species responsible of atmospheric acidity and of photochemical oxidants. Wet and dry deposition processes are applied in a large number of areas and there is a need for analytical methods which can yield more accurate and reliable results.

Measurements of species which are present at ppb-ppt level in the complex atmosphere system and the evaluation of temporal and geographical distribution are a challenging analytical problem. Two main approaches can be taken to achieve this aim: the use of optical and electronic instruments, which might be able to yield concentration of a certain species in real time or the adoption of enrichment processes to obtain measurable amounts of the selected species: both approaches have been followed.

Research activity in the field of instrumental analysis has been oriented to the application of fluorescence generated by laser irradiation which can be used for the detection of nitric acid, by surface chemiluminescence due to the reaction with various laser dyes of ozone for its determination in the troposphere and stratosphere and to the development of monitors based on infrared tunable diode lasers and derivative detection. These lines of research seem to be very promising.

Investigations on enrichment processes making use of diffusion samplers (denuders) have been carried out in various research centers of the European Community (FRG, Italy, The Netherlands, Denmark) and in Sweden and in this area remarkable results have ben obtained. Denuders have proven to be indispesable tools for the analysis of reactive compunds such as NH_3, HCl, HF, HNO_3, HNO_2, H_2SO_4, NH_4NO_3 and $(NH_4)_2\ SO_4$.

The extensive research activity carried out in various centers brought to overcome some limitations related to these devices, which have been originally conceived as low volume samplers. The development of the annular geometry where the sample flows 20 to 100 times higher compared to simple denuders as well of the thermo-denuder where a fractionated compound is converted to a easily to quantify product and measured with high sensitivity, enlarged the field of application of these devices and extended the possibility of application.

A workshop on new procedures developed to measure atmospheric acidity has been held in Rome in June 1985 and in order to assess the reliability of various methods a field intercomparison study was organized. This workshop was held at the Schauinsland-EMEP Station (Freiburg-FRG) in October 1985 where five measuring methods have been compared during a 5-day experiment. The workshop showed that the sampling through denuders permits to measure gaseous and aerosolic atmospheric species at the trace levels. The various methods yield a sufficient agreement on sulphate, ammonium and sulfur dioxide while noticeable differences for other species have been observed.

There is no doubt about the outstanding performances, which can be obtained with the denuders but it is believed that minor operational procedures might exert a noticeable impact and further research is needed to optimize the analytical procedures.

Through the use of denuders it has been shown that it is possible the determination of nitrous acid, a compound which has been identified very recently in the gas phase by differential optical absorption spectroscopy in the ultraviolet region. It seems this species has been overlooked in atmospheric chemistry whereas the possibility of determining it quite accurately can supply useful information on a scientific and practical stand point.

A critical evaluation of the procedure carried out to determine both nitric and nitrous acid might yield additional information upon the oxidizing power of the atmosphere.

Noticeable results have been obtained in the determination of hydrogen peroxide in gas and condensed phase and upon its role in solid precipitation.

Interesting results have been obtained in the determination of species related to photochemical smog. Peroxy-acetyl nitrate (PAN) can be determined without any enrichment in a semi-automatic analyzer by using a gaschromatograph equipped with an electron capture detector and a suitable column to operate the fractionation of various oxidants. It seems that further improvement of this procedure might permit to determine routinously this pollutant strictly correlated to photochemical smog.

A number of investigations have been carried out for the determination of compounds of a specific environmental interest such as polichlorinated dibenzo dioxins and dibenzo furans in outdoor air and nitrated and oxygenated polyaromatic hydrocarbons in urban areas and of carbonyl compounds in air and in dispersed liquid phase. These investigations are of a great importance also for the understanding of environmental problems.

The reported activity summarizes the large amount of work which has been carried in this working party and the high standard of the results which contribute to a better knowledge of the atmospheric environment.

DETECTION OF GASEOUS NITRIC ACID BY OH(A→X)-FLUORESCENCE GENERATED BY ArF-LASER IRRADIATION

Th. Papenbrock and F. Stuhl
Physikalische Chemie I, Ruhr-Universität Bochum
D-4630 Bochum 1, Federal Republic of Germany

Summary
This work represents a status report on the feasibility to detect gaseous HNO_3 in air at low mixing ratios by ArF excimer laser irradiation at 193 nm. The basis of this detection method is the two-photon-dissociation of HNO_3 to generate electronically excited $OH(A^2\Sigma^+)$ radicals which fluoresce at around 308 nm. The detailed mechanism of the formation of OH(A) has been studied previously showing that intermediates are formed having a pressure dependent lifetime. The apparatus described here serves to establish (and improve) the sensitivity of detection and the detection limit. Furthermore, interferences by other atmospheric constituents are being studied.

1. INTRODUCTION

Nitric acid (HNO_3) is one of the acidic constituents of the atmosphere of the Earth. Some of the formation processes previously discussed (1) are

$$OH + NO_2 + M \rightarrow HNO_3 + M \qquad (I)$$

$$N_2O_5 + H_2O \rightarrow 2\ HNO_3 \qquad (II)$$

$$NO_3 + CH_2O\ (C_2H_5CO) \rightarrow HNO_3 + HCO\ (CH_3CO) \qquad (III)$$

Nitric acid can hence be considered to be a product of reaction steps terminating atmospheric HO_x- and NO_x-chains. Since HNO_3 is efficiently rained out from the atmosphere, its mixing ratio is not expected to be large. In fact, a number of studies indicate values of about 1 ppbv (2,3) with maximum concentration at noontime (2).

Several detection methods have been used to monitor gaseous nitric acid in air. Among them are (a) conversion to NO (possibly via NO_2) with subsequent chemiluminescent detection of NO (4), (b) thermo (5) and diffusion denuders (6), (c) filter techniques (2a,3,7), (d) Fourier transform spectroscopy (8), and (e) IR diode laser absorption (2b,c,9). Recently, several investigations have been performed to compare a number of these monitoring techniques (2). None of these methods seems to be well established for the direct, fast, reliable, and sensitive detection of HNO_3. Because of the important role of HNO_3 in atmospheric chemistry, it seems to be worthwhile to further develop these existing methods and, additionally, to search for new ones.

We have recently observed strong $OH(A^2\Sigma^+ \rightarrow X^2\Pi)$-fluorescence being generated when HNO_3 is irradiated by the unfocussed beam of an ArF excimer laser (10). So far, this OH(A) emission has been found to be unique for HNO_3, since it has not been observed for a number of other atmospheric constituents containing OH. (For HONO, only very weak emission has been detected (11) and for H_2O, for example, none (10a).)

In the present paper, we will summarize the basic photochemical processes leading to the generation of excited OH(A) radicals. We will further describe the apparatus in detail which is used to study the feasibility of the detection of HNO_3 in air. First experiments are reported in which calibrated gas mixtures in the ppbv-range are used and possible interferences are investigated.

2. EXPERIMENTAL

For the present feasibility study, gas mixtures containing small amounts of nitric acid were irradiated by the short light pulses of an ArF excimer laser. OH(A→X)-fluorescence resulting from the two-photon-dissociation of HNO_3 was detected by a photomultiplier at right angles to the laser beam. After calibration, the signal from the photomultiplier represents a measure of the mixing ratio of gaseous HNO_3.

A schematic diagram of the apparatus used is shown in Fig. 1. The excimer laser (Lambda Physik, EMG 101) produced an uv light pulse at 193.3 nm of about 15 ns duration and typically a fluence of 40 mJ cm^{-2}. The laser intensity was continuously monitored by a pyroelectric detector (Gentec ED 500). Since previous experiments (10) suffered from stray light, much care was taken to avoid scattered light from both the laser and the windows of the photolysis cell. For this reason the cross section of the laser beam was reduced to 1 cm^2 by a prism. Furthermore, several apertures were introduced into the laser beam as shown in Fig. 2.

The photolysis cell which is shown in Fig. 2 consisted of pyrex glass and contained the apertures made of brass. The optical windows were made of quartz. The major dimensions of the cell were 92 cm in length and 14 cm in diameter. It could be evacuated by a turbomolecular pump when low total pressures were required. The pressure in the cell was measured by capacitance pressure transducers (MKS model 221). Usually, air at several mbar premixed with traces of HNO_3 was introduced into the photolysis cell. The incoming flow of this mixture was directed towards the photolysis volume observed by the photomultiplier. During an experiment, the cell was evacuated by a rotary pump.

Fluorescence from the OH($A^2\Sigma^+ \rightarrow X^2\Pi$)-transition was detected at 308.1 ± 2.2 nm through an interference filter by a photomultiplier (EMI 9813QB). In a few experiments, the filter was replaced by a 0.5 m-monochromator (Minuteman 305), in order to disperse the fluorescence at a resolution of 0.05 nm. The signal from the photomultiplier (operated at 1.2 kV) was fed to either a transient recorder (Biomation 8100) or a gated integrator (PAR 162/165). Both instruments were connected to a computer (Zenith Z 110) to accumulate and evaluate the data and to control the experiment. Typically, the signals from 250 laser shots were accumulated to obtain a data point. This procedure lasted 125 s at a laser repetition frequency of 2 Hz.

By means of a storage oscilloscope or the transient recorder, OH(A) decays were registered as demonstrated in Fig. 3 by

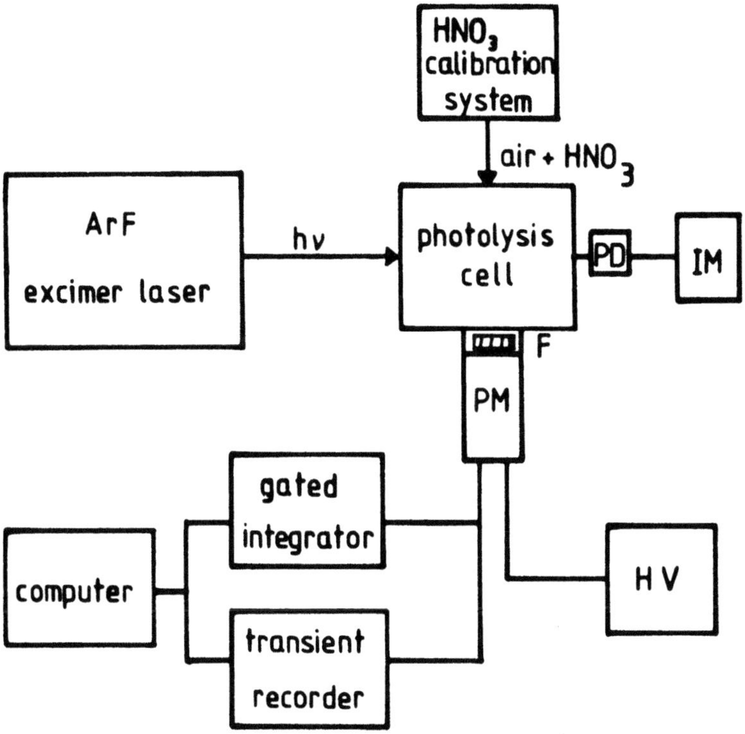

Figure 1 Schematic diagram of the apparatus. PD = pyroelectric detector, IM = intensity meter, PM = photomultiplier, HV = high voltage power supply, F = optical filter.

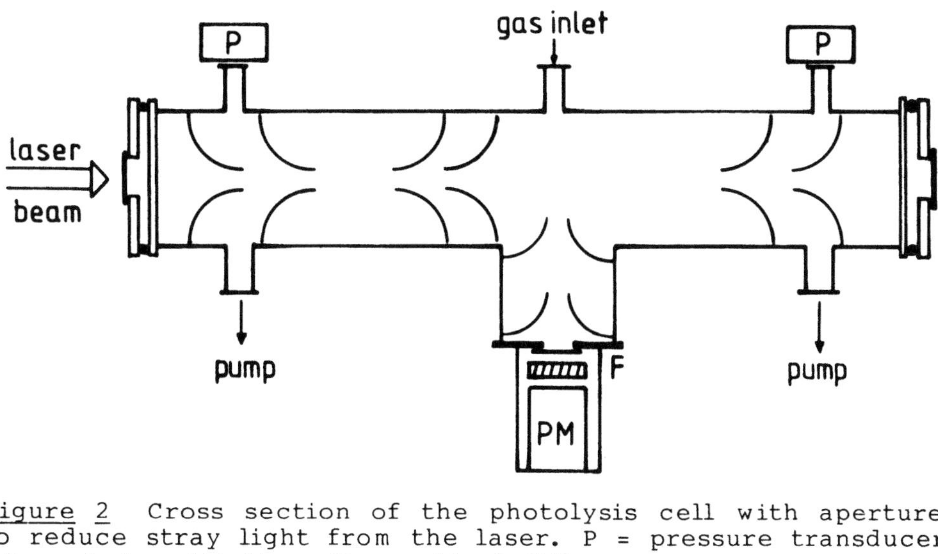

Figure 2 Cross section of the photolysis cell with apertures to reduce stray light from the laser. P = pressure transducer, PM = photomultiplier, F = optical filter.

the oscilloscope trace of a single laser shot. This detailed information was found to be very useful in the early stages of the experiments, in order to discriminate scattered laser light from the desired OH(A → X)-fluorescence. It hence facilitated to position the gate of the integrator as indicated in Fig. 3. The OH(A) decay curves allow us, moreover, to extrapolate the fluorescence intensity at time zero, the time the laser was fired.

The introduction of well known mixing ratios of HNO_3 into the photolysis cell was not found to be easy. Several methods were attempted to admix HNO_3 to dry air: (a) diffusion of HNO_3 through a capillary, (b) saturation of flowing air with HNO_3 vapor at different temperatures, and (c) permeation of HNO_3 into a teflon tube. In most experiments, method (c) was used. In the present way of handling method (a), the reproducibility of the mixing ratios was found to be inferior to that obtained by the other two methods. Method (b) permits one to calculate the mixing ratios via the known vapor pressure of HNO_3. Therefore, this method was initially preferred. However, because of the long glass tubes used for the saturator and the additional mixing stages, ad- and desorption of HNO_3 hampered a reasonably fast adjustment of the desired values of the mixing ratio.

The calibration of the mixing ratios was performed by method (b) and/or by titration. Titration allowed calibration only at relatively large mixing ratios ($>$340 ppb). Smaller values were obtained by further dilution of the premixed flow. The values thus calculated from the premixed and admixed flows were additionally confirmed by quantitative conversion of HNO_3 to NO_2 on hot quartz surfaces and by subsequent conversion of NO_2 to NO by $FeSO_4$. The final NO product was then monitored by a NO/O_3-detector (4a,12). Both compressed and synthetic air were used as diluents. The compressed air was dried and purified by charcoal and cotton filters before use. The titration was done with a well defined amount of NaOH solution diluted with distilled water (typically 100 µl of 0.1 n NaOH + 80 ml H_2O). Alternatively, the flowing gas mixture was bubbled through distilled H_2O and the dissolved HNO_3 was determined by titration. To monitor the progress of the titration, a pH-meter (Knick) was used. Care was taken not to dissolve CO_2 from ambient air during long titration times (up to 25 h). Therefore, the solution was covered with a layer of slowly flowing Ar.

3. RESULTS AND DISCUSSION

First we would like to briefly summarize the knowledge on the photochemical process generating the excited OH(A)-radicals. It has been previously observed (10,13) that strong $OH(A^2\Sigma^+ \rightarrow X^2\Pi)$-emission is generated upon photolysing HNO_3 by an ArF-laser. The rotational distribution found for the excited OH(A) corresponds to a rotational temperature of about 3750 K. The OH(A) emission is generated by absorption of two laser photons. It has been recently demonstrated with two ArF lasers in a pump-probe arrangement that a long-living intermediate is formed by the first absorbed laser photon. This intermediate species is subsequently photolyzed by the second photon to yield the excited OH(A) radicals. Furthermore, it has been

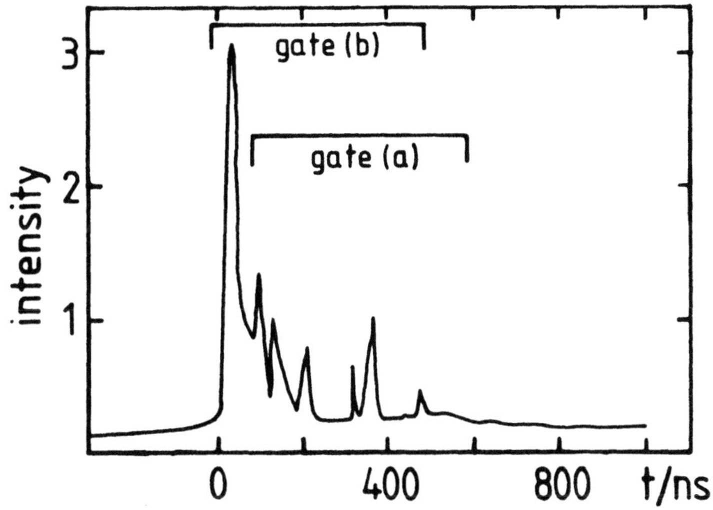

Figure 3 Time dependence of the stray light and OH(A ⟶ X)-fluorescence signal for 10 ppb HNO_3 in 8 mbar synthetic air for a single shot experiment. The indicated gate positions were used for the integrator.

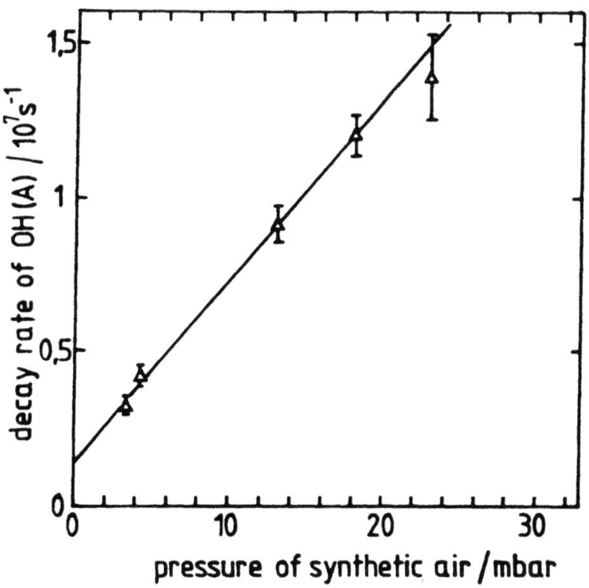

Figure 4 Decay rates of the OH(A ⟶ X)-fluorescence signal as a function of pressure of synthetic air. The mixing ratio of HNO_3 was 180 ppb. The error limits represent three times the standard deviation.

observed that the intermediate species is efficiently quenched by added gases such as O_2 and N_2. The intermediate could not be identified directly, but, on the basis of energy consideration, it was proposed to be HONO, possibly in its triplet state (13).

Based on the number of photons registered by the photomultiplier at relatively high mixing ratios but low laser fluence, the detection sensitivity was estimated to be 1 detected photon per ppb and laser shot (14). This value is in rough agreement with the strength of the OH(A) decay signal shown in Fig. 3 for 10 ppbv HNO_3. Accumulation of such decays by the transient recorder and computer furnished precise lifetime measurements of OH(A). As the lifetime of the OH(A) fluorescence signal determines position and length of the gate of the integrator (see Fig. 3), we have measured the lifetimes in dependence on the pressure of synthetic air. The resulting decay rates, τ^{-1}, for OH(A) are presented in Fig. 4 for a constant mixing ratio of about 180 ppbv. In this mixture, the mole fraction of HNO_3 is so low that the parent molecule, HNO_3, does not participate in the quenching process and the resulting rate constant of $(2.46 \pm 0.4) \times 10^{-11}\ cm^3\ s^{-1}$ is entirely due to the quenching by air. It should be noted that this value is in reasonable agreement with literature values (15) and that it is specific for the rotational distribution of OH(A) generated in these experiments (13).

In spite of the apertures, some stray light from the laser pulse still is observed, mainly during the first 20 ns of the OH(A) fluorescence. Therefore, the position of the gate has to be set very carefully as indicated in Fig. 3. Usually position (a) was used which opened the gate 20 ns after the intensity maximum of the laser pulse was reached. On the other hand, gate position (b) covering the entire OH(A) decay was applied for the investigation of the stray light signal. It is evident from Figs. 3 and 4, that high pressures of air eventually shorten the decay of the fluorescence signal such that the emission occurs before gate (a) opens. To avoid this situation, sufficiently low total pressures in the range of several mbar were used.

For most of the experiments, calibration method (c) was applied. The calibration of the tubes obtained by titration are given in Fig. 5. The rate of permeation was found to be a linear function of the length of the teflon tube. In these runs, tubes with 1.6 mm outer diameter and a wall thickness of 0.3 mm were used. Lowering the temperature of the liquid HNO_3 from room temperature to 252 K, dramatically decreased the rate of permeation. In order to start with low HNO_3 mixing ratios, in most of the experiments, the length of the teflon tube was 43 cm and the temperature of the liquid HNO_3 was 252 K. The resulting premixed flows were then further diluted as desired.

First calibration curves were obtained for this detection method with a permeation tube as described above. The results are shown in Figs. 6 and 7 for large and small ppbv values, respectively. It should be noted that the intensity scales in both figures are not comparable since in Fig. 6 (for high mixing ratios) the fluorescence signal was measured by the transient recorder and represents the initial intensity at time zero. For the data points in Fig. 7, the signal was much weaker

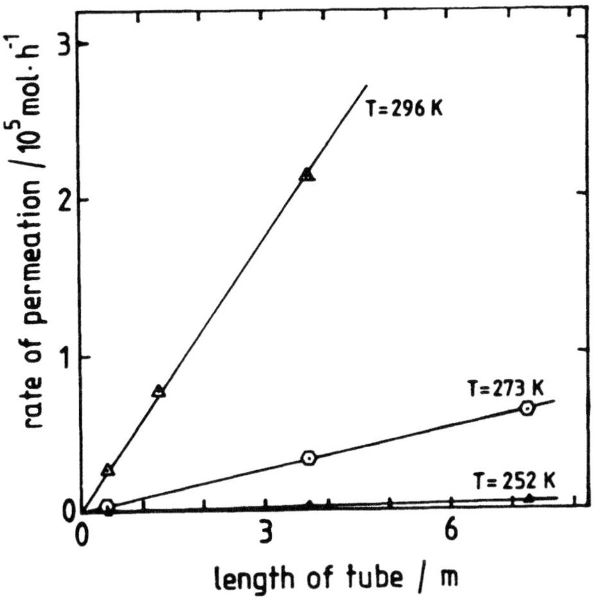

Figure 5 Rate of permeation of HNO_3 into a teflon tube of 1.6 mm outer and 1.0 mm inner diameter as a function of tube length and temperature.

and the gated integrator was used in position (a) (see Fig. 3). A data point in Fig. 6 required 250 decay curves to be accumulated and averaged which took 170 s for each point. With the gated integrator (Fig. 7) the signals from 105 shots were averaged by the gated integrator and the computer. In this case, it took about 100 s to obtain a data point. The adjustment of a new mixing ratio in the photolysis cell was found to be complete in half an hour. Therefore, it was waited this long before a calibration point was taken. Some data points in Fig. 7 were obtained with compressed air. These points were found to be consistently higher by about 6 ppbv than those measured with synthetic air.

In contrast to a previous preliminary calibration (10), the curves of Fig. 6 and Fig. 7 were found to be linear. This improvement is probably due to the much lower stray light level in the present photolysis cell. The 3 σ-detection limit was calculated to be 0.8 ppbv which is a factor of 5 lower than previously measured in preliminary experiments (10). It should be noted that this limit is in accord with the calculated detection sensitivity. In the present experiments, this detection limit was achieved without elaborate detection geometry. For example, no lenses or mirrors were used to collect the fluorescence light and the viewing angle of the photomultiplier was restricted by two apertures. Therefore, the total light intensity received by the photomultiplier was much lower than previously (10).

The intercepts in Fig. 7 correspond to signals equivalent to 7.5 and 13.5 ppb. The cause of these "background signals" is

Figure 6 Fluorescence intensity as a function of the HNO_3 mixing ratio in the range from 10 to 500 ppbv. The error limits represent three times the standard deviation. The total pressure in the photolysis cell was 4.4 mbar. The intensity is represented by the initial fluorescence intensity at time zero as measured by the transient recorder. The signals from 250 laser shots at 43 mJ cm^{-2} were averaged to obtain a data point.

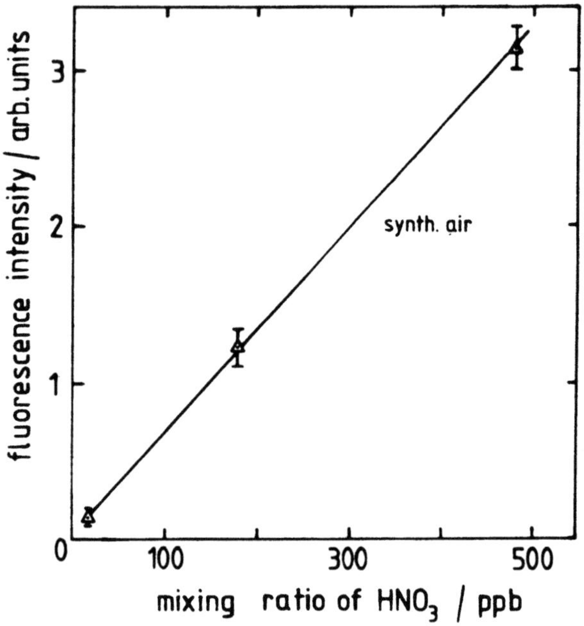

Figure 7 Fluorescence intensity as a function of the HNO_3 mixing ratio in the range from 2.3 to 12.5 ppbv. The error limits represent three times the standard deviation. The total pressure in the photolysis cell was 8 mbar. The intensity was measured by the gated integrator. The gate was opened 20 ns after the maximum of the intensity of the excimer laser; it was 500 ns long. The data points are averages from 105 laser shots at 33 mJ cm^{-2}.

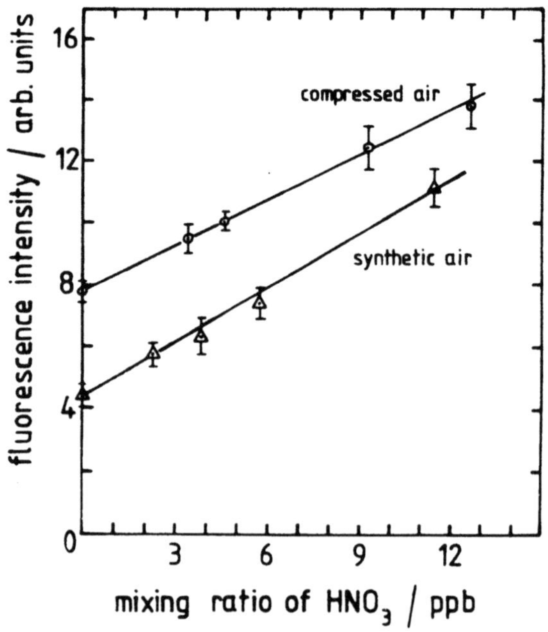

not evident. Most likely, these intercepts are due to one or several of the following effects: (a) There was still some stray light generated from the laser light. (b) The synthetic air contained some HNO_3 and the compressed air contained even more, although it was fed through filters. (c) HNO_3 could not be completely removed from the photolysis cell, since it is very efficiently adsorbed. (d) Synthetic air and even more so compressed air contained constituents yielding emission at 308 nm. Several experiments indicated that at least part of the intercept is caused by (a) and (c). Future work will be devoted to study the "background signal" and to improve the detection limit.

Possible interferences were investigated for several OH-containing constituents of the atmosphere. H_2O, H_2O_2, CH_3OH and C_2H_5OH have been added to the photolysis system and did not give any emission (10). Since excited HONO has been proposed to be the intermediate species in the photolysis of HNO_3 (13), the OH(A) generation in the ArF-laser photolysis of HONO has been studied in detail in a separate investigation (11). It has been observed that OH(A) is formed in the 193 nm photolysis of HONO with a low quantum yield of 10^{-5} by the absorption of one photon. For the laser fluxes used here, it has been estimated that the OH(A) fluorescence intensity from the one-photon dissociation of HONO is about three orders of magnitude smaller than that from the two-photon dissociation of HNO_3, provided equal concentrations of the acids are photolyzed (11). For the atmosphere, in fact, similar concentrations of HNO_3 and HONO (16) have been reported and hence interference by HONO can be expected to be negligible. Further studies of interfering effects are planned.

ACKNOWLEDGEMENT

We thank Profs. D. Klockow and D. Kley for useful discussions and gratefully acknowledge financial support by the BMFT Grant No. 07045741.

REFERENCES

(1) J.G. Calvert and W.R. Stockwell, Environ. Sci. Technol 17, 428 A (1983)

(2) (a) C.W. Spicer, J.E. Howes, Jr., T.A. Bishop and L.H. Arnold, Atmospheric Environment 16, 1487 (1982); (b) J.G. Walega, D. H. Stedman, R.E. Shetter, G.I. Mackoy, T. Igucki and H.I. Schiff, Environ. Sci. Technol. 18, 823 (1984); (c) K.G. Anlauf, J.W. Bottenheim, K.A. Brice, P. Fellin, H.A. Wiebe, H.I. SchifF, G.I. Mackay, R.S. Braman, and R. Gilbert, Atmospheric Environment 19, 1859 (1985).

(3) E.E. Lewin, R.G. DePena, and J.P. Shimshock, Atmospheric Environment 20, 59 (1986) and references therein.

(4) (a) T.J. Kelly, D.H. Stedman, J.A. Ritter, and R.B. Harvey, J. Geophys. Res. 85, 7417 (1980); (b) D.W. Joseph and C.W. Spicer, Anal. Chem. 50, 1400 (1978).

(5) (a) R. Nießer and D. Klockow, J. Arosol. Sci. 13, 175 (1982); (b) J. Slanina, L.V. Lamoen-Doornenbal, W.A. Lingerak, W. Meilof, D. Klockow, and R. Nießner, Int. J. Environ. Anal. Chem. 9, 59 (1981).

(6) (a) D.J. Eatough, V.F. White, L.D. Hansen, N.L. Eatough,

and E.C. Ellis, Anal. Chem., 57, 743 (1985); (b) J. Forrest, D.J. Spandau, R.L. Tanner, and L. Newman, Atmospheric Environment 16, 1473 (1982).
(7) (a) T. Okita, S. Morimoto, M. Izawa and S. Konno, Atmospheric Environment 10, 1085 (1976); (b) J. Forrest, R.L. Tanner, D. Spandau, T.D.'Ottavio and L. Newman, Atmospheric Environment 14, 137 (1980); (c) B.R. Appel, S.M. Wall, Y. Tokiwa, and M. Haik, Atmospheric Environment 14, 549 (1980); (d) R.S. Braman, T.J. Shelley, and W.A. McClenny, Anal. Chem. 54, 358 (1982).
(8) P.L. Hanst, Appl. Opt. 17, 1360 (1978).
(9) H.I. Schiff, D.R. Hastie, G.I. Mackay, T. Igucki, and B.A. Ridley, Environ. Sci. Technol. 17, 352 A (1983).
(10) (a) Th. Papenbrock, H.K. Haak, and F. Stuhl, Ber. Bunsenges. Phys. Chem. 88, 675 (1984); (b) H.K. Haak and F. Stuhl, J. Photochem. 17, 69 (1981).
(11) R.D. Kenner, R. Rohrer, and F. Stuhl, J. Phys. Chem. to be published, May 1986.
(12) These experiments were performed by M. Woyde.
(13) R.D. Kenner, F. Rohrer, Th. Papenbrock, and F. Stuhl, J. Phys. Chem. 90, 1294 (1986).
(14) This estimate was performed by F. Rohrer.
(15) K. Schofield, J. Chem. Phys. Ref. Data 8, 733 (1979).
(16) C. Keller, Dissertation, Universität zu Köln (1984), A. Sjödin, M. Frem, Atmospheric Environment, 19, 985 (1985), G.W. Harris, W.P.L. Winer, J.N. Pitts, Jr., U. Platt, and D. Perner, Environ. Sci. Technol. 16, 414 (1982).

PREPARATION OF STANDARD ATMOSPHERES OF NITROGEN ACID COMPOUNDS: THE NO_2 PERMEATION TUBE

I. ALLEGRINI, P. BUTTINI, V. DI PALO and M. POSSANZINI
Istituto sull'Inquinamento Atmosferico del C.N.R.
Area della Ricerca di Roma
Via Salaria Km. 29,300 -C.P. 10
00016 Monterotondo Stazione (Roma)

Summary

The growing interest towards nitrogen acid compounds (HNO_3 HNO_2, NO_2) involved in acid deposition stimulates the development of proper analytical methods for atmospheric trace levels measurements. The precision and accuracy of such methods should be tested by means of highly reliable standard atmospheres containing the species of interest. Within the framework of a program on the certification and standardization of analytical techniques for air pollution, it was found that a NO_2 permeation tube can also be regarded as a constant source of HNO_3 and HNO_2. These species are present on the "green grade" NO_2 cylinders to be used for the preparation of the tubes. The individual species can be selectively removed by using differently coated annular denuders, so that pure NO_2 and mixtures of HNO_2-NO_2 and HNO_3-HNO_2- NO_2 can be obtained. A detailed description of the experimental set-up is presented, including results obtained using different operating conditions.

1. INTRODUCTION

Nitrogenous compounds play a very important role in the definition of the chemical processes occurring in the atmosphere. NO_2 easily reacts with the hydroxyl radicals to yield nitric acid, whereas surface-catalyzed reactions involving nitric oxide and water lead to the formation of nitrous acid (1). Removal of HNO_3 from the atmosphere by either wet or dry deposition contributes appreciably to the acidification of the environment. Nitrous acid has been recognized as an important initiator of photochemical smog through the generation of OH radicals.

In order to fully understand the nitrogen chemistry in the atmosphere , a number of analytical techniques have been deviced for the measurements of these two nitrogen acids; they include instruments based on spectrochemical systems (long path diffe rential optical absorption spectroscopy and Fourier transform infrared spectroscopy with tunable diode laser (2,3) and wet-chemical procedures. Among these, denuder techniques coupled to ion chromatographic analysis have been proved to be a successful way to obtain reliable field data (4). Unfortunately, the comparability of HNO_2 and HNO_3 concentration data seems to be very poor, owing to the fact that the harmonization is still not yet reached. This prevents any interpretation from data obtained in different locations and using different techniques. Thus, in order to understand the mechanism governing formation and removal of HNO_2 and HNO_3, precise and accurate calibration techniques are required.

Although, in principle, several techniques could be employed for the preparation of standard atmospheres, the permeation based devices offer the proper way to achieve the goal and a series of different approaches have been attempted.

Since their introduction, the FEP-Teflon permeation tube has been proved to be a very useful tool for calibration of NO_2 measuring procedures, including specific analyzers and a number of modifications in the basic tube construction and operating conditions have been suggested (5,6). Performances and characteristics of NO_2 tubes have been described by Hughes et al (7), who investigated the effects of humidity and temperature on the emission rate and long term stability. Great care must be exercized to avoid any reaction between NO_2 and water at the inside and outside walls of the tube; otherwise NO_2 might be lost as a result of NO_2 conversion to HNO_3 and a decrease in permeation rate and lifetime might occur. For these reasons NO_2 must be dried by passing it, as a liquid, over a suitable dessiccant before filling the tube and dry nitrogen must be used as carrier gas.

If, however, a NO_2 tube contains water in a significant percentage (2-3 vol %), the characteristic brown colour changes to green due to the presence of impurities of N_2O_3 and HNO_3 resulting from the reactions

$$3NO_2 + H_2O \rightleftharpoons 2HNO_3 + NO$$
$$NO + NO_2 \rightleftharpoons N_2O_3 \overset{H_2O}{\rightleftharpoons} 2HNO_2$$

Under these conditions a permeation tube filled with "green grade" NO_2 can be regarded as a self containing source of HNO_3 and HNO_2 and could provide dynamic mixtures of NO_2, HNO_3 and HNO_2.

The purpose of this paper is the evaluation of the NO_2 permeation tube as a standard device for the preparation of gaseous mixtures of the three nitrogenous species and the application of the annular denuder technique to measure the respective permeation rates and selectively isolate HNO_2 and HNO_3 from NO_2.

2. EXPERIMENTAL

Apparatus

In order to prepare a NO_2 permeation tube, the bulb of a 5 ml transfer pipette was sealed at one end and the tube on the other side was cut and fire-polished at about 2,5 cm from the bulb. A lenght of FEP Teflon tubing (5mm i.d. x 1 mm wall thickness) was then forced over the open end. The device was than filled with liquid NO_2 delivered, through a narrow glass tube, by a "green grade" NO_2 small cylinder. The operation does not require any careful control to prevent absorption of ambient water vapor by the nitrogen dioxide. Then, a short glass tubing closed at the top was inserted in the Teflon tube leaving a permeation window about 0,5 cm long.

Test atmospheres of NO_2,HNO_2 and HNO_3 were prepared by attaching a mixing chamber and a dilution system with clean dry air to the outlet of a thermostated condenser containing the permeation tube (Fig. 1). The nitrogen carrier flowrate through the tube was maintained at approximately 70-80 ml/min and the flow of diluent air was adjusted to obtain a total flowrate of 1 l/min through the collecting system. The latter consisted of three annular denuders, each 10 cm in length, 10 mm and 13 mm in annular diameters. Annular walls were sandblasted to increase the loading capacity and also obtain a more homogeneous coating. The first two denuders were coated with Na_2CO_3 in a water-methanol (1:1 v/v) solution to collect HNO_2 and HNO_3. The second denuder was used to correct for the NO_2 interference in the measurement of the above species (8). The third denuder was coated with 5% NaOH and 10% guaiacol (o-methoxyphenol) in methanol to chemisorb NO_2, which was found to be quantitatively converted to nitrite by reaction with phenoxide ions (9).

HNO_3 was alternatively collected on a sodium chloride-coated denuder, through which NO_2 and HNO_2 proceed unaffected.

A Dionex ion chromatograph mod. QIC was used to analyze the extractive solutions of the denuders. The separator was a 4 x 250 mm HPIC-AS3 column filled with low capacity anion exchange resin.

Procedure

Wall coating procedure has been reported elsewhere (8). After sampling for time periods varying from 3 to 18 hours, each denuder was extracted with 10 ml deionized water. Then, aliquots of the extracts were injected into the microloop of the ion chromatograph. The eluent system was 3 mM $NaHCO_3$/2,4 mM Na_2CO_3 solution mixture. By referring to the calibration graphs based on the peak heights of nitrite and nitrate standards, the amounts (ug) of NO_2,HNO_2 and HNO_3 sorbed were determined and divided by sampling duration to obtain the respective permeation rates.

3 RESULTS AND DISCUSSION

After the filling of the tube, it takes a few days in order to reach a constant permeation rate (within $\pm$ 3%) of NO_2 at 27°C.

This was verified by mixing the NO_2 output of the tube with dry clean air to a known dilution and monitoring the mixture with a NO_x chemiluminescent analyzer (Environment, mod AC 30 M). Likewise, conditioning of about two-three days was required to attain the equilibrium stage after temperature changes to 37°C and 45°C.

Typical ion chromatograms related to the extracts of the three annular denuders connected in series are shown in Fig. 2. The first chromatograms shows peaks due to nitrite and nitrate. The former is the sum of HNO_2 and the fraction of NO_2 converted to nitrite on the denuder walls, while the latter is the sum of HNO_3 and the fraction of NO_2 which is partly oxidized to nitrate when reacting on alkaline substrates. The geometry of the denuders and the flow rate are such that a collection efficiency larger than 99% for the reactive species is assured; thus, HNO_2 and HNO_3 will be completely absorbed on the first denuder. The positive interference from the uptake of NO_2 can be evaluated by means of the second denuder which only collects NO_2 mostly as nitrite. Results from the experiments and previous findings in dry clean air, show that the uptake of NO_2 is 0,5-0,6%. According to the general theory of the denuders, such a small efficiency means that the absolute amounts of NO_2 collected in the first and second denuder will be almost the same and, subtracting the signals of nitrite and nitrate of the second denuder to those found on the first, a reliable measurement of HNO_2 and HNO_3 can be performed.

Although the method has been proved to be suitable for the direct measurement of HNO_2 and HNO_3, its differential nature should be taken into account, thus its accuracy and precision is consistently lower than that obtained by the classic usage of denuders. Previous results in ambient atmosphere show that the reproducibility, as measured in parallel runs, is about 10 % for HNO_2 and 7% for HNO_3 while for the measurement of NH_3 and SO_2 which usually do not require any correction, the reproducibility is better than 4% (8). In addition, ambient concentrations of HNO_2 and HNO_3 are much lower than that of NO_2, thus interference from NO_2 could be significant.

The third chromatogram, relative to the guaiacol coated denuder, displays the nitrite peak only, as no nitrate is originated from NO_2 absorbed on such a reducing sorbent layer. The use of this novel denuder open several possibilities in the measurement of ambient concentrations of NO_2. As it is well known, commercial chemiluminescent analyzers also measure other nitrogenous compounds, including HNO_2 and HNO_3, while the luminol-based chemiluminescent technique is still affected by PAN interference. In remote locations, where the concentration of NO_2 is comparable to that of other nitrogenous interfering species, erratic results through instrumental methods might be obtained. Classical wet chemical procedures have inherent disadvantages such as flow rate limitations, relative humidity effects and strong interferences from other pollutants, while the use of alkaline guaiacol as denuder coating eliminates these problems and provides a direct and simple way for the analysis of ambient NO_2.

By using the three denuders in series, it is possible to compute the permeation rate of NO_2,HNO_3,HNO_2 from the permeation tube at different temperatures as shown in Table I. It should be stressed that the permeation rates of HNO_3 and HNO_2 are two orders of magnitude lower than that of NO_2. This is not a drawback because the output from the permeation tube gives concentration ratios similar to those obtained in moderately polluted atmospheres. The coefficent of variation of the permeation tube rates were found to be very different for the three species as they ranged from 3-4% for NO_2 to 7-10% and 14-21% for HNO_3 and HNO_2, respectively, in the 27-45°C temperature interval.

The relative deviation of the NO_2 emission rates is a clear indication of the reliability of the analytical method employed while the scatter observed for nitrous and nitric acid cannot be explained on the basis of the existing data. There are several factors which could cause the poor reproducibility. First of all, it should be recalled that the measurement of HNO_2 and HNO_3 is differential; thus, if the NO_2 deposition with formation of nitrite and nitrate on the second denuder is not fully controlled, errors might arise. In addition, the absolute amounts of acid are very low and errors due to the extraction and analysis of the denuders might also contribute to the degree of precision. Another important parameter to be taken into account, is the final humidity of the carrier gas, which alters the apparent permeation rate of HNO_3 and HNO_2. The humidity of the carrier gas has been kept at a very low level by using pure N_2 cylinders and molecular sieve traps in series. The effect of temperature on the rates can be observed in Fig. 3, where plots of the log of permeation rates of NO_2,HNO_3 and HNO_2 versus the reciprocal of absolute temperature are shown. The slopes of the straight lines drawn through the data points are quite different, indicating an average increase, for a 1°C temperature increment, of about 8% for NO_2, 4% for HNO_3 and 6% for HNO_2 emissions rates.

The performances of a " green grade" NO_2 permeation tube have been exploited to investigate the response of a commercial chemiluminescent NO-NO_2 analyzer to HNO_3 and HNO_2.

Fig.4 shows a typical experiment where the permeation tube kept at 37°C and coupled with selective denuders, has been used as calibration source. The signal A corresponds to the output of the NO_2 permeation tube mixed with dilution air such that the final concentration, read on the instrument, was 996±5 ppb. The signal C has been obtained by inserting a NaCl coated denuder which removes HNO_3 only and the new measured concentration was 987±5ppb. The signal B is typical of two Na_2CO_3 coated denuders in series which completely remove HNO_2 and HNO_3 plus about 1% of the incoming NO_2. In this case the reading decreased to 966±5ppb. This indicates that the analyzer response follows, within the experimental errors, the permeation rates experimentally found for the reactive gases.

4 CONCLUSIONS

The characteristic of the "green grade" NO_2 permeation tube for the preparation of standard atmospheres of reactive nitrogen compounds, allows the calibration and testing of field instruments and apparatuses intended for the monitoring of these species in remote locations. A proper combination of different annular denuders is able to prepare NO_2,NO_2+ HNO_2+ HNO_3 and NO_2 + HNO_2 mixtures at concentration levels similar to those encountered in receptor areas where such measurements are usually performed. Although the errors in the measurement of the permeation rates of HNO_2 and HNO_3 are still not below 10%, the technique is very valuable because the generation of constant and known mixtures of HNO_2 and HNO_3 is a very difficult task.

The permeation tube, coupled to a similar permeation device for SO_2, could be properly set in a small field audit calibrator to be conveniently used in monitoring networks.

REFERENCES

(1) PLATT,U.,The origin of HNO_2 and HNO_3 in the atmosphere. In "Chemistry of multiphase atmospheric systems",NATO ASI Series,W. Jaeschke,ed.,Springer-Verlag,Berlin,1986,pp 299-319

(2) PERNER,D. and PLATT,U. Geophys. Res. Lett. 6,917,1979.

(3) TUAZON,E.C.,WINER,A.M.,GRAHAM,R.A.,EASTON,R.R.,and PITTS,J. N.,Jr.,Atmos. Environ. 12,865,1978.

(4) DE SANTIS,F.,FEBO,A.,PERRINO,C.,POSSANZINI,M.,and LIBERTI A. Simultaneous measurements of HNO_3,HNO_2,HCl and SO_2 in air by means of high efficiency annular denuders.Proc. on Advancements in Air Pollution Monitoring Equipment and Procedures Friburg,2-6 June 1985,pp 68-75.

(5) HUYGEN,C.,Atmos. Environ. 5,55,1971.

(6) LINQVIST,F.,and LANTING,R.W.,Atmos. Environ. 6,943,1972.

(7) HUGHES, et al. Anal. Chem. 49,1823,1977.

(8) ALLEGRINI,I., et al. Annular denuder method for sampling gaseous and aerosolic materials: application to the determination of species relevant to acid deposition. Submitted to Environ. Sci. Technol.

(9) BUTTINI,P.,DI PALO,V.,and POSSANZINI,M.,Coupling of denuder and ion chromatographic techniques for NO_2 trace level determination in air.Submitted to Sci. Tot. Environ.

TABLE I

Permeation rates of HNO_2, HNO_3 and NO_2 and their variation with temperature.

MEASURED RATES (µg/min)

DATE 86	T= 27°C HNO_2	HNO_3	NO_2	DATE 86	T= 37°C HNO_2	HNO_3	NO_2	DATE 86	T= 45°C HNO_2	HNO_3	NO_2
22-4	0,011	0,023	1,23	19-5	0,028	0,031	2,96	3-6	0,044	0,046	5,24
28-4	0,016	0,020	1,32	20-5	0,031	0,029	3,01	5-6	0,035	0,047	5,40
30-4	0,015	0,018	1,24	22-5	0,024	0,032	3,01	6-6	0,030	0,041	5,45
2-5	0,012	0,019	1,28	23-5	0,020	0,031	3,02	10-6	0,039	0,038	5,65
7-5	0,010	0,024	1,26	26-5	0,023	0,025	2,90	11-6	0,038	0,040	5,52
12-5	0,010	0,021	1,26	28-5	0,018	0,030	2,81				
13-5	0,009	0,024	1,33	30-5	0,024	0,032	2,86				
14-5	0,014	0,020	1,20								
Mean	0,012	0,021	1,26		0,024	0,030	2,94		0,037	0,042	5,45
Rel.std. dev.	21,4%	10,8%	4,6%		18,5%	6,7%	2,8%		13,9%	9,3%	3,0%

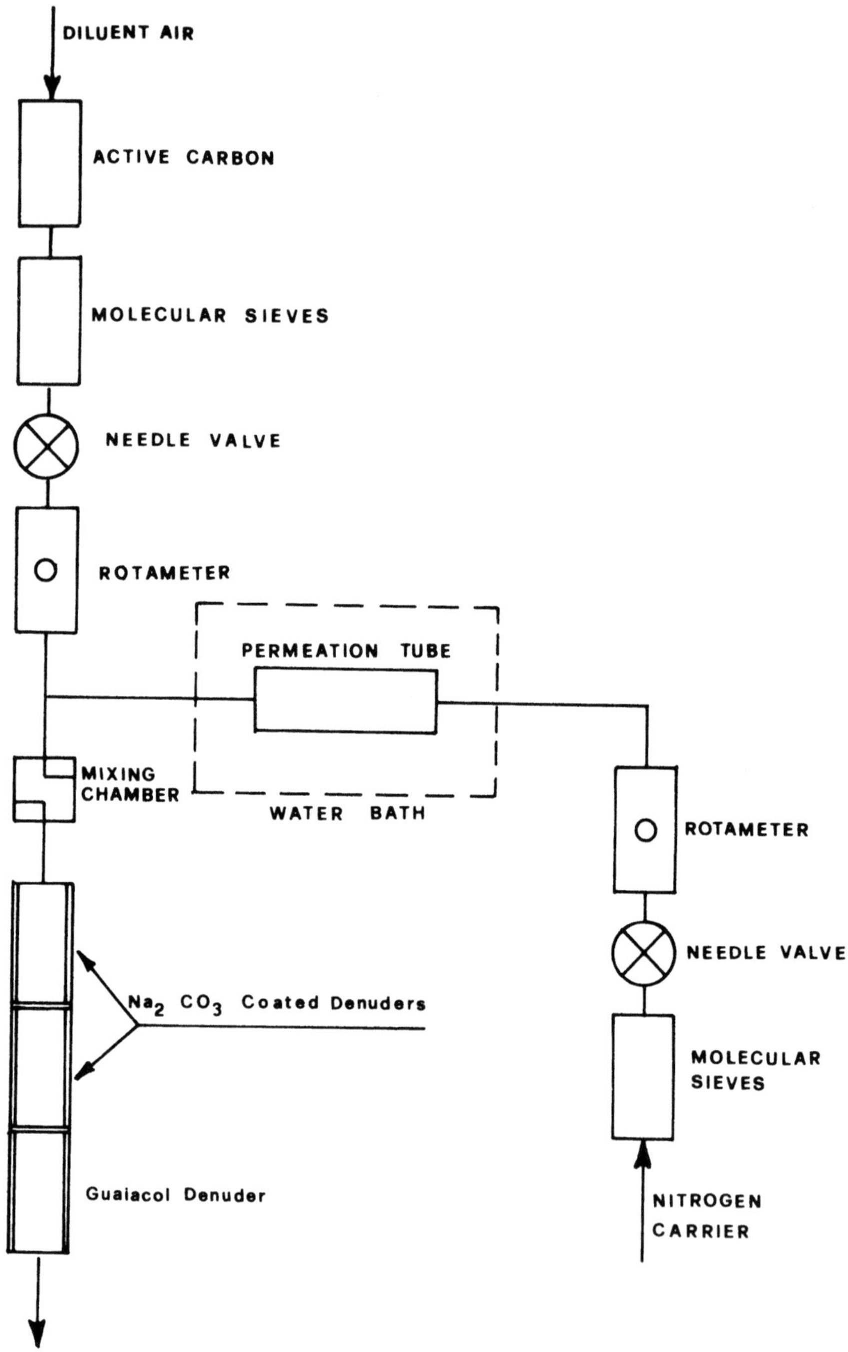

Fig. 1
Apparatus for preparing and sampling test mixtures of NO_2, HNO_2 and HNO_3 generated by a "green grade" NO_2 permeation tube.

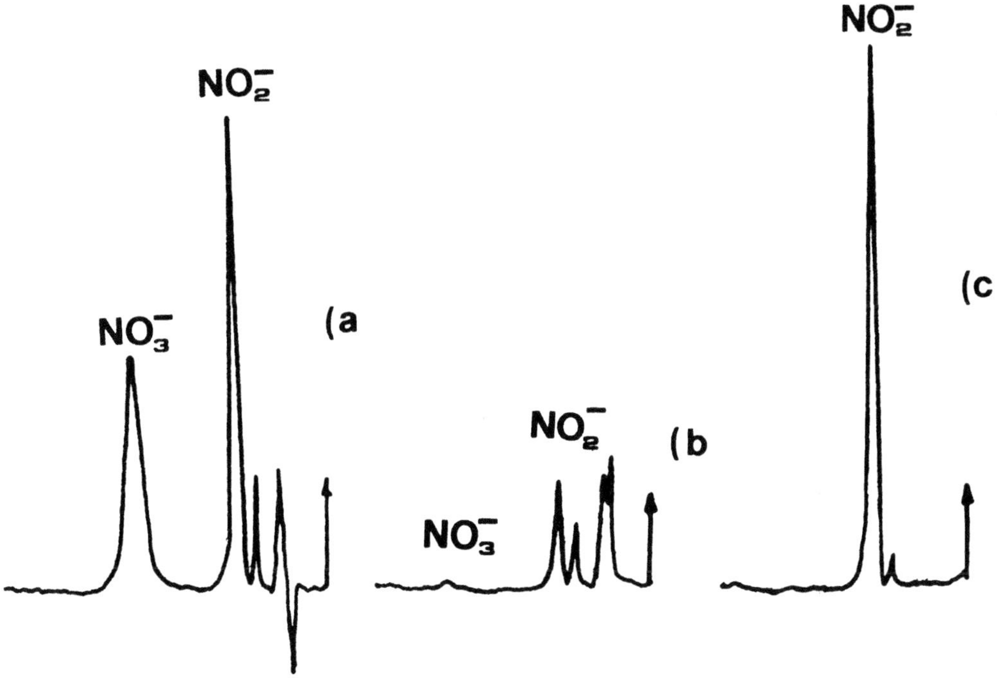

Fig. 2

Ion chromatograms of denuder extracts after collection of the effluent from the NO_2 permeation tube.

a) NO_2^- and NO_3^- peaks result from the absorption of nitrous acid, nitric acid and interfering nitrogen dioxide (1st sodium carbonate-coated denuder).

b) NO_2^- and NO_3^- peaks indicate the positive NO_2 interference to be subtracted in the measurement of HNO_2 and HNO_3 (2nd sodium carbonate coated denuder).

c) NO_2^- peak is a measure of NO_2 quantitatively sorbed on the guaiacol-coated denuder (3rd denuder).

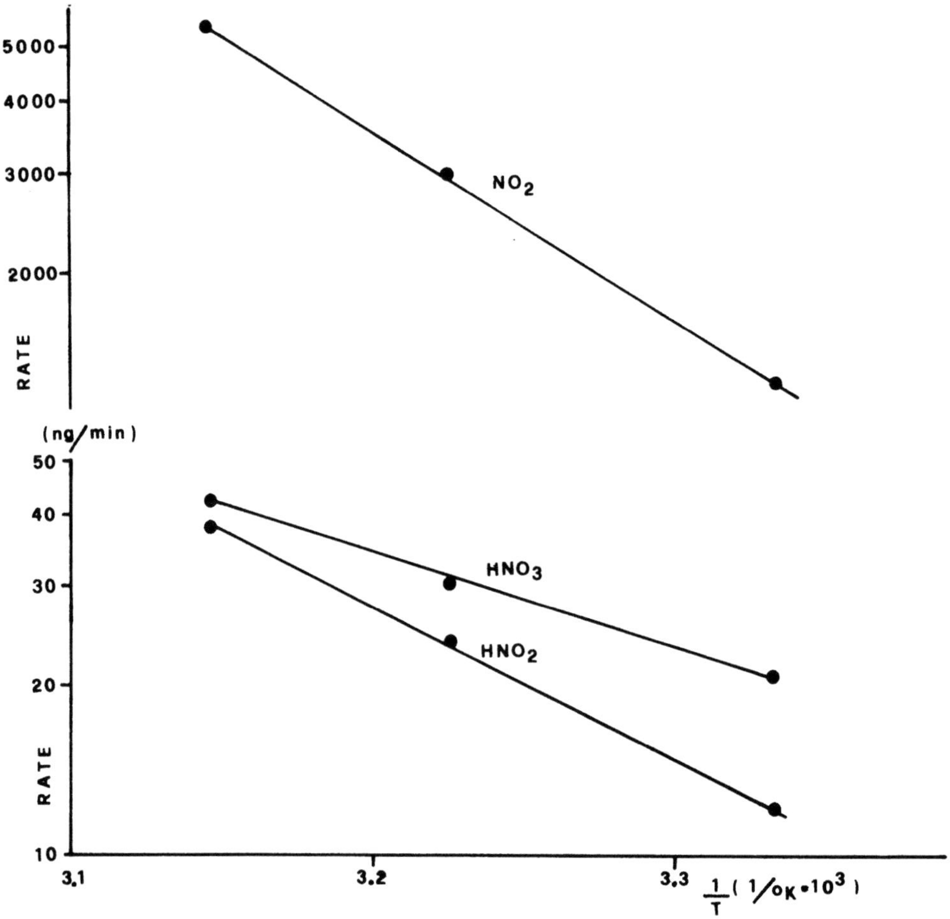

Fig.3 Temperature-permeation rate relationships for NO_2, HNO_2 and HNO_3.

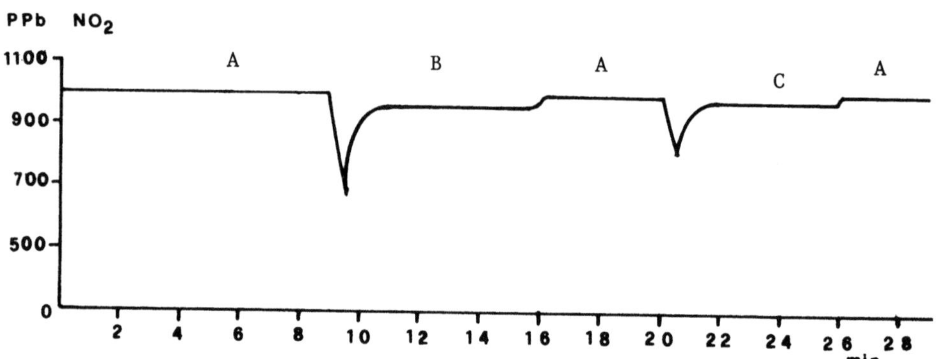

Fig.4 Response of a chemiluminescent NO_x analyzer to HNO_2 and HNO_3.
A: Signal obtained from the permeation tube (NO_2, HNO_2 and HNO_3).
B and C: Decreased signals after insertion of a Na_2CO_3 and NaCl-coated denuder, respectively.

AUTOMATED DENUDER SYSTEMS

J.Slanina, A.M. van Wensveen, C.A.M.Schoonebeek and P.I.Voors

Netherlands Energy Research Foundation ECN,
P.O. Box 1, Petten, The Netherlands

SUMMARY

The selectivity of denuder systems, for the analysis of e.g. strong acids in air, is generally very good, but drawbacks are that the methods are labour intensive and that relatively long sampling times are needed. To mitigate these problems two types of automated denuder systems have been developed. The first type is based on the thermo-denuder concept. The compound of interest is sampled and the denuder is heated to convert the compound in an easily to measure product. The detection limits for e.g. sulfuric acid or SO_2 are in the order of 0.01 PPB for a sampling period of 1 hour. The second kind consists of an annular denuder which is contineously wetted by a minimal amount of water. The water is analysed off-line and the concentrations in air of NH_3, HNO_3, HCl, HF, H_2O_2 and other readily in water soluble compounds can be measured simultaneously. Detection limits vary from 0.01 PPB for NH_3 and SO_2 to 0.01 PPB for H_2O_2 for sampling period of 1 hour.

Introduction.

Denuders have proven to be indispensable tools for the analysis of reactive compounds such as NH_3, HCl, HF, HNO_3, H_2SO_4 NH_4NO_3, $(NH_4)_2SO_4$ and H_2O_2 in air 1,2) . Generally the selectivity of denudersystems is better than obtainable by means of the traditional filter- or impinger methods 3).

Denuder systems have two drawbacks however.
Denuders are inherent low volume samplers so either long sampling periods are necessary or relative high detection limits must be accepted.
Futhermore it must be noted that the traditional manual denuder methods are rather labour intensive. These characteristics of the manual denuder methods have restricted their use and make them unattractive for e.g. dry deposition measurements and the characterisation of concentrations of acids in plumes. A number of approaches are possible to remedy these flaws. An obvious one is to develop automated denuder systems. If a denuder can be combined with a very sensitive detector system the problems of speed and detection limits can be mitigated.

High volume denuder systems, as developed by Liberti and coworkers [4)] and at ECN [5)], could be very usefull as the sample flows in these systems is 20 to 100 times higher compared to simple

denuder systems.
Furhtermore it would be extremely helpfull for many areas of environmental research if multicomponent denuder systems would be available.
So the research on denuder systems at ECN has been directed to two different approaches: The first is based on the thermo-denuder concept. The compound of interest is sampled in a denuder. After sampling the denuder is heated and the compound is converted to a easily to quantify product and measured with high sensitivity.
The second approach is based on an annular denuder which is wetted by a minimal amount of water. All readily in water soluble substances can be sampled by such a system, inluding SO_2, NH_3, HCl, HF, H_2O_2 and a number of organic compounds. Analysis can either be performed in-situ, or off-line in the laboratory.

Thermo-denuder systems.

The first automated thermo-denuder system has been developed for the determination of sulfuric acid and ammoniumsulfate in air [6]. The system is based on copper/copperoxide coated denuders. As figure 1 indicates, the system consists of two parallel sytems, each consisting of two denuders, placed in series. Initially sampling is started by means of the first set of denuders, thermostating the first denuder at 120 °C to capture sulfuric acid and the second at 300 °C so ammoniumsulfate is decomposed and the resulting sulfuric acid is absorbed by the coating of the denuder. Next sampling is performed by the second set of denuders under identical conditions as described for the first set. The first two denuders are analysed by heating them, sequentially, to a temperature of 800 °C to decompose the coppersulfate, formed under sampling conditions, to SO_2. Measurement of the liberated SO_2 is performed by means of a Flame Photometric Analyser or by means of a Pulsed Fluoresence Analyser. The detection limit of the first detection system is lower, but the reliability of the second is better.
The second set is analysed in the same way as described for the first set of denuders.
Detection limits in the order of 0.01 microgram m^{-3} are easily attained by this system, for sampling periods of one hour. Organic sulfur compounds and SO_2 can interfere so a active carbon coated denuder is inserted in the sampling line to trap these interfering substances.

The same type of copper/copperoxide coated denuders is employed for the determination of SO_2. Sampling is carried out at ambient temperature in this case. The detection, performed by heating the denuder tube to 800 °C and measuring the liberated SO_2 is performed in the same way as described for the case of sulfuric acid and ammonium sulfate.
A comparison with coated filter-packs and impingers is given in table 1. Regression analysis confirms the performance of the method as a slope of 0.98, a intercept of 0.4 microgram m^{-3} and a correlation coefficient of 0.999 is found comparing the results of the denuder system with those obtained by filterpacks.
The advantages of the method are the very low detection limit, in

the order of 0.01 PPB mixing ratio and the high precision. The latter quality is of importance in the case of dry deposition measurement by means of the gradient method. The existing methods for SO_2 are not precise enough at low SO_2 concentrations for gradient measurements. A system, consisting of two sets containing three SO_2 denuders each, has been developed for this purpose (fig.2). The first set is employed to sample the SO_2 concentrations at three diffferent heighths and is analysed by heating the denuders. While the first set is analysed the second set is sampling in the same way as described for the first set. The precision obtained by three denuders at the same height and for a sampling time of fifteen minutes is given in tabel 2.

Currently thermo-denuder systems for the determination of NH^3 and ammonium in air and for the dermination of HNO_3 and ammoniumnitrate are under development.

Wet denuder system.

The wet denuder system is basically a annular denuder coated by a water layer. The construction is given in figure 3. The annular denuder consists of two glass tubes. The inner tube has a outer diameter of 40 mm, the outer tube an inner diameter of 45 mm, leaving an annulus of 2.5 mm. The tubes are kept in place by teflon stops, equiped with spacers at the ends of the tube. The front plug is tapered to avoid turbulence at the entrance of the tube. A layer of water is present in the annulus to wet the active surfaces of the denuder. The solution is pumped in counter-flow mode into the system by means of a small hole at the end of the tube and extracted at the entrance. Evaporation must be accounted for in such a way that always sufficient solution is present to wet the walls of the denuder.
Two different modes of operation have been tested:

In the batch-wise mode the tube is filled with so much solution that sufficient is left at the end of the sampling period. The temperature and the relative humidity of the sampled air is measured and an estimation is made of the losses by evaporation. The minimal amount of water (16 ml for a 30 cm denuder) plus the estimated loss is pumped in the denuder at the start of the sampling period. After the sampling period the denuder is tilted by means of a pneumatic cylinder and the water is pumped out. The tube is washed out by two portions of 16 ml, ensuring that less than 1% of the sampled amount is left. In the continuous mode water is pumped in at such a rate that a specified flow can be extracted at the outlet (the difference between the input and the output is caused by evaporation). The output flow is directed to a sample changer which monitors the flow by counting droplets and changes sampling tubes after a specified number of drops. In the continuous mode it is not possible to calculate the concentration is the air sample directly, but a mixing tank model based on well known theory can be used for this purpose. Chemical interactions of the chemical species, which are sampled, must be avoided because off-line analysis is a much simpler way to operate this type of system, compared to the extensive in-situ analytical methods necessary in

case of interactions of the sampled compounds. Three problems can arise: Losses of SO_2 can occur if the pH of the absorbing solution is lowered below pH 5 by the absorption of acid compounds. Sulfite, the product of absorption of SO_2, will react with H_2O_2, even at pH 7. Ozone will generate artifact H_2O_2 in the wet denuder system. The first problem is solved by adding monoethanolamine as a buffer to stabilise the pH of the absorption solution between pH 6 and 7. Under these condition the reaction of sulfite with hydrogenperoxide is so slow that stabilization in the denuder system is not necessary. PHOPAA is added to the sampling tubes before starting the system, and the resulting adduct is very stable even for a period of two weeks and can be measured very sensitively by means of a fluorescence detector [7].
The denuder system is equipped with a cyclone (50% cut-off at 5 micron particles) to avoid interference by gravitational losses of large particles in the denuder system. If a small amount of NO is added to the sample stream in de the cyclone (1 liter per minute containing a mixing ratio of 1 PPB) no generation of H_2O_2 in the system by ozone is observed, due to the rapid reaction of NO with ozone [8].
The sampling efficiency of a wet denuder system with a length of 30 cm has been tested for sample flows between 20 and 50 liter per minute. The results are given in table 3 and compared to the theoretical efficiency for annular denuder systems [9]. As the efficiency is near the theoretical value, a sampling rate of 50 liter per minute is generally chosen and the losses are corrected using the theoretical efficiencies as calculated for the different species in air.

The detection limits are dependant on the sampling rate, the flowrate of the absorbing solution through the tube and the analytical possibilities for trace analysis of the species of interest. In the case of sampling during one hour at a rate of 50 l min-1, a flow of 30 ml per hour of the absorbing solution and a detection limit of 20 nanogram per milliliter for most species (mainly caused by contamination problems) the detection limits are about 0.2 microgram per m^3 for NH_3, HNO_3, HCl, HF and SO_2. In the case of H_3O_3 the detection limit is in the order of 0.01 microgram m^3 as the detection limit for the analytical method is typically in the order of 1 nanogram per milliliter.

We hope to apply this wet denuder system in studies concerning the uptake of pollutants by fog and dew, below-cloud scavenging processes and related subjects.

Literature.

1) Slanina J., v. Lamoen-Doornenbal J., Lingerak W.A., Meilof W., Klockow D. and Niessner R.,
Int. J. Environ. Anal. Chem. 9, (1981), 59

2) Slanina J.,
V.D.I. Berichte 429 (1982), pp 177-182

3) Niessner R.,
EIN NEUER WEG ZUR BESTIMMUNG STARKE SAEUREN UND IHRER SALZE IN DER ATMOSPHAERE.
Thesis, Universitaet Dortmund, 1981.

4) de Santis F., Febo A., Possaninzi M. and Liberti A.,
Annali di chimica, 74 (1984) 135

5) Slanina J.,
The use of filterpacks and denuder systems for the monitoring of reactive and acid species.
In proceedings of workshop on Environmental Monitoring for Architectural Conservation, Rome, 1984, in press.

6) Slanina J., Schoonebeek C.A.M., Klockow D., Niessner R.,
Anal. Chem. 57 (1985) 1955.

7) Lazrus A.L., Kok G.L., Gitlin S.N., Lind J.A., MacLaren S.E.,
Anal. Chem. 57 (1985), 917

8) Tanner R.L.,
Personal communication

9) Possanzini M., Febo A., Liberti A.,
Atmospheric Environment, 17 (1983) 2605

TABLE 1

COMPARISON RESULTS FOR TEST ATMOSFERES OF DENUDERS, FILTERPACKS AND IMPINGERS IN MICROGRAM PER M³ SO²

AUTOM.DENUDER SYSTEM	FILTERPACK	IMPINGER
30.9	31.2	-
32.4	32.0	36.6
2.7	3.0	-
24.0	25.0	25.4
23.4	21.5	-
2.4	2.5	2.5
10.4	10.6	11.9
10.7	9.8	-
12.3	12.5	12.1
57.1	56.9	-
56.5	58.5	-
28.2	27.9	-

TABLE 2
PRECISION OBTAINED WITH THREE IDENTICAL SO^2 DENUDER SYSTEMS.

MIXING RATIO (PPB)	NUMBER OF DETERMINATIONS	REL. STAND. DEVIAT. IN %
0.8	8	5
1.3	10	3.6
2.8	14	2.2
5.1	6	2.3
6.9	8	3
15.8	26	2.2

TABLE 3.
COMPARISON MEASURED AND THEORETICAL EFFICIENCY OF THE WET-DENUDER SYSTEM.

FLOW RATE IN L PER MIN.	MEASURED EFF. IN %	THEOR. EFF. IN %
25	98	96
41	90	88
60	85	77

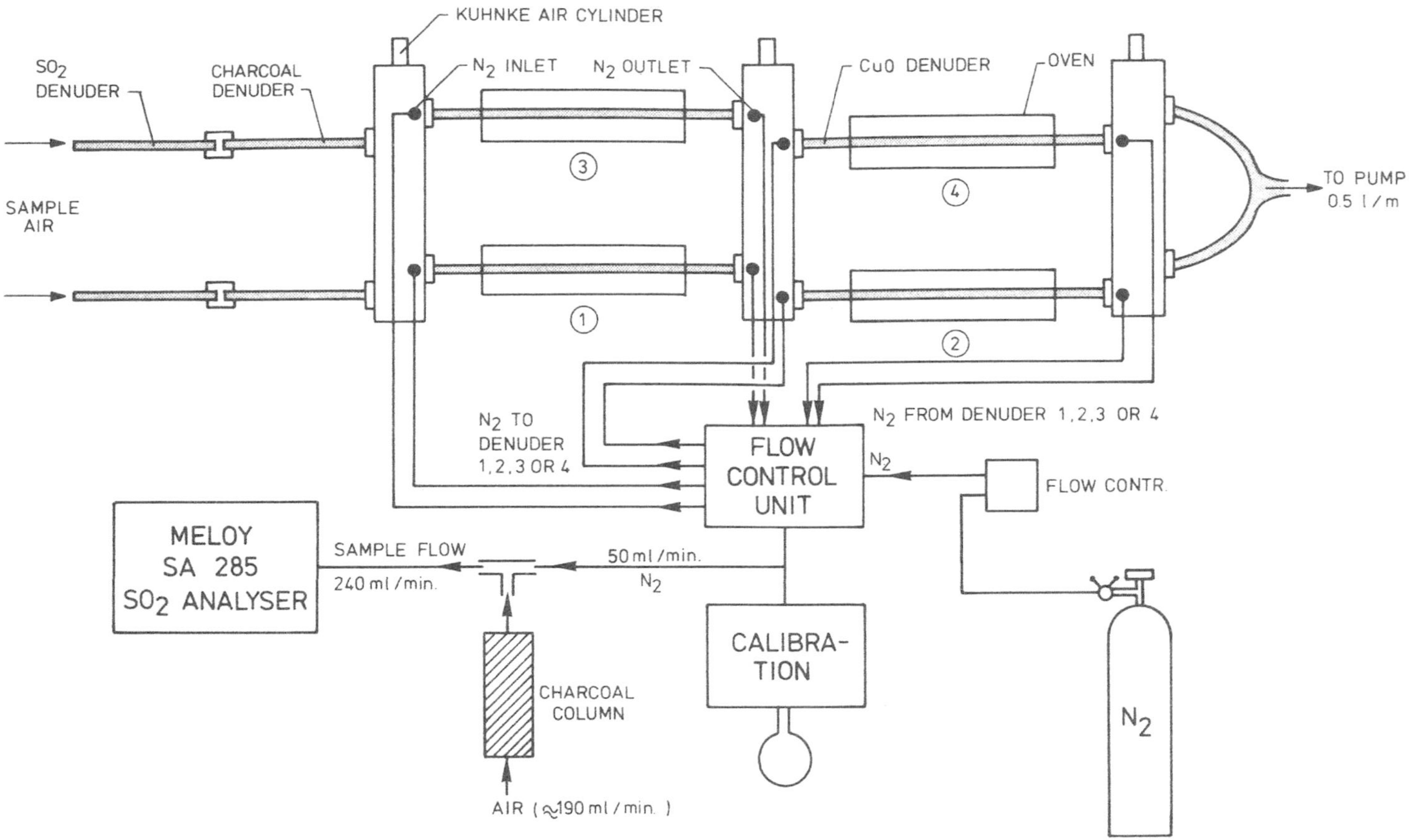

FIG.1: GENERAL SET-UP

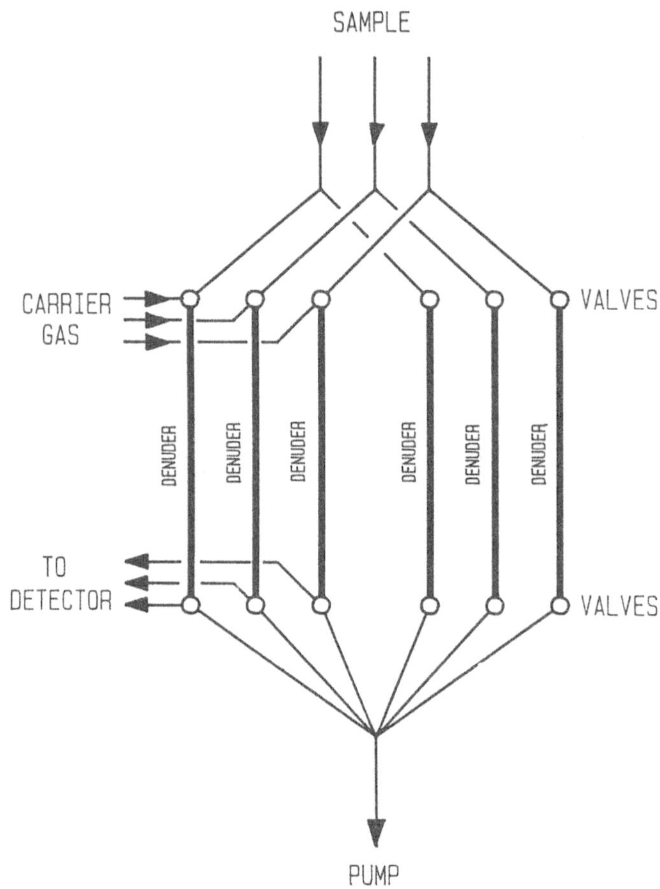

FIGURE 2

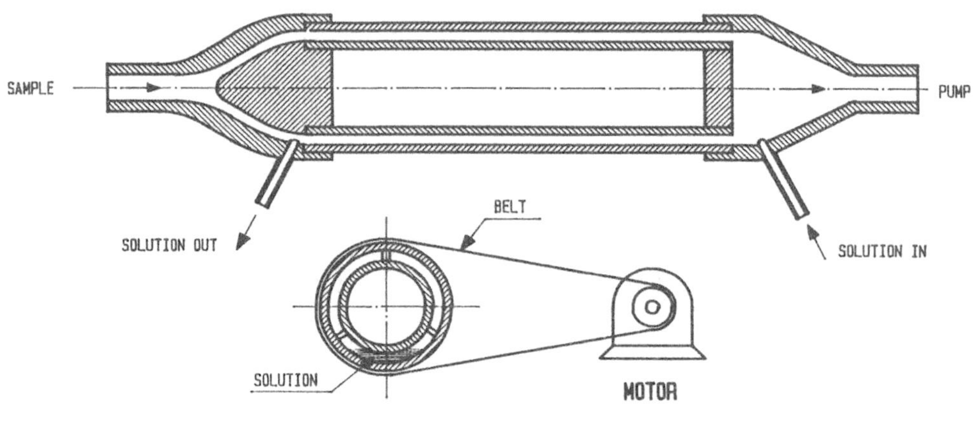

FIGURE 3 ANNULAR DENUDER

A BATTERY-POWERED LIGHT-WEIGHT OZONE ANALYZER FOR USE IN THE TROPOSPHERE AND STRATOSPHERE

S. SAHAND, W. SPEUSER and U. SCHURATH
Institut für Physikalische Chemie, Universität Bonn, Wegelerstr. 12, D-5300 Bonn 1

Summary

A novel light-weight ozone analyzer for stratospheric soundings up to 45 km altitude has been developed and tested in the laboratory. Surface chemiluminescence due to the reaction of ozone with various laser dyes (preferentially coumarin 47) adsorbed on silica gel plates is measured with a cheap side-window photomultiplier. Extremely high air flow rates past the chemiluminescent surface could be achieved down to very low pressures with a miniature electric fan, thus eliminating the flow rate dependence of the signal, while maintaining the highest possible sensitivity of the detector. The instrument has been calibrated under simulated stratospheric pressure conditions in an evacuable smog chamber, using long path uv absorption at 254 nm as reference method. The sonde will soon be field tested in collaboration with the Meteorological Observatory of DWD on Hohenpeißenberg.

1. INTRODUCTION

Ozone concentrations in ambient air have been measured by wet chemical analysis, particularly by the buffered KI method (1,2), by uv absorption at 253.7 nm (3,4), long path ir absorption (1), gas phase chemiluminescence with ethene (5), gas phase chemiluminescence with NO (6), dry surface chemiluminescence of adsorbed dyes (7,8) chemiluminescence of oil-coated paper (9), and very recently by wet surface chemiluminescence of dye solutions (10). Commercial instruments for tropospheric ozone monitoring networks are based on ethene chemiluminescence, or on uv absorption. They are heavy-weight and expensive, and cannot be used under stratospheric conditions. With the advent of inexpensive light-weight electrochemical ozone sondes (11,12), balloon soundings of the troposphere have become feasible on a routine basis, and have provided invaluable information on long term ozone trends (13,14).

Electrochemical sondes have the advantage, at least in principle, of yielding absolute ozone concentrations, although heterogeneous losses, reduced pumping efficiency at low pressures, and background current problems have been found to reduce their overall accuracy (15-17). An upper limit of about 32 km is imposed on the useful altitude range of wet chemical instruments by the vapor pressure of water, ca. 4.6 torr at the triple point. The altitude limitation can be overcome by measuring surface chemiluminescence of ozone with rhodamine B adsorbed on silica gel with a photomultiplier. A light-weight ozone sonde based on this principle was originally developed by Regener (7,18), and successfully compared with an electrochemical sonde (19), but due to problems of long-term stability and calibration the Regener sonde has been ousted by the much easier-to-handle electrochemical sondes. Only for rocket-borne measurements of ozone profiles in the upper stratosphere and mesosphere has a Regener-type instrument been successfully used, which is, however, too expensive for routine soundings with balloons (20,21).

Model calculations assessing the impact of anthropogenic halocarbon injections on stratospheric ozone predict that the largest relative change of the ozone mixing ratio should occur around 40 km altitude (22). This altitude range is inaccessible to electrochemical sondes, and quite difficult to measure with ground-based LIDAR systems, which are still in an experimental stage (23).

The following paper reports on a light-weight ozone analyzer suited for balloon soundings up to 45 km altitude (ca. 1 torr total pressure). The prototype is developed in cooperation with the Meteorological Observatory of DWD on Hohenpeißenberg, where it will be launched in tandem with a standard electrochemical sonde. Overlapping profiles will be used to validate the sensitivity of the prototype sonde which has been measured under simulated stratospheric pressure conditions in the laboratory.

2. DESIGN OF THE PROTOTYPE SONDE

One of the disadvantages of surface chemiluminescence detectors is their complex dependence on air flow and total pressure (21,24), which must be measured wery accurately for data reduction. However, it has been reported that above a critical flow rate the chemiluminescence intensity becomes proportional to the absolute ozone concentration in air, independent of pressure and flow rate (8,21). Extremely high pumping speeds can be achieved nowadays with miniaturized light-weight electrical fans, provided that the aerodynamic resistance of the sonde, particularly of the light traps, is sufficiently low. The performance of commercial fans at very low pressures is not known, but we anticipated that they would be more suitable than conventional piston or bellows pumps for maintaining the critical flow rate which ensures constant absolute sensitivity of the sonde over the entire pressure range of 760 torr to approximately 1 torr.

While the design of a chemiluminescence detector for laboratory conditions presents no particular problems, severe limitations on power consumption, price and total weight had to be taken into consideration in constructing the prototype sonde. After a series of laboratory tests concerning the properties of materials and sub-units, such as photomultiplier, high voltage supply, electric fan, and shape and dimensions of the sonde body and light traps, a prototype sonde was designed, figure 1. The sonde body (not shown in figure 1) was made of black PVC. Air is aspired by a miniature fan through 550 mm length of corrugated teflon tubing (25 mm o.d., ca. 21 mm i.d.). A black plastic mantle was heat-shrunken on the teflon tube, which was bent to form a light trap, as shown in figure 1. The air exits from the sonde through an internally blackened curl of corrugated aluminium tubing (600 mm lenght, 40 mm o.d.). The air is forced past a 18 x 18 mm^2 chemiluminescence plate which is mounted 3 mm in front of a side window photomultiplier, in order to maximize the flow speed of the air parallel to the chemiluminescent surface. The photomultiplier socket, miniature high voltage power supply, and signal processing electronics which couple the photomultiplier output to the radiosonde, are encapsuled in a gas-tight PVC box. Under field conditions the chemiluminescence plate can be contacted with a polyethylene flask containing enough liquid water to maintain the temperature of the chemiluminescent surface near 0 °C, if necessary.

3. THE AIR PUMPING SYSTEM

A small vacuum chamber was constructed to test the miniature fans, air flow resistance of the sonde body with the light traps, etc., at reduced pressure. The chamber consisted of two glass cylinders of ca. 10 liter volume each. The test object could be installed in the flange separating the two half-chambers, as depicted in figure 2 for a typical test configuration. The total pressure p, and the differential pressure Δp between the two half-chambers, was measured with MKS Baratrons. Air could be admitted to the upper half-chamber via calibrated flow meters, while the lower half-chamber could be pumped with a 60 m^3/h roughing pump and a Roots pump.

Figure 3 shows the relative static pressure difference $\Delta p/p$ which could be maintained between the two half-chambers by various miniature fans at their nominal supply voltages, as function of total pressure. The fan denoted 361 L (supplier: Micronel) was selected for the prototype sonde. The power consumption at 12 V was found to decrease from 1 W at ambient pressure to a constant level of 0.15 W below 20 torr.

In order to evaluate the volume flow rate through the sonde in its typical configuration throughout the pressure range of 760 - 1 torr, two sets of measurements had to be carried out in sequence: 1. the conductance of the sonde was determined as function of pressure by measuring the pressure drop due to the sonde at given flow rates through the test chamber in the configuration shown in figure 2, but with the fan turned off and covered with a gas-tight lid. 2. the relative differential pressure $\Delta p/p$ maintained by the fan in the test configuration of figure 2 at zero external flow through the test chamber was measured as function of total pressure p. At all pressures the measured $\Delta p/p$ was nearly a factor 2 lower than $\Delta p/p$ in figure 3 which pertains to static conditions (zero conductance). Combination of the results of 1. and 2. yielded the volume flow rate through the prototype sonde as function of total pressure, which is plotted in figure 4. Although the pressure dependence is quite significant, the miniumum flow rate of 4 liter/minute below 2 torr is still far in excess of flow rates which can be achieved with conventional piston or bellows pumps under comparable conditions (e.g. 18). We note in passing that the flow through the sonde as function of $\Delta p/p$ does not obey Poiseuille's law, indicating a change of flow conditions between high and low pressure.

4. DYE TESTING

Although a number of dyes have been reported to emit chemiluminescence when exposed to ozone (8,10), no systematic studies of sensitivity, dye stability, effect of additives etc. seem to have been performed. Rhodamine B, which has been most extensively used for solid surface detection of ozone, fluoresces preferentially in the red spectral region. A modern equivalent of the red sensitive photomultiplier used by Regener in his original sondes would be forbiddingly expensive for routine soundings. One of the cheapest photomultipliers (Hamamatsu 931 B) has a side window and an opaque S-4 photocathode which is most sensitive in the 400 to 500 nm spectral range (cf. the photocathode responsivity curve in figure 5, upper trace). We have therefore tested a number of Lambdachrome laser dyes which fluoresce at shorter wavelengths than rhodamine B.

In a screening test of several solid adsorbents, thin layer chromatographic plates (E. Merck, 0.2 mm silica gel 60 on glass, silanized, without fluorescence indicator) were selected. The dyes were dissolved in methanol or dioxane, as recommended by the supplier. The plates were either dipped into the solution, or wetted by means of a 100 μl syringe, then dried at 60 ^{o}C in vacuo. The dye concentrations were optimized by recording the chemiluminescence intensity of the adsorbed dyes at a given ozone concentration under reproducible conditions for several hours, in parallel with a Bendix ozone analyzer as reference instrument. For most of the dyes concentrations of 2.5 g/liter were adequate, while higher concentrations reduced the fluorescence yield by energy transfer quenching (25). Rhodamine B and rhodamine 6G, which exhibit energy transfer quenching at even lower concentrations, were used at concentrations of 0.3 and 0.07 g/liter, respectively, while the dye BiBuQ was dissolved in dioxane at a concentration of 0.63 g/liter. The dye DPS was insoluble at 6 g/liter in dioxane, and was used as the saturated solution.

Apart from the pure dye solutions, solutions with 4 g/liter gallic acid added as antioxidant were also tested. Gallic acid has been reported to enhance the chemiluminescence yield, and also to increase the lifetime of the dyes which are more or less rapidly destroyed by exposure to ozone (26). In the upper half of figure 5 the relative output of various laser dyes, when excited in solution with a nitrogen laser, is compared with the photocathode sensitivity of the Hamamatsu 931 B photomultiplier used in our sonde. The wavelenght dependence of the output maxima was taken as a measure of the fluorescence maxima, although the fluorescence spectra are typically broader, and sometimes red shifted with respect to the laser output curves (27). The lower half of figure 5 shows a bar graph of the sensitivities (in nA/ppb ozone at ambient pressure, photomultiplier supply voltage - 830 V, dark current at room temperature 0.12 nA), measured in laboratory air with different dyes adsorbed on silica gel as described, both with and without gallic acid additive. The comparison was carried out with the ozone sonde in its standard configuretion, figure 1. Before measuring the sensitivities shown in figure 5, the plates had been pre-ozonized overnight in a rapid flow of laboratory air containing approximately 100 ppb ozone.

According to figure 5 and other tests, coumarin 47 is most suitable for detecting ozone by surface chemiluminescence, although some or the other dyes are also acceptable when activated with gallic acid. However, more important than the high sensitivity of the pale yellow chemiluminescence plates prepared with coumarin 47 (which is much more than sufficient for the purpose of this work) is their excellent stability. No systematic sensitivity losses of coumarin 47 plates were noted, neither after storage times of several months in a desiccator or simply in laboratory air, nor after repeated testing at ozone levels up to several hundred ppb, over periods of several hours.

5. EFFECTS OF HUMIDITY AND FLOW RATE

It is well known that humidity enhances the chemiluminescence efficiency of dyes adsorbed on silica gel (8), although no acceptable explanation has been found so far. The humidity effect was confirmed in this work. While interference by water vapor does not pose a problem in the dry stratosphere, a preliminary attempt at reducing the water vapor effect was made by coating the dried plates with a thin layer of silicon oil, or by treating it with a commercial hydrophobing emulsion. Table 1 shows that the effect could not be eliminated. However, the sensitivity of the plates was significantly reduced by the coating. Several other hydrophobing agents and coatings were applied to the dried plates after deposition of the dyes, but without success. It was noted, however, that the silicon oil treatment had a favorable effect on the relaxation time from high sensitivity in humid air to low sensitivity in dry synthetic air.

The sensitivity reduction observed after coating the plates with silicon oil is thought to be due to the inert oil film which reduces the reaction rate of ozone with the adsorbed dye (i.e. ozone molecules have to diffuse through the oil film in order to react with the dye). The chemiluminescence intensity is governed by the dynamic equilibrium established by (a) the supply of ozone molecules from the gas phase to the surface by turbulent and molecular diffusion, (b) the rate of diffusion of ozone molecules through the inert oil film, and (c) the rate of reaction with the solid surface. The slowness of step (b) is held responsible for the observed loss of sensitivity. However, it also reduces the concentration gradient of ozone in the gas phase close to the surface, thus reducing the critical flow rate which must be exceeded in order to establish constant sensitivity, independent of the pumping speed through the sonde. This is particularly important at low total pressures, cf. figure 4. Indeed, while a reduction of flow rate by a factor of 3 normally resulted in a 10 % decrease of the sensitivity, no such change in sensitivity was observed when coumarin 47 adsorbed on silica gel was coated with silicon oil.

The time constant of the chemiluminescence could be estimated by holding a Pen Ray lamp in front of the intake of the sonde, and switching this photochemical ozone source on and off while observing the photomultiplier signal with a fast Keithley 610 C Solid State Electrometer. The time constant was considerably shorter than one second. This compares very favorably with the substantially longer time constants of electrochemical ozonesondes.

6. LABORATORY TESTS AT SIMULATED STRATOSPHERIC PRESSURES

The pressure dependence of the air flow rate through the sonde has been already discussed in section 3, cf. figure 4. A realistic sensitivity test of the sonde as function of total pressure was carried out in a 425 liter evacuable smog chamber which has been described elsewhere (28). The entire prototype sonde containing a coumarin 47/gallic acid plate coated with silicon oil, was placed inside the smog chamber, which was evacuated, flushed, and then filled with dry synthetic air to the desired total pressure. Absolute ozone concentrations in the smog chamber were measured by long path uv absorption (407 cm folded optical path) at 254 nm, using a stabilized deuterium lamp, monochromator, and solar blind photomultiplier. Ozone in oxygen from an ozonizer could be injected in small increments into the chamber. Two of the calibration curves, which were obtained in 755 torr and 1 torr synthetic air, are plotted in figure 6, B and C. Sensitivities of 0.83 nA/10^{10} molecules cm^{-3} ozone were found at both pressures. Slightly higher sensitivities were determined at intermediate pressures (1.0 and 1.05 nA/10^{10} molecules cm^{-3} ozone at 10 and 50 torr total pressure, respectively). This weak pressure dependence was confirmed independently by injecting ca. 1.5×10^{13} cm^{-3} ozone into the smog chamber at a total pressure of 1 torr, then increasing the total pressure within a few minutes to a final value of 755 torr. During the pressure change, the signal from the sonde and the total pressure were recorded simultaneously. The chemiluminescence intensity showed little pressure dependence, passing through a weak maximum in the 10 - 50 torr range. This procudure is, however, less reliable than a point-by-point calibration at different pressures, due to possible concurrent losses of ozone during the pressure change.

After several hours of sonde testing in dry synthetic air, the smog chamber was evacuated and filled with laboratory air. The calibration curve A, which was obtained immediately afterwards, is shown for comparison with the dry air calibration curves B and C in figure 6. Apart from confirming the humidity effect, the calibration curve in laboratory air also shows that there is no memory effect when changing from dry synthetic air to humid laboratory air.

7. SUMMARY AND OUTLOOK

We have constructed and tested in the laboratory the prototype of a light-weight ozone analyzer suitable for balloon soundings of the stratosphere. The instrument measures surface chemiluminescence due to the reaction of ozone with coumarin 47 and gallic acid deposited on a silica gel plate and coated with silicon oil. The sensitivity is nearly pressure independent in the range of 760 - 1 torr total pressure, and more than adequate for ozone detection up to 45 km altitude. A time constant of less than 1 second has been estimated. The weight of the prototype, including 10 gas-tight lithium batteries for several hours operation is about 700 g. Depending on the influence of temperature on the chemiluminescence reaction, which remains to be determined, a slightly higher operational weight of the finalized sonde must be envisaged. A preliminary field test of the sonde will be carried out at Hohenpeißenberg Observatory in the near future.

Further laboratory work is needed in order to understand and eliminate the humidity effect. While hydrophobic treatments of the silica gel plates after dye deposition and drying were uneffective, mixing of a coumarin 47 solution with a hydrophobing emulsion ("Silicon-Imprägnierungsmittel" of unknown composition, Carl Roth KG, Cat.-No. 0027) before deposition and drying was found to reduce the humidity effect significantly.

Table 1: Sensitivity ratios of various chemiluminescent dyes in dry synthetic air and humid laboratory air (ca. 7 torr water vapor pressure). All plates activated with gallic acid. Column A: plates coated with silicon oil ("Rotisilon A", Carl Roth KG, Cat. No. 0831). Column B: plates treated with hydrophobing emulsion and dried ("Silicon-Imprägnierungsmittel" of unknown composition, Carl Roth KG, Cat. No. 0027).

Dye	Sensitivity ratios in dry and humid air	
	Column A	Column B
Rhodamin B	0.46	0.46
Rhodamin 6G	0.47	0.45
Coumarin 153	0.57	0.50
Coumarin 307	0.55	0.53
Coumarin 47	0.50	0.46 (0.40) [a]
BiBuQ	0.66	0.50

(a) Coumarin 47 without gallic acid

ACKNOWLEDGEMENTS

This work is supported by the Bundesminister für Forschung und Technologie. We gratefully acknowledge the continuing support and expertise by our collaborators of Hohenpeißenberg Observatory, and the skills of our electronics workshop headed by B. Fassbender.

REFERENCES

1 J.N. Pitts, J.M. McAfee, W.D. Long, A.M. Winer, Environ. Sci. Technol. 10, 787 (1976)
2 D.L. Flamme, Environ. Sci. Technol. 11, 978 (1977)
3 M.H. Proffitt, R.J. McLaughlin, Rev. Sci. Instr. 54, 1719 (1983) and references cited therein

4 K.H. Becker, A. Heindrichs, U. Schurath, Staub-Reinhalt. Luft 35, 326 (1975)
5 G.W. Nederbragt, A. Van der Horst, J. Van Duijn, Nature 206, 87 (1965)
6 A. Fontijn, A.J. Sabadell, R.J. Ronco, Anal. Chem. 42, 575 (1970)
7 V.H. Regener, J. Geophys. Res. 65, 3975 (1960)
8 J.A. Hodgeson, K.J. Krost, A.E. O'Keeffe, R.K. Stevens, Anal. Chem. 42, 1795 (1970)
9 F. Chisaka, S. Yanaglhara, Anal. Chem. 54, 1015 (1982)
10 J.D. Ray, D.H. Stedman, G.J. Wendel, Anal. Chem. 58, 598 (1986)
11 A.W. Brewer, J.R. Milford, Proc. Roy. Soc. London A 256, 470 (1960)
12 W.D. Komhyr, Tech. Rep. ERL 200-APCL 18, NOAA, Boulder, Colorado, 1971; Ann. Geophys. 25, 221 (1969)
13 W.Attmannspacher, R. Hartmannsgruber, P. Lang, Meteorol. Rdsch. 37, 193 (1984)
14 J. Logan, ESF Workshop Dubrovnik 11 - 17 May 1986
15 D.C. Thornton, N. Niazy, J. Geophys. Res. 87, 8943 (1982)
16 M.P. McCormick, T.J. Swissler, E. Hilsenrath, A.J. Krueger, M.T. Osborn, J. Geophys. Res. 89, 5315 (1984)
17 R.A. Barnes, A.R. Bandy, A.L. Torres, J. Geophys. Res. 90, 7881 (1985)
18 V.H. Regener, J. Geophys. Res. 69, 3795 (1964)
19 W.S. Hering, H.U. Dütsch, J. Geopyhs. Res. 70, 5483 (1965)
20 E. Hilsenrath, L. Seiden, P. Goodman, J. Geophys. Res. 74, 6873 (1969)
21 E. Hilsenrath, P.T. Kirschner, Rev. Sci. Instr. 51, 1381 (1980)
22 NASA, Ozone Assessment Report, in press
23 J. Werner, K.W. Rothe, H. Walther, in "Atmospheric Ozone", D. ReidelPublishing Co., Dordrecht, 1985, p. 446
24 E.H. Steinberger, J. Sivan, J. Neumann, J. Geopyhs. Res. 72, 4524 (1967)
25 Y. Liang, P.F. Moy, J.A. Poole, A.M. Ponte Goncalves, J. Phys. Chem. 88, 2451 (1984)
26 D. Bersis, E. Vassiliou, Analyst 91, 499 (1966)
27 "Laser Dyes - Properties of Organic Compounds for Dye Lasers", Mitsuo Maeda, Academic Press, Tokyo, 1984
28 K. Henrich, H. Lippmann, U. Schurath, W. Wendler, pp. 218-227 in: Physico-Chemical Behaviour of Atmospheric Pollutants, 2nd Europ. Sympos., Varese 1981. Proceedings edited by B. Versino and H. Ott, D. Reidel Publishing Co., Dordrecht, 1982

FIGURE CAPTIONS

Figure 1: Exploded view of prototype sonde (PVC body not shown). 1, photomultiplier; 2, chemiluminescent surface; 3, electric fan; 4, encapuled high voltage supply and electronics; 5, air intake light trap (inside: corrugated teflon); 6, exhaust light trap (corrugated aluminium); 7, water reservoir in thermal contact with chemiluminescent surface (optional, see text).

Figure 2: Vacuum test chamber. In the test configuration shown the pressure difference Δp between intake and exhaust of the sonde, sustained by the electric fan under operational conditions, could be measured as function of total pressure.

Figure 3: Static differential pressures $\Delta p/p$ maintained by commercial electric fans, as function of total pressure, at nominal voltages.

Figure 4: Volume flow rate through the prototype sonde, as function of total pressure.

Figure 5: Upper: photocathode responsivity of 931 B photomultipliers, and relative laser output of Lambdachrome dyes (see below) in solution, as function of wavelength. Lower: sensitivities measured with Lambdachrome dyes deposited on silica gel, as described in text, without gallic acid (black bars), and activated with gallic acid.

Figure 6: Calibration curves of the prototype sonde. A, at 755 torr in laboratory air; B and C, in dry synthetic air at 755 and 1 torr, respectively.

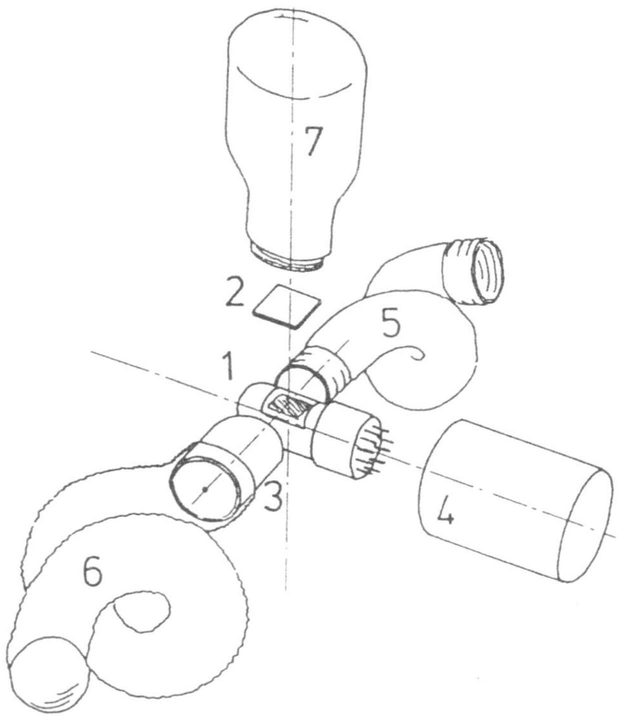

Figure 1

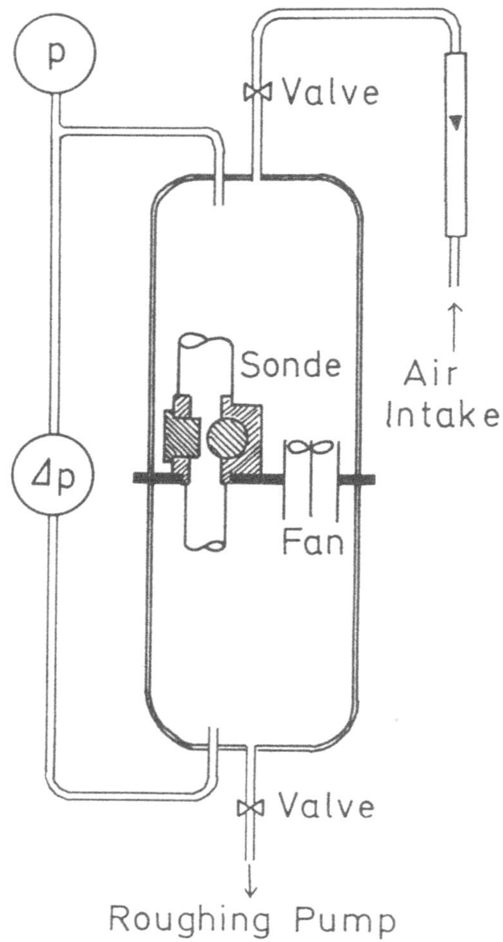

Figure 2

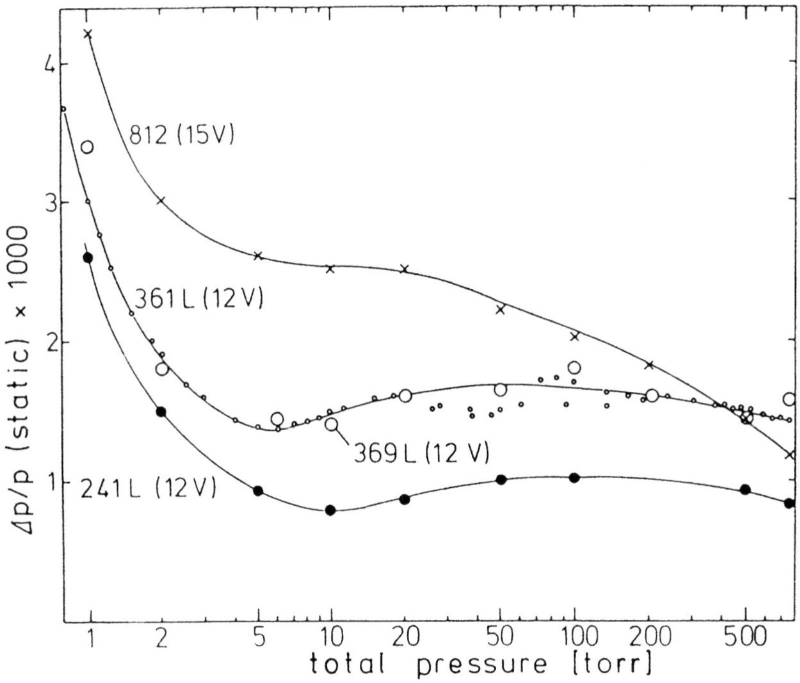

Figure 3

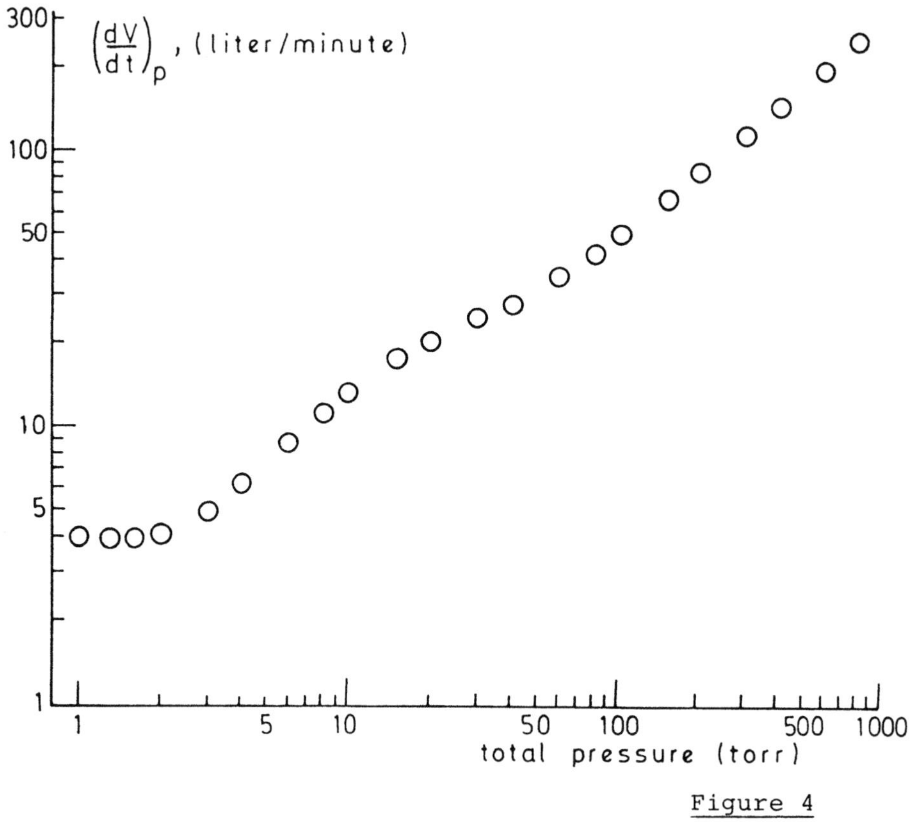

Figure 4

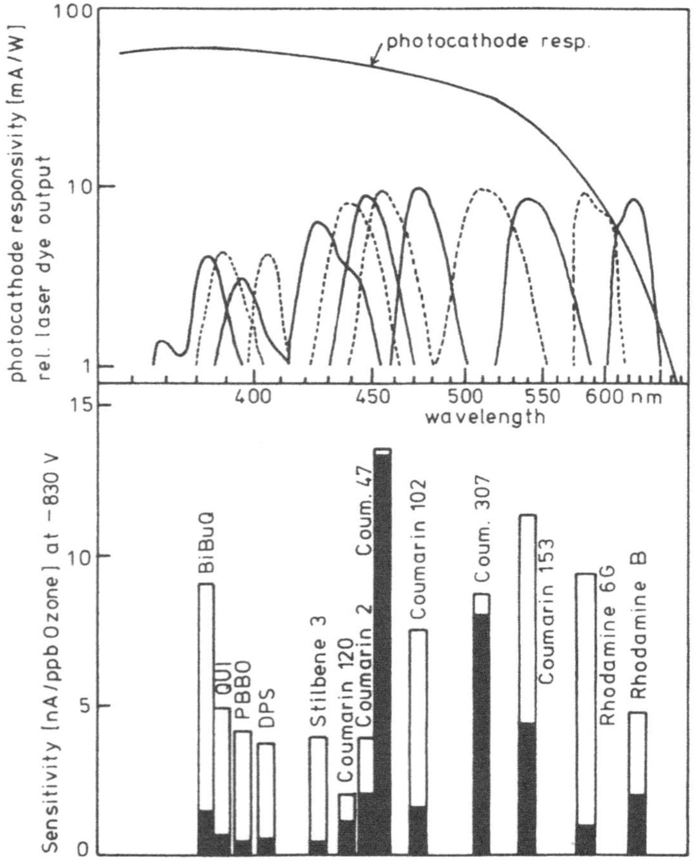

Figure 5

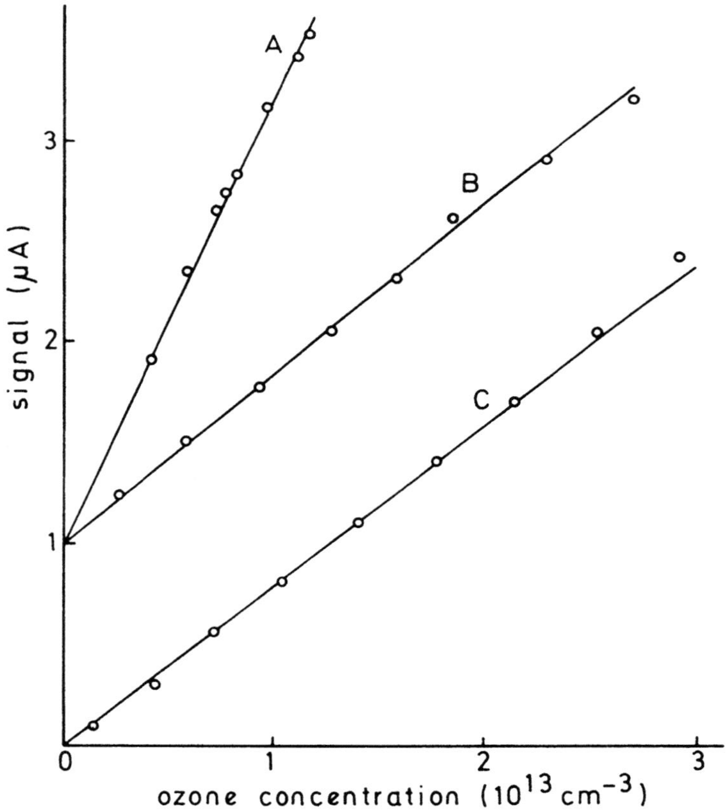

Figure 6

H_2O_2 IN SOLID PRECIPITATION

Albrecht Neftel and Andreas Sigg
Physics Institute, University of Bern, CH-3012 Bern, Switzerland

Peter Jacob
Chemistry Department, University of Dortmund, D-4600 Dortmund, FRG

Abstract

H_2O_2 concentrations up to 20 µeq/l were found in snow from alpine regions. To extract useful information for atmospheric chemistry models, the snow concentration has to be translated into atmospheric concentrations. To do so, two questions have to be answered first: 1) The stability of H_2O_2 in snow and ice, 2) the scavenging ratio c_{air}/c_{snow} and its dependence upon the formation of the snow.

To investigate the first question, we measured during winter 1985/1986 the H_2O_2 in snow profiles at the experimental snow field of the Snow and Avalanche Research Institute, Weissfluhjoch-Davos, Switzerland, situated at an altitude of 2500 m.a.s.l. The results show a degradation of H_2O_2 in the snow with increasing age. The degradation is linked to the exposure of the snow to sunlight. The second question has only recently been adressed by measuring the H_2O_2 concentration in the gas phase, fog droplets and precipitation during a winter precipitation event. The H_2O_2 concentration in the fog droplets and the precipitation is determined by both, the gasphase concentration and chemical reactions occuring in the cloud droplets.

1.) Introduction

In the past few years increasing interest is drawn to atmospheric hydrogen peroxide. H_2O_2 is the most important species for the oxidation of SO_2 in cloud droplets (Penkett 1979, Calvert et al. 1985). H_2O_2 is also an important species within the odd hydrogen cycle in atmospheric chemistry, it acts as a source for OH radicals and as a sink for HO_2 radicals. The recombination of two HO_2 radicals is the most important source of H_2O_2. Reliable data on atmospheric H_2O_2 are an important tool to verify model calculations.
We found that H_2O_2 is one of the dominant trace species in polar ice samples (Neftel et al. 1984). It can be detected in ice samples aged by more than 10,000 years. Ice cores from

polar locations offer an unique chance to evaluate the load of atmospheric impurities far back in time. In order to be able to link the measured concentrations in the ice cores to the atmospheric concentration at the time of precipitation formation, both the behaviour in snow and ice and the scavenging ratio $c_{air}/c_{prec.}$ must be known. Both questions are adressed in this paper. The behaviour of H_2O_2 after deposition was analysed in a seasonal snow cover located at the Weissfluhjoch at an elevation of 2540 meter above sea level (m.a.s.l.), situated in the eastern part of Switzerland. Three profiles were taken through the whole snow cover during the winter 1985/1986.
The investigations of the scavenging behaviour of H_2O_2 are embedded in a larger project of the laboratory of atmospheric physics at the ETH Zürich studying the scavenging behaviour of aerosols (Schumann et al, 1986).

Analytical procedure

To determine the H_2O_2 concentration in aqueous solution, we used the chemiluminescencent reaction between hydrogen peroxide and bis-(2.4.6 - trichlorphenyl)oxalate (TCPO) in a borate buffer medium of pH 8 (Klockow and Jacob, 1986). By using segmented flow technique the sample size could be reduced to 200µl. Gaseous H_2O_2 in ambient air is determined by means of a cryogenic sampling technique. Ambient air is drawn through a glass cylinder kept between -40°C and -50°C with a flow rate of 1.4m^3/h. Ca. 40% of the atmospheric water vapor condensates at the cylinder surface. Within 10% error the ratio of H_2O_2 to H_2O molecules is the same in the collected ice as in the air drawn through the cylinder (Jacob et al 1986). The ambient H_2O_2 air concentration is determined as

$$c_{H2o2\ air} = (c_{H2O2}/c_{H2O})_{cond} * c_{H2O\ air}$$

The water concentration in the air is given by the relative humdity and the temperature.

Sampling procedure

The snow studies were done at the experimental snow field of the Swiss Snow and Avalanche Research Institute, Davos-Weissfluhjoch.
Profiles through the snow cover are taken routinely each second week from the first snowfall until end of june, in order to supervise the metamorphosis of the snow for avalanche prediction. Grain size, hardness, density and temperature profiles are measured. The surface is marked with a thread, so that corresponding layers can be identified in subsequent profiles. The samples were taken with a polycarbonate plate fixed on a mast with meter measure, in order to provide vertically equally spaced samples. Snow samples of 2 to 3 cm thickness were taken with polyethylene beakers and brought frozen to the laboratory. Snow samples stored at -20°C in the dark did not change their H_2O_2 concentration during the time period of our investigation. Profiles through the snow cover

were taken the 30th of january, the 28th of february and the 15th of april.
From the 15th of december to the end of january each morning a surface snow sample was taken and stored immediatly in the dark at -20°C. Aliquots could be analysed for H_2O_2. These samples do not represent the continuous snow record of this time period, because only a surface layer of approx. 3cm was sampled.

The scavenging behaviour of H_2O_2 during winter precipitation events was studied at the Rigi (1600m.a.s.l.), a mountain which rises 1200m above the Swiss midlands, between the lake of Lucerne and lake of Zug. During most of the winter precipitation events the station is located within the precipitating cloud.
The ambient air was first sucked through a fog water collector that collects fog droplets with diameter greater than 10um (Däumer 1984). The small sample size of 200µl allowed a time resolution of one hour for air and cloud droplets measurements. Continuous measurements of temperature and relative humdity were provided by by the ETH group. The precipitation was collected near the station on a polethylene shell. All analysis were performed directly at the station.

Results and Discussion

Fig.1 shows the temperature profiles in the snow cover. Due to geothermal heat flux the bottom is maintained near zero degree during the whole winter . The temperature decreases then with increasing snow height to the mean air temperature, which increases with the length of the days. The diurnal temperature variations are strongly damped in the snow cover. During the examined time period the snow never reached zero degree celcius, except for a few days in april where thin melt layers were formed at and near the surface. The strong temperature gradients present in such seasonal snow covers enhances recristallisation processes.
Fig. 2 shows the H_2O_2 concentration in the three profiles as function of the accumulated precipitation in mm water equivalent. Three distinct peaks are visible in all three profiles at 190mm, 280mm and 420 mm water equivalent. It is clearly visible that the H_2O_2 concentration decreases with increasing storage time in the snow cover. The structure in the profiles is conserved and no broadening of the peaks is visible. This confirms the results of a pilot study carried out during winter 84/85 (Neftel 1985).
The triangles in fig. 2a shows the concentration of the surface snow samples. Unfortunatly it cannot be reconstructed at which time they are situated in the corresponding precipitation event. The surface samples corresponding to the peaks show equal or higher values than found in the first profile. We have therefore to assume that the H_2O_2 concentrations recorded in the first profile are already lowered compared to the original precipitation.
Fig. 3 shows the development of the total accumulation during the winter 85/86. Comparing the H_2O_2 concentration with the

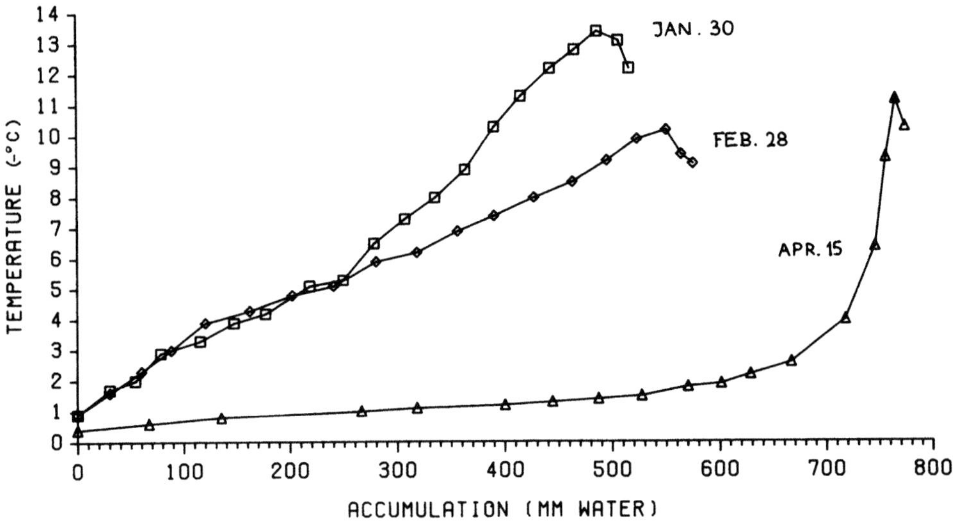

Fig. 1 Temperature distribution in the three profiles.

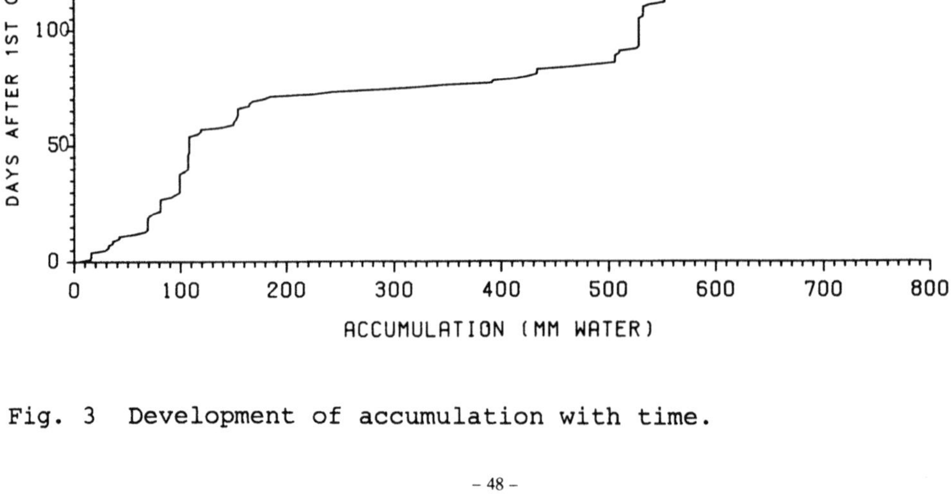

Fig. 3 Development of accumulation with time.

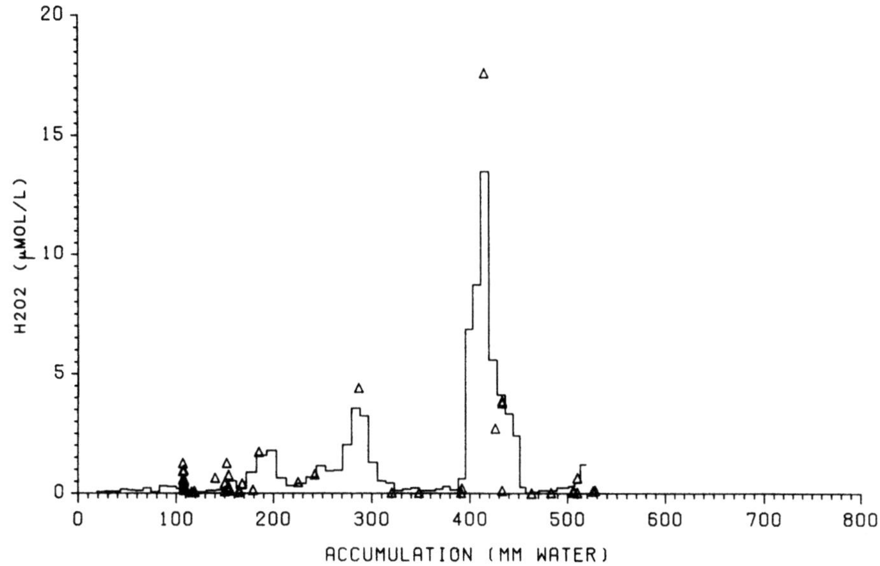

Fig. 2a H_2O_2 concentration in the profil of the 30th of january. The surface snow samples are plotted with triangles at the corresponding accumulation.

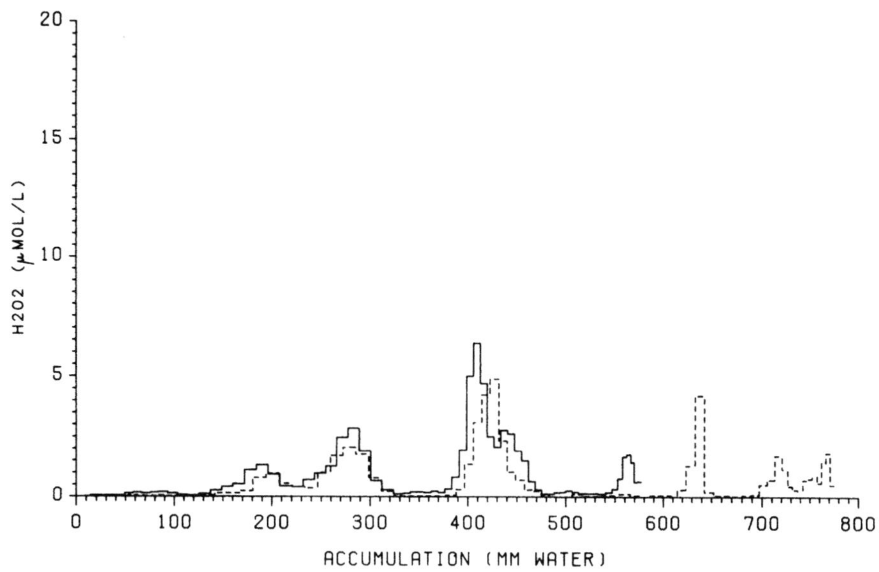

Fig. 2b H_2O_2 concentration in the profile of the 28th february and 15th of april 1986.

precipitation intensity shows no correlation. The surface samples at the other hand exhibit a decreasing trend on subsequent days without precititation. What are possible causes for the observed degradation of H_2O_2 ?
We discuss three possibilities:

a) Photolytical decay
b) Chemical reaction or catalytic destruction
c) Recristallisation leading to a loss of adsorbed H_2O_2

To a)
The measured degradation of H_2O_2 is related to the amount of light. Subsequent measurements of the surface snow samples without new precipitation in between exhibit decreasing concentrations. The peak at 564mm water equivalent recorded in the second profile, which corresponds to the snow fallen in the last week of february, vanished entirely in the third. During the first two weeks of march, almost no precipitation was recorded, so that the exposure of the surface snow to sunlight was large. High H_2O_2 concentrations are always associated with high precipitation rates, where most of the snow is rapidly protected from the sunlight.
We favorize the photolytical decay being responsible for the largest part of the degradation of H_2O_2 in snow. Measurements of the photolytical decay rate are only available for the gas and liquid phase, where similar values have been obtained (Graedel and Weschler 1981). A reasonable mean for a 24 hour mean value for the gas phase is $2*10^{-6}s^{-1}$ for a clear day (Chameides and Davis, 1983; Jason report 1985) corresponding to a halflife of 4 days. The dissociation energy of the reaction:

$$H_2O_2 \quad -> \quad 2\ OH$$

is 200kJ/mol, thus the threshold wavelength for the photolytical dissociation of H_2O_2 in two OH radicals is 580nm. In the ice the rate will be most probably lower due to an enhanced recombination rate of the produced OH radicals. Thus a halflife of 4 days can be regarded as a lower limit for snow receiving full radiation. OH radicals are possibly rather mobile in the ice and can either diffuse into the air or react with other species in the ice. Strong radiation heats the snow surface to temperatures near 0^oC, so that liquid water may be present, where such reactions could occur. There is a number of species present in the ice not measured in this study, such as aldehydes, which at least in the liquid phase have high rate constants for reactions with H_2O_2.

In fresh snow the light is exponentially attenuated. The extinction coefficient depends upon the wavelength and is minimal for blue light. At 1m depth the light is attenuated to 1% and is concentrated in a narrow interval around 440nm (Schwertfeger and Weller 1977). Photolytical disintegration of H_2O_2 can therefore only be efficient in the first half meter of a snow cover. But in greater snow depths the H_2O_2

concentration still decreases slowly. This indicates that also other mechanisms contribute to the decrease.
It can be expected that the H_2O_2 concentration in the atmosphere increases with increasing solar radiation (Logan et al, 1981). Nevertheless the highest concentrations are found in january precipitation. A restriction has to be made. The vertical exchange of air masses increases with increasing solar radiation. More species consuming H_2O_2 such as SO_2 or competing with HO_2 as NO can be brought at higher elevations. This can also be seen from the increasing $SO_4^{2-}$and NO_3^- concentrations in the snow (Neftel, this conference).
On clear days in late winter and spring the snow surface reaches temperatures near zero degree celcius due to the increased global radiation as it can be seen in fig.4. In such conditions the photolytical decay is enhanced and, as mentioned above, the produced OH radicals have the possibility to react either in the liquid water contained in the snow or eventually in the liquid like layer at the surface of the snow crystals (see below). Thus, initially present high H_2O_2 concentrations in the snow cannot be maintained.

To b)
At each surface of ice crystals exists a thin layer of an amorphous, liquid like structure. The thickness increases exponentially going toward 0°C. Golecki and Jaccard (1978) measured at -10°C a thickness of 40nm and at -1°C 80nm. It can be speculated that similar chemical reactions occur in this disordered structure as in the liquid phase but with much smaller rate constants due to the lower mobility of the molecules and the lower temperature. This would enable e.g. the oxidation of S(4) to S(6) by H_2O_2 and the catalytic destruction of H_2O_2 on surfaces of dust particles. The former should be accompanied by a corresponding increase of the sulfate concentration and should be enhanced at higher temperatures. This could not be observed as the data compiled in table 1 show. The rate of degradation is not related to the temperature in the snow cover. Nevertheless catalytic destruction of H_2O_2 on surfaces, especially on transition metals is very effective. This mechanism might be responsible for the decreasing H_2O_2 content at greater depths (>0.5m) where no light is avaible for the photolytical decay.

To c)
Temperature gradients in the snow cover enhance evaporation and recondensation processes leading to a considerable loss of the specific surface in few days. In the case of gases which are insoluble in the ice, the recristallisation process leads to a complete loss of the adsorbed gases, as it was measured e.g. for CO_2 (Schwander 1980). This is not observed for H_2O_2. At the time of the measurements the surface snow samples were completely recristallised and still exhibited higher concentrations than the corresponding layers in the snow field. We have measured for H_2O_2 a surprisingly high partition coefficient c_{water}/c_{ice} of 0.014 (unpublished data). H_2O_2 is therefore rather easily built into the ice matrix.

TABLE 1

Accum. mm w.equ.	Profile No	Depth below surface cm	age (days)	H_2O_2 conc. ppB (w/w)	Degrad.rel. to 1st profile	Degrad.rel. to 2nd profile	Sulphate conc. (SO_4^{2-}-S) ppB (w/w)	Temp. of the snow layer at the sampling date
415	1	45	11	458	1	-	70	-11.3
409	2	65	40	216	0.47	1	69	-7.6
427	3	112	85	166	0.37	0.79	78	-1.3
284	1	93	16	121	1	-	34	-6.5
283	2	98	55	97	0.8	1	28	-5.9
279	3	150	100	70	0.58	0.72	25	-1.1
198	1	120	19	61	1	-	39	-4.4
190	2	122	58	45	0.75	1	52	-4.7
201	3	164	103	33	0.55	0.75	46	-1.0

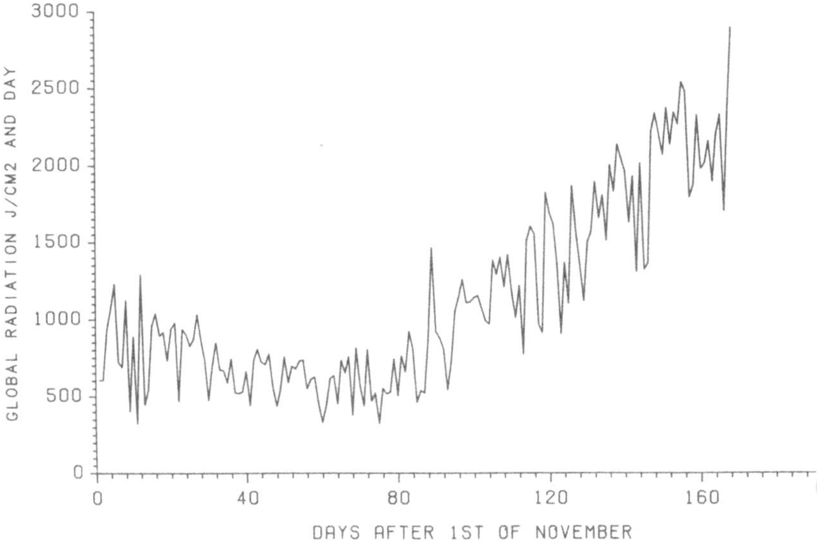

Fig. 4 Global radiation at the Weissfluhjoch.

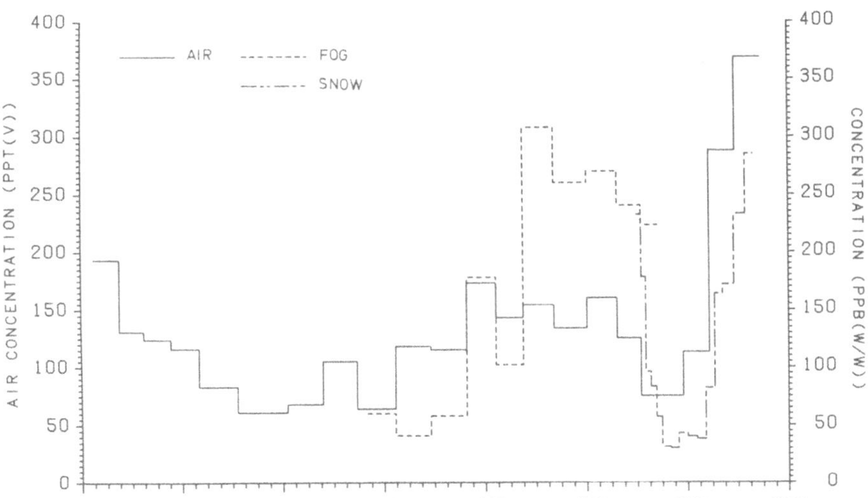

Fig. 5 H_2O_2 concentration of the gasphase, fog and precipitation at the Rigi during 3nd and 4th of april 1986.

We conclude therefore that during recondensation processes the H_2O_2 is not changed. The mixing length of the recristallisation process cannot be larger than a few cm, otherwise the peaks would have spread out, going from one profile to the next.

Our measurements showed clearly that H_2O_2 is not a conservative tracer in a seasonal snow cover of mid-latitudes. The concentrations found in aged snow covers does not reflect directly the original concentrations. The largest contribution to the degradation comes most probably from photolytical decay. This process must be investigated in more details in future studies in order to be able to reconstruct original concentrations e.g. in polar ice cores. The large variability that we found in fresh surface samples points toward a large variability in the precipitation, which must be linked to similar variations in the H_2O_2 concentration of the precipitating clouds.

Last winter we started a project to study the relation between the atmospheric H_2O_2 concentration and H_2O_2 in the solid precipitation. The investigations were made at Rigi-Staffel, at an elevation of 1600 m.a.s.l. The Rigi, located in Central Switzerland towers up 1000m over the Swiss Midlands. The station is located at the north west face, which drops steeply to the lake of Lucerne. For alpine conditions precipitation fronts reach the Rigi mountain relatively undisturbed. Fig.5 shows the measured H_2O_2 concentration in air, fog and precipitation during the recorded event at the 3nd and 4th of april.
Theoretically the ratio of the air concentration to the precipitation concentration (scavenging ratio) can be caracterized by two extreme cases:
a) liquid droplets in thermodynamical equilibrium with the gas phase. The scavenging ratio is given by the Henry constant.
b) snow flakes entirely grown over the gas phase. H_2O_2 molecules are trapped by co-condensation (Jacob et al. 1986). The same ratio of H_2O_2 to H_2O is adjusted in the solid phase as in the gas phase. Rimed snow flakes are between the two cases. The difference between the two types of incorporation of H_2O_2 is large and renders difficult the interpretation of measured concentrations in the precipitation.
The concentration measured in the cloud droplets is clearly below the equilibrium concentration calculated as

$$c_{fog} = H * c_{atm}$$

$$H = 9.7*10^4*\exp[6600(1/T-1/298)] \text{ M/atm.}$$

(Martin and Damschen 1981)

There are two possible explanation for this discrepancy:
a) an overestimation of the gas phase concentration due to small droplets (<10µm) passing the fog collector and trapped in the cryogenic sampling system.
b) decrease of the H_2O_2 concentration in the cloud water during the sampling period of one hour.

It is possible that H_2O_2 partially decays already catalytically during the collection process. To minimize this danger the whole fog collector was constructed of teflon. H_2O_2 in the droplets is also consumed for the S(4) oxidation. Kelly et al (1985) reported that as long as SO_2 concentrations exceed H_2O_2 in the gasphase, only low concentrations of H_2O_2 were observed. SO_2 concentrations recored at the Jungfraujoch (3500 m.a.s.l.) are generally above 1.0µg/m^3 (NABEL), thus higher than the H_2O_2 concentration measured at the Rigi. The precipitation started with a light rain fall, which turned later in graupel snow, haevily rimed snow and at the end light rimed snow. Because the H_2O_2 concentration measured in the first fraction of the precipitation is equal to the concentration previously measured in the fog, we assume that the H_2O_2 in the collected fog water did not decrease substantially during the sampling time. The fast decrease in the concentration can be partially explained by an increased portion of the precipitation grown over the gas phase.

Large changes on a small time scale were registered versus the end of the campaign, both in the air and precipitation concentration. Astonishingly these changes were not associated with similar changes in one of the other parameters measured by the ETH group (aerosol concentration and size distribution, temperature, humdity). It is very important to monitore in future studies the SO_2 concentration in the air and the sulfur species in the precipitation in order to estimate the H_2O_2 consumption in the droplets.

Acknowledgments

We thank Prof C. Jaccard and H. Beck from the Swiss Snow and Avalanche Research Institute for the excellent support, Prof. H. Oeschger for the opportunity to do this work and his continuous interest. We thank Prof. A. Waldvogel, Bettina Zinder and Thomas Schumann, all at the laboratory for atmopsheric physics at the ETH Zürich, for the excellent support during the Rigi campaign. Claudia Mäder, Ursula Michel, Bruno Gisler and David Kistler, all at the EAWAG, helped with the measurements. Yves Page was so kind to provide us the surface snow samples. The project was supported by COST action 611.

References

Calvert J.G, Lazrus A., Kok L., Heikes B.G., Walega J.G., Lind J and Cantrell C.A. (1985)
Chemical mechanisms of acid generation in the troposphere
Nature 317 27-35

Chameides W.L. and Davis D.D (1982)
The free radical chemistry of cloud droplets and its impact upon the composition of rain.
Geophys.Res 87, 4863

Däumer B. (1984)
Entwicklung, Eichung und Erprobung eines Nebelsammlers
Diplomarbeit. Universität Dortmund unpubl.

Golecki I and Jaccard C. (1978)
Intrinsic surface disorder in ice near the melting point.
Journal of Physics C: Solid state physics 11 4229-4237

Graedel T.E. and Weschler C.J. (1981)
Chemistry within aqueous atmospheric aerosols and raindrops.
Reviews of Geophysics and Space Physics 19 503-539

Jacob P., Tavares T.M. and Klockow D. (1986)
Methodology for determination of gaseous hydrogen peroxide in ambient air.
(in press) Fres. Zeitschrift. f.Chemie

JASON (1985)
Acid deposition
The mitre corporation 1820 Dolley madison Boulevard McLean, Virginia 22102, Chapter 10

Kelly T.J., Daum P.H. and Schwartz S.E. (1985)
Measurements of peroxides in cloudwater and rain.
Journal of Geophys. Res. 90 7861-7871

Klockow D. and Jacob P. (1986)
The preoxyoxalate chemiluminescence and its application to the determination of hydrogen peroxide in precipitation in
Chemistry of multiphase atmospheric systems (W. Jaeschke Ed.)
NATO ASI Vol G6 117-128

Martin L.R. and Damschen D.E. (1981)
Aqueous oxidation of sulfur dioxide by hydrogen peroxide at low pH.
Atmospheric Environment 15 1615-1621

Nabel (1984)
Messresultate des nationalen Beobachtungnetzes für Luftfremdstoffe.
Bundesamt für Umweltschutz, Bern , Switzerland

Neftel A., Jacob P. and Klockow D. (1984)
Measurements of hydrogen peroxide in polar ice samples.
Nature 311 43-45

Neftel A. (1985)
Snow as recording device for metastable compounds in the atmosphere: The example of H_2O_2 .
Proceedings of a workshop within concerted action COST 611
"Physico-chemical behaviour of atmospheric Pollutants" RIVM, Bilthoven, Netherland 64-69

Penkett S.A., Jones B.M.R., Brice K.A. and Eggleton A.E.J. (1979)
The importance of atmospheric ozone and hydrogen peroxide in cloud and rainwater.
Atmospheric Environment 13 123-137

Schumann T., Zeinder B. and Waldvogel A. (1986)
Measurements of aerosol and hydrometeor concentrations and their chemical composition during winter precipitation along a mountain slope.
2nd International Aerosol conference, Berlin Sept. 1986

Schwander J. (1980)
CO_2 - Gehalt von Schnee- Firn und Eisproben.
Lizentiatsarbeit Physikalisches Institut, Universität Bern, Switzerland.

Schwartz S.E. (1986)
Mass-transport considerations pertinent to aqueous phase reactions of gases in liquid-water clouds in Chemistry of multiphase atmospheric systems (Ed. W. Jaeschke)
NATO ASI Series Vol G6 Springer Verlag.

Schwerdtfeger P. and Weller G.E. (1977)
Radiative heat transfer processes in snow and ice.
Meteorological studies at Plateau Station, Antarctica
Antarctica Res. Serie Vol 25 American Geophysical Union, Washington DC 35-39

DETECTION OF NITRATED AND OXYGENATED POLYAROMATIC HYDROCARBONS (PAH) IN SUSPENDED PARTICULATE MATTER SAMPLED IN URBAN AREAS AND THEIR RELATION WITH ANTHROPOGENIC EMISSION AND PHOTOCHEMICAL SMOG FORMATION

A.CECINATO,E.BRANCALEONI,C.DI PALO,R.DRAISCI and P.CICCIOLI
Istituto sull'Inquinamento Atmosferico del CNR - Area della Ricerca di Roma
Via Salaria km 29,300,C.P. 10, 00016 Monterotondo Scalo - ITALY

Summary

A method for the evaluation of nitrated and oxygenated polyaromatic hydrocarbons (nitro - PAH,oxy - PAH) in suspended particulate matter is described. It is based on the sequential use of low and high resolution liquid chromatographic techniques for separating the soluble organic fraction (SOF) into components characterized by different polarity. Nitro - PAH have been detected in the moderately - polar fraction using HRGC combined with specific detection (thermoionic and electron capture detection). Positive identification has been achieved by capillary GC combined with mass spectrometry working either in the scan mode or in the selected ion detection (SID) mode. Oxy - PAH have been identified most by GC -MS. The results obtained on different samples collected in a urban area, indicated that, in addition to 1-nitropyrene emitted from Diesel exhaust,2-nitropyrene and 2-nitrofluoranthene can be present in urban samples. The possible origin of these components is discussed by comparing the results obtained on urban samples collected in different conditions,with those obtained by analyzing anthropogenic emissions.

1. INTRODUCTION

The discovery that particulate matter,emitted from certain anthropogenic sources,is strongly mutagenic in the Ames assay carried out without microsomal activation,has provided compelling evidence that other genotoxic species than polyaromatic hydrocarbons (PAH) are present in the soluble organic fraction (SOF) extracted from environmental samples. The results of detailed investigations (1 - 4),carried out on Diesel particulate matter,has confirmed that directly acting mutagenic activity has to be attributed most to PAH containing polar groups in the molecule. In particular,the presence of moderately - polar PAH, containing nitro groups attached to the benzenic rings,makes a significant contribution to the direct mutagenic activity of the sample. Interest on nitro - PAH in general ,and particular nitropyrene/fluoranthene isomers,has been increased by recent findings that these compounds can induce sarcomas (5) and mammalian tumors in rats (6). The concern about possible health effects caused by moderately - polar PAH,and specifically nitro-PAH, has been strengthened by the observation (7) that other nitro-PAH than those present in Diesel emission can be found in suspended particulate matter sampled in urban areas. It has been suggested (7) that some nitro-PAH,specifically 2-nitrofluoranthene and 2-nitropyrene,can be formed as a consequence of gas with particles reactions taking place when photochemical smog develops in the atmosphere. The fact that photochemical smog,through the increase of nitric and nitrous acid or nitrogen and oxygen containing radicals,can increase the mutagenic activity of suspended particulate matter has called for detailed investigations aimed to specifically detect these nitrated components. In this work,a method for detecting moderately - polar PAH at trace levels is described. It consists in the sequential use of low and high resolution liquid chromatography for the fractionation of the SOF. Components,characterized by different polarity,are separated and enriched during the liquid chromatographic processes. Selective detection of nitro-PAH is carried out by connecting HR capillary columns to an electron capture detector (ECD),a thermoionic detector (TID) and a mass spectrometer (MS). Mass spectrometry is used either for the identification of individual components through their mass spectra or for the quali/quantitative evaluation of specific isomers by working in the selected ion detection (SID) mode. Oxygenated PAH are identified by GC-MS employed in the scan mode. The full procedure has been applied to dust samples collected in downtown Rome when different conditions were observed in the city. The results confirmed that 2-nitrofluoranthene and 2-nitropyrene can be actually present in particulate matter and they are not originated by Diesel emission .While the content of 1-nitropyrene is strickly related to the emission strength of automobile exhaust,the amount of 2-nitrofluoranthene/pyrene isomers is not related to this source. The possible ways through which these nitrated compounds can be immitted into the atmospheric environment are evaluated by comparing the content of nitro-PAH found in different urban samples with that measured in particulate matter emitted by different anthropogenic sources.

2. EXPERIMENTAL

Atmospheric dust samples were all collected on the roof of a building (25 m from the ground) located near the historical centre of Rome where heavy traffic can develop during rushing hours. Particles were collected on a glass fibre filter (20 x 25 cm) connected to an high - volume Staplex pump assembled in a sampling shelter built according to the EPA recommendations (8). The sampling site was placed in elevated position in order to reduce,as much as possible ,the collection of large particles suspended by human activities and to avoid the exposure of the filter to direct diesel emission. This choise was necessary because the sampling shelter was unable to discriminate particles having a diameter smaller than 100 μm. A detailed description of the collection methods previously used for sampling Diesel soot and industrial emission particulate is reported in ref. 9. All samples were extracted for 6 h in a Soxhlet apparatus using 50 mL of a benzene:methanol mixture, capable to remove efficiently both polar and non-polar constituents from particulate matter. After evaporation to dryness (T = 30°C) with a flow of pure N_2,the residue was passed over an Alumina bed for removing fine particles and high molecular weight components polymerized in the elution extract. Aliquots of the solution were injected into a semipreparative column (320x 8 mm),filled with basic Alumina 70 to 150 mesh (Woelm,Carlo Erba,Milano,Italy). Prior to the analysis,the column was activated at 230°C under a flow of ultrapure He for at least 2 h. The elution of compounds,having different polarity,was performed with n-hexane,CH_2Cl_2,CH_3CN and a mixture of CH_3CN + CH_3OH (3:1),used isocratically in that order. The liquid chromatograph (Violet,Rome,Italy) was equipped with a four way valve for the solvent delivery and connected to a UV absorption detector set at 254 nm. The four fractions collected were dried, dissolved in proper solvents and analyzed by GC to check the efficiency of separation. The CH_3CN eluate,containing the bulk of moderately-polar PAH,was further fractionated on a 250 x 4 mm HPLC column,packed with 10 μm silica particles (Erbasil,Carlo Erba). Before each analysis,the column was cleaned with CH_2Cl_2 and conditioned with n-hexane. The HPLC separation was carried out in gradient elution by adding CH_2Cl_2 to n-hexane at a rate of 2% v/v per minute. Three subfractions were collected during this process and aliquots used for identification and quantification of moderately-polar PAH by GC and

GC/MS. GC separations were carried out on a Dani HR 6800 gas chromatograph (Dani Spa,Monza,Italy),equipped with a programmed temperature vaporizer (PTV) for the sample injection. The gas chromatographic apparatus was connected either to an ECD or a TID for the identification of nitro-PAH in the HPLC subfraction. The ECD (a Carlo Erba model HT 25) was used in the modulated frequency mode with a pulse width of 5 μsec. A flow of pure N_2 (30 mL/min) was added to the carrier gas (He) for generating the thermal electrons necessary for the electron - capture process. When the column was connected to the thermoionic detector (a DANI model CT-NPD 68/20),the temperature of the pearl was kept at 700°C and small flow rates of air and H_2 were added to the column outlet in order to optimize the response towards nitrogen containing compounds and reduce the detection of other organic components. GC/MS analyses were all performed on a DANI 3900 gas chromatograph,directly connected to a VG 70-70 F double - focusing mass spectrometer,equipped with a VG 2350 Data System (VG Analytical,Altrincham,U.K.). When the determinations were carried out in SID,the resolution of the instrument was set at a value of 2,000 measured at 10% valley. All GC and GC/MS separations were carried out on a 12 m fused silica capillary column coated with DB-5 and supplied by J&W Scientific Co. (Rancho Cordova,CA,USA). The inner diameter of the column was 0.32 mm and the film thickness 1μm.

3. RESULTS AND DISCUSSION

In Figure 1a an example of the chromatographic profile obtained when the SOF,extracted from urban particulate matter,was subjected to the prefractionation process, carried out on the Alumina column, is reported. Figures 1b,c, d,and e show,instead,the HRGC profiles obtained when each fraction of Figure 1a was injected into the capillary column connected to a FID. Chromatograms have been scaled in sensitivity in order to see better the different distribution of organic components present in each fraction . Preliminary tests, carried out with reference compounds added to the sample prior the separation, confirmed that alkanes,PAH,moderately-PAH and polar compounds are eluted most in the n-hexane,CH_2Cl_2,CH_3CN and CH_3CN + CH_3OH fractions respectively. While the identity of the constituents present in the former two fractions was easily attributed on the basis of their retention times,fraction 3 and 4 were still to complex and the sensitivity and specificity of the FID so low that no attempt to identify nitro or oxy-PAH could be accomplished with success. Since previous investigations (2,9),carried out on different emission samples and dust collected in rural areas,have shown that a substantial amount of the most polar fraction (CH_3CN + CH_3OH eluate) usually contain long-chain polar compounds,members of different homologous series but all characterized by a low aromatic content,we focused our attention on fraction 3 (CH_3CN eluate) where most part of nitro and oxy-PAH are eluted. Similarly to what has been experienced in the separation of Diesel soot, positive identification and quantitative determination of nitro and oxy-PAH can be better accomplished if the moderately-polar fraction is further separated on a HPLC silica column working in gradient elution. Through this step it is , in fact,possible to obtain subfractions sufficiently enriched to be subjected to GC and GC/MS determinations. The HPLC profile of the CH_3CN fraction,relative to the same sample shown in Figure 1,is reported in Figure 2a. In this Figure are also indicated the number of subfractions collected at the column outlet. Since our main task was the quantitative evaluation of the most active direct mutagens,particulare care was paid to separate the bulk of nitro-PAH from the rest of the sample. This task was achieved by selecting the HPLC column and the gradient used in such a way that nitrated compounds,with molecular weight ranging from 1-nitronaphtalene to 6-nitrobenz(a)pyrene,could be all collected

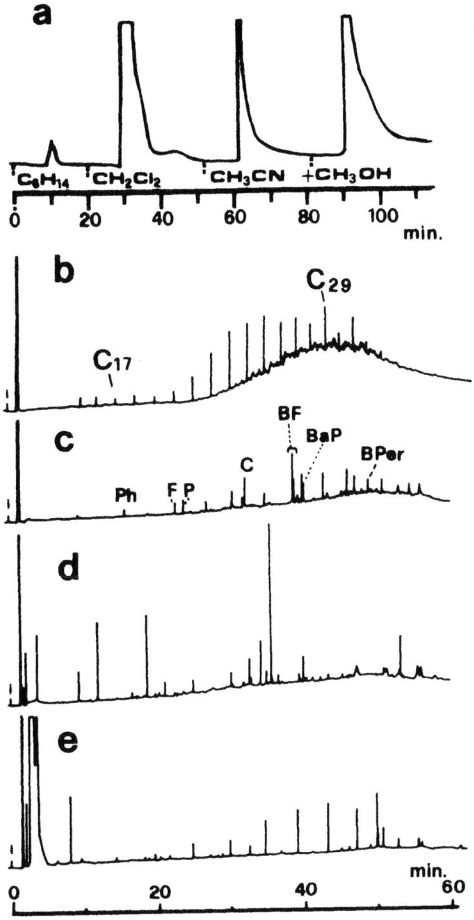

Figure 1

a) Low resolution liquid chromatographic separation of the soluble organic fraction extracted from urban particulate matter collected in downtown Rome.
b) GC profile obtained by analyzing the C_6H_{14} LRLC fraction(alkanes)by capillary chromatography connected to a FID.
c) GC profile recorded during the separation of the fraction (CH_2Cl_2 eluate) collected from the Alumina column. Ph:phenanthrene,F:fluoranthene,P:pyrene,C:chrysene,BF:benzofluoranthenes, BaP:benzo(a)pyrene,BPer:benzperylene.
d) GC-FID profile of the CH_3CN eluate containing moderately-polar PAH (nitro + oxy-PAH).
e) GC-FID profile of the CH_3CN+CH_3OH eluate containing polar compounds.

in the same subfraction with little interference from oxy - PAH or other nitroarenes. This subfraction (numbered as II in Figure 2a) was collected when the content of CH_2Cl_2 in the eluent was changed from 31 to 56% v/v. Once fractionated,nitro -PAH present in subfraction II were determined by following three indipendent approaches,each one providing different pieces of information. The first step was to inject small aliquots of subfraction II into the capillary column connected to an ECD. Since this detector is exstremely sensitive to nitro - PAH, and particularly to nitropyrene/fluoranthene isomers, the occurrence of intense peaks,having a retention equal or close to that of pure nitrated components,gives an important,preliminary information about the presence and levels of these direct mutagens in the sample. However,the ECD is sensitive to any electron - absorbing species and its response must be taken only as an indication that needs further confirmation before to draw any conclusion. Additional evidence on the presence of nitro - PAH can be gained by injecting aliquots of the same subfraction into the same capillary column connected to an alkali flame (2) or a thermoionic detector. Although such detectors are,by far, less sensitive than the ECD, their selectivity is sufficient to tell us whether electron-capturing species contain nitrogen atoms in the molecule and ,consequently, are likely to be nitro-PAH . Figures 2b and c show the GC profiles recorded when subfraction II was analyzed by capillary chromatography connected, alternatively, to an ECD or a TID. Arrows in these Figures are used to indicate the retention of 1-nitropyrene and 3-nitrofluoranthene isomers measured by injecting pure standards. If we focus our attention on that part of the chromatogram where the most genotoxic components are eluted (upper part), we can say that both detectors give consistent results as they suggest that other nitro-PAH, in addition to 1-nitropyrene, were present in this urban sample. In order to confirm that these additional components were, actually, nitro-PAH, subfraction II was subjected to the GC-MS analysis. This determination was accomplished by using the MS in the scan mode. This approach was preferred because it allowed a better check of the identity of the individual components by simply recalling, from the computer memory, the full spectrum recorded during the various scans. Figure 2d shows the trace obtained when the current, generated by integrating the signals of the ions recorded at m/z 247, 217, 201, 200 and 189, was plotted against the number of scans. These m/z values were relative to the fragments M^+, $(M-NO)^+$, $(M-NO_2)^+$, $(M-HNO_2)^+$ and $(M-CNO_2)^+$, typically observed in the mass spectra of nitropyrene/fluoranthene isomers. The GC-MS provided striking evidence that, in addition to 1-nitropyrene, two more nitropyrene/fluoranthene isomers were present in subfraction II. As can be seen, one compound is eluted before 1-nitropyrene whereas the other is eluted just after this component. Since differences in the mass spectra, collected during the GC-MS run, are unlikely to give sufficient information on the position of the nitro-group and spatial distribution of benzenic rings, the retention order was the only way to attempt an identification. This task was facilitated by the fact that, on capillary columns coated with slightly polar phases, 1-nitropyrene is eluted after all nitrofluoranthene isomers and it is always the first-eluted mononitropyrene isomer (10). By keeping in mind these considerations, it was easy to identify the last-eluted component of Figure 2d as 2-nitropyrene. More difficult was instead, to establish which nitrofluoranthene isomer was corresponding to the peak eluted before 1-nitropyrene. The problem was arising from the fact that 2-nitrofluoranthene and 3-nitrofluoranthene, both having a retention very close to the unidentified peak, were difficult to separate on our column. This difficulty was overcome by spiking the real sample with known amounts of 3-nitrofluoranthene isomer. Through this approach we found that the peak width of the unidentified compound of Figure 2d increased and, more importantly, different ratios were observed between the intensity of the molecular ion and that relative to the losses of NO and NO_2. On the light of these observations, we were able to establish that, very likely, the compound investigated was the 2-nitrofluoranthene isomer.

The detection of both 2-nitrofluoranthene/pyrene isomers in urban samples was extremely important as fully confirmed the observations made by Pitts et al. (7) who have suggested that such components can be formed by reaction of NO_2 with pyrene and fluoranthene when the concentration of OH radicals is increased by photochemical reactions. In order to confirm the results shown in Figure 3 and to be sure that the presence of these unexpected components was not arising from the long exposure (4 days) of our sample to gaseous pollutants, two more samples were collected in the same site . In this case the sampling time was reduced to 24 h and suspended particulate matter collected in different conditions. The first 24 h sample (numbered as no. 2) was collected when the traffic was heavy, the heating appliances were switched on and the sky was partially cloudy. That day (April 10 1986) the photochemical smog was not particularly high and near to the levels measured in late winter-early spring in our regions. The second 24 h sample (numbered as no. 3) was collected later (April 25, 1986) when the traffic was

Figure 2

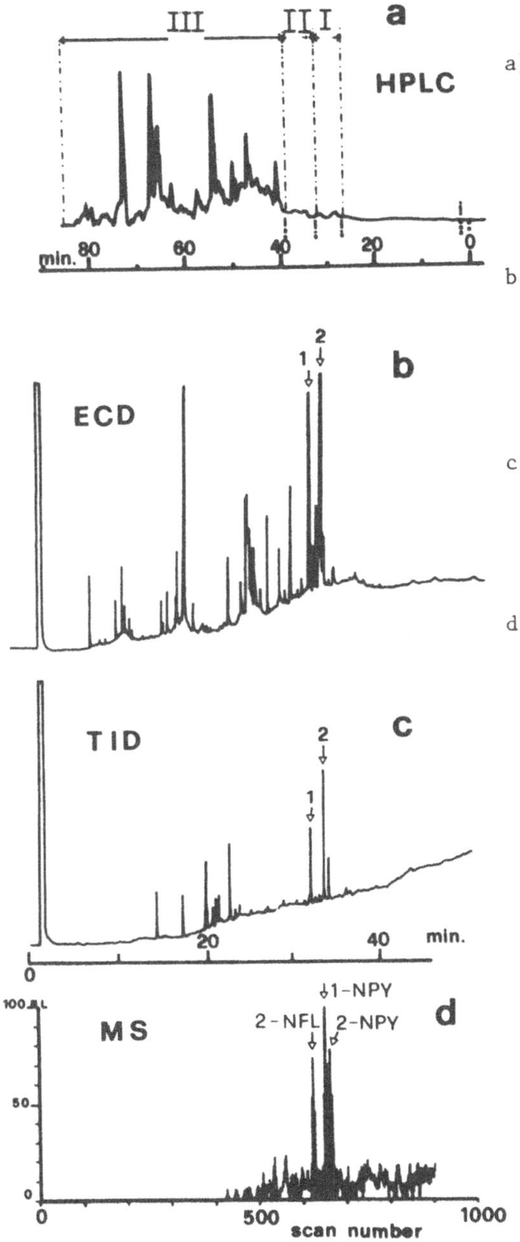

a) HPLC profile obtained during the fractionation of the CH_3CN eluate shown in Figure 1d and containing moderately-polar PAH. Number in romanic are used to indicate the various subfractions collected.The liquid fractionation was carried out on sample no.1 of Table I.

b) HRGC profile obtained by injecting subfraction II of Figure 2a into the capillary column connected to the ECD. Numbers 1 and 2 refer,respectively to the retention of 3-nitrofluoranthene and 1-nitropyrene measured with pure compounds.

c) HRGC profile of the same subfraction shown in Figure 2b obtained when the capillary column was connected the a thermoionic detector. Numbers 1 and 2 have the same meaning as in Figure 2b.

d) GC-MS trace recorded during the analysis of the same subfraction II shown in the previous two Figures. For peak identification and the method used for recording nitro-PAH see text.

not particularly intense, the heating appliances were switched-off by law but the sun was shining. It is important to notice that in the late afternoon of the previous day ozone reached very high values (100 ppb) and suspended particulate matter had enough time to react with photochemical pollutants to generate nitro-PAH. In Figure 3b and c are reported, respectively the SID traces recorded at m/z 247 (M^+ of nitropyrene/fluoranthene isomers) when subfraction II of sample no.2 and 3 were analyzed by GC-MS. For sake of comparison, in the same Figure (3a) is reported the profile obtained by analyzing a light-duty Diesel exhaust. In order to provide a comprehensive picture of the composition of the various urban samples, in Table I are reported the volume sampled for each filter (including that sampled for 4 days and numbered as sample no.1), the amount of dust collected and the content of alkanes together with that of PAH, moderately-polar PAH, nitro-PAH and their possible precursors. By comparing the results reported in Figure 2 and 3 and in Table I, the following conclusions can be drawn:

1) Diesel soot does not contain other nitropyrene /fluoranthene isomers than 1-nitropyrene in appreciable amounts.
2) The content of 1-nitropyrene in urban samples truly reflects the impact of Diesel emission on the environment whereas the amount of 2-nitrofluoranthene/pyrene isomers is not related to this emission source.
3) The content of 2-nitrofluoranthene/pyrene isomers appears to be indipendent from the amount of dust collected and from the esposure of the filter to gaseous species. Dust sampled for 4 days shows, in fact , the same content of 2-nitrofluoranthene/pyrene isomers than that exposed for 24 h providing that similar environmental conditions and emission rates are generated within the city.
4) The content of 2-nitrofluoranthene/pyrene isomers does not seem to be influenced by the content of their possible precursors and it is indipendent from the total amount of PAH present in the sample. This aspect is better seen when sample no.2 is directly compared with sample no.3.
5) The presence of 2-nitrofluoranthene/pyrene isomers seems not to be related to the photochemical smog conditions established in the city.

This last observation is supported by our previous results obtained by analyzing urban and suburban samples (10) collected in different seasons. We found that urban and suburban samples collected during winter, when the wheather was cold and rain was falling, were both containing 2-nitrofluoranthene/pyrene isomers that, at that time, we were unable to positively identify. The only difference between the dust collected in the two sites was in the content of 1-nitropyrene that was lacking in the suburban sample. Suburban samples collected during summer season, when ozone reached mixing ratios higher than 100 ppb, did not show, instead, the presence of these components. Since the only difference between the urban 24 h samples reported in Figure 3b and c was due to the fact that heating appliances were shut-off during sampling no.3, it seems more reasonable to believe that, in contrast with the hypothesis made by Pitts, 2-nitrofluoranthene/pyrene isomers can be generated by combustion sources . This hypothesis is partly supported by two facts. The first is that some industrial emission sources actually do contain 2-nitrofluoranthene/pyrene isomers. An example, of such a source is shown in Figure 4 where the GC-MS traces, recorded during the analysis of particulate matter emitted from the stack of an industrial plant producing carbon electrodes, is reported. The second fact is that nitration of PAH should give, as major product, 1-nitropyrene whereas the formation of 2-nitrofluoranthene/pyrene isomers should take place to a lesser extent.

At the present time, we are not in the position to fully support our statements with sufficient experimental data and it is fair to say that the question whether photochemical smog can increase the mutagenic properties of particulate matter is, in our opinion, still open. We would like to stress

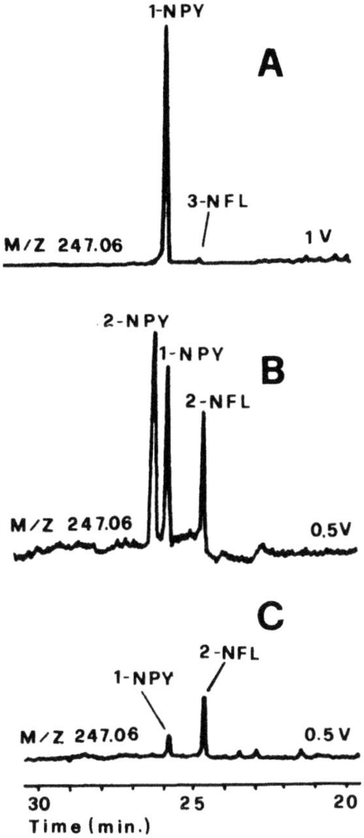

Figure 3

a) GC-MS profile recorded during the SID determination of nitro-PAH present in a Diesel soot extract.
b) GC-MS profile recorded during the SID determination of nitro-PAH present in the urban sample indi̲cated as no.2 in Table I.
c) GC-MS profile recorded during the SID determination of nitro-PAH present in the urban sample indi̲cated as no.3 in Table I.

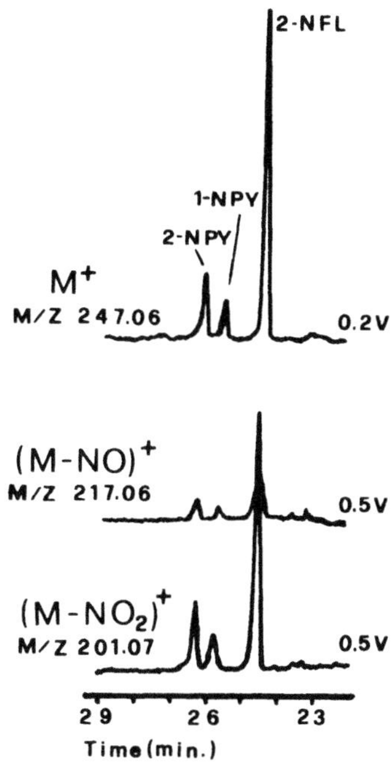

Figure 4

SID profiles showing the presence of 2-nitrofluoranthene/pyrene iso̲mers in the particulate matter sam̲pled form the stack of an industri̅al plant producing carbon electrodes.

TABLE I

Experimental conditions and chemical features of urban dust samples collected in downtown Rome

1 - A : SAMPLING CONDITIONS:

Sample 1 : sky: sunny; domestic heating : on; traffic: medium; $[O_3]$: ~60 ppbv; sampled volume: 4,000 m^3; total s.p.m.: 70 $\mu g/m^3$.

Sample 2 : sky: part. cloudy; domestic heating : on; traffic: heavy; $[O_3]$: ~40 ppbv; sampled volume: 1,200 m^3; total s.p.m.: 208 $\mu g/m^3$.

Sample 3 : sky: sunny; domestic heating: shut off; traffic: low; $[O_3]$: ~70 ppbv, sampled volume: 1,200 m^3; total s.p.m.: 75 $\mu g/m^3$.

1 - B : ORGANIC CONTENT OF THE VARIOUS SAMPLES COLLECTED:

Sample:	Aliphatics mg/g	PAH mg/g	mod.pol.PAH mg/g	polar fract. mg/g	Fluoranthene, µg/g	Pyrene µg/g	2-NF µg/g	1-NP µg/g	2-NP µg/g
1	19.0	4.8	.9	23.8	4.0	4.6	.36	.90	.70
2	34.4	7.9	9.8	96.2	4.0	4.4	.39	1.0	1.1
3	23.3	8.7	21.5	31.7	2.3	2.3	.41	.25	n.d.

here the crucial importance of establishing unambiguously the real source from which 2-nitrofluoranthene/pyrene isomers are generated. Because if, in contrast with our opinion, the main source of these components turns-out to be photochemical reactions, the impact of photochemical smog on human health can be more dramatic than that arising from gaseous pollutants. Therefore more laboratory studies should be devoted to find out whether, in the presence of OH radicals, HNO_3, NO_2, or other nitrating agents, pyrene and fluoranthene deposited on particulate matter can be actually converted into nitrated derivatives.

As far as oxy-PAH is concerned, all samples investigated were found to contain the same components. The most aboundant oxy-PAH found in subfraction III are listed in table II. Since these components can be either formed during Diesel and oil combustion or by photochemical reactions, it is difficult to rely on them for establishing whether and to what extent urban dust has been subjected to photochemical oxidation. We can only say that, consistently with our hypothesis, samples numbered as 1 and 2 were characterized by an higher content of oxy-PAH which were identified most as keto-PAH. The fact sample no.3 was characterized by an higher content of moderately-polar PAH but by a lower content of oxy-PAH is due to the fact that, owing to the time of sampling, natural emission from plants was higher. This effect was demostrated by the presence, in subfraction III, of some moderately-polar PAH typically found in trees emission.

TABLE II

Most abundant OXY-PAH detected in subfraction III of urban dust collected in downtown Rome.

No.	Compound	M.W.
1	Anthracene-9,10-dione	208
2	Methyl(anthracen/phenanthren)-x,y-diones (up to two)	222
3	Benz(anthracen/fluoren)-x-one *	230
4	Methylbenz(anthracen/fluoren)-x-ones (up to three)	244
5	Benz(anthracen/phenanthren)-x,y-diones (up to three)	258
6	6H-Benz(cd)pyren-6-one*	254

* Most abundant compounds, their concentration was in the range of 10 μg/g

AKNOWLEDGEMENTS

We are indebted to the MS Service of the Area della Ricerca di Roma who provided the mass-spectrometer and to A.Brachetti for technical assistance.

REFERENCES

(1) D. SCHUETZLE, T.L. RILEY, T.J. PRATER, T.M. HARVEY, D.F. HUNT; Anal. Chem.

54,265,(1982).
(2) A. LIBERTI, P. CICCIOLI, E.BRANCALEONI, C.DI PALO, A. CECINATO; HRGC&CC, 7,601,(1984).
(3) I.T. SALMEEN, A.M. PERO, R. ZATOR, D. SCHUETZLE, T.L. RILEY; Env.Sci.Tech. 18,375,(1984).
(4) J.S. SIAK,T.L. CHAN, P.S. LEE; Environ.Int. 5,243,(1981).
(5) M. HIROSE, M.S. LEE, C.Y. WANG, C.M. KING; Cancer.Res. 44,1158,(1984).
(6) H. OHGAKI, C. NEGISHI, K. WAKABAYASHI, K. KUSAMA, S. SATO, T. SUGIMURA; Carcinogenesis 5,583,(1984).
(7) J.N. PITTS Jr., J.A. SWEETMAN, B. ZIELINSKA, A.M. WINER, R. ATKINSON; Atmos.Env. 19,1601,(1985).
(8) National Primary and Secondary Ambient Air Standards,Federal Register; 36,no. 84,Part.II,Friday,April 30,1971,Appendix B,pag.8191-8194.
(9) P. CICCIOLI, E. BRANCALEONI, A. CECINATO, C.DI PALO, P.BUTTINI, A.LIBERTI; J. Chromatog. 351,451,(1986).
(10) T. RAMDAHL, J.A. SWEETMAN, B. ZIELINSKA, R. ATKINSON, A.M. WINER, J.N. PITTS Jr.. HRGC&CC 8,849,(1985).

FIELD INTERCOMPARISON OF SAMPLING AND ANALYTICAL METHODS FOR SO2 AND SO4 IN AMBIENT AIR

A pilot project in EMEP among 9 European countries
LANGENBRUGGE, 7 NOVEMBER 1985 TO 24 JANUARY 1986

K. NODOP and J. E. HANSSEN
Norwegian Institute for Air Research
P.B.130, N - 2001 Lillestrøm, NORWAY

Summary This field intercomparison of sampling and anlytical methods for sulphur dioxide and particulate sulphate in ambient air is the first one in EMEP. It has been conducted as a pilot project at the EMEP site Langenbrügge, Federal Republic of Germany, starting on 7 November 1985 giving a set of 79 measurements altogheter. Nine countries participated with a total of 12 methods. The set up of the instruments was done by the participants and the collected samples were sent to the national laboratories for chemical analysis. Sulphur dioxide was collected on impregnated filters and two different absorption solutions. Particulate sulphate was collected on filters. The design of the samplers varied from 8-port automatic instruments to those needing a daily change. The analytical methods used by the participants included Thorin, Sulfonazo III, West and Gaeke, ion chromatography and X-ray fluorescence. A quality assurance programme accompanied with the comparison.

The daily mean concentration of SO_2 was in the range of 1 to 137 μgSm^{-3} and of SO_4 0.3 to 18 μgSm^{-3}. Generally the agreement between the methods is better at higher concentrations than at lower ones. Up to ca 5 μgSm^{-3} for both SO_2 and SO_4, the standard deviation is always greater than 20 % of the composite mean. The correlation among the sulphur dioxide methods is better than for particulate sulphate. The mean concentration for the whole period varies from 12.7 to 18.4 for SO_2 and 2.5 to 3.7 for SO_4 (μgSm^{-3}). The observed differences in concentrations can mainly be explained by differences in the samplers and their design (e.g. inlet dimension), but in some cases the analytical handling caused the differences.

1. INTRODUCTION

The 'Co-operative programme for monitoring and evaluation of the long-range transmission of air pollutants in Europe' (EMEP) is based upon national networks with their own methods for sampling and chemical analysis, even though recommendations are worked out by the Chemical Co-ordinating Centre

(CCC). High quality of data is essential to fulfil the tasks of the EMEP project. To ensure comparable results intercomparisons of sampling equipment and analytical methods have to be conducted. Annual laboratory intercomparisons of analytical methods have been performed by the CCC since the start of EMEP, but no intercomparisons of sampling methods have been made due to lack of money, although the importance was stressed by the Steering Body of EMEP long ago.

However, as a pilot project within EMEP a field intercomparison of methods for measuring sulphur dioxide and particulate sulphate in ambient air was able to be conducted with the support of the Ministry of Interior and the Umweltbundesamt of the Federal Republic of Germany, and in co-operation with the CCC. To keep the first intercomparison to a managable size, a small number of countries were invited to participate to cover the methods being used within EMEP. The following nine countries joined the intercomparison: Czechoslovakia, Denmark, Federal Republic of Germany, Finland, France, Hungary, Norway, Sweden and United Kingdom.

The results of the Langenbrügge field intercomparison including detailed information will be published as EMEP report (EMEP/CCC - 2/86).

2. GENERAL PERFORMANCE

The field intercomparison started on the 4 November 1985 with a meeting of the experts from the participating countries to discuss the proposed programme and the set up of the instruments. Participants were asked to bring along exactly the same method and equipment which are in operation at their own EMEP stations. The sampling time was 24 hours with a time switch at 10 GMT. During the exercise the staff at the station changed the daily samples and sent them to the participating laboratories for routine analysis. Fresh sampling supplies were also sent to the station by mail. The measuring period started on the 7 November 1985 and continued until the 25 January 1986 giving a set of 79 measurements altogether.

A quality assurance programme accompanied with the field intercomparison. Questionaires were sent out to the participants to get detailed information on their sampling and analytical methods used. The laboratories were asked to perform duplicate analysis of ca. 10% of the samples by either the same or a different analytical method. An intercomparion of analytical methods with help of synthetic samples on own material was conducted after the measurement period. In addition the results of ninth intercomparisons of analytical methods within EMEP are used. During the field intercomparion the operation of the instruments was checked regularly.

The results of the measurements by the participants were forwarded to the CCC for data evaluation. A draft report was circulated among the participants and the results of the intercomparison were discussed on the ´Workshop on Data Quality Assurance´ in Freiburg, Federal Republic of Germany, 2 to 4 July 1986.

3. LOCATION AND ARRANGEMENT

The field intercomparison took place at the EMEP station D 02 Langenbrügge (52°48′N, 10°454′E, 74 m asl) in the Federal Republic of Germany. Langenbrügge was chosen because it had shown a wide range of concentrations at that time of the year and it met all technical requirements.

A small wooden house next to the station building was used for housing the instruments. All the sampling inlets were arranged within a distance of 2 m on one wall, about 1.5 m above the ground. The Danish equipment was the only one which could stay outside, and it was placed in front of the commonly used wall. Inside, all instruments were placed on a shelf next to the wall. The set up of the equipment, inside and outside, is shown in Pictures 1 and 2 and the whole station Langenbrügge in Picture 3.

The German measurements are part of the routine programme at the station, therefore no extra instruments were set up. They are connected to one manifold with the air inlet 1.6 m above the roof of the building (Picture 3). The manifold has a diameter of 50 cm^2 and an average air velocity of 2.1 ms^{-1}.

4. SAMPLING METHODS AND EQUIPMENT

In the EMEP project sulphur dioxide in ambient air is measured mainly by methods making use of impregnated filters or absorption solutions. In the comparison 3 countries were participating with potassium hydroxide (KOH) impregnated filters, 2 using tetracloromercurate (TCM) and 4 using hydrogen peroxide (H_2O_2) as absorbing media. Particulate sulphate was collected on an upstream filter in the sampling line. Detailed information on each method is summarized in Table 1 specifying e.g. the measuring and analytical methods, flow rate, exposed filter area and face velocity. The tubing length used was determined by the set up and may not be the same as normally used.

The participants and their methods are identified by their country names. The first nine methods listed in Table 1 are those used in the EMEP network by the participants. In addition some participants wished to compare also other methods or the same method in parallel. In these cases the country code has a 2 as suffix. Altogether, 13 methods measuring SO_2 and 12 methods measuring SO_4. The results from a UV fluorescent measurements of SO_2 (ENVIRONNEMENT AF 20M) have been included (D2). The main emphasis on data evaluation and assessment concentrates on the methods used in EMEP. These are CS1, DK, N1, D1, H, F1, S, SF and UK for SO_2 and SO_4, with the exception of Hungarian SO_4 results (see Table 1).

The systems set up at Langenbrügge varied from 8-port automatic instruments to those needing a daily change at exactly 10 GMT. Starting with the filter methods, three participants used open-face filter holders, containing first the particulate matter collecting filter (inert) and then the impregnated filter. They were mounted in an inverted funnel serving both as inlet and as rain protection. The main difference between them is the flow rate used (6 to 58 m^3 per day). The amount of KOH used for impregnation differed also. The samples were changed daily.

Picture 1. EMEP Station Langenbrügge

Picture 2.

Picture 3.

Table 1: Measuring methods for SO_2 and SO_4 used by the participants in the field intercomparison								
Participants	Sulphur dioxide measuring method absorbing media and agent Impregnated filter (IF) Absorption solution Vol. cm^3		Flow rate		Exposed Filter Area Ø mm	Face Velocity cm/s	Analytical methods for	
			m^3/d	l/min			SO2 and	SO4
CS1	IF	0.5 M KOH	6-8	4-7	28	14	Thorin	Thorin
DK	IF	0.5 M KOH (2M+glycerol) *1	58	40	42	48	Sulfonazo III	Sulfonazo III
N1	IF	1. M KOH	18	13 *2	40	17	I C	I C
D1	20	0.2 M TCM	1	0.7	22	3	West Geake	XRF
H	50	0.4 M TCM	1.5	1	20	5	West Geake	XRF *3
S	75	0.3 % H2O2 + HCl	2	1.4	25	5	I C	XRF
SF	100	0.3 % H2O2 + HClO4	3	2.1	22	9	Thorin	XRF
UK	70	0.45% H2O2	2	1.4	25	5	I C	XRF
F1	70	0.3 % H2O2 + HClO4	2-3	1-2	27	7	Thorin	I C *4
CS2	IF	0.5 M KOH	6-8	4-7	28	14	Thorin	Thorin
N2	100	0.3 % H2O2 + HClO4	3.6	2.5	41	3	Thorin	I C
F2	70	0.3 % H2O2 + HClO4	2	1.4	40	2	Thorin	I C *4
D2	-	ENVIRONNEMENT AF 20 M	-	-	-	-	-	-

*1 at own EMEP stations 0.5M is used, both impregnations were used in the comparison.
*2 at own EMEP stations a flow rate of 24 m^3/day is used.
*3 at own EMEP stations a separate system is used to measure SO_4: Teflon filter and the analytical method is IDA.
*4 at own EMEP stations SO_4 is analysed with XRF.

Among the participants using absorption solutions the flow rate varied between 1 and 3 m^3 per day. The wash bottles were arranged on a circular track or in line connected to a manifold controlled by a time switch. The filters for collecting particulate sulphate were mounted inside.

The actual air volume was usually measured with a dry gasmeter, and readings were provided each day. One participant used the same sample volume each day in calculations. For another system the air volume was determinded from before-and -after flow rate measurements. Most of the systems had flow control devices in the form of a critical orifice or simply a tubing clamp.

5. ANALYTICAL METHODS

All chemical analyse were carried out by the participating national laboratories. Sulphur dioxide was determined by the wet chemical methods Thorin, Sulfonazo III and West and Gaeke using the filter extract or the absorption solutions. Particulate sulphate was measured as total sulphur by the X-ray fluorescence methods or as water soluble sulphate extracted from the filters by wet chemical methods. SO_2 and SO_4 were also determined by ion chromatography. The methods used by the participants are given in Table 1. Most of the laboratories also provided duplicate analysis of the samples by the same or a different method.

6. RESULTS AND DATA PROCESSING

Altogether 79 daily samples of each component were collected by each participant during the measuring period. The results were checked immediately upon their arrival, and the participants were asked to reanalyse samples which were suspect. If errors were detected, the data set was corrected. Concentrations which lay below the detection limit are reported as half the detection limit. Due to the discussions on the EMEP wokshop one participant detected an error in his calculations and the results of one participant were excluded because of malfunction of the sampler. Otherwise the original data sets have been kept unchanged. A mean concentration was calculated from all the individual measurements for each 24-h sampling period. Since a priori no measurement set can be regarded as the true one, the calculated mean concentration was used in the data evaluation as the reference. The mean concentrations may be biased by a data set which is considerably higher or lower than all the others. Values which were than 2 times the standard deviation away from the mean were excluded and the mean recalculated. This procedure was repeated once more.

The mean concentrations of SO_2 and SO_4 were in the range of 1 to 137 μgSm^{-3} and 0.3 to 18 μgSm^{-3}, respectively. The highest concentrations measured by any participant were 195 μgSm^{-3} of SO_2 and 28 μgSm^{-3} of SO_4. The mean concentrations and the standard deviations for the whole period are shown in Figure 1.

To easily identify days on which the general agreement was bad to give them a closer check, the ratio of the SD to the mean was plotted versus the mean concentrations (Figure 2). Also the ratio of the concentrations measured by each participant and the mean is shown in Figure 3. Generally the agreement between the participants is better at higher concentrations than at lower ones. Up to ca 5 μgSm^{-3} for both SO_2 and SO_4, the standard deviation is always greater than 20 % of the mean.

Regression analyse were performed between the results of each participant and the composite means. The classical regression line was chosen to describe the relationship with the mean as the independent variable. Calculations were made for SO_2 and SO_4 covering the whole concentration range (Table 2). The correlation coefficient, the ratio of the mean of each method to the mean of all methods and the standard deviation of the regression line are listed as well. The correlation coefficient varies between 0.96 and 1.00 for SO_2 and 0.95 and 0.99 for SO_4. The correlation among the sulphur dioxide methods is better than for particulate sulphate. The concentration means for the whole measuring period ranges from 12.7 to 18.4 μgSm^{-3} of SO_2. If this is compared to the composite, the greatest deviation is 21 %. The corresponding results for SO_4 are 2.5 to 3.7 μgSm^{-3} and 26 %.

The flow rates of all samplers were cross-checked with one wet flow rate meter. The control showed that flow rates measured and the flow rates given by the participants were not always identical. The air volume measured by the different gasmeters varied within 5 %. The flow rate was also measured in different parts of the sampling lines. Some instruments were not free of leakages. One instrument of the multiday design showed different flow rates between the ports.

Figure 1:

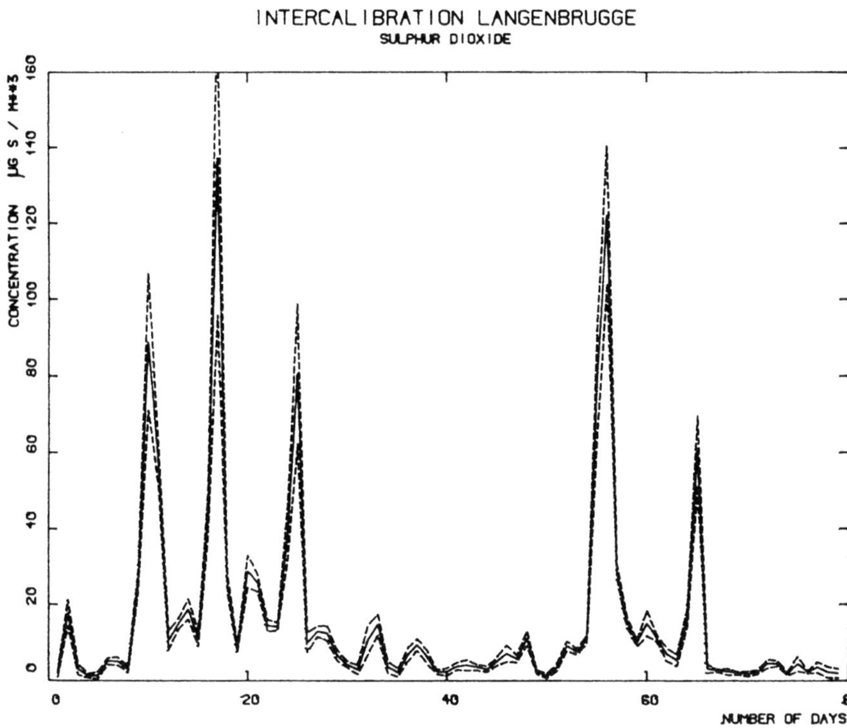
INTERCALIBRATION LANGENBRUGGE
SULPHUR DIOXIDE
CONCENTRATION µG S / M**3
0
20
40
60
80
100
120
140
160
0
20
40
60
80
NUMBER OF DAYS

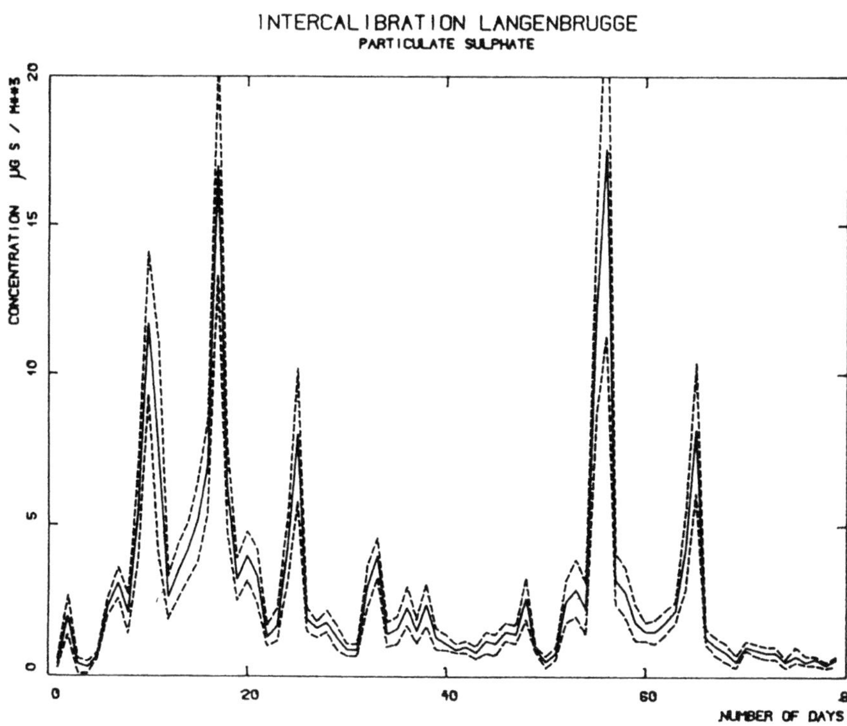
INTERCALIBRATION LANGENBRUGGE
PARTICULATE SULPHATE
CONCENTRATION µG S / M**3
0
5
10
15
20
0
20
40
60
80
NUMBER OF DAYS

Figure 2:

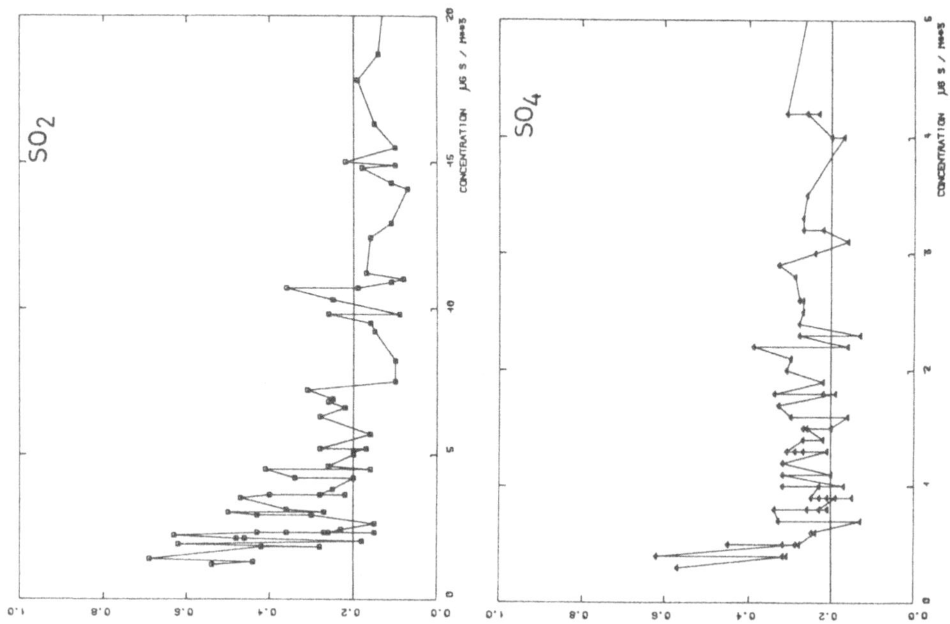

Figure 3:

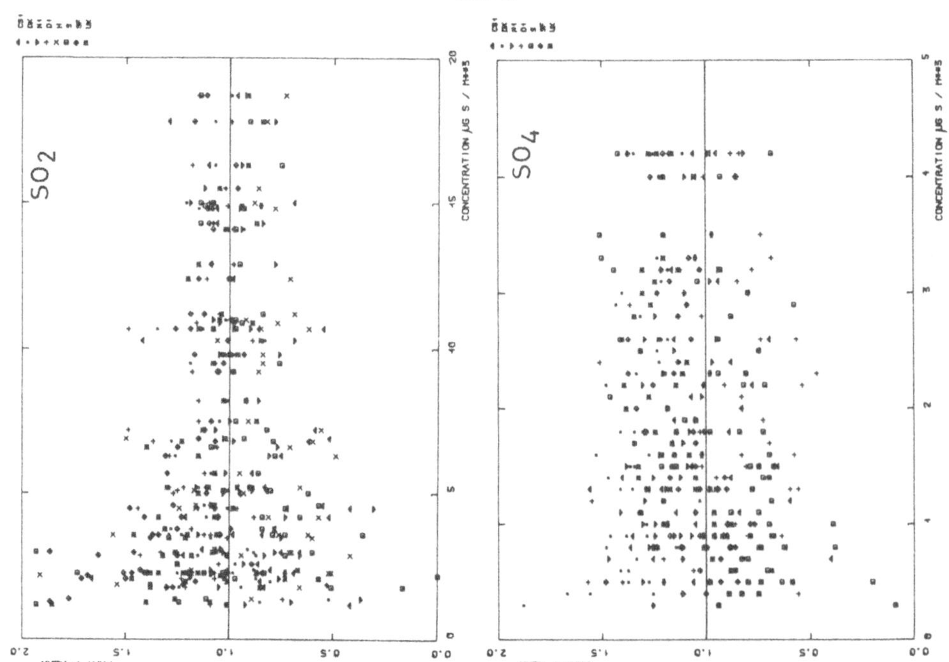

The data evaluation of the results from the comparison of analytical methods has not been finished yet, because there is still data missing from some participants. For the present it can be stated that some participants give systematically lower or higher results than the theoretical values. The deviation is in the order of 5 to 10 % and at higher concentrations better than at lower concentrations. The precision of the analytical methods used is in the order of 3 % with some exceptions. Again the precision is also better at higher concentrations than at lower ones.

Table 2: Statistics for the whole period at LangenbrUgge

SO2	MEAN	SD	PART	SD	R	N	P/M	A	B	SD
CS1	15.5	24.4	12.7	16.4	.96	60.	.82	.65	2.66	4.64
CS2	14.2	23.4	11.8	16.2	.96	65.	.83	.67	2.27	4.33
DK	16.7	26.4	18.4	31.2	.99	75.	1.10	1.16	-1.08	5.32
N1	16.0	25.7	16.9	30.9	.99	73.	1.06	1.20	-2.18	3.56
D1	16.6	26.3	17.1	25.1	.99	76.	1.03	.95	1.40	3.23
H	16.0	27.3	12.7	20.8	.99	68.	.79	.76	.58	2.96
S	16.7	26.8	18.0	32.6	.99	70.	1.07	1.20	-2.12	5.29
SF	16.6	26.6	18.4	28.7	1.00	74.	1.11	1.08	.53	2.20
UK	16.5	26.1	16.8	26.4	1.00	77.	1.02	1.01	.12	2.43
N2	16.7	26.2	16.4	25.2	1.00	76.	.98	.96	.44	2.00
F2	22.7	31.7	20.2	29.1	.99	47.	.89	.91	-.44	4.44
D2	16.7	26.2	19.6	33.0	1.00	76.	1.17	1.26	-1.42	2.54
SO4	MEAN	SD	PART	SD	R	N	P/M	A	B	SD
CS1	2.8	2.9	2.8	3.1	.99	63.	1.01	1.05	-.11	.43
CS2	2.7	2.9	2.7	2.9	.99	65.	1.01	1.01	.01	.50
DK	2.9	3.4	3.7	4.3	.99	73.	1.26	1.24	.07	.50
N1	2.9	3.4	3.2	3.7	.99	75.	1.10	1.09	.05	.49
D1	2.8	3.0	2.6	3.0	.98	72.	.95	.97	-.08	.62
H	3.1	3.5	2.1	2.1	.95	70.	.69	.57	.39	.68
S	2.4	2.8	2.5	3.5	.95	66.	1.05	1.17	-.29	1.15
SF	2.8	3.4	3.4	4.8	.97	77.	1.21	1.39	-.51	1.24
UK	2.9	3.4	3.2	3.5	.99	76.	1.10	1.01	.25	.46
N2	2.9	3.4	2.6	2.9	1.00	76.	.91	.85	.19	.22
F2	3.1	3.2	2.1	2.1	.98	61.	.67	.65	.07	.39

7. CONCLUSION

This measuring method intercomparison study was the first held within EMEP. The main objective was to evaluate the precision by which the selected pollutants can be determined. The first step in this study was to compare the reported concentrations as measured by the participants. Some remarks have to be made before interpreting the results.

The concentration range measured in the intercomparison is typical for Langenbrügge at that time of the year, but this may not be the case at the background stations in the other participating countries. Therefore only certain daily concentrations (e.g. low concentrations) should be taken into account when assessing the different methods. Since no data set can be regarded as the true one, the composite mean concentrations had been used as the reference, but this mean may be biased.

This intercomparison was conducted as a side-by-side parallel run of samplers to compare the concentrations obtained. It was successful in that it enabled an assessment of the results, because both the concentration range and the number of methods and samplers were sufficiently large.

The intercomparison showed that there are not negligible concentration differences between the methods used by the participating countries. When talking about methods in the sense that one distinguishes between the absorbing media (e.g. impregnated filters, absorption solution for SO_2), no method gives generally higher or lower results than the others. It seems to be rather the case that differences in the samplers and their design can account for most of the observed differences in concentrations. The flow rate and inlet dimension especially can be named here considering particulate sulphate. The concentration differences are partly caused by the anlytical handling of the samples. The general maintainance and periodical checks of the instrument are as important as all the other aspects of handling of samples.

8. ACKNOWLEDGEMENTS

One outhor (K. Nodop) is working at the CCC on a contract from the Institute for Meteorology and Geophysics, University of Frankfurt, fundend by the Umweltbundesamt of the Federal Republic of Germany. The authors wish to thank the staff at Langenbrügge for their extra work with changing samples every day; also all participants and participating laboratories for their co-operation, especially those providing a lot of additional information and duplicate analyse. Also the financial support of the WMO to two participants is gratefully acknowledged.

FIELD MEASUREMENTS ON THE USE OF DENUDERS FOR THE EVALUATION OF ACID DEPOSITION.

An intercomparison experiment carried out within the frame of COST Project 611.

Istituto sull'Inquinamento Atmosferico del C.N.R.
Area della Ricerca di Roma
Via Salaria Km 29,300 - C.P. 10
00016 Monterotondo Stazione (Roma) ITALY

Summary

A field comparison study on significant chemical species for acid deposition has been carried out with the aim to evaluate results obtained by using different techniques, namely annular and tubular denuders, thermal denuders, filter packs and fluorescent method for SO_2. Results have been satisfactory for SO_2 in gas phase, SO_4^{2-} and NH_4^+ in particulate matter, while noticeable differences have ben observed for NH_3, HNO_3 and NO_3^-. The use of different devices and flow rates and the very small concentration of analytes might partially explain the differences in the results. Since a greater number of date should be collected for more reliable statistic analysis, a second experiment should be planned to definitively evaluate the proposed techniques and suggest a reference method to be used within CEC for a better harmonization of measurements related to acid deposition.

The CEC-Workshop on the Measurement of Atmospheric Acidity (Rome, 25-26 June 1985) and the ECE-Workshop on the Advancements in Air Pollution Monitoring Equipments and Procedures (Freiburg, FRG, 2-6 June, 1985) emphasized the importance of field comparison studies on significant chemical species in the acid deposition problem. On the basis of the conclusive remarks of these Workshops a collaborative program for the measurements of selected species involved in acid deposition was agreed with the aim of comparing the results obtained by using different techniques, namely annular, tubular and thermal denuders, filter pack and instrumental methods.

The first intercomparison study was carried out at the Schauinsland-EMEP Station of the Unweltbundesamt on October 1985, using the facilities of the existing laboratory.

Nighttime and daytime levels of the following species have been measured:

SO_2, HNO_3 HNO_2, HCl, NH_3 in gas phase

$SO_4^=$, NO_3^-, Cl^-, NH_4^+, $H^+(H_2SO_4)$ in particulate matter

Five groups from Institutes of Germany, Italy, Norway and Sweden partecipated

to the experiments and carried out the sampling.
Analysis of filter and denuder extracts was performed by the EMEP-Station. Anion analysis was made by means of a Dionex Ion Chromatograph while ammonium ion was measured by indophenol blue method and/or ion selective electrode.

The comparison study was performed over a continuous 5-day period from 21 to 25 October 1985.
The sampling and analysis methods are summarized in Table I, which includes the names of the organizations performing the measurements and an organization-method code for subsequent reference. The methods employed are described in detail in several publications and reports and a reference list is given at the end.

After a preliminary 4-h sampling test on 22nd October it was observed that the relevant species were present at very low concentrations. This would cause some detection problems for tubular denuders which operated at low air flowrates. Therefore it was decided to extend the sampling time to 12 h. Only during one day (Oct. 23) 4-h sampling intervals were selected. The sampling schedule, including sampling intervals, is shown in Table II. A fluorescent SO_2 analyzer was operated over the entire test period and the data averaged for comparison with the integrated procedures.

RESULTS

Fig. 1 shows bar graphs of all results from the integrating procedures measuring gaseous and particulate species. The length of the horizontal bars corresponds to the duration of sampling. It can be seen that the different methods often yielded conflicting results for gaseous species, whereas, with few exceptions, they came into closer agreement in measuring ionic species in particle phase. Undoubtedly a large part of the scatter is due to the very low concentrations being measured. Indeed, the observed concentrations of NH_3, HNO_3 and HNO_2 were within the range 0.2-0.5 $\mu g/m^3$, and those of SO_2 within 2-3 $\mu g/m^3$. As a consequence, sample concentration was in some cases near the minimum detection level of the analytical method.

From Fig. 1 it also appears evident that the diurnal variation of the measured pollutants was not so significant to show two distinct daytime and nighttime concentration patterns. However, daytime NH_3 and SO_2 concentrations about 20-30% larger than at night were generally measured by all methods; whereas, in some instances, HNO_3 showed an opposite and unexpected trend. This was observed especially by SERI-CD method, which surprisingly yielded a nighttime HNO_3 maximum of 4.6 $\mu g/m^3$, while nighttime values not exceeding 0.5; 1.0 and 1.5 $\mu g/m^3$ were observed by the other methods.

STATISTICAL ESTIMATES

The approach followed for the intercomparison of the different measuring methods involved the assessment of reference concentration values. After examining the data variability, the geometric mean (GM) of the observations as measure of central tendency in each sampling block was adopted. Hence, the relative average deviation (RAD) from GM as a measure of variation for each method was calculated.

A least-square linear correlation was then employed to regress the values of any method against the corresponding GM in all sampling runs. In Fig. 2 scattergrams of the observed data vs GM are shown. All concentrations were expressed in nmole/m^3.

The behaviour of each method can be characterized by four parameters associated with the regression analysis: the slope (b) of the regression line, which measures the bias proportional to the concentration level; the y-intercept (a), which measures the constant bias of a method; the correlation coefficient (r), which measures the degree of correspondence of each method to the reference data;

the standard error of estimate of y on x (Sy, x) which estimates the scatter about the fitted line and serves as a measure of precision.

Two lines, drawn parallel to the regression line and separated vertically by + Sy, x from it, would include about 68% of the sample points.

The vertical bars associated with every data point indicate the geometrical standard deviation (Sg) from the geometric mean of the concentrations observed in each sampling run.

Table III shows the descriptive statistics obtained from regression fits. In addition to the estimated slope and intercept and the related standard deviations (S_b and S_a), an indication is given wheter the 95% confidence intervals of b and a contain 1 and 0, respectively.

Sulphur dioxide

The geometric mean of SO_2 in all 12-h sampling intervals ranged from 28 to 54 nmole/m^3. IIA-AD method yielded lower SO_2 concentrtions (about a factor of two) than the SERI-CD method. However the relative average deviation from GM was consistently lower for the former (15%)than for the latter (38%). In addition the RAD of IIA method was very close to values obtained by the NIAR-FP method (16%) and the fluorescent analyzer (FLA) (14%). The above discrepancy was also outlined by the parameters of the correlation graphs. IIA-AD data were sufficiently correlated with the geometric mean and produced intercept and slope which were not statistically different from 0 and 1. SERI-CD method yielded a lower $S_{y,\ x}$ value associated with the regression line and a better correlation coefficient, but it gaves a slope considerably less than 1.0, as well as FLA did. Moreover, b and a values of NIAR-FP method were outside the 95% confidence interval around 1.0 and 0.0, respectively.

Ammonia

IIA-AD method measured NH_3 concentratins larger of a factor varying from two to four compared to those of the other methods, in the GM range 0.2 - 0.5 ug/m^3. As a consequence, a high RAD (93%) characterized this method. Particularly, IIA-AD results were about four times larger than those measured by the SERI-CD method (RAD 37%). UD and UMS methods exhibited good agreement on the NH_3 concentrations over the full range, with values fluctuating around the reference concentration (RAD was 17% for UD method and 15% for UMS method).

Considering the correlation statistics, it should be noted that UMS-TD method yielded the best values of b (0.75), a (4 nmole/m^3), $S_{y,\ x}$ (2 nmole/m^3) and r (0.96). SERI-TD and UD-TD methods also showed a good correlation against GM, whereas IIA-AD method produced an almost lower r value (0.69) and a slope wery different from 1.0.

Nitric acid

In the GM range 0.2 - 1.5 μg/m^3, SERI-CD method measured HNO_3 levels about a factor of three larger than the reference values, whereas IIA-AD method yielded, in turn, results three times lower.

These notable differences could not be explained by the possible NO_2 and particulate nitrate interferences which could affect SERI-CD method. In this procedure no use is made of a preseparator to exclude coarse particles and no differential technique is applied to take into account the above interferents. However, the combination of these two interfering effects should be less than 5-10%. Therefore, the source of the observed discrepancy is, for the time being, unknown.

Although all methods were well correlated with GM, they showed a significant proportional bias relative to the reference data set. The UD-TD method only was highly correlated with the geometric mean of the concentrations and yielded a regression line with slope and intercept values not statistically different from those of a 1:1 correspondence between the two data sets.

Nitrous acid

Both SERI-CD and IIA-AD methods measured atmospheric HNO_2 from the difference of nitrite amounts found on two Na_2CO_3-denuders connected in series. SERI-CD method always measured very low HNO_2 concentrations (< 0.1 μg/m^3) while HNO_2 levels in the 0.1 - 0.5 μg/m^3 range were observed with the IIA-AD method.

Particulate ammonium, sulfate and nitrate

The measurement methods showed good agreement on the sulfate and ammonium concentrations in aerosols, whereas they gave spreaded results on nitrate particulate. It should be outlined that IIA-AD, UD-TD and UMS-TD methods employed a cyclone or a virtual impactor having a cut size of 2.5 μm at the respective working conditions.

The last two methods measured only nitrate associated to NH_4^+ as they used a denuder heated at 410 °K to quantitatively dissociate ammonium nitrate and collect the evaporated HNO_3.

Fig. 3 shows that both IIA-AD and SERI-TD method found a very good correlation between ($SO_4^=$) + (NO_3^-) and (NH_4^+), consistent with the existence of $(NH_4)_2SO_4$ and NH_4NO_3. This was furtherly verified by the multiple linear regression analysis which gaves the following relationships for the two methods:

for IIA-AD $NH_4^+ = 2.26\ SO_4^= + 1.04\ NO_3^- - 17$ (r = 0.98, n = 10)

for SERI-CD $NH_4^+ = 1.97\ SO_4^= + 0.94\ NO_3^- + 8$ (r = 0.99, n = 8)

The above results clearly demonstrated that no sulfate or nitrate containing coarse particles were present in the sampling location, as very similar coefficients were obtained with or without particle pre-separator. In addition, NH_4Cl on aerosol samples was found, by thermodenuder technique, to be absent or present at negligible levels.

Average nitrate concentrations, according to the different methods, ranged from 24 to 65 nmol/m^3 and the gaseous to particulate nitrate ratios from 0.17 to 0.55. IIA-AD method measured nitrate about a factor of three lower than GM data, which, however, were strongly influenced by the SERI-CD values, whose mean was three times higher than the GM.

An average particulate nitrate less of 15% from the Teflon filter was measured by the IIA-AD method, which includes a back-up nylon filter to recover HNO_3 released from the Teflon filter.

This was not an unexpected result, since greater losses (30-40%) have been observed in other locations under high ambient temperatures and lower relative humidities than those encountered during this field comparison. The daytime and nighttime temperature ranges were 5-10°C and -5 - 5°C, respectively and the entire measuring period was characterized by stable atmospheric conditions.

Reproducibility tests

The precision of the SERI-CD method was evaluated in duplicated measurements for each species. The analyses of seven paired sampled (ten for NH_4^+) gave coefficients of variation of 7.8%, 41.2%, 22.3%, 2.9%, 8.1% and 4.3 for SO_2, NH_3, HNO_3, $SO_4^=$, NH_4^+ and NO_3^- over the concentration ranges shown in Table III. With the IIA-CD method only two parallel samples were taken, indicating average differences of 7.1% for $SO_4^{=}$, 30% for HNO_3, 13.9%for NH_3, 5.1%for NH_4^+, 9.7%for SO_2, 18.6%for NO_3.

CONCLUSIONS

Five measuring methods have been compared during a 5-day field experiment in Schauinsland, FRG. Four analytical procedures consisted in the sampling through denuders, which were adopted with different geometry and different flow rates. The workshop showed that the sampling through denuders permits to measure gaseous and aerosolic atmospheric species at the trace levels. The procedures used require however, further investigations, as the accuracy is not satisfactory. The methods yield a sufficient agreement on particulate sulfate, ammonium and sulfur dioxide, while noticeable differences in measuring ammonia, nitric acid and particulate nitrate have been observed.

It is difficult to find a reasonable explanation upon the differences which have been found in the comparison of these results. Definitely, since the analytical work involved very small concentrations, minor operational procedures might have a noticeable impact.

As the measurement of the species which have been the object of the experimental workshop seems to be of relevant importance, a second experiment should be planned in order to test the above methods even at larger concentrations that those measured during the first field intercomparison. Moreover, a greater number of data should be collected for more reliable statistic analysis.

ACKNOWLEDGEMENTS

We are indebted to the Unweltbundesamt which has provided helpful assistance for experimental work and hospitality.

REFERENCES

ALLEGRINI, I., DE SANTIS, F., DI PALO, V., PERRINO, C. and LIBERTI, A., (1985). Determination of ammonia and ammonium ion in air by means of high-efficiency annular denuder annular denuders. In Advancements in Air Pollution Monitoring Equipment and Procedures Proc., Freiburg, FRG, June 2-6, pp. 76-84.

ALLEGRINI, I., DE SANTIS, F., DI PALO, V. and LIBERTI, A. (1984). Measurement of particulate and gaseous ammonia at a suburban area by means of diffusion tubes. J. Aerosol. Sci., 15, 466-471.

DE SANTIS, F., FEBO, A., POSSANZINI, M. and LIBERTI, A. (1985). Simultaneous measurements of nitric acid, nitrous acid, hydrogen chloride and sulfur dioxide in air by means of high-efficiency annular denuders. In advancements in Air Pollution Monitoring Equipment and Procedures Proc.. Freiburg, FRG, June 2-6, pp. 68-75.

FERM, M. (1979). Method for determination of atmospheric ammonia. Atmos. Environ., 13, 1385-1393.

FERM, M., FUGLSANG, K., HANSSEN J.-E., LATTILA, H. and SODERLUNG, R. (1985). Field intercomparison of measurements of total ammonium, nitrate and sulfur concentrations. In Advancements in Air Pollution Monitoring Equipment and Procedures Proc.. Freiburg, FRG, June 2-6, pp. 52-66.

FERM, M. and SJODIN, A. (1985). A sodium carbonate coated denuder for determination of nitrous acid in the atmosphere. Atmos. Environ., 19, 979-983.

FERM, M. (1986). A Na_2CO_3 coated denuder and filter for determination of gaseous HNO_3 and particulate NO_3^- in the atmosphere. Atmos. Environ., (in press).
KLOCKOW, D. and TECKENTRUP, A. (1980). Sampling and analysis of sulfur dioxide and particulate sulfate in air using impregnated filters. Intern. J. Environ. Anal. Chem. 8, 137-148.
LEVIN, E. and ZACHAK-CHRISTIANSEN, B. (1977). Efficiency of 0.5 N KOH impregnated filters for SO_2 collection. Atmos. Environ., 11, 861-862.
NIESSNER, R. and KLOCKOW, D. (1980). A thermoanalytical approach to speciation of atmospheric strong acids. Intern. J. Environ. Anal. Chem. 8, 163175.
POSSANZINI, M., FEBO, A. and LIBERTI, A. (1983). New design of a high-performance denuder for the sampling of atmospheric pollutants. Atmos. Environ., 17, 2605-2610.
SLANINA, J. et al. (1981). Application of a thermodenuder analyser to the determination of H_2SO_4, HNO_3 and NH_3 in air. Intern. J. Environ. Anal. Chem., 9, 59-70.

Table 1
Operators and measuring techniques.

COUNTRY: ITALY   INSTITUTE: ISTITUTO SULL'INQUINAMENTO ATMOSFERICO C.N.R.

System	Position in System	Components	Coating			Filter Dimensions		Temperature	Flow rate	Preseparation		Back up Filter	drying Procedure		Elution		Analytical Methods
			Sink	Matrix	Vol.	L cm	ø mm	tube C°	l/h	Typ/Art	Cutoff µm	kind	Gas	hand-ling	Volume ml	Medium	
Annular Denuder Method A D M	1st Denuder 2nd Denuder	HNO_2 HNO_3 SO_2 Interferents (NO_2, PAN, particles)	Na_2CO_3 1% Glycerol 1%	H_2O/MeOH 40/60		22	30 d_1 33 d_2	20 °C	900	Cyclone	2,5	Teflon (particulate) Nylon (HNO_3 released by Teflon filter)	N_2	hand-made	10	H_2O	IC
A D M	3rd Denuder	NH_3	Citric Acid 1%	MeOH		13	30 d_1 33 d_2	20 °C	900	Cyclone	2,5	4 th Denuder coated with citric acid (NH_3 released by Teflon filter)	N_2	"	10	H_2O	Indophenol

COUNTRY: NORWAY INSTITUTE: NORWEGIAN INSTITUTE FOR AIR RESEARCH

CODE: NIAR-FP

System	Position in System	Components	Coating			Filter Dimensions		Temperature tube C°	Flow rate l/h	Preseparation		Back up Filter kind	drying Procedure		Elution		Analytical Methods
			Sink	Matrix	Vol.	L cm	Ø mm			Typ/Art	Cutoff µm		Gas	handling	Volume ml	Medium	
Filter	2 1	$SO_4^{=}$ SO_2	 KOH 1 M	- water	- 300 ul		47 47	ambient	900	18 cm 14 cm	Face velocity 22.6 cm/s	-		over night	25 25	H_2O H_2O	IC
Filter	1	HNO_3/NO_3^-	NaOH 1 %	methanol	300 ul		47	ambient	900 (1300)	funnel	cutoff depends	-		air teight box with	depends on concen-tration and sampling time	H_2O	IC
Filter	1	NH_3/NH_4^+	oxalic acid 3 %	methanol	300 ul		47	ambient	900 (1300)			-		Silicia Gel		H_2O	Indophenol Orion electrode 95 - 10
							effectiv 41										

COUNTRY: SWEDEN INSTITUTE: SWEDISH ENVIRONMENTAL RESEARCH INSTITUTE

CODE: SERI-CD

System	Position in System	Components	Coating			Filter Dimensions		Temperature	Flow rate	Preseparation		Back up	drying Procedure		Elution		Analytical Methods
			Sink	Matrix	Vol.	L cm	ø mm	tube C°	l/h	Typ/Art	Cutoff µm	Filter kind	Gas	hand-ling	Volume ml	Medium	
Denuder+ coated filter	1 2	NH_4 NH_4^+	oxalic acid	1 % in methanol	the Denuder is filled	50	3	5 - 10 °C above ambient	180	funnel	depends on windspeed	W 40 coated in the same way as denuder	air	in-situ	2	H_2O	gas sensitive electrode
Denuder+ coated filter	1 2	NH_3 NH_4^+	citric acid	1 % in methanol	"	50	3	"	180	"	"	W 40 coated in the same way as denuder	air	in-situ	2	H_2O	"
two equal denuders in series than coated filter	1 2	HNO_2 HNO_3 NO_3^- SO_2 SO_4^{--}	Na_2CO_3	1%NaOh in methanol	"	75	4	"	120	NO		"	air	in-situ	2-5	H_2O	IC

COUNTRY: FEDERAL REPUBLIC OF GERMANY INSTITUTE: UNIVERSITÄT DORTMUND

CODE: UD-TD

System	Position in System	Components	Coating			Filter Dimensions		Temperature tube C°	Flow rate l/h	Preseparation		Back up Filter kind	drying Procedure		Elution		Analytical Methods
			Sink	Matrix	Vol.	L cm	ø mm			Typ/Art	Cutoff µm		Gas	handling	Volume ml	Medium	
Denuder	1	HCl HNO_3	NaF	0,2% by weight in H_2O/ 2-PrOH 1:1	1 ml	90	6	ambient	120	virtual impactor	2,5		N_2	hand-made	2	H_2O	IC
Denuder	2	Scrubber for H_2S and SO_2	TOM		0,4 ml	35	6	ambient	120	"	2,5		N_2	"			
Denuder	3	H_2SO_4 NH_4NO_3 NH_4Cl^-	NaF	0,2% by weight in H_2O/ 2-PrOH 1:1	1 ml	90	6	136	120	"	2,5		N_2	"	2	H_2O	IC
Denuder		NH_3	citric acid	0,2 % by weight in 2-PrOH	1 ml	90	6	ambient	120	virtual impactor	2,5		N_2	hand-made	2	H_2O	Indophenol

COUNTRY: FEDERAL REPUBLIC OF GERMANY INSTITUTE: UNWELTBUNDESAMT MESSSTATION SCHAUINSLAND

CODE: UMS-TD

System	Position in System	Components	Coating			Filter Dimensions		Temperature	Flow rate	Preseparation		Back up Filter	drying Procedure		Elution		Analytical Methods
			Sink	Matrix	Vol.	L cm	Ø mm	tube C°	l/h	Typ/Art	Cutoff µm	kind	Gas	hand-ling	Volume ml	Medium	
Denuder	1	HCl HNO_3	NaF	0,2 % weight in H_2O Isopropa-nol 1 : 1	1 ml	90	6	ambient Air	120	virtual Impactor	2,5	-	N_2	hand-made	2	H_20	IC
Denuder	2	Scrubber for H_2S SO_2	K_2CO_3	0,5 % weight in methanol	1 ml	35	6	ambient Air	120	"	2,5		N_2	"	-	-	-
Denuder	3	H_2SO_4 NH_4Cl NH_4NO_3	NaF	0,2 % weight in H_2= Isopropa-nol 1 : 1	1 ml	90	6	136	120	"	2,5	-	N_2	"	2	H_2O	IC
Denuder	1	NH_3	citric acid	0,1 weight %	1 ml	90	6	ambient Air	120	virtual Impactor	2,5	-	N_2	hand-made	2	H_2O	Indophenol
Filter		SO_4	PTFE														

TABLE II - SAMPLING SCHEDULE/SAMPLING INTERVALS (h)

Run	October 21	October 22	October 23	October 24	October 25
1	14.00-18.00				
2	19.00 —	07.00			
3		08.00-19.00			
4		20.00 -	07.00		
5			08.00-12.00		
6			12.30-16.00		
7			16.30-19.00		
8			20.00 -	08.00	
9				08.30-19.00	
10				20.30 -	08.00

TABLE III - DESCRIPTIVE STATISTICS OF THE REGRESSION FITS FOR OBSERVED DATA vs GEOMETRIC MEAN OF CONCENTRATIONS ($nmole/m^3$).

SULFUR DIOXIDE

Y variable	GM	GM	GM	GM
X variable	IIA-AD	SERI-CD	NIAR-FP	FLA
No. of pairs	8	7	5	8
Average Y (Range)	42(28-54)	45(33-54)	40(28-54)	44(28-54)
Average X (Range)	37(26-48)	69(51-86)	34(19-51)	43(26-59)
Relative average deviation from GM (RAD%)	15.6	37.8	16.1	14.2
Correlation coefficient (r)	0.80	0.95	0.96	0.65
Slope ($b \pm s_b$)	1.05±0.32 (yes)	0.68±0.10 (no)	0.68±0.11 (no)	0.61±0.29 (yes)
Intercept ($a \pm s_a$)	3 ± 12 (yes)	-2 ± 7 (yes)	16 ± 4 (no)	18 ± 13 (yes)
Standard error of estimate ($s_{y,x}$)	6	3	3	8

* Parenthesis indicates whether the 95% confidence interval for the slope contains 1.0

** Parenthesis indicates whether the 95% confidence interval for teh intercept contains 0.0

TABLE III - DESCRIPTIVE STATISTICS OF THE REGRESSION FITS FOR OBSERVED DATA vs GEOMETRIC MEAN OF CONCENTRATIONS (nmole/m^3).

AMMONIA

Y variable	GM	GM	GM	GM
X variable	JIA-AD	SERI-CD	UD-FP	UMS-TD
No. of pairs	8	8	7	6
Average Y (Range)	18(9-25)	18(9-25)	17(9-24)	17(9-25)
Average X (Range)	36(22-61)	9(5-13)	18(9-26)	17(5-26)
Relative average deviation from GM (RAD%)	92.8	37.0	17.5	14.7
Correlation coefficient (r)	0.69	0.85	0.77	0.96
Slope ($b \pm s_b$)	0.29±0.12 (yes)	1.63±0.40 (yes)	0.65±0.24 (yes)	0.75±0.10 (yes)
Intercept ($a \pm s_a$)	7 ± 5 (yes)	4 ± 4 (yes)	5 ± 5 (yes)	4 ± 2 (yes)
Standard error of estimate ($s_{y,x}$)	4	3	4	2

* Parenthesis indicates whether the 95% confidence interval for the slope contains 1.0

** Parenthesis indicates whether the 95% confidence interval for teh intercept contains 0.0

TABLE III - DESCRIPTIVE STATISTICS OF THE REGRESSION FITS FOR OBSERVED DATA vs GEOMETRIC MEAN OF CONCENTRATIONS (nmole/m^3).

NITRIC ACID

Y variable	GM	GM	GM	GM	GM
X variable	IIA-AD	SERI-CD	NIAR-FP	UD-TD	UMS-TD
No. of pairs	8	7	4	7	7
Average Y (Range)	12(3-25)	13(3-25)	8(4-13)	12(3-25)	13(3-25)
Average X (Range)	4(0-8)	36(8-74)	16(10-25)	10(3-19)	18(4-45)
Relative average deviation from GM (RAD%)	41.3	125.3	72.3	19.0	32.2
Correlation coefficient (r)	0.73	0.97	0.97	0.91	0.94
Slope ($b \pm s_b$)	2.40±0.93 (yes)	0.29±0.03 (no)	0.54±0.08 (no)	1.31±0.25 (yes)	0.54±0.09 (no)
Intercept ($a \pm s_a$)	1 ± 4 (yes)	2 ± 1 (yes)	0 ± 1 (yes)	-1 ± 3 (yes)	3 ± 2 (yes)
Standard error of estimate ($s_{y,x}$)	6	2	1	4	3

* Parenthesis indicates whether the 95% confidence interval for the slope contains 1.0

** Parenthesis indicates whether the 95% confidence interval for teh intercept contains 0.0

TABLE III - DESCRIPTIVE STATISTICS OF THE REGRESSION FITS FOR OBSERVED DATA vs GEOMETRIC MEAN OF CONCENTRATIONS (nmole/m^3).

AMMONIUM

Y variable	GM	GM
X variable	IIA-AD	SERI-CD
No. of pairs	10	10
Average Y (Range)	190 (85-293)	190 (85-293)
Average X (Range)	178 (79-301)	202 (92-288)
Relative average deviation from GM (RAD%)	6.6	6.6
Correlation coefficient (r)	0.99	0.99
Slope ($b \pm s_b$)	0.96 ± 0.04 (yes)	0.99 ± 0.05 (yes)
Intercept ($a \pm s_a$)	19 ± 9 (yes)	-12 ± 12 (yes)
Standard error of estimate ($s_{y,x}$)	11	12

* Parenthesis indicates whether the 95% confidence interval for the slope contains 1.0

** Parenthesis indicates whether the 95% confidence interval for teh intercept contains 0.0

TABLE III - DESCRIPTIVE STATISTICS OF THE REGRESSION FITS FOR OBSERVED DATA vs GEOMETRIC MEAN OF CONCENTRATIONS ($nmole/m^3$).

SULFATE

Y variable	GM	GM	GM
X variable	IIA-AD	SERI-CD	NIAR-FP
No. of pairs	10	9	6
Average Y (Range)	56(13-86)	54(13-86)	50(30-76)
Average X (Range)	57(29-84)	58(32-90)	44(24-68)
Relative average deviation from GM (RAD%)	5.6	11.1	11.2
Correlation coefficient (r)	0.97	0.98	0.98
Slope ($b \pm s_b$)	1.09 ± 0.09 (yes)	0.91 ± 0.06 (yes)	0.99 ± 0.99 (yes)
Intercept ($a \pm s_a$)	- 7 ± 6 (yes)	1 ± 4 (yes)	6 ± 5 (yes)
Standard error of estimate ($s_{y,x}$)	6	5	4

* Parenthesis indicates whether the 95% confidence interval for the slope contains 1.0

** Parenthesis indicates whether the 95% confidence interval for teh intercept contains 0.0

TABLE III - DESCRIPTIVE STATISTICS OF THE REGRESSION FITS FOR OBSERVED DATA vs GEOMETRIC MEAN OF CONCENTRATIONS ($nmole/m^3$).

NITRATE

Y variable	GM	GM	GM	GM	GM
X variable	IIA-AD	SERI-CD	NIAR-FP	UD-TD	UMS-TD
No. of pairs	8	8	5	7	7
Average Y (Range)	43(3-86)	43(3-86)	35(27-43)	38(3-86)	47(3-86)
Average X (Range)	24(6-56)	65(3-117)	40(7-56)	46(1-75)	65(2-112)
Relative average deviation from GM (RAD%)	61.7	43.7	49.2	32.6	44.5
Correlation coefficient (r)	0.92	0.92	- 0.67	0.84	0.91
Slope ($b \pm s_b$)	1.37±0.23 (yes)	0.63±0.11 (no)	-0.27±0.21 (no)	0.76±0.22 (yes)	0.64±0.12 (yes)
Intercept ($a \pm s_a$)	10 ± 7 (yes)	3 ± 8 (yes)	45 ± 8 (no)	3 ± 11 (yes)	2 ± 10 (yes)
Standard error of estimate ($s_{y,x}$)	12	12	8	15	13

* Parenthesis indicates whether the 95% confidence interval for the slope contains 1.0

** Parenthesis indicates whether the 95% confidence interval for teh intercept contains 0.0

Fig. 1 - Concentration levels measured by the different methods.

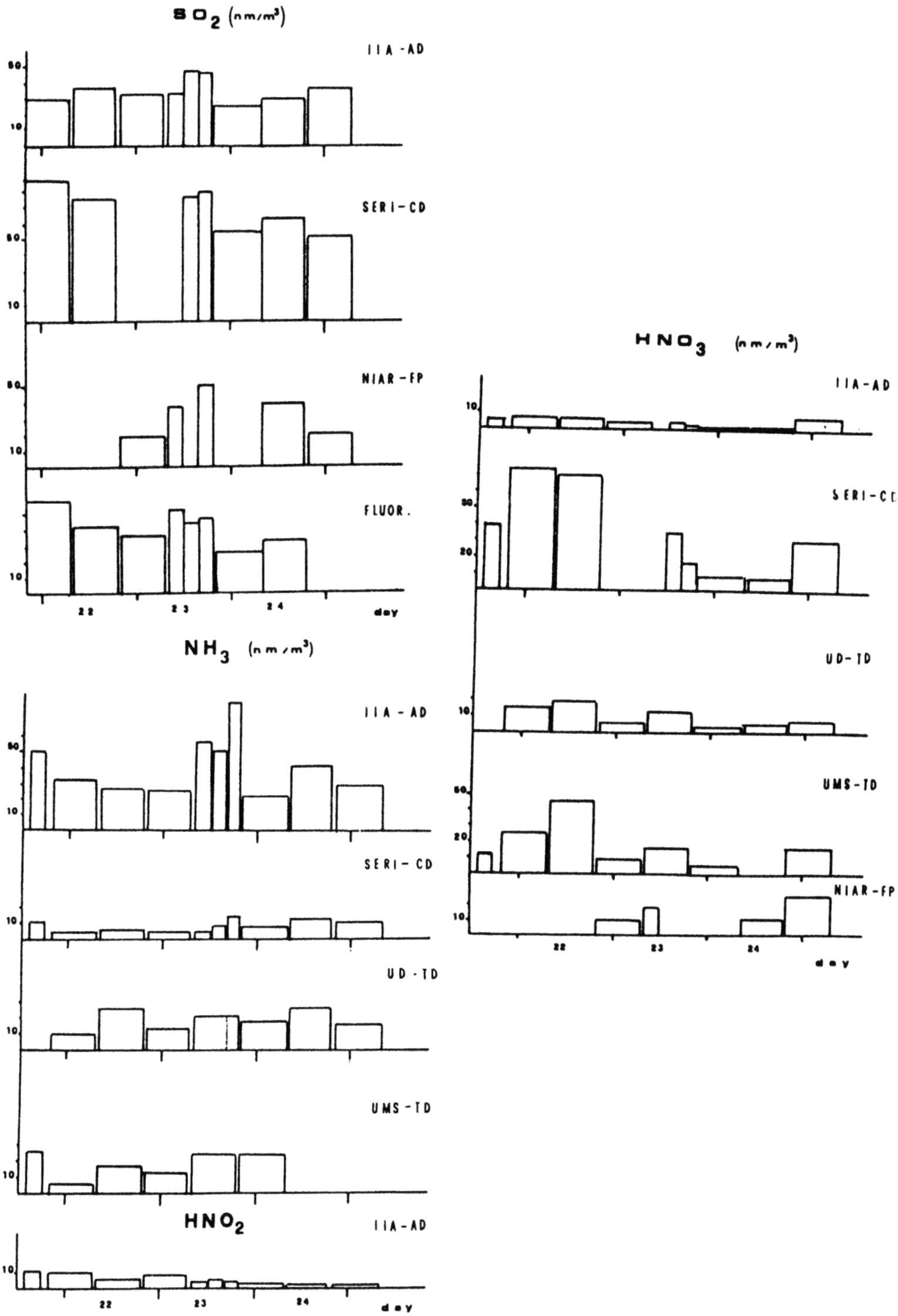

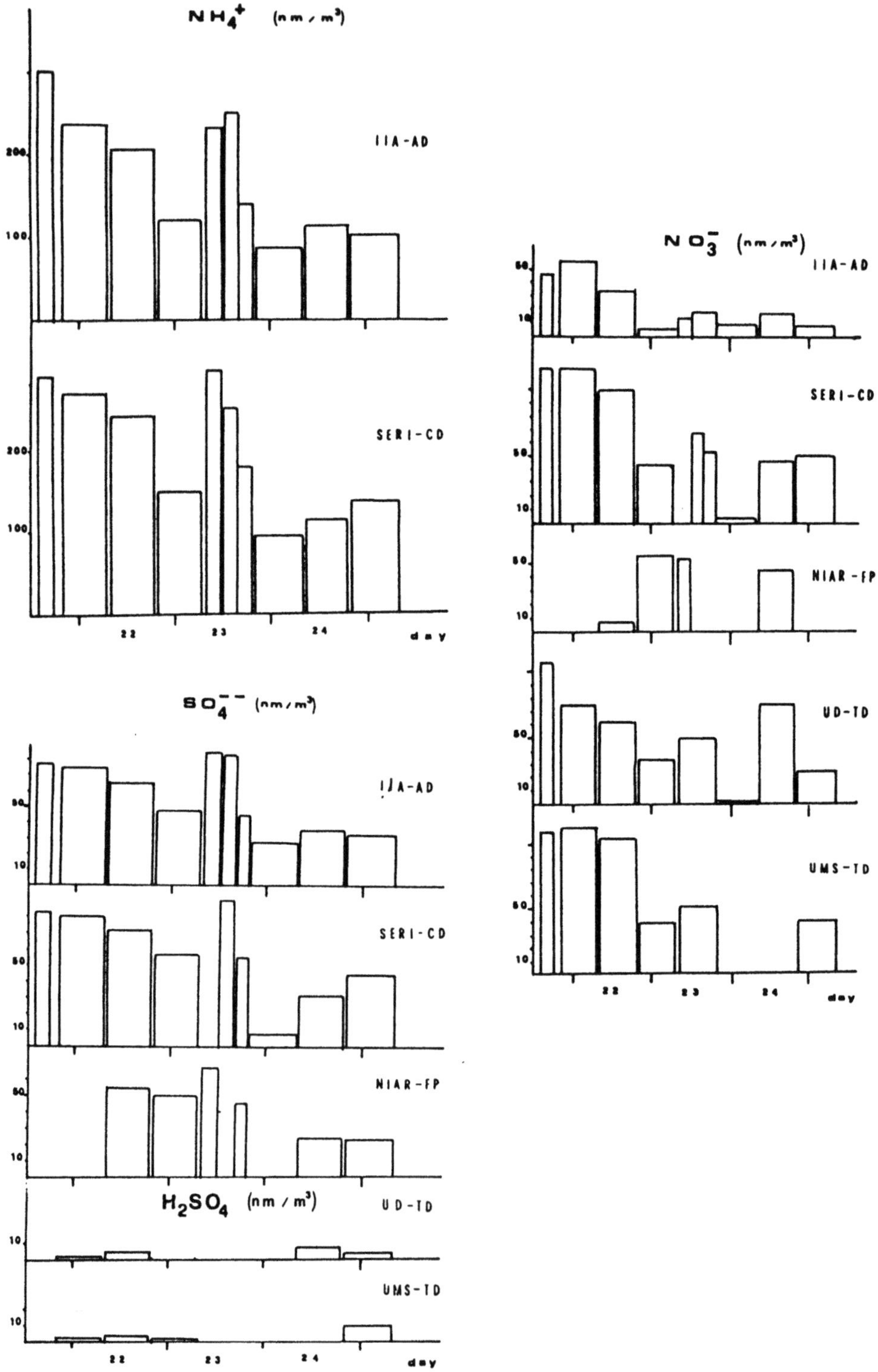
NH_4^+ (nm/m³)
IIA-AD
SERI-CD
200
100
22
23
24
day
NO_3^- (nm/m³)
IIA-AD
SERI-CD
NIAR-FP
UD-TD
UMS-TD
50
10
SO_4^{--} (nm/m³)
IJA-AD
SERI-CD
NIAR-FP
H_2SO_4 (nm/m³)
UD-TD
UMS-TD

Fig. 2 - Regressions of the different methods vs. the geometrical mean.

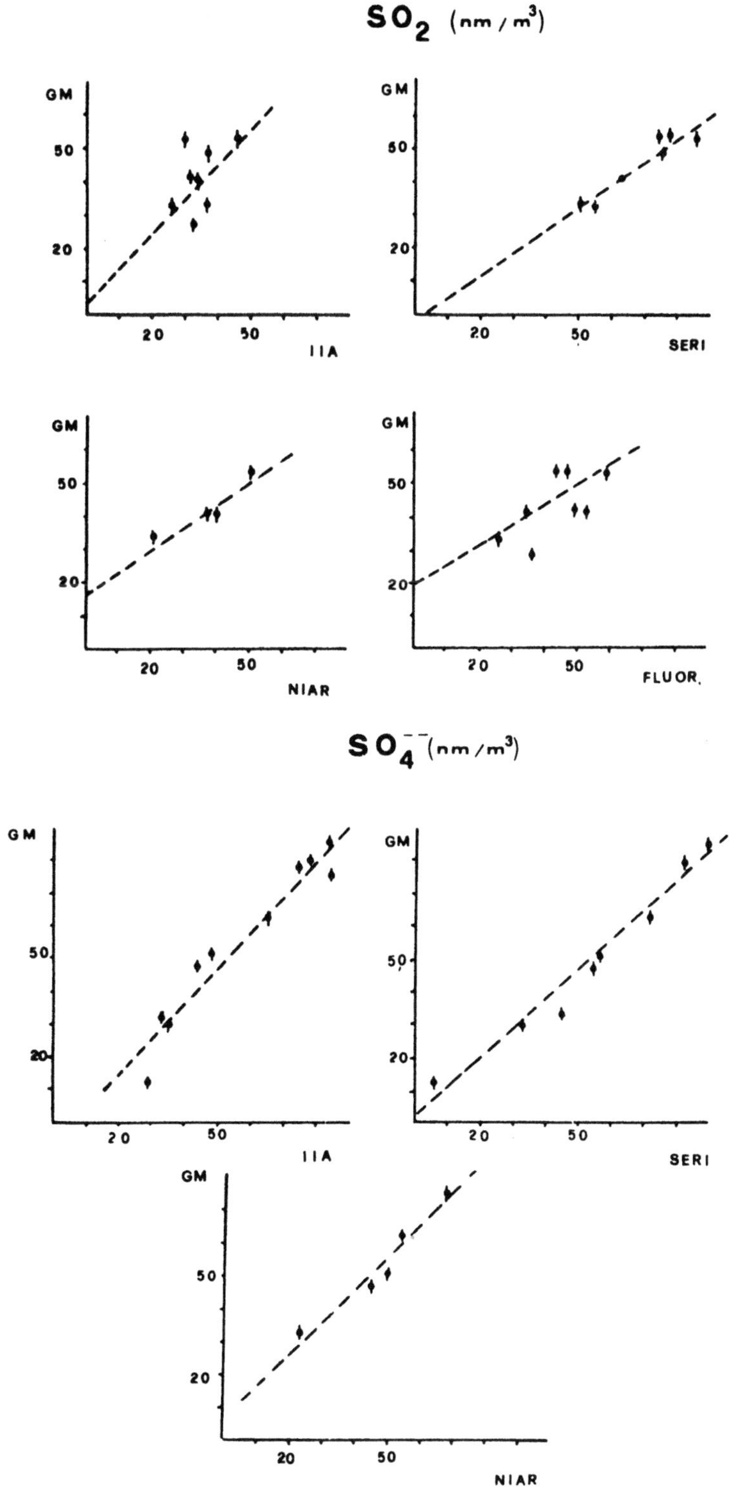

NH_3 (nm / m³)

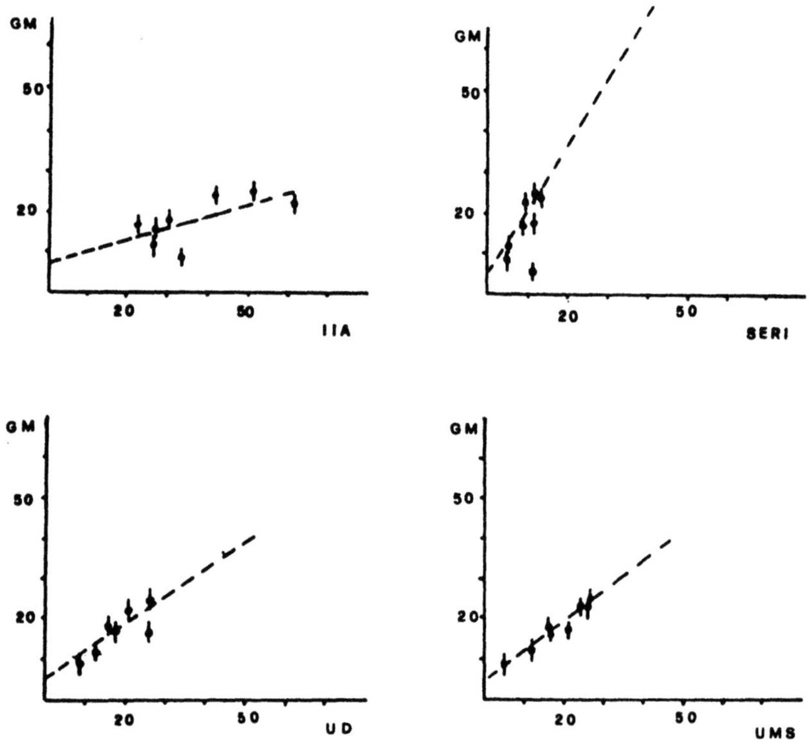

NH_4^+ (nm / m³)

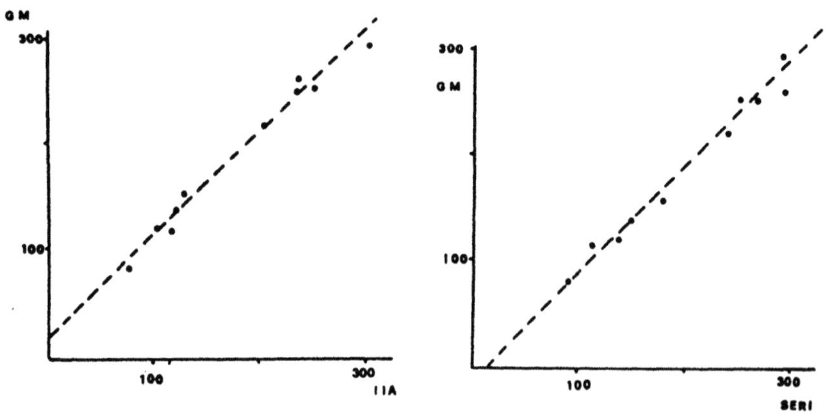

HNO_3 (nm / m^3)

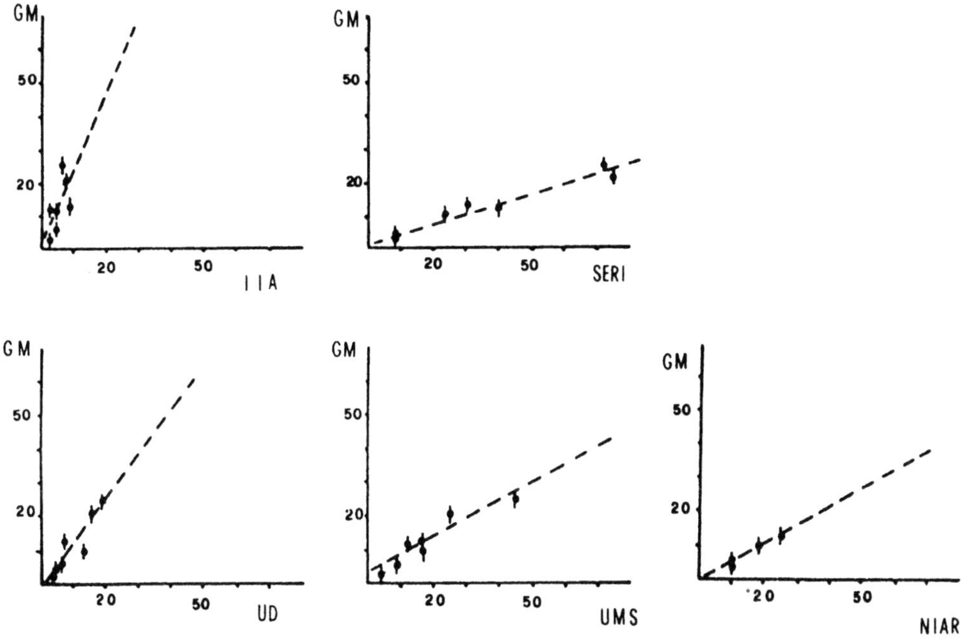

NO_3^- (nm / m^3)

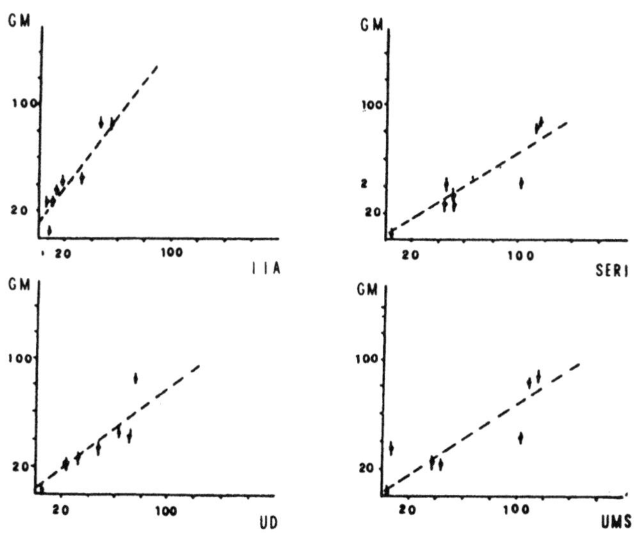

Fig. 3 - Correlation between $(SO_4^{=}) + (NO_3^{-})$ and (NH_4^{+}).

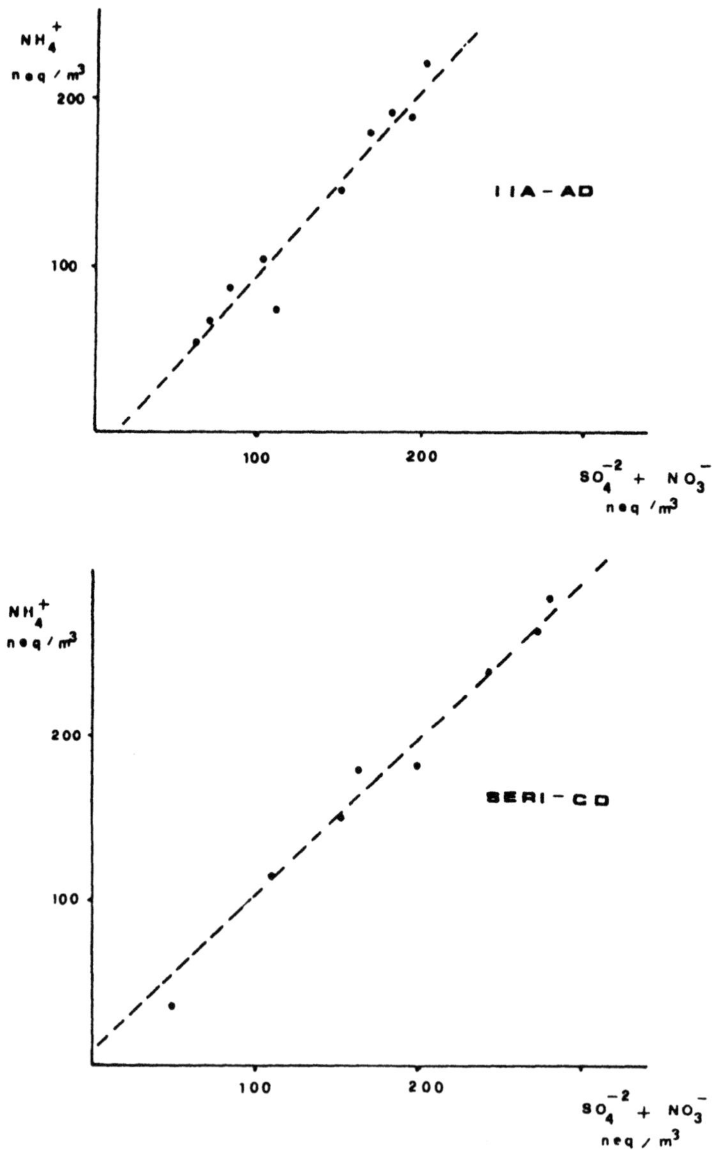

USE OF AIR - SO_2 MIXTURES FOR INTERCOMPARISON OF SO_2 MONITORS IN THE EEC

M. PAYRISSAT, H. RAU and G. SERRINI
Commission of the European Communities
Joint Research Centre - Ispra Establishment
21020 Ispra (Va) - Italy

Abstract

For an intercomparison programme among 29 continuous SO_2 analysers in eight EC Member States, cylinders of air-SO_2 mixtures were used as comparison standards. Two or three samples of two batches for each participating Member State were measured successively by the Central Laboratory (CL) of the JRC Ispra, calibrating with the TCM reference method, as referred in the EC Directive No. 80/779 EEC, and by one or more National Laboratories (NL) using their own calibration methods. Regular checks of the mixtures carried out by the CL showed an exponential decay of SO_2 concentration in the cylinders. In the two batches (190 and 40 ppb SO_2, respectively), mean losses of 0.48 and 3.7% per month were found with coefficients of variation of 28 and 48%. The CL values were corrected for the observed SO_2 decay and a statistical analysis performed between $\overline{NL}$ and $\overline{CLm}$, $\overline{CLm}$ being the corrected CL concentrations at the measuring month of NL. In the average, 68% of the 186 $\overline{NL}$-$\overline{CLm}$ cross schecks showed comparability (± 6.3%) versus the CL reference standard. This standard, dynamically generated by permeation, showed a quite good agreement with the TCM method. Considering on the one hand reference laboratory instruments (RI) and, on the other, network instruments (NI), the range of comparability included 73% RI and only 40% LI cross checks. The calibration procedures are discussed in so far as the main quality parameter is concerned.

1. INTRODUCTION

In the framework of the implementation of the EC Directive No.80/779/EEC setting limit and guide values for sulphur dioxide and suspended particulates, JRC Ispra, on behalf of DG XI, acts as a Central Laboratory (CL) with the task to clear harmonization in this area for the Member States. The Quality Assurance Programme (QAP), performed during 15 months since November 1984 (in fact, an intercomparison exercise for measurements of SO_2, black smoke and suspended particulate material) was conducted by the CL in cooperation with National Coordinating Organizations (NCO), nominated by the Member States (Table I).

The SO_2 QAP here presented consisted of intercomparisons between continuous methods and the tetrachloromercurate-pararosaniline (TCM method) as laid down in the EC Directive (1) to which the continuous methods must show a "satisfactory correlation" as reported in article 10.1. Actually, the comparison of different calibration methods was implemented with transfer standards of unknown SO_2 concentrations for the participants. These were air-SO_2 mixtures filled in pretreated aluminium and steel cylinders and measured by the CL with a continuous analyser, regularly calibra-

Table I. Organization of the EC Quality Assurance Programmes (Directive 80/779/EEC)

CEC. Bruxelles — **DGXI** — Proposal to the Member States (K.H. Zierock)

▽

JRC Ispra — **Central Laboratory, CL** — Programme coordinator (M. Payrissat)

▽

National Coordinating Organisations, NCO

Member State	Organisation or Laboratory	National coordinator (*)
(*) B	Institut d'Hygiène et d'Epidémiologie	G. Verduyn
(*) DK	National Agency of Environmental Protection (Risö Nat. Lab.)	L. Gründahl
(*) FRG	Pilotstation des UBA (Offenbach)	W. Rudolf
(*) F	Ministère de l'Environnement	M. Legrand
(*) GR	Ministry of Environment	J. Viras
L	Administration de l'Environment	T. Weber
IR	National Institute for Physical Planning	P.J. Flanagan
(*) I	Istituto Superiore di Sanità	M.A. Bertolaccini
(*) NL	Rijksinstituut voor Volksgezondheid en Milieuhygiene (RIVM Lab.)	H.K. Van de Viel
(*) UK	Warren Spring Laboratory	G. Mc. Innes

(*) Participating to the SO_2 quality programme

▷

National Laboratories, NL

National tests of Sulfur dioxide

Black smoke and Suspended particulates

ted against the TCM reference method. The needed stability checks of the standards, in a previous exercise conducted by l'Institut d'Hygiène et d'Epidémiologie of Brussels and CEN Mol (2,3), were carried out during the ring test by a portable calibration device (4). In the present exercise, where each Member State received different cylinders, the stability checks have been carried out in the CL before and after the measurements of each national laboratory (NL). Although the results are still under examination and discussion with the participants, the question whether the NL-CL cross check procedure used in the exercise allows a correct intercomparison is considered hereafter.

2. TESTS, INSTRUMENTATION AND CALIBRATION PROCEDURES

2.1 The test procedures proposed by the CL

Two representative levels of air-SO_2 mixtures, prepared by two European manufacturers (Air Liquide and British Oxygen Company), were checked by the CL and shipped for analysis to the participants. These two batches were foreseen at nominal concentrations of 200 and 50 ppb, respectively, in conjunction with a mixture of SO_2-free air containing 320 ppm CO_2 for the zero checks (3). For the implementation, the CL delivered special stainless steel regulators, able to minimize the losses of SO_2, and an instruction manual of measuring procedures (5) defining an overflow rate (1 l/min) and measuring sequence time (20 min). In this manual, the CL proposed to dintinguish between a reference instrument (RI), well maintained and calibrated in the laboratory, and a network instrument (NI). Only the participants performing the tests with RI were asked to carry out one prior 2-point calibration (zero/span) using their routine laboratory standard.

The programme included national A-tests and international B-tests. The following Member States: United Kingdom, The Netherlands, Denmark, Greece and Belgium, participated in A-tests characterized by only cross checks in which different cylinders were successively measured by the CL and each NCO. The remaining participants (Italy, Federal Republic of Germany and France) carried out A-tests as national round robin tests for which the cylinders, sent to the respective NCO, were circulated to na-

TABLE II. Cross checks of different sets of mixtures between the CL and NCO - Comparison test of the means at 5% risk probability threshold.

MEMBER STATE	Measured Concentrations							Calculated Values				Comparison NL↔CLm at 5% risk
	NCO AT MONTH m			CL AT MONTH m±n				k ($month^{-1}$)	CLm at month m	NCO-CLm at month m	t exp	
	m	ppb	st. dev.	n	ppb	n	ppb					
UK	11.84	31	2.3	— 1	34	+ 4	33	0.014	34 ppb	—3 ppb	1.85	NCO = CL
	2.85	21	2	— 4	27	+ 4	21	0.048	22	— 1	1.03	NCO = CL
	11.84	202	3.5	— 1	201	+ 7	197	0.0027	200	+ 2	0.77	NCO = CL
	2.85	186	5.6	— 4	200	+ 4	192	0.0045	196	—10	2.766	NCO ≃ CL
NL	11.84	46	1	— 1	45	+ 7	44	≃0	45	+ 1	1.29	NCO = CL
	2.85	20	1	— 4	27	+ 4	13	0.074	21	— 1	1.29	NCO = CL
	11.84	199	3.1	— 5	204	+ 7	191	0.0055	198	+ 1	0.42	NCO = CL
	2.85	189	2.7	— 1	193	+ 4	186	0.0043	190	— 1	0.45	NCO = CL
(*) DK	11.84	34	1.2	— 1	36	+ 7	24	0.050	34	0	0	NCO = CL
	2.85	42	1.5	— 1	45	+ 4	44	≃0	44.5	— 2.5	2.48	NCO = CL
	11.84	199	4	— 1	199	—	n.v	n.v	199	≃ 0	≃0	NCO = CL
	2.85	189	2.5	— 4	199	+ 4	191	0.0055	195	— 6	2.769	NCO ≃ CL
(*) GR	1.85	20	0	— 3	28	+12	18	0.030	26	— 6	n.v	NCO ◄ CL
	1.85	34	2	— 3	39	+12	25	0.0310	35	— 1	1.19	NCO = CL
	1.85	200	2.9	— 3	202	+ 1	201	0.0038	200	0	0	NCO = CL
	1.85	196	4	— 3	201	+ 5	194	0.0044	198.5	— 2.5	0.89	NCO = CL
(*) B	**8.85	25	0.6	—10	31	+ 5	22	0.023	25	0	0	NCO = CL
	3.85	17	1.2	— 2	31	+ 3	24	0.050	28	—11	12.7	NCO ◄ CL
	**8.85	183	2.1	—14	203	+ 5	18.7	0.0053	188	— 5	2.47	NCO = CL
	3.85	172	5	— 2	196	+ 3	185	0.0072	189	—17	5.14	NCO ◄ CL

(*) Network instrument
(**) After re-check

tional laboratories (NL). For an estimate of the reproducibility, two rounds (A1 and A2) were carried out. All the Member States participated also in 3 international B-tests in which the cylinders were consecutively measured by 3 NCOs.

2.2 The instrumentation and the calibration procedures

The laboratories having performed only cross schecks with the CL (UK, NL, DK, GR, B) used an UV-fluorescent (UV-F) monitor, except Belgium, which used a flame photometric detector (FPD). All instruments were RI instruments, except the Danish (NI calibrated in the laboratory) and the Belgian (settled in mobile laboratory). Concerning the primary standards for the calibrations, the NCOs mentioned above declared to use a multipoint calibration with commercial or home-made (NL) permeation tubes, except UK and B which implemented the hydrogen peroxide method. Except NL it seems that all NCOs performed the prior calibration with a transfer standard.

Italy participated with one RI (FPD) at ISS-Roma, and four NI (conductimetric or coulometric) in Rome, Milan, Torin, Bologne. At ISS, the primary standard was a permeation tube gravimetrically calibrated. The NI were calibrated by 1 or 2 points (one above 200 ppb) by means of permeation devices incorporated in the analysers or static dilution system (Rome).

Germany participated with 14 RI in 12 different laboratories. The primary standards were permeation devices, TCM method or static injection. The permeation rates were used as given in the certification (Saarbrücken, Oppenheim), checked gravimetrically (Wiesbaden, Essen) or against TCM method (Wiesbaden, München, Itzehohe, Karlsruhe, Hamburg). A home-made primary standard by static injection (UBA, Offenbach) was used to prepare 3 independent standards. The remaining laboratories used the TCM method by means of transfer standards checked each 3-4 months.

France participated with 4 RI: 3 UV-F (IRCHA, Marseille, Douai) and 1 coulometric (Saint Etienne). All laboratories used permeation devices (home-made tube in St. Etienne), checked gravimetrically (IRCHA, St. Etienne). One gas cylinder was used as transfer standard (Marseille) for the prior calibrations.

The Central Laboratory used 2 UV-F monitors (one for cross check and emergency), regularly calibrated with two calibrators at two dynamic dilution levels: one 3 channel "Tracor" loaded by 2 "Metronics" permeation tubes and one 3 channel ATEC built by SCK-Mol, loaded by 3 home-made tubes. The output concentrations were determined by weighing regularly the 5 permeation tubes (± 0.01 mg) and measuring the flow rates with a gas meter at 0.5% (Tracor) and CFM (ATEC). With the zero and 5 different permeation tubes in the range 0 - 250 ppb, a 6 point calibration curve was obtained to calibrate the monitors. This curve was cross checked with the TCM reference method mentioned in the Directive, using the improved procedures (6). The measured calibration factor (SO_2 concentration in the sample solution per absorbance unit) using sodium bisulphite and permeation gas mixtures, was found within 3% difference. The routine cross checks, carried out by transfer standards, were two air-SO_2 mixtures regularly checked by TCM.

Many logistic problems were encountered (delay of cylinder delivery, probable lack of pre-stability tests by one manufacturer, delay of shipping to some laboratories, not replacement of CL staff, etc.), so that the programme lasted more than was foreseen.

3. RESULTS AND DISCUSSION

3.1 Comprehensive measurements

The final figure requested actually was the mean of three measurements carried out in two consecutive days (5). For each SO_2 level, the required measurements were the differences between two readings: the EEC span and the EEC zero level obtained after 20 minutes.

A large number of participants observed that they did not attain stable readings after 20 minutes. Several parameters seemed to be responsible for this phenomenon, e.g. the SO_2 losses in the sampling line, maybe into the regulator. Consequently, the observed concentration was slightly lower than the real concentration of the bulk of the mixture.

The standard deviations of the three measurements were given by the participants in the average 0.6 - 2 ppb at 40 ppb level and 2 - 4 ppb at 190 ppb level. The mean concentrations were then within ± 3 ppb and ± 7 ppb, respectively, at 95% confidence level.

The results have been reported in a previous work (7) on eight tables, including - as shown in the example of Table II - two main headings titled "measured concentrations" on the left-hand side and "calculated values" on the right-hand side. Under the column "measured concentrations" we have mentioned for each mixture two CL measurements: one carried out before and one after the national test. For cross checks carried out with time intervals of less than 3 - 4 months, a fair agreement NCO-CL was generally found (difference within ± 4 ppb). At increased time interval, the differences became greater, mostly negative versus prior CL checks and positive versus later CL checks. The SO_2 concentrations in the cylinders were then unstable, as we noticed during 20 months by 3/4 month checks since June 1984.

Exponential CL(m) decay curves, $\overline{CLm}=\overline{CLo}e^{-km}$ versus the measuring month m fitted for most mixtures with correlation coefficients $r \geq 0.96$. For the high level (nominal concentration 200 ppb in October 1984), a mean decay constant $\bar{k}$ equal to 0.0048 $month^{-1}$ was found with a coefficient variation CV = 28%. For the low level (average initial concentration 40 ppb in November 1984), higher decay constants ($\bar{k}$ = 0.037 $month^{-1}$) were found with CV = 48%, indicating lower stability and higher fluctuations of the batch.

The stability of SO_2 in the cylinders has been questioned also in a recent exercise, carried out by BCR, CEC Brussels, with "Air Liquide" cylinders at 90 ppb. The impurities of the gas matrix and the interaction at the smoothed or coated surfaces (absorption centers for SO_2 molecules) would be responsible of different SO_2 losses and k avlues.

Figures 1 and 2 show the measurements of different air-SO_2 mixtures with RI instruments carried out in Germany and France. Figure 3 shows the measurements of other cylinders with NI instruments. In all the figures, the CL(m) decays have been represented in dotted lines.

3.2 Statistical evaluation and discussion

The intercomparisons NL versus CL, for all the 186 performed cross checks, have been attempted by means of the comparison test of the means (t test). The actual $\overline{NL}$ values and the corresponding $\overline{CLm}$ values (CL concentration calculated at the measuring month m of NL) allowed to obtain t_{exp} (Table II):

$$t_{exp} = \frac{\overline{NL}-\overline{CLm}}{\sqrt{\frac{s^2_{\overline{NL}} + s^2_{\overline{CL}}}{3}}} \quad (s^2\text{: variances within laboratory})$$

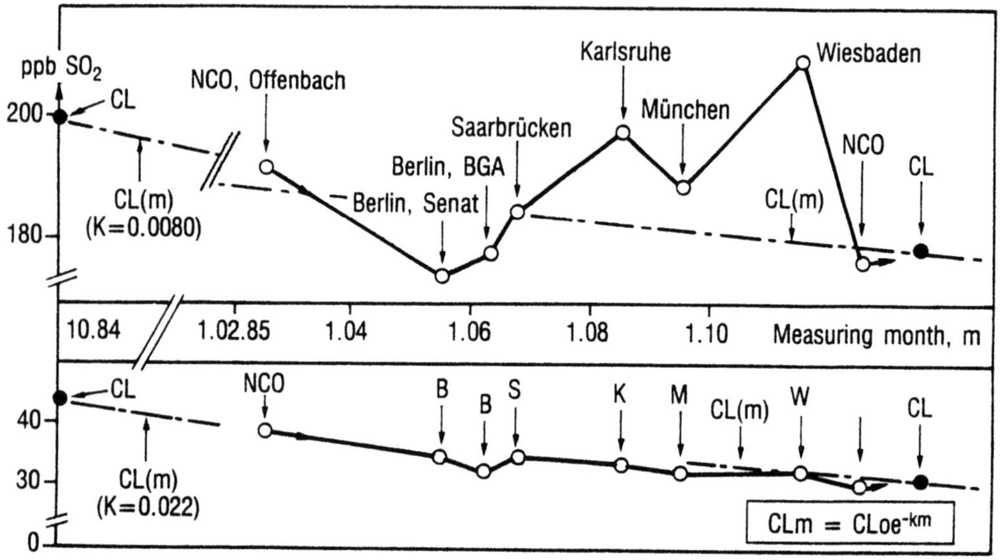

Fig. 1. Check of reference laboratory instruments - National test with Germany, 2nd set of laboratories. Filled circles (●): CL measurements; empty circles (O): NL measurements; Dotted lines: CL(m) regressions versus time.

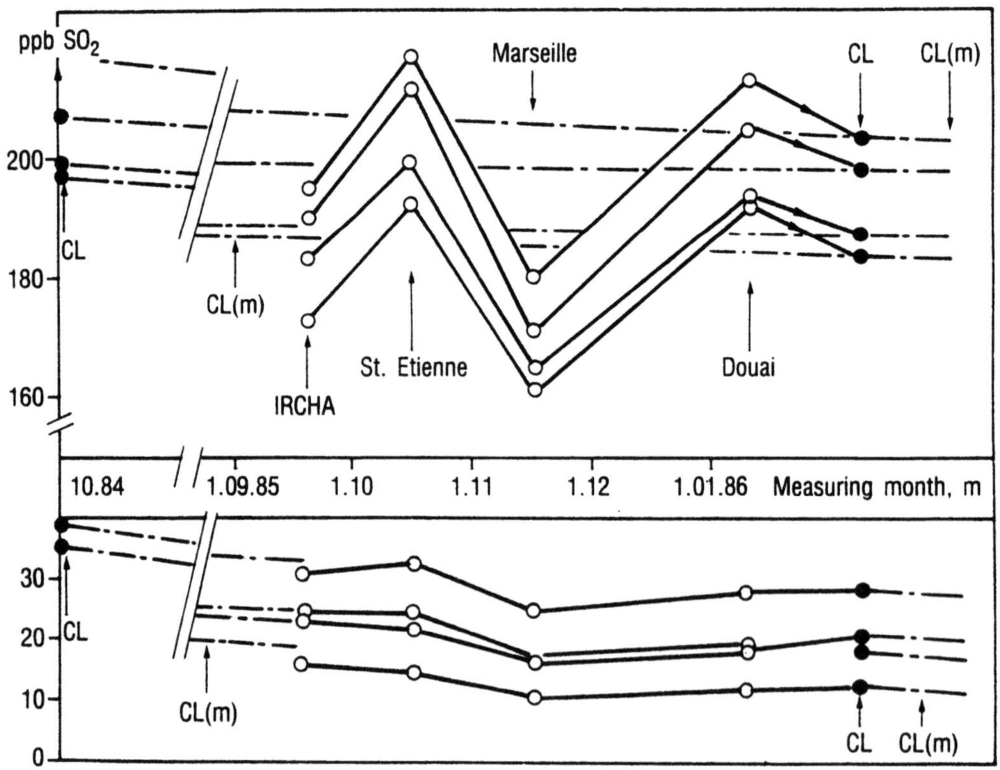

Fig. 2. Check of reference laboratory instruments - National test with France.

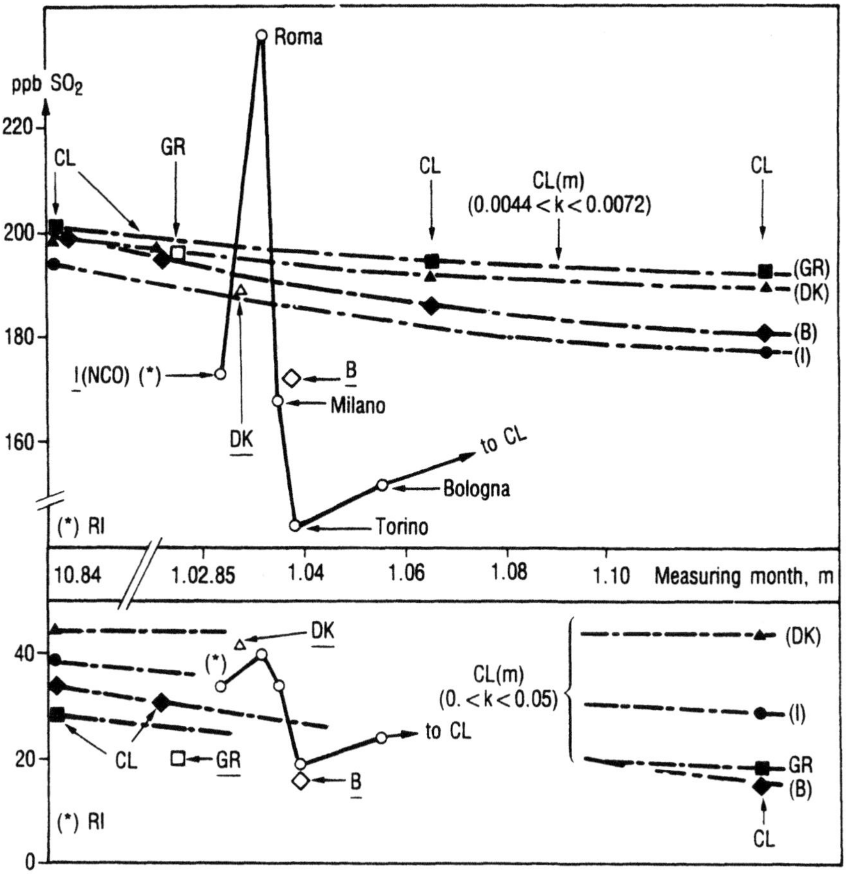

Fig. 3. Check of network instruments - CL-NL cross checks with Greece, Denmark, Belgium - National test with Italy.

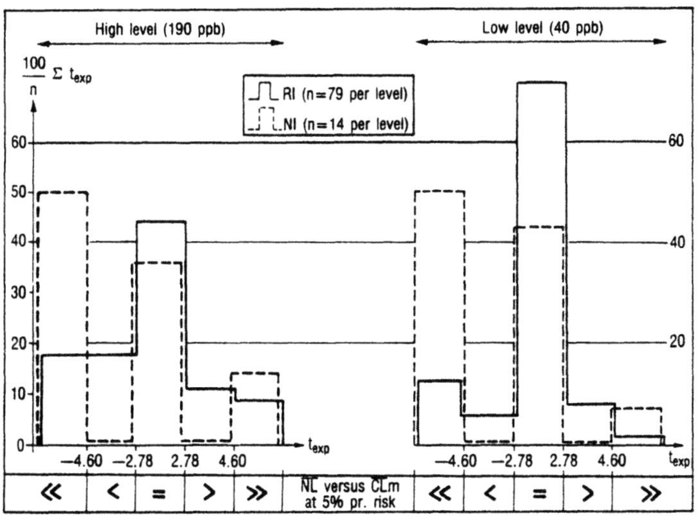

Fig. 4. Comparison test of the means, $\overline{NL}$ versus $\overline{CLm}$. Distribution of the t_{exp} value for Reference and Network instruments.

The actual NCO and NL variances and averaged CL variances, taken as 0.81 and 7.84 for low and high level have been considered for the calculations.

As shown in the last column of Table II, we have distinguished, for 4 degrees of freedom:

- the $\overline{NL}$ means not significantly different from $\overline{CLm}$ at 5% risk probability ($t_{exp} \leqq 2.776$) by the sign "=";
- the NL means significantly larger "≫" or lower "≪" than $\overline{CLm}$ corresponding to $t_{exp} \geqq 4.60$ (threshold value for 1% risk probability);
- the intermediate means ($2.776 < t_{exp} < 4.60$), not found in Table II, either lower "<" or higher ">" than $\overline{CLm}$.

Our purpose was to know whether the range of comparability in this way defined allows to individualize the most important quality parameters. From the distributions of the t_{exp} values (Figure 4), it can be argued that the medians of the RI distributions show the same comparability with the CL whatever the primary calibration standard may be (RI distributions quite gaussian). In addition, comparing RI and NI distributions, it can be concluded that the NI values are biased for the lack of calibration maintenance (a prior 2 point calibration was not required for NI).

Regarding the RI intercomparisons for each single laboratory (Table III), a certain scattering of Δ% values is observed because, in this case, of the relatively low number of observed cross checks. In the individual $\bar{\Delta}$% values, calculated versus $\overline{CLm}$ and illustrated in Figure 5, we have reported the comparability range δ, defined according to the statistical test used in this work.

The interval of comparability δ, at 95% confidence interval, was calculated for $t_{exp} = 2.776$, extracting Δ% from (1) and assuming $\overline{CLm}$ equals 190 and 40 ppb. Due to the lower coefficient of variation at high level, the above interval was more stringent at 190 ppb (δ = 3.5%; 43% LI and 36% NI cross checks) than 40 ppb level (δ = 6.3%; 72% LI and 44% NI , even if a 6.3% "averaged accuracy" at both levels shows a similar stringency : 73% NI and 40% NI cross checks). Out of the above mentioned δ range, a larger proportion of biased values is found at high concentration, due to the deviation of the calibration curves under comparison. In fact, comparing a given NL calibration curve versus a CL calibration curve having a different slope, the proportion of biased values increases with the concentration level.

As already known, the TCM method and permeation system can achieve accuracies of the order of 4 and 2% (8). If the majority of the $\bar{\Delta}$% fluctuations already mentioned (here and there the 0% line inside the δ range) can be explained by random errors such as differences of permeation temperatures, tube weighings, flowrate measurements, SO_2 absorptions (caused also by the CL), it is difficult to explain by these arguments the largest deviations observed for the remaining 27% RI.

4. CONCLUSIONS

As confirmed by the present study, the air-SO_2 mixtures in pretreated cylinders, at concentration ranges between 40 and 200 ppb, can be used for the intercomparison of SO_2 analysers, providing the control of the factors influencing the measurements reliability: concentration stability and implementation procedures. Due to the SO_2 concentration losses in the cylinders (up to 3.7% per month), it is advisable to reduce at maximum the duration of the exercises. To this end, for community-wide exercises, the multiple cross check procedure versus a pilot laboratory, as presented, could be an advantage compared to the ring test procedure. However, to

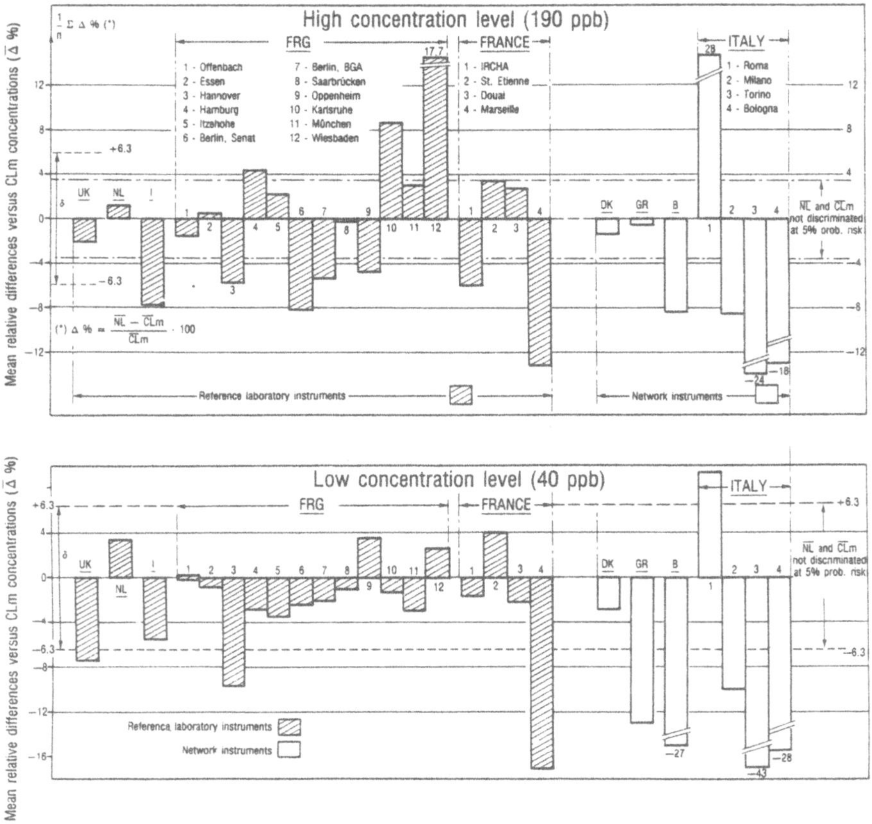

Fig. 5. Mean relative differences between National Laboratories and Central Laboratory corrected from the concentration variability of the cylinders versus time.

draw valuable conclusions, the exercise needs to be repeated.

The most important quality parameters here identified are related to the implementation of the calibration primary standards. From the 27% of Reference instruments out of the comparability range and still under examination, we do not know to what extent other parameters influencing the quality of the measurements could be identified.

If the calibration procedures are the most important parameters for harmonization of measuring methods, some problems remain for the implementation of air-SO_2 mixtures as calibration standards. Defined methods and procedures for mixture preparation and certification are still needed to use these gaseous reference materials as first order reference material (9).

ACKNOWLEDGEMENTS

The authors wish to thank the Directorate General for the Environment Consumer Protection and Nuclear Safety, CEC, DG XI, Brussels, who supported this work, the authorized staff of the participating Member States, the J. Büsing contribution during a too short stay at the CL, and H.J. Peperstraete for helpful redaction advise.

TABLE III. Intercomparison results versus the Central Laboratory, after decay corrections of the mixtures

Member State	Laboratory (NL)	Measurement method	n Nr. of crosschecks per level	Mean relative difference $\overline{NL}$ versus $\overline{CLm}$			
				Low level		High level	
				$\bar{\Delta}$ % (*)	stand. error (**)	$\bar{\Delta}$ % (*)	stand. error (**)
			RI INSTRUMENTS				
UK	WSL, NCO	UV-F	2	− 6.7	2.1	− 2.1	3.1
NL	RIVM, NCO	UV-F	3	+ 2.7	4.5	+ 1.2	1.2
I	ISS, NCO	FPD	3	− 5.6	0.7	− 7.8	0.7
D	UBA, Offenbach	UV-F	8	0	1.8	− 1.6	0.4
	Essen	UV-F	3	− 0.9	2.3	+ 0.5	0.3
	Hannover	FPD	2	− 9.7	1.1	− 5.7	1.5
	Hamburg	UV-F	2	− 2.8	2.8	+ 4.4	0.2
	Itzehoe (1)	UV-F	2	− 1.3	4.2	+ 3.1	0.5
	Itzehoe (2)	FPD	2	− 5.6	2.7	+ 1.3	0.2
	Itzehoe (1) + (2)		4	− 3.5	2.4	+ 2.2	0.6
	Berlin, Senat	Conduct.	2	− 2.3	5.6	− 8.2	0.3
	Berlin, BGA	UV-F	2	− 2	8.9	− 5.3	0.5
	Saarbrücken	UV-F	2	− 1	4.4	− 1.1	0.5
	Oppenheim	FPD	1	+ 3.6	n.v.	− 4.8	n.v.
	Karlsruhe (1)	UV-F	2	+ 1.1	6.6	+ 6.7	1.4
	Karlsruhe (2)	Conduct.	2	− 3.6	7.5	+ 10.5	0.3
	Karlsruhe (1) + (2)		4	− 1.3	4.3	+ 8.6	1.2
	München	UV-F	2	− 2.9	2.9	+ 3	0.3
	Wiesbaden	Conduct.	2	+ 2.7	2.7	+ 17.7	1.1
F	IRCHA	UV-F	8	− 1.6	2.3	− 5.8	0.8
	Saint-Etienne	Coulom.	8	+ 4.1	2.8	+ 3.5	1.2
	Douai	UV-F	8	− 2.1	2.7	+ 2.8	0.8
	Marseille	UV-F	6	−17	2.8	−13.1	0.5
			NI INSTRUMENTS				
DK	RISÖ, NCO	UV-F	2	− 2.8	2.8	− 1.5	1.5
GR	NCO	UV-F	2	−13	10	− 0.6	0.9
B	NCO	FPD	3	−27	8.7	− 8.4	0.6
IK	Roma, Network	FPD	2	+ 9.5	3.3	+28	0.3
	Milano, Network	Coulom.	2	− 9.5	6.7	− 8.4	1.7
	Torino, Network	Coulom.	2	−43	0.2	−24	1.2
	Bologna, Network	Coulom.	2	−28	1.5	−18	0.7

(*) $\frac{1}{n} \Sigma \frac{\overline{NL} - \overline{CLm}}{\overline{CLm}} \cdot 100$

(**) $s/\sqrt{n}$ (s : stand. deviation on Δ % values)

REFERENCES

(1) Journal Officiel des Communautés Européennes, No. L 229, pp.48, "Méthode de référence pour l'analyse de l'anhydride sulfureux", Office des Publications Officielles des Communautés Européennes, BP 1003, Luxembourg (30.8.1980).

(2) MUYLLE, E.P., PEPERSTRAETE, H.J., "Harmonisation of methods for measurements of SO_2", CEC Study Contract ENV/223/74-FR. Institut d'Hygène et d'Epidémiologie, Bruxelles (10.81).

(3) MUYLLE, E.P., PEPERSTRAETE, H.J., "Harmonisation of methods for mea-

surements of SO_2", Report EUR 8052 EN, Office for Official Publications of the European Communities (1982).
(4) ISO - Standard 6349, "Gas analysis preparation of calibration gas mixtures - permeation method" (1979).
(5) PAYRISSAT, M. and ZIEROCK, K.H., "The organization of the first European Quality Assurance Programme", Proc. of the EC Workshop on Monitoring Sulphur Dioxide, Black Smoke and Suspended Particulates, RIVM, Bilthoven, 5-7 November 1984.
(6) PAYRISSAT, M., RAU, H. and SERRINI, G., "Improvement of the EC reference method for the determination of sulphur dioxide", Proc. of the EC Workshop on Monitoring Sulphur DIoxide, Black Smoke and Suspended Particulates, RIVM, Bilthoven, 5-7 November 1984.
(7) PAYRISSAT, M., "First Quality Assurance Programme - test procedures, data collection and evaluation", Preliminary final report presented at the meeting of Governmental Experts, Brussels, 26.5.1986.
(8) VAN DE WIEL, H.J., UITERWIJK, J.W. and REGTS, T.A., "Dynamic generation of test atmospheres - quantitative aspects", Proc. of the Int. Workshop on Air Pollution Reference Measurement Methods and Systems, Bilthoven, 1977, p.41-51.
(9) PEPERSTRAETE, H.J., "Gaseous reference materials - certification - traceability", Proc. of the 2nd Eur. Symp. on Physico Chemical Behaviour of Atmospheric Pollutants, Varese, October 1981, p.68-75.

DETERMINATION OF POLYCHLORINATED DIBENZO-P-DIOXINS AND DIBENZOFURANS IN OUTDOOR AIR

P. KIRSCHMER and M. BUCK
Landesanstalt für Immissionsschutz des Landes Nordrhein-Westfalen, Essen

Summary

Polychlorinated dibenzo-p-dioxins (PCDD) and Dibenzofurans (PCDF) were determined in 50 outdoor samples collected in Nordrhine-Westphalia at different sites in areas of different land use and emission structures. It was found, that higher chlorinated PCDD/PCDF are collected on glass fiber filter and that lower chlorinated PCDD/PCDF (Tetra- and Penta-CDD/CDF) are absorbed on polyurethane foam. There is no indication that 2,3,7,8-Isomers of PCDD/PCDF are predominantly decomposed by photochemical reactions in outdoor air. 2,3,7,8-TCDF was found at 14 sites but no 2,3,7,8-TCDD could be detected. The mean concentration for the sum of PCDD (4-8) and PCDF (4-8) was 3,2 pg/m³ and 5,2 pg/m³ respectively in the Rhine-Ruhr-District. The detection limit for TCDD was in the order of 0,1-0,2 pg/m³ and for higher chlorinated PCDD/PCDF (5-8) in the order of 0,005-0,015 pg/m³.

Zusammenfassung

Polychlordibenzo-p-dioxine (PCDD) und Polydibenzofurane (PCDF) wurden in 50 Immissionsproben von Nordrhein-Westfalen an verschiedenen Meßstellen bestimmt. Die Meßstellen lagen in Gebieten mit unterschiedlicher Flächennutzung und Emissionsstruktur. Die Messungen ergaben, daß höher chlorierte PCDD/PCDF auf dem Glasfaserfilter zurückgehalten werden, und daß niedriger chlorierte PCDD/PCDF (Tetra- und Penta-CDD/CDF) auf dem Polyurethanschaum absorbiert werden. Ein bevorzugter Abbau der in 2,3,7,8-Stellung chlorierten PCDD/PCDF durch photochemische Reaktionen konnte nicht festgestellt werden. 2,3,7,8-TCDF wurde an 14 Meßstellen bestimmt, während 2,3,7,8-TCDD nicht nachgewiesen werden konnte. Im Rhein-Ruhr-Gebiet betrug die mittlere Konzentration für die Summe der PCDD(4-8) 3,2 pg/m³ und für PCDF(4-8) 5,2 pg/m³. Die Nachweisgrenze für TCDD lag bei 0,1-0,2 pg/m³; für die höher chlorierten PCDD/PCDF (5-8) wurde eine Nachweisgrenze von 0,005-0,015 pg/m³ erreicht.

1. EINLEITUNG

In den vergangenen 4 Jahrzehnten ereigneten sich mehrere industrielle Unfälle, bei denen Polychlorodibenzo-p-dioxine (PCDD) freigesetzt wurden. 1977 und 1978 fanden zwei unabhängige Forschergruppen (1, 2) heraus, daß auch in Flugstäuben von Hausmüllverbrennungsanlagen PCDD enthalten sind. In der Folgezeit wurde in vielen weiteren Proben, wie z.B. Deponien, Ruß von Bränden, Sedimenten, Fischen und in Muttermilch, PCDD und PCDF nachgewiesen. Diese Analysenergebnisse werfen die Frage auf, auf welchen Wegen die PCDD/PCDF in die entsprechenden Matrices gelangen. Für die Abgase von Müllverbrennungsanlagen und die Abgase anderer Verbrennungen ist der Luftpfad, der Weg auf dem Spurenstoffe sowohl in das Wasser, den Boden wie auch in biologische Matrices gelangen können, sofern in der Luft selbst keine Umwandlung stattfindet. Um die Bedeutung des Luftpfades für die menschliche Gesundheit beurteilen zu können ist die Kenntnis der Konzentration von PCDD/PCDF im Immissionsbereich wichtig. Ziel der Untersuchung war es, einen ersten Überblick über die Immissionen von PCDD/PCDF in NRW zu erhalten. Dazu wurde ein Meßverfahren eingesetzt, das geeignet war Konzentrationen von 2,3,7,8-TCDD bis zu 0,1 pg/m³ und Konzentrationen von höher chlorierten PCDD/PCDF von 0,015 pg/m³ und weniger nachzuweisen. Diese Nachweisgrenze ist erforderlich, um Konzentrationen im Bereich des Richtwertes für 2,3,7,8-TCDD von 3 pg/m³, sicher zu erfassen.

2. DURCHFÜHRUNG DER MESSUNGEN

2.1. Meßstellen und Meßzeiten

Die Meßstellen wurden in Gebiete mit unterschiedlicher Flächennutzung und Emissionsstruktur gelegt (siehe Tab. I und Abb. I). Die Meßwerte sollten zeigen, ob die Strukturunterschiede in den verschiedenen Meßgebieten eine unterschiedliche Immissionsbelastung ergeben.

An jeder Meßstelle wurden drei Messungen über je drei Tage durchgeführt, um eine minimale statistische Absicherung zu erhalten. Diese räumlich-zeitliche Stichprobenmessung erlaubt es die Gesamtbelastung des Gebietes zu erfassen und ermöglicht es Schwerpunkte innerhalb dieses Gebietes zu erkennen.

Tab. I: Beschreibung der Meßstellen

Meßstelle	Flächennutzung und Emissionsstrukturen
Egge, Eifel, Selm	Land- und Forstwirtschaft
St. Augustin, Schwalmtal	Hausmülldeponie, Sondermülldeponie
Essen-Mitte	Nahbereich der Kfz-Verkehrsemission
Solingen, Essen-Katernberg, Leverkusen	Nahbereich von Müllverbrennungsanlagen
Dortmund-Derne	Nahbereich einer Kokerei
Bottrop, Oberhausen, Köln-Vogelsang, Duisburg-Walsum, Bochum-Gerthe, Essen-Altendorf	Belastungsgebiet des Rhein-Ruhr-Gebietes, mit unterschiedlichen Emissionsquellen im Nahbereich jeder einzelnen Meßstelle

2.2. Probenahme

Es wurden Proben von 1000 m³ gesammelt. Die Partikel wurden auf einem Glasfaserfilter zurückgehalten. Gasförmige Anteile der PCDD/PCDF wurden auf einem Polyurethanschaum zurückgehalten. Das Glasfaserfilter und der Polyurethanschaum wurden einzeln analysiert.

2.3. Analyse

Einzelheiten des Analysenverfahrens sind in (3) beschrieben. Im nachfolgenden wird der Analysengang nur kurz skizziert: Die Glasfaserfilter und die Polyurethanschäume wurden mit Toluol im Soxhlet extrahiert. Die Reinigung der Extrakte erfolgte in zwei Schritten: Im ersten Schritt wurden die PCDD/PCDF mit Toluol/Diethylether (9:1) von einer Florisil PR/Na_2SO_4/K_2CO_3-Säule eluiert. Im zweiten Schritt erfolgte die Trennung mit HPLC (Nucleosil 5 µm, NO_2, Chrompack).

Die Identifizierung und Quantifizierung erfolgte gaschromatographisch auf einer SP 2330 Trennkapillare und massenspektrometrisch mit Elektronenstoß-Ionisation (EI) und mit Negativ-Ionen Chemischer Ionisation (NICI).

3. ERGEBNISSE UND DISKUSSION

Die Analyse der Partikelfilter und des Polyurethanschaumfilters ergab für die einzelnen Chlorierungsgrade der PCDD/PCDF unterschiedliche Meßwerte. OCDD/OCDF und Hepta-CDD/CDF befanden sich überwiegend auf dem Partikelfilter. Tetra- und Penta-CDD/CDF wurden überwiegend von dem Polyurethanschaumfilter absorbiert und nur zu einem geringen Teil von dem Partikelfilter. Hexa-CDD/CDF waren im Partikelfilter und im Polyurethanschaumfilter zu finden. Dieses Ergebnis belegt die Notwendigkeit eines Absorptionsfilters für die Gasphase.

2,3,7,8-TCDD konnte an keiner Meßstelle nachgewiesen werden (NWG: 0,1 pg/m^3). An fast allen Meßstellen sind die Meßwerte für PCDF höher als für PCDD. In den Abb. II-IV ist zu erkennen, daß in dem urbanen, industrialisierten Rhein-Ruhr-Gebiet die PCDD- und PCDF-Immissionen um ein vielfaches höher liegen als in den quellfernen Gebieten des Egge-Gebirges und der Eifel. Das Auftreten der sehr geringen PCDD und PCDF-Konzentrationen in den quellfernen Gebieten dürfte in erster Linie die Folge von Transporten aus industriellen oder urbanen Ballungsgebieten sein. Es ist davon auszugehen, daß in anderen Ballungsgebieten sowie auch in quellfernen Gebieten ähnliche Belastungsverhältnisse herrschen. Dies zeigen auch PCDD/PCDF-Immissionsmessungen aus Hamburg (4). Die PCDD/PCDF-Immissionen im Ballungsgebiet dürften auf Verbrennungsvorgänge in der vielfältigsten Form zurückzuführen sein. Diese Annahme wird bestätigt durch den Vergleich der Isomeren-Verteilung von PCDD- und PCDF-Immissionen mit Emissionsmessungen an Müllverbrennungsanlagen. In beiden Fällen erhält man sehr ähnliche Isomeren-Verteilungen für die PCDD und PCDF (3).

Die Annahme eines schnellen Abbaues durch UV-Strahlung der 2,3,7,8-chlorsubstituierten Dibenzo-p-dioxine bzw. Dibenzofurane kann durch die Meßergebnisse nicht bestätigt werden (Tab. II). Der Anteil der 2,3,7,8-chlorsubstituierten DD/DF ist bezogen auf die Konzentration der Gesamt-PCDD/PCDF bzw. die Summe der PCDD/PCDF eines entsprechenden Chlorierungsgrades (4-8) nicht reduziert. Daraus läßt sich schließen, daß ein UV-Abbau sofern er stattfindet, nur relativ langsam erfolgen kann und zu keiner nennenswerten Reduzierung der 2,3,7,8-Isomere beiträgt.

Gemessen an einem praktizierten Richtwert in NRW von 3 pg/m^3 für 2,3,7,8-TCDD sind die gefundenen PCDD/PCDF-Immissionen gering. Da für andere PCDD und PCDF Isomere keine Richtwerte existieren, ist eine Bewertung der gefundenen Immissionen zum jetzigen Zeitpunkt noch nicht möglich. Eine ausführliche Darstellung der Ergebnisse wird in (5) veröffentlicht.

Abb. I: Lage der Meßstellen in NRW

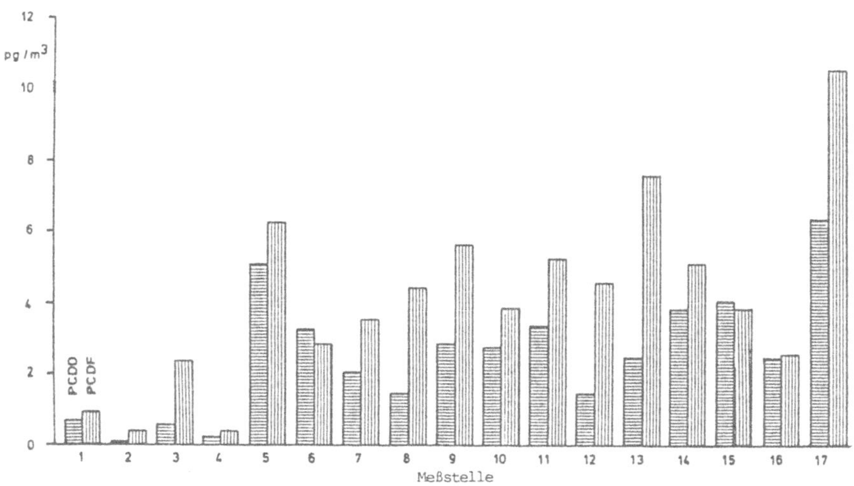

Abb. II: Vergleich der Summe PCDD mit der Summe PCDF an den einzelnen Meßstellen
1: Egge, 2: Eifel, 3: Selm, 4: St. Augustin, 5 und 6: Schwalmtal, 7: Essen-Mitte, 8: Solingen, 9: Essen-Katernberg, 10: Leverkusen, 11: Bottrop, 12: Dortmund-Derne, 13: Oberhausen-Osterfeld, 14: Duisburg-Walsum, 15: Bochum-Gerthe, 16: Köln-Vogelsang, 17: Essen-Altendorf

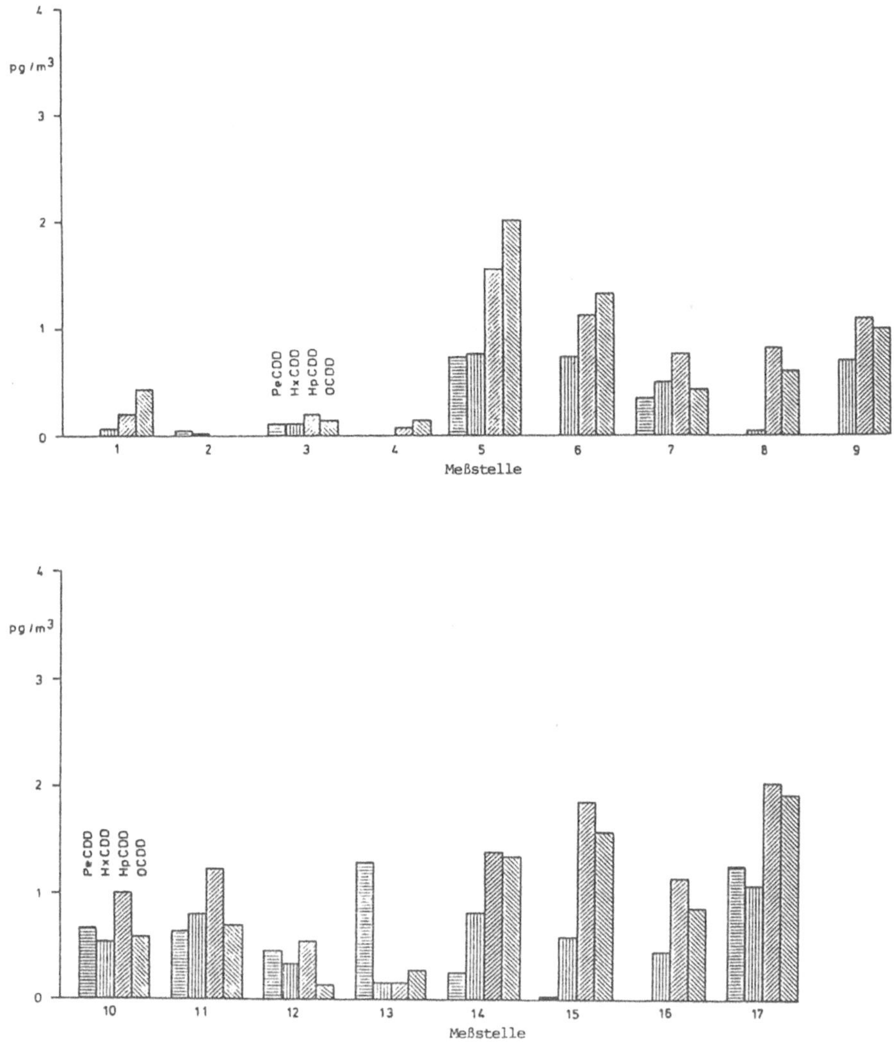

Abb. III: Vergleich der Konzentrationen für die unterschiedlichen Chlorierungsgrade (5-8) für PCDD an den einzelnen Meßstellen. Nummerierung der Meßstellen siehe Abb. II

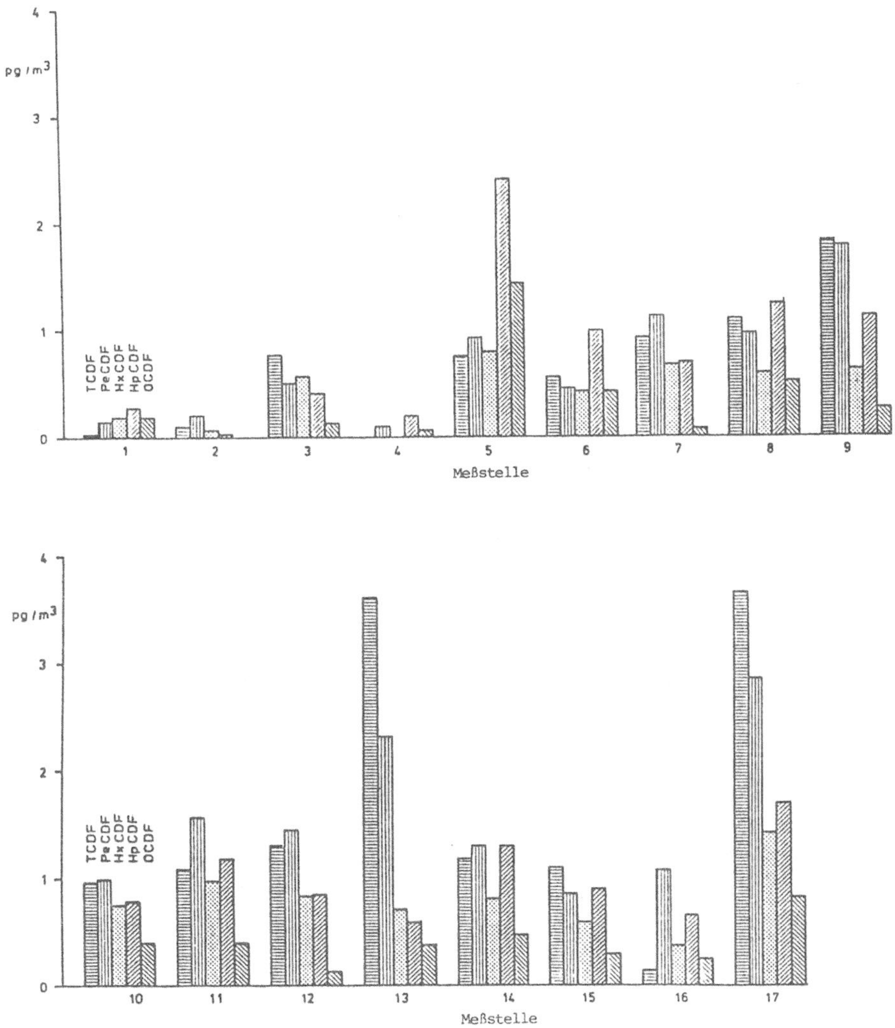

Abb. IV: Vergleich der Konzentrationen für die unterschiedlichen Chlorierungsgrade (4-8) für PCDF an den einzelnen Meßstellen. Nummerierung der Meßstellen siehe Abb. II

Tab. II: Mittelwerte und Bereiche der PCDD/PCDF-Immissionskonzentrationen im Rhein-Ruhr-Gebiet [pg/m³] gemessen an 10 Meßstellen in der Zeit vom 21.06.1985 - 13.10.1985

PCDD	Mittelwerte	Bereich	PCDF	Mittelwerte	Bereich
2,3,7,8-	*	*	2,3,7,8-	0,09	* - 0,20
1,2,3,7,8-	0,01	* - 0,06	1,2,3,7,8-		
			1,2,3,4,8-	0,15	0,04 - 0,35
1,2,3,4,7,8-	0,03	* - 0,08	2,3,4,7,8-	0,10	0,04 - 0,31
1,2,3,6,7,8-	0,06	* - 0,28	1,2,3,4,7,8-	0,13	0,03 - 0,48
1,2,3,7,8,9-	0,04	* - 0,08	1,2,3,6,7,8-	0,08	0,02 - 0,25
			2,3,4,6,7,8-	0,12	0,03 - 0,50
Σ TCDD	*	*	Σ TCDF	1,59	* - 4,42
Σ PeCDD	0,50	* - 1,66	Σ PeCDF	1,53	0,44 - 3,53
Σ HxCDD	0,62	0,15 - 2,16	Σ HxCDF	0,79	0,15 - 2,24
Σ HpCDD	1,16	* - 4,61	Σ HpCDF	0,99	0,16 - 2,66
OCDD	0,89	* - 3,71	OCDF	0,35	* - 1,23
Σ PCDD	3,16	0,15 -10,09	Σ PCDF	5,25	1,50 -13,83

* = ≤ Nachweisgrenze

LITERATUR

(1) OLIE, L., VERMEULEN, P.L., HUTZINGER, O. (1977). Chlorodibenzo-p-dioxins and Chlorodibenzofurans are Trace Components of Fly Ash and Flue Gas of some Municipial Incinerators in the Netherlands. Chemosphere 6, S. 455.

(2) BUSER, H.R., BOSSHARDT, H.-P., RAPPE, C. (1978). Identification of Polychlorinated Dibenzo-p-dioxin Isomeres Found in Fly Ash. Chemosphere 7, S. 165.

(3) OEHME, M., MANØ, S., MIKALSEN, A., KIRSCHMER, P. (1986). Quantitative Method for the Determination of Femtogram Amounts of Polychlorinated Dibenzo-p-dioxins and Dibenzofurans in Outdoor Air. Chemosphere 15, S. 607.

(4) BRUCKMANN, P., HACKHE, K.H. Immissionsmessungen von Dioxinen und Dibenzofuranen in Hamburg an Orten mit unterschiedlicher Luftqualität. Vortrag auf dem 21. Meßtechnischen Kolloquium der Ländereinrichtungen für Immissionsschutz, Mai 1986 in Wiesbaden.

(5) BUCK, M., KIRSCHMER, P. (1986). Immissionsmessungen Polychlorierter Dibenzo-p-dioxine und Dibenzofurane in Nordrhein-Westfalen, LIS-Berichte, Essen, in Vorbereitung.

MEASUREMENT OF ATMOSPHERIC NITROUS AND NITRIC ACID BY MEANS OF ANNULAR DENUDERS

A. Febo, F. De Santis, C. Perrino

Istituto sull'Inquinamento Atmosferico del C.N.R.
Area della Ricerca di Roma
Via Salaria Km 29,300 - C.P. 10
00016 Monterotondo Stazione (Roma) ITALY

Summary

A new sampling set-up for the determination of nitrous and nitric acid, consisting of a train of annular denuders, is described. The first denuder, coated with sodium chloride, collects nitric acid. The second and third denuders, coated with Na_2CO_3, are used for the determination of nitrous acid. The purpose of the two denuders set in series is to collect HNO_2 and to account for the interferences of species which may give nitrite ion as well as of the action of oxidants which during the sampling cause at variable extent nitrite to nitrate conversion. By subtracting the total amount of nitrite and nitrate, expressd as nitrite, found on the second denuder to that found on the first it is possible to perform a correct determination of nitrous acid.

INTRODUCTION

Accurate measurements of ambient concentrations of nitrous and nitric acids are of great importance to the understanding of the role of these species in the acid deposition phenomenon as well as to the validation of atmospheric photochemical reaction schemes.

Several sampling and analytical techniques have been so far developed, based on the use of spectrochemical techniques, filter packs and denuder systems. Tunable diode laser (1) and long-path differential U.V. (2) adsorption techniques have been proven to provide a reliable method for the measurement of HNO_3 and HNO_2 concentrations respectively, but their cost and maintenance are often prohibitive for most air quality studies. Filter pack methods, on the other hand, are recognized to suffer from both negative an positive artifacts, such as reaction of HNO_3 with basic particulate (3-4) and evolution of HNO_3 from co-collected ammonium nitrate (5).

The more successful approach to prevent sampling artifacts is the use of diffusion denuders which allow to selectively collect gaseous species leaving the particulate matter unaffected.

Denuders have been used as preseparators to enhance the selectivity of non-selective methods, as it is the case of the denuder difference method for nitric acid (6), or as collectors for both nitrous and nitric acid by using a sodium carbonate adsorbing layer (7-9).

For a correct use of sodium carbonate denuders the following points should be verified:

- Na_2CO_3 is a perfect sink for HNO_2 and HNO_3;
- No other gaseous species deposited on the Na_2CO_3 denuder with a good efficiency

give rise to the formation of the same ions;
- On Na_2CO_3, HNO_3 yields only NO_3^- and HNO_2 yields only NO_2^-.

The first point has been verified, and the sampling efficiency of HNO_2 and HNO_3 on Na_2CO_3 has been shown to be that predicted by the theory (9). As regards the second point it has been found that NO_2 and PAN on Na_2CO_3 yields nitrites, which interfere with the measurement of HNO_2. Moreover, particulate nitrate might affect the determination of HNO_3. However, the efficiency for the collection of these interfering species is small (1-3%) and their mass distribution along the denuder is fairly constant (10). Therefore, by arranging two identical denuders in series the amount of relatively unreactive interferents collected in the first denuder will be approximately equal to that found in the second denuder. By subtracting the amount of nitrite found in the second denuder to that found in the first one it is possible to obtain the amount due to nitrous acid (9). This sampling train also allows to take into account the small interference due to particulate matter.

The third point was also verified in laboratory experiments and the reaction products were proven to be stable (9).

Field trials have shown, however, that a transformation of nitrite to nitrate during the sampling may occur. This aspect has been extensively studied and a procedure for the correct determination of nitrous and nitric acid, which takes into account also the oxidation phenomena, is proposed.

EXPERIMENTAL

Glass annular denuders with sandblasted annulus walls, 22 cm long, 33 mm o.d. and 30 mm i.d., as previously described (11), have been used. The operating flowrate was 12 l/min. Three different coating solutions have been used: 1% Na_2CO_3 in 1:1 water methanol; 1% Na_2CO_3 + 1% glycerine in 1:1 water-methanol; 0.5% NaCl in 1:9 water-methanol.

Walls coating was obtained by pouring few milliliters of the solution into the annular space, sealing both ends and rotating the denuder to ensure complete wetting of the surface. The excess solution was then drained out and the walls dried with a gentle flow of clean air. After sampling the denuder were leached with at least 10 milliliters of distilled water and the solution analyzed by IC.

Artificial HNO_2 atmospheres have been obtained by means of a device based on room temperature oxalic acid sublimation onto solid sodium nitrite (13).

RESULTS AND DISCUSSION

If a sodium carbonate denuder, of the dimensions reported in the experimental section, is operated at the flowrate of 12 l/min, a collection efficiency of about 100% for HNO_2 and HNO_3 is expected. As a consequence, if two denuders are operated in series, the amount of nitrate collected on the second denuder should be only due to the small interference of particulate matter, while the amount of nitrite should be due to the interference of NO_2 and PAN. The unexpectedly high nitrate amount found on some occasions on the second denuder, which was not justified by a particularly high particulate nitrate concentration, suggested the hypothesis of an oxidation of nitrite to nitrate during the sampling.

The phenomenon, not observed in laboratory trials, in which purified air was used, seemed to occur mainly during daytime and to be related to the occurrence of photochemical smog episodes in the sampled atmosphere.

The hypothesis of a nitrite to nitrate conversion due to the action of an oxidant atmosphere was tested in a laboratory experiment. Artificial atmospheres of about 150 ppb of nitrous acid were flowed through a sodium carbonate annular denuder. Only the nitrite peak was observed in the IC analysis. After being loaded with nitrous acid, a similar denuder was exposed for one hour to an artificial atmosphere containing about 100 ppb of ozone. The analysis showed a decrease of the nitrite peak and the appareance of a nitrate peak. Replicate tests showed a conversion of about 30% in these experimental conditions. If the denuder was exposed to ozone for a longer sampling period or to an higher concentration, the nitrate peak increases as the nitrite decreases and finally disappears.

To confirm the hypothesis of an oxidation process occurring during the sampling, the same experiment was performed by using annular denuders coated with sodium carbonate and glycerine, which is supposed to prevent oxidation. As a result, a nitrite to nitrate conversion was still observed, but to a less extent (about 20%).

On a sodium carbonate denuder, as a consequence of the effect determined by an oxidizing atmosphere, it's impossible to discriminate the contribution due to nitric acid from that due to the conversion to nitrate of the nitrite coming from HNO_2, NO_2, PAN. In order to accurately measure nitrous and nitric acid in real atmospheres, which might have oxidizing properties, a sampling train consisting of a sodium chloride followed by two sodium carbonate denuders has been set up.

Laboratory trials showed that sodium chloride is a perfect sink for nitric acid and that it has no effect on the sampling of HNO_2, NO_2 etc.. The use of sodium fluoride as coating layer, proposed by some Authors (12), was discarded as at the working condition effective for the collection of nitric acid it showed a significant efficiency in the sampling of nitrous acid.

The two denuders set in series allow the correct determination of nitrous acid by taking into account both the interferences of NO_2, PAN etc. and the action of oxidants. As the nitric acid has been removed by the sodium chloride denuder, the nitrate collected on the sodium carbonate denuders, apart from a small interference of particulate nitrate, stems from the oxidation of nitrite coming from both nitrous acid and the interferents (NO_2, PAN, etc.). Thus, by expressing the nitrate as nitrite and by subtracting the total amount found on the second denuder to that found on the first one, a correct measurement of nitrous acid may be performed.

If the concentration of particulate nitrate in the area under investigation is supposed to be much higher than the concentration of gaseous nitric acid, two sodium chloride denuders should be used, in order to take into account the interference due to particulate matter.

The described set-up has been tested in a real atmosphere by running two side-by-side sampling trains for 24 hours at a flow rate of 12 l/min. One train employed one NaCl and two Na_2CO_3 denuders, while in the other one, Na_2CO_3 plus glycerine was used instead of Na_2CO_3 only. The results of three experiments of this type are shown in table I.

It can be observed that the amounts of nitrate collected on the NaCl denuders of the two sampling lines are quite similar. As a result of the oxidation of nitrite, carbonate denuder on both lines show fairly high amounts of nitrate. The conversion of nitrite to nitrate clearly depends upon the oxidizing power of the atmosphere, which results not just from ozone alone, but from the entire spectrum of oxidants present in ambient air.

The conversion also depends on the characteristics of the adsorbing layer: when glycerine is added to sodium carbonate, the nitrite to nitrate conversion is much smaller. Nevertheless, by expressing nitrate as nitrite and by subtracting the amount found on the second denuder to that found on the first one, the same concentration of nitrous acid is obtained on both sampling lines.

For each sampling, an oxidation factor, which depends on the chemical nature of the adsorbing layer, the oxidizing power of the atmosphere and the distribution of the collected species on the denuder, may be introduced. The meaning of this factor, which could possibly be of notable interest, is currently under investigation.

TABLE I: Measurement of HNO_2 and HNO_3 at Montelibretti (Rome).

Date 1986	Type of coating	ION CONTENT (μg)						ATMOSPHERIC CONCENTRATION	
		Denuder NO_3^-	NaCl NO_2^-	Denuder 1 NO_3^-	Denuder 1 NO_2^-	Denuder 2 NO_3^-	Denuder 2 NO_2^-	HNO_2 (μg/m^3)	HNO_3 (μg/m^3)
9/10	C/G	14.5	-	6.5 →	12.3 4.8 17.1	3.6 →	3.3 2.7 6.0	0.67	0.88
	C	14.5	-	20.5 →	8.0 15.3 23.3	13.8 →	2.3 10.3 12.6	0.65	0.88
9/11	C/G	20.5	-	7.8 →	16.5 5.8 22.3	4.2 →	2.5 3.1 5.6	0.98	1.32
	C	20.7	-	16.4 →	9.0 12.3 21.3	5.7 →	1.0 4.3 5.3	1.00	1.33
9/12	C/G	20.0	-	7.1 →	17.3 5.1 22.4	3.4 →	2.7 2.5 5.2	1.11	1.29
	C	21.0	-	19.3 →	7.5 14.1 21.6	5.3 →	- 3.9 3.9	1.11	1.39

CONCLUSIONS

The occurrence of a nitrite to nitrate conversion on sodium carbonate coated denuders in the presence of an oxidizing atmosphere has been investigated. It has been shown that in these conditions a positive interference in the measurement of nitric acid and a negative interference in the measurement of nitrous acid occur.

This effect should be taken into account whenever sodium carbonate denuders are employed for analyzing nitrous and nitric acid. The accuracy of the results so far reported in the literature (7-9) might be affected at a variable extent by the occurrence of this artifact.

The denuder set-up described in this work provides a reliable technique for the measurement of nitrous and nitric acid which keeps into account the conversion of nitrite coming from both nitrous acid and interferent species as NO_2 and PAN.

REFERENCES

1) Schiff H.I., Hastie D.R., Mackay G.I., Iguchi T., Ridley B.A.. Environ. Sci. Technol., 17, 352A (1983).
2) Platt U., Perner D., Patz H.W.. J. Geophys. Res., 84, 6329 (1979).
3) Appel B.R., Tokiwa Y., Haik M., Kothny E.L.. Atmos. Environ., 18, 409 (1984).
4) Klockow D.. V.D.I. Berichte 429, 165 (1982).
5) Appel B.R., Tokiwa Y., Haik M.. Atmos. Environ., 15, 283 (1981).
6) Shaw R.W.Jr., Stevens R.K., Bowermaster J.. Atmos. Environ., 16, 845 (1982).
7) Ferm M., Sjödin A.. Atmos. Environ., 19, 979 (1985).
8) Ferm M.. Atmos. Environ., 20, 1193 (1986).
9) De Santis F., Febo A., Perrino C., Possanzini M., Liberti A. in "Advancements in Air Pollution Monitoring Equipment and Procedures"; Proceedings of the ECE Workshop, Freiburg, F.R.G., June 2-6, 1985 pp. 68-75.
10) Febo A., Allegrini I.. In "Measurement of Atmospheric Acidity"; Proceedings of the CEC Workshop, Montelibretti (Rome), 25-26 June 1985.
11) Possanzini M., Febo A., Liberti A.. Atmos. Environ., 17, 2605 (1983).
12) Niessner R., Klockow D.. Intern. J. Environ. Anal. Chem., 8, 163 (1980).
13) Braman R.S., De La Cantera M.A.. Anal. Chem., 58, 1533-1537 (1986).

EVALUATION OF ANALYTICAL METHODS FOR THE DETERMINATION OF DEGRADATION PRODUCTS OF CHLORINATED HYDROCARBONS

Determination of Chloroacetate and Dichloroacetate in Liquid Samples

G. Fuchs, K. Bächmann
Technische Hochschule Darmstadt
Fachbereich Anorganische Chemie und Kernchemie
D-6100 Darmstadt

SUMMARY

An ionchromatographic method was developed for the analysis of mono- and dichloroacetate in precipitation samples. The method is based on an enrichment and matrix stripping cycle.

Concentration values of 0.3 - 2.3 ng ml^{-1} and 0.4 - 4.4 ng ml^{-1} for mono- and dichloroacetate were measured in four precipitation samples, respectively.

1. INTRODUCTION

Chlorinated hydrocarbons are mainly decomposed in the atmosphere by a reaction with OH radicals. Some of the theoretically possible reaction products were found in smog chamber experiments (1, 2), but these experiments were performed with much higher initial mixing ratios of chlorinated hydrocarbons than generally found in the atmosphere. Besides phosgene and hydrogen chloride, chlorinated derivatives of acetic acid were detected in these experiments. Chlorinated derivatives of acetic acid should form monochloro-, dichloro-, and trichloroacetate in liquid phases due to hydrolysis. However, these chlorinated acetates should be enriched in liquid cloud droplets or in rain drops due to wash-out effects. Besides the wet deposition, a high dry deposition rate of these species must be assumed. In order to confirm the occurence of these possible degradation products in the atmosphere, we developed an ion chromatographic method for the determination of chlorinated acetates in precipitation.

2. EXPERIMENTAL

Ion chromatography with a conductivity detector was used for the analysis of chloroacetate and dichloroacetate. The experimental conditions necessary to realize the ion chromatographic separations and determination of monochloroacetate are listed in tab. 1.

Table 1: Experimental Conditions

Ion chromatographic device: DIONEX 2000 i
Separation column: AS 4, Precolumn: AG 4, with inline filter
Detector: conductivity with suppressor system AFS 1,
regenerant 25 mM H_2SO_4, 2 ml min^{-1}
Eluent: 1 mM $NaHCO_3$, 2 ml min^{-1}
Sample volume: 100 µl

The low concentrated eluent must be produced from deionized and degased water. Fig. 1 presents a separation of 4 µg ml^{-1} fluoride, acetate, formate, monochloroacetate, and chloride, respectively. No difficulties occur in the determination of monochloroacetate in this concentration range

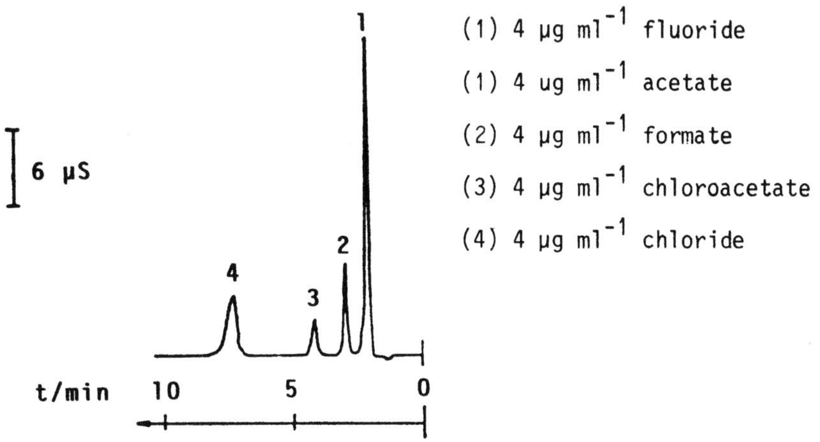

Fig. 1: Separation of fluoride, acetate, formate, chloroacetate and chloride, concentration: 4 µg ml^{-1} each

A separation of 80 ng ml^{-1} monochloroacetate besides an excess of 4 µg ml^{-1} fluoride, acetate, formate, and chloride is shown in Fig. 2. The detection limit of monochloroacetate besides the above mentioned excess of other anions is estimated to 20 ng ml^{-1}.

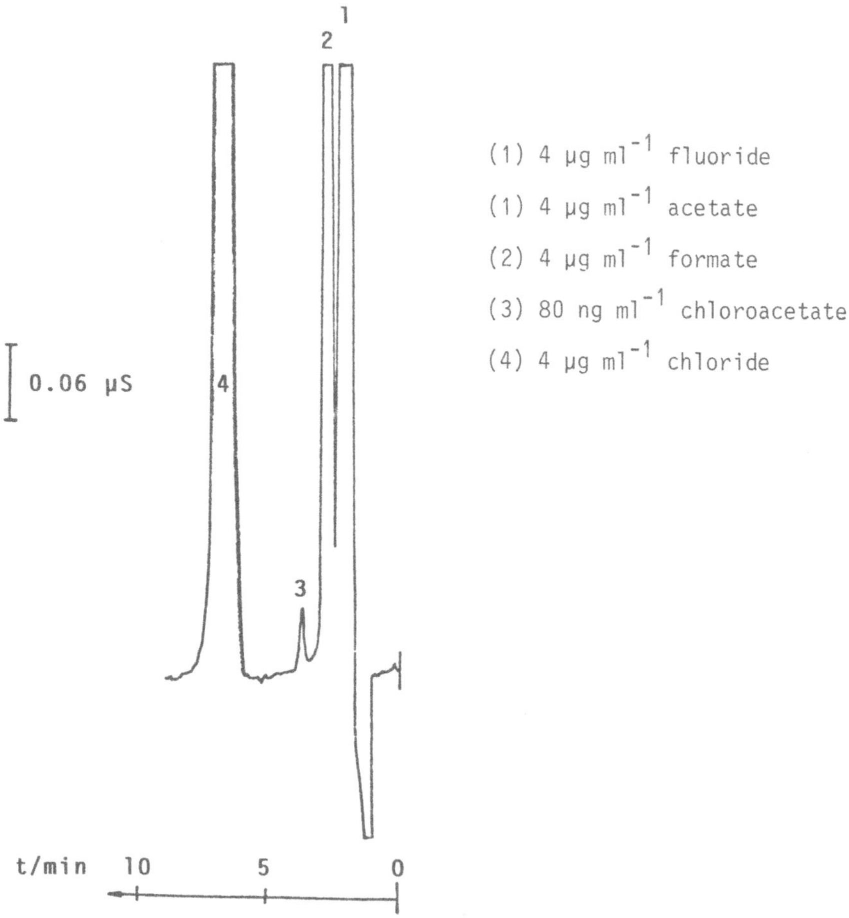

Fig. 2: Determination of 80 ng ml^{-1} chloroacetate besides an excess of 4 µg ml^{-1} fluoride, acetate, formate and chloride

It was not possible to determine monochloroacetate or dichloroacetate in a rain water sample by direct injection of a 100 µl sample. A special preconcentration cycle was developed. A preconcentration column was used instead of the 100 µl sample loop.

Figures 3 - 5 show the steps necessary to get a detectable chloroacetate signal. Figure 3 presents a separation of 5 ml concentrated standard solution (1 ng ml^{-1} chloroacetate besides an excess of fluoride, acetate, formate, and chloride). No chloroacetate signal was detected. The capacity of the preconcentration column was too low to concentrate a larger volume of standard solution due to the large amount of other anions besides chloroacetate.

During this separation, 2 ml were sampled downstream the conductivity detector at the time of the chloroacetate elution (region 3). This procedure was repeated nine times. The total volume sampled downstream the detector was 10 x 2 ml = 20 ml. In this 20 ml sample, chloroacetate was enriched relative to the excess of other ions.

These 20 ml were concentrated and separated as shown in figure 4. A chloroacetate signal was obtained on the tailing of the formate peak. Again, 2 ml were sampled downstream the detector (peak 3) concentrated on the preconcentration column and separated. Figure 5 presents the chromatogram of these 2 ml concentrate. This chloroacetate signal corresponds to a concentration of 1 ng ml^{-1} besides a thousandfold excess of other ions. The total volume concentrated was 50 ml (10 concentrations).

This preconcentration/sampling technique was applied to the analysis of several rain and snow water samples. One example is given in the figures 6 - 8.

A similar procedure is used for the analysis of dichloroacetate. Since nitrite and dichloroacetate have the same retention times under the above mentioned conditions, no specific signal is obtained. Therefore, nitrite must be transformed to nitrate by a selective oxidation before the last separation step is performed. 25 µl 30 % H_2O_2 / ml sample is added and the mixture kept at a temperature of 90°C for 30 minutes. Under these conditions, nitrite was oxidized to form nitrate, the dichloroacetate signal showed no effect.

Calibration was carried out by concentration of 5 ml standard solution.

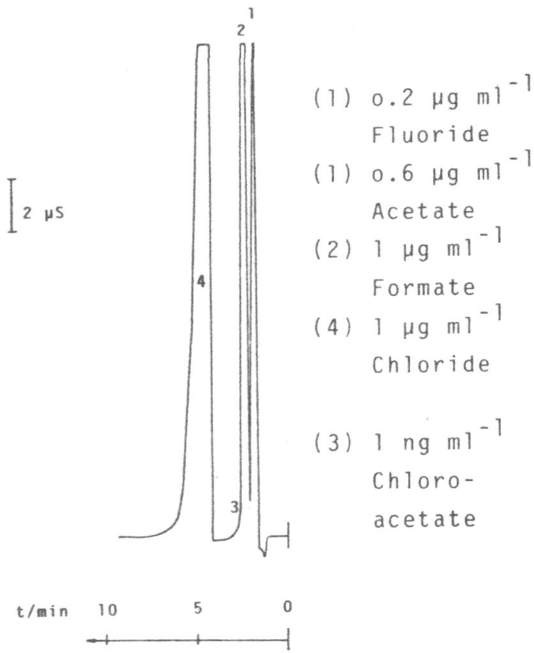

Fig.3: Separation of 5 ml concentrated standard solution

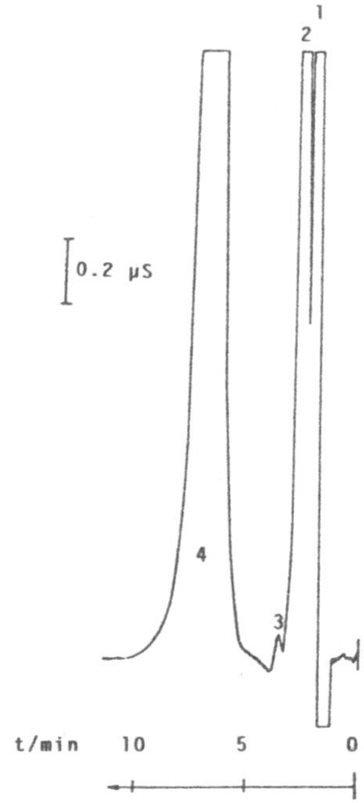

Fig.4: 10 x 2 ml sampled downstream the detector at the time of the chloro- acetate elution, concentra- ted and separated

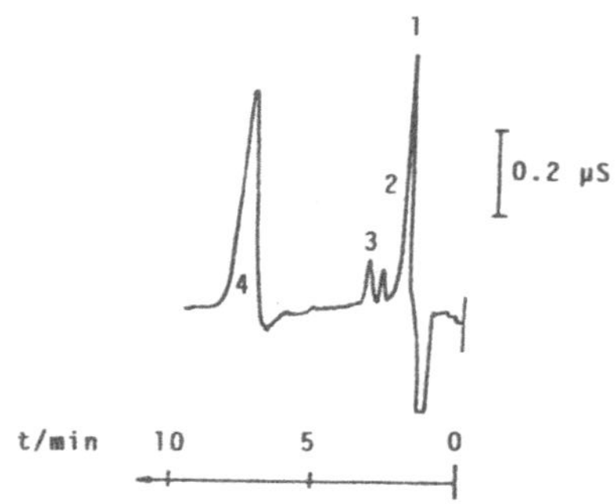

Fig.5: 2 ml sampled during chromatogram Fig.4, concen- trated and separated

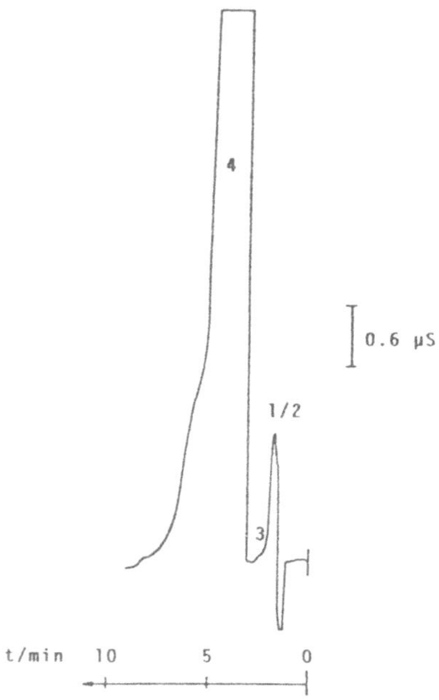

Fig.6: Separation of 5 ml concentrated rain water sample

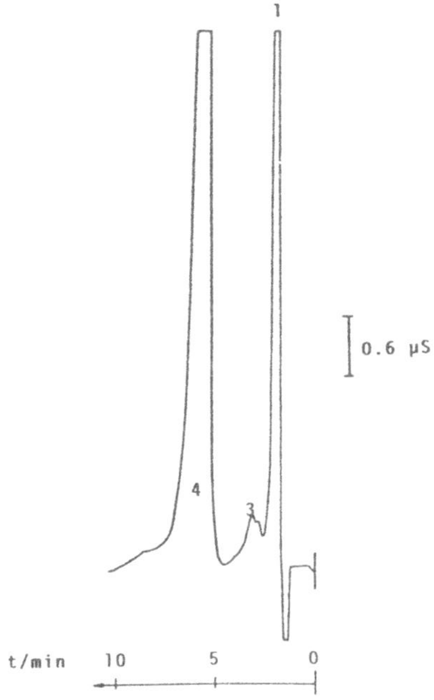

Fig.7: 10 x 2 ml sampled downsteam the detector at the time of the chloro-acetate elution, concentrated and separated

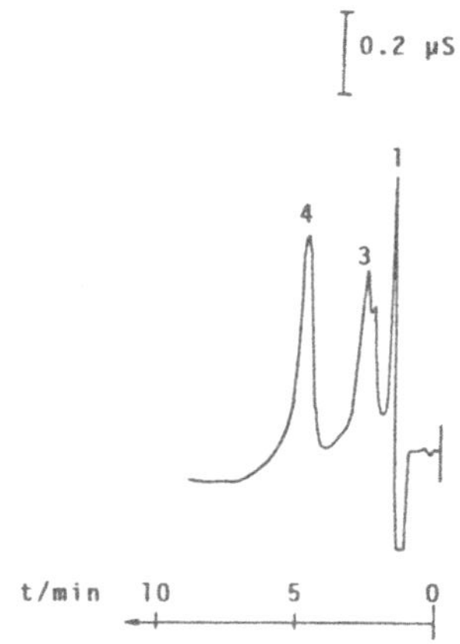

Fig.8: 2 ml sampled during chromatogram Fig.4, concentrated and separated

3. RESULTS

Using the above described technique we analysed three different rain water and one snow water sample. The concentrations of chloro- and dichloroacetate are listed in table 2. For comparison, the chloride, nitrate and sulfate concentration values determined by conventional ion chromatography are mentioned.

Tab. 2: Concentrations of chloro- and dichloroacetate, chloride, nitrate and sulfate in rain and snow water samples (ng ml^{-1})

	Rain (14.4.86)	Rain (18.4.86)	Rain (24.4.86)	Snow (12.2.86)
Chloroacetate	0.3	1.3	2.3	0.5
Dichlororacetate	4.4	1.1	0.4	not determined
Chloride	2066	1722	1316	1272
Nitrate	4766	3404	5583	2727
Sulfate	5642	3284	5895	21106

4. DISCUSSION

Using the above described method, it is possible to determine chloro- and dichloroacetate concentrations in rain- and snow water samples besides a more than thousandfold excess of chloride, nitrate, and sulfate. The concentration values are in the low ppb-range.

REFERENCES

1. C.R. Pearson, C_1 and C_2 Halocarbons, in: O. Hutzinger (ed.) The Handbook of Environmental Chemistry, Vol. 3, part B Anthropogenic Compounds, Berlin 1982, p. 69 ff
2. B.W.Gay, P.L. Hanst, J.H. Bufalini, R.C. Noonan, Env. Sci. & Tech. 10, 58 (1976)

SAMPLING OF ALDEHYDES AND CARBONYL COMPOUNDS IN AIR AND THEIR DETERMINATION BY LIQUID CHROMATOGRAPHIC TECHNIQUES

P. CICCIOLI, R. DRAISCI, A. CECINATO and A. LIBERTI
Istituto sull'Inquinamento Atmosferico del CNR- Area della Ricerca di Roma
Via Salaria km 29,300, C.P. 10, 00016 Monterotondo Scalo - ITALY

Summary

A simple method for the collection of low boiling aldehydes and ketones in air has been developed. It is based on the use of small traps filled with Carbopack B adsorbent coated with suitable amounts of 2.4-Dinitrophenylhydrazine (DNPH) and H_3PO_4. Aldehydes and ketones are enriched by passing 100 to 200 L of air through the trap. After sampling, the cartridges are extracted with small volumes of CH_3CN and aliquots subjected to HPLC fractionation.Specific detection of DNPH derivatives is carried out with a UV-visible adsorbance detector set at 365 nm or by an amperometric detector working in the oxidation mode. The methods allows the determination of acetaldehyde, formaldehyde, propionaldehyde and acetone at sub-ppb levels. Typical profiles of carbonyl compounds collected in a Diesel exhaust, in urban air and in a rural area are presented. Diurnal variations of formaldehyde, acetaldehyde and acetone,occurring during a two-days photochemical smog event observed in a rural area,are reported together with the trends of ozone, PAN and toluene. The results are critically discussed.

1. INTRODUCTION

Carbonyl compounds in general, and particularly aldehydes, are important air pollutants as they take part actively to the photochemical smog formation (1). In urban areas substantial amounts of aldehydes are released into the the atmospheric environment as a consequence of automobile exhaust or industrial emission. While almost 30% of formaldehyde found in urban air is directly produced by internal combustion of gasoline or Diesel fuel, most part of all carbonyl compounds is formed by photochemical oxidation of hydrocarbons immitted in the atmosphere by anthropogenic activities. In forested or rural areas aldehydes and ketones are formed by photochemical degradation of terpene hydrocarbons emitted from plants. Among the various aldehydic components, acetaldehydeis a product of particular interest to atmospheric chemistry as it is the precursor of peroxyacetyl nitrate (PAN) which exhibits adverse effects on plants growth. Moreover, photoxidation of acetaldehyde by HO_2 radicals (2) can give rise to the formation of CH_3COOH and HCOOH which, in turn, can contribute to increase the acidity of the atmosphere. Another important carbonyl compound is acetone as it can be produced by degradation of both anthropogenic and natural hydrocarbons and particularly isoprene emitted from trees. In spite of the importance of carbonyl compounds in determining the levels of photochemical pollution in rural and urban areas, their monitoring is not as frequent as that of other photochemical pollutants. The reason of this lack of data has to be attributed most to the fact that the techniques presently available are either too expensive for routine work to be carried out in the field (optical methods) or they require special care in order to maintain the quality of the sample collected (impinger methods). To overcome some of these limitations and improve the collection efficiency, solid supports coated with suitable reagents have been proposed for the collection of aldehydes (3-5). In particular, small cartridges filled with chromatographic adsorbents (4,5),coated with 2.4 Dinitrophenylhydrazine (DNPH), appear to be extremely promising for field monitoring of carbonyl compounds as they require little training, are easy to handle and do not require expensive equipment for the collection of sample and its analysis.

In this work a simple method, based on the use of very small cartridges filled with Carbopack B coated with DNPH and H_3PO_4, is described. It allows the quantitative collection of formaldehyde,acetaldehyde,acrolein,propionaldehyde and acetone and makes possible their determination at sub-ppb levels when volumes of air ranging from 100 to 200 L are passed through the adsorption trap. DNPH derivatives, recovered by liquid extraction, are separated and quantitated by HPLC. Specific detection is carried out by two different methods. With the first method, DNPH derivatives are detected by the change in absorbance of light occurring in the UV region near to visible (365 nm). With the second method, dinitrophenylhydrazones present in the HPLC eluate are oxidized in a electrochemical cell and the products, dicharged at the working electrode, are measured as the change in current occurring with respect to a reference electrode (voltammetric detection). Practical applications of this procedure are reported. They include the determination of carbonyl compounds in a automobile exhaust pipe, in a urban site and in a rural area not directly exposed to anthropogenic emission sources. Daily trends of formaldehyde, acetaldehyde and acetone, measured in a rural site during a two-days photochemical smog episode, are also presented together with the variations of PAN, ozone and toluene. The results obtained are critically discussed.

2. EXPERIMENTAL

The cartridges used for sample collection were all made by glass tubings having an internal diameter of 0.5 cm. The total length of the trap was 5cm. Carbopack B, a graphitized carbon black having a specific surface area of 90 m^2/g, was used as solid sorbent. The material, supplied by Supelco (Bellefonte,PA,USA), was sieved in order to obtain a particle size ranging between 20 and 40 mesh. The cartridges were filled with different amounts of adsorbent depending upon the type of measurements to be performed. For the sampling in rural areas, traps were filled with 20 mg of dry packing. Cartridges used in urban areas or in emission sources were, instead, filled with 40 mg of dry sorbent. A 100 mesh,stainless steel screen was placed at the trap inlet whereas a glass wool plug was inserted at the outlet to hold the sorbent in the cartridges. The sorbent was coated by passing through the traps 1mL of a CH_3CN solution containing 1mg/mL of DNPH and 1mg/mL H_3PO_4. After coating, the cartridges were dried using a flow of purified N_2, capped at both ends and ready to use. The recovery of DNPH derivatives was obtained by passing suitable amounts of CH_3CN though the cartridge. The volume of eluent used was dependent upon the amount of sorbent present into the trap. 4 to 5 mL of eluent were sufficient to ensure the complete recovery of DNPH derivatives from small traps normally used in rural areas. Longer traps, used in urban areas or emission sources, required an amount of CH_3CN ranging between 9 and 10 mL. The air sampling was performed on a home-made sampling device allowing a constant flow rate through the cartridge. The total volume as well as the sampling rate could be electronically controlled. The apparatus allowed the sampling rate to be changed from 0.5 to 5 L/min whereas the total volume could be preset from 1 L to 1 m^3. After extraction, 20 μL of the solution were injected into a 45mm x 25cm Erbasil-ODS (Carlo Erba Spa,Milan,Italy) reversed phase column filled with 10μm silica particles. A Jasco (Japan Spectroscopic Co,Tokio,Japan) BIP model liquid chromatograph was used for the solvent delivery. The column outlet was connected either to a Jasco Uvidec 100 V UV-visible absorbance detector or to an ELDEC 103 (Violet,Rome,Italy) voltammetric detector equipped with a glassy-carbon electrode. Depending upon the detector used, two different solutions were used as mobile phase for separating DNPH derivatives. When detection was accomplished by the UV-visible absorbance detector, the mobile phase was a solution of CH_3CN+H_2O(61:39 v/v). When the HPLC column was connected to the electrochemical detector, the eluent was a solution of CH_3CN+H_2O (55:45 v/v) containing 0.01 M H_3PO_4 and 0.01 M KNO_3. The addition of salt and acid was necessary to permit the oxidation and dicharge of DNPH and their derivatives to the glassy-carbon electrode. In all cases the flow rate of the eluent was 0.75 mL/min.

Standard atmospheres containing known amounts of aldehydes were generated by using permeation tubes. The permeation tube for formaldehyde was prepared by filling a PTFE tube (5 mm i.d., 10 cm in length),having a thickness of 0.5 mm, with paraformaldehyde. A permeation rate of 0.2 μg/min was obtained by keeping the permeation tube at a temperature of 84°C. Permeation tubes for acetaldehyde and propionaldehyde were obtained by filling the same PTFE tube with pure,liquid compounds supplied by Carlo Erba Spa. The surface of the PTFE exposed to the permeation process was 4 cm. By keeping the temperature at 31°C, the permeation rates were 0.5 and 0.35 μg/min for acetaldehyde and propionaldehyde respectively. In order to obtain homogeneous atmospheres, the dilution gas, emerging from the condensers containing the permeation tubes, was sent into a mixing chamber (20L in volume)maintained under continuous stirring before to be adsorbed into the cartridges.

3. RESULTS AND DISCUSSION

3.1. OPTIMIZATION OF THE HPLC PARAMETERS

Before to evaluate the performances of cartridges filled with Carbopack B adsorbent, efforts were devoted to optimize the chromatographic conditions requested for separating and detecting carbonyl compounds as dinitrophenylhydrazones. This preliminary evaluation was made by using standard solutions containing known amounts of formaldehyde,acetaldehyde,acrolein,acetone and propionaldehyde. The results obtained when the column was connected to the UV-visible absorbance detector suggested that the separation of acrolein,acetone and propionaldehyde was only partly accomplished but the improvements obtained by increasing the water content in the eluent above 39% v/v were not fully compensated by the substantial increase in retention caused by the hydrophobic properties of the mobile phase. Therefore, a compromise between resolution and analysis time was reached by selecting a composition of the mobile phase equal to 61% v/v of CH_3CN and 39% v/v H_2O. Since the addition of salts and acids to the eluent (necessary for electrochemical detection) increases the solubility of DNPH derivatives in the liquid phase, the composition of the mobile phase was changed when the HPLC column was connected to the voltammetric detector. Changes were made in such a way that a retention and resolution similar to that measured when the column was attached to the UV-visible absorbance detector was achieved. The aim was reached by using a composition of the mobile phase corresponding to 55% v/v of CH_3CN and 45% v/v of H_2O. Once established the chromatographic conditions, the next step was the optimization of the detectors performances. Experiments, carried out at 254 and in the region ranging between 355 and 365 nm, suggested that, by setting the absorbance detector in the latter region, the response of DNPH derivatives was increased at least of a factor two. The UV spectra, carried out on standard solutions of DNPH derivatives of low boiling aldehydes and acetone, confirmed that higher sensitivity can be gained by working in the near-visible UV region. Figure 1a shows the spectrum of acetaldehyde recorded during these preliminary experiments. In routine work, the wavelength selected was 365 nm for a better detection of all compounds but values ranging between 355 and 375 nm could be also used for a better detection of specific components.

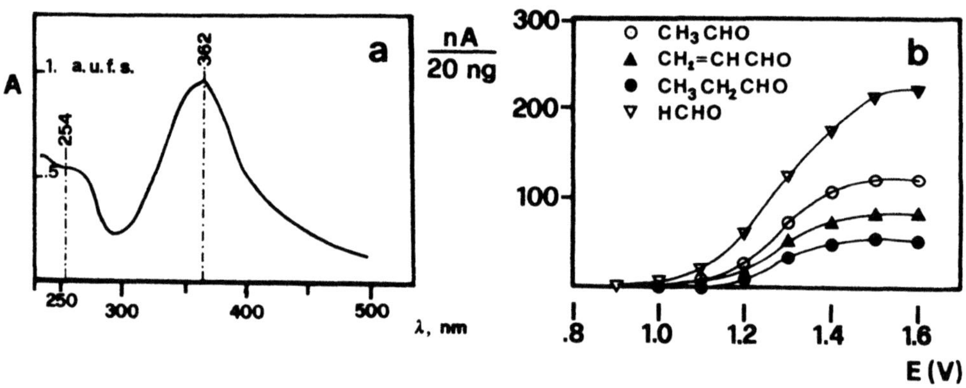

Figure 1 - a) UV absorbance spectrum of acetaldehyde.
b) Voltammograms of different aldehydic compounds carried out on the electrochemical detector.

To evaluate the performances of the electrochemical detector, voltammograms were generated by measuring the current recorded when the voltage applied to the working electrode was changed from 0.8 to 1.7 V. Current measurements were made when the same standard solution was injected into the HPLC column. The results obtained are shown in Figure 1b. The trend shown in this Figure suggested that the highest response is seen when 1.7 V are applied to the cell. In practical work, however, a voltage of 1.4 V was found to give a more sensitive response because of a better signal-to-noise ratio. A comparison between the performances of the two detectors for the analysis of DNPH derivatives is reported in Figure 2. By looking at the results shown in this Figure we can say that UV absorbance detection is slightly more sensitive than voltammetric detection and stable conditions are easy to achieve yet to maintain. In spite of the drawbacks associated with the use of the electrochemical detector, its use can be of a certain utility to confirm the results of the UV-visible absorbance detector especially when interfering compounds are present in the extract of real samples.

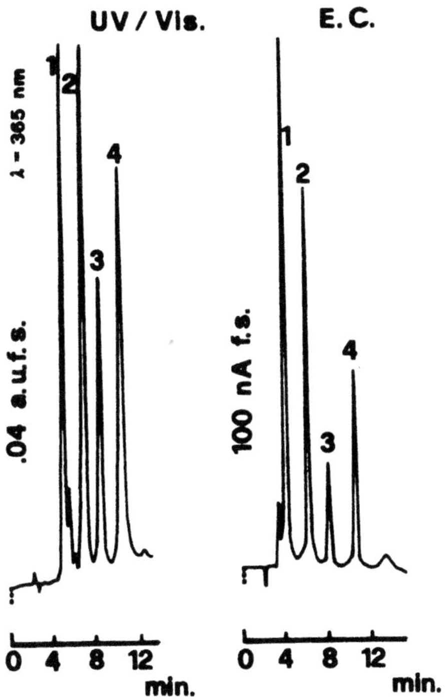

Figure 2 - Chromatographic profiles obtained when the same standard solution, injected into the HPLC column, was detected by a UV-visible absorbance and a voltammetric detector. The recorder attenuation was set at values where the same signal-to-noise ratio was measured at the amplifier outlet.
1. DNPH peak, 2. formaldehyde-2.4 DNPH 3. acetaldehyde- 2.4 DNPH- 4. propionaldehyde-2.4-DNPH. The amounts injected were 6 ng for formaldehyde, 3 ng for acetaldehyde and 7 ng for propionaldehyde respectively. For the experimental conditions see the text.

3. 2 EVALUATION OF THE ADSORPTION EFFICIENCY AND CAPACITY OF TRAPS FILLED WITH CARBOPACK B

In order to gain information about the adsorption efficiency and capacity of traps filled with Carbopack B adsorbent, experiments were undertaken using synthetic atmospheres containing known concentrations of aldehydes. Since the collection efficiency of traps was compared with that given by liquid solutions where the gas stream was bubbled, low flow rates (200 mL/min) were used during the experiments. The results,obtained by passing 10 L of air containing 450 ppb of formaldehyde, 800 ppb of acetaldehyde and 600 ppb of propionaldehyde, indicated that the adsorption efficiency of traps was slightly higher (3 to 5%) than that measured by using the impinger method and always higher than 98% of the sample passed through the trap. Measurements of the capacity, made by following an approach similar to that already described, indicated that 100 L of air containing 40 ppb of aldehydes can be passed through the adsorbent before the reagent present on the smaller traps is consumed. A value almost double can be collected on traps containing 40mg of Carbopack B. The capability of the cartridges to retain DNPH derivativatives was also measured by adsorbing 20 μL of a standard solution containing 60 ng of formaldehyde, 30 ng of acetaldehyde and 128 ng of propionaldehyde into the adsorbent and by measuring the recovery when 200 L of air were passed through the cartridge at a flow rate of 2 L/min. The results of these experiments showed that no losses of DNPH were occurring in these conditions. Further experiments, carried out by adding water up to a value of the relative humidity close to 90%, indicated that, in contrast with other methods, the efficiency was not reduced by the presence of water in the trap. The storage stability was evaluated by keeping the traps at room temperature and in the dark for three days after sample collection. The results of the HPLC analysis showed that no substantial losses occurred in these conditions ($<$ 2% of the total sample).

3.3 PRACTICAL APPLICATIONS OF CARBOPACK B TRAPS

Traps packed with Carbopack B coated with DNPH and H_3PO_4 were used to evaluate the content of aldehydes and carbonyl compounds present in a light-duty Diesel exhaust pipe, in a urban atmosphere and in a rural area not directly exposed to anthropogenic emission. Figure 3a, b and c shows the chromatographic profiles obtained when 0.4 L of Diesel exhaust, 100 L of urban air and 150 L of a rural atmosphere were collected into the traps filled with Carbopack B adsorbent. The chromatograms were recorded when the HPLC column was connected to the UV-visible absorbance detector. Figure 3d, e and f show the HPLC profiles of the same samples when the voltammetric detector was used for detecting DNPH derivatives.

The capability of the method to detect small variations of carbonyl compounds occurring during the day has been tested by performing a two-days monitoring campaign in a rural area located 30 km NE of Rome. The results obtained are shown in Figure 4 together with the profiles of toluene, PAN and ozone. During the first day the sun was shining and the temperature reached 27°C around noon. The second day the sky was partially cloudy and the temperature was 24°C. The meteorological conditions were, in the second day, not proper for the development of photochemical smog. In both cases the prevalent wind was blowing from SW and therefore pollutants,developed within the urban area, could be transported in our sampling site.

By looking at the results shown in Figure 4, it is possible to conclude that all trends are consistent with the chemical reactivity of the species investigated. The sudden increase of the mixing ratios of PAN and ozone, observed the first day around 3p.m., is , in fact, concurrent with a slight decrease in the mixing ratio of acetaldehyde. The second day, when the production of photochemical smog was lower, due to the decreased intensity of the solar radiation, the content of aldehydes, and particularly acetaldehyde, was higher and reached a maximum value around 3 p.m.. By contrast

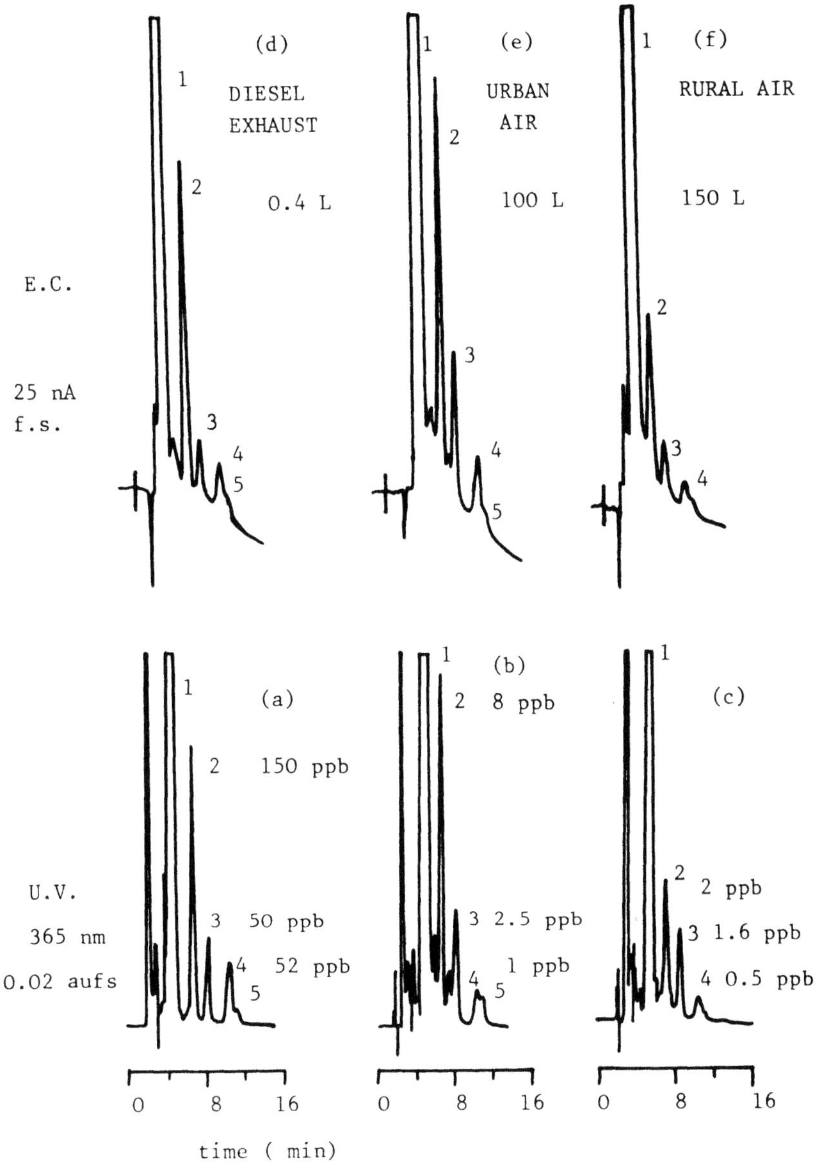

Figure 3 - HPLC separations of aldehydes and carbonyl compounds collected with Carbopack B traps in the exhaust pipe of a light-duty Diesel engine, in a urban atmosphere with low traffic and in arural area not directly exposed to anthropogenic emission. 1. DNPH; 2. formaldehyde-DNPH; 3. acetaldehyde-DNPH; 4. acetone-DNPH; 5. propionaldehyde-DNPH. In the Diesel exhaust sample peak no. 4 contained also acrolein not resolved from acetone.

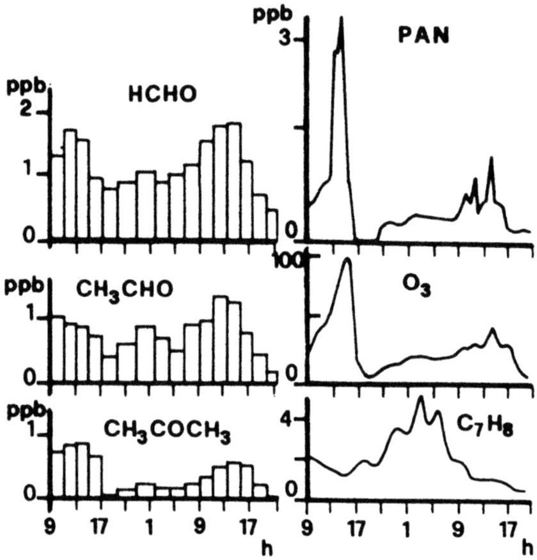

Figure 4- Diurnal variations of acetaldehyde, formaldehyde, acetone, ozone, PAN and toluene observed in a rural site located 30 km NE of Rome (20-21 May 1986).

the mixing ratios of PAN and ozone were close to the normal values observed in late winter-early spring in our region. Consistently,the mixing ratio of acetone, which is generated by degradation of both natural and anthropogenic hydrocarbons diffusing or emitted near the site, was higher the first day when photochemical reactions were taking place to a major extent. The trend of formaldehyde is somehow similar to that of acetaldehyde but a distinct maximum, centered one hour before that of PAN and ozone, can be observed also in the first day of measurements. It is likely that the emission rate of this compound was,at that time, slightly higher than the rate of degradation due to photochemical processes. It is important to notice that,in both days, transport from the city was the prevalent effect determining the content of aldehydes and photochemical oxidants in our site as the mixing ratio of toluene was, during daylight, rather low.

REFERENCES

(1) R.A. COX , Proceedings of the First European Symposium on "Physico-Chemical Behaviour of Atmospheric Pollutants" Ispra,16-18 Oct. 1979,B.Versino & H.Ott Edrs.,CEC Publ.,1980, pp.91-109

(2) G.K. MOORTGAT, J.P. BURROWS, G.S. TYNDALL,W. SCHNEIDER,R.A. COX, Paper presented at the Worshop on "Chemistry Related to Tropospheric Ozone"

organized within the framework of Working Party 2 "Chemical and Photo chemical Reactions" of the Concerted Action "Physico-Chemical Behaviour of Atmospheric Pollutants", COST 611, Cologne 12-13 Nov. 1985, CEC Publ., p.73.

(3) U.SCHMIDT and D.C. LOWE, Proceedings of the Second European Symposium on "Physico-Chemical Behaviour of Atmospheric Pollutants",Varese, 29 Sept.-1Oct.1981, B.Versino & H.Ott Edrs, D.Riedel Publ.Co.,1982, pp. 377-386.

(4) K. KUWATA,M. UEBORI, H. YAMASAKI,Y. KUGE,Y. KISO, Anal.Chem., 55, 2013,(1983).

(5) F. LIPARI, S.J. SWARIN, Environ. Sci. Technol., 19,70,(1985).

THE USE OF A FILTER PACK FOR THE MONITORING ORGANIC AND INORGANIC ACID SPECIES IN AIR.

D. Brocco, A. Franco, A. Liberti and R. Tappa

Istituto sull'Inquinamento Atmosferico del C.N.R.
Area della Ricerca di Roma
Via Salaria Km 29,300 - C.P. 10
00016 Monterotondo Stazione (Roma) ITALY

Summary

This paper describes a filter pack for the measurement of volatile organic acids (HCOOH, CH_3COOH), total nitrate (HNO_3 + NO_3^-), total ammonium (NH_3 + NH_4^+), nitrous acid, sulfur dioxide and sulfate. The filter pack consists of four filter coupled in a series: a filter for particulate matter, two KOH - and a $HClO_4$ - impregnated filter respectively for acid and basic components. The organic and inorganic acid species in the filter water extracts were directly analyzed by ion chromatography. The multiple filter pack was compared with an annular denuder system in field measurements. Atmospheric acid species concentrations have been measured in rural and urban areas.

1. INTRODUCTION

The determination of organic acids, in addition to inorganic acid species more commonly measured, is of a great importance for a complete understanding of the nature of acidity in the acid deposition.

At present, only a limited number of sampling and analytical methods for determination of organic acids have been suggested, based on the use of spectrochemical techniques (1), liquid chromatography (2) and gas chromatography (3). The methods reported are expansive and unpratical for routine monitoring and furthermore do not detect other gaseous acid species.

This paper presents a filter pack - ion chromatography method for the sampling and the simultaneous determination of volatile organic acids and inorganic acid species in the atmosphere. This filter pack procedure was compared to an annular denuder system, sampling inorganic acid species.

2. EXPERIMENTAL

Filter Pack

The filter pack consists of the combination of three filters which have different chemical characteristics to obtain a selective absorption of gaseous and aerosolic species. The three filters were mounted in series in a multiple filter holder. The first filter was a 0.8 um pore size cellulose acetate filter for collection atmospheric particles. The second filter was a Whatman 41 filter impregnated with an aqueous solution of 1 N KOH to collect gaseous organic and inorganic acid species. A second KOH-impregnated filter was used to account for the absorption of NO and NO_2

interferences. The third filter was a Whatman 41 filter impregnated with an aqueous solution of 1% perchloric acid for trapping gaseous ammonia. The filter pack system was protected from rain by an inverted funnel and the outside wall was covered with an aluminium foil.

The sampled filters were extracted with 10 ml of deionized water in 20 ml glass bottle. The bottle was shaken for few minutes and then left standing for analysis.

Particulate water-soluble containing nitrate, sulphate and ammonium were determined from the first filter. Gaseous organic and inorganic acid compounds were measured from the second impregnated filters and the ammonia content was obtained from the third filter.

A ion chromatography (IC) method for the simultaneous determination of organic and inorganic anions from a single injection has been developed. A typical chromatogram showing the separation of acid species is reported in Fig. 1.

The instrument conditions for IC analyses were:

a) separator system: two guard-columns HPIC-AG-3 (50 x 4 mm i.d.) and analytical column HPIC-AS3 (250 x 4 mm i.d.)
b) anion fiber suppressor: AFS-1
c) eluent: 2 mM $NaHCO_3$ - 1.6 mM Na_2CO_3; flow rate: 0,50 ml/min
d) regenerant for AFS-1: 0.025 N H_2SO_4; flow rate: 1.50 ml/min
e) injection volume = 50 ul.

A Dionex ion chromatography (QIC Ionchrom Analyzer) equipped with a conductivity detector was used.

The separator columns and the anion fiber suppressor have been supplied from Dionex. The IC method does not detect organic acid larger than C_2 which might be present in the atmosphere. The use of ion exclusion chromatography system (ICE) coupled with IC for the determination of all volatile organic acids is reported (4).

Annular denuder system (ADS)

The denuder system (5) consists of three annular denuders for the collection of gaseous acid species and ammonia and a set of filters down stream of the denuder for the collection of particles and ammonium nitrate evaporated in the sampling. Acid species are recovered on the first Na_2CO_3 coated annular denuder. A second Na_2CO_3 coated annular is used to account for the absorption of interferences. A third annular denuder coated with 1% citric acid is used to collect NH_3. Aerosol particles are collected on a teflon filter and a nylon back-up filter is used for trapping HNO_3 set free from the dissociation of NH_4NO_3. A third 1% $HClO_4$ impregnated Whatman 41 filter is set for collection of ammonia coming from the dissociation of NH_4NO_3.

After the sampling the denuders and the filters were extracted with 10 ml deionized water. Analysis of filters and denuders were carried out by ion chromatography.

The air monitoring by filter pack and annular denuder system was performed daily for 18 m^3 air volume sampled. The flow rate was of 12,5 l/min.

The experiments have been carried out at the centre of Rome (urban station) and at Montelibretti, situated about 30 Km east of Rome (rural station). The last station operated within the EMEP project.

3. RESULTS AND DISCUSSION

The atmospheric concentrations of organic acids and inorganic acid species obtained with parallel measurements at rural and urban stations by filter pack are collected in Table I.

Acetic and formic acids were the most abundant organic species. The propionic acid was found in some samples in not determinable concentrations. In order to examine the KOH-impregnated filter efficiency to collect organic acids a second KOH-filter was used. Recoveries of HCOOH and CH_3COOH on the first KOH-filter were 96-98% at 24 h (18 m^3). Mean amounts of these acids on the particulate filter were less than 2% of total recoveried acids. These results indicate that HCOOH and CH_3COOH in the atmosphere are mostly present in the vapor phase.

The measurements of HNO_2, HNO_3 and NH_3 by filter collection are affected by various artifacts as positive as well negative interferences. To evaluate these artifacts the results obtained by filter pack method were compared with those measured with ADS. The comparative results are collected in Table II.

As it can be seen from results, no remarkable differences were observed for HNO_2.

As for the annular denuder, the nitrite collected on the KOH-filter is affected by NO and NO_2 interference. To take into account this contribution we can suppose that, for the large difference in absorption of NO and NO_2 on KOH-filter in comparison with nitrous acid, the amount of the interferents on two KOH-filters set in series is fairly costant. If we use two filters, by subtracting the amount of nitrite found on the second filter to that collected on the first, it is possible to get the amount due to nitrous acid.

Nitric acid concentration obtained by filter pack were definitely lower than those obtained by denuder system. It seems that the loss of HNO_3 due to absorption on particulate matter and on surfaces in filter holder is the dominant artifact at very low levels of HNO_3. This nitrate positive artifact some time can be balanced by the dissociation of NH_4NO_3 collected on the particle filter (6, 7).

Nitrogen dioxide also might interfere, since a little fraction of NO_2 trapped by the KOH-filter will be converted to nitrate. This interference can be taken into account by subtracting the nitrate amount found on the second KOH-filter to that collected on the first. However, the filter pack yields reliable measurements for total nitrate (HNO_3 + NO_3^-).

As far as it concerns the determination of NH_3 and NH_4^+ noticeable differences were observed from the values obtained by filter pack and the ones measured by ADS, but a good agreement exists between the sum NH_3 + NH_4^+ obtained with the two procedures.

The data relative to SO_2 and $SO_4^=$ concentrations measured with the two procedures show a good agreement and no interferences have been found.

In conclusion a filter pack of four filters coupled in series, a particle filter, two KOH-and a $HClO_4$-impregnated filter, is a reliable device for the sampling of volatile organic acids, total nitrate (NO_3^- + HNO_3), total ammonium (NH_4^+ + NH_3), particulate sulfate and gaseous nitrous acid and sulfur dioxide. This system also provide results for total suspended particulate and other particulate components such Na^+, K^+, Ca^{++}, Mg^{++} and heavy metals.

The simplicity of the sampling equipment required makes this method suitable for monitoring networks.

REFERENCES

(1) HANST, P.L., WONG, N.W., BRAGIN, J. (1982), Atmos. Environ. 16, 969-981.

(2) VARIOLATO, S., PFATTLI, P., ZITTING, A. (1983), J. Chromatogr. 258, 207-211.

(3) KAWAMARA, K., LAI-LING NG and KAPLAN, J. (1985), Environ. Sci. Technol. 19, 1086-1089.

(4) BROCCO, D., TAPPA, R. (1986), Determination of organic and inorganic acid species in the atmosphere and rain water by ion chromatography, J. Chromatogr., in press.

(5) ALLEGRINI, I., DE SANTIS, F., FEBO, A., LIBERTI, A., PERRINO, C. and POSSANZINI, M. (1986), Annular denuder method for sampling gaseous and aerosolic materials. Submitted to Environ. Sci. Technol.

(6) APPEL, B.R., WALL, S.M., TOKIWA, Y. and HAIK, M. (1979), Atmos. Environ. 13, 319-325.

(7) APPEL, B.R., WALL, S.M. and HAIK, M. (1980), Atmos. Environ. 14, 549-554.

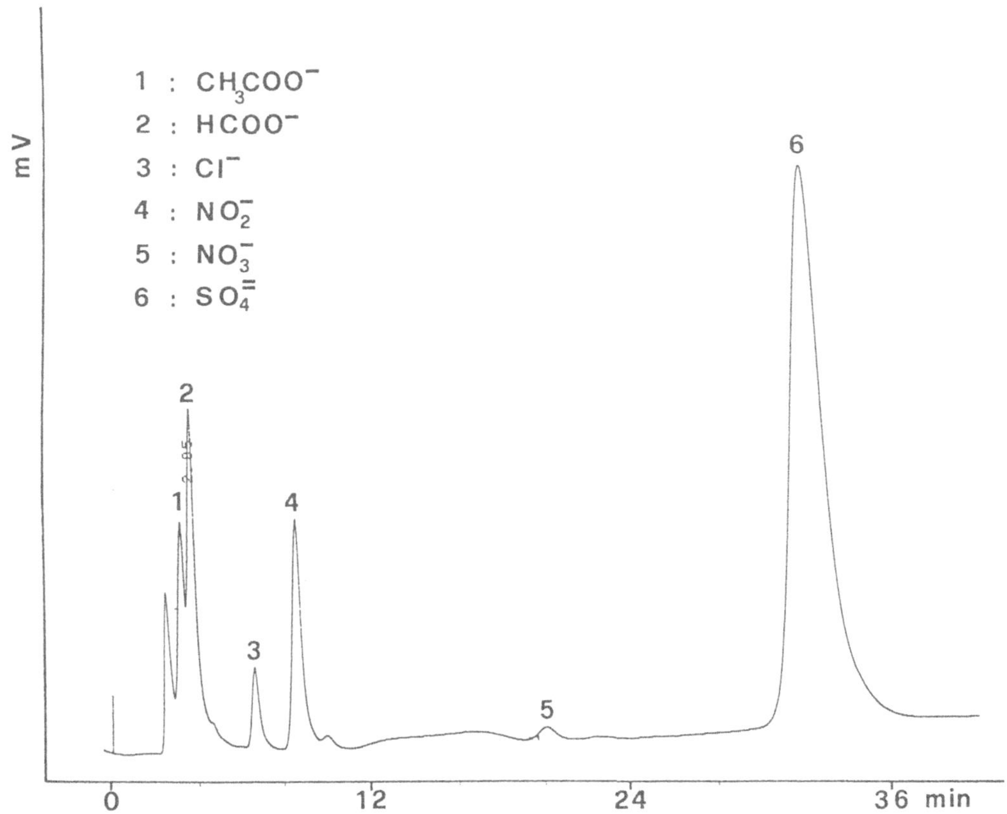

Fig.I - Chromatogram of a KOH-impregnated filter extract.

TABLE I

Atmospheric concentrations, in $\mu g/m^3$, of organic and inorganic acid species measured at rural (M) and urban (R) areas (March '86).

Day	Gas Phase												Aerosolic Phase					
	HCOOH		CH_3COOH		HNO_2		HNO_3		SO_2		NH_3		NO_3^-		$SO_4^=$		NH_4^+	
	M	R	M	R	M	R	M	R	M	R	M	R	M	R	M	R	M	R
19	2,3	7,2	7,6	26,5	0,9	5,1	1,5	25,6	6,9	24,8	3,4	2,7	10,6	7,9	12,5	15,5	4,6	6,2
20	0,5	6,7	0,4	27,5	0,6	4,0	0,5	20,1	4,9	21,6	4,4	3,1	8,4	3,9	10,3	17,1	5,6	6,0
21	2,0	9,0	7,0	38,7	0,4	9,9	0,8	27,7	10,6	51,0	4,8	4,2	15,7	10,6	18,1	15,9	6,5	6,2
22	3,9	5,4	14,2	20,2	0,7	4,4	0,4	3,7	5,5	43,6	2,5	2,2	9,6	3,7	11,9	8,6	3,7	3,5
23	3,8	5,0	14,8	20,4	0,5	3,5	0,5	13,8	5,9	13,6	2,3	1,5	9,2	2,3	11,7	11,4	3,5	3,2
25	3,6	5,7	20,9	21,4	0,4	6,1	0,4	3,9	3,7	49,6	3,1	3,6	2,4	3,2	5,8	8,0	1,6	1,5
26	2,9	4,5	17,2	18,5	0,4	4,2	0,4	11,4	3,9	32,5	2,2	2,9	2,3	3,7	3,0	6,3	0,7	0,1
27	1,2	3,4	5,5	14,4	0,8	3,2	0,5	6,8	3,0	29,4	2,4	2,1	4,5	2,8	2,0	4,3	1,1	1,0

TABLE II

Comparative data, in $\mu g/m^3$, relative to the sampling by ADS (D) and Filter Pack (F).
Rural area (May '86)

Day	Gas Phase								Aerosolic Phase					
	HNO_2		HNO_3		SO_2		NH_3		NO_3^-		$SO_4^=$		NH_4^+	
	D	F	D	F	D	F	D	F	D	F	D	F	D	F
6	0,5	1,2	0,8	0,5	9,2	10,2	0,6	1,3	9,5	9,9	9,1	11,0	-	5,1
7	0,5	0,8	1,4	1,4	5,9	6,0	1,7	1,8	3,8	7,3	11,6	13,8	4,9	4,7
8	0,5	0,7	1,1	0,2	5,9	5,3	1,7	1,7	3,3	5,9	10,1	10,8	6,7	4,6
9	0,5	0,8	1,1	0,2	2,8	2,9	2,4	2,3	2,3	4,0	8,4	9,4	4,6	2,3
13	1,1	0,8	1,3	0,1	2,9	2,9	3,4	3,8	3,6	6,2	7,3	7,4	3,5	2,9
14	0,8	0,7	1,4	0,2	3,7	3,8	3,5	4,3	4,0	6,7	10,0	10,2	4,5	3,6
15	0,7	0,7	1,2	0,2	3,2	2,8	2,9	3,8	2,3	4,1	5,0	5,1	2,9	1,8
16	0,7	0,5	2,9	0,1	3,9	3,3	2,9	4,1	4,6	6,2	10,4	10,2	3,6	2,0

Measurements 6-9/3 have been carried out by using only a KOH-filter in F.

IR-TUNABLE DIODE LASER TECHNIQUES FOR TRACE GAS DETECTION AT PPBV LEVEL

F. Cappellani, G. Melandrone and G. Restelli
Commission of the European Communities
Joint Research Centre - Ispra Establishment
21020 Ispra (Varese) - Italy

Summary

Infrared tunable diode lasers have been used in recent years as fast and sensitive detectors of trace gases at the ppbv level in ambient air or in reaction chambers. Their sensitivity, however, is frequently limited by noise of various origins, arising from the laser source or from the system used for measurement of the spectral absorptions. In this paper, post detection data handling techniques for improvement of the signal to noise ratio in second harmonic detection are discussed with preliminary experimental results.

1. INTRODUCTION

Photochemical oxidant formation and acid deposition are important issues in tropospheric chemistry where gases and radicals present at trace level in air are recognized to play a central role. It is then of paramount importance to have the capability to measure these species both in ambient air and in gas chambers where simulation experiments, aimed at the understanding of atmospheric, physical and chemical, processes are conducted.

The achievement of the combination of high specificity and sensitivity below the ppbv level with conventional chemical techniques is a difficult and sometimes impossible target, especially in the case of reactive species: the measurement of radicals and of the complex family of nitrogenous species are typical examples of this difficulty.

Further problems are encountered if real - or nearly real - time measurements are needed. For these reasons, optical techniques have been introduced; among them long path absorption with infrared tunable diode lasers (TDL) has been proposed (1) and recently used for sub-ppbv detection of trace species "in field" (2) and of radicals (3) in smog chamber experiments, both directed to studies of atmospheric chemistry.

The achievement of adequate sensitivities in a trade-off with speed of measurement is still a matter of development. Best results up to now have been obtained with the use of harmonic detection, a technique which has been demonstrated to be capable of measuring infrared absorptions as low as 10^{-5} cm^{-1} (4). In this technique, only changes in laser power associated with a change in laser wavelength are detected and the noise associated with temporal fluctuations of the laser intensity is nearly completely eliminated. The enhancement in the signal to noise ratio and, hence, in sensitivity is generallly limited by different factors depending both upon the laser source and the optical system used.

In the following, the operation of a high sensitivity TDL monitor is presented where main emphasis has been upon post-detection data handling techniques as a means of improving the ultimate detection limit.

2. NOISE REDUCTION TECHNIQUES

The basic advantage of the tunable diode laser resides in its extremely narrow emission linewidth: 3-30 MHz using closed cycle refrigerators and much narrower with cryogenic coolant Dewar systems. This characteristic permits the analysis of absorption lines down to the Doppler regime.

Operation with the gas in the cell in the Doppler or Voigt regime has the consequence of substantially increasing the separation between nearby absorbing lines so increasing the specificity. It has been also demonstrated that spectral absorption analysis of the gas at low pressure leads to a definitive improvement in detection sensitivity in spite of the reduced optical thickness (concentration x path length) (4).

In the harmonic detection scheme, the TDL is current modulated and tuned across an absorption line of the gas at pressures in the 15-40 Torr range. The absorption signal is generally detected and suitably processed at the second harmonic.

If the modulation amplitude is sufficiently small, the output signal is proportional to the second derivative of the absorption line. It has been shown, however, that the signal is maximized at a modulation amplitude about twice the halfwidth of the absoprtion line (5): in this case the signal is proportional to the second Fourier component of the lineshape.

The analysis of isolated absorption lines in the Voigt regime requires a modulation amplitude which is a few percent of the total TDL drive current, so that variations of the laser emission intensity with the modulation current are minimized. This is an important advantage with respect to analysis of the gas at atmospheric pressure which would require much larger modulation amplitudes, since fluctuations in this variation result in a "background signal" which limits the signal to noise ratio (6).

It has been observed (7) that the ultimate limit to the sensitivity is set by complex interference patterns originating by scanning the TDL across the Fabry-Perot etalons resulting from different elements of the optical system. This type of noise appears to be more or less evident with the TDL used; sometimes related to its far field pattern.

A prerequisite to any effort is to reduce this "optical noise" by a careful design of the optical system and of the IR beam alignment. In addition to this step, which is not discussed here, attempts have been made to evaluate the capabilities offered by signal processing techniques such as correlation analysis and Fourier transform methods. Background phasing and subtraction techniques have also been considered.

Cross Correlation

Correlation analysis provides information about the coherence within a spectrum or between two spectra (8). The use of this technique has been proposed recently for application to the TDL detection of weak absorption signals; preliminary experiments in laboratory conditions on the detection of CF_2Cl_2 have given promising results (9).

The correlation function is obtained by evaluating the time-averaged or the time-integrated product of the two spectra as a function of their relative displacement. In practice, the approach taken is to cross correlate to spectrum obtained from the multipass cell with a practically noise-free spectrum of the analyte gas in a simultaneously recorded reference gas cell. Air and reference gas at the same temperature and pressure are analysed simultaneously by splitting the laser beam in an asymmetric (85-15) beam splitter into two branches; the most intense going through the air sample and the other through the reference cell.

The cross correlation technique is expected to be most effective with a spectrum which is deteriorated by random noise and is nearly ineffective for the elimination of a purely periodical noise.

Fourier Transform Techniques

Very frequently the optical noise appears in the second derivative spectrum as sinusoidal interference fringes with a more or less regular period; the 2f absorption signal is severely distorted and small signals disappear and escape any evaluation.

The possibility of applying Fourier Transform (FT) techniques depends on the fact that the 2f absorption linewidth as resulting from the molecular lineshape and the modulation parameter is much greater or smaller than the noise fringe period.

In this case it should be possible to convert the digitized 2f spectrum to the frequency domain by direct FT. After locating the frequency peak (one or more) corresponding to the periodic noise and putting them to zero, an inverse FT should give back the suitably cleaned original spectrum. At this stage, a cross correlation with the reference spectrum could be further used to increase the S/N ratio.

Unfortunately, however, the experimental evidence is that at least part of the optical fringe noise has a period close to the absorption linewidth. This occurs for gases in the Voigt (10-40 Torr) regime and etalons in the optical systems with a length in air of the order of 1 metre (see, for instance, the base path of most commercial multiple reflection gas White cells).

In principle, the sinuoidal nature of the interference fringes in comparison with the non sinusoidal shape of the absorption signal might be used to separate the respective contributions in the FT spectrum even for the worst case. In practice, however, for weak absorption lines which are merged into the noise, this procedure is impractical.

Examples of FT analysis and speculations on the potential use of the information obtained are given in Refs. (9, 10).

Background Noise Subtraction

The hypothesis that the TDL noise is correlated in subsequent scans or in the two beams into which the laser beam is divided by a splitter has been suggested as a means for suppressing the laser noise by pre-subtraction or by simultaneous subtraction of background noise from the spectrum signal (11-13). In this study, background subtraction has been performed by recording first the spectrum with the flowing air and then the spectrum obtained after evacuating the gas cell.

It has been observed that operating in a short time sequence the shape of the background spectrum (interface fringe pattern) is nearly the same in the two spectra except for a small shift of the order of some tenths of the fringe period. By suitable computer phasing of the two spectra it has been possible to subtract a considerable part of the background noise and then to correlate the resulting spectrum with the reference one.

This procedure may pose serious problems depending on spurious effects due to gas release from the cell walls when evacuating of flowing the cell with an intert gas.

3. EXPERIMENTS AND RESULTS

The experimental set-up is shown schematically in Fig. 1. The TDL is cooled by a closed cycle refrigerator in a mount modified (14) to minimize laser frequency emission jitters due to mechanical vibrations induced by the refrigerator.

Before entering the multipass gas cell, the beam is splitted and one branch used for simultaneous detection of the reference gas line with a 10 cm long gas cell; this spectrum is used for cross-correlation measurements. Alternatively, the laser beam is directed to a monochromator for laser emission calibration and mode analysis.

The main beam is directed through a multiple reflection White gas cell, 1.4 m base, 64 m total beam path, where the gas mixture (or air) to be analysed is continuously drawn at a pressure in the range 15-40 Torr. The TDL drive current is repeatedly scanned through one or more absorption features with a 1-2 kHz sine modulation, superimposed on the diode current, of amplitude equivalent to a tuning equal to about one absorption line Voigt halfwidth. Scan times in the range of a few seconds were used (64 scans).

The signals detected from the lock-in amplifiers tuned to detect the second harmonic signals of the frequency modulated absorptions are sent to a data acquisition and control system which allows scanning of the spectral region of interest, averaging of a prefixed number of scans and digital storing of the data to perform computations on the spectra recorded.

Spectral averaging resulted in a sizeable improvement of the signal to noise ratio; however, using a number of scans in excess of 64-128 was ineffective, probably because of slight drifts in the laser emission.

The experiments were conducted by flowing ambient air through the multipass cell and analysing absorption lines of N_2O in the spectral range 1290-1310 cm^{-1}. The N_2O concentration in air was assumed constant (300 ppbv + 10%) to evaluate detection sensivities. Selection of N_2O absorption line within the wide range of strengths available (0.15 x 10^{-18} - 10^{-21} cm^{+1} $molec^{-1}$) was used to simulate signals due to a lower concentration of the trace gas. Preliminary results are presented in the following.

Fig. 2 shows the 2f spectrum of the 1312.081 cm^{-1} N_2O line of intensity equal to 0.7×10^{-20} cm^{+1} $molec^{-1}$ (bottom), together with the cross correlation function resulting from input of the reference spectrum shown. In this preliminary stage, the spectral signal manipulations were done using a DATA 6000 processor. As can be seen from Fig. 2, an improvement in S/N is evident. The cross-correlation function allows a better definition of the peak at zero displacement which, when normalized to the autocorrelation function of the reference spectrum, should allow us to quantify the concentration of the analyte in the gas micture (9).

If the noise is essentially periodic it appears that the FT procedure previously described is an effective alternative to the cross-correlation analysis. Its use is even preferable if problems related to the increased complexity of the optical system and of the signal calibration relative to the cross-correlation procedure are considered.

The effect of FT analysis is shown in Fig. 3 where the atmospheric N_2O line at 1240.814 cm^{-1} is extracted from the spectrum recorded in the multipass cell (21 Torr) strongly distorted by a periodical noise.

Frequently however, the character of the interference fringe pattern with a period comparable to the linewidth of the signal makes it useless to perform cross-correlation analysis or to implement an FT procedure. In this case, a striking improvement is obtained by background subtraction since the fringes are repeated in position and strength for the spectra of the cell with air and of the cell in vacuum. Fig. 4 showns an example of this subtraction: the N_2O doublet around 1293.16 cm^{-1} (strength: 0.15×10^{-21} cm^{+1} $molec^{-1}$) is clearly evident after the subtraction while undetectable in the original spectrum. Attempts to avoid the need of evacuating the gas cell by subtracting the background from a splitted part of the beam after suitable nmormalization of the two signals (12) were quite unsuccessful. The result was attributed to the predominance of an optical interference noise over laser modulation noise. The optical noise is, obviously, uncorrelated in the two beams because of the asymmetry of the optical layout.

REFERENCES

(1) HINKLEY et al. (1976). Techniques for detection of molecular pollutants by absorption of laser radiation, in E.D. Hinkley (Ed.) Laser Monitoring of the Atmosphere, Springer-Verlag, Berlin, 237-295.

(2) ANLAUF, K.G. et al. (1985). A comparison of three methods for measurement of atmospheric nitric acid and aerosol nitrate and ammonium. Atmos. Environ. 19, 325-333.

(3) COX, R.A. and JENKIN, M.E. (1986). Kinetic studies of HO_2 reactions using diode laser spectroscopy in chemistry related to tropospheric ozone, COST 611 WP2 Workshop, Cologne, 12-13 November 1985, Extended Abstracts, 70-72.

(4) REID, J. et al. (1978). High sensitivity pollution detection employing tunable diode lasers. Appl. Opt. 17, 300-307.

(5) REID, J. and LABRIE, D. (1981). Second-harmonic detection with tunable diode laser - comparison of experiment and theory. Appl. Phys. 32B, 203-210.

(6) CASSIDY, D.T. and REID, J. (1982). Atmospheric pressure monitoring of trace gases using tunable diode lasers. Appl. Opt. 21, 1185.

(7) HASTIE, D.R. et al. (1983). Tunable diode laser systems for measuring trace gases in tropospheric air. Envir. Sci. Techn. 17, 352 A.

(8) see, for example, LAM, R.B. et al. (1982). Cross correlation signal/noise enhancement with application to quantitative gas chromatography/Fourier transform spectrometry. Anal. Chem. 54, 1927-1931.

(9) SASAVITSKI, I.I. et al. (1984). IR tunable diode lasers for spectral gas analysis. J. of Technical Physics (USSR) 54, 1542-1551.

(10) SANO et al. (1983). Analytical description of tunable diode laser derivative spectrometry. Japanese J. Appl. Phy. 22, 1883-1888.

(11) CASSIDY, D.T. and REID, J. (1982). High sensitivity detection of trace gases using sweep integration and tunable diode lasers. Appl. Opt. 21, 2527-2530.

(12) SACHSE, G.W. et al. (1983). Airborne tunable diode laser system for trace gas measurements. Proc. SPIE 438, 99-104.

(13) REID, J. et al. (1985). High sensitivity detection of trace gases at atmospheric pressure using tunable diode lasers. Opt. Quantum Electron. 17, 31-39.

(14) MELANDRONE, G. et al. (1985). Frequency jitter from mechanical vibrations in a diode laser mounted on a closed cycle refrigerator. Appl. Spectrosc. 39, 63-68.

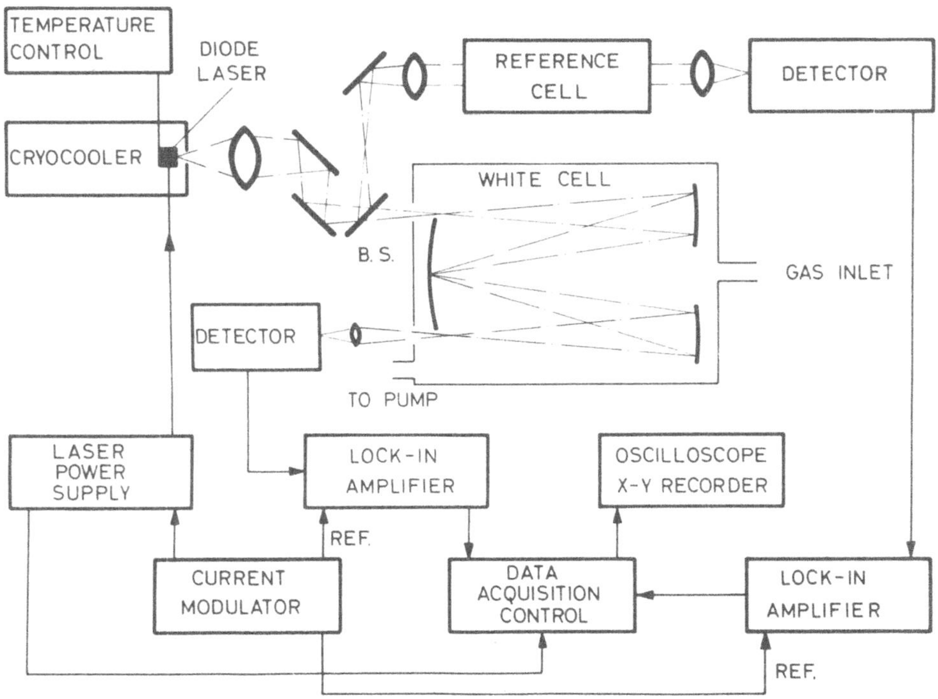

Fig. 1 - Layout of the experimental set-up.

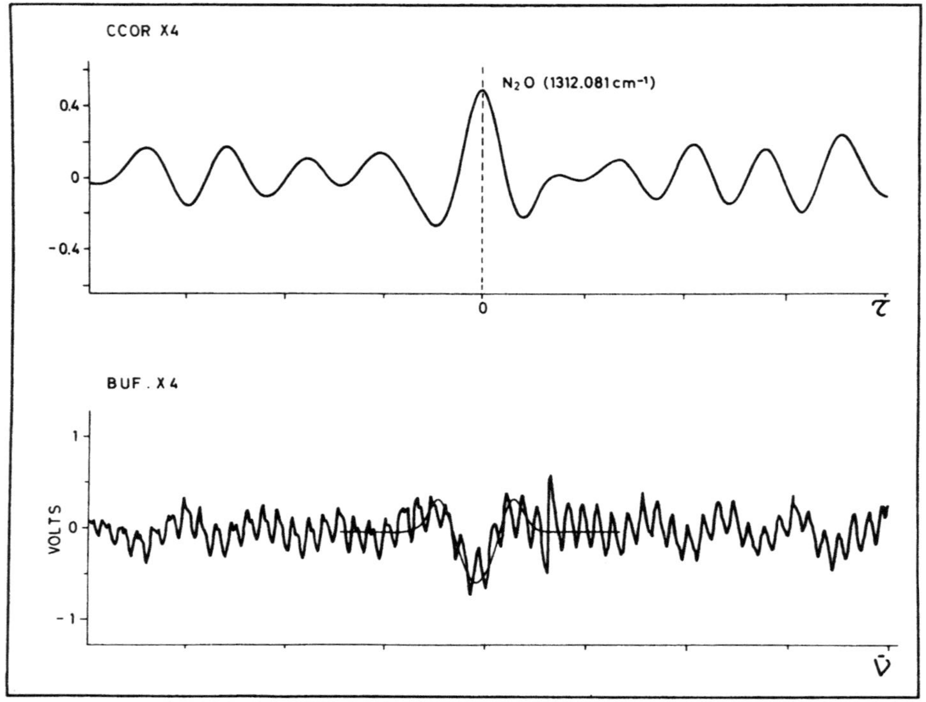

Fig. 2. N_2O line at 1312.081 cm^{-1} detected in the White cell at 15 torr air flux (bottom) and cross correlation (up).

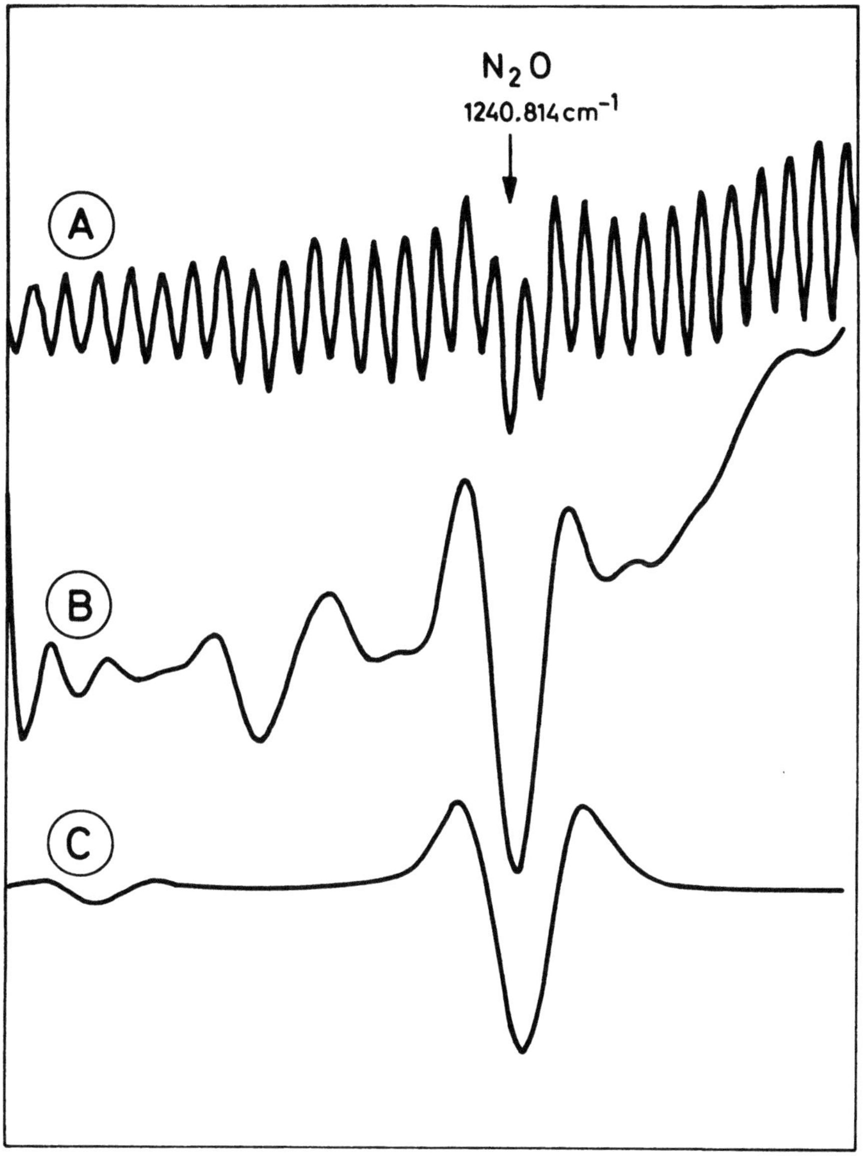

Fig. 3 - N_2O line at 1240.814 cm^{-1} recorded in the ambient air spectrum flowing at 21 Torr in a 64 m multipass cell (A). Same spectrum after application of FT analysis (B). Reference N_2O spectrum (C).

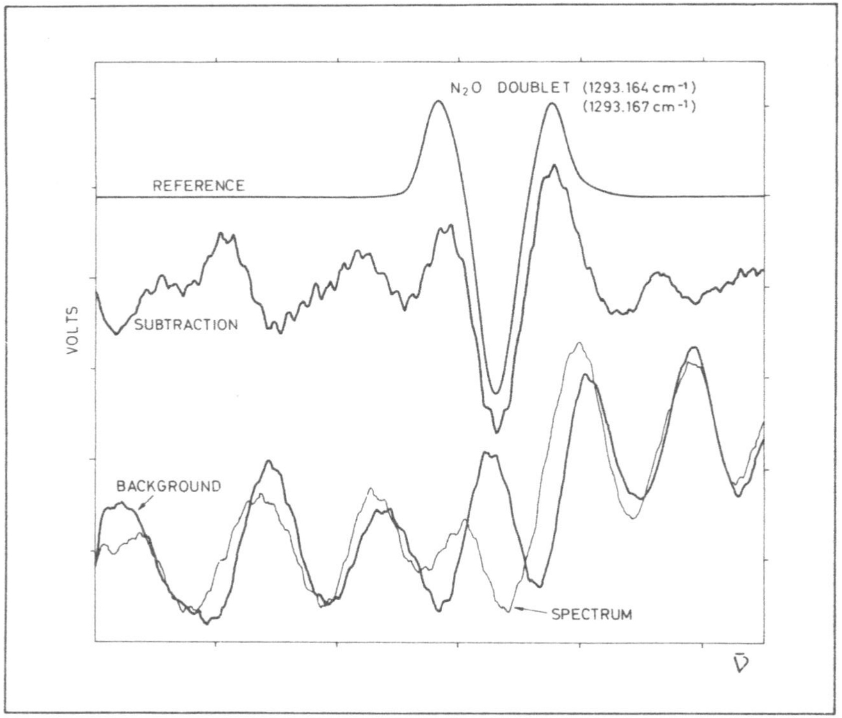

Fig. 4 - N_2O doublet at 1293.16 cm^{-1} arising by subtracting the noise spectrum obtained by emtying the cell.

MEASUREMENT OF AIRBORNE HYDROCARBONS BY HEADSPACE-CHROMATOGRAPHY

J. MÜLLER and F. RIEDEL
Federal Environmental Agency, Pilotstation Frankfurt
D - 6050 Offenbach, Frankfurter Strasse 135

Summary

A method was developed in order to measure airborne hydrocarbons in a simple and efficient way.

Sampling was carried out with charcoal tubes in which the hydrocarbon compounds are adsorbed. The loaded charcoal is dissolved in benzyl alcohol and the solution analyzed in a head-space chromatograph. After separation in a carbopack-column the substances are split and simultaneously are detected in a FID (aliphates and aromates) and ECD (halogenated hydrocarbons).

During several months measurements in urban and remote areas were undertaken. At remote stations the hydrocarbon concentrations roughly were five times lower. However, carbon tetrachloride due to natural sources homogeneously is distributed. Also, 1,1,1-trichloroethane was found to be enriched in remote areas which is induced by its long residence time. The atmospheric behaviour of the different compounds is regarded in respect to the reaction rates.

Day and night concentration variances can be influenced by wet removal processes. The effectiveness depends on the individual solubility of each compound.

1. INTRODUCTION

Due to man-made activities the atmospheric burden of many hydrocarbon compounds in the last decades considerably increased (1,2). Halocarbons used as solvents, dry cleaners and propellants are enriched in the atmosphere on account of their long residence times. Therefore, hydrocarbons ubiquitously are found on the northern and southern hemisphere (9). Several compounds play an important role in atmospheric chemistry (3) and may contribute to forest damages (4).

Therefore, it is necessary to observe the concentrations in industrialized and remote areas. By knowledge of concentration gradients, diurnal and seasonal variations information about the source distribution and residence times of the individual compounds can be obtained.

In order to get a profound and quick overlook there is need for efficient but not too sophisticated measuring methods (5). Sampling should be possible anywhere in the field and analysis be done in a laboratory.

2. SAMPLING AND ANALYSIS

Sampling is carried out by sucking ambient air through a little tube filled with active charcoal (Fig. 1). The airborne hydrocarbon compounds are adsorbed on the charcoal surface. At ambient air temperature the light aliphates pass the sampling system. However, by cooling down the device light compounds can also be captured. In urban air collection of about 0.2 m^3 is needed for analysis. In remote areas for equivalent sensitivity 1 m^3 air is sucked through the sampling tube.

The tubes can be sent by post to the laboratory. The loaded charcoal is put into the vessel of a headspace-chromatograph. After addition of benzyl alcohol the vessels are closed and by shaking the hydrocarbon compounds are desorbed and homogeneously distributed in the solution. This method is also used for personal environment sampling (6).

By heating up to 120° C the hydrocarbons preferentially evaporate and are enriched in the headspace of the vessel. After some hours equilibrium between liquid phase and gasphase is attained. Then, by automatic needle a certain volume is taken out of the vessel´s headspace and is injected into the gaschromatograph (Perkin Elmer F 45). The substances are separated in a carbopack column and then split into equal quantities (Fig. 2). The aliphates and aromates are detected in a FID-detector and the halocarbons simultaneously in an ECD-detector. Two chromatograms are plotted and the peaks are integrated.

Calibration is carried out by adding quantitatively hydrocarbon compounds to benzyl alcohol which furtheron is handled in the same way as ambient air samples. In Fig. 3 the respective chromatograms are represented.

3. RESULTS AND DISCUSSION

Sampling of ambient air was carried out at three stations: in urban air at Frankfurt (M) and at the remote stations in Deuselbach (Hunsrück in 500 m height) and Schauinsland (Schwarzwald, 1100 m height). The total concentration of hydrocarbons in urban areas roughly is five times higher then in remote areas. This gradient in concentration is in the same order of magnitude as observed for other pollutants like SO_2 , NOx or heavy metals like lead or iron.

The amount of anthropogenic sources roughly is correlated to the population density. This mostly determines the spacial distribution of emitted primary pollutants. In the case of secondary products due to atmospheric reactions, however, the spacial patterns may become different.

In Fig. 4 the mean values of the different compounds measured between Oct. 85 and June 86 are represented. It can be noticed that every compound according to its residence time and the distribution of its sources has an individual behaviour. The concentration gradients for halocarbons are less distinct than for aliphates and aromates. In urban air, due to local sources larger standard deviations are abserved. In addition to sources by evaporation during industrial applications most hydrocarbon compounds are also emitted by fuel combustion processes (7, 8).

In Fig. 5 the percentage profiles at the three stations are plotted. In this plot the concentrations of the measured 9 hydrocarbon compounds are added and the sum put 100 %. On the figure can be noticed that the halocarbon compounds at the remote stations become relatively enriched.

In Fig. 6 the monthly mean values and standard deviations of carbon tetrachloride are represented. Among urban and remote stations only slight concentration differences are noticed. This fact has also been found by other investigators (8, 9). Even between the northern and southern hemisphere no significant concentration differences were observed (9). This either is due to an extreme long residence time or natural $CC\ell_4$ which may be generated by airchemical reactions.

In Fig. 7 the monthly concentrations of 1,1,1-trichlorethane are plotted which is beside $CC\ell_4$ the most enriched halocarbon compound. The concentrations at the two remote stations are roughly the same inspite of the heigh difference. During the sampling period urban and remote concentrations vary in a different way.

Due to man-made sources, the standard deviations in urban air in comparison to the remote stations demonstrate large deviations. For trichlorethane no natural sources are known. Therefore, the relative enrichment in remote areas according to Fig. 5 is caused by the long residence time. In some papers a sufficiently long residence time is estimated inducing that at least 15 % of the atmospheric burden escapes into the stratosphere and interacts with the ozone layer in a manner similar to fluorocarbons (10,11).

In Fig. 8 the relative standard deviations of the monthly mean values are plotted in dependance of the reaction rates with OH-radicals (12, 13). This is known to be the dominant removal process in the atmosphere. These reaction rates, theoretically, can be bound to the individual ionisation energies (13).

Substances constantly released into the atmosphere which have high reaction rates also underly high seasonal concentration variances according to the disposability of OH-radicals. Substances less reactive have less seasonal concentration variances. On Fig. 8 this tendency can be clearly noticed. The relative standard deviations S [%] in a rough approximation are inversely proportional to the residence times τ :

$$S[\%] \sim \frac{1}{\tau}$$

1,1,1-Trichlorethane which has the lowest reaction rate with OH-radicals and therefore, among the measured hydrocarbons has the longest residence time also demonstrates the lowest seasonal variances.

In Fig. 9 the relative enrichment factors (quotients of Fig. 5) between Schauinsland (remote) and Frankfurt (urban) are plotted versus the reaction rates. This plotting is reasonable as far as compounds of similar source distribution are taken into account. This is the case for most anthropogenic pollutants whose sources are distributed according to the population density and which are not generated by secondary atmospheric reactions. In Fig. 9 can be seen that the enrichment of anthopogenic compounds in remote areas is clearly related to the individual reaction rates. Thus, by field measurements of the enrichment factors and application of Fig. 9 the reaction rates of primary hydrocarbon pollutants in the atmosphere can be predicted.

Furtheron, day and night concentrations were measured in Schauinsland. In Fig. 10 the corresponding concentrations are plotted. The halocarbons have less distinct differences in the mean values.

Due to the lack of OH-radicals during night heterogenous reaction processes are responsible for the atmospheric removal. A slight solubility in water induces scavenging by droplets or by water films on aerosol surfaces.

REFERENCES

(1) Crutzen, P.J. (1979) Ann. Rev. Earth Planet Sci. 7 : 443

(2) Pearson, C.R. (1982) in O. Hutzinger (ed.) Handbook Environmental Chemistry-Antropogenic Compounds, Vol 3/B, Springer, Berlin

(3) Lovelock, J.E. (1977), Nature 267: 32

(4) UBA-Texte 17(1986), c/o UBA, Berlin

(5) Müller, J., F. Riedel (1984) c/o Reidel Publ. "Third European Syposium: Physico-Chemical Behavior of Atmospheric Pollutants" pp. 373 - 377

(6) Kolb, B., P. Pospisil (1978), Applied Chromatography No. 33, c/o Perkin-Elmer, Überlingen

(7) Ballschmiter, K., P. Mayer (1983), Chemosphere, Vol. 12, No. 6, pp. 879-885

(8) Neuber, E., H.-W. Georgii, J. Müller (1981) Staub-Reinhalt. Luft 41, pp. 91-97

(9) Tille, K.J.W., K. Bächmann (1986) c/o Physico-Chemical Behaviour of Atm. Poll., Stresa (Italy): printed CEC, Luxenbourg

(10) Singh, H.B., L.J. Salas, R.E. Stiles (1982), Environ. Sci. Technol., Vol. 16, No. 12, pp. 872-879

(11) Singh, H.B., L.J. Salas, H. Shigeishi, E. Scribner (1979) Env. Sci. Technol., 203, pp. 899-903

(12) Güsten, H., L. Klasinc, D. Marić (1984) J. Atm. Chem. 2, pp. 83 - 83

(13) Gaffney, J.S. and Levine, S.Z. (1979), Int. J. Chem. Kinet. 11, pp. 1197 - 1209

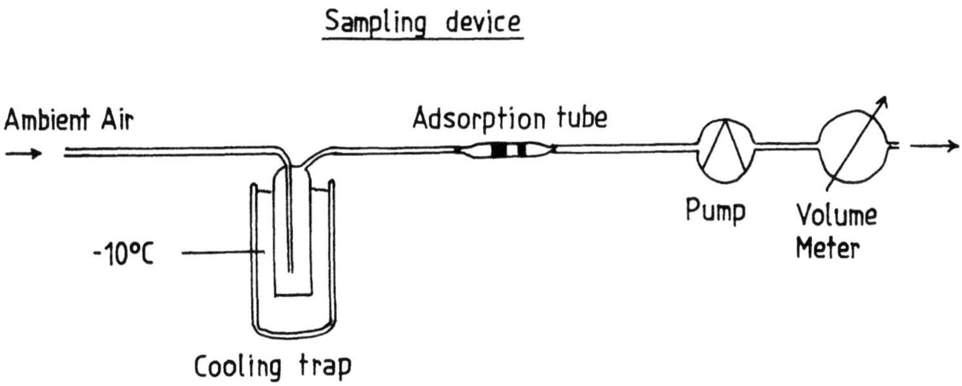

FIGURE 1: Sampling device

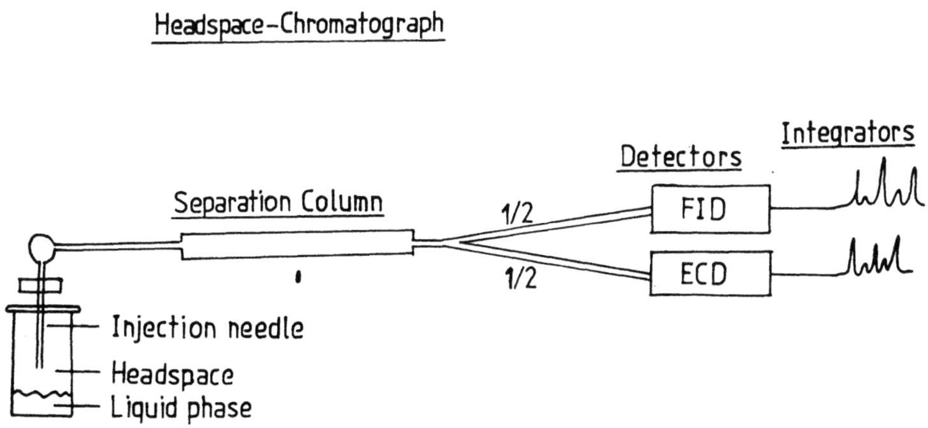

FIGURE 2: Analysis device

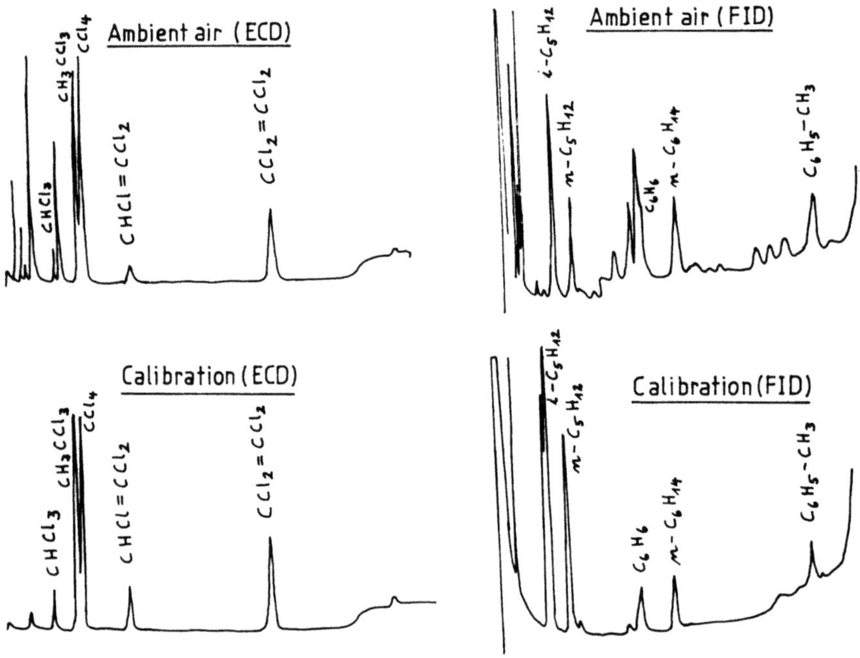

FIGURE 3: Chromatograms of FID and ECD

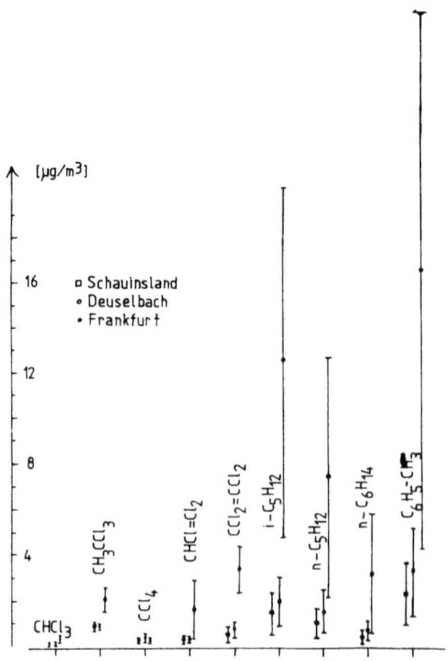

FIGURE 4: Concentrations in urbain air (Frankfurt) and at remote stations (Deuselbach, Schauinsland)

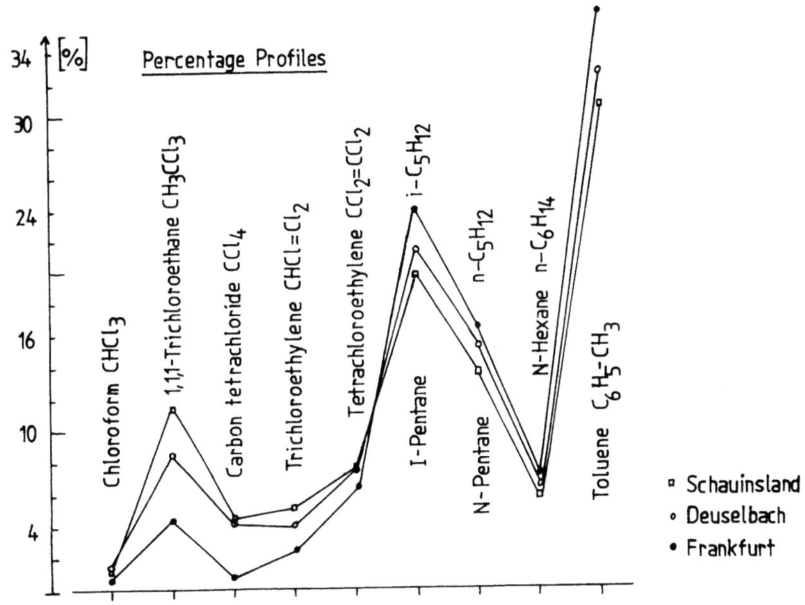

FIGURE 5: Percentage profiles (relative fraction of each compound) at the three stations.

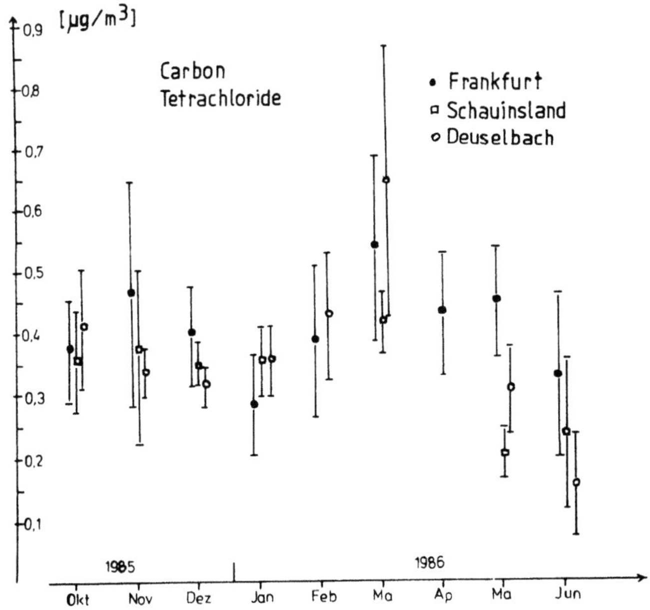

FIGURE 6: Monthly variations and standard deviations of carbon tetrachloride

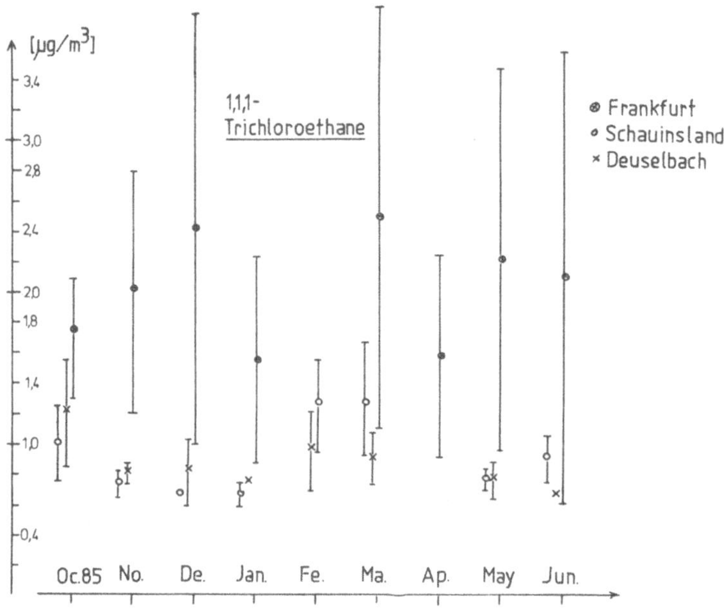

FIGURE 7: Monthly variations and standard deviations of 1,1,1-Trichloroethane

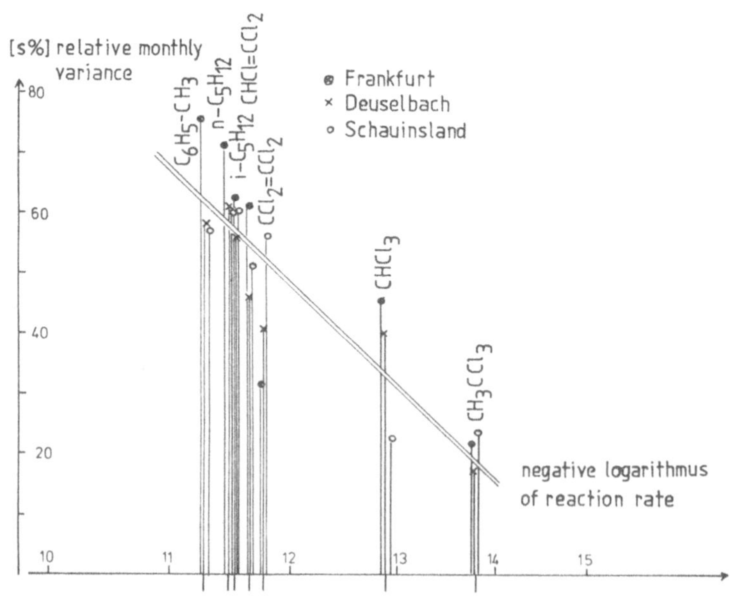

FIGURE 8: Seasonal standard deviations versus reaction rates at the three stations.

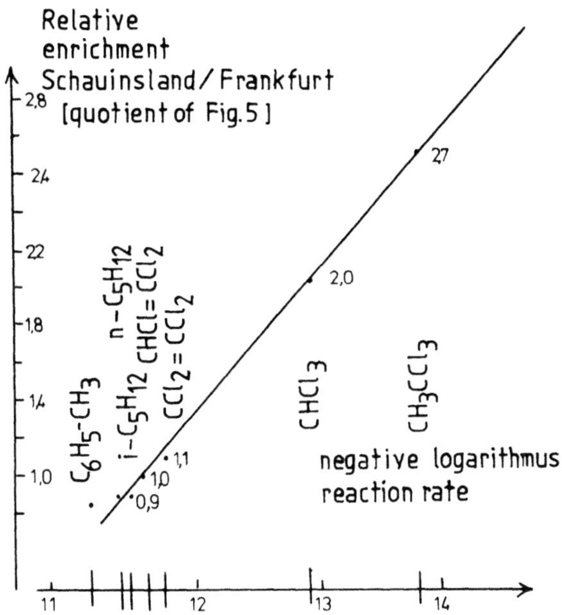

FIGURE 9: Relative enrichment (quotients of Fig. 5) between Schauinsland (remote) and Frankfurt (urban).

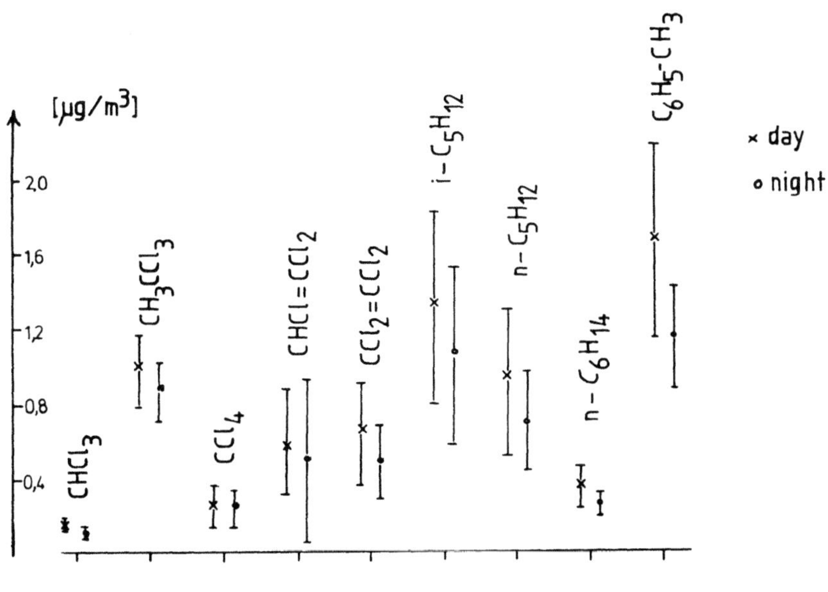

FIGURE 10: Day and night concentrations and standard deviations at the remote station Schauinsland.

A STUDY OF ATMOSPHERIC DEPOSITION CHEMISTRY IN PENINSULAR AND INSULAR ITALY

E. Ioannilli, M. Bettinelli, A. Franciotti and G.L. Ziliani
ENEL-DCO Unità Laboratorio di Piacenza

ABSTRACT

A major monitoring effort on atmospheric deposition chemistry over all Italy is being established cooperatively by ENEL and the Ministery of Agriculture.
As a preliminary result, a first annual set of data on wet and dry deposition for five remote stations in peninsular and insular Italy and one in Po Valley has became available; in this paper these results are presented and discussed.
Acid rain events, quite frequent in the Po Valley, also happen with less frequency also in the southern and insular sectors; the minimum pH values do not reach the thresholds for acute effects to the vegetation, indicated by the current literature; annual average wet deposition is alcaline, not acid in southern Italy; total (wet plus dry) deposition has been alcaline, except in the Po Valley station.
A distinct geographical pattern of the equivalent ratio non marine SO_4/NO_3 in the wet-only precipitation, shows up across the Italy; it ranges from average values less than 2 in the Po Valley up to average values higher than 7 in the South, with many events higher than 20. Considering that the ratio of the precursors in the emissions of large combustion sources in Italy is between 6:1 to 10:1, this fact indicates that a substantial fraction of rain acidity in the North must come from domestic heating and transports.
A thorough meteorological and chemical study, together with an extension of the monitoring network, is necessary to fully document and interprete the apparent spatial and temporal patterns.

1. INTRODUCTION

It is generally recognized that the "acid rain" issue has had on the italian energy development an impact heavier than that in other EEC Countries.
This is because the urgent national need to overcome the weakness of the present energetic structure by shifting rapidly from oil to coal, had to cope with the ecologists struggle against new sulphur emission sources.
For these reasons the Italian Electricity Board (ENEL) decided a direct involvement in the "acid rain" research, in order to give his contribution to clarify the real situation in Italy and so help the Authorities to make the best policy choices, in the national interest.

Besides other interrelated tasks [(1)], a major effort in monitoring the chemical characteristics of dry and wet deposition in the whole Country has been planned. A sampling network, based on 60 stations equipped with wet-dry automatic collectors, is being established in cooperation with the Ministery of Agricolture; in these stations the deposition chemistry together with the health of forests will be monitored for at least three years.
As preliminary result of this program a substantial set of data for five remote stations in peninsular and insular Italy and one in the Po Valley has become available. Given the scarcity of data on this subject, particularly for the southern sector of Italy, we deemed it useful to present in this paper the results of the first year of operation of these stations (October '84 - Sept. '85); also considering that they represent the first consistent set of deposition chemistry data, obtained with the modern wet-dry automatic collectors available in Italy.
These collectors, differently from the old "bulk samplers" allow from one hand to determine the acidity concentration of pure rain samples, to evaluate the possibility of acute effects; from the other hand, by integrating the dry and wet samples, we can evaluate the overall deposition to the ground, and so its long term potential for ecosystem acidification.
Therefore they supply all the necessary information for evaluating the possibility of negative environmental effects of atmospheric deposition on natural ecosystems.

2. EXPERIMENTAL

Stations location

The stations are distributed as follows: one in the Po Valley (Piacenza, 60 m a.s.l.), three in the peninsular (Vallombrosa, 840 m a.s.l.; Suvereto, 50 m a.s.l.; Rossano Calabro, 1200 m a.s.l.) and two in the insular (Termini Imerese, 1000 m a.s.l.; Villasor, 30 m a.s.l.) portion of the Country. All stations but Piacenza are located in rural or forestal districts about 20 km apart from any significant industrial and/or urban emission source. Piacenza station is located in a rural zone in the periphery of a small village, in the densely populated and industrialized Po Valley.

Sampling methods

In all stations dry and wet deposition is collected by using

---

(1) Document: Contributo dell'ENEL alla conoscenza di alcune importanti problematiche ambientali- Aprile '85 ENEL-Rome - Via G.B. Martini 3 -

an automatic wet-dry sampler. A conductivity sensor detects the occurence of precipitation and activates a motor which shifts a cover from the wet collector to the dry collector. The standard collection vessels are formed from 0.20 cm tick, high density polyethylene and have a capacity of 13 liters (I.D. 28.6 cm). A polyurethane foam pad, enclosed in polyethylene film, is affixed to the underside of the cover and ensures a tight seal which reduces evaporative losses from the sample and prevents the leakage of particulate material into the "wet" collector during dry periods.

Analytical methods

A quality assurance procedure is established for all sampling and analysis steps. Wet samples are taken weekly and weighted; a 250 ml subsample is poored in polyetilene bottles and stored in refrigerator at 4°C; if the wet sample is larger than 250 ml, the pH and conductivity are measured in the remaining sample in the local laboratory; dry collectors are removed monthly and carefully closed. These samples are sent to the Central Laboratory for later chemical analysis, together with a "sample registration sheet" in which all relevant data and observations are recorded. The sampling frequency is a compromise between the data meaningfullness and the operating cost, considered that we deemed it advisable to analyze all samples in the same Central Laboratory. Doing this way, we aknowledge that short lived components, such as organic acids, are lost by mycrobial activity, as no biocide addition is practiced (Keene and Galloway, 1984).

The analytical methods employed are presented in the following table, together with the associated precision (coefficient of variation from repeated measurements mode in several days) (Bettinelli et al., 1985).

Chemical species	Methodology	Coefficient of variation %
H^+	potentiometry (Ross's electrode)	5.0
acidity	Gran's titration	1.5
alkalinity	Gran's titration	2.5
conductivity	conductimetry	2.0
Na	atomic emission spectrometry	4.0
K	atomic emission spectromety	4.0
Mg	atomic absorption spectrometry	3.0
Ca	atomic absorption spectrometry	2.5

Chemical species	Methodology	Coefficient of variation %
NH_4	molecular absorption spectrometry, ion chromatography	2.8
Cl	ion chromatography	3.0
NO_3	ion chromatography	3.5
SO_4	ion chromatography	5.0

For dry deposition analysis 500 ml of distillate water is added in the dry collector, and the dry deposit is removed from the walls by a brush; the sample is then shaken for 24 hours in a mechanical shaker. The liquid extract is filtered and stored at 4°C, until analyzed. On this extract the same analyses as for wet sample are performed.

3. DEPOSITION CHEMISTRY PATTERNS - RESULTS AND DISCUSSION

3.1 Data base

Detailed data and summary statistics of the concentration of several chemical parameters measured in the wet samples, are available on request.

The following tables show the annual weighted average concentrations of the main ions and of the strong acidity (μeq/l), of the conductivity (μS/cm) and the median of pH, either for wet as for dry deposition (500 cm^3 leachate of monthly sample).

Wet only deposition

Chemical parameter / Station	Ca	Mg	Na	K	NH_4	NO_3	SO_4	Cl	Strong acidity	pH med.	Cond.
Piacenza 60 m a.s.l.- Rural	28	8	14	2	63	47	86	16	34	4,6	29
Suvereto (LI) 50 m a.s.l.- Rural	42	28	94	6	17	20	74	114	20	4,6	32
Vallombrosa (FI) 840 m a.s.l.- Remote	26	10	29	3	21	19	54	32	23	4,6	20
Rossano Calabro (CZ) 1200 m a.s.l.- Remote	30	10	26	5	7	4	32	34	- 7	5,3	13
Termini Imerese (PA) 1000 m a.s.l.- Remote	79	21	39	5	13	8	53	47	-50	6,3	19
Villasor (CA) 30 m a.s.l.- Rural	78	47	168	7	18	11	76	207	-25	6,2	41

Dry deposition

Chemical parameter / Station	Ca	Mg	Na	K	NH_4	NO_3	SO_4	Cl	Strong Acidity	pH med.	Cond.
Piacenza 60 m a.s.l.- Rural	493	93	106	26	426	212	756	107	- 70	6,2	150
Suvereto (LI) 50 m a.s.l.- Rural	433	69	213	18	9	84	192	223	- 242	7,1	80
Rossano Calabro (CZ) 1200 m a.s.l.- Remote	271	70	132	28	18	43	154	149	- 173	6,8	58
Termini Imerese (PA) 1000 m a.s.l.- Remote	691	78	266	31	7	62	257	301	- 453	7,2	113
Villasor (CA) 30 m a.s.l.- Rural	838	183	488	32	3	63	364	586	- 532	7,2	152

Acidity deposition patterns

The "wet only" deposition, in the station in the Po Valley and on those of the hilly areas of Tuscania is, on the average, relatively acid; at the contrary, on the stations in Calabria (Sila), Sicily and south Sardinia, it is alcaline.
The dry deposition is always alcaline, particularly in the stations in Sicily and Sardinia.
Figure 1 shows the seasonal average input of wet, dry and "total" acidity (or alcalinity) for the period Oct. 1984 - Sept. 1985 (geq/ha•season). A general seasonal pattern of the wet acidity deposition emerges for all stations: in cold months the rain is relatively more acid than in springtime and summer; the annual average wet deposition ranges from acidity to alkalinity, depending on the zone. We observe that the integrated amount of deposition (wet plus dry) over long periods is alcaline (even if the wet component is consistently acid) except in the Piacenza station.

Ionic composition patterns

Looking into the ionic average composition, we can note important difference among the various stations.
The Piacenza station, placed in a rural district in the Po Valley and so very different from all others, shows either in wet as in dry deposition, ammonium and nitrate concentrations distinctly higher than all other stations.
The ammonium concentration (exceedingly high in dry deposition), is clearly linked to the release of ammonia from the agricultural areas all around the station, heavily fertilized (500 kg N/ha yr) by urea, ammonium nitrate, manure, etc.; also the emission of the nitrate precursors is certainly more intense in the Po Valley in respect to other zones, because of the heavier domestic heating, vehicular

traffic and industrial development.

The sulphate is higher than in other stations, but not so much in wet deposition, as it is in the dry deposition: this can be due to the dry deposit of SO_2, whose average concentration in the area is around 40 $\mu g/m^3$, well above that in the other stations. Villasor and Suvereto stations show the greatest influence of the marine component, as is shown by the high concentrations of sodium, magnesium and chloride, in about the same proportions as in the sea water; this component is also readily apparent in the other stations, included that in the Po Valley.

Looking at the composition of the dry deposit, we observe that in the Po Valley station it is dominated by a terrigenous component (Ca, Mg, K); in the others stations the marine aerosol prevails; those alcaline components become more or less sulphated and nitrated according to the gaseous pollution existing at the station; of course also secondary particulate dry deposition play a role. A significant contribution to the calcium content can be due also to the following mechanism: when the southern circulations are associated to a deep central mediterranean trough, the air masses coming from Africa transport saharian dusts, which contain more than 10% $CaCO_3$. This phenomenon, which affect the dry as well as the wet deposition, is very intense in the islands and in the western coast of Italy, but it is significant also in the Po Valley.

Patterns of the ratio SO_4/NO_3

Figure 2 shows the average value of the equivalent ratio: non marine SO_4/NO_3 (the sodium has been considered all marine for this calculation) in the wet and dry deposition.

In wet deposition this ratio shows a distinct pattern across the Country: it ranges from values around 2 in the north, to values around 3 in the centre, up to values higher than 5 in the south and islands; looking at the ratio is a complex response to the emissions of sulphate and nitrate precursors, the availability of oxidants and the source-receptor time-distance: so it is difficult to give a general interpretation of the above observations. We can notice that the ratio of the emissions of sulphate and nitrate precursors is certainly lower in the northern than in the southern part of the Country, because of the heavier domestic heating and vehicular traffic in the former; also the presence in the South of large volcanoes (Etna, Stromboli, Vulcano) which were active during the study period, emitting more than 10^3 t/d of sulphur, could explain the extreme values (>> 20) of

the ratio in all southern stations. Extensive studies are urgently needed to assess the trajectories of air masses during precipitation in the South and to establish the role of the kinetic of oxidation in this respect.
The same ratio in the dry deposition does not show such a clear pattern; the nitrate seems to be relatively more abundant, in respect to the sulphate, in the dry than in the wet deposition, in all station except that in the Po Valley, where the opposit happens.
The Po Valley situation could be explained considering the gaseous dry deposition (which is far more active for SO_2 than for NO_x), as indicated above; for the other stations, as nitrate does not came from natural particulate sources, we must think that secondary nitrate particles reach the South by long range transport in dry days, more than in rainy days. Deeper meteorological investigations are necessary to give a reasonable interpretation of these facts.
In any case, given that the ratio in the emissions of large combustion sources (80% using oil in Italy) can vary between 6:1 to 10:1, finding lower values in the wet or dry deposit seems to indicate that a large contribution to the acidity deposition, particularly in the North, must be associated to emission sources with high content of nitrate precursors, typically the domestic heating (using methane or light distillates) and transports.

4. CONCLUSIONS

Summarizing from the above informations, we can tentatively draw the following conclusions:

- Acid precipitation events, quite frequent in the Po Valley also happen, with less frequency, in peninsular and insular Italy; it is important to note that minimum pH values do not reach the thresholds for acute effects to the vegetation indicated by the current literature, at least for the acid rain alone.
- The integrated wet deposition in the Po Valley near Piacenza and in the hilly areas of Tuscania is acid in cold seasons, but tends to be neutral or even alcaline in spring and summer; in southern Italy and islands it is normally alcaline.

  Total deposition (wet plus dry) is normally largely alcaline (except in the Po Valley station); this can be due to the input of marine aerosols, saharian dust and terrigenous calcareous component.

  In this situation no long term acidification processes, driven directly by the atmospheric deposition, are envisaged.

- The equivalent ratio SO_4/NO_3 in the wet deposition ranges from average values less than 2 in the Po Valley up to 7 in Calabria; considering that the same ratio in the emission of large oil combustion sources in Italy varies from 6 to 10, a consistent contribution to rain acidity, at least in the North, must came from sources emitting mostly nitrate precursors, such as domestic heating and trasport.

From the preliminary results given above (which will be greatly extended in the near future), it does not seem that the acid rain proper could make any problem for natural ecosystems in peninsular and insular Italy, as no acute nor long term effects can be envisaged.

Of course, the possibility of synergisms between acid rains and other conventional and not conventional pollutants, cannot be ruled out; the planned studies must well try to clarify this problem. But at the present status of the research and of the evidence available for the italian situation, we think that the limited resources available for the prevention of atmospheric pollution must be spent primarily for reducing hydrocarbon emissions from mobile and stationary sources. The expected advantages are the improvement of in town air quality, the reduction of phytotoxic ozone on forests downwind of cities, the slowing down of SO_2 oxidation and so decreasing the acid rains proper.

On the contrary, enforcing desulphurization for the new coal power plants (burning less than 1% sulphur) would only slow down the modernization of the italian power plant park, requested by the National Energy Plan, with the consequence of more foreign expenses for fuel, more diseconomics and, paradoxically, more atmospheric pollution (due to old oil plants), without any foreseable environmental improuvement.

REFERENCES

Keene W.C., Galloway J.N. (1984): Organic Acidity in precipitation of North America - Atm. Env. II, 2491-2497

Keene W.C., Galloway J.N. and Holden J.D. Jr. (1983): Measurements of weak organic acidity in precipitation from remote area of the world - J. Geophys. Res. 88, 5122-5130

Keene W.C. and Galloway J.N. (1985): Gran's Titration: inherent errors in measuring the acidity of precipitation - Atm. Env. 19, 199-202

Brosset C. (1980): A method for evaluating the acid-base balance in natural waters - Wat. An. Soil Pollut. 14, 251-265

Lee Y.H. (1980): The linear plot: a new way of interpreting titration data, used in determination of weak acids in lake water - Wat. An. Soil Pollut. 10, 457-469

Bettinelli M., Ioannilli E. and Ziliani G.L. (1985): Acidity measurements by different methods for a precipitation monitoring network - in "Measurement of Atmospheric Acidity - Proceedings of workshop organized within the framework of the working Party 1 - Detection, Identification and Analysis of Air Pollutants of the Concerted Action Cost 611 "Physico-Chemical Behaviour of atmospheric Pollutants" Area della Ricerca di Roma, Montelibretti (Italy), 25-26 June 1985

Jervis T.R. (1979): Rainfall acidity natural variance and subsequent time-dependence of pH - Atm. Env. 13, 160

Fig.1 SEASONAL PATTERN OF WET AND DRY ACIDITY DEPOSITION IN SOME STATIONS

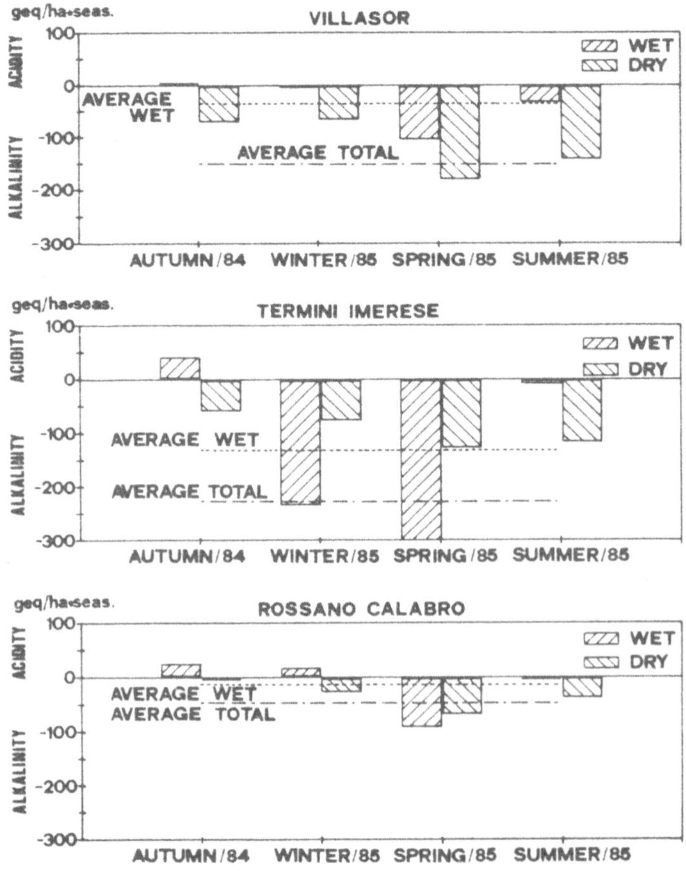

Fig.1 SEASONAL PATTERN OF WET AND DRY ACIDITY DEPOSITION IN SOME STATIONS

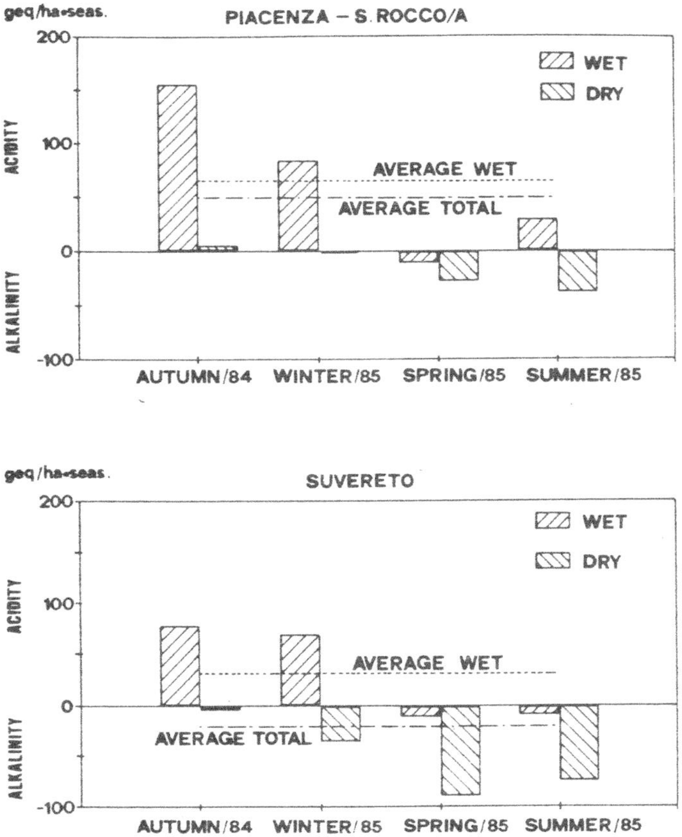

Fig.2 AVERAGE VALUE OF THE EQUIVALENT RATIO: NON MARINE SO4/NO3 IN WET AND DRY DEPOSITION

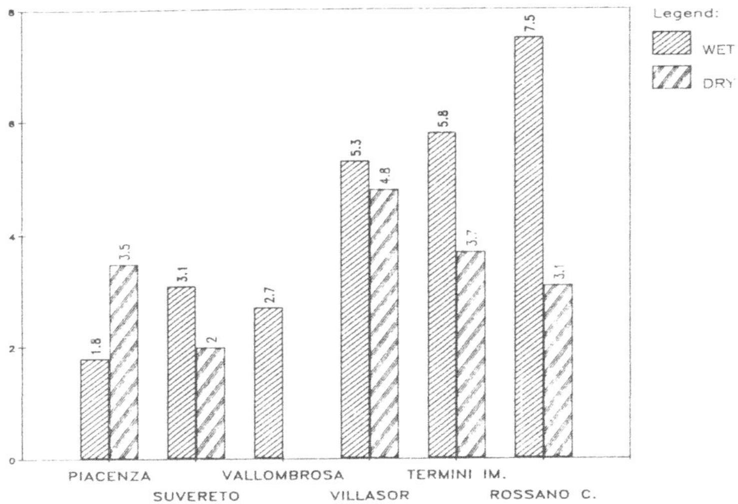

CHEMISTRY OF ATMOSPHERIC DEPOSITIONS AT L. TOGGIA (2160 m a.s.l.), WITH EMPHASIS ON THE SAMPLING PROCEDURE

R. Mosello and G.A. Tartari
C.N.R. - Istituto Italiano di Idrobiologia, Pallanza

Summary

The chemistry of bulk deposition at L. Toggia (2160 m a.s.l., Val d'Ossola, NW Italy) was studied for the five year period 1981-85, with monthly (1981-82) and weekly (1983-85) samples. The mean annual pH values (4.84-5.59) indicate the absence of strong acidity; 25% of the 1983-85 samples showed pH below 4.8 and 10% fall in the range 4.12-4.52. The minimum monthly values of pH (4.8-4.9) are measured in summer, in coincidence with the maximum sulphate and nitrate concentrati ons (64-75 and 18-21 ueq/l, respectively) and notwithstanding the maximum ammonia concentration (25-36 ueq/l). Ammonia is the second cation in importance, after calcium; this suggests that ammonia oxidation, both in the soil and in freshwater, may play an important role in the acidification process at high altitudes in the Alps. The analyses of snow cores, collected before the beginning of snowmelt, may give useful information on the chemistry of bulk deposition during the snow period, which may be as long as eight months. The snow pack chemistry shows spatial differences, so that the analyses of at least five cores, collected in different undisturbed sites of the studied area, is advisable. Six comparisons between the snow cores and bulk deposition chemistry, carried out in 1984, 1985 and 1986, do not show differences more marked than those found within the snow cores, determined both from the spatial variability of the chemistry of snow pack and analytical errors. The best agreement was for pH, conductivity and nitrate, the worst for calcium, sulphate and sodium.

1. INTRODUCTION

The chemical characteristics of atmospheric deposition at different altitudes on the Alpine chain is of interest for several reasons. A first point concerns the effects on freshwater. A previous study (1) considering 320 alpine lakes, mainly located at altitudes ranging from 1800 to 2700 m a.s.l., did not detect any strongly acidic lakes (pH below 4.7), but showed that about 50% of the sampled waterbodies are low buffered, with total alkalinity (T.A.) values equal to or below 0.2 meq/l. In smaller alpine areas, with unfavourable geological conditions, the situation may be worse: in Canton Ticino a group of 56 lakes, not included in the 320 lakes cited above, showed T.A. equal to or below 0.2 meq/l in 86% of the cases, while in 18% of the lakes the T.A. was not detectable. Secondly, there is the effect of atmospheric pollution on vegetation. A recent study (2) carried out in Switzerland, indicates a higher percentage of tree damage in mountain areas (42%) as compared with the plain (27%). The same study, as well as many other papers (see, e.g. 3,4), stress the basic importance of vegetation not only from a general ecological viewpoint, but also in ensuring the stability of mountain slopes and in the prevention of phenomena such as landslides and avalanches. A third point concerns the studies on the transboundary transport of pollutants: to what extent do the Alps act as a barrier?

During 1981, in connection with studies dealing with freshwater acidification, a sampling of atmospheric deposition was carried out at three different sites located in the western Italian Alps (Piemonte, Val d'Ossola, Pennine-Lepontine Alps): L. Cavalli (1500 m a.s.l.), L. Devero (1870 m) and L. Toggia (2160 m) (5). In the following years studies on the chemistry of deposition at high altitude were limited to the station of L. Toggia, where there is a hydroelectric dam with personnel present for the whole year. At the same time, because of the need to obtain information on deposition chemistry in remote sites, where continuous sampling is not possible, a research was started with the aim of assessing the reliability of snow cores as indicative of deposition during the period from September-October to April-May. Though less detailed than the above method, this sampling makes possible an evaluation of the acid input to water bodies during the period of snow melt, when acute acid stress may occurr (6,7). Snow pack investigations have already been used as a method of obtaining an estimate of the chemical composition of precipitation in temperate regions (8, 9, 10). Studies on Alpine snow samples have been aimed both at getting information on the chemistry of atmospheric deposition (11, 12, 13, 14) and at studying paleoprecipitation chemistry (15).

The aim of this paper is twofold:

1) to present and discuss the chemistry of deposition at L. Toggia during the five years period 1981-1985;
2) to discuss to reliability of snow core chemistry as representative of deposition during the winter period. This will be made on the basis of a comparison with the chemical results obtained from weekly deposition samples collected at L. Toggia and snow cores collected in the same area.

2. SAMPLING AND METHODS

Bulk depositions were collected monthly from January 1981 to September 1982 and weekly from October 1982 to December 1985, using polyethylene bottles with a funnel. The snow was collected in a polyethylene cylinder 28 cm in diameter and 60 cm high. Both types of collectors were located at a height of 1.5 m, above the ground. The mean time lag between sample collection and chemical analysis was 15 days during the monthly and 7 days during the weekly samplings. Snow cores were collected using a plexiglass cylinder, 50 cm long and 5.1 cm in diameter; the sampling are carried out by making a hole in the snow down to the ground; the sample is taken by inserting the cylinder vertically into the snow, closing off the bottom end with a flat piece of polyethylene and removing the cylinder sideways. The entire column was then collected in 50 cm fractions, and the whole stored in a 3 l polyethylene bottle. All the material used for the removal and the transportation of the samples was tested in order to verify the absence of contamination. The cores were sampled in two sites at 250 m E-SE and 1600 m NW from the bulk collector; in the following discussion the two samples will be indicated as A and B. Six samplings were performed, respectively in April 1984; February, March and April 1985, February and May 1986.

Concerning the analytical methods in the period 1981-83 sulphate was determined by turbidimetry, nitrate by salicylate, chloride by ion selective electrode; during 1984-85 these anions were measured by a ion chromatograph. Comparison between the different methods showed a good agreement for nitrate and chloride, while the measurements of sulphate in the first three years were less reliable. Ammonia was measured by blue indophenol, calcium, magnesium, sodium and potassium by A.A.S. The pHmeter calibration (pHM84, Radiometer) is made with two buffers, covering the range of measurements, and gives a precision of 0.03 u, while accuracy, because of the problems involved in this kind of measure, is thought to be 0.2 u. Details on the precision, lower detection and quantification limits of the analytical methods used are presented elsewhere (16).

3. RESULTS AND DISCUSSION

Tab. 1 shows the amount of precipitation during the five years of the study, the mean volume weighted concentrations, of the main ions and the mean pH and conductivity values. The highest volume of precipitation was measured during 1981 (1945 mm), while the lowest is relative to 1985 (738 mm); the historical mean of the period 1951-1985 is 1263 mm. pH mean values, ranging between 4.84 and 5.59, were obtained by weighting the hydrogen ion concentrations on the volume of the single precipitation event;

this procedure is correct only in the absence of events presenting a detectable total alkalinity (17, 18). As such cases are detected at L. Toggia, for a better description of the pH variation, Tab. 2 reports a comparison of the mean values with other values of statistical relevance. The mean values always fall in the 25-75% le range and are lower by 0.1-0.4 pH units than the 50%le. Returning to the ionic concentrations of Tab. 1, calcium and ammonia, ranging between 21-47 and 13-19 ueq/l respectively, are the most important cations, while sulphate (33-59 ueq/l) is the main anion. The ratio sulphate/nitrate is between 2.5 and 3.9. The total ionic concentrations range between 122 and 169 ueq/l , with a conductivity variation of 10.1-12.6 uS/cm at 18 °C. The ionic balance shows differences from 3 to 15% of the total solute concentrations, generally with an excess of cations; this excess is not explained only by the hydrogen ion, calculated from the mean volume weighted pH. Indeed, this is only 2 to 5 ueq/l higher than the hydrogen ion concentration calculated from the 50%le pH. The bicarbonate ion could be more important; indeed, if we calculate the net acidity on the basis of the yearly values:

$$NA = H^{+} - HCO_3^{-} = SO_4^{--} + NO_3^{-} + Cl^{-} - (NH_4^{+} + Ca^{++} + Mg^{++} + Na^{+} + K^{+})$$

we obtain negative values, i.e. positive alkalinity, in three cases (-12, -6, -11 ueq/l respectively for 1981, 1983 and 1984) and positive values in the remaining two years (+16 and +2 ueq/l respectively for 1982 and 1985).

In order to verify the presence of seasonal variations of pH, ammonia, nitrate and sulphate, the monthly values of the single years, the monthly and yearly values of the whole five year period were calculated as volume weighted means (Fig. 1). The monthly values of the single years show a marked variability, higher than the seasonal variations presented by the five-year monthly means. pH shows its highest values (5.2-5.6 u) during winter and its lowest values (4.8-4.9 u) during spring and summer; there is a peak (5.5 u) in August. The other three ions show higher values during summer and the minima in autumn and winter. An attempt at an explanation of these variations may be made by considering separately ammonia and the other three ionic species. A significant fraction of ammonia cames from the bacterial degradation of organic matter and animal excrements in the soil; these biological processes increase with increasing temperature (19, 20) and are weak or absent during winter time, under the snow cover. To explain the seasonal variation of the other three ions we must consider the stratification of the atmosfere During winter time the presence of thermal stratification does not permit the polluted masses of air in the Po Valley to reach the altitude of the sampling station (2160 m); this happens during summer, when stratification is absent, with a consequent increase

Tab. 1 - Volume of precipitation, pH, conductivity (uS/cm at 18 'C) and mean volume weighted annual concentrations (ueq/l) in bulk deposition at L. Toggia.

Year	mm	pH	H	NH4	Ca	Mg	Na	K	Cat.	SO4	NO3	Cl	An.	Ions	Cond.
1981	1945	5.59	3	19	47	5	7	3	84	40	13	16	69	153	12.6
1982	1317	4.94	11	13	39	7	9	3	82	59	15	13	87	169	12.1
1983	1215	4.84	14	18	34	7	7	3	83	35	13	15	63	146	12.1
1984	1087	5.31	5	16	38	7	7	2	75	33	14	12	59	134	10.1
1985	738	4.91	12	15	21	6	10	2	66	37	13	6	56	122	10.5

Tab. 2 - Mean pH calculated from the volume weighted hydrogen ion concentrations compared with the distribution of the pH values (%le = percentile, n = number of data).

Year	mean	min.	25%le	50%le	75%le	max.	n
1981	5.59	4.9	5.4	6.0	6.1	6.5	11
1982	4.94	4.42	4.78	5.11	5.38	5.97	19
1983	4.84	4.12	4.55	5.12	5.91	6.96	32
1984	5.31	4.34	4.82	5.65	5.93	7.07	30
1985	4.91	4.29	4.80	5.05	5.39	6.25	30

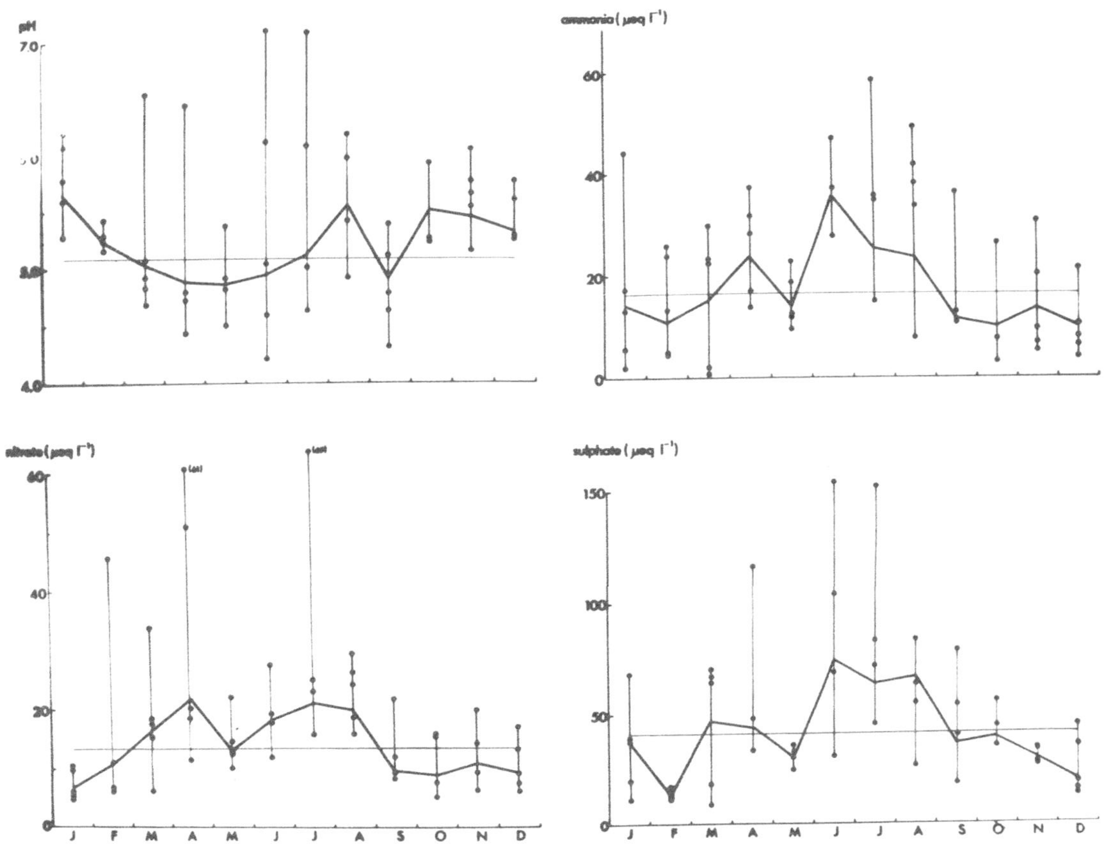

Fig. 1 - Monthly values for the single years (dots), monthly mean values (full line) and yearly mean value (horizontal line) for the whole period of study.

in the transport of sulphate, nitrate and hydrogen ion. The higher pH values of July and August are probably determined by dust deriving from a limited automobile traffic on an unsurfaced road not far from the sampling station. The disturbance factor is of minor intensity in September and absent in the remaining period of the year.

Before discussing the relationship between bulk precipitation and snow core chemistry, it is important to assess the spatial variability that the snow pack may present. Tab. 3 shows the volume of melt water and the chemical composition of three cores collected in a limited area (about 100 m^2) apparently undisturbed and with uniform wind exposure. The water volume shows a variation of 10% with a range between 530 and 625 mm. pH (5.01-5.33) and conductivity (5.0-5.2 uS/cm) show quite a good agreement, while the range of variation is larger for calcium, sulphate, nitrate and sodium. These variations are only partially explained by the precision of the analytical methods (Tab. 3). Though the possibility of accidental pollution of the sample, both in the field or in the laboratory, can not be excluded, we think that the results indicate a spatial variability of the chemistry of snowpack. The most likely causes are the presence of windborne dust and mechanical rearrangement of the snow determined by wind. A spatial variability was observed also in an other study (21), on the basis of a higher number of samples collected in a high-altitude Scottish catchment. Also the comparison between the samples collected at sites A and B shows differences in the concentrations, high for calcium and sulphate, lower for the other ions. Tab. 4 shows the results of two of the six samplings carried out, while the following discussion refers to the whole set of data. The pH values are in good agreement, with differences lower than 0.3 u. While the differences in concentrations between the two sites are not systematic, the volumes of meltwater of cores B are 20-60% higher than those of cores A.

The concentration of bulk precipitation to be compared with the snow cores was calculated as volume weighted concentrations of the weekly samples, considering the period of time from the first snowfall to the day of the core sampling. The comparison shows that the water equivalents of the snow column are systematically higher (30-300%) than the volumes of precipitation. The highest difference was measured in the second sampling of 1986 (Tab. 4) after a heavy snowfall which occurred in April. Both an inefficient measurement of precipitation volumes in the case of snow events and the action of wind on the snowpack may be responsible for the observed differences. Concerning the chemical characteristics, in five of the six cases examined precipitation pH agrees quite well with those of the snow cores, with differences of 0.1-0.2 u; no explanations are available for the highest difference (0.5 u, second comparison). The differences are more distinct in the case of conductivity, calcium and sulphate; however,

Tab. 3 - Chemical characteristics of three snow cores collected in an area of 100 m2, compared with the precision (P) of the analytical methods in the same range of concentration. Units: snow height and water equivalent in cm; others as in Tab. 1.

	A1	A2	A3	mean	S.D.	P
Snow depth	212	192	245	216	27	-
Water equivalent	53	63	55	57	5	-
pH	5.21	5.01	5.33	5.18	0.16	0.03
Hydrogen ion	6	10	5	8	3	2
Ammonia	2	8	2	4	3	0.3
Calcium	22	8	21	17	8	5
Magnesium	3	2	2	2	1	3
Sodium	10	2	10	7	5	3
Potassium	1	0	1	1	1	3
Cations	44	30	40	39	7	-
Sulphate	12	7	14	11	4	0.5
Nitrate	10	2	3	5	4	0.2
Chloride	8	8	8	8	0	0.1
Anions	30	17	25	24	7	-
Ions	74	47	65	63	14	-
Conductivity	5.0	5.2	5.1	5.1	0.1	0.5

Tab. 4 - Chemical characteristics of the snow cores collected at sites A and B, compared with the mean concentrations of bulk deposition (P). Units as in Tab. 3.

Period	1.10.84 - 12.02.85			18.11.85 - 8.05.86		
	A	B	P	A	B	P
Snow depth	130	180	-	400	450	-
Water equivalent	37	56	30	156	186	49
pH	5.01	4.97	5.49	4.97	5.22	4.99
Hydrogen ion	10	11	3	11	6	10
Ammonia	3	5	4	10	7	9
Calcium	21	11	40	9	12	10
Magnesium	0	0	5	4	2	3
Sodium	3	8	9	3	2	5
Potassium	1	1	2	0	0	1
Cations	37	36	63	37	30	38
Sulphate	8	13	23	13	11	9
Nitrate	9	9	9	10	9	13
Chloride	2	2	8	3	3	3
Anions	19	24	40	27	23	25
Ions	56	60	103	64	53	63
Conductivity	5.2	6.5	7.4	6.7	5.0	8.2

they are not systematic and are of the same order as those present among the snow cores.

4. CONCLUSIONS

Though the mean annual pH values of atmospheric precipitation at L. Toggia (4.84-5.59, period 1981-85) indicate the absence of strong acidity, some events show an acidic character, with sulphate and nitrate as the main anion. The weekly samples collected in 1983-85 show a 50%le and 25%le pH value of 5.29 and 4.80 respectively, and 10% of the measurements range between 4.12 and 4.52. The seasonal variations of pH and sulphate and nitrate concentrations indicate a stronger influence of pollution coming from the plain during summer; the minimum pH monthly values (4.8-4.9) are measured in June-September, though in this period the highest concentration of ammonia is also present. On an annual basis ammonia is the second cation in importance, after calcium. This suggests that ammonia oxidation, both in the soil and in fresh water, may play an important role in the acidification process at high altitude in the Alps. A further topic to be studied in detail in the near future is the episodicity of the hydrogen ion inputs to soil and freshwater, both during spring, when the pollutants are released more intensely in the first phase of snowmelt, and during autumn, when heavy precipitation is frequent.

The analyses of snow cores, collected before the beginning of snowmelt, may provide good information on the chemistry of bulk deposition in the snow period, which in this area may be as long as eight months. As the chemical characteristics of snow pack show a spatial variability, low for pH and conductivity higher for calcium and sulphate, the collection of more samples (at least five) is recommended. Because of the very low solute content of the meltwater, every care must be taken to avoid both the contamination of the samples and analytical errors. The sampling technique used by the authors, consisting of the excavation of the snow pack down to the ground and the sideways extracton of the cores, appears to be reliable but may give some practical trouble, considering that the depth of the snow cover may reach 3-4 m. A comparison between the chemistry of snow cores and of bulk precipitation does not show differences more marked than those found within the snow cores, determined both from the spatial variability of the chemistry of snow pack and the errors associated with the analytical techniques. On the basis of six comparisons, carried out in 1984, 1985 and 1986, the chemical variables that showed the best fit were pH, conductivity and nitrate, while calcium, sulphate and sodium showed the worst agreement, without any regular pattern. An important point to stress is the difference between the water equivalent of the snow cores and the volume of precipitation measured with the recording gauges; in this case the former measures were systematically higher than the latter. A discussion of the problems connected with this topic is not part of the aim of the present paper. However the authors stress that the correct measurements of the volume of precipitation is of capital importance in studies on acidification carried out at high altitude, because of its implication in the estimation of the ionic deposition.

ACKNOWLEDGMENTS

We are grateful to the ENEL personnel of the Ossola Valley for their help in sampling the bulk precipitation. Our thanks are also due to Mr. A. Ferrari and B. Sulis of this Institute for their contribution in the field and laboratory work.

REFERENCES

(1) MOSELLO, R. (1984). Hydrochemistry of high altitude alpine lakes. Schweiz.Z.Hydrol., 46 : 86-99.

(2) UFFICIO FEDERALE DELLE FORESTE SVIZZERE (1985). Risultati del l'inventario Sanasilva 1985 dei danni ai boschi. Berna e Bir - mensdorf : 48 pp.

(3) DEPARTEMENT FEDERAL DE L'INTERIEUR SUISSE (1984). Dépérissement des forets et pollution de l'air. Berne : 120 pp.

(4) BACH, W. (1985). Forest dieback: extent of damages and control strategies. Experientia, 41 : 1095-1104.

(5) MOSELLO, R. and G. TARTARI (1982). Chemistry of the precipitation in the L. Maggiore watershed (N. Italy). Mem.Ist.Ital. Idrobiol., 40 : 163-180.

(6) JOHANNESSEN, M. and A. HENRIKSEN (1978). Chemistry of snow meltwater: changes in concentration during melting. Wat.Resour. Res., 14 : 615-619.

(7) JEFFRIES, D.S., C.M. COX and P.J. DILLON. (1979). Depression of pH in lakes and streams in Central Ontario during snowmelt. J.Fish.Res.Bd.Can., 36 : 640-646.

(8) HENRIKSEN, A. (1972). Quantitative chemical analysis of snow. Vatten, 28 : 409-412.

(9) ELGMORK, K., A. HAGEN and A. LANGELAND (1973). Polluted snow in southern Norway during the winters 1968-71. Envir.Pollut., 4 : 41-52.

(10) WRIGHT, R.F. and H. DOVLAND (1978). Regional surveys of the chemistry of the snowpack in Norway, late winter 1973, 1974, 1975 and 1976. Atmosf.Environ., 12 : 1755-1768.

(11) STUMM, W. and G. RIGHETTI (1982). Tessiner Bergseen: saurer Regen, saure Traufe. Neue Zürcher Zeitung, 231 : 31.

(12) ZOBRIST, J. (1983). Die Belastung der schweizerischen Gewässer durch Niederschläge. VDI Berichte, 500 : 159-164.

(13) SMIDT, S. (1985). Depositionsmessungen in verschiedenen Höhenlagen. Allgem.Forstzeitung, 96 Jq., 10 : 270-271.

(14) PSENNER, R. and U. NICKUS (in press). Snow chemistry of a glacier in the Central Alps. Zeit.Glaciol.und Glacial Geol.

(15) SCHOTTERER, U., H. OESCHGER, D. WAGENBACH and K.O. MUENNICH. (1984). Information on paleoprecipitation on a high altitude glacier (Monte Rosa, Switzerland).

(16) MOSELLO, R. (in prep.) Effects of acid precipitation on lake ecosystems in north-western Italy. Final Report CEE-CNR ENV.878.I (S).

(17) REUSS, J.O. (1975). Chemical/biological relationships relevant to ecological effects of acid rainfall. EPA-660/3-75-032 : 46 pp.

(18) LAU, Y.K. (1982). Precipitation chemistry. Alberta Environment, Canada : 75 pp.

(19) SOEDERLUND, R. and B.H. SVENSSON (1976). The global nitrogen cycle. Ecological Bull., 22 : 23-73.

(20) LENHARD, U. and G. GRAVENHORST. (1980). Evaluation of ammonia fluxes into the free atmosphere over Western Germany. Tellus, 32 : 48-55.

(21) BRIMBLECOMB, P., M. TRANTER, P.W. ABRAHMS, I. BLACKWOOD, T.D. DAVIES and C.E. VINCENT. (1985). Relocation and preferential elution of acidic solute through the snowpack of a small, remote, high-altitude Scottish catchment. Annals of Glaciology, 7 : 141-147.

ANALYSIS OF CARBONYL COMPOUNDS IN THE ATMOSPHERIC LIQUID PHASE

M.C. FACCHINI and G. CHIAVARI
Istituto Chimico "G. Ciamician", Università di Bologna
Via Selmi 2, 40126 Bologna (Italy)

S. FUZZI
Istituto FISBAT - C.N.R.
Via de'Castagnoli 1, 40126 Bologna (Italy)

Summary

Two high-performance liquid chromatographic methods for the analysis of carbonyl compounds in the atmospheric dispersed liquid phase (fog, clouds and precipitation) are presented: a) determination of formaldehyde as lutidine derivative; b) speciation of a wide range of carbonyls as 2,4-dinitrophenylhydrazones. Electrochemical detection was employed for both procedures. The proposed methods are discussed in terms of the requirements for atmospheric sample analysis. Particular care is taken in the assessment of the sample conservation procedure, when the analysis is not immediately carried out in the field. Preliminary results of fog and rain sample analysis are presented.

1. INTRODUCTION

A large number of organic compounds have been identified as constituents of the atmospheric dispersed liquid phase: clouds, fog, precipitation (1). Among them, the importance of carbonyl compounds in aqueous atmospheric chemistry has been underlined by several authors. These compounds are involved in a variety of chemical reactions such as: a) formation of adducts with several compounds (bisulphite, alchools, amines), thus acting as oxidation inhibitors (2,3,4); b) reaction with OH and HO_2 radicals to produce organic acids (1,5,6). Aldehydes can also be accumulated in the dispersed liquid phase, due to the high water solubility; in their hydrated forms (gem-diols) they are less sensitive to photodecomposition (5) and therefore clouds and fog, subjected to a reduced insolation, are an ideal reservoir for these compounds (4). Formaldehyde is the most abundant carbonyl compound in the atmosphere, but other carbonyls have been detected in atmospheric liquid samples (3,7,8). Good field data are needed as an input to physico-chemical models in order to completely understand the role of carbonyls in the chemistry of the atmospheric liquid phase.

Two analytical methods are presented in this paper; both of them use the HPLC technique and employ electrochemical detection (ElCD):

a) HPLC - ElCD detection of formaldehyde as lutidine derivative;

b) HPLC - ElCD speciation of several carbonyl compounds as 2,4-dinitrophenylhydrazones.

The first procedure is specific for formaldehyde, even in the presence of abnormally high concentrations of other aldehydes, and it is fairly simple and rapid. The second one is to be preferred when a speciation of different carbonyl compounds is required (for example input data for a model). The two methods are analyzed in terms of:

i) sensitivity and precision;

ii) interferences;

iii) sample handling and analytical procedure.

A comparison, using the same guideline, is also provided with analogous procedures reported by other authors. Finally, experimental field data on carbonyl compound concentration in fog and rain are reported.

2. EXPERIMENTAL PROCEDURES

2.1. HPLC - ElCD detection of formaldehyde as lutidine derivative

A widely used technique for formaldehyde determination in atmospheric liquid samples is the Nash spectrophotometric method (9). This procedure is based on the reaction between formaldehyde, acetylacetone and ammonia:

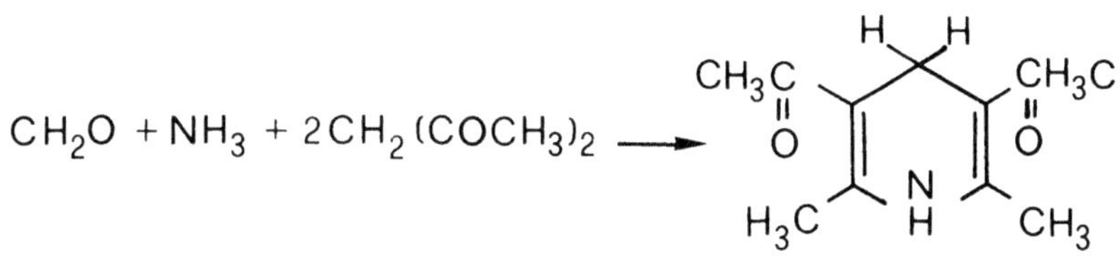

(1)

The final product of this reaction is spectrophotometrically examined at 412 nm. Several data have been reported in literature on the analysis of cloud, fog, rain samples and atmospheric aerosol (4,10,11). However, some interferences due to other aldehydes are possible (9). We have recently set up a HPLC method for formaldehyde determination based on reaction (1), which avoids any possible interference of other aldehydes, even if these are present in abnormally high concentration (12). Furthermore, detection is performed by ElCD, instead of spectrophotometrically. Electrochemical detection improves, in the first place, the detection limit of the analytical technique: 1.5 μM as compared to 5.5 μM for the spectrophotometric method. On the other hand, the specificity of this procedure for formaldehyde is due to both the chromatographic separation and the fact that only formaldehyde is electrochemically active at a potential lower than + 0.8 V.

It is well known that bisulphite-aldehyde addition compounds are formed in the atmospheric liquid phase (4); this fact introduces an error in the formaldehyde determination. The addition of I_2 has been proposed to eliminate this interference (13). We found that the addition of I_2 does not af-

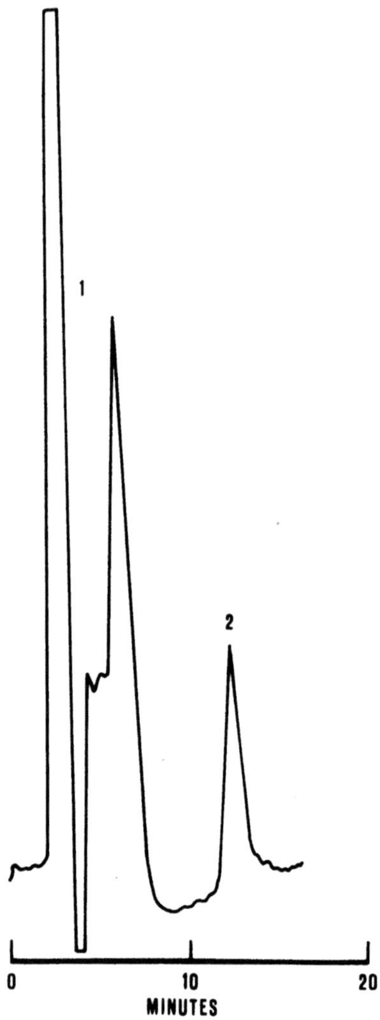

Fig. 1 - Formaldehyde analysis as lutidine derivative. HSO_3^- present was oxidized by addition of I_2.
1) excess reagent + I_2;
2) formaldehyde, 6.7 μM.

fect the chromatographic response of the formaldehyde-lutidine compound, as shown in Fig. 1.

A careful study was performed on the handling and conservation of the samples from field studies and on the reproducibility of the results. It is not always possible to analyze the samples immediately after collection, and formaldehyde losses from atmospheric liquid samples were reported at a rate of 15-20% per day (10). We thus chose to perform the derivatization of the samples directly in the field, after I_2 addition to oxidize HSO_3^-. The treated samples, stored in the dark at 4°C were then stable for up to four days; after that period, a decrease in the chromatographic response was observed. Glass vials must be preferred for storage of the samples, since an increase of formaldehyde concentration has been observed in samples stored in polyethylene bottles, probably due to a HCHO release from the plastic surface (14). The samples are then injected into the chromatographic column without any further manipulation, avoiding possible contamination.

2.2. HPLC - ElCD speciation of several carbonyl compounds as 2,4-dinitrophenylhydrazones

Results on the speciation of a wide range of carbonyl compounds have been reported in literature, employing a HPLC technique based on the derivatization to 2,4-dinitrophenylhydrazones:

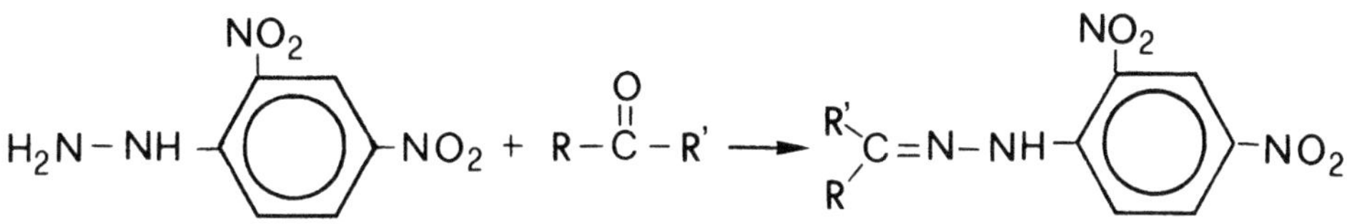

(2)

and spectrophotometric detection at 360 nm (3,7).

ElCD detection has recently been proposed (15,16) as an improvement of this HPLC technique. A gain in sensitivity is reported of up to 50 times (15) Some carbonyl compounds are present in atmospheric liquid samples in concentrations of fractions of micromoles, and preconcentration procedures are usually applied before analysis of real samples (3). The higher sensitivity achieved employing ElCD detection permits the analysis of these samples without any preconcentration, which might cause contamination and increase the time of analysis. The derivatization procedure is not affected by HSO_3^- concentrations up to 100 μM.

Eight carbonyl compounds were investigated in our study: formaldehyde, acetaldehyde, acrolein, propionaldehyde, aceton, n-butyraldehyde, iso-valeraldehyde, benzaldehyde. Several problems were found in applying this technique to the analysis of real samples. The above reported loss of formaldehyde from the sample solutions was observed for the other low molecular weight carbonyls, in particular acetaldehyde and acrolein. It is therefore important to derivatize the samples immediately after collection. In order to obtain a quantitative conversion of the carbonyls into the final products, derivatization was carried out in a two-phase system by adding dinitrophenylhydrazine (DNPH) solution and iso-octane (17,18). Carbonyls and DNPH reagent are water soluble and the reaction takes place in the aqueous phase; as the reaction proceeds, the partitioning of the derivatives into the organic phase shifts the equilibrium towards the formation of the dinitrophenylhydrazones. This procedure is particularly important for the derivatization yield of C_3 carbonyls. As above reported, plastic bottles were excluded for the storage of the samples before analysis; only glass vials were used. After

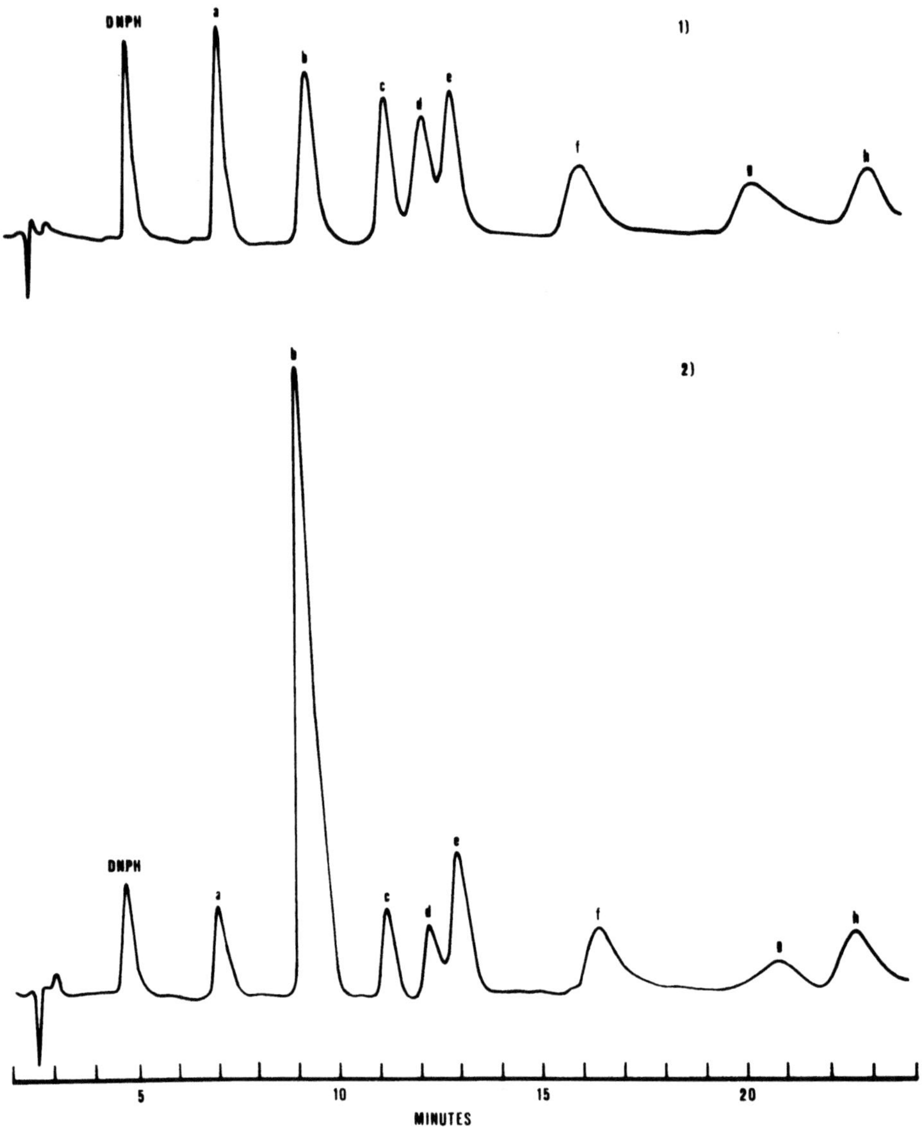

Fig. 2 - Analysis of a standard containing several carbonyl compounds (concentration ca. 2.5 μM), after derivatization to 2,4-dinitro phenylhydrazones: a) formaldehyde, b) acetaldehyde, c) acrolein, d) propionaldehyde, e) aceton, f) n-butyraldehyde, g) iso-valeraldehyde, h) benzaldehyde; DNPH = reagent.
1) immediate analysis after derivatization in the two-phase system;
2) after 72 hours, samples stored without separation of the phases.
Notice the abnormal increase of the acetaldehyde peak and the poor reproducibility of the results for the other carbonyls.

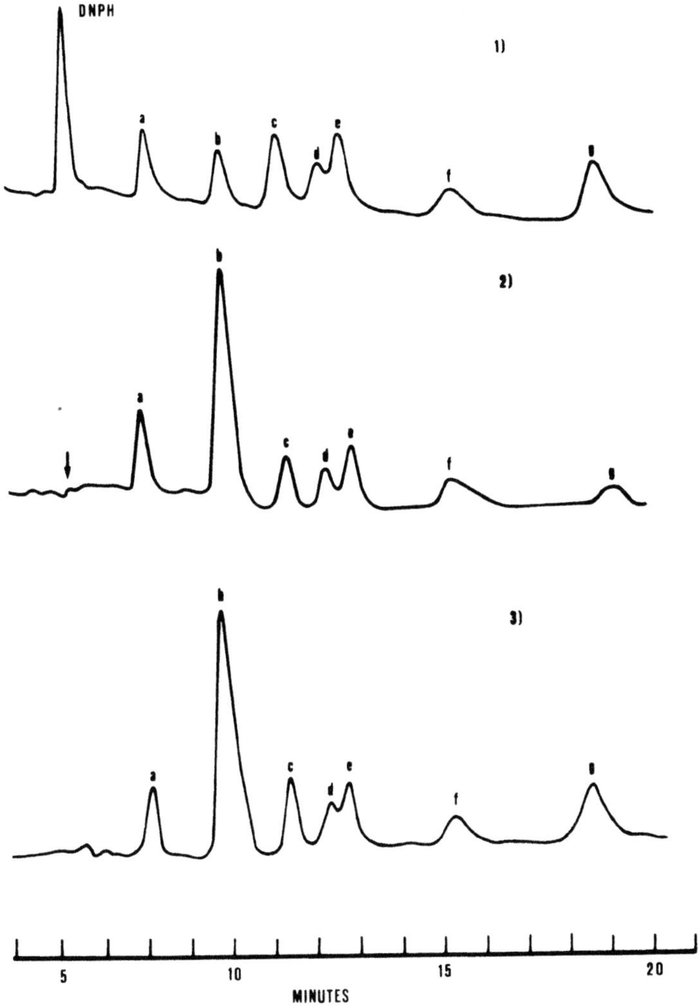

Fig. 3 - Same as Fig. 2, but with separation of the organic phase from the aqueous one:

1) immediate analysis after derivatization;
2) after 24 hours;
3) after 48 hours.

Concentration of carbonyls ca. 7.5 μM. Notice the disappaerence of the DNPH peak after 24 hours. The increase of the acetaldehyde peak is, in this case, much less pronounced than in the previous one. Moreover, the reproducibility of the results for the other carbonyls is very satisfactory. Benzaldehyde was not added to the mixture.

the derivatization reaction is completed (ca. 40 minutes), the organic phase is separated from the aqueous one. In fact, problems in reproducibility of results were observed when the samples were stored in contact with the aqueous phase. In this case, an abnormal increase of the acetaldehyde peak was also observed within 1 to 3 days (Fig. 2). The samples conserved in the organic phase exibit a good reproducibility over a period of 72 hours for all carbonyls except acetaldehyde, whose increase is, however, much less pronounced than in the case of samples conserved in the two-phase system (Fig. 3). On the other hand, the acetaldehyde blank signal shows the same rate of increase as the samples, so that a linear, reproducible response is possible for this compound by blank subtraction. We have no explanation for this behaviour of acetaldehyde; however, similar evidences were reported by other authors (14).

2.3. Reproducibility of the analytical procedures

This subject is common to both procedures reported in this paper. Repeated calibrations were performed employing standards obtained by microscale derivatization reactions, which simulate real samples both in concentration range and available sample volume. The concentration range of atmospheric interest was investigated (0-100 μM); a correlation factor better than 0.96 was obtained for calibration plots over the entire range. Reproducibility of more than 90% was also achieved. These satisfactory results were obtained by applying the complete analytical procedure also to the blanks. Blank values were then subtracted from the corresponding chromatographic peaks.

3. ANALYSIS OF CARBONYLS IN FOG AND RAIN SAMPLES

Analysis of carbonyls in fog and rain samples was performed using both procedures. Speciation of different carbonyl compounds was obtained by derivatization as 2,4-dinitrophenylhydrazones. An example of a fog sample analysis is reported in Fig. 4. All investigated carbonyls were found in this sample, four additional unknown peaks were also detected. Table 1 reports the average concentration of different carbonyls detected in fog water samples collected in the Po Valley, Italy (19). On the contrary, only formaldehyde was detected in rain samples collected in the same area (mean concentration = 2.05 μM).

Some of the samples were analyzed for formaldehyde using both 2,4-dinitrophenylhydrazone and lutidine derivatization procedures. The results obtained in the two cases agreed within the analytical error.

4. CONCLUSION

The two methods of carbonyl compounds analysis reported herein were examined taking into account all implications concerning their use during field operations. In particular, since it is not often possible to perform analysis immediately after sampling, particular care was devoted to the assesment of the conservation procedures.

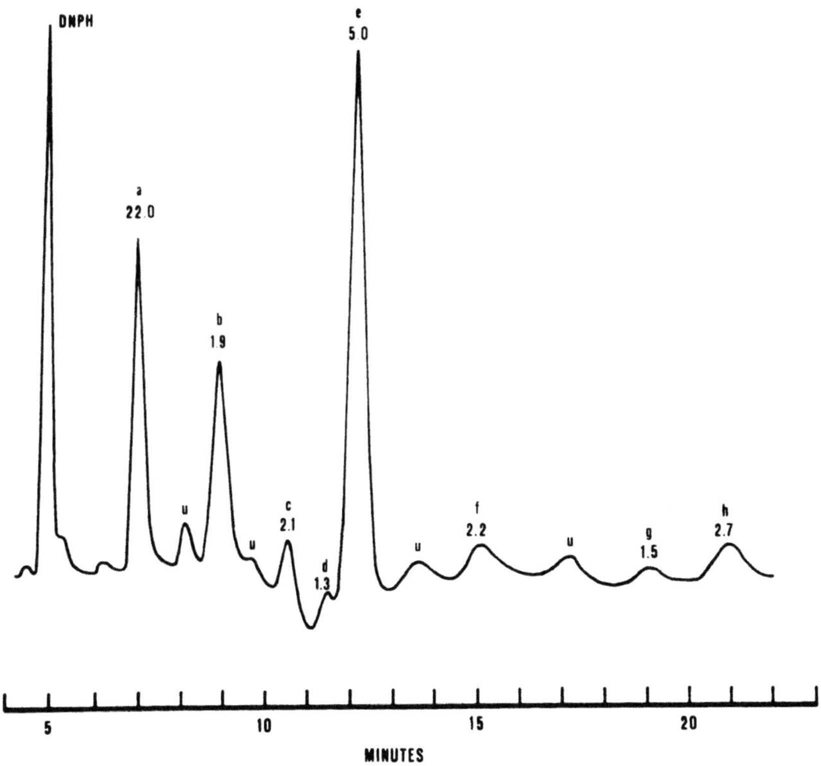

Fig. 4 - HPLC analysis of a fog sample after derivatization of the carbonyls to 2,4-dinitrophenylhydrazones. For the identification of the peaks refer to Fig. 2. Four unknown peaks (u) also appear. Concentration of the identified carbonyls is reported in μM units.

carbonyl compound	mean concentration (μM)	analytical error (μM)
formaldehyde	12.7	0.90
acetaldehyde	0.7	0.24
acrolein	0.4	0.25
propionaldehyde	0.3	0.10
aceton	1.7	0.19
n-butyraldehyde	0.5	0.45
iso-valeraldehyde	0.3	0.09
benzaldehyde	0.3	0.18

Table 1 - Mean concentration of carbonyl compounds detected in fog water samples collected in the Po Valley, Italy. Analytical error is also reported.

The analysis of formaldehyde as lutidine derivative proposed in this paper is a simple and accurate analytical technique; it represents an improvement with respect to the spectrophotometric method as regards selectivity and detection limit. In the case of carbonyl compound speciation by the 2,4-dinitrophenylhydrazones technique, it should be underlined that the higher sensitivity achieved employing ElCD permits the analysis of samples without preconcentration steps, which could introduce contamination of the samples and considerably increase the time of analysis.

Acknowledgements - This work was partially supported by Progetto Strategico "Clima e Ambiente dell'Area Mediterranea" of the Italian National Research Council. G. Nardini and G. Orsi are gratefully acknowledged for their help in the collection of fog samples

REFERENCES

(1) GRAEDEL T.E. and WESCHLER C.J. (1981) Chemistry within aqueous atmospheric aerosols and raindrops. Rev. Geophys. Space Phys., 19, 503-539.

(2) ADEWUYI Y.G., CHO S.Y., TSAY R.P. and CARMICHAEL G.R. (1984) Importance of formaldehyde in cloud chemistry. Atmos. Environ., 18, 2413-2420.

(3) GROSJEAN D. and WRIGHT B. (1983) Carbonyls in urban fog, ice fog, cloud water and rainwater. Atmos. Environ., 17, 2093-2096.

(4) MUNGER J.W., JACOB D.J. and HOFFMANN M.R. (1984) The occurrence of bisulphite-aldehyde addition products in fog and cloudwater. J. Atmos. Chem., 1, 335-350.
(5) CHAMEIDES W.L. and DAVIS D.D. (1983) Aqueous-phase source of formic acid in clouds. Nature, 304, 427-429.
(6) CALVERT J.G., LAZRUS A., KOK G.L., HEIKES B.G., WALEGA J.G., LIND J. and CANTRELL C.A. (1985) Chemical mechanisms of acid generation in the troposphere. Nature, 317, 27-35.
(7) STEINBERG S. and KAPLAN I.R. (1984) The determination of low molecular weight aldehydes in rain, fog and mist by reversed phase liquid chromatography of the 2,4-dinitrophenylhydrazone derivatives. Intern. J. Environ. Anal. Chem., 18, 253-266.
(8) FACCHINI M.C., CHIAVARI G. and FUZZI S. (1986) An improved method for carbonyl compound speciation in the atmospheric liquid phase. Chemosphere (in press).
(9) NASH T. (1953) The colorimetric estimation of formaldehyde by means of the Hantzsch reaction. Biochem. J., 55, 416-421.
(10) KLIPPEL W. and WARNECK P. (1978) Formaldehyde in rain water and on the atmospheric aerosol. Geophys. Res. Lett., 5, 177-179.
(11) ZAFIRIOU O.C., ALFORD J., HERRERA M., PELTZER E.T., GAGOSIAN R.B. and S.C. LIU (1980) Formaldehyde in remote marine air and rain: flux measurements and estimates. Geophys. Res. Lett., 7, 341-344.
(12) CHIAVARI G., FACCHINI M.C. and FUZZI S. (1985) Determination of formaldehyde as its lutidine derivative in the atmospheric liquid phase by high-performance liquid chromatography. J. Chromatogr., 333, 262-268.
(13) SMITH R.V. and ERHARDT P.W. (1975) Nash determination of formaldehyde in the presence of bisulphite. Anal. Chem., 47, 2462-2464.
(14) LOWE D.C., SCHMIDT U., EHHALT D.H., FRISCHKORN C.B. and NUERNBERG H.W. (1980) Determination of formaldehyde in clean air. Environ. Sci. Tech. 15, 819-823.
(15) JACOB W.A. and KISSINGER P.T. (1982) Determination of carbonyl 2,4-dinitrophenylhydrazones by liquid chromatography/electrochemistry. J. Liq. Chromatogr., 5, 669-676.
(16) CHIAVARI G. and BERGAMINI C. (1985) High-performance liquid chromatography of carbonyl compounds as 2,4-dinitrophenylhydrazones with electrochemical detection. J. Chromatogr., 318, 427-432.
(17) SELIM S. (1977) Separation and quantitative determination of traces of carbonyl compounds as their 2,4-dinitrophenylhydrazones by high-pressure liquid chromatography. J. Chromatogr., 136, 271-277.
(18) FUNG K. and GROSJEAN D. (1981) Determination of nanogram amounts of carbonyl as 2,4-dinitrophenylhydrazones by high-performance liquid chromatography. Anal. Chem., 53, 168-171.
(19) FUZZI S. (Editor) Heterogeneous Atmospheric Chemistry Project. FISBAT-C.N.R. Report, February 1986, Bologna, Italy.

WORKING PARTY 2

CHEMICAL AND PHOTOCHEMICAL REACTIONS;
MECHANISMS AND RATES

Chairman: R.A. COX

Photo-oxidation of sulphur containing compounds

Kinetic studies of oxidation reactions of sulphur compounds: reaction SO_2 + OH in the presence of O_2 and reaction IO + DMS

A comparison of models for the oxidation of acetylene at 300 K

Rate constants for reaction of OH radicals with acetates and glycols in the gas phase

OH-Dynamics during hydrocarbon oxidation in air

Aqueous phase chemistry of biogenic sulphur compounds

The contribution of hydrogen peroxide to atmospheric liquid phase chemistry

A study of the HO_2 + CH_3CHO reaction in the photolysis of CH_3CHO, and its consequences for atmospheric chemistry

Laboratory studies of the nitrate radical

Quelques aspects de la chimie des aldehydes dans la basse troposphere

Kinetics of the formation of nitrous acid from the thermal reaction of nitrogen dioxide and water vapour

FT-IR laboratory studies of kinetics of the NO_x/O_3 system of tropospheric interest

OH-Initiated oxidation of P-xylene under atmospheric conditions

Products and kinetics of the OH initiated oxidation of SO_2, CH_3SH, DMS, DMDS and DMSO

Absolute rate constant for the reaction of OH with SO_2 in the presence of water at atmospheric pressure

The role of transition metal ions in atmospheric droplet chemistry

SUMMARY BY THE CHAIRMAN

R.A. COX
The Harwell Laboratory, Harwell, UK

Current status of Working Party 2

There are currently 25 projects registered in Working Party 2, directed towards obtaining an understainding of the chemical transformation of atmospheric trace gases, both in the gas phase and in acqueous droplets and aerosols. 18 projects relate to the study of rate constants and mechanisms of gas phase reactions, whilst the remainder address chemistry involving particulates and droplets. Most of the studies are entirely laboratory based although close collaboration with field studies is a feature of a few of the projects.

Research results reported at the 4th European Symposium on Physico Chemical behaviour of Atmospheric Pollutants

The three main topics reported on at the meeting were: (a) Atmospheric sulphur chemistry, (b) NO_x chemistry and (c) Hydrocarbon oxidation reactions. The results of the activities in these areas are summarised below.

(a) Atmospheric sulphur chemistry

The main objective of the work has been to obtain quantitative information on the rates of atmospheric oxidation of sulphur compounds from man made sources (mainly SO_2) and natural sources (organic sulphur compounds such as dimethyl sulphide, DMS). Oxidation processes in the gas phase and the liquid phase have been studied.

For the gas phase, the approach has been to establish the rate constants and reactivity patterns of the reactions with sulphur compounds of OH radicals, NO_3 radicals and also the completely novel oxidising radical IO, which is believed to be present in marine atmospheres. The major gas phase oxidation process for SO_2 is oxidation by OH and the work of Martin et.al. (F) and Nielsen et.al.(DK) has provided important information on the primary reaction:

$$OH + SO_2 + M \longrightarrow HSO_3 + M \qquad (1)$$

and on the secondary oxidation of the HSO_3 radical to give SO_3:

$$HSO_3 + O_2 \longrightarrow SO_3 + HO_2 \qquad (2)$$

These data, together with previous knowledge, now allow a fairly clear understanding of the mechanism for the homogeneous oxidation of SO_2 although there remain some minor discrepancies in the rate constant. The production of HO_2 in reaction (2) when followed by the well known reaction with NO:

$$HO_2 + NO \rightarrow NO_2 + OH \quad (3)$$

leads to prompt regeneration of OH under typical lower atmospheric conditions. Therefore the presence of SO_2 does not directly suppress the concentrations of OH and HO_2 radicals by interfering with the radical loss reactions, as had been widely assumed in earlier atmospheric models.

The reactions of OH with a range of organic sulphur compounds has been studied jointly by Nielsen and Pagsberg (DK) and Treacy et.al. (IRL) and also by Barnes et.al. (D). A clear pattern of structure/reactivity is emerging and some of the earlier discrepanicies in the rate studies have been resolved by observing the effects of NO and O_2 concentration on the reaction rates. The kinetics of the reaction of IO radicals with DMS has been investigated by Martin et.al. who observe a rapid reaction producing dimenthylsulphoxide, DMSO:

$$IO + CH_3SCH_3 \rightarrow I + CH_3SOCH_3 \quad (4)$$

Based on estimates of the IO concentrations in the sunlit marine boundary layer, it appears that reaction (4) could be a major sink for DMS.

Further information on the products of DMS oxidation have been reported; Barnes et.al. find up to 70% yield of SO_2 from DMS oxidation by OH in a 'low NO_x' atmosphere, compared to much lower conversion to SO_2 when NO_x is present. They also find a very rapid reaction rate for OH with DMSO.

A very complex picture is emerging for the oxidation of organic sulphur compounds in the environment, and the details are far from resolved. It is clear that there is a significant flux of DMS from the sea but it is not known what the oxidation products are and how much they contribute to atmospheric sulphate burdens. Thus the impact of marine sulphur on atmospheric acidity cannot be assessed at the present time.

Oxidation of sulphur dioxide and its hydrolysis products in the liquid phase has also been studied. Evidence from field studies (Barmbauer et.al., D) appears to confirm the earlier suggestion, based on laboratory work, that hydrogen peroxide in solution is a major oxidising agent for dissolved S^{iv} species. A reassessment by Cocks and McElroy (UK) also indicates that SO_2 oxidation in droplets catalysed by transition metal ions (FE^{3+} and Mn^{2+}) may be important, at least for the concentrations of these ions in UK rain.

(b) <u>NO_x chemistry</u> A number of aspects of NO_x chemistry have been investigated within the COST 611 programme. The reactions of N_2O_5 and NO_2 with water vapour have been investigated using infra red spectroscopy to follow reactants and products. An upper limit for the homogeneous reaction:

$$N_2O_5 + H_2O \rightarrow 2HNO_3 \quad (5)$$

implies a very slow gas phase reaction. The reaction probably occurs rapidly on droplets and aerosols, however. A detailed study of the kinetics of the heterogeneous reaction

$$NO_2 + H_2O \rightarrow HNO_2 + HNO_3 \quad (6)$$

is also reported (Jenkin and Cox, UK). If the laboratory results are scaled to account for typical conditions in the atmospheric boundary layer a small source of HNO_2 is predicted. The reaction rates were insufficient to account for all field observations of HNO_2 unless surface characteristics in the outside environment are generally more favourable for reaction, as is indeed possible.

Two laboratory studies of reactions of the nitrate radical, NO_3 have been reported. Hjorth et.al. (JRC) have obtained an upper limit for the rate constant for the reaction

$$NO_3 + CO \longrightarrow NO_2 + CO_2 \qquad (7)$$

The reaction is too slow to provide a significant pathway for either NO_3 or CO in the troposphere. New measurements of the absolute absorption cross section for NO_3 in its visible absorption band at 662 nm are reported by Wayne et.al., which confirm the 'high' values which now seem to be emerging from studies in the USA. Measurements are also reported for the rate constant for the reaction of NO_3 with HO_2:

$$NO_3 + HO_2 \longrightarrow HNO_3 + O_2 \qquad (8a)$$
$$\longrightarrow OH + NO_2 + O_2 \qquad (8b)$$

The reaction is relatively rapid (k_8 = (3 to 8) x 10^{-12} cm^3 molecule $^{-1}$ s^{-1} at 308 K) but assessment of its significance in the atmosphere requires knowledge of the product channels and the temperature dependence of k_8.

(c) Hydrocarbon Oxidation Chemistry

The mechanism for the oxidation of hydrocarbons and other volatile organic compounds in the atmosphere are very complex and studies involving both experimental measurements and computer modelling have been undertaken in order to provide better understanding of these processes. Two experimental studies have been reported in which the reactive free radical species OH and HO_2 have been directly monitored during photo-oxidation of selected organics, C_2H_4 (Schmit et.al.,D) and CH_3CHO (Moortgat et.al.,D). These studies have given new insight into the reaction mechanism of these oxidations. New kinetic data for the reactions

$$HOCH_2CH_2OO + NO \longrightarrow HOCH_2CH_2O + NO_2 \qquad (9)$$

and

$$CH_3OO + HO_2 \longrightarrow CH_3OOH + O_2 \qquad (10)$$

have also been obtained from these studies.

Rate constants for the reaction of OH radicals with a series of oxygenated organics have also been determined (Hartmann et.al.,D). These data allow the atmospheric lifetimes of several industrially important organics to be estimated as well as providing useful structure-reactivity relationships which can be used as a basis for estimating the reactivity of other organics towards OH radicals.

The problems of modelling the complex chemistry of hydrocarbon oxidation are also illustrated by the work of Langhans et.al. (B) and Moortgat et.al. (D).

Conclusions and future directions

Considerable progress continues to be made in establishing a reliable data base for understanding physico chemical behaviour of trace atmospheric gases. The problems of importance in the current major issues of atmospheric pollution - acid deposition and photochemical oxidants - are mostly covered by the activities in the concerted action. It is anticipated that the assessments and recommendations from the COST 611 Task Forces dedicated to these major issues, will help to maintain this important focus for the laboratory and field work concerned with chemical transformations.

The study of reaction kinetics and mechanisms today involves application of a variety of novel, spectroscopy based techniques, particularly for the detection and measurements of trace concentrations of chemically reactive species. The development of these techniques, which is one of the activities of working party (1) goes hand in hand with the kinetics studies in working party (2) and relates closely to applications for in-situ field measurements in the atmosphere.

In the future, and with the aid of the new techniques becoming available for both gas phase and condensed phase systems, a major improvement in our konwledge of the chemical processes involved in atmospheric chemistry can be forseen, through collaborative research undertaken in the laboratories of the European countries.

PHOTO-OXIDATION OF SULPHUR CONTAINING COMPOUNDS+)

O.J. NIELSEN
Chemistry Department, Risø National Laboratory, DK-4000 Roskilde, Denmark

J. TREACY, L. NELSON and H. SIDEBOTTOM
Department of Chemistry, University College Dublin, Dublin, Ireland

Summary

The reactions of OH with dimethyldisulfide (DMDS), dimethylsulfide (DMS), and diethylsulfide (DES) have been investigated applying both a photolytic relative technique and an absolute method using pulse radiolysis combined with kinetic spectroscopy. Furthermore, the reactions of OH with di-tert-butyldisulfide (DTBDS), ethylpropylsulfide (EPS), and dipropylsulfide (DPS) were studied using the absolute method. All measurements were done at room temperature and atmospheric pressure. The relative method using methyl nitrite as OH source was found to give unreliable OH rate constants with organic sulfides. The following absolute OH rate constants were obtained:

$$k(OH + DMDS) = (300 \pm 30) \times 10^{-12}$$
$$k(OH + DTBDS) = (41 \pm 4) \times 10^{-12}$$
$$k(OH + DMS) = (3.5 \pm 4) \times 10^{-12}$$
$$k(OH + DES) = (4.5 \pm 5) \times 10^{-12}$$
$$k(OH + EPS) = (4.9 \pm 5) \times 10^{-12}$$
$$k(OH + DPS) = (5.2 \pm 5) \times 10^{-12}$$

(units are $cm^3 molecule^{-1} s^{-1}$)

The data for the sulfides are consistent with the concept of group rate constants. The relative measured rate constants for the reaction of OH with DMS and DES were found to increase with increasing concentrations of added NO. These results provide some indication of the reasons for the difference in reported rate constants for these reactions. Still some discrepancies remain to be resolved.

1. INTRODUCTION

On a global scale biogenic and anthropogenic emissions of sulphur compounds are thought to be equal (1,2). In order that the contribution of anthropogenic emissions to acid deposition can be more precisely defined it is clear that an understanding of the natural sulphur cycle is desirable. Reduced sulphur compounds such as H_2S, CS_2, CH_3SCH_3, and CH_3SSCH_3 are considered to be the most important biogenic sources of sulphur in the troposphere (3).

+) This work was presented at a European Science Foundation workshop on Tropospheric Ozone Chemistry in Dubrovnik 11.-17. May 1986.

Oxidation of these compounds is initiated by reactions with OH radicals, however, detailed reaction mechanisms remain obscure. Numerous experimental studies have been reported concerning the kinetics of these reactions, however, there is often conflict between rate constant data determined by using absolute methods at low pressure and relative techniques under simulated atmospheric conditions.

This work is concerned with the determination of rate constants for reactions of OH radicals with a number of reduced sulphur compounds at room temperature and atmospheric pressure using both an absolute and a relative rate method. Our results provide some indication for the difference in reported rate data for these reactions.

2. EXPERIMENTAL

Relative rate method

OH radicals are generated by photolysis ($\lambda > 290$ nm) of methyl nitrite in air:

(1) $$CH_3ONO + h\nu \rightarrow CH_3O + NO$$
(2) $$CH_3O + O_2 \rightarrow CH_2O + HO_2$$
(3) $$HO_2 + NO \rightarrow OH + NO_2$$

Experiments were carried out at 298 ± 2 K and 730-750 torr in a 50-L FEP Teflon cylindrical reaction chamber. The chamber was surrounded by 20 fluorescent lamps (10 blacklamps, Philips TL 20/08 and 10 sunlamps, Philips TL 20/09) giving a photolytic half-life for CH_3ONO of about 30 minutes. All quantitative analyses were made by GC. Series of experiments were carried out in which CH_3ONO/NO/sulphur compound/reference compound/air mixtures were photolysed ($[CH_3ONO]_o$ = 2-10 ppm; $[NO]_o$ = 0-20 ppm; $[S]_o$ = 1-10 ppm; $[RH]_o$ = 1-10 ppm). The OH radicals generated via reactions (1)-(3) react with the sulphur compound and the reference hydrocarbon:

(4) $$OH + S \rightarrow products$$
(5) $$OH + RH \rightarrow products$$

Providing the sulphur compound and the hydrocarbon are consumed only by reaction with OH radicals k_4 can be determined:

(I) $$\ln([S]_o/[S]_t) = (k_4/k_5)\ln([RH]_o/[RH]_t)$$

Absolute method

The experimental set-up is shown schematically in Figure 1, and has been described in detail previously (4,5). The field emission accelerator (FEBETRON 705B) provides single pulses of 2-MeV electrons with a pulse duration of 30 ns and a maximum corrent of 3000 amperes. The gas samples are contained in a 1-L stainless steel irradiation cell mounted directly onto the accelerator. A set of internal aluminized spherical mirrors provide optical path lengths of 40, 80, 120 or 160 cm by multiple reflections. Most experiments were carried out using an optical path length of 120 cm. Gas mixtures were prepared by admitting one component at a time and reading the corresponding partial pressure with a MKS Baratron model 170 absolute electronic membrane manometer with a resolution of 10^{-5} bar. The analyzing light

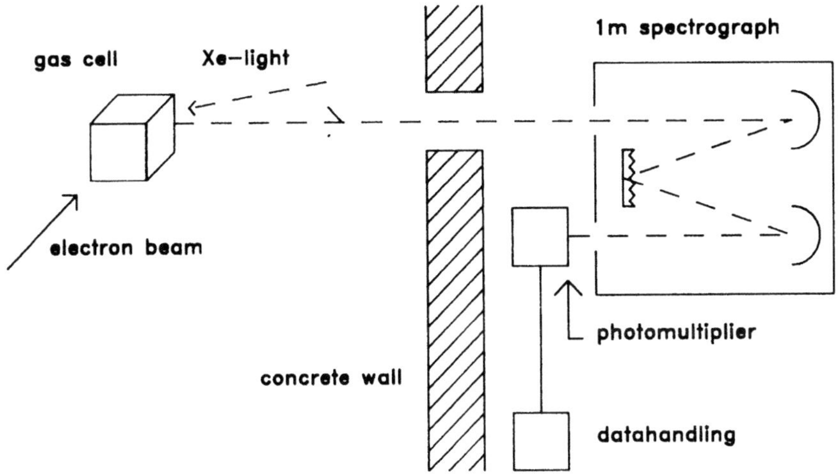

Figure 1. Schematic diagram of the experimental set-up.

source is a pulsed 150-W high-pressure Xe-lamp. The analyzing light is dispersed in a Hilger and Watts 1-meter monochromator with a 1200 grooves/mm grating blazed at 300 nm. The light intensity in the plane of the wavelength band-pass selected by the exit slit is monitored using a Hamamatsu R955 photomultiplier and the output signals are digitized in a transient recorder. The experimental data are transferred to a large central computer facility for further processing and analysis.

When irradiating mixtures of 15 torr H_2O and Ar to 1 atm initially excited Ar-atoms, $Ar^{\bullet}$, are formed. OH radicals are produced through reaction (6):

(6) $$Ar^{\bullet} + H_2O \rightarrow OH + H + Ar.$$

Formation of OH is extremely rapid compared to the time-scale of the OH decay. On adding an excess of a compound, S, to the system pseudo-first-order kinetics are observed due to:

(7) $$OH + S \rightarrow \text{products}.$$

k_7 is determined from a plot of reciprocal OH half-lives, t, versus S concentrations with a slope equal to $\ln(2)/k_7$ (see Figure 2). OH concentration was followed by monitoring the transient absorption around 309 nm.

3. RESULTS AND DISCUSSION

RSSR + OH

In Table 1 the present kinetic results are compared with the previous determinations. Since the reaction of OH with dialkyl sulfides is significantly faster than reactions involving hydrogen abstraction the reaction presumably involves addition to the S-S bond. This has also been confirmed in a product study (6). The absolute value for k(OH + DMDS) determined in

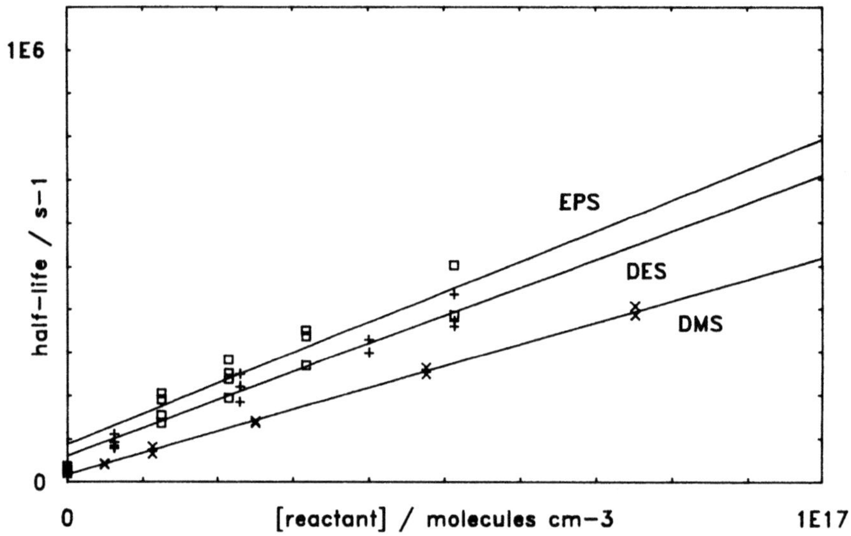

Figure 2. Absolute measurement of OH + dialkylsulphides.

this study is seen to be slightly higher than the other 3 values in Table 1. The reaction of OH with DTBDS has not been studied before. The lower value of k(OH + DTBDS) compared to that of k(OH + DMDS) must be due to steric hindrance by the bulky t-butyl groups causing a large reduction in the OH addition to the S-S bond.

RSR' + OH

The rate constants for the reaction of OH with DMS and DES measured relative to that of c-C_6H_{12} were found to increase with increasing concentrations of added NO. In Table 2 the present kinetic results are compared with the previous determinations except for the reactions OH + EPS and OH + DPS which have not been studied before.

Values of k(OH + DMS) and k(OH + DES) from the relative method derived from the initial rate slope in Figure 3, conversion less than 10%, are higher than the values obtained in our absolute measurements using pulse radiolysis.

Curvature of the plots in Figure 3 has previously been reported by Cox and Sheppard (7) and by Atkinson et al. (8). Atkinson et al. (8) explained this curvature as a consequence of NO_3 formation under conditions where O_3 generation was significant. This was confirmed in the present work by photolyzing NO_2 in DMS/c-C_6H_{12}/air mixtures. The DMS was observed to decay quite rapidly while the c-C_6H_{12} concentration remained virtually constant. Furthermore, DMS showed a high decay rate in O_3/NO_2/air mixtures in the dark. The above explanation cannot be used to rationalize the apparent dependence of the rate constant for reaction of OH with DMS and DES on the NO concentration. High values of added NO would suppres O_3 formation and hence the presence of the NO_3 radical cannot be responsible for the curvature in

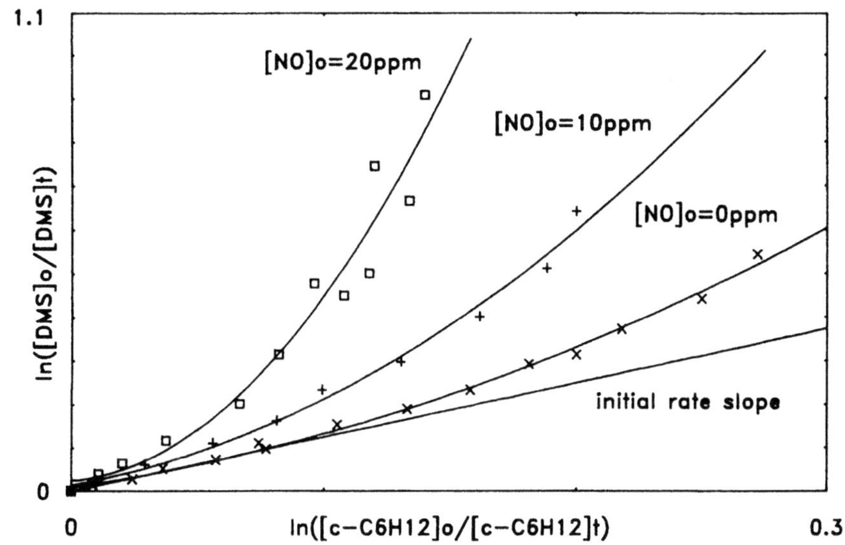

Figure 3. Relative measurement of HO + DMS.

Figure 3. Barnes et al. (9) have recently found a similar dependence of the OH + CH_3SH rate constant on NO concentration. We concur with their suggestion that a reaction product is involved in the additional removal of sulphides in static photolysis systems involving NO_x species. The CH_3S radical has been proposed as an important intermediate in the oxidation of DMS (8, 9, 10, 11).

In order to test whether CH_3S radicals play a role in the decompostion of DMS in the presence of NO_x and O_2, DMDS was used as a photolytic source of CH_3S radicals during our experiments. The results showed that photolysis of DMDS in the presence of DMS in either pure N_2 or air did not result in additional loss of DMS. However, when the photolysis was carried out in air with small amounts of added NO a significant decrease in DMS occurred. Further increases in DMS decay was observed with increasing addition of NO. Hence, it is concluded that the presence of O_2 and NO_x in the system will lead to unreliable OH rate constant data for organic sulphides. When CH_3S is formed in the presence of O_2 and NO the subsequent reactions are expected:

(8) $$CH_3S + O_2 \rightarrow CH_3SO_2$$
(9) $$CH_3SO_2 + NO \rightarrow CH_3SNO + NO_2$$

The important question seems to be how methyl thionitrite (CH_3SNO) attacks DMS and DES.

Values of k(OH + DMS) seem to fall in one group around 10 x 10^{-12} and another around 4 x 10^{-12} $cm^3 molecule^{-1} s^{-1}$. Our absolute value of k(OH + DMS) seems to support the low value as seen in Table 2. It is generally agreed that the reaction of OH with DMS is both addition and H-atom abstraction. Work is in progress of determining the branching ratio by monitoring the formation of CH_3 radicals at 216 nm.

Our absolute value for k(OH + DES) is a factor of 3 lower than the two previous measurements (12,13). Martin et al. (13) clearly had a problem with the coating of their reactor and details of the study by Wine et al. is not yet available in the literature. When the value of k(OH + DMS) is around either 4 or 10 x 10^{-12} $cm^3 molecule^{-1} s^{-1}$ it seems difficult to find a chemical argument for k(OH + DES) to be between 12 and 15 x 10^{-12} $cm^3 molecule^{-1} s^{-1}$. This discrepancy initiated this study of the reactions of OH with EPS and DPS which had not been studied before. The values were found to be (4.9 ± 0.5) and (5.2 ± 0.5) x 10^{-12} $cm^3 molecule^{-1} s^{-1}$ for EPS and DPS, respectively.

Going from DMS via DES and EPS to DPS is successive addition of two, one, and one CH_2-group. If we had been dealing with alkanes this should add 0.8 x 10^{-12} $cm^3 molecule^{-1} s^{-1}$ per CH_2-group to the total OH rate constant according to Carter and Atkinson's concept of group rate constants (14). In this case of the alkyl sulfides it looks like this value is around 0.5 x 10^{-12} $cm^3 molecule^{-1} s^{-1}$, which is in acceptable agreement as this number should approach the alkane case as the alkyl chains in the sulfides grow longer.

ACKNOWLEDGEMENT

Thanks are due to Dr. A. Holm, University of Copenhagen, for the sample of EPS. Also thanks are due to O. Jørgensen and S. Petersen for purifying the sample of DPS, H. Egsgaard for checking the purity of all samples, J. Munk and P. Genske for technical assistance, and P. Pagsberg for valuable discussions (all from the Chemistry Department at Risø).

REFERENCES

(1) C.F. CULLIS and M.M. HIRSCHLER, Atm.Environ., 14, 1263 (1980)
(2) T.E. GRAEDEL, Rev.Geophys. Space Phys., 15, 421 (1977)
(3) N.D. SCE and M.K.W. KO, Atm.Environ., 14, 1223 (1980)
(4) K.B. HANSEN, R. WILBRANDT, and P. PAGSBERG, Rev.Sci.Instrum., 50, 1532 (1972)
(5) O.J. NIELSEN, Risø-R-480 (1984)
(6) S. HATAKEYAMA and H. AKIMOTO, J.Phys.Chem., 87, 2387 (1983)
(7) R.A. COX and D. SHEPPARD, Nature, 284, 330 (1980)
(8) R. ATKINSON, J.N. PITTS, JR., and S.M. ASCHMANN, J.Phys.Chem. 88, 1584 (1984)
(9) I. BARNES, V. BASTIAN, K.H. BECKER, E.H. FINK and W. NELSEN, J.Atm. Chem., to be published (1986)
(10) H. NIKI, P.D. MAKER, C.M. SAVAGE and L.P. BRIETENBACH, Int.J.Chem. Kinet., 15, 647 (1983)
(11) D. GROSJEAN, Environ.Sci.Technol., 18, 450 (19084)
(12) P.H. WINE; priv.comm.
(13) D. MARTIN, J.L. JOURDAIN, and G. LeBRAS, Int.J.Chem.Kinet. 17, 1247 (1985)
(14) W.P.L. CARTER and R. ATKINSON , J.Atm.Chem. 3, 377 (1985)
(15) M.J. KURYLO, Chem.Phys.Lett. 58, 238 (1978)
(16) R. ATKINSON, R. PERRY, and J.N. PITTS, JR., Chem.Phys.Lett. 54, 14 (1978)
(17) P.H. WINE, N.M. KREUTTER, C.A. GUMP, and A.R. RAVISHANKARA, J.Phys. Chem. 85, 2660 (1981)
(18) T.J. WALLINGTON, Int.J.Chem.Kinet. (in press)
(19) R. ATKINSON, J.N. PITTS, JR., and S. ASCHMANN, J.Phys.Chem. 88, 1584 (1984)

Table 1: Summary of reported room temperature rate constants for the reactions OH+R, R= DMDS and DTBDS.

R	$cm^3 mole^{-1} s^{-1}$	method	pressure Torr	M	reference
DMDS	$(198 \pm 18) \times 10^{-12}$	absolute	5-200	Ar	(17)
	$(300 \pm 30) \times 10^{-12}$	absolute	760	Ar	This work
	$(220 \pm 80) \times 10^{-12}$	relative	760	air	(7)
	$(200 \pm 20) \times 10^{-12}$	relative	740	air	This work
DTBDS	$(41 \pm 4) \times 10^{-12}$	absolute	760	Ar	This work

Table 2: Summary of reported room temperature rate constants for the reactions OH+R, R= DMS, DES, EPS and DPS.

R	$cm^3 mole^{-1} s^{-1}$	method	pressure Torr	M	reference
DMS	$(8.3 \pm 0.9) \times 10^{-12}$	absolute	10-50	Ar/SF_6	(15)
	$(9.8 \pm 1.2) \times 10^{-12}$	absolute	50-98	Ar	(16)
	$(4.3 \pm 0.6) \times 10^{-12}$	absolute	50	Ar	(17)
	$(3.2 \pm 0.2) \times 10^{-12}$	absolute	0.5	He	(13)
	$(3.6 \pm 0.2) \times 10^{-12}$	absolute	?	?	(18)
	$(3.5 \pm 0.4) \times 10^{-12}$	absolute	760	Ar	This work
	$(9.1 \pm 1.4) \times 10^{-12}$	relative	760	air	(7)
	$(10.0 \pm 1.0) \times 10^{-12}$	relative	735	air	(19)
	$(9.3 \pm 1.8) \times 10^{-12}$	relative	740	air	This work
DES	$(15.0 \pm 2.3) \times 10^{-12}$	absolute	?	?	(12)
	$(12.0 \pm 1.4) \times 10^{-12}$	absolute	0.5	He	(13)
	$(4.5 \pm 0.5) \times 10^{-12}$	absolute	760	Ar	This work
	$(23.0 \pm 2.0) \times 10^{-12}$	relative	740	air	This work
EPS	$(4.9 \pm 0.5) \times 10^{-12}$	absolute	760	Ar	This work
DPS	$(5.2 \pm 0.5) \times 10^{-12}$	absolute	760	Ar	This work

KINETIC STUDIES OF OXIDATION REACTIONS OF SULFUR COMPOUNDS : REACTION SO_2 + OH IN THE PRESENCE OF O_2 AND REACTION IO + DMS

D. MARTIN, J.L. JOURDAIN, G. LAVERDET and G. LE BRAS
Centre de Recherches sur la Chimie de la Combustion et des Hautes Températures - CNRS - 45071 ORLEANS Cédex 02 - FRANCE

Summary

The reactions of atmospheric interest, SO_2 + OH in the presence of O_2 and IO + DMS, have been studied using the discharge flow method with EPR or mass spectrometric analysis of the species. The following rate constants were measured at room temperature :

$HOSO_2 + O_2 \rightarrow HO_2 + SO_3$ (1) $\quad k_1 = (3.5 \pm 1) \times 10^{-13}$

IO + DMS $\rightarrow$ products (4) $\quad k_4 = (1.5 \pm 0.5) \times 10^{-11}$

(units are cm^3 $molecule^{-1}$ s^{-1})
The value obtained for k_2 confirms the recent hypothesis concerning the HO_x conservation in the gas phase oxidation of SO_2 in the atmosphere. The value obtained for k_4 indicates that IO is a potential candidate for DMS oxidation in marine atmosphere.

1. INTRODUCTION

The title reactions have been studied in relation with their potential role in the atmospheric cycle of sulfur and in acid deposition. The reaction of SO_2 with OH is the main gas phase oxidation path of SO_2 in the atmosphere. Recently STOCKWELL and CALVERT (1) have found some evidence for occurrence of a reaction between the adduct formed $HOSO_2$ and O_2. They proposed the following mechanism :

$$HOSO_2 + O_2 \rightarrow SO_3 + HO_2 \quad (1)$$

The rate constant of this reaction has already been measured by MARGITAN in a flash photolysis study (2). In the present paper we report a determination of k_1 using the discharge flow method which is complementary of the flash photolysis technique.

The IO + DMS reaction has been suggested to explain the low residence time measured for DMS in coastal areas (CARLIER and MOUVIER (3, 4)). The residence time measured by these authors, 0.3 - 3 hours, is much lower than the value of 44 hours which is calculated if the OH + DMS reaction is assumed to be the only daytime sink for DMS. The reaction IO + DMS had never been studied before and the data reported in this paper and those obtained in a concerted photolysis of BARNES et al (5) are the first available.

2. EXPERIMENTAL

The title reactions have been studied using the discharge flow method with EPR or mass spectrometric analysis of the species. The experimental set up have already been described (6, 7). The reactors, made of quartz, were coated with halocarbon wax. The total pressure in the reactors was around 1 torr for the IO + DMS study and 2 torr for the SO_2 + OH + O_2 one.

In the SO_2 + OH study, OH, produced by the H + NO_2 reaction, were monitored in the gas phase by EPR. SO_2 was flowed with or without O_2 through a moveable axial tube. NO was used to convert HO_2, produced via reaction (1), into OH via reaction NO + HO_2 $\rightarrow$ NO_2 + OH. NO could be added either with SO_2 and O_2 or at a fixed position upstream from the EPR cavity.

In the IO + DMS experiments, IO was produced from the reaction O + I_2 $\rightarrow$ IO + I ($k = 1.38 \times 10^{-10}$ at 298K (8)). O atoms were flowed directly into the reactor and I_2 through the external tube of two axial concentric injectors. The O + I_2 reaction was completed upstream of the end of the central injector, through which CH_3SCH_3 was flowed. I_2 was used in excess over O atoms to prevent any reaction between O atoms and CH_3SCH_3. The IO radical was detected by mass spectrometry at its parent peak (m/e = 143) but some contribution of reaction products was also observed at this peak. Therefore, the IO radical was indirectly monitored at the peak m/e = 46 (NO_2^+) after conversion of NO into NO_2 via reaction IO + NO $\rightarrow$ I + NO_2. NO was introduced into the reactor 7 cm upstream of the sampling cone of the mass spectrometer.

3. RESULTS AND DISCUSSION

a) Reaction SO_2 + OH in the presence of O_2

An extensive study of the SO_2 + OH reaction was first carried out in the absence of O_2 in order to determine the kinetic parameters needed to make the mathematical simulation of the similar chemical system in the presence of O_2. The results obtained, which are described in detail in a recent paper (9), have led to the following expression of the bimolecular rate constant for the reaction of SO_2 with OH, in the presence of He and SO_2, at room temperature and in the pressure range 1 - 6.4 torr :

$$k_2 = (7.8 \pm 0.2) \times 10^{-32} \times |He| + (1.33 \pm 0.4) \times 10^{-30} \times |SO_2| + (2.3 \pm 0.2) \times 10^{-15} \text{ cm}^3 \text{ molecule}^{-1} \text{ s}^{-1}.$$

In the presence of O_2 and NO the following main reactions occurred

$$OH + SO_2 \rightarrow HOSO_2 \quad (2)$$

$$HOSO_2 + O_2 \rightarrow HO_2 + SO_3 \quad (1)$$

$$HO_2 + NO \rightarrow NO_2 + OH \quad (3)$$

The expected decrease of OH consumption, by adding O_2 and NO in the reactor, is shown in the figure below.

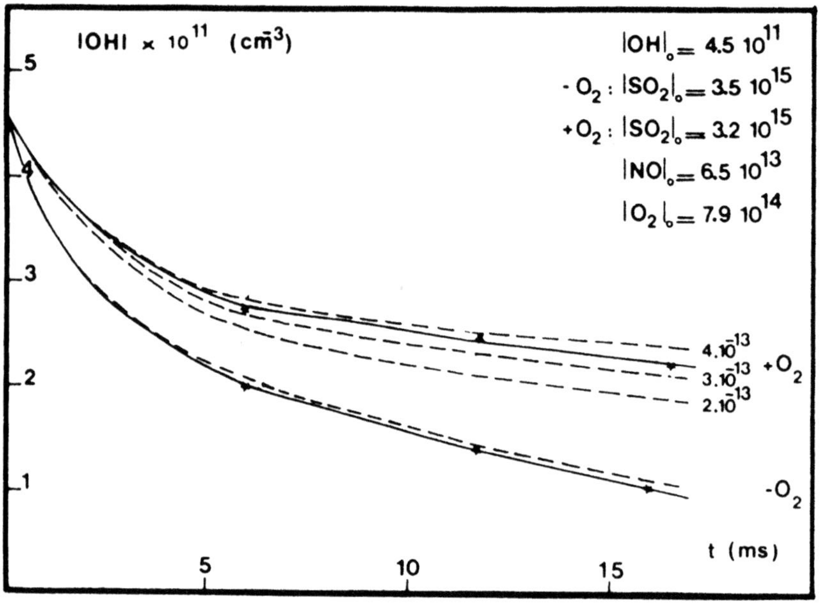

The experimental decay curves of OH in the presence or absence of O_2 were fitted by a computer program including all the possible reactions, in order to get the rate constant of reaction (1). A good fit for all the experiments was obtained by considering a first order loss rate of 50 s^{-1} for $HOSO_2$, which was attributed to a wall reaction. The simulation of all the experiments, with O_2 concentrations ranging from 3.5×10^{14} to 1.5×10^{15} cm^{-3}, led to the following value of k_2 :

$$k_2 = (3.5 \pm 1) \times 10^{-13} \text{ cm}^3 \text{ molecule}^{-1} \text{ s}^{-1} \text{ at } 298K.$$

This value agrees well with the flash photolysis determination of MARGITAN (2) ($(4 \pm 2) \times 10^{-13}$) and also with a recent discharge flow measurement of C.J. HOWARD (10), $k_2 = (4.4 \pm 0.9) \times 10^{-13}$, done during the course of our work.

For atmospheric application, our value of k_2 leads to confirm the very high rate of reaction (2) in the atmosphere, where the gas phase oxidation of SO_2 can be represented by the reaction : $SO_2 + OH + O_2 \rightarrow SO_3 + HO_2$. The inclusion of this reaction in the atmospheric models leads to a more linear response of H_2SO_4 concentration changes with changes in SO_2 emissions, compared to the previous situation where the global reaction $SO_2 + OH \rightarrow SO_4^{--}$ was considered.

b) Reaction IO + DMS (4)

This reaction was studied under the second order conditions with similar concentrations of IO and DMS since the pseudo-first order conditions (excess of DMS over IO) could not be applied for the following reasons : i) the monitoring of IO decay at the parent peak m/e = 143 was not valid due to the contribution of reaction products at this peak, ii) the indirect monitoring of IO decay at the NO_2^+ peak (NO + IO $\rightarrow$ NO_2 + I) was impossible, due to a large contribution of CH_3SCH_3 at mass m/e = 46 in the presence of a high excess of DMS. In the second order conditions used, the rate constant k_4 was determined from the measurement of CH_3SCH_3 reached at a fixed reaction time and making a mathematical simulation of the chemical system considered. In addition to reaction (4), the self combination IO + IO $\rightarrow$ products (5) and the wall reaction of IO were included in the simulation. The rate constants of these two reactions were measured in separate experiments and the values obtained were respectively $k_5 = (3 \pm 0.5) \times 10^{-11}$ cm^3 $molecule^{-1}$ s^{-1} and $k_w = (40 \pm 10)$ s^{-1} at 298K. The validity of the second order procedure used was checked on the similar reaction O + CH_3CH_3 $\rightarrow$ products, the rate constant of which being presently reasonably known.

For the IO + DMS reaction, the values of experimental and calculated ratios : $(|CH_3SCH_3|_o - |CH_3SCH_3|)/|CH_3SCH_3|_o$ are reported in the table below for typical initial CH_3SCH_3 concentrations used and at a fixed reaction time of 30 ms. The adjustable parameter k_4 was varied from 10^{-11} to 5×10^{-11} and the best fit obtained from 13 experiments corresponded to the following value :

$$k_4 = (1.5 \pm 0.5) \times 10^{-11}\ cm^3\ molecule^{-1}\ s^{-1} \text{ at } 298K.$$

A mass spectrometric analysis of the reaction products led to the qualitative detection of the peak at mass m/e = 78 which corresponds to the parent peak of DMSO, $CH_3S(O)CH_3$. DMSO is expected to be produced via :

$$CH_3SCH_3 + IO \rightarrow CH_3S(O)CH_3 + I$$

$\|CH_3SCH_3\|_o$ 10^{12} cm^{-3}	$(\|CH_3SCH_3\|_o - \|CH_3SCH_3\|)/\|CH_3SCH_3\|_o$ experimental	calculated k_4 : 10^{-11}	2×10^{-11}	3×10^{-11}	5×10^{-11}
18.20	0.03	0.05	0.06	0.07	0.08
13.10	0.07	0.07	0.09	0.10	0.11
8.34	0.08	0.08	0.09	0.14	0.15
4.17	0.13	0.11	0.17	0.21	0.25
1.83	0.17	0.12	0.21	0.27	0.38

The present data on the IO + DMS reaction (11) can be compared with those of Barnes et al (5) obtained simultaneously from a continuous photolysis of $I_2/NO_2/CH_3SCH_3/N_2$ mixtures at atmospheric pressure, using FTIR analysis of the species. Their value of k_4, $(3 \pm 1) \times 10^{-11}$ at 296K, is in good agreement with our determination. They have also detected

$CH_3S(O)CH_3$ as a reaction product and in addition they were able to measure the branching ratio, found to be unity, for the channel yielding $CH_3S(O)CH_3$ and I atoms.

Concerning the atmospheric role of the IO + DMS reaction, the impact on both iodine and CH_3SCH_3 chemistry can be considered. The tropospheric chemistry of iodine compounds has been discussed by CHAMEIDES and DAVIS (12) and more recently by JENKIN et al (13). The reactions IO + hν $\rightarrow$ I + O (5) and IO + NO $\rightarrow$ I + NO_2 (6) are involved in O_3 depletion cycles by iodine species since I formed reacts with O_3 via : I + O_3 $\rightarrow$ IO + O_2. If reaction (4) produces I atoms, as our experiments suggest, this reaction would be an additional conversion route of IO into I. Its relative importance can be estimated by comparing the rates of reactions (4), (5) and (6). These rates are respectively 0.2(4), 0.2 and 0.06 s^{-1} if typical concentrations of 1.25×10^{10} (13) and 3×10^9 cm^{-3} (14) are taken for NO and CH_3SCH_3 (in marine atmosphere). The rate constants k_4 and k_6 were taken equal to 1.5×10^{-11} (this work) and 1.67×10^{-11} (8). Under these conditions, reaction (4) would contribute to 13 % in the conversion of IO into I in marine atmosphere. Therefore, reaction (4) cannot be, a priori, considered to be negligible in marine tropospheric iodine chemistry.

About the effect of reaction (2) on the CH_3SCH_3 atmospheric lifetime, the rate of reaction (4) has to be compared to that of reaction OH + CH_3SCH_3 (7) which is the main daytime atmospheric reaction of CH_3SCH_3. The rate ratio for reactions (2) and (7) is k_2 $|IO|/k_7$ $|OH|$. Computed IO concentrations between 10^8 and 10^9 cm^{-3} in marine atmosphere have been reported (12). Even if such concentrations are highly uncertain at the present time, due to lack of kinetic data for several key reactions, a rough estimation of the rate ratio k_2 $|IO|/k_7$ $|OH|$ can be made. Using the following values : $k_2 = 1.5 \times 10^{-11}$ (this work), $k_7 = 6.3 \times 10^{-12}$ (15), $|IO| = 10^8$ cm^{-3} and $|OH| = 10^6$ cm^{-3} (typical value), we obtain k_2 $|IO|/k_7$ $|OH| = 238$. Consequently, the IO + CH_3SCH_3 reaction would be the main sink for CH_3SCH_3 in marine atmosphere and would correspond to an atmospheric lifetime $\tau = 1/k_2$ $|IO| = 0.18$ hour for CH_3SCH_3. This result would support the low atmospheric lifetime measured by CARLIER and MOUVIER (3, 4) in marine atmosphere. However this conclusion remains to be confirmed after having determined more precisely the atmospheric IO concentrations. Besides, reaction (4), which has been shown to produce $CH_3S(O)CH_3$, could be a noticeable source of this compound recently measured at levels 1-5 10^7 cm^{-3} in marine atmophere (16).

REFERENCES

1 - STOCKWELL W. R. and CALVERT J. G. Atm. Env. 17, (1983), 2231.

2 - MARGITAN J. J. J. Phys. Chem. 88, (1984), 3314.

3 - CARLIER P. Quadriennal Ozone Symposium IAMAP - Kalithéa (Grèce) Sept. 1984.

4 - CARLIER P. and MOUVIER G. Atmos. Chem. in press.

5 - BARNES I., BECKER K. H., CARLIER P. and MOUVIER G. Submitted for publication.

6 - JOURDAIN J. L., LE BRAS G. and COMBOURIEU J. J. Phys. Chem. 85, (1981), 655.

7 - POULET G., LE BRAS G. and COMBOURIEU J. J. Chem. Phys. 69, (1978), 767.

8 - RAY G. W. and WATSON R. T. J. Phys. Chem. 85, (1981), 2955.
9 - MARTIN D., JOURDAIN J. L. and LE BRAS G. J. Phys. Chem. in press.
10 - HOWARD C. J. Communation at the seventeenth Symposium on free radicals, Granby. Colorado. Aug. 18-23, (1985).
11 - MARTIN D., JOURDAIN J. L., LAVERDET G. and LE BRAS G. Submitted for publication.
12 - CHAMEIDES W. L. and DAVIS D. D. J. Geophys. Res. 85 C12, (1980), 7383.
13 - JENKIN M. E., COX R. A. and CANDELAND D. E. J. Atmos. Chemistry 2, (1985), 359.
14 - ANDREAE M. O. et al. J. Geophys. Res. 90 D7, (1985), 12891.
15 - ATKINSON R. Chem. Review in press.
16 - HARVEY G. R. and LANG R. F. Geophys. Res. Lett 13, (1986), 49.

A COMPARISON OF MODELS FOR THE OXIDATION OF ACETYLENE AT 300 K

C. VINCKIER and J. PEETERS

Laboratory for Analytical and Inorganic Chemistry
Department of Chemistry
K.U.Leuven, Belgium

Summary

The use of kinetic sub-models in the simulation of the physico-chemical behaviour of atmospheric pollutants is a widely accepted approach. These sub-models are generated primarily in laboratory kinetic studies in order to reduce the complexity of the overall reacting system. For the particular case of the degradation of hydrocarbons, it will be shown that one not only needs the concentration profiles of the stable species formed, but also that the qualitative and quantitative determination of the reactive intermediates such as atoms or radicals is required for the validation of the proposed kinetic model.

INTRODUCTION

Numerical modelling and simulation is a widely accepted method for the description and elucidation of complex physico-chemical systems. Especially in the field of air pollution, modelling has been used to predict the behaviour and fate of air pollutants when transported between their emission source and various points in space. Also the modelling of "episodes" where the concentration of a pollutant varies with time in a certain region, is highly important. Elaborate models have been presented in the past on E.E.C. or NATO sponsored Symposia or Colloquia (1,2) and in general they contain two major parts : 1) Transport of Air Pollutants, which largely depends on the various types of pollutant sources, the complexity of the terrain, meteorological factors, etc. 2) Physico-chemical Transformation of Air Pollutants in which the chemical kinetics plays a crucial role in pollutant formation and (or) degradation processes.

In this paper we will only discuss some aspects of the latter part which in itself is already so complicated that one has developed sub-models in order to follow or predict the air pollutant concentration under various conditions such as day or night regimes, dry or wet climate, polluted areas near industrial plants or rural zones, etc... However the validation of these models in the field or in real conditions is a demanding and almost impossible task due to the strong coupling between transport and transformation in the troposphere. In order to eliminate the complications due to transport phenomena it became standard practice to develop, refine and validate chemical transformation sub-models by laboratory simulation of atmospheric chemical processes in so-called "smog chambers" (3). These systems have the advantage that one can start in relatively simple and well characterized conditions with one or two hydrocarbons added; once the reaction mechanism is understood, one can proceed to more complex systems. Only a reaction mechanism with the appropriate rate parameters is now required to model the observed concentration-time behaviour of the various pollutants. However, because of the large reaction schemes involved (often more than 100 reactions), such systems are still too complicated to gain accurate kinetic information on individual chemical reactions. Therefore separate laboratory experiments have been designed in which only a very limited number of chemical reactions occur and in which the determination of kinetic parameters is straightforward. It will be shown in this paper that even in such well-controlled conditions it is necessary to gather detailed information not only on the stable reaction products formed but also on the reactive intermediates such as atoms or radials in order to assess the validity of the proposed kinetic model and to obtain the associated kinetic coefficients.

RESULTS

The role of hydrocarbons RH is essential in the photochemical oxidant cycle, but unfortunately there still exists a lack of kinetic information on e.g. the primary adduct formation $R\dot{H}OH$ for unsaturated compounds and on the subsequent reactions thereof with O_2 (4). On the other hand the oxidation of acetylene by atomic oxygen although of little direct relevance to atmospheric conditions, is an example of a kinetic system which, at the current state of knowledge, is well suited for model validation. Indeed, beside a large variety of concentration-time data available (5,6,7), also the kinetic parameters of a number of elementary reactions involved are already known (8,9,10). The concentration of stable products such as CO, C_3H_4, C_4H_2, C_6H_2, CH_2O, CO_2 and H_2 have been determined and their profiles have been simulated by a numerical model including 27 reactions (6). However, as is also the case quite often for atmospheric transformations, the problems in describing the C_2H_2 degradation already starts in the first step; the total rate constant k_{tot} of C_2H_2 + O is well established (11) but the discussion on the relative importance of the two major reaction channels is still open :

$$C_2H_2 + O \rightarrow CH_2 + CO \qquad \text{1a}$$

$$C_2H_2 + O \rightarrow HCCO + H \qquad \text{1b}$$

While both intermediates CH_2 and HCCO have been detected by means of Molecular Beam Sampling-Mass Spectrometry (MB-MS) (8,9,12), some authors judged channel 1a to be the most important by far (95 %) (13) and others completely neglected the reaction chain starting with HCCO in their kinetic model calculations (6). When the calculated concentration profiles were compared with the experimental results, the agreement seemed to be satisfactory, except at short reaction times where mixing phenomena were held responsible for deviations between calculated and measured profiles (6). The overall picture however led to the conclusion that for a broad variety of experimental conditions, the deviations were within the experimental scatter so that the proposed kinetic model with associated rate constants was considered to be valid. However our laboratory studies of the same C_2H_2-O system have shown that beside CH_2, the radical HCCO is also a primary product of reaction 1 (9). Even more important, our recent mass spectrometric calibration of the HCCO radical (14) together with the determination of the rate constants of its destruction reactions revealed that reaction channel 1b accounts for about 60 $\pm$ 20 % of the total C_2H_2 + O consumption rate. So it was completely puzzling to us how a good fit between measured and calculated profiles of stable species could be obtained with a mechanism that neglects the dominant first-step route. This brought us to the basic question in kinetic modelling about the uniqueness of a mechanism that simulates accurately the formation of stable products (15, 16). It is our opinion that at quasi-stationary concentrations of the intermediates, the question must be answered in the negative. Indeed, after a few trials one succeeds in establishing two different hypothetical reaction mechanisms resulting in the same concentration profiles for all stable species, as examplified by :

Mechanism A

$$C_2H_2 + O \rightarrow CH_2 + CO \qquad 1$$

$$CH_2 + O \rightarrow CO + 2H \qquad 2$$

$$CH_2 + H \rightarrow CH + H_2 \qquad 3$$

$$CH + O_2 \rightarrow CO_2 + H \qquad 4$$

$$CH + C_2H_2 \rightarrow C_3H_2 + H \qquad 5$$

$$CH + O \rightarrow CO + H \qquad 6$$

Mechanism B

$$C_2H_2 + O \rightarrow HCCO + H \qquad 1'$$

$$HCCO + O \rightarrow 2CO + H \qquad 2'$$

$$HCCO + H \rightarrow CH_2 + CO \qquad 3'$$

$$CH_2 + O_2 \rightarrow H_2 + CO_2 \qquad 4'$$

$$CH_2 + C_2H_2 \rightarrow C_3H_2 + H_2 \qquad 5'$$

$$CH_2 + O \rightarrow CO + H_2 \qquad 6'$$

When the reactive intermediates CH_2, CH, C_3H_2 respectively HCCO, CH_2, C_3H_2 are removed immediately after their formation by the very fast destruction reactions (2-6), respectively (2'-6'), they will be present in negligibly low concentrations and the stationary-state principle may then be invoked for these species. When these stationary concentrations are used in the expressions for the net chemical reactions rates of CO, H_2, CO_2, O and H, one finds with $k_1[C_2H_2][O] = U_A$, respectively $k_{1'}[C_2H_2][O]^2 = U_B^2$:

Mechanism A

$$\frac{d[CO]}{dt} = U_A \left\{1 + \frac{k_2[O]}{k_2[O]+k_3[H]} + \frac{k_3[H]}{(k_2[O]+k_3[H])} \times \frac{k_6[O]}{(k_4[O_2]+k_5[C_2H_2]+k_6[O])}\right\}$$

$$\frac{d[H_2]}{dt} = U_A \times \frac{k_3[H]}{k_2[O]+k_3[H]}$$

$$\frac{d[CO_2]}{dt} = U_A \left\{\frac{k_3[H]}{(k_2[O]+k_3[H])} \times \frac{k_4[O_2]}{(k_4[O_2]+k_5[C_2H_2]+k_6[O])}\right\}$$

$$\frac{d[O]}{dt} = -U_A \left\{ 1 + \frac{k_2[O]}{k_2[O]+k_3[H]} + \frac{k_3[H]}{(k_2[O]+k_3[H])} \times \frac{k_6[O]}{(k_4[O_2]+k_5[C_2H_2]+k_6[O])} \right\}$$

$$\frac{d[H]}{dt} = U_A \left\{ \frac{2\,k_2[O]}{k_2[O]+k_3[H]} - \frac{k_3[H]}{k_2[O]+k_3[H]} (1 - 1) \right\} = U_A \times \frac{2\,k_2[O]}{k_2[O]+k_3[H]}$$

Mechanism B

$$\frac{d[CO]}{dt} = U_B \left\{ \frac{2\,k_{2'}[O]}{k_{2'}[O]+k_{3'}[H]} + \frac{k_{3'}[H]}{k_{2'}[O]+k_{3'}[H]} + \right.$$

$$\left. \frac{k_{3'}[H]}{(k_{2'}[O]+k_{3'}[H])} \times \frac{k_{6'}[O]}{(k_{4'}[O_2]+k_{5'}[C_2H_6]+k_{6'}[O])} \right\}$$

$$= U_B \left\{ 1 + \frac{k_{2'}[O]}{k_{2'}[O]+k_{3'}[H]} + \frac{k_{3'}[H]}{(k_{2'}[O]+k_{3'}[H])} \times \frac{k_{6'}[O]}{(k_{4'}[O]+k_{5'}[C_2H_2]+k_{6'}[O])} \right\}$$

$$\frac{d[H_2]}{dt} = U_B \times \frac{k_{3'}[H]}{k_{2'}[O]+k_{3'}[H]}$$

$$\frac{d[CO_2]}{dt} = U_B \left\{ \frac{k_{3'}[H]}{(k_{2'}[O]+k_{3'}[H])} \times \frac{k_{4'}[O_2]}{(k_{4'}[O_2]+k_{5'}[C_2H_2]+k_{6'}[O])} \right\}$$

$$\frac{d[O]}{dt} = -U_B \left\{ 1 + \frac{k_{2'}[O]}{k_{2'}[O]+k_{3'}[H]} + \frac{k_{3'}[H]}{(k_{2'}[O]+k_{3'}[H])} \times \frac{k_{6'}[O]}{(k_{4'}[O]+k_{5'}[C_2H_2]+k_{6'}[O])} \right\}$$

$$\frac{d[H]}{dt} = U_B \left\{ 1 + \frac{k_{2'}[O]}{k_{2'}[O]+k_{3'}[H]} - \frac{k_{3'}[O]}{k_{2'}[O]+k_{3'}[H]} \right\} = U_B \times \frac{2\,k_{2'}[O]}{k_{2'}[O]+k_{3'}[H]}$$

The above equations with $k_1 = k_{1'}$ or $U_A = U_B$ show that both mechanisms A and B will result in the same concentration profiles for CO, H_2, CO_2, O and H provided that

$$\frac{k_3}{k_2} = \frac{k_{3'}}{k_{2'}} \; ; \; \frac{k_4}{k_6} = \frac{k_{4'}}{k_{6'}} \; ; \; \frac{k_5}{k_6} = \frac{k_{5'}}{k_{6'}}$$

regardless of the initial mixture composition. The formation of various hydrocarbons during this oxidation process can then start from the radical C_3H_2, the formation rate U_f of which in mechanism A is given by :

$$U_{f,C_3H_2} = U_A \left\{ \frac{k_3[H]}{(k_2[O]+k_3[H])} \times \frac{k_5[C_2H_2]}{(k_4[O_2]+k_5[C_2H_2]+k_6[O])} \right\}$$

and in mechanism B by :

$$U_{f,C_3H_2} = U_B \left\{ \frac{k_{3'}[H]}{(k_{2'}[O]+k_{3'}[H])} \times \frac{k_{5'}[C_2H_2]}{(k_{4'}[O_2]+k_{5'}[C_2H_2]+k_{6'}[O])} \right\}$$

Both formation rates are identical when the conditions mentioned above are fulfilled. Subsequent reactions of C_3H_2 such as

$$C_3H_2 + H \rightarrow C_3H_3 \quad 7$$

$$C_3H_3 + H + M \rightarrow C_3H_4 + M \quad 8$$

$$C_3H_3 + O \rightarrow C_2H + CH_2O \quad 9$$

$$C_2H + C_2H_2 \rightarrow C_4H_2 + H \quad 10$$

lead of course to the same formation rates of C_3H_4, CH_2O and C_4H_2 in both mechanisms since the reactions are exactly the same. Moreover it is easily demonstrated that at the given rate constant ratios, all the above conclusions remain valid for any mixture of mechanisms A and B, $U = \alpha U_A + (1-\alpha)U_B$, subject to the sole additional condition that $\alpha k_1 + (1-\alpha)k_{1'} = k_{tot}$. Mechanisms A and B separately ($\alpha = 0$, $\alpha = 1$) or together ($0 < \alpha < 1$) result in exactly identical net reaction rates for CO, H_2, CO_2, O, H, C_3H_4, CH_2O and C_4H_2. It shoud be pointed out that when the concentration of the intermediates CH_2, HCCO and CH are negligibly low due to fast destruction, one doesn't even have to invoke "slight experimental scatter" to prove that the two mechanisms and all mixtures thereof are indistinguishable. It might look puzzling that the fate of CH_2 is different in mechanism A from B, but this could be explained by recognizing that CH_2 is not necessarily formed in the same electronic state in the two mechanisms. Indeed, reaction 3' probably leads to CH_2 in its singlet electronically excited state with a reactivity towards O_2 and C_2H_2 that is known to be much higher than that of the triplet ground state CH_2 (10, 17, 18). Actually recent experiments carried out at our laboratory have proven that a large fraction of methylene formed in the acetylene oxidation is present as $^1CH_2$ (14). Since the reactivity of CH and $^1CH_2$ towards O_2 and C_2H_2 is comparable, the conditions for both mechanisms to be identical even become realistic. Anyway, the hypothetical kinetic schemes elaborated above clearly demonstrate that the concentration-time behaviour of stable compounds, even for a wide variety of initial mixture compositions, do not provide a sufficient base for the validation of a given model. One cannot exclude a priori that these concentration profiles are fully compatible with two or more completely different mechanisms, at least when the intermediates have reached their stationary concentrations. The latter requirement is absolutely relevant to atmospheric pollution chemistry where most of the reactive intermediates such as OH, RO, RO_2 do indeed fulfill the quasi-stationary state conditions.

In conclusion one may state that for the validation of kinetic models one not only needs the concentration profiles of stable species but first and foremost full and accurate information on the nature and concentration profiles of the reactive intermediates involved. This of course requires highly sophisticated experimental techniques such as long path absorption, LIF, LMR and Molecular Beam Sampling-Mass Spectrometry. Additionally it is clear that the confidence in the kinetic model will grow when the reaction rate constants and the product distribution of the elementary reactions involved are known by separate and direct kinetic experiments. The importance of the foregoing in the field of air pollution modelling lies in the fact that certain kinetic models for hydrocarbon chemistry may perfectly fit the laboratory results, but may go wrong in a real environment due to unexpected radical reactions of the intermediates with atmospheric trace species and even aerosols.

CONCLUSION

Due to the complexity of transport and transformation processes of the air pollutants in the environment, many laboratory experiments have been undertaken in order to elucidate the reaction mechanisms responsible for the transformation of these pollutants. Simulation of these kinetic experiments is a wide-spread technique in which calculated concentration profiles are compared with the experimental profiles. It has been shown in this work that modelling of the stable species formed does not guarantee the validity of the model since one does not have a definite proof of the uniqueness of the model. One needs to follow, beside the stable species, the concentration profiles of the reactive intermediates such as atoms and radicals as well. These experimental data together with reaction rate constants of the elementary reactions measured in independent experiments will strongly enhance the confidence in the kinetic model. However one should keep in mind that these kinetic models, deduced from laboratory experiments can only be used in an overall environmental model when the physico-chemical characteristics of the polluted atmosphere are comparable to those of the experiment. When the chemical composition of the atmosphere becomes too different from the laboratory experimental conditions, unexpected radial reactions can occur so that the kinetic model developed for laboratory experiments may not be applied for the simulation of the chemistry in polluted air.

ACKNOWLEDGEMENTS

Both the authors are indebted to the National Fund for Scientific Research, Belgium (NFWO) and to the Fund for Joint Basic Research (FKFO) for their financial support. Also a E.E.C. research grant in the frame of the "European Stimulation Action" is gratefully acknowledged.

REFERENCES

(1) European Symposia on "Physico-Chemical Behaviour of Atmospheric Pollutants" (3 issiues).
(2) Nato International Technical Meetings on "Air Pollution Modelling and its Application" (15 issues).
(3) Carter W.P.L., Lloyd A.C., Sprung J.L. and Pitts J.N., Int. J. of Chem. Kinetics, 9, 45, 1979.
(4) Falls A.H., McRae G.J. and Seinfeld J.H., Int. J. of Chem. Kinetics, 11, 1137, 1977.
(5) Jones I.T.N. and K.D. Bayes, 14th Symposium (International) on Combustion, 277, 1973.
(6) Homann K.H. and Schweinfurth H., Ber. Bunsenges. Phys. Chem., 85, 569, 1981.
(7) Vinckier C. and Debruyn W., 17th Symposium (International) on Combustion, 623, 1979.
(8) Vinckier C. and Debruyn W., J. Phys. Chem., 83, 2057, 1979.
(9) Vinckier C., Schaekers M. and Peeters J., J. Phys. Chem., 89, 508, 1985.
(10) Böhland T., Temps F. and Wagner H.G., Ber. Bunsenges. Phys. Chem., 88, 455, 1984 and 88, 1222, 1984 and 89, 432, 1985.

(11) Herron J.T. and Huie R.E., J. Phys. Chem. Ref. Data, 2, 467, 1974.
(12) Many references on this topic are given in ref. 9 of this paper.
(13) Blumenberg B. Hoyermann K. and Sievert R., 16th Symposium (International) on Combustion, 841, 1977.
(14) Peeters J., Schaekers M. and Vinckier C., submitted for publication in J. Phys. Chem..
(15) Noyes R.M., J. Phys. Chem., 81, 2315, 1977.
(16) Frenklach M., Chapt. 7, Combustion Chemistry, W.C. Gardiner Ed., Springer-Verlag, N.Y., 1984.
(17) Langford A.O., Petek H. and Moore C.B., J. Chem. Phys., 78, 6650, 1983.
(18) Canosa-Mass C.E., Ellis M., Monty Frey H. and Walsh R., Int. J. Chem. Kin., 16, 1103, 1984.
(19) Duewer W.H., J. Phys. Chem., 81, 2483, 1977.

RATE CONSTANTS FOR REACTION OF OH RADICALS WITH ACETATES AND GLYCOLS IN THE GAS PHASE

D. HARTMANN, A. GEDRA*, D. RHÄSA, and R. ZELLNER
Institut für Physikalische Chemie, Universität Göttingen
3400 Göttingen, FRG

Summary

Rate constants for the gas phase reaction of OH radicals with (1) n-butyl-acetate, (2) 2-ethoxy-ethanol, (3) 2-butoxy-ethanol, and (4) 1-acetoxy-2-ethoxy-ethane between T = 298 - 500 K have been measured using a combined laser photolysis/resonance fluorescence (LPRF) technique. The results are:

$$k_1(T) = (3.1 \pm 0.7)\ 10^{-11} \exp(-594 \pm 126/T)\ cm^3/s$$

$$k_2(T) = (1.8 \pm 0.4)\ 10^{-11} \exp(-120 \pm 30/T)\ cm^3/s$$

$$k_3(T) = (1.4 \pm 0.3)\ 10^{-11}\ cm^3/s$$

$$k_4(T) = (3.6 \pm 0.8)\ 10^{-12} \exp(383 \pm 80/T)\ cm^3/s$$

The rate constants at 298 K have been analysed in terms of individual C-H bond contributions using estimated thermochemical data and empirical group reactivity correlations. It is found that the primary reaction is dominated in all cases by H atom abstraction from a CH_2 group with the absolute reactivity depending on the α-substituent and decreasing in the order $-O-CH_2- > HO-CH_2- > -CH_2-CH_2- \sim CH_3-CH_2- > -C(O)O-CH_2-$. Based on these findings and on generally accepted oxidation schemes a simplified mechanism is suggested for the atmospheric degradation of glycols. Assuming an average tropospheric OH concentration of $5 \times 10^5\ cm^{-3}$ the atmospheric lifetimes are estimated to be 5.5 days and 1.7 days for compounds (1) and (2, 3, 4), respectively.

1. INTRODUCTION

Homogeneous gas phase reactions are the dominant degradation pathways of volatile organic compounds. Due to its reactivity and concentration level ($\sim 5 \times 10^5\ cm^{-3}$ in the global annual tropospheric average [1]) the OH radical is considered to be the most important oxidant. Consequently, the corresponding rate coefficients, k_{OH}, have become the accepted

---

*) Permanent address: Central Research Institute for Chemistry, Budapest, Hungary

quantities to characterize the atmospheric degradability of volatile organic material.

The status of rate measurements for reactions of the OH radicals with organic compounds is in general extremely comprehensive and well documented in a number of current reviews [2 - 5]. Moreover, based on the available data set of k_{OH}-values a number of empirical methods have been devised to predict rate coefficients from thermochemical data and individual group additivity rules [4, 6]. However, there is an apparent lack of rate data for reactions of OH with acetates and glycols. These compounds have extensive use as solvents for dyes, rubber material, plastics, and oils [7] and are being released, for instance, during any handling and application of paint.

In the present work rate constants for the reactions of OH radicals with n-butyl-acetate, 2-ethoxy-ethanol, 2-butoxy-ethanol, and 1-acetoxy-2-ethoxy-ethane have been measured using a combined laser photolysis/resonance fluorescence technique. The data are being used for comparison with predictive reactivity rules and to deduce atmospheric residence times and degradation routes for these compounds.

2. EXPERIMENTAL

The experiments described in this work have been performed in a combined laser photolysis/resonance fluorescence (LPRF) system. A schematic representation of the experimental set-up is provided in Fig. 1. Since the LPRF-technique has been described fully in previous publications from this laboratory [8 - 12] only the major aspects will be summarized here. OH radicals are generated by excimer laser (λ = 193 nm) photolysis of HNO_3 at initial concentrations of $\sim 1 \times 10^{12}$ cm^{-3}. The subsequent decay of OH in the presence of excess reagent is monitored by time resolved A-X resonance fluorescence excited by a microwave discharge of H_2O in Ar. The optical signal is detected by a photomultiplier after passing through a narrow band interference filter. The analogue output from the PMT is then preamplified and fed into a signal averager coupled to a microcomputer for processing of the digital information.

The reaction cell is operated under slow flow conditions (linear flow velocities 5 - 30 cm/s) in order to renew the gas mixture in the excitation volume between consecutive laser pulses. Known reagent concentrations are provided by passing He at atmospheric pressure through a saturator kept at 50 - 80 ^{o}C. Direct measurements of the reagent flow rate by a gravimetric technique agreed well with the one calculated on the basis of the vapour pressure curves.

A series of preliminary experiments on rate measurements of k_1 and k_2 were performed in a conventional flash photolysis system combined with resonance absorption for detection of OH. This method, however, produced rate coefficients that were up to a factor of 2 (in case of n-butyl-acetate) larger than those from the laser photolysis experiments. A subsequent analysis of the UV spectra of n-butyl-acetate and 2-ethoxy-ethanol revealed that this effect was most likely due to the formation of photofragments from these reagents in the broad

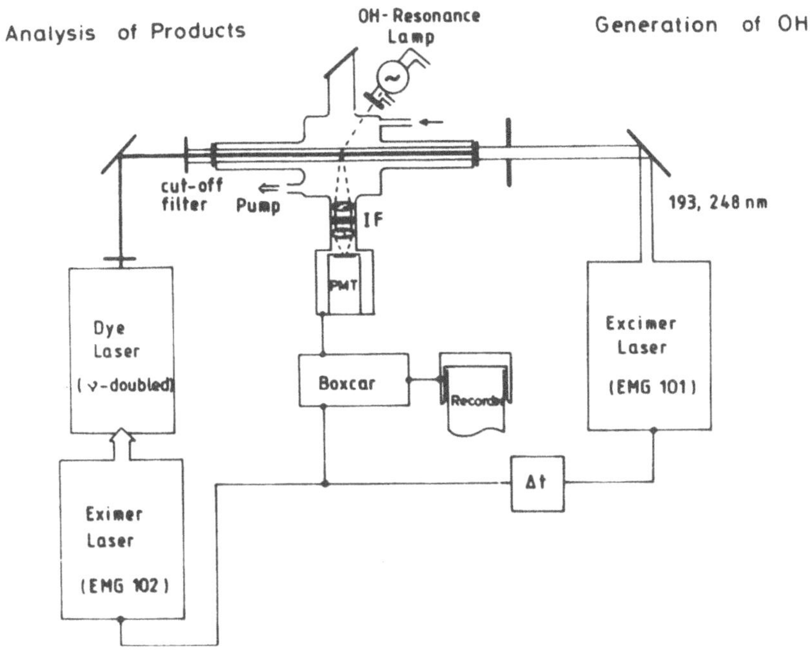

Fig. 1: Schematic representation of laser photolysis/resonance fluorescence experiment

band photolysis by a convential flash lamp at a short wavelength cut-off of 165 nm (spectrosil quartz).

The reagents used had the following purities: n-butyl-acetate (Fluka, > 99 % p. a.), 2-ethoxy-ethanol (Hoechst, "reinst"), 2-butoxy-ethanol (Fluka, > 98 %, purum), 1-acetoxy-2-ethoxy-ethane (Hoechst, "reinst"). They were used without further purification.

3. RESULTS AND DISCUSSION

3.1. Rate Coefficient Measurements

All kinetic experiments to determine the rate coefficient for OH consumption were carried out under pseudo-first order conditions with the reagent concentration in at least 10 fold excess over OH. A representative example of an OH fluorescence intensity trace together with its digitized logarithmic display is shown in Fig. 2. As can be seen the decay of OH is exponential over at least one order of magnitude in concentration. In order to derive the second order rate coefficients, the dependence of $k_{1st} = -\Delta \ln[OH]/\Delta t$ on reagent concentration has been determined. The results are summarized in Figs. 3a-d where experiments for each compound at different temperatures are shown. As can be seen, there is a good linear dependence of k_{1st} on reagent concentration ($[A]$) which can be described by $k_{1st} = k_A [A] + k'$ and where the intercept k' accounts for the loss of OH by reaction with HNO_3 and by transport and

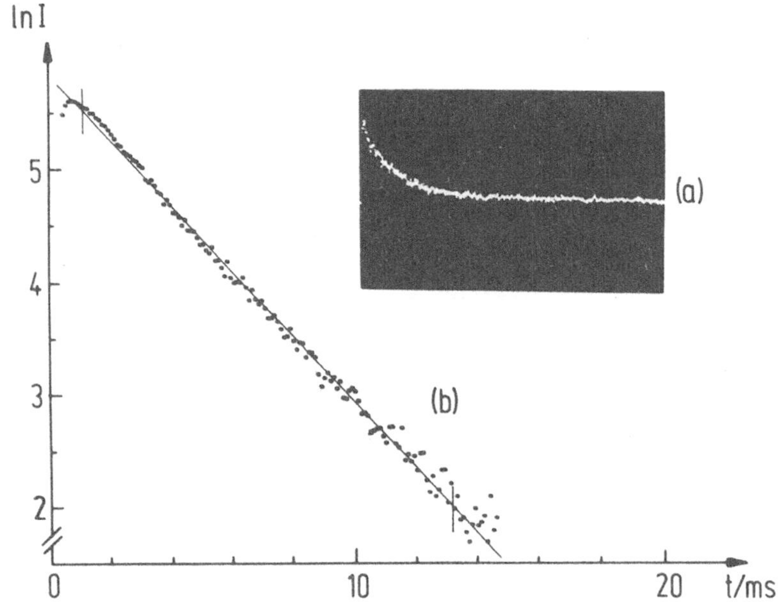

Fig. 2: Multichannel analyzer oscilloscope trace (a) and corresponding logarithmic display (b) of OH decay in the presence of 1-acetoxy-2-ethoxy-ethane; p = 14 mbar, T = 401 K, $[C_6H_{12}O_3] = 1.5 \times 10^{13}\ cm^{-3}$

flow [8, 9]. The second order rate coefficients k_A obtained from these plots are collected in Table I.

Table I: Summary of measured rate coefficients for reaction of OH with acetates and glycols

Reagent	T/K	$10^{11} \cdot k_A/cm^3\ s^{-1}$
(1) n-butyl-acetate	298	0.43 ± 0.08
$CH_3C(O)OCH_2CH_2CH_2CH_3$	400	0.68 ± 0.13
	516	1.0 ± 0.2
(2) 2-ethoxy-ethanol	298	1.2 ± 0.3
$CH_3CH_2OCH_2CH_2OH$	485	1.4 ± 0.3
(3) 2-butoxy-ethanol	298	1.4 ± 0.3
$CH_3CH_2CH_2CH_2OCH_2CH_2OH$	505	1.4 ± 0.3
(4) 1-acetoxy-2-ethoxy-ethane	298	1.3 ± 0.2
$CH_3C(O)OCH_2CH_2OCH_2CH_3$	401	0.9 ± 0.2
	506	0.8 ± 0.2

All data were taken at total pressures of 15 ± 2 mbar of He. If represented in Arrhenius form the following expressions are

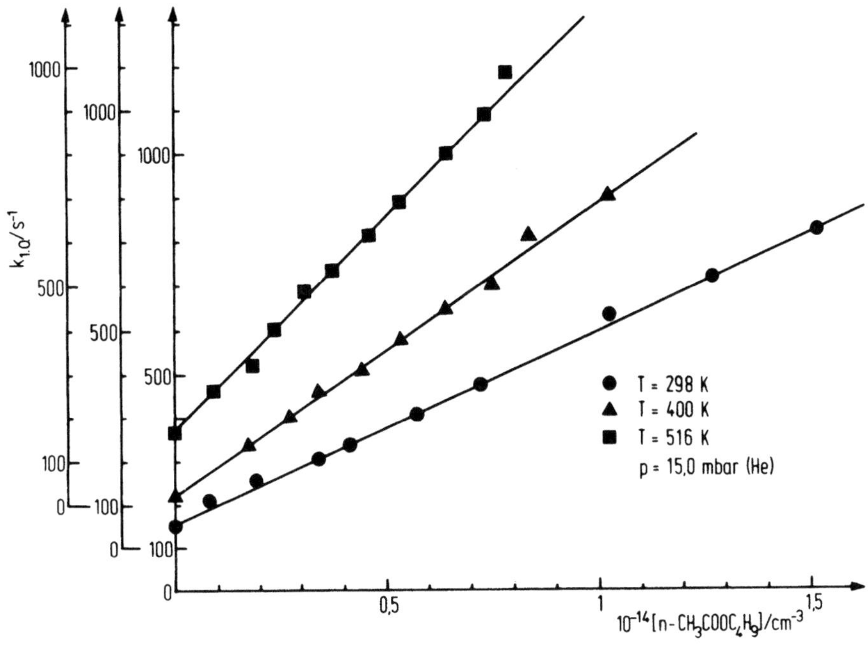

Fig. 3a: Dependence of first-order rate coefficients for OH decay on the n-butyl-acetate concentration at different temperatures

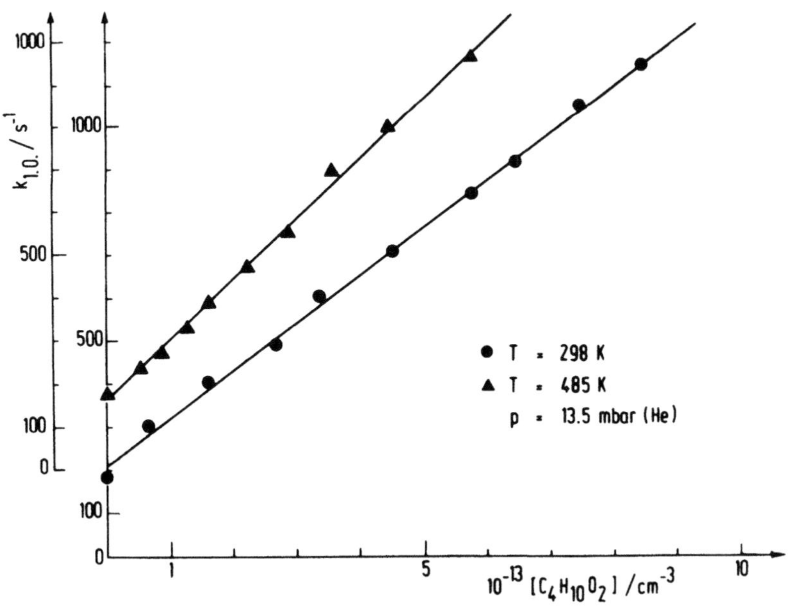

Fig. 3b: Dependence of first-order rate coefficient on the 2-ethoxy-ethanol concentration

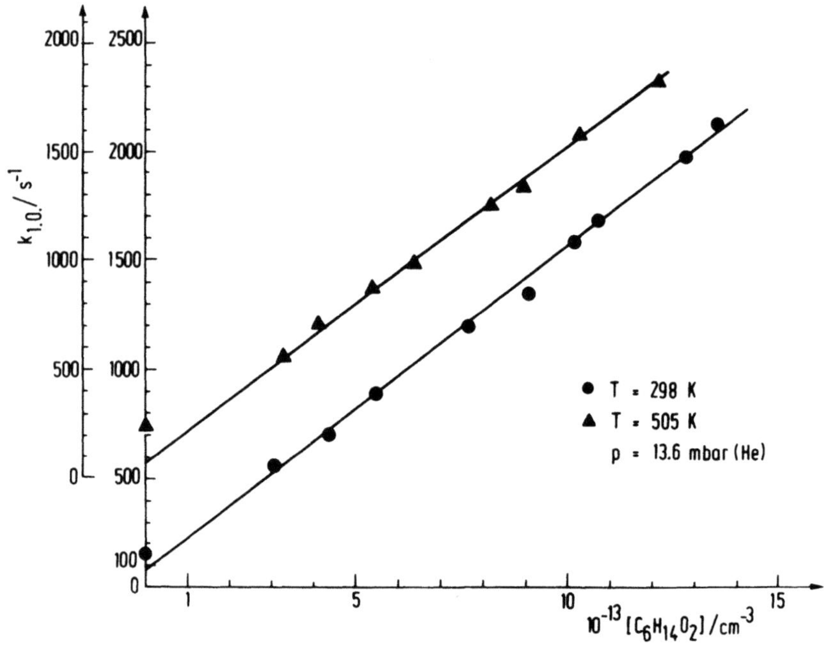

Fig. 3c: Dependence of first-order rate coefficient on the 2-butoxy-ethanol concentration

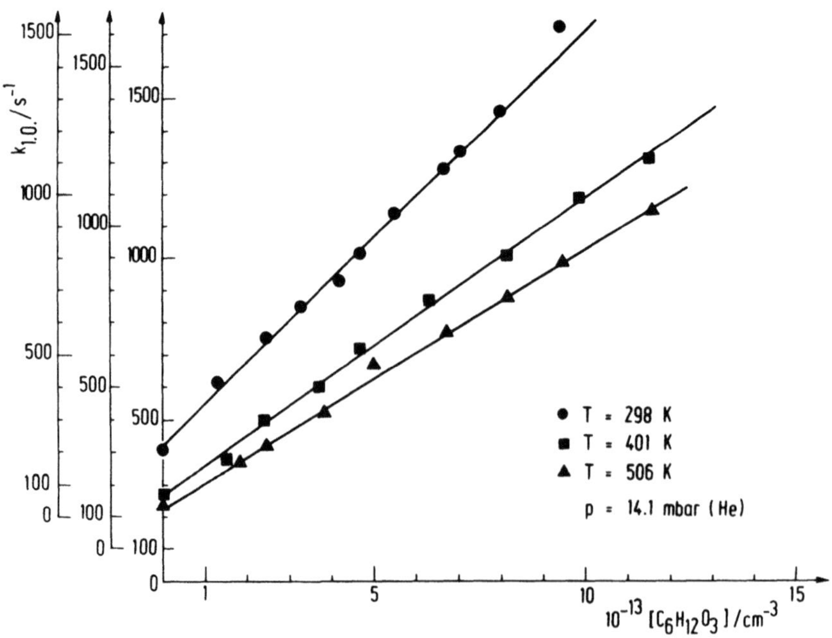

Fig. 3d: Dependence of first-order rate coefficient on the concentration of 1-acetoxy-2-ethoxy-ethane

obtained:

$$k_1(T) = (3.1 \pm 0.7)\ 10^{-11}\ \exp(-594 \pm 126/T)\ cm^3/s$$

$$k_2(T) = (1.8 \pm 0.4)\ 10^{-11}\ \exp(-120 \pm 30/T)\ cm^3/s$$

$$k_3(T) = (1.4 \pm 0.3)\ 10^{-11}\ cm^3/s$$

$$k_4(T) = (3.6 \pm 0.8)\ 10^{-12}\ \exp(383 \pm 80/T)\ cm^3/s$$

To our knowledge none of these reactions has been studied before so that no data are available with which our results can be directly compared. There are, however, a number of studies of OH reactions with simpler acetates (methyl-acetate (MA), ethyl-acetate (EA), n-propyl-acetate (nPA), and sec.-butyl-acetate (sBA)). A comparison of the rate coefficients ($k_{MA} = 1.7 \times 10^{-13}$ [13], $k_{EA} = 1.7 \times 10^{-12}$ [13, 14], $k_{nPA} = 4.2 \times 10^{-12}$ [15], $k_{sBA} = 5.4 \times 10^{-12}$ [15] in units of cm^3/s) shows that there is a strong increase in reactivity from methyl- to ethyl-acetate followed by a much more gradual increase upon substitution by larger groups. Our result for n-butyl-acetate ($k_{nBA} = (4.3 \pm 0.8)\ 10^{-12}\ cm^3/s$) is essential identical to that for n-propyl-acetate. The larger rate coefficient for sec-butyl-acetate is comprehensible on the basis of a slightly lower C-H bond strength in this substituent.

The rate coefficients at 298 K for the other compounds studied are essentially identical ($k \simeq (1.3 \pm 0.1) \times 10^{-11}\ cm^3/s$) and about a factor of 3 larger than that for n-butyl-acetate. This larger reactivity is apparently imposed by the ether group ($-CH_2-O-CH_2-$) which is the common element of all these compounds. Whereas the temperature dependence, however, is slightly positive or zero for the ethylene glycol derivatives 2-ethoxy-ethanol and 2-butoxy-ethanol, respectively, we find a slightly negative temperature dependence for the reaction of OH with 1-acetoxy-2-ethoxy-ethane. The temperature coefficient reported corresponds to $E_4 = -3.2$ kJ/mol which is equivalent to a T^{-1} dependence.

3.2. Comparison with Predictions of Rate Coefficients Based on Thermochemical Data

Reactions of OH with saturated hydrocarbons including acetates and glycols are assumed to proceed by simple H atom abstraction. As a consequence the overall rate coefficient may be considered as the sum of the individual C-H bond contributions, i. e. $k_{OH} = \sum N_i \cdot k_i$. In here k_i is the specific rate coefficient for a particular C-H bond (allowing for the nature of α- and β-substituents) and N_i is the number of equivalent H-atoms. For the present purpose we have adopted two different methods for the prediction of k_i's recently described in the literature. In the method developed by Heicklen [6] k_i is calculated from a collision theory rate coefficient expression with the reaction cross section and activation energy deduced empirically from the reactivities of simple hydrocarbons and the corresponding C-H bond strengths, respectively. The empirical method proposed by Atkinson [4] assumes characteristic group reactivities for primary, secondary, and tertiary C atoms and allows for the effects of β-substituents. The results obtained from an application of both methods to the reactions considered here are summarized in Table 2.

Table 2: Reaction enthalpies, relative reactivities of individual C-H bonds and absolute overall rate coefficients as predicted by empirical methods

Reagent	Primary Radical a)	ΔH_R kJmol b)	Rel.Rate % H c)	Rel.Rate % A d)	$10^{11}\cdot k_{298}$ H c)	$10^{11}\cdot k_{298}$ A d)
	$\dot{C}H_2CO_2CH_2CH_2CH_2CH_3$	- 85	6	-		
$CH_3CO_2CH_2CH_2$	$CH_3CO_2\dot{C}HCH_2CH_2CH_3$	- 97	20	36		
$-CH_2CH_3$	$CH_3CO_2CH_2\dot{C}HCH_2CH_3$	-102	34	34	0.17	0.42
	$CH_3CO_2CH_2CH_2\dot{C}HCH_3$	-102	34	26		
	$CH_3CO_2CH_2CH_2CH_2\dot{C}H_2$	- 85	6	4		
	$\dot{C}H_2CH_2OCH_2CH_2OH$	- 85	0.5	1.3		
$CH_3CH_2OCH_2$	$CH_3\dot{C}HOCH_2CH_2OH$	-122	43	33		
$-CH_2OH$	$CH_3CH_2O\dot{C}HCH_2OH$	-122	43	43	2.1	1.7
	$CH_3CH_2OCH_2\dot{C}HOH$	-114	13.5	22		
	$CH_3CH_2OCH_2CH_2\dot{O}$	- 64	-	0.7		
	$\dot{C}H_2CH_2CH_2CH_2OCH_2CH_2OH$	- 85	0.4	1		
	$CH_3\dot{C}HCH_2CH_2OCH_2CH_2OH$	-102	3	5		
$CH_3CH_2CH_2CH_2$	$CH_3CH_2\dot{C}HCH_2OCH_2CH_2OH$	-102	3	6.5		
$-OCH_2CH_2OH$	$CH_3CH_2CH_2\dot{C}HOCH_2CH_2OH$	-122	40	35	2.2	2.1
	$CH_3CH_2CH_2CH_2O\dot{C}HCH_2OH$	-122	40	35		
	$CH_3CH_2CH_2CH_2OCH_2\dot{C}HOH$	-114	13.6	17		
	$CH_3CH_2CH_2CH_2OCH_2CH_2\dot{O}$	- 64	-	0.5		
	$\dot{C}H_2CO_2CH_2CH_2OCH_2CH_3$	- 85	0.5	-		
	$CH_3CO_2\dot{C}HCH_2OCH_2CH_3$	- 97	2	10		
$CH_3CO_2CH_2CH_2$	$CH_3CO_2CH_2\dot{C}HOCH_2CH_3$	-122	48.5	50	1.8	1.4
$-OCH_2CH_3$	$CH_3CO_2CH_2CH_2O\dot{C}HCH_3$	-122	48.5	39		
	$CH_3CO_2CH_2CH_2OCH_2\dot{C}H_2$	- 85	0.5	1		

a) Primary radical formed in reaction with OH by H-atom abstraction
b) Reaction enthalpies calculated from bond dissociation energies as estimated by the method of Heicklen [6]
c) according to the method of Heicklen [6]
d) according to the method of Atkinson [4]

As can be seen the highest reactivity is associated with the H-atom of the $-O-CH_2$-moiety which has a bond energy of

377 kJ/mol [6]. Moreover, comparing both relative and absolute reactivities the following scale of decreasing reactivity can be deduced: $-O-CH_2- > HO-CH_2- > -CH_2-CH_2- \sim CH_3-CH_2- > -C(O)O-CH_2-$. H-atom abstraction from terminal OH or CH_3 groups is relatively unimportant.

It is interesting to compare the predictions of the absolute rate coefficients with those found experimentally. There is excellent agreement between the Atkinson method and experiment for compounds (1) and (4). For compounds (2) and (3) the rate coefficient is overestimated. The Heicklen method is in general further off the experimental result with an overestimate at (2), (3), and (4) and an underestimate at compound (1). As a whole, however, the predictive methods are in agreement with experiment by better than a factor of 2.

3.3. Atmospheric Degradation

The present study does not provide direct insight into the atmospheric degradation route of acetates and glycols. However, based on the above evidence of the nature of the primary radicals formed in reaction with OH and accepted general principles of atmospheric hydrocarbon oxidation [5] certain conclusion may be drawn with respect to the subsequent oxidation. As an example the proposed atmospheric degradation scheme of the glycol 2-ethoxy-ethanol is presented in Fig. 4.

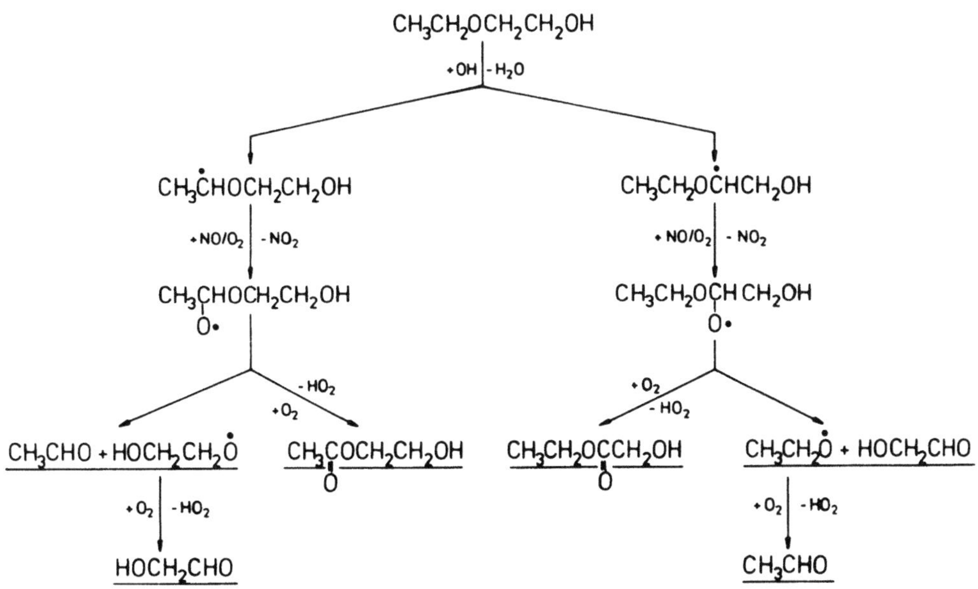

Fig. 4: Proposed atmospheric degradation route of 2-ethoxy-ethanol

Two different routes are considered which according to Table 2 occur with equal probability. In each case the primary radical is expected to be rapidly transferred into the corresponding oxy radical in a sequence of reactions involving O_2 and NO. Once formed oxy radicals react in at least two different ways: The reaction with O_2 leads to the formation of esters. The alternative is the unimolecular decomposition to yield an aldehyd and a terminal oxy radical which in reaction with O_2 produces a second aldehyd. Hence, as a net result, the $-CH_2-O-CH_2-$ moiety of the initial glycol is in part oxydized to yield esters and in part fragmented to yield two different aldehyds. Since aldehyds react rapidly with OH ($k(OH+CH_3CHO) = 1.6 \times 10^{-11}$ cm^3/s [16, 17]) they will be readily further degraded. Esters on the other hand are less reactive than glycols (see above) and will consequently accumulate.

The atmospheric lifetime of 2-ethoxy-ethanol may be calculated from the expression $\tau = (k_{OH} [OH]_{ss})^{-1}$, which for an average tropospheric OH concentration of 5×10^5 cm^{-3} yields ~ 1.7 days. With this short lifetime the diurnal variation of OH must be taken into account. Since OH concentrations of ~ 3×10^6 cm^{-3} [18] are predicted for the daytime atmosphere, the lifetime of 2-ethoxy-ethanol may be well below one day. The same applies to 2-butoxy-ethanol and 1-acetoxy-2-ethoxy-ethane. Only n-butyl-acetate is more persistent with an estimated lifetime of ~ 5.5 days. It should be noted finally, that none of these compounds have absorptions above 240 nm and hence cannot be degraded photolytically in the troposphere.

ACKNOWLEDGEMENT

Support of this work by BMFT, Hoechst AG and "Fonds der chemischen Industrie" is gratefully acknowledged. One of us (R. Z.) thanks DFG for a "Heisenberg-Stipendium".

REFERENCES

(1) P.J. Crutzen, in "Atmospheric Chemistry", E.D. Goldberg (Ed.), Springer 1982.

(2) R. Atkinson, K.R. Darnall, A.C. Lloyd, A.M. Winer, J.N. Pitts jr., Adv. Photochem. 11, 375 (1979).

(3) R. Atkinson, A.C. Lloyd, J. Phys. Chem. Ref. Data, 13, 315 (1984).

(4) R. Atkinson, Chem. Reviews, 1986, in press.

(5) H.Gg. Wagner, R. Zellner; Erdöl, Kohle, Erdgas, Petrochemie, 27, 212 (1984).

(6) J. Heicklen, Int. J. Chem. Kin. 13, 651 (1981).

(7) "Ullmanns Enzyklopädie der Technischen Chemie", Vol. 16, Verlag Chemie, Weinheim (1978).

(8) K. Lorenz, R. Zellner, Ber. Bunsenges. phys. Chem. 87, 629 (1983).

(9) R. Zellner, K. Lorenz, J. Phys. Chem. 88, 984 (1984).

(10) J. Hägele, K. Lorenz, D. Rhäsa, R. Zellner, Ber. Bunsenges. phys. Chem. 87, 1023 (1983).

(11) B. Fritz, K. Lorenz, W. Steinert, R. Zellner; in "Physico-chemical Behaviour of Atmospheric Pollutants", p. 192, B. Versino, H. Ott (Eds.), Reidel (1982).

(12) K. Lorenz, D. Rhäsa, R. Zellner, in "Physico-chemical Behaviour of Atmospheric Pollutants", p. 158, B. Versino,

G. Angeletti (Eds.), Reidel (1984).
(13) I.M. Campbell, P.E. Packinson, Chem. Phys. Lett., 53, 585 (1978).
(14) C. Zetzsch, private communication.
(15) A.M. Winer, A.C. Lloyd, K.R. Darnall, R. Atkinson, J.N. Pitts jr., Chem. Phys. Lett., 51, 221 (1977).
(16) R. Atkinson, J.N. Pitts jr., J. Chem. Phys. 68, 3581 (1978).
(17) R. Zellner, K. Lorenz, D. Rhäsa, in "Methods of the ecotoxicological evaluation of chemicals. Photochemical degradation in the gasphase", report Kernforschungsanlage Jülich, Jül-Spez-279 (1984).
(18) R. Zellner, K.J. Niemitz, in "Formation, Distribution, and transformation of atmospheric pollutants", DECHEMA, in press (1986).

OH-Dynamics during Hydrocarbon Oxidation in Air

V.H. Schmidt, Yuhui Su, K.H. Becker, and K. Brockmann
Physikalische Chemie/FB 9
Bergische Universität - GH Wuppertal
D-5600 Wuppertal 1, FRG

Summary

The VOC oxidation initiated by OH reactions in 1 atm of air has been analysed by measuring the OH concentration-time profiles with and without NO present in a pulsed laser experiment. It has been shown that important reaction parameters of the radical chain can be determined in comparison to current chemical models. More generally it is discussed, under what experimental conditions the study of the chain reactions during the VOC oxidation propagated by OH + VOC, RO_2 + NO, RO + O_2, and HO_2 + NO can improve the present knowledge in a qualitative and quantitative manner.

1. Introduction

Reactions of hydroxy radicals with volatile organic constituents of the troposphere are part of complex chain reactions controlled by NO_x and atmospheric oxygen. Thus it is important not only to study the primary OH reactions under careful separation from other reactions (e.g. those of the products), but also to investigate the influence of NO_x and of oxygen on the chain propagation and termination. Many studies on tropospheric chain reactions were carried out by relative rate measurements and product analyses. Recently it has been shown that lasers are important tools to perform absolute rate constant measurements of OH reactions under tropospheric conditions, i.e. in 1 atm of air /Schmidt et al., 1984; Wahner and Zetzsch, 1985/. In these studies OH radicals are produced by laser flash photolysis (LP) with UV lasers and the OH concentration is followed by a second laser either by laser-induced fluorescence (LIF) /Schmidt et al., 1984/ or long-path laser absorption (LPA) /Wahner and Zetzsch, 1985/.

For the OH reactions of acetylene and other alkynes it was found by LP/LIF measurements that regeneration of OH already occurs if oxygen is present /Schmidt et al., 1984, 1985b/. It could be shown, that the main reaction path proceeds via reaction of the acetylene-OH adduct with oxygen, with one product being OH. This chain system consists of two radicals in the main reaction path and is a rather unusual example of a tropospheric OH chain, because only the alkyne and oxygen reactions propagate the chain. In most other cases of oxidation of volatile trace constituents HO_2 is involved as an intermediate and NO takes part in a chain propagating reaction with HO_2 and/or other peroxy radicals, as is the case for hydrocarbons such as alkenes and alkanes. Simple examples of tropospheric chain reactions with NO involved are the reactions of OH + SO_2 and OH + H_2, where three radicals constitute the chain (OH, OH-SO_2 adduct, HO_2 and OH, H, HO_2).

With respect to the NO oxidation, the problem of the kinetics and of the product formation of OH chain reactions with organic compounds is closely related to the tropospheric ozone formation. It is of great importance to know the kinetics and yields of RO_2 and HO_2 in such systems which in the troposphere consist of a broad spectrum of different reagents including the secondary products of the hydrocarbon oxidation. The aim of this study is to demonstrate that the LP/LIF method may be used to simulate the VOC oxidation chain. This should result in a more general description of those parameters of hydrocarbon oxidation systems, which are most important for atmospheric modeling.

Under pseudo-first-order conditions the observed OH decay curves are exponential and the rate constant can be derived in a rather simple manner. If the OH radicals in a chain reaction are monitored, the decay curves are at least bi-exponential or consist of a superposition of even more different time-dependent exponential terms /Matsen and

Franklin, 1950/. The analysis of such decay curves composed of several exponential terms can provide not only informations about the starting OH reaction but also on the subsequent radical transformations without direct observation of the other radicals in the chain /Schmidt et al., 1984/.

Results will be reported on measurements of the OH chain reactions during the oxidation of ethene and 1-butene. Already in the case of the simplest alkene, ethene, six radicals take part in the chain, which makes it impossible to describe the full system in closed form. A reduction to a 4-radical system for lower alkenes and different ways to analyse the reaction system will be discussed.

2. Experimental

The experimental set-up has already been described previously /Schmidt et al., 1984/. The experiments were improved by using a reaction cell which can be cooled to -50°C, a digital data aquisition system and a higher time resolution of better than 0.5 µs.

A gas mixture of hydrogen peroxyde, buffer gas (air or inert gases) and reactants is slowly passed through the reaction cell. OH radicals (about $10^{11}/cm^3$) are produced by excimer laser photolysis of H_2O_2. The OH concentration is monitored as a function of the reaction time by LIF using a frequency-doubled dye laser tuned to a line of the rotationally resolved (1,0) band of the OH ($A^2\Sigma^+ \leftarrow X^2\Pi$) system at 282 nm. The fluorescence is filtered by a small double monochromator at 308 nm ((0,0) band) and detected by an UV photomultiplier. The pulsed signal is sampled by a boxcar integrator and the digitized data are transferred to a microcomputer system.

All reactions were studied in an excess of the reactants resulting in pseudo-first-order kinetics for the decay of OH. By properly choosing the reactant concentrations, OH decay times of a few µs up to 20 ms could be attained. The decays of the OH concentrations were measured over 2 to 3 orders of magnitude by shifting the time delay between the excimer laser and the dye laser pulses. The concentrations of the different gases were calculated from the total pressure in the reaction cell and from the total and the individual gas flow rates measured by calibrated flow meters. All measurements were performed at room temperature (295 $\pm$2K).

3. Results

Measurements of OH reactions with the three lowest alkenes, ethene, propene, and 1-butene were performed in air or inert buffer gases (N_2, Ar) and both, with and without the presence of NO. Without NO pure exponential decay curves were obtained indicating that no reproduction of OH occurs (Fig. 1, curve a). Rate constant measurements of OH with ethene and propene at different total pressures and with different buffer gases /Schmidt, 1985b/ are in good agreement with the literature values /Atkinson, 1985/.

If NO is added to the reactant gas mixture, strong deviations from the exponential curve shape are observed for C_2H_4 as well as for i-C_4H_8 (Fig. 1b, c and Fig 2). At rather low NO partial pressures two parts of the curves can be distinguished. The first part is exponential in the time range from 10 to 100 μs depending on the alkene concentration and OH rate constant. The second part towards longer reaction times is comparativly flat and rises in amplitude when the NO partial pressure is increased. A more or less pronounced maximum of the OH concentration as a function of time is observed in the second part, which disappeares with increasing NO partial pressure.

4. Discussion

The observed non-exponential OH decay curves were expected from experiences with other OH chain reactions. But the shapes of the curves are different compared to the earlier results obtained with the same experimental set-up. In the reaction systems of OH + C_2H_2, OH + H_2, OH + SO_2, and OH + CS_2 the second part towards longer reaction times can be ascribed to a second, slow exponential decay /Schmidt et al., 1984, 1985a/. The decay curves were fully described by a sum of two time-dependent exponential terms. With the alkene oxidation system, especially the intermediate time regime is complicated due to a local concentration maximum, although the curves seem to go over into an exponential decay at long reaction times.

Fig. 3 shows the reaction scheme for OH + C_2H_4 taken from the literature /Atkinson and Lloyd, 1984; Atkinson, 1985/. Radical-radical and radical-product reactions are not included, because all reactions were performed under pseudo-first-order conditions. The OH is sequentially transformed to ß-hydroxy ethyl, ß-hydroxy ethyl peroxy, ß-hydroxy ethoxy and HO_2, partly to CH_2OH, and then back to OH. The chain propagating steps are reactions with ethene, oxygen, and NO or thermal decomposition. Chain termination is mainly expected by the reaction of OH + NO. For ethene all chain propagating reactions (1) - (7) are reasonably well known except reaction (3), which has an estimated rate constant of $\sim 7 \times 10^{-12}$ cm^3/s /Atkinson and Lloyd, 1984/.

Since six radicals are propagating the chain, a full description of this system requires the solution of a coupled system of six linear first-order differential equations with constant coefficients, therefore, an explicit description of the time dependence of the radical concentrations cannot be given. The system can be reduced to four equations, if it is taken into account that under tropospheric conditions the reaction of the ethyl radical with O_2 is several orders of magnitude faster than the other reactions, and if the parallel pathways of the decomposition of the hydroxy ethoxy radical and its reaction with O_2 are combined to only one reaction step described by the sum of the rate constants. This is possible, because the subsequent reaction of CH_2OH with oxygen is much faster than the steps (4) and (5).

The solution of the resulting differential equation system is given for example by Walter /1986/ and is discussed with respect to pseudo-first-order kinetics by Matsen and Franklin /1950/. The solution of the differential equation system results in an "Eigenwert" problem. Under the assumption that all eigenvalues are non-degenerate, the general solution for OH is:

$$(OH) = \sum_{i=1}^{4} A_i \exp(\lambda_i t) . \qquad (1)$$

The closed forms of the expressions for the coefficients A_i and λ_i are still very complex and will not be given here. The λ_i can be complex (with a real and an imaginary quantity), resulting in oscillating terms in equation (1). However, analysis of the solution for chain reactions of three radicals shows that these oscillations are very slow compared to the damping term (real part of the complex λ_i parameters) and cannot be observed under usual experimental conditions. These oscillations cannot be the reason for the local maxima in the experimentally observed OH concentration-time profiles.

In order to describe the experimental decay curves, computer simulations of the decay curves of fig. 4 according to the model of fig. 3 were performed. If literature rate constants (tab. I) are used for the simulations, reasonable agreement is only achieved for the first part of the decay curves. Better agreement is obtained, when the regeneration rate of OH is made faster, which leads to a lifting of the whole second part of the curves. With this alteration the experimentally observed OH concentration-time profile can be satisfactorily simulated, however, only with low NO concentrations in the range of 0.010 mbar. At higher NO concentrations in the order of 0.070 mbar an additional loss process has to be introduced in to the simulation treatment which brings the second part of the curves into a steeper form closer to the experimental data. The rate determining reactions in the model shown in fig. 3 are the two NO reactions. The reaction rate constant of HO_2 with NO is well known, but a faster OH regeneration rate would occur with a higher value of the rate constant of step (3), which is not well known up to now. Additional radical losses out of the chain (termination reactions) can be localized anywhere in the chain. At low NO concentrations a value of k_3 = $1.7 \times 10^{-11} cm^3/s$ yields the best approximation to the experimental data, without any additional loss process. At higher NO concentrations a loss rate of 12,000 s^{-1} has to be added for better approximation as shown in fig. 5. The known loss processes such as NO + OH have already been included in the model.

The nature of the additional loss reaction could be the addition reaction of the hydroxy peroxy radical with NO, for which up to now no unambiguous evidence has been presented /Atkinson, 1985/. A rate constant of about 5×10^{-12} cm^3/s is estimated from the above first order rate constant and the given NO concentration. However, the results obtained at low NO concentrations where no additional loss rates are required imply that the formation of nitrates is negligible in the case of ethene. The yield of radical transformation very likely is near unity for the hydroxy peroxy radical reaction with NO. If the loss reaction is attributed to the $HOCH_2CH_2O$ radical, it has to be taken into account that its concentration is about 5 times lower than that of $HOCH_2CH_2OO$ under the applied experimental conditions and, therefore, the loss rate would be 5 times higher (60,000 s^{-1}), requiring a bimolecular rate constant of about $3 \times 10^{-11} cm^3/s$. For this reaction no rate constant is known in the literature. It can be compared with that of the reaction CH_3O + NO for which values of $2.08 \times 10^{-11} cm^3/s$ /Sanders et al., 1980/ and $2.5 \times 10^{-11} cm^3/s$ /Lorenz et al., 1984/ have been estimated.

5. Conclusions

It has been demonstrated that tropospheric chain reactions can be studied in a quantitative way by time-resolved measurements of the OH radical concentration in LP/LIF experiments. Due to the fact that only one radical could be observed, deviations from current models can be observed in general, but the cause of such deviations cannot easily be attributed to a particular reaction step. However, variations of the O_2 and NO concentrations could already give hints what particular step would be most likely responsible for the deviations.

Direct observation of at least two radicals of the chain, OH and preferentially RO_2 radicals, would improve the quantitative analysis of the reaction systems. With such improvements including the measurement of the NO_2 formation rate it should be possible to determine most reaction parameters needed for modeling the VOC oxidation process also in the case of higher hydrocarbons.

In the case of ethene the analysis of the OH dynamics at different NO concentrations give strong indications that the rate constant of the reaction $HOCH_2CH_2OO$ + NO is significantly higher than previously assumed. A value of $1.7 \times 10^{-11} cm^3/S$ is derived from the best approximation between model calculations and experimental results. In addition, this reaction should form the related alkoxy radical which forms HO_2 by its O_2 reaction with a yield near unity. A loss term would have been recognized in the model calculation also at low NO concentrations.

The loss term at higher NO concentrations can be attributed to the reaction $HOCH_2CH_2O$ + NO which competes with the reaction $HOCH_2CH_2O + O_2$. The model predicts a rate constant of about $3 \times 10^{-11} cm^3/S$ similar to the value estimated for the reaction CH_3O+NO.

The treatment of the propene system has not yet been finished.

6. References

R. Atkinson und A. Lloyd, J. Phys. Ref. Data, 13, 315 (1984)

R. Atkinson, Chem. Rev. 85, 69 (1985)

D.L. Baulch, R.A. Cox, P.J. Crutzen, R.F. Hampson Jr., J.A. Kerr, J.Troe, and R.T Watson, J. Phys. Chem. Ref. Data, 13, 1259 (1984)

K. Lorenz, D. Rhäsa, and R. Zellner, Proceedings of the Third Europ. Symp. "Physico-Chemical Behaviour of Atmospheric Pollutants", pp. 223, B. Versino und G. Angeletti (edit.), Reidel Publ. Company, Dordrecht 1984

J.J. Margitan, J. Phys. Chem. 88, 3314 (1984)

F.A. Matsen and J.L. Franklin, J. Am. Chem. Soc. 72, 3337 (1950)

N. Sanders, J.E. Butler, L.R. Pasternack and J.R. McDonald, Chem. Phys. 48, 203 (1980)

V. Schmidt, G.Y. Zhu, K.H. Becker and E.H. Fink, Proceedings of the Third Europ. Symp. "Physico-Chemical Behaviour of Atmospheric Pollutants", pp. 177, B. Versino und G. Angeletti (edit.), Reidel Publ. Company, Dordrecht 1984

V. Schmidt, G.Y. Zhu, K.H. Becker and E.H. Fink, Ber. Bunsenges. Phys. Chem. 89, 321 (1985a)

V. Schmidt, Thesis, Wuppertal, (1985b)

W. Walter, Gewöhnliche Differentialgleichungen, Springer-Verlag, Berlin 1986

A. Wahner and C. Zetzsch, Ber. Bunsenges. Phys. Chem. 89, 323 (1985)

Reaction in 1 bar of air at 295 K	Rate constant (cm^3/s)	Partial pressure of reactant (mbar)	First order rate constant ($10^3 s^{-1}$)
1) $OH + C_2H_4$	8.1×10^{-12}	0.80	160
2) $OHC_2H_4 + O_2$	"fast"	208	"fast"
3) $HOCH_2CH_2OO + NO$	7×10^{-12}	0.089	15
4) $HOCH_2CH_2O + O_2$	7×10^{-15}	208	37
5) $HOCH_2CH_2O$ decomp. to products		-	140
6) $CH_2OH + O_2$	2×10^{-12}	208	10000
7) $HO_2 + NO$	8.3×10^{-12}	0.089	18
8) $OH + NO$	5.0×10^{-12}	0.089	11

Table I: Bimolecular and pseudo-first-order rate constants of the chain propagating steps (1) - (7) and chain terminating reaction (8) of the ethene oxidation /Atkinson and Lloyd, 1984; Atkinson, 1985/. The rates are calculated with the reactant concentrations of the measurement of fig. 1b.

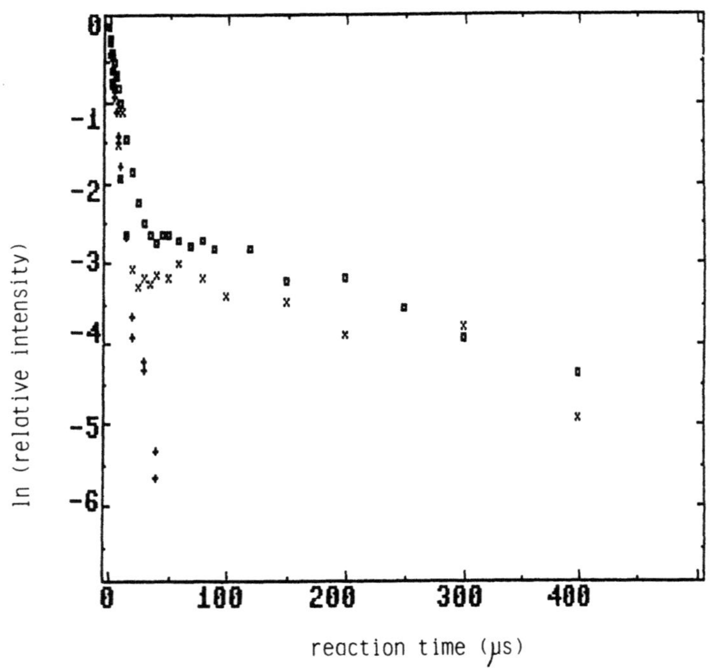

Fig. 1: OH concentration vs. reaction time at a total pressure of 1 bar of synthetic air with partial pressures of 1a) (+): 0.80 mbar of C_2H_4 and no NO; 1b) (x): 0.80 mbar C_2H_4 and 0.089 mbar of NO; 1c) ( ): 0.40 mbar of C_2H_4 and 0.089 mbar of NO.

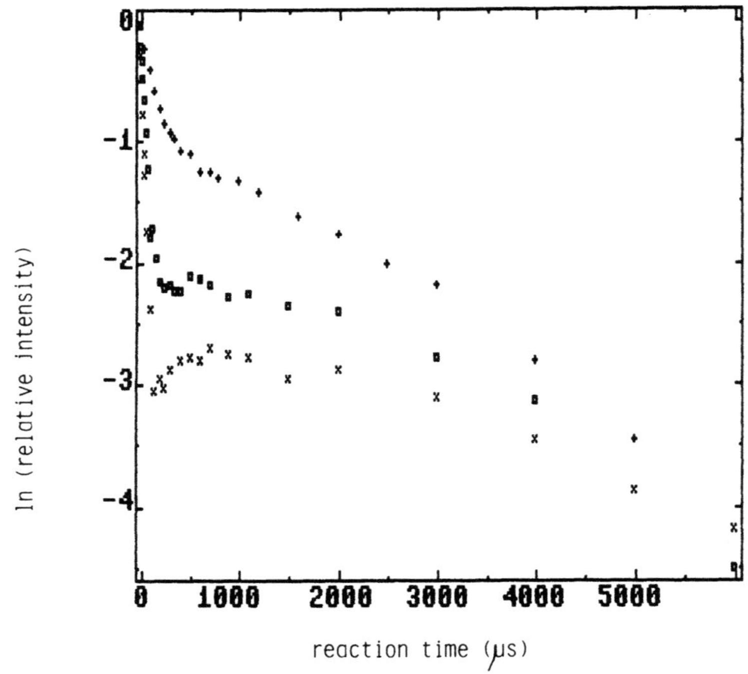

Fig. 2: OH concentration vs. reaction time at a total pressure of 0.13 bar of synthetic air with partial pressures of 0.01 mbar of NO and a) (x): 0.04 mbar of 1-C_4H_8, b) ( ): 0.02 mbar 1-C_4H_8 and, c) (+): 0.004 mbar of 1-C_4H_8.

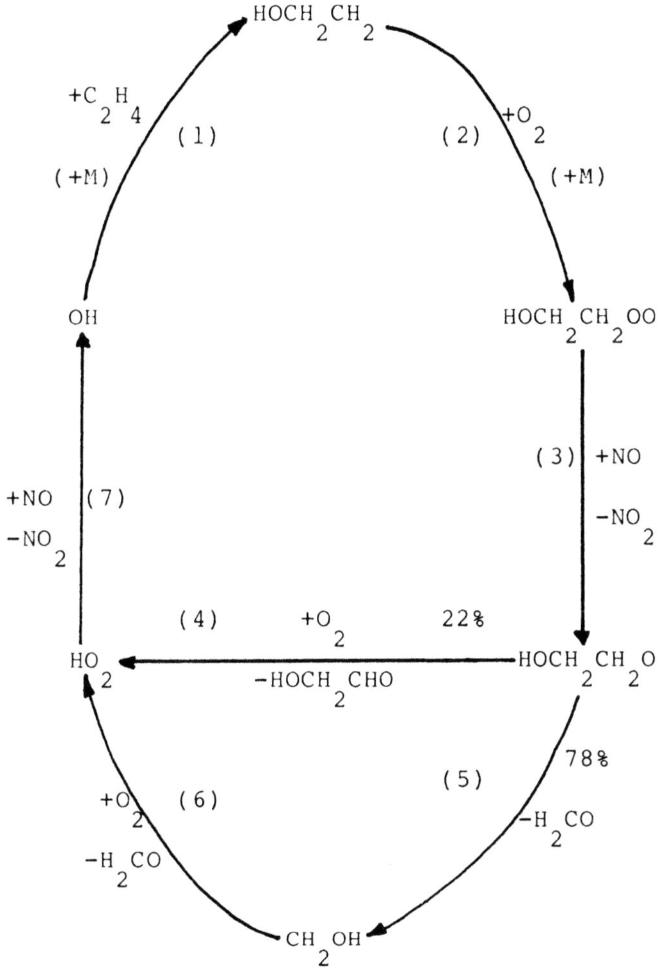

Fig. 3: Reaction scheme for the system OH + C_2H_4 in the presence of oxygen and NO according to Atkinson and Lloyds /1984/ and Atkinson /1985/.

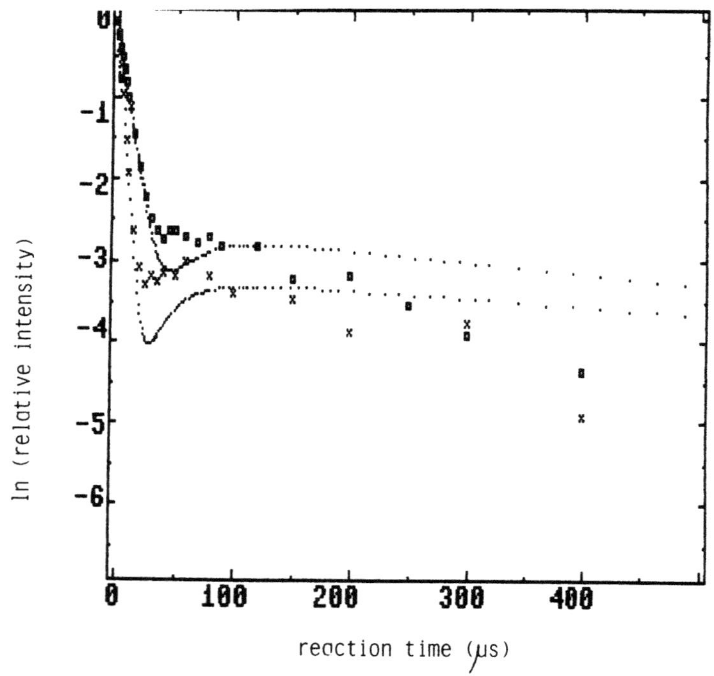

Fig. 4: OH concentration vs. reaction time for the system OH + C_2H_4; comparison of the experimental results (curve 1b) (x) and 1c) ( )) with computer simulations (dotted lines) according to the model of fig. 3 and the rate constants of tab. I.

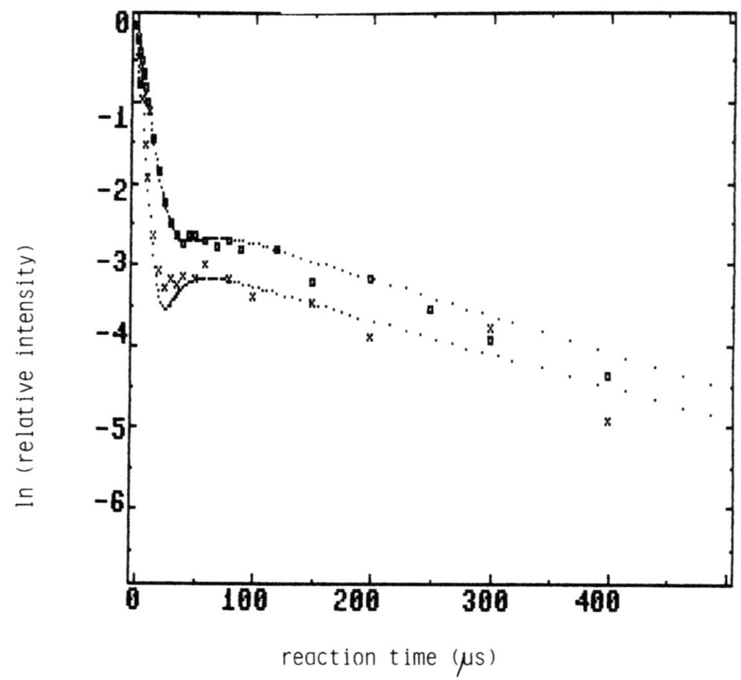

Fig. 5: OH concentration vs. reaction time for the reaction of OH + C_2H_4; comparison of the experimental results (curve 1b) (x) and 1c) ( )) with computer simulations (dotted lines) according to the model of fig. 3 with k_3 = 1.7 x 10^{-11} cm^3/s and an assumed additional loss of CH_2CH_2OO of 12,000 s^{-1}; the values for the other constants (tab. I) are taken from the literature.

AQUEOUS PHASE CHEMISTRY OF BIOGENIC SULPHUR COMPOUNDS

Peter Brimblecombe and David Shooter
School of Environmental Sciences
University of East Anglia
Norwich NR4 7TJ UK

Summary

Biological reduced sulphur compounds (CH_3SCH_3, H_2S, COS and CS_2) are rather insoluble and their most significant aqueous phase chemistry is likely to occur in seawater. Dimethyl-sulphide (DMS) appears to undergo photosensitized oxidation in sunlit seawater (2.4×10^{-5} s^{-1}) and thermal oxidation that is some three orders of magnitude slower. The gas phase oxidation products of DMS are more soluble in water and are likely to have an important aerosol chemistry.

In remote areas biological activity may represent the principal source of sulphur derived acidity in rainfall. Volcanic emissions are also likely to be significant sources of sulphur in the atmosphere. Volcanoes are often erratic and widely, although not randomly, distributed point sources of atmospheric materials. Of the 20 ± 10 Tg(S) a^{-1} emitted by volcanoes two thirds is likely to be in the form of reduced sulphur gases such as H_2S and SO_2 (Moller, 1984). The global reduced sulphur emission from biological sources has been estimated as 70 ± 10 Tg(S) a^{-1} of which 60% is as dimethyl-sulphide(DMS), 30% as H_2S and the remaining 10% as CS_2 and COS (Moller 1984). The emission fluxes of the various biological sulphur compounds are under constant revision (for instance Bates and Gammon, 1985; Ferek et al 1986) because of the limited spatial extent of measurements and the problem of seasonal variation (Turner and Liss, 1985). While most attention is paid to these sulphides other compounds such as dimethyl-disulphide have been detected in the ocean and compounds such methane thiol are likely to be present in the atmosphere as they are far from uncommon products of biological activity (Jocelyn, 1972).

1. SOLUBILITY

The compounds above, which represent the principal biological sources of reduced sulphur compounds, all have relatively low solubilities in water. The Henry's law constants for these compounds are listed, where they are known, in Table 1. It is clear that a rather low solubility is required if there is to be a large flux of these trace compounds into the atmosphere. This is obviously true where the oceans are postulated as a source, but also important when considering a soil source except where they are extremely dry, but in those cases biological activity is probably small. In contrast to this behaviour the rather high solubility of a compound such as dimethyl-sulphoxide (DMSO) means that we expect its flux from the oceans to the atmosphere to be rather small even though it is likely to be present in seawater at concentrations much in excess of that of DMS (Andreae,

1980). Harvey and Lang (1986) argue that the low atmospheric concentrations they find for DMSO (2-4 ng m^{-3}) suggest that there is little transfer from the sea surface into the atmosphere.

Table 1. Some properties of biogenic reduced sulphur compounds in the atmosphere.

	DMS	COS	CS_2	H_2S
Henry's law constant mol l^{-1} atm^{-1}	0.56	0.023		0.014
Atmospheric concentration ppt	30	570	61	75
Flux to atmosphere Tg(S) a^{-1}	15-30	1-1.7	3.5-14	19-38
Flux removed in rain Tg(S) a^{-1}	0.0003	0.0002		
First order rate const. to remove 10% in rain s^{-1}	0.62	0.045		

Notes: Henry's law constants at 15°C from Dacey et al (1984) and Rasmussen et al (1982). Fluxes and concentrations are all from Moller (1984) although there are more recent values his are a self consistent set.

The fact that low solubility compounds make up the most significant part of the flux of reduced sulphur compounds from the oceans means that chemistry in atmospheric droplets is not going to be a dominant feature of their removal processes. An exception to this would be cases where the rate of reaction in solution is very fast. The global wet removal can be gauged by assuming the total annual precipitation (0.5×10^{18} kg) is in equilibrium with the compound in the atmosphere. As seen in Table 1 this contributes only insignificantly to the removal rate. We can also estimate the importance of reactions taking place in droplets, but need to account for the fact that water spends relatively little of the time in the atmosphere as a liquid. Assuming that the total annual precipitation amount remains as liquid water for about five hours the contribution of solution phase reactions to the global budget can also be calculated. Table 1 gives the first order rate constant required if reactions in droplets are to contribute 10% to the flux of the gas from the atmosphere. It can be seen that unless the rate constant for the reactions are very fast the solution phase reactions cannot contribute significantly to the overall removal process.

A number of solution phase reactions of the biological reduced sulphur compounds are well know. Carbonyl sulphide, for instance, hydrolyses in water:

$$H_2O + COS \longrightarrow H_2S + CO_2$$

at a rate of $5.4 \times 10^{-6}\ s^{-1}$, too slow for the uncatalyzed hydrolysis in droplets to be important for the annual flux. It is clear that radicals react with some small sulphur molecules very rapidly. For instance

$$OH + CH_3SH \longrightarrow RS + H_2O$$

has a rate constant $3.1 \times 10^8\ l\ mol^{-1}s^{-1}$ at pH 7 (Packer, 1974). However as the concentration of the OH radical in atmospheric droplets is probably less than 10^{-12} mol l^{-1} the contribution of such radical reactions to the total flux would be small. A remaining possibility is that photochemical processes in the droplet phase could be important. For example in the gas phase an experimental photo-oxidation rate of $4.5\ 10^{-5}\ s^{-1}$ has been determined for CS_2 (Wine et al, 1981). However the rapid photo-oxidation required to make significant contribution to solution photochemistry would probably need to be photosensitized in order to be make the droplet chemistry fast enough to be important.

One of the reasons that aqueous chemistry in droplets appears so ineffective arises because atmospheric water exists as vapour most of the time. If the reactions in seawater were to be considered then the reaction rate can be integrated over the whole year, so considerably slower reaction rates could influence the flux. This being so we have started to study reduced sulphur reactions both in seawater and atmospheric droplets.

2. PHOTO-OXIDATION OF AQUEOUS DIMETHYL-SULPHIDE

The aqueous photo-oxygenation of DMS, COS and CS_2 are worthy of investigation because of the influence of these reactions on the geochemical fluxes; most particularly that in seawater. The flux of DMS has been extensively studied so our investigation began with this compound, although carbonyl sulphide is also of great interest because while some have argued that the oceans can act as a source for COS (Ferek and Andreae, 1984) there have been suggestions that the oceans are a sink (Johnson, 1981).

Photo-oxidation of aqueous DMS at near-visible wavelengths was investigated by analysing its concentrations during irradiation of previously air equilibrated solutions over time. Where the DMS concentrations were high (4.0×10^{-4} mol l^{-1}) the loss of oxygen reacting with the DMS could be monitored using an oxygen electrode. The chromatographic method adopted for determining DMS was sensitive enough for experiments to be run using the DMS naturally present in seawater (full details of the irradiation experiments may be found in Brimblecombe and Shooter (1986)).

Irradiation of DMS solutions with a sodium lamp showed no significant loss of the sulphide however oxidation was rapid in the presence of photosensitizers such as methylene blue (at 5.3×10^{-6} mol l^{-1}). With the photosensitizer the reaction proved to be first order in DMS concentration in both distilled water and seawater. In the high concentration experiments where oxygen loss could be measured it appeared that two molecules of DMS were oxidized by every molecule of bimolecular oxygen. This was consistent with, although didn't prove that DMS photo-oxidation yeilds dimethylsulphoxide the usual product of dialkylsulphide oxidations in solution. Rose bengal and anthroquinone also proved to be photosensitizers for DMS oxidation.

A further series of experiments were undertaken in seawater where the only photosensitizers were those naturally present. Both light from a sodium lamp and sunlight brought about photo-oxygenation of DMS. It seems that in both samples of coastal seawater and that collected well out in the North Sea there is sufficient natural photosensitizer to photo-oxygenate DMS in sunlight. The first order rate constant for this reaction in summer sunlight near 50° N latitude appears to be about $2.4 \times 10^{-5}\ s^{-1}$. This rate is rapid enough to cause significant oxidation of DMS in surface waters, but even if reasonable quantities of this photosensitizer were transferred to atmospheric droplets photo-oxidation would not be an important pathway for the removal of DMS from the atmosphere. The rate of photosensitized oxidation of DMS in the oceans is such that it implies a residence time of less than a day. If the photo-oxidation takes place over a significant part of the surface water column containing DMS it could represent a loss process that is larger than transfer to the atmosphere.

Although we have assumed DMSO to be the product of these oxidation reactions, when DMS is irradiated with ultra-violet light from a mercury lamp (a few minutes 10cm from a 1000W Hanovia Type 6751 mercury arc) we find a range of products which include not only DMSO, but also dimethyl-sulphone (DMSO2), MSA, formate and sulphate. Work is currently underway to see if photosensitized irradiation in sunlight can yeild a similarly wide range of products.

3. THERMAL OXIDATION OF AQUEOUS DIMETHYL-SULPHIDE

Dimethyl-sulphide is generally regarded as stable to oxidation by molecular oxygen at ambient temperatures. Recent work has shown that aryl sulphides oxidize reasonably rapidly at high temperatures ($100^{o}C$) and oxygen partial pressures (Correa and Riley, 1985). In addition to this it is possible that the presence of metal ions might well catalyze oxidation by oxygen. The high stability of DMS to oxidation by molecular oxygen in aqueous solution required that most of the experiments were conducted at high temperature to reduce reaction times to manageable periods.

The DMS ($\sim 3 \times 10^{-7}$ mol l^{-1}) for the kinetic runs was made up in saline solution (0.7 mol dm^{-3} sodium chloride BDH AR or Sepramar grade adjusted to the chosen pH with sodium hydroxide or hydrochloric acid) or filtered (0.22 μm Millipore) coastal seawater. Aliquots (6 cm^{3}) of the DMS solutions were sealed in glass ampoules with a small head space ($\sim 0.5\ cm^{3}$) to prevent loss of DMS to the gas phase. These were left in the dark in ovens at a range of temperatures. The graph of the logarithm of DMS concentration against time curved upwards. The reason for the curvature in first order plots may be inhibition by reaction products frequently shown by organosulphides under autoxidation. Apparent first order rate constants were determined from the initial slope, as it was there that the inhibition would be smallest.

The reaction is sensitive to pH and the apparent first order rate constant at $80^{o}C$ is shown as a function of pH in Fig. 1. Kinetic studies of the oxidation of DMS in saline solution at pH 7.8 were carried out over a range of temperatures between 20 and 120 ^{o}C. The

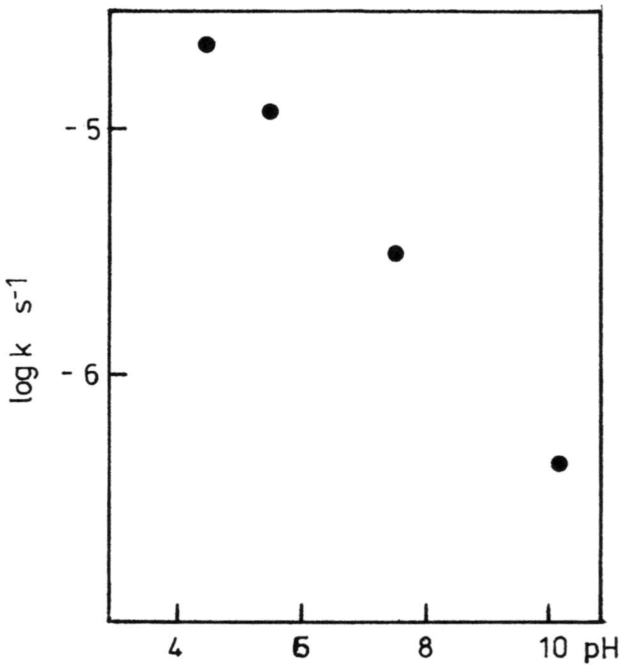

Fig.1 First order rate constant for the thermal oxidation of DMS as a function of pH.

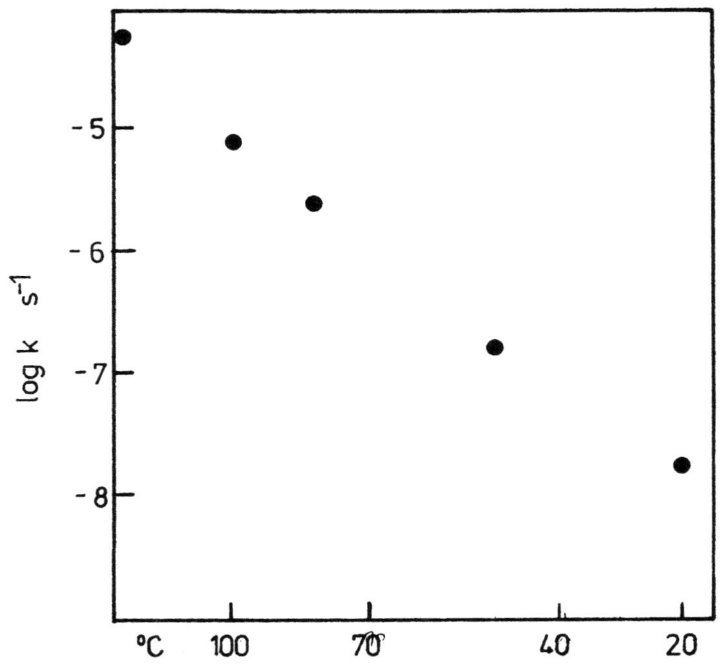

Fig.2 First order rate constant for the thermal oxidation of DMS as a function of temperature.

Arrhenius plot from this work is shown in Fig. 2. It should be emphasised that this work is based on first order initial rates and that the point at 20 ^{o}C was obtained with some difficulty, because of the long reaction times involved. The reaction was followed for six months which represents only a fraction of a half life. The activation energy extracted from the data is 80 kJ mol^{-1} K^{-1} which can be compared with 88.6 kJ mol^{-1} K^{-1} for the oxidation of the benzyl methyl sulphide. In addition to this preliminary experiments showed that the presence of copper(II) (10^{-4} mol l^{-1}) gave a ten-fold increase in rate compared with pure sodium chloride solutions at 80 ^{o}C and pH 5.6. Reactions in the sodium chloride solution show rates that are nearly an order of magnitude slower than those in seawater. While the thermal reaction proves to be extremely slow with a half life of a number of years at ambient temperatures, this is not a long time in terms of the lifetime of deep ocean waters. DMS found in old waters would seem to require a recent source.

4. REACTIONS OF DIMETHYL-SULPHIDE OXIDATION PRODUCTS

Among the products of gas phase photo-oxidation of DMS are a number of compounds which are much more soluble than the reduced precursors. For instance possible oxidation products of DMS (Grosjean and Lewis, 1982; Grosjean, 1984)

Production of MSA:

$$CH_3SCH_3 + OH \longrightarrow CH_3SOH + CH_3$$

$$CH_3SOH + O_2 \longrightarrow CH_3SO_3H$$

Production of DMSO:

$$CH_3SCH_3 + OH \longrightarrow CH_3SOHCH_3$$

$$CH_3SOHCH_3 + O_2 \longrightarrow CH_3SOCH_3 + HO_2$$

The Henry's law constants for typical oxidation products of DMS are listed in Table 2. These compounds are soluble enough to partition into the aerosol phase in the atmosphere. There are a few measurements of these compounds in rainwater (Andreae, 1980; Harvey and Lang, 1986; Saltzman et al, 1983). Assuming that the isolated measurements can be taken as indicative of precipitation concentrations as a whole we can estimate global wet deposition. As we can see they are very much larger than those predicted for the less soluble gases and it is possible that these could represent much more significant fractions of the flux of reduced sulphur compounds.

Table 2. Some products of the oxidation of biogenic sulphur compounds in the atmosphere.

	DMSO	DMSO2	MSA
Henry's law constant mol l^{-1} atm^{-1}	$\sim 10^5$	large	6.5×10^{13}*
Concentration in rain mol l^{-1}	1.5×10^{-8}	6×10^{-8}	3×10^{-7}
Flux removed in rain Tg(S) a^{-1}	0.2	1	5

* units for the MSA constant are $mol^2 kg^{-2} atm^{-1}$ from Clegg and Brimblecombe (1985). The value for DMSO has been estimated on the basis of vapour pressure and that Raoult's law is obeyed.

There are few studies of likely reactions of these compounds in atmospheric aerosol droplets. In particular it would be interesting to establish whether these compounds are further oxidized through to products such as the sulphate anion. Saltzman et al (1983) showed that methane sulphonic acid reacted with OH radicals in solution.

Photochemical experiments similar to the ones described above using photosensitizers and DMS at high concentration were run with DMSO. There proved to be no detectable loss of oxygen even in the presence of high concentrations of methylene blue (5×10^{-6} mol l^{-1}). This suggested that DMSO resists photosensitized oxidation. However Harvey and Lang (1986) have recently suggested that DMSO may undergo disproportionation in solution:

$$2CH_3SOCH_3 \longrightarrow CH_3SCH_3 + CH_3SO_2CH_3$$

They point out that such disproportionation is likely to be promoted by light, metal ions and oxygen. Although it is important to note that stoichiometrically the reaction requires two DMSO molecules and might not be rapid at low DMSO concentrations. Nevertheless the possibility of recycling DMS by this process needs to be investigated. The dimethyl-sulphone (DMSO2) produced in such a reaction is likely to be more resistant to further oxidation than DMSO.

The ultimate oxidation products of the sulphur in the compounds under discussion should be sulphuric acid. It may be that oxidation never proceeds this far, but methane sulphonic acid is likely intermediates. In marine aerosols strong acids are able to displace less soluble acids from solution. In terms of mass the principal species degassed will be hydrogen chloride, but the greater insolubility of hydrogen fluoride means that it will be lost from marine aerosols more readily than HCl. Traces of acids should lead to the marine aerosol being heavily depleted in fluorine.

5. CONCLUSIONS

The reactions investigated so far have all been relatively slow and more relevant to seawater than aerosol droplet chemistry. Further investigations will be concerned with: (i) the photochemical oxidation and disproportionation of DMSO (ii) photosensitization of the hydrolysis and oxidation of COS and (iii) the existence of photosensitization in natural aerosols.

REFERENCES

Andreae, M.O., 1980. Dimethylsulfoxide in marine and freshwaters. Limnol. Oceanogr., 25: 1054-63.

Andreae, M.O. and Barnard, W.R., 1984. The marine chemistry of dimethylsulfide., Mar. Chem., 14: 267-79.

Andreae, M. O. and Raemdonck, H., 1983. Dimethylsulfide in the surface ocean and marine atmosphere: a global view. Science., 221: 744-7.

Barnard, W.R., Andreae, M.O. and Watkins, W.E., 1982. The flux of dimethylsulfide from then oceans to the atmosphere. J. Geophys. Res., 87: 8787-93.

Bates, T.S. and Gammon, R.H., 1985. Oceanic dimethylsulphide and the global atmospheric sulphur cycle, Trans. Am. Geophys. Union, 66, 1309 (abstract).

Brimblecombe, P. and Shooter, D. 1986. Photo-oxidation of dimethylsulphide in aqueous solution, Marine Chemistry (in press).

Clegg, S.L. and Brimblecombe, P. 1985. The solubility of methanesulphonic acid and its implications for atmospheric chemistry, Env. Tech. Letts., 6:269-78.

Cline, J.D. and Bates, T.S., 1983. Dimethyl sulphide in the equatorial Pacific Ocean: a natural source of sulfur to the atmosphere., Geophys. Res. Lett., 10:949-52.

Correa, P. E. and Riley, D. P., 1985. Highly selective direct oxidation of thioethers to sulfoxides using molecular oxygen. J. Org. Chem., 50: 1787-8.

Dacey, J.W.H., Wakeham, S.G. and Howes, B.L., 1984. Henry's law constants for dimethylsulfide in freshwater and seawater. Geophys. Res. Lett., 11:991-994.

Ferek, R.J. and Andreae, M.O. 1984. Photochemical production of carbonyl sulphide in marine surface waters, Nature, 307:148-50.

Ferek, R.J., Chatfield, R.B. and Andreae, M.O. 1986. Vertical distribution of dimethylsulphide in the marine atmosphere, Nature, 320:514-16.

Grosjean, D., 1984. Photooxidation of methyl sulphide, ethyl sulphide and methanethiol, Env. Sci. & Tech., 18:460-68.

Grosjean, D. and Lewis, R., 1982. Atmospheric photooxidation of methyl sulfide. Geophys. Res. Lett., 9: 1203-6.

Harvey, G.R. and Lang, R.F., 1986. Dimethylsulphoxide and dimethylsulphone in the marine atmosphere, Geophys. Res. Lett., 13, 49-51.

Hatakeyama, S. Okuda, M. and Akimoto, H. 1982. Formation of sulfur dioxide and methanesulfonic acid in the photooxidation of dimethyl sulfide in the air. Geophys. Res. Lett., 9:583-6.

Jocelyn, P.C., 1972. The Biochemistry of the SH Group, Academic Press, London.

Johnson, J.E. 1981. The lifetime of carbonyl sulfide in the troposphere, Geophys. Res. Lett., 8:938-40.

Möller, D., 1984. On the global natural sulfur emission. Atmos. Environ., 18: 29-39.

Packer, J.E., 1974. The radiation chemistry of thiols, in Patai, S., the Chemistry of the Thiol Group, Vol. II, J. Wiley & Sons, London.

Rasmussen, R.A., Hoyt, S.D. and Khalil, M.A.K. 1982. Atmospheric carbonyl sulphide (OCS): techniques for measurement in air and water, Chemosphere, 11:869-75.

Saltzman, E.S., Savoie, D.L., Zika, R.G. and Prospero, J.M. Methane sulphonic acid in the marine atmosphere, J. Geophys. Res., 88:10897-902.

Turner, S.M. and Liss, P.S., 1985. Measurements of various sulphur gases in a coastal marine environment., J. Atmos. Chem., 2:223-32.

Wine, P.H., Chameides, W.L. and Ravishankara, A.R. 1981. Potential role of CS_2 photooxidation in tropospheric sulphur chemistry, Geophys. Res. Lett., 8:543-46.

FIELD INVESTIGATIONS ON THE VARIABILITY OF H_2O_2 CONCENTRATION IN ATMOSPHERIC GAS PHASE

P. Jacob, A. Bambauer, T. M. Tavares* and D. Klockow

University of Dortmund, Department of Chemistry, D-4600 Dortmund 50, FRG

*Instituto de Química, Universidade Federal da Bahia, RB-40000 Salvador, BA, Brasil

Summary

Concentrations of H_2O_2 in atmospheric gas and liquid phase were determined consecutively in several periods and events, respectively, during the past two years. Gaseous samples were taken by a cryogenic technique, and rain by wet only collection. The samples were analyzed by means of the peroxyoxalate chemiluminescence method.

Following specific research goals, field measurements showed the following: a. Evident yearly and diurnal variations of gaseous H_2O_2 concentrations. b. Decrease of H_2O_2 gas phase concentrations during cloud formation and rain. c. Frequently higher H_2O_2 gas phase concentrations in the forest compared to outside. d. Evident negative correlations between atmospheric gaseous SO_2 and H_2O_2 in rain.

INTRODUCTION

The increasing interest in H_2O_2 in the atmosphere is caused by its important role in liquid phase oxidation of S(IV) (1, 2) and its relation to atmospheric radical chemistry (3). Furthermore recent investigations suggest that H_2O_2 itself might contribute to plant damages (4). Therefore extensive measurements of H_2O_2 in atmospheric gas and liquid phase were undertaken during a period of 2 years, aiming to find correlations with meteorological parameters and other gaseous compounds at different locations.

EXPERIMENTAL

Sampling sites were the Campus of the University of Dortmund, a city situated at the eastern border of the highly industrialized Ruhr area, and a site near Schmallenberg (450 m asl) in the Sauerland, a hilly forested region with little industrial activity, about 80 km in the southeast of Dortmund.

Gaseous H_2O_2 was trapped together with water vapor in a -45 °C cold glass tube, by pulling air through it with a rate of about 1 m^3 h^{-1} (5). After a sampling time of 30 to 60 min. in a height of 1.5 m above ground the sample was melted and H_2O_2 immediately determined by a chemiluminescent technique using bis(2,4,6-trichlorophenyl)oxalate and perylene as a sensitizer in a flow injection system (6). Rainwater was collected with a borosilicate glass funnel (19 cm diameter) in polypropylene vessels.

RESULTS AND DISCUSSION

a) Yearly and diurnal variations

Gas phase production of H_2O_2 requires UV-sunlight below 320 nm (7), which is significantly lower in intensity during winter compared to summer (8). Thus model calculations show a clear yearly variation for HO_2 as a precursor and H_2O_2 (7), with summer/winter ratios up to 100. The same behavior could be observed in precipitation as the main sink for H_2O_2 (6), and very pronounced in arctic snow (9).

Figure 1 shows the gaseous H_2O_2 level during a sunny period in November and December 1984 with concentrations up to 33 pptv, which are much lower compared to a clear sky summer situation in May, June and July 1985/86 (Fig. 2- 5), where concentrations up to 700 pptv could be found in the late afternoon. In long lasting periods without rain H_2O_2

accumulates in the atmosphere, as to be seen by increasing peak concentrations (Fig. 2), and decreases significantly during rain events.

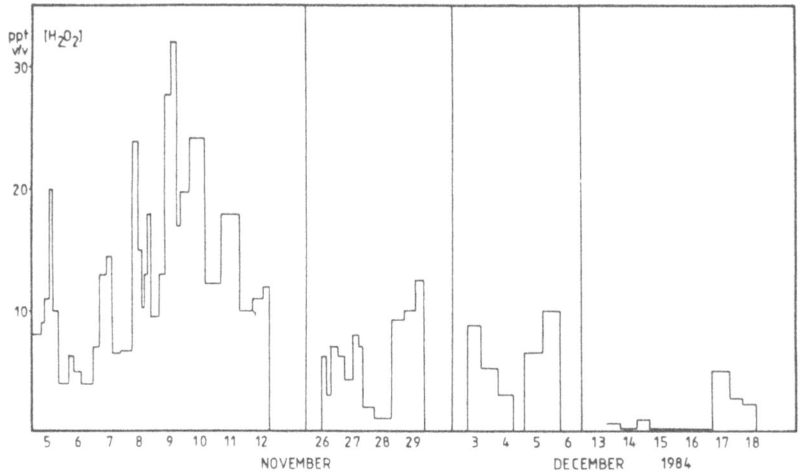

Fig. 1: H_2O_2 concentrations in ambient air; Campus of the University of Dortmund (1.5 m above ground), November/December 1984

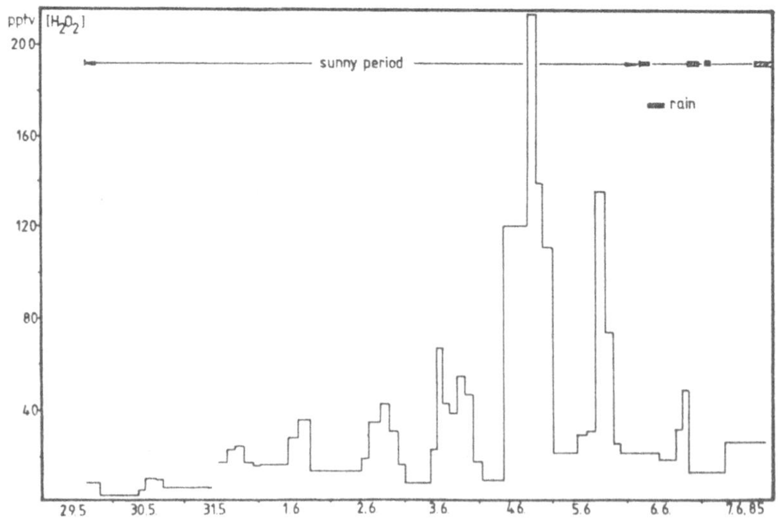

Fig. 2: H_2O_2 concentrations in ambient air; Campus of the University of Dortmund (1.5 m above ground), May/June 1985

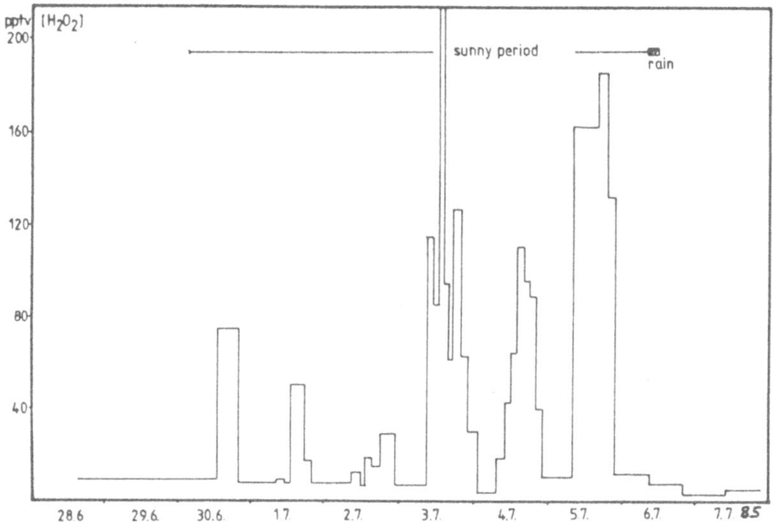

Fig. 3: H_2O_2 concentrations in ambient air; Campus of the University of Dortmund (1.5 m above ground), June/ July 1985

Late in the evening the H_2O_2 concentrations usually drop down very fast to concentrations close to the detection limit, most likely because of deposition with dew. In several occasions the dew was carefully micropipetted from plants and examined but, with one exception, H_2O_2 could not be detected, most probably as a a consequence of a fast depletion through catalytic decomposition by plant surfaces or atmospheric deposits on plants.

b. Influence of cloud formation and rain on gas phase H_2O_2 concentration

An evident diurnal variation, with high H_2O_2 maxima in the late afternoon, could only be observed on cloudless days. The appearence of clouds lowers the H_2O_2 concentration and leads to a less pronounced peak-shape (cf. Figs. 4 and 5). This is to be expected (10), because the solar flux is reduced by clouds and cloud droplets are an effective sink for H_2O_2. Falling rain decreases the H_2O_2 concentration in the gas phase below the cloud basis as can be taken from the results of consecutive measurements of gas phase H_2O_2 during two different 10 days periods (Figs. 2 and 3).

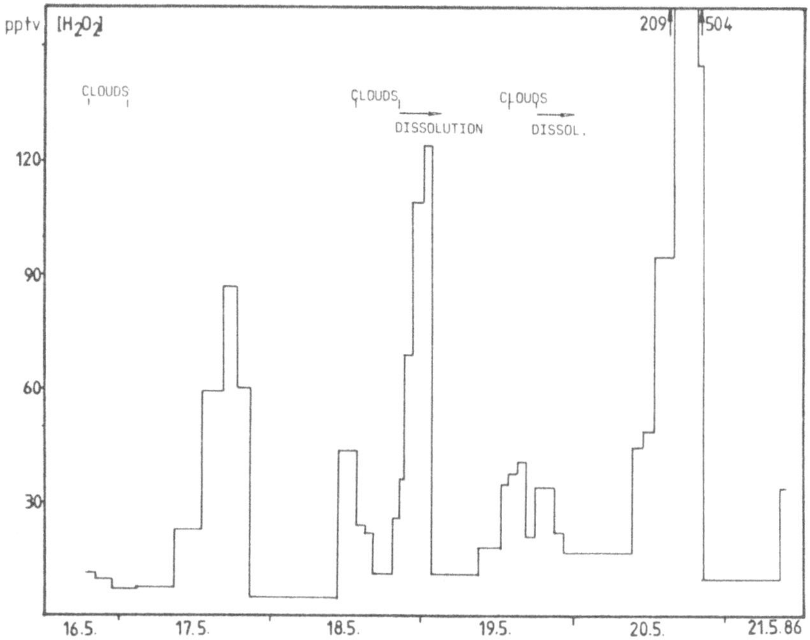

Fig. 4: H_2O_2 concentrations in ambient air; Campus of the University of Dortmund (1.5 m above ground), Mai 1986

The distribution of H_2O_2 between gas and liquid phase during rain events was investigated, and the data obtained already reported somewhere else (5). It was found that in the beginning of the precipitation event the H_2O_2 concentration in the liquid phase was much higher than expected from the measured gas phase mixing ratio and Henry's Law constant, but decreased rapidly, until the relation was the opposite, that is higher concentrations in the gas phase than gas/liquid equilibrium would predict. The latter oberservation could also be made during longer lasting precipitation events when no changing of air masses occured.

These results show that near the ground gas/liquid equilibration of H_2O_2 is not completed. The reason is that raining clouds are some hundred to some thousand meters above the sampling height of 1.5 m. This means that if the

H_2O_2 concentrations in or just below the clouds differ from those close to the ground - and recent measurements have shown this (11) - and if the phase transition is not fast enough, one cannot expect an equilibrium to be established. Another factor causing deviation from equilibrium is reaction of liquid phase H_2O_2 with reducing species already present in the cloud elements or scavenged by falling rain drops from the atmosphere below the cloud basis. Even generation of H_2O_2 in raindrops by autoxidation and photochemical processes is possible (12). In order to obtain comprehensive information about gas/liquid equilibria simultaneous measurements of H_2O_2 in gas and liquid phase within clouds are required.

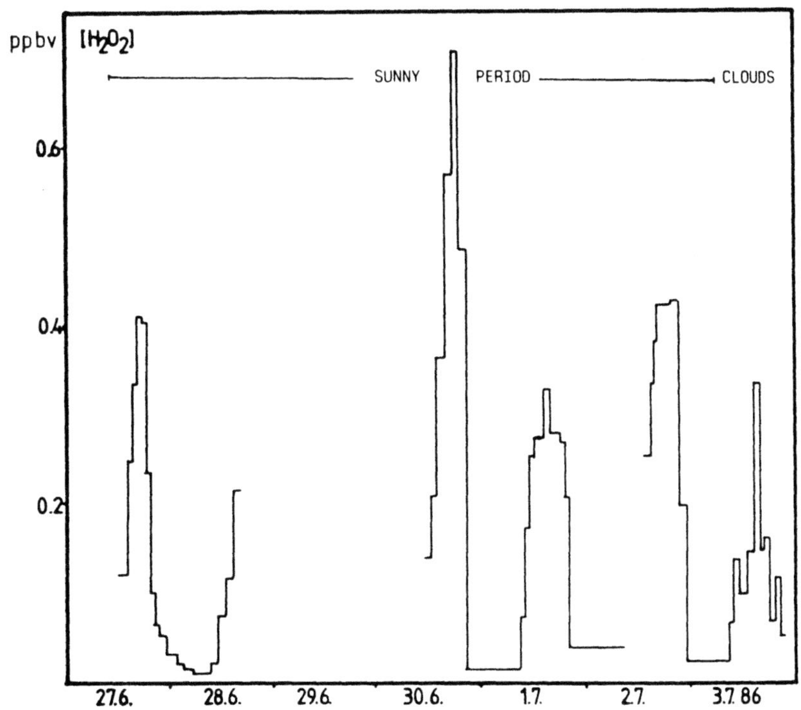

Fig. 5: H_2O_2 concentrations in ambient air; Campus of the University of Dortmund (1.5 m above ground), June/ July 1986

c. H_2O_2 measurements in a forest

A possible direct influence of H_2O_2 on plants has already been suggested (4). For comparison, gaseous H_2O_2 was sampled at two locations of the Sauerland site, about 40 m apart from each other, inside and outside a forest stand (conifers). As can be seen in Figs. 6 and 7 the concentrations in the forest are frequently considerably higher than outside, with similar diurnal variations. Most likely part of the H_2O_2 detected in the forest is a product of reactions between biogenic alkenes (e. g. terpenes) and ozone (13).

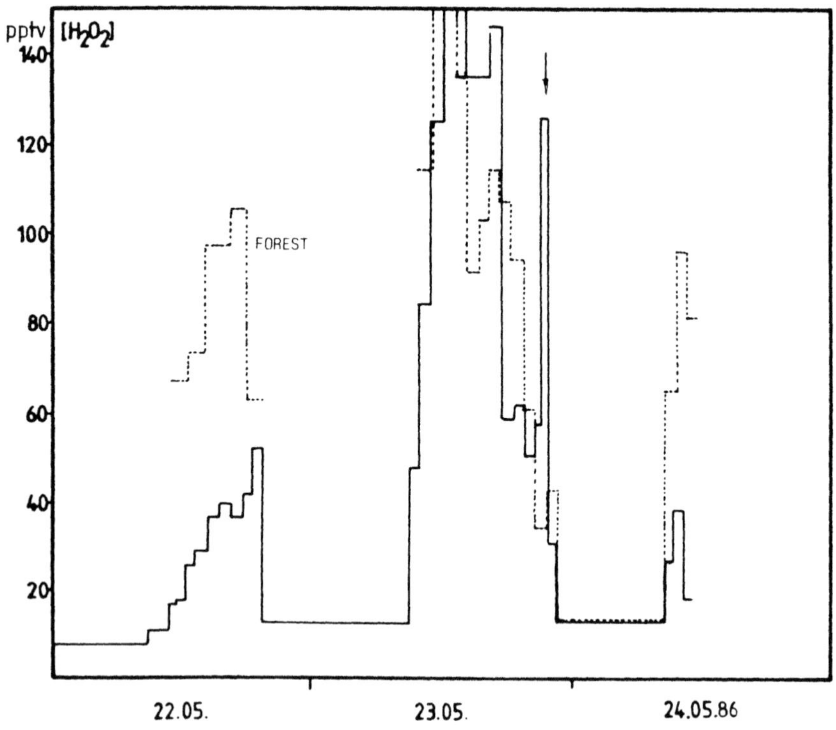

Fig. 6: H_2O_2 concentrations inside and outside a forest stand in the Sauerland (1.5 m above ground), May 1986

On May 18, 1986, in Dortmund (Fig. 4), and May 23, 1986, outside the forest in the Sauerland (Fig. 6), a sharp H_2O_2 peak could be observed in the late evening. This rapid increase in H_2O_2 gas phase concentration was associated with the event of cloud dissolution, and through the evaporation

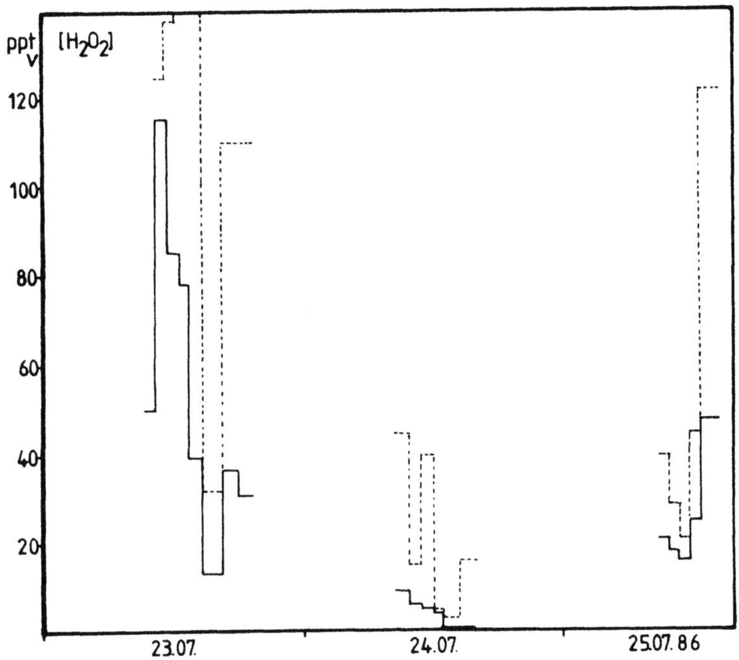

Fig. 7: H_2O_2 concentrations inside and outside a forest stand in the Sauerland (1.5 m above ground), July 1986

of cloud droplets H_2O_2 could have been released to the gas phase. The same peak could not be observed inside the forest, most probably because of the "filtering" property of the canopy. The rapid transport of the H_2O_2 to the ground from the altitude where the clouds were before evaporation could have been initiated through nocturnal vertical air movements operative at the sampling sites due to boundary layer disturbances (14).

d. Correlations between H_2O_2 and other reactive atmospheric trace gases

During a 3 days period, liquid and gas phase H_2O_2, as well as airborne SO_2, NO_2, NO and O_3 were continuously measured at the open field station (outside the forest) in the Sauerland.

Fig. 8 shows the H_2O_2 content in rain together with the gaseous SO_2 mixing ratio. It can be seen that there is a negative correlation of gaseous SO_2 and H_2O_2 in rain, which can be explained by the fast reaction of the two compounds in the liquid phase. No correlation was found between gas phase H_2O_2 and SO_2, as to be expected from the non-interaction of these two compounds in the gaseous phase.

Fig. 9 shows the H_2O_2 content in the gas phase together with gaseous NO_2 and O_3 concentrations at ground level before, during and after three rain events. The interpretation of these results is as follows:

The main source of gaseous H_2O_2 is the recombination of two HO_2 radicals ($k=3.1\times10^{-12}$ cm³ molecule^{-1}s^{-1} (15)). NO is a competitor in that it consumes HO_2 radicals according to:

$$NO + HO_2 \rightarrow NO_2 + OH \quad (1) \quad (k=8\times10^{-12}\ \text{cm}^3\ \text{molecule}^{-1}\text{s}^{-1}\ (15))$$

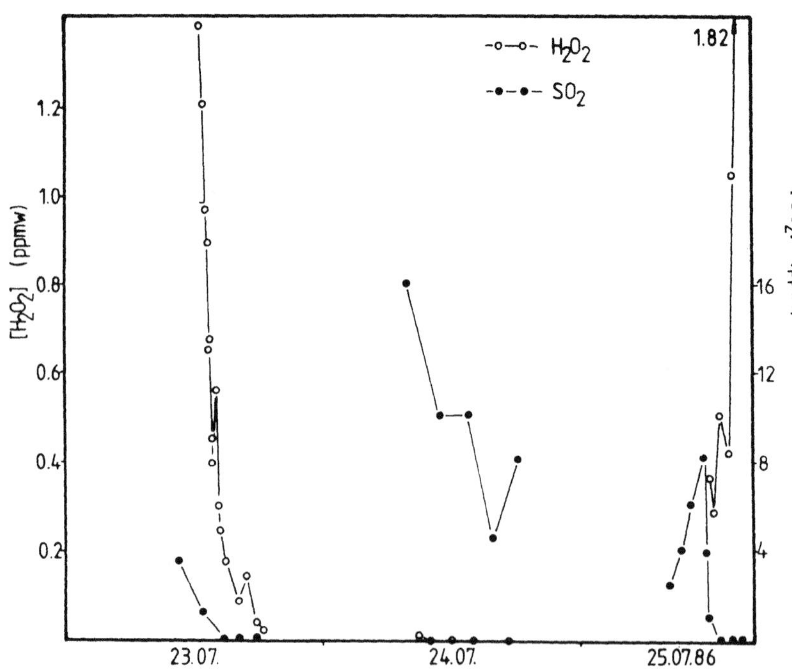

Fig. 8: H_2O_2 concentrations in rain (1.5 m above ground) and gaseous SO_2 concentration (8 m above ground); Sauerland station, open field site

Furthermore NO is the main reacting partner of O_3, also producing NO_2:

$$NO + O_3 \rightarrow NO_2 + O_2 \quad (2) \quad (k=1.8\times10^{-14}\ cm^3\ molecule^{-1}s^{-1}\ (10))$$

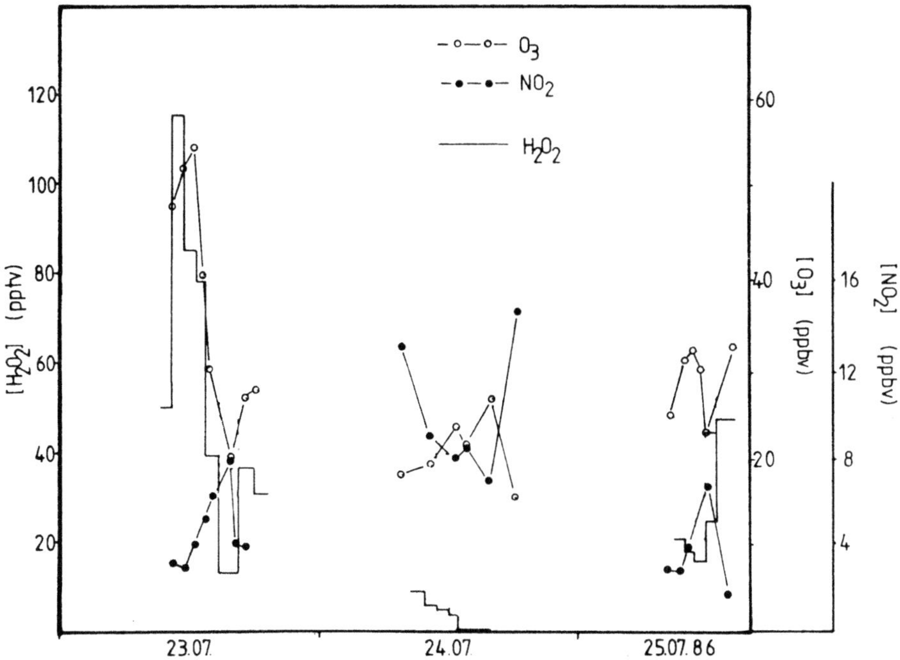

Fig. 9: Concentrations of H_2O_2 (1.5 m above ground), NO_2 and O_3 (8 m above ground) in the gas phase; Sauerland station, open field site

Accordingly NO levels were always found to be lower than 4 ppbv, whereas NO_2 as a product of reactions (1) and (2) was present in significantly higher mixing ratios and was negatively correlated to O_3. The behavior of gas phase H_2O_2 is influenced by two factors, namely the gas phase chemistry just mentioned and its uptake by rain and therefore more difficult to explain.

Before the rain event of July 23 started (Figs. 8 and 9), gaseous H_2O_2 was high in the presence of significant amounts of SO_2. As the frontal rain began, a wash out process followed by liquid phase reaction of dissolved S(IV)

with H_2O_2 took place, decreasing rapidly the concentrations of gas phase H_2O_2 and SO_2.

The precipitation event of July 24, a local convective storm and not a new front, took place in an atmospheric environment rich in SO_2 and consequently led to a total depletion of the already low gas phase H_2O_2.

The event of July 25, result of a new front, showed an increase in H_2O_2 concentration both in the liquid and gaseous phase. The course of the O_3 and NO_2 concentration indicates that some NO was present before it started to rain, such limiting H_2O_2 production. With the new front the situation changed leading to higher H_2O_2 concentrations in the gas phase (Fig. 9). In the liquid phase this increase became obvious only after removal of SO_2 through wash out and conversion in rain (Fig. 8).

CONCLUSIONS

From the given results of field experiments the following aspects should be pointed out:

- Inside a forest stand concentrations of gaseous H_2O_2 are frequently significantly higher than in the open field.
- Evaporating clouds may release H_2O_2 into the gas phase.
- Atmospheric H_2O_2 and SO_2 react very fast via the liquid phase (rain).

Further investigations should include the determination of terpenes and organic peroxides and of reducing species like sulfite and nitrite in the liquid phase.

ACKNOWLEDGEMENTS

This work was partly supported by the Minister of Science and Research of the State of Nordrhein-Westfalen and partly by the Federal Environmental Protection Agency, Berlin.

One of us (T. M. T.) gratefully acknowledges travel support by Deutscher Akademischer Austauschdienst.

REFERENCES

(1) PENKETT, S. A., JONES, B. M. R., BRICE, K. A., EGGLETON, A. E. J. (1979). The Importance of Atmospheric O_3 and H_2O_2 in Oxidizing SO_2 in Cloud and Rainwater. Atmos. Environ. 13, 123 - 137

(2) MARTIN, L. R., DAMSCHEN, D. E. (1982). Aqueous Oxidation of Sulfur Dioxide by Hydrogen Peroxide at Low pH. Atmos. Environ. 15, 1615 - 1621

(3) CHAMEIDES, W. L., DAVIS, D. D. (1982). The Free Radical Chemistry of Cloud Droplets and its Impact upon the Composition of Rain. J. Geophys. Res. 87, 4863 - 4877

(4) MASUCH, G., KETTRUP, A., MALLANT, R. K. A. M., SLANINA, J. (1986). Effects of H_2O_2 Containing Acid Fog on Young Trees; Intern. J. Environ.Anal. Chem. 27, 183 - 213

(5) JACOB, P., TAVARES, T. M., KLOCKOW, D. (1986). Methodology for the Determination of Gaseous Hydrogen Peroxide in Ambient Air. Fres. Z. Anal. Chem. 325, 359 - 364

(6) KLOCKOW, D., JACOB, P. (1986). The Peroxyoxalate Chemiluminescence and its Application to the Determination of Hydrogen Peroxide in Precipitation. In: "Chemistry of Multiphase Atmospheric System" (W. Jaeschke, Ed.) Springer Verlag, Berlin - Heidelberg - New York - Tokyo, p. 117 - 130

(7) LOGAN, J. A., PRATHER, M. J., WOFSY, S. C., McELROY, M. B. (1981). Tropospheric Chemistry: A Global Perspecive; J. Geophys. Res. 86, 7210 - 7254

(8) MÖLLER, F. (1973). Einführung in die Meteorologie, Bd. II; B.I. Wissenschaftsverlag, Mannheim - Wien - Zürich, p. 9 - 48

(9) JACOB, P., NEFTEL, A., KLOCKOW, D. (1986). Die Peroxyoxalat-Chemilumineszenz und ihre Anwendung zur Bestimmung von Wasserstoffperoxid in Niederschlägen, Eisbohrkernen und Außenluft. In: VDI-Berichte Nr. 608 "Meßtechnik in der Luftreinhaltung", VDI-Verlag, Düsseldorf, in press

(10) THOMPSON, A. M., CICERONE; R. J. (1982). Clouds and Wet Removal as Causes of Variability in Trace-Gas Composition of the Marine Troposphere; J. Geophys. Res. 87, 8811 - 8826

(11) CALVERT, J. G., KOK, G. L., HEIKES; B. G., WALEGA, J. G., LIND, J., CANTRELL, C. A. (1985). Chemical Mechanisms of Acid Generation in the Troposphere; Nature 317, 27 - 35

(12) KLOCKOW, D., JACOB, P., BECKER, K.-H. (1986). Zur Rolle des Wasserstoffperoxids in der atmosphärischen Flüssigphasenchemie. Statuskolloquium des Ministers für Umwelt, Raumordnung und Landwirtschaft NRW, Düsseldorf, 29./30.10.1986

(13) GÄB, S., HELLPOINTNER, E., TURNER, W. V., KORTE, F. (1985). Hydroxymethylhydroperoxide and Bis(Hydroxymethyl)Peroxide from Gas-Phase Ozonolysis of Naturally Occurring Alkenes; Nature 316, 535 - 536

(14) WINKLER, P. (1980). Störungen der nächtlichen Grenzschicht; Meteorol. Rdsch. 33, 90 - 94

(15) FISHMAN, J., SOLOMON, S., CRUTZEN, P. J. (1979). Observational and Theoretical Evidence of a Significant In-Situ Photochemical Source of Tropospheric Ozone; Tellus 31, 432 - 446

A STUDY OF THE HO_2 + CH_3CHO REACTION IN THE PHOTOLYSIS OF CH_3CHO, AND ITS CONSEQUENCES FOR ATMOSPHERIC CHEMISTRY.

G.K. MOORTGAT, J.P. BURROWS, W. SCHNEIDER, G.S. TYNDALL and R.A.COX+
Max-Planck-Inst. für Chemie, Air Chemistry Division, D-6500 Mainz, F.R.G.
+Combustion Center, Harwell Laboratory, Didcot, Oxon. U.K.

Abstract

A novel apparatus has been built to study the behaviour of reactants, products and intermediates (radicals) during and after photolysis. The apparatus consists of a "double multi-path" spectrometer, combining both ir- and uv-absorption spectrometry with additional capability of modulated photolysis for transient detection. Photo-oxidation of CH_3CHO gave CO, CO_2, H_2O, CH_2O, $HCOOH$, CH_3OH, CH_3OOH, CH_3COOH and CH_3COOOH as identifiable products by long path FTIR spectrometry. The effect of variation the O_2- and CH_3CHO-concentration upon the different photolysis products were studied. Absorption-time profiles at selected wavelengths in the 210-275 nm region were recorded in the modulated photolysis of CH_3CHO in 700 torr air. Transient absorptions were assigned to peroxy radicals and kinetic analysis of the data assisted by computer simulation resulted in a value $k_{10} = 4.8 \times 10^{-12}$ $cm^3.molecule^{-1}.s^{-1}$ for the reaction $HO_2 + CH_3O_2 \longrightarrow$ Products. This value is somewhat higher than the value $k_{10} = 3.5 \times 10^{-12}$ $cm^3.molecule^{-1}.s^{-1}$ obtained from the computer simulations of the photo-oxidation products measured by long path FTIR spectrometry. The formation of CH_3COOH in the gas phase oxidation of CH_3CHO is believed to occur via the addition of HO_2 to CH_3CHO. However on the basis of our experimental parameters, this reaction is not a major source of acetic acid in the atmosphere. A complete mechanism of the photo-oxidation is included.

1.0 INTRODUCTION

Acetaldehyde is of particular interest to atmospheric chemistry since it is formed as a product of the reaction of O_3 and/or OH radicals with natural non-methane hydrocarbons, especially higher olefins such as isoprene and terpenes (1,2). The photo-oxidation of hydrocarbons in photochemical smog produces acetaldehyde as a major intermediate product (3,4). Acetaldehyde is also directly emitted as product from incomplete combustion processes in diesel engines, aircraft exhausts, power plants and waste burning (5).

The removal of acetaldehyde in the atmosphere occurs primarily by two routes. One is the reaction with OH radicals, which abstract the aldehydic hydrogen, and leads to the formation of acetyl radicals. The acetyl radical then adds oxygen to form peroxyacetyl, and is the major precursor to peroxyacetylnitrate (PAN), an air pollutant having important physiological effects (6). The second route is the photolysis of CH_3CHO. Three different decomposition pathways are known after photoexcitation of the CH_3CHO molecule:

$$CH_3CHO + h\nu \longrightarrow CH_3 + HCO \qquad (1)$$
$$\longrightarrow CH_4 + CO \qquad (2)$$
$$\longrightarrow CH_3CO + H \qquad (3)$$

The quantum yields of the three photolysis channels have recently been determined by several groups (7-10). The determination of the intermediate and final products of the photo-oxidation of acetaldehyde and the elucidation of the mechanism of the photolysis in air remains a field to be investigated.

Using a modulated broad band photolysis apparatus, Cox et al. (11) performed a preliminary study of the photolysis of CH_3CHO in O_2. Although CH_3O_2 was detected and its rate of removal measured, any HO_2 formed in the system was lower than the detection limit. In an earlier photolysis study of CH_3CHO in air at 313 nm, Weaver et al. (12) observed an increase of the quantum yield of the products CO_2, CH_3COOOH and CH_3OH with increasing partial pressure of CH_3CHO. Recent investigations by Meyrahn et al. (9-10) also reveal CO_2 quantum yields larger than expected from pathway 3 alone, and increasing rapidly with CH_3CHO concentration, indicative that a chain reaction is taking place. CO_2 formation must be initiated by abstraction of the aldehydic hydrogen of CH_3CHO by a free radical, such as HO_2, CH_3O, CH_3O_2. etc.

In a recent long path FTIR spectrometric study by Moortgat and McQuigg (13) the products of the photo-oxidation of small concentrations of CH_3CHO in air (50 ppm) at 700 torr were determined: CO, CO_2, HCHO, CH_3OH, HCOOH, CH_3OOH and CH_3COOH. The formation of the products CO, HCHO, CH_3OH and CH_3OOH can be explained by the reactions 4 to 11 listed in Table 1. Recent studies of the photo-oxidation of CH_2O (14,15) have shown that the formation of HCOOH occurs via a chain reaction initiated by the reaction of CH_2O with HO_2 with a rate constant near 1×10^{-13} $cm^3.molecule^{-1}.s^{-1}$ (15-17). Moortgat and McQuigg (13) postulated that the formation of CH_3COOH occurs analogously by the reaction of HO_2 + CH_3CHO, but at a much slower rate. Barnes et al. (18) observed an overall slow effective decay of HO_2 in the presence of CH_3CHO $k_{eff} = 1 \times 10^{-17}$ $cm^3.molecule^{-1}.s^{-1}$.

In the present study we report the determination of the final products by long path FTIR spectroscopy as a function of various initial CH_3CHO and O_2 concentrations. The photolysis set-up was modified by incorporation of a long path uv-monitoring device as well as a modulated photolysis system, allowing the detection of peroxy radicals. In addition, computer simulation of the observed photo-oxidation products was performed to unravel the different elementary processes. The aim of the study was to establish the mechanism of acetaldehyde photo-oxidation, in particular to investigate if the HO_2

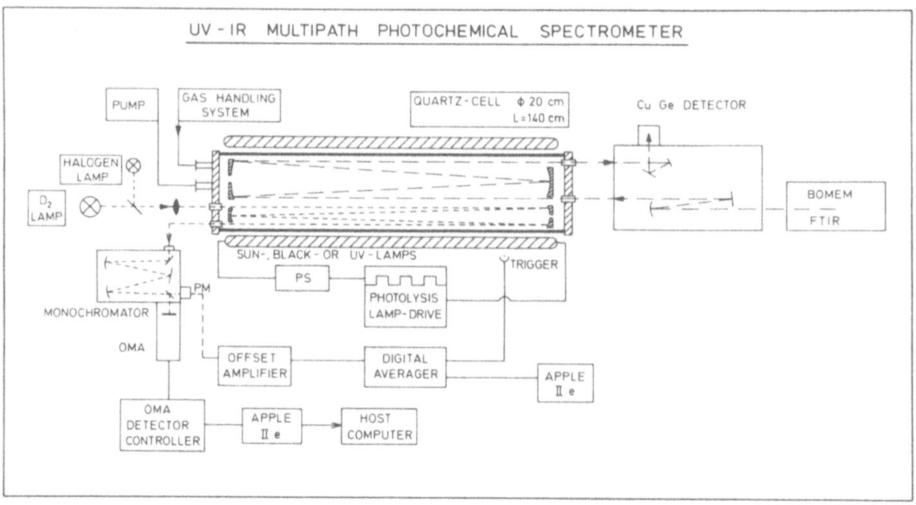

Fig 1. Schematic diagram of the "double multi-path" spectrometer.

radical, arising initially from the reactions 4 and 11, reacts with CH_3CHO leading to formation of CH_3COOH. CH_3CHO is likely to be a precursor of CH_3COOH which is a component of natural precipitation and contributes to its acidity (19).

2. EXPERIMENTAL.

A novel apparatus has been built at the MPI in Mainz to study the behaviour of reactants, products and intermediates (radicals) during and after the photolysis. The apparatus consists of a "double multi-path" spectrometer, combining both ir- and uv-absorption spectroscopy with the additional capability of modulated photolysis for transient detection. A schematic diagram is shown in Figure 1. The photolysis experiments were performed in a 46 liter (1.40 m length, 20 cm diameter) quartz cell, equipped with two independent sets of White-optic mirror arrangements. The cell is surrounded with 16 photochemical lamps (type Sunlamp, TL40W/12 Philips, spectral distribution 275-390 nm with maximum emission at 305 nm), 6 of which can be operated in square wave modulated fashion.

The uv source is a 30 Watt D_2 lamp and after multipassing the reaction volume, the monitoring beam is focussed on the entrance slit of a 0.25 m Spex monochromator, and detected on a PM-tube or an optical multichannel analyser (OMA), EG&G Model 1461, using a model 1421 Silicon Diode Array detector. Absorption-time profiles obtained during the modulated photolysis are fed into a signal averager (EG&G Model 4203) and processed by an Apple IIe microcomputer. For the ir measurements a Bomem DA03.01 Fourier transform infrared (FTIR) spectrometer was employed. A Ge/KBr beamsplitter and closed cycle liquid helium cooled Cu-doped Germanium detector were used to obtain spectra from 400-4000 cm^{-1} with a resolution of 1 cm^{-1}.

A conventional greaseless gas handling system was employed. The gas mixtures were all made up using high purity air, N_2 or O_2 from l'Air Liquide. All experiments were performed at a total pressure of 700 torr and 25 ^0C. Reference calibration curves for CO, CO_2, CH_3OH, HCOOH, CH_2O and CH_3COOH were obtained from dilute (10-30 ppm) mixtures of the pure compounds in 700 torr synthetic air at 25°C. The chemicals CH_3CHO, CH_3OH, HCOOH and CH_3COOH (all Merck) were purified by vacuum distillation. CH_2O was obtained by heating paraformaldehyde under vacuum, and collecting the monomer at -78 C. CH_3COOOH was prepared by reacting 30 % H_2O_2 with acetic anhydride and contained some acetic acid as impurity (20). The pure CH_3OOH sample was prepared by Dr. Blom (University of Giessen). Reference spectra for CH_3OOH and CH_3COOOH were obtained using authentic samples, and compared with published data (21,22). The use of higher concentration CH_3CHO caused part of the infrared spectrum to be completely saturated, and made the determination of certain expected product like CH_2O and H_2O_2 totally impossible.

3 Results and Discussion

3.1 End-product analysis using FTIR spectroscopy.

Photo-oxidation of CH_3CHO-air mixtures gave CO, CO_2, H_2O, CH_2O, HCOOH, CH_3OH, CH_3OOH, CH_3COOH and CH_3COOOH as identifiable products. The effect of variation the O_2-and the CH_3CHO-concentration upon the different photolysis products was studied. In Figure 2, the ratio R_X/R_{CO} for the different products CH_3OH, CO_2, HCOOH, CH_3OOH, CH_3COOH and CH_3COOOH are displayed against the initial CH_3CHO (left side) or O_2 (right side) concentration. The effects of the variation of CH_3CHO and O_2 upon the product yields suggest that there are three types of products behaviour: Type A, CH_3OOH, whose rate of formation is independent of both CH_3CHO and O_2; Type B, Ch_3COOH and HCOOH, whose rates of formation are CH_3CHO-dependent but O_2-independent; and Type C, CO_2

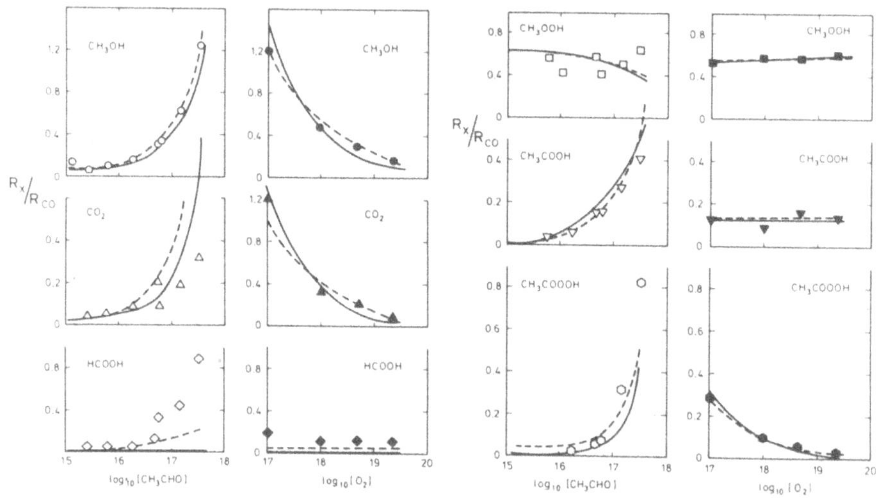

Fig.2 Effect of the variation of [CH_3CHO] and [O_2] upon the rate of formation of the products in the photolysis of CH_3CHO. The curves represent the computer simulated for two models (A ——, B ---) as described in the text.

CH_3OH and CH_3COOOH, whose rates of formation are both CH_3CHO and O_2 dependent. Such a behaviour suggests three types of mechanistic implications.

3.2 Mechanism of Photo-oxidation of CH_3CHO.

The major pathway in the photodissociation of CH_3CHO in the wavelength region used in the experiments is pathway 1. The primary radical products, CH_3 and HCO, react rapidly with O_2 to form CH_3O_2 and HO_2, radicals via reactions 4 and 5. Channels 3 and 4 account for only ~3% and 4.4 % respectively of the photolysis products (7-10). The mechanism of product formation in the CH_3CHO photo-oxidation therefore involves reactions of CH_3O_2 and HO_2, and the radical and molecular species derived from these primary peroxy radicals.

A mechanism which can account for the observed behaviour is presented schematically in Figure 3. It comprises three separate components, which interact with each other through certain coupling elementary processes. Section A, which involves mutual and cross reactions of the peroxy radicals dominates product formation at low CH_3CHO concentrations. Section B involves reactions following the addition of HO_2 to CH_3CHO. The principle characteristic product is acetic acid and the importance of the mechanism increases with CH_3CHO concentration. Section C centres round the CH_3CO_3 radical formed by H-abstraction from CH_3CHO followed by reaction 28. The characteristic products are CO_2 and peracetic acid, CH_3COOOH, both of which are formed in linear chain reactions, which increase in importance at high CH_3CHO concentration.

A) Low acetaldehyde, Mechanism A.

The dominant product in the low CH_3CHO regime is CH_3OOH, formed in reaction 10:

$$HO_2 + CH_3O_2 \longrightarrow CH_3OOH + O_2 \qquad 10$$

This reaction competes with the self reactions 6 to 9 of CH_3O_2 and HO_2. The CH_3O formed in reaction 7 reacts predominantly with O_2 at low CH_3CHO to yield formaldehyde and HO_2 via reaction 11. Under these conditions none of the above products are expected to show a quantum yield dependence on CH_3CHO or O_2. The relative yields of CH_3OOH, CH_3OH, CH_2O, CH_3COOCH_3 and H_2O_2 provi-

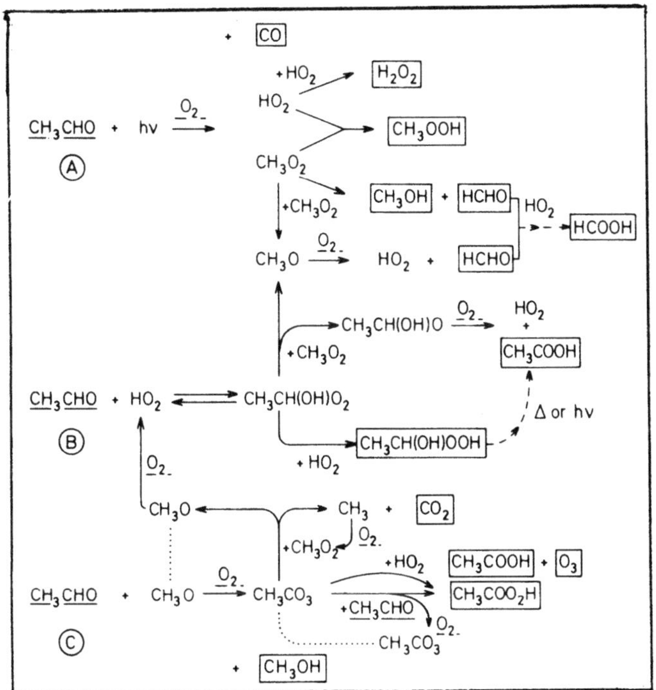

Fig.3 Schematic diagram of the mechanism of the acetaldehyde photo-oxidation.

de information on the relative rates of the above processes. Unfortunately, only the the yields CH_3OOH and CH_3OH could be measured under our experimental conditions.

B. Acetic Acid formation, Mechanism B.

The proposed mechanism for formation of CH_3COOH is initiated by the addition reaction 20 of HO_2 with CH_3CHO to form a complex $CH_3CH(O)OOH$, which rearranges by intermolecular H-transfer to $CH_3CH(OH)O_2$ (In Table 1, this radicalis written as $HAcHO_2$). This formulation is based on the analogous mechanism for the HO_2 + CH_2O reaction for which there is a substantial amount of evidence from kinetics and product analysis studies (14-17). The expected fate of $CH_3CH(OH)O_2$ is reaction with other peroxy radicals, particularly CH_3O_2 and HO_2 which are produced in the primary photolysis

$$CH_3CH(OH)O_2 + CH_3O_2 \longrightarrow CH_3CH(OH)O + CH_3O + O_2 \qquad 23$$

$$CH_3CH(OH)O_2 + HO_2 \longrightarrow CH_3CH(OH)OOH + O_2 \qquad 22$$

The oxy radical $CH_3CH(OH)O$ (or HAcHO, see Table 1) formed in 23 reacts readily with O_2 to form acetic acid:

$$CH_3CH(OH)O + O_2 \longrightarrow CH_3COOH + HO_2 \qquad 24$$

The peroxide $CH_3CH(OH)OOH$ (or $HAcHO_2H$, see Table 1) is expected to be photolysed or decompose heterogeneously on the wall to form a secondary source of CH_3COOH: $CH_3CH(OH)OOH$ + wall, hv $\longrightarrow$ $CH_3COOH + H_2O$ 25

No experimental information is available for kinetic parameters of any of the elementary steps proposed in mechanism B. Although the overall rate coefficients for reactions 22 to 24 may be estimated from analogy with other peroxy radical reactions the relative importance of the different product channels cannot be estimated a priori. It is assumed that CH_3COOH is produced only by reaction 23 and that, if the equilibrium between HO_2 and $CH_3CH(OH)O_2$ in reaction 20 is maintained, the rate of formation of CH_3COOH, R_{AcOH}, is given by $R_{AcOH} = k_{23} \cdot K_{20}^{*} [CH_3CHO][HO_2][CH_3O_2]$. (i)

The radical concentration product $[HO_2][CH_3O_2]$ is to a first approximation independent of $[O_2]$ and lineraly dependent on $[CH_3CHO]$ as indicated from the effect of these variables on the yield of CH_3OOH produced in reaction 10.

Thus equation (i) predicts the observed effect of $[O_2]$ and $[CH_3CHO]$ on the yield of CH_3COOH (see Figure 2). The relative yields of CH_3COOH and CH_3OOH at low $[CH_3CHO]$ give $k_{23}.K_{20}^*/k_{10} \approx 5 \times 10^{-18}$ $cm^3.molecule^{-1}$.

C. Formation of CO_2 and CH_3COOOH, Mechanism C.

The proposed mechanism for CO_2 formation, which was originally proposed by Weaver and Heicklen (12) is initiated by the abstraction of H from CH_3CHO by methoxy radicals:

$$CH_3O + CH_3CHO \longrightarrow CH_3OH + CH_3CO \qquad 27$$

The CH_3CO radical rapidly forms CH_3CO_3 which reacts rapidly with CH_3O_2:

$$CH_3CO_3 + CH_3O_2 \longrightarrow CH_3 + CO_2 + CH_3O + O_2 \qquad 32$$

Since CH_3 reacts with O_2 to form CH_3O_2 again, the reaction is effectively decomposition of CH_3CO_3 into CH_3O and CO_2, catalysed by CH_3O_2. The chain reaction is terminated by the reaction of CH_3O with O_2 and therefore the yields of CO_2 and the enhanced CH_3OH yield at high $[CH_3CHO]$ are expected to be strongly dependent on O_2, as observed (see Figure 2). The acetylperoxy radicals may also abtract H from CH_3CHO to form peracetic acid, reaction 31

$$CH_3CO_3 + CH_3CHO \longrightarrow CH_3COOOH + CH_3CO \qquad 31$$

This chain process for CH_3COOOH formation only becomes important at high $[CH_3CHO]$ and also shows an O_2 dependence, arising from the dependence of the $[CH_3CO_3]$ on the relative rates of CH_3O reaction with CH_3CHO and O_2. Termination of the reaction chains involving CH_3CO_3 can also occur via reaction with HO_2 :

$$CH_3CO_3 + HO_2 \longrightarrow CH_3COOOH + O_2 \qquad 33$$

$$CH_3CO_3 + HO_2 \longrightarrow CH_3COOH + O_3 \qquad 34$$

The relative importance of these two channels has been recently established by product analysis studies (23). The additional source of CH_3COOH from reaction 34 would be expected to increase with decreasing O_2, as a result of the $[O_2]$ dependence of $[CH_3CO_3]$. The absence of any significant O_2 dependence of CH_3COOH (Figure 2) suggests that reaction 34 is not an important source of CH_3COOH, at least at acetaldehyde concentrations of 5×10^{16} molecule cm^{-3}. The other major product detected in this work was formic acid, which is believed to arise exclusively as a secondary product of the reaction of HO_2 with formaldehyde, following a similar mechanism to that formulated above for acetic acid formation.

4.0 Computer simulations

The products obtained from the photo-oxidation of acetaldehyde using various initial O_2 and CH_3CHO concentrations, as shown in Figure 2, were computer-simulated using the FACSIMILE computer programme available at Harwell. Table 1 shows a list of 50 elementary reactions which describe the proposed mechanism in Figure 2 of CH_3CHO in air.

Appropriate rate constants for most reactions were selected from recent compilated tables (24-26). Recent studies of the photo-oxidation of CH_2O have shown that the formation of HCOOH occurs via a chain reaction of CH_2O with HO_2, as indicated in the reaction sequence 15 to 19. Several values for k_{15} and the associated reverse reaction rate constant k_{-15} have been measured (15-17). In a recent preliminary flash photolysis study (27) of CH_2O the equilibrium constant for the reaction $HO_2 + CH_2O \rightleftharpoons OOCH_2OH$ was estimated to be near 4×10^{-17} $cm^3.molecule^{-1}$. Using Veyret's value (16) $k_{15} = 7.5 \times 10^{-14}$ the rate constant $k_{-15} = 750$ s^{-1} was used in the computer simulations. The rate constants associated with mechanism B, for the formation of CH_3COOH are estimated from analogeous HCOOH reaction mechanism in the photo-oxidation of formaldehyde. Assuming an equilibrium constant $K_{20} = 1 \times 10^{-17}$ $cm^3.molecule^{-1}$, the value $k_{20} = 1 \times 10^{-15}$ $cm^3.molec^{-1}.s^{-1}$ and $k_{-20} = 100$ s^{-1} were used. Since the mechanism includes the reactions between the various peroxy radicals, initially the selected values for the reaction rate constants were assumed from analogous radical-radical reactions.

The initial photoconversion rates for CH_3CHO were obtained experimentally from the yield of CO formation. The photolysis rate for the other photolabile compounds, such as CH_2O, H_2O_2 and CH_3OOH were calculated, relative to the CH_3CHO photolysis rate, from the known absorption cross-sections, quantum yields and the spectral distribution of the light source. The photolysis rate for $CH_3CH(OH)OOH$ was estimated from the photodecomposition rate of the analogous species HO_2CH_2OH (15).

Two approaches were used in the simulation of the products. First a selected set of rate constants were used on a trial basis in the simulation program and concentration time profiles calculated for the different experimental photolysis mixtures. It became evident that the simultaneous simulations of the main photolysis products such as CH_3OH, CH_3OOH, CO_2, CH_3COOH and CH_3COOOH were mainly dependent on the adjustment of the reaction rate constants of a few reactions, namely reaction 10, 23, 27 and 32. The second approach was then to use the fitting routine of the FACSIMILE programme whereby the reaction rate constants for the four cited reactions were automatically adjusted to obtain the best fit through the experimental data.

Computer fitting resulted in the following rate constants at 300 K: $k_{10} = (3.5 \pm 1.0) \times 10^{-12}$, $k_{23} = (9.0 \pm 4.0) \times 10^{-13}$, $k_{32} = (1.5 \pm 0.5) \times 10^{-11}$ and $\bar{k}_{27} = (1.5 \pm 1.0) \times 10^{-13}$ all in units of $cm^3.molec^{-1}.s^{-1}$. The computer generated results are compared with the experimental data in Figure 2. The agreement between the computer simulated and experimentally observed products trends gives good support for the validity of the proposed mechanism in Figure 3. As can be seen in Figure 2, the results obtained with the reaction rate constants used in model A (full line) simulate fairly well the effect of the variation of the O_2 upon the all the products except HCOOH, but less well the effect of the variation of the acetaldehyde. Indeed the values obtained for the formation of the HCOOH are extremely low, and do not match our experimental results. Model A underestimates the CH_3COOOH yields but produces too much CH_3COOH and CO_2 at the high CH_3CHO concentrations. In model B an attempt was made to simulate the HCOOH yield by modifying the reaction mechanism for the HCOOH formation: reaction 16 was altered into

$$OOCH_2OH + HO_2 \longrightarrow HCOOH + H_2O + O_2 \qquad 16B$$

and the rate constant changed to $k_{15} = 1.1 \times 10^{-13}$ $cm^3.molecule^{-1}.s^{-1}$ and $k_{-15} = 30\ s^{-1}$. In addition, reaction 35 was added to the list of reactions. Model B contains also minor changes of k_{10}, k_{27} and k_{32}. The results of the simulations using the input data listed in model B are displayed in Figure 2 (dotted line). There is definitely an improvement of the simulation of the variation of the acetaldehyde concentration. But there is a systematic discrepancy at the higher acetaldehyde concentrations in the CO_2 and CH_3COOOH products yields. Also the existing HCOOH mechanism does not seem to be conform with the experimental data.

In conclusion, the agreement between computer simulation and the observed trends in the experimental data seem to confirm the proposed mechanism. It is possible that there are unidentified errors of detail either in the rate constants or the mechanism in the two models employed here. On the other hand, a certain degree of experimental error is associated with the results obtained by the FTIR analysis, certainly at the higher acetaldehyde concentrations, where larger part of the spectrum is obscured.

5.0 Modulated Photolysis of CH_3CHO

Measurements of the absorption-time profiles during modulated photolysis of CH_3CHO-Air mixtures (total pressure = 700 torr) were made using photolysis periods of 6 s - 10 s. Transient absorptions showed a characteristic rise and fall during the alternating photolysis and dark periods. At the low CH_3CHO concentrations where mechanism A is operating, absorption at 250 nm was expected to be the dominated by CH_3O_2 radicals whereas at 220 nm absorp-

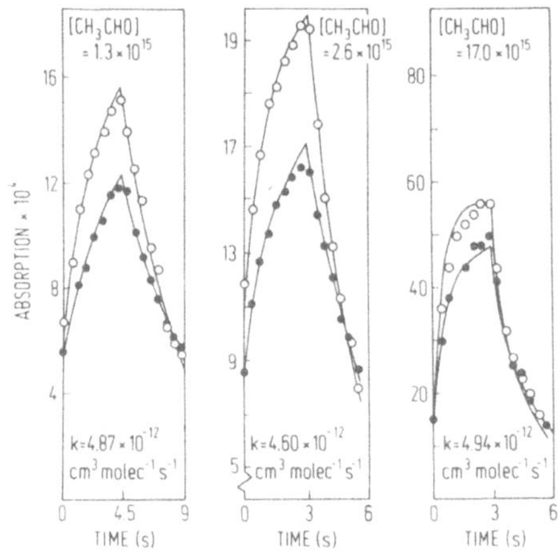

Fig.4 Computer fits to transients absorptions at 250 nm (o) and 220 nm (o) in modulated photolysis of CH_3CHO-Air mixtures: P_T = 700 torr T= 298 K.

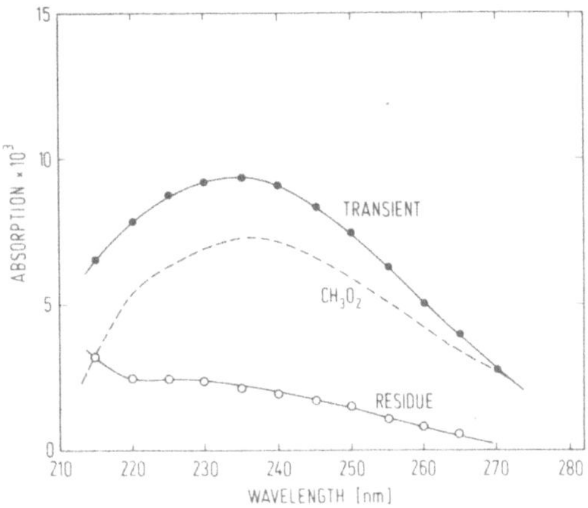

Fig. 5 Transient absorption spectrum in the modulated photolysis of CH_3CHO (2 torr) in Air (700 torr) at 303 K. Spectral slit: 5 nm. Also shown is the CH_3O_2 absorption spectrum.

tion components due to CH_3O_2 and HO_2 are expected. The data in are in accord with this since absorption at 220 nm was greater than at 250 nm as expected if HO_2 was present. On the basis of knowledge of the rates of the reactions of CH_3O_2 and HO_2 we expect that the reaction of CH_3O_2 with HO_2 would occur in competition with the mutual reaction of HO_2 and CH_3O_3 (reactions 6 and 9) Consequently, the decay kinetics of the radical absorption is complicated and not amenable to simple analysis to obtain a value of k_{10}. The rate constant k_{10} could be obtained from the analysis of the steady state concentrations of CH_3O_2 and HO_2 during the photolysis. These can be estimated from the maximum absorptions reached during the modulation cycle, with addition of the estimated residual absorption at the end of the cycle, using the known cross-sections of HO_2 and CH_3O_2 at 220 nm and 250 nm. Using residual absorptions estimated assuming second order decay kinetics i.e. (linear absorbance)$^{-1}$ vs time plots an average value of k_{10} = (6.7$\pm$ 2.2) x 10^{-12} cm^3. $molecule^{-1}.s^{-1}$ was obtained from the modulated photolysis of low [CH_3CHO] in the range (1.3 -17.0) x 10^{15} molecule.cm^{-3}. This value is probably an upper limit value, since in most cases, the radical absorption barely reached the steady state during the light-on part of the photolysis cycle and therefore the steady states were probably underestimated. Considerable uncertainty also arises from the uncertainty in the absorption cross section values.

The rate constant for reaction 10 was also determined by computer fitting of the two modulation waveforms at 220 and 250 nm. The absorption data are very sensitive to the variable k_{10}, and also the absorption cross sections for the two radicals at the two wavelengths. HO_2 was taken from the JPL-publucation (24) (σ 220 nm = 4.0 x 10^{-18} cm^2.$molecule^{-1}$; σ250 nm = 0.5 x 10^{-18} cm^2 molecule $^{-1}$) and σCH_3O_2 was determined in separate experiments in this study, σCH_3O_2 = 3.8 x 10^{-18} cm^2.$molecule^{-1}$ at 250 nm. As is seen in Figure 4, the "best fit" of k_{10} from analysis of the time-resolved absorption profiles of CH_3O_2 and HO_2 in the modulated photolysis of CH_3CHO-air mixtures at 700 torr and 300 K was near 4.8 x 10^{-12} cm^3.$molecule^{-1}$.s^{-1}. This value agrees with previously reported values from molecular modulation (29) and flash photolysis studies (30) taking into account uncertainties especially due to the absorption coefficient for CH_3O_2. This value is however somewahat higher than that derived for k_{10}, based on the yields of CH_3OOH observed in the FTIR product studies: $k_{10} = (3.5 \pm 1.0) \times 10^{-12}$ $cm^3.molecule^{-1}.s^{-1}$

The wavelength dependence of the radical absorptions near steady state was conducted at high $[CH_3CHO]$ (6.4×10^{16} $molecule.cm^{-3}$) to establish the spectrum of the combined transient absorption in the range 215-270 nm. A similar spectrum for CH_3O_2 was also obtained as part of this study in the modulated photolysis of mixtures containing Cl_2 (1.60×10^{15} molecule cm^{-3}), CH_4 (6.4×10^{-18} $molecule.cm^{-3}$) in air at 700 torr. These are compared on Figure 5. The transient spectrum in CH_3CHO photolysis is similar to CH_3O_2 but shows stronger absorption at shorter wavelengths. A residual spectrum was obtained by subtraction of the CH_3O_2 spectrum assuming that all the absorption at 270 nm is due to CH_3O_2. The remaining spectrum is completely different to that of HO_2 and shows some similarities to the spectrum reported for CH_3CO_3, i.e. a shoulder near 240 nm with increasing absorption at 220 nm (28). However an assignment cannot be made in the absence of knowledge of the spectrum of $CH_3CH(OH)O_2$, which is likely to be similar to CH_3O_2.

6.0 Atmospheric Implications of Results.

The formation of acetic acid in the gas phase oxidation of atmospheric acetaldehyde can occur by the addition of HO_2 to acetaldehyde to form the $CH_3CH(OH)O_2$ radical, which under atmospheric conditions will react with either NO or the most abundant peroxy radical, HO_2:

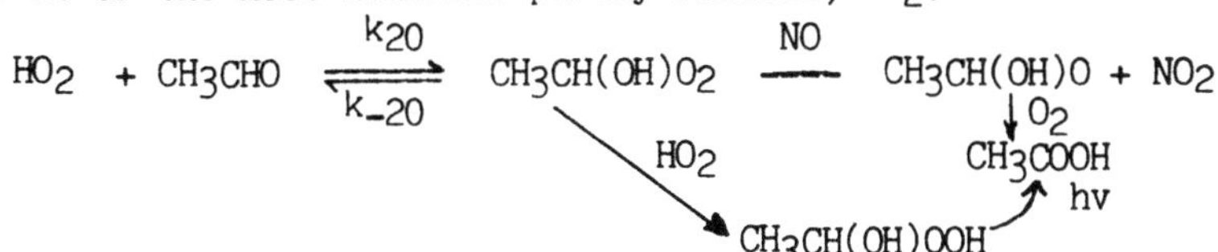

In the lower atmosphere and particularly in the continental boundary layer, the NO route is likely to dominate over HO_2, which would only be important in the background troposphere. These considerations are based on knowledge of reactivity of other peroxy radicals with HO_2 and NO. The experimental data indicate a value of $K_{20}^* \simeq 1 \times 10^{-17}$ $cm^3.molecule.s^{-1}$ which, with the upper limit of $k_{20} = 1 \times 10^{-15}$ $cm^3.molecule^{-1}.s^{-1}$, gives $k_{-20} = 100$ s^{-1} at 300 K. The rate of $CH_3CH(OH)O_2$ with respect to reaction with NO and HO_2 is much shorter ($\sim 10^{-2}$ s^{-1} in background air, ~ 1 s^{-1} in the continental boundary layer air). Moreover the rate of formation of the $CH_3CH(OH)O_2$ radical will be very slow under daytime atmospheric conditions (k_{20} $[HO_2] = 10^{-7}$ s^{-1}) and the equilibrium reaction (20) will not be established. We conclude therefore on the basis of our experimental observations and a consideration of other estimated parameters that the reaction of HO_2 with CH_3CHO is not a major source of acetic acid in the atmosphere.

The reaction of CH_3O_2 with HO_2 is a key reaction in the background troposphere. Its importance lies in the competion between HO_2 and NO for the oxidation of CH_3O_2, which determines the balance between O_3 production and destruction cycles in the oxidation of atmospheric CH_4. The presents results indicate that the channel for CH_3OOH formation

$CH_3O_2 + HO_2 \longrightarrow CH_3OOH + O_2$ with $k_{10} = 3.7 \times 10^{-12}$ $cm^3.molecule^{-1}.s^{-1}$ is somewhat slower than the value 6.5×10^{-12} $cm^3.molecule^{-1}.s^{-1}$ currently recommended for calculations of atmospheric chemistry (29). This implies a more efficient production of O_3 from CH_4 oxidation in the troposphere. The possibility also exists for an additional channel producing formaldehyde: $CH_3O_2 + HO_2 \longrightarrow CH_2O + H_2O + O_2$.

References

1) Finlayson B.J., Pitts J.N., Jr. and Atkinson R., J. Am. Chem. Soc., 96, 5357 (1974).
2) Atkinson R., Darnall K.R., Lloyd A.C., Winer A.M., and Pitts J.N., Jr., Adv. in Photochem., 11, 375 (1979).
3) Calvert J.G.," The Chemistry of the polluted troposphere" in "Chemistry of the unpolluted and polluted troposphere", Nato Advanced Study Inst. Series, Ed. H.W. Georgii and W.Jaeschke, 425 (1982).
4) Altschuller A.P., Atm. Env., 17, 2383 (1983).
5) Gelinas R.J. and Skewes-Cox, P.D., J. Phys. Chem., 81, 2468 (1977).
6) P.J. Crutzen, "The major biogeochemical cycles and their interactions, Ed. B. Bolin, R.B. Cook, Scope, 67 (1983).
7) Horowitz A. and Calvert J.G., J. Phys. Chem., 86, 3105 (1982).
8) Horowitz A. Kerschner G.J. and Calvert J.G., J. Phys. Chem., 86, 3094 (1982).
9) Meyrahn H., Moortgat G.K. and Warneck P., Proc. 5th Biann. Coll. of SFB-73, Verlag Fried. Viehweg und Sohn, 65 (1982).
10) Meyrahn H., Ph.D. Thesis, Mainz (1984).
11) Cox R.A., Burrows J.B., Chant S. and Patrick R., Cost 61a bis meeting, "Physico-chemical Behaviour of Atmospheric Pollutants", Working Group 2 Discussions Meeting 10th and 11th July 1980, at AERE, Harwell, U.K.
12) Weaver J., Meagher J. and Heicklen J., J. Photochem., 6, 111 (1976/77).
13) Moortgat G.K. and McQuigg R.D., "Physico-chemical behaviour of atmos pheric Pollutants", Ed. B. Versino and G. Angeletti, D. Reidel Publ. Co., 194 (1984).
14) Su F., Calvert J.G., Shaw J.H., Niki H., Maker P.D., Savage C.M. and Breitenbach L.P., Chem. Phys. Lett., 65, 221 (1979).
15) Su F., Calvert J.C. and Shaw J.H., J. Phys. Chem., 83, 3185 (1979).
16) Veyret B., Rayez J.-Cl. and Lesclaux R., J. Phys. Chem., 86, 3424 (1982).
17) Barnes I., Becker K.H., Fink E.H., Reimer A., Zabel F., Chem. Phys. Lett., 115, 1 (1985)
18) Barnes I., Becker K.H., Fink E.H., and Zabel F., Private Communication.
19) Likens G.E. and Galloway J.N., Tellus, 35B, 16 (1983).
20) Nielsen T., Hansen A.M. and Thomsen E.L., Atmosph. Environm., 16, 2447 (1982).
21) Kan C.S., Calvert J.G. and J.H. Shaw, J. Phys. Chem., 84, 3411 (1980).
22) Niki H., P.D. Maker, C.M. Savage and L.P. Breitenbach, J. Phys. Chem., 87, 2190 (1983).
23) Niki H., Maker P.D., Savage C.M. and Breitenbach L.P., J. Phys. Chem., 588 (1985).
24) J.P.L. Publication no.7, 85-37 (1985)
25) Niki H., Maker P.D., Savage C.M. and Breitenbach L.P., J. Phys. Chem., 85, 877 (1981).
26) Parkes D.A., Intern. J. Chem. Kin., 9, 451 (1977).
27) Veyret B., McAdam K. and Lesclaux R., 2nd French-German Workshop on Tropospheric Chemistry, St. Hughes de Biviers, France, 11-13 March, 1986.
28) Addison M.C., Burrows J.P., Cox R.A. and Patrick R., Chem. Phys. Lett., 73, 283 (1980).
29) Cox R.A. and Tyndall G.S., Chem. Phys. Lett., 65, 357 (1979).
30) McAdam K., Veyret B. and Lesclaux R., Chem. Phys. Lett., in Press (1986)

<table>
<tr><th colspan="2">Table I. Reaction</th><th colspan="2">Rate Constant</th></tr>
<tr><th></th><th></th><th>Model A</th><th>Model B</th></tr>
<tr><td>1</td><td>$CH_3CHO + h\nu \rightarrow CH_3 + HCO$</td><td colspan="2">4.0 E-05</td></tr>
<tr><td>2</td><td>$\rightarrow CH_4 + CO$</td><td colspan="2">5.0 E-07</td></tr>
<tr><td>3</td><td>$\rightarrow CH_3CO + H$</td><td colspan="2">1.2 E-06</td></tr>
<tr><td>4</td><td>$HCO + O_2 \rightarrow HO_2 + CO$</td><td colspan="2">5.5 E-12</td></tr>
<tr><td>5</td><td>$CH_3 + O_2 + M \rightarrow CH_3O_2 + M$</td><td colspan="2">3.9 E-32</td></tr>
<tr><td>6</td><td>$CH_3O_2 + CH_3O_2 \rightarrow CH_3OH + CH_2O + O_2$</td><td colspan="2">2.2 E-13</td></tr>
<tr><td>7</td><td>$\rightarrow 2\ CH_3O + O_2$</td><td colspan="2">1.2 E-13</td></tr>
<tr><td>8</td><td>$\rightarrow CH_3OOCH_3 + O_2$</td><td colspan="2">3.0 E-14</td></tr>
<tr><td>9</td><td>$HO_2 + HO_2 \rightarrow H_2O_2 + O_2$</td><td colspan="2">2.8 E-12</td></tr>
<tr><td>10</td><td>$HO_2 + CH_3O_2 \rightarrow CH_3OOH + O_2$</td><td>3.5 E-12</td><td>4.0 E-13</td></tr>
<tr><td>11</td><td>$CH_3O + O_2 \rightarrow CH_2O + HO_2$</td><td colspan="2">1.3 E-15</td></tr>
<tr><td>12</td><td>$CH_2O + h\nu \rightarrow HCO + H$</td><td colspan="2">6.7 E-05</td></tr>
<tr><td>13</td><td>$\rightarrow H_2 + CO$</td><td colspan="2">3.3 E-05</td></tr>
<tr><td>14</td><td>$H + O_2 + M \rightarrow HO_2 + M$</td><td colspan="2">5.5 E-32</td></tr>
<tr><td>15</td><td>$HO_2 + CH_2O \rightarrow OOCH_2OH$</td><td>7.5 E-14</td><td>1.1 E-13</td></tr>
<tr><td>-15</td><td>$OOCH_2OH \rightarrow HO_2 + CH_2O$</td><td>750</td><td>30</td></tr>
<tr><td>16A</td><td>$OOCH_2OH + HO_2 \rightarrow HO_2CH_2OH + HO_2$</td><td>6.5 E-12</td><td>--</td></tr>
<tr><td>16B</td><td>$\rightarrow HCOOH + H_2O + O_2$</td><td>--</td><td>6.5 E-12</td></tr>
<tr><td>17</td><td>$2\ OOCH_2OH \rightarrow 2\ OCH_2OH + O_2$</td><td colspan="2">2.3 E-13</td></tr>
<tr><td>18</td><td>$OCH_2OH + O_2 \rightarrow HO_2 + HCOOH$</td><td colspan="2">3.5 E-14</td></tr>
<tr><td>19</td><td>$HO_2CH_2OH + h\nu,wall \rightarrow H_2O + HCOOH$</td><td colspan="2">2.6 E-03</td></tr>
<tr><td>20</td><td>$HO_2 + CH_3CHO \rightarrow HAcHO_2$</td><td colspan="2">1.0 E-15</td></tr>
<tr><td>-20</td><td>$HAcHO_2 \rightarrow HO_2 + AcH$</td><td colspan="2">100</td></tr>
<tr><td>21</td><td>$2\ HAcHO_2 \rightarrow 2\ HAcHO + O_2$</td><td colspan="2">2.3 E-13</td></tr>
<tr><td>22</td><td>$HAcHO_2 + HO_2 \rightarrow HAcHO_2H + O_2$</td><td colspan="2">6.5 E-12</td></tr>
<tr><td>23</td><td>$HAcHO_2 + CH_3O_2 \rightarrow HAcHO + CH_3O + O_2$</td><td colspan="2">9.0 E-13</td></tr>
<tr><td>24</td><td>$HAcHO + O_2 \rightarrow CH_3COOH + HO_2$</td><td colspan="2">3.5 E-14</td></tr>
<tr><td>25</td><td>$HAcHO_2H + h\nu,wall \rightarrow CH_3COOH + H_2O$</td><td colspan="2">2.7 E-03</td></tr>
<tr><td>26</td><td>$CH_3O_2 + CH_3CHO \rightarrow CH_3OOH + CH_3CO$</td><td colspan="2">8.1 E-20</td></tr>
<tr><td>27</td><td>$CH_3O + CH_3CHO \rightarrow CH_3CO + CH_3OH$</td><td>6.0 E-14</td><td>1.5 E-13</td></tr>
<tr><td>28</td><td>$CH_3CO + O_2 \rightarrow CH_3CO_3$</td><td colspan="2">2.0 E-12</td></tr>
<tr><td>29</td><td>$2\ CH_3CO_3 \rightarrow 2\ CH_3CO_2 + O_2$</td><td colspan="2">2.5 E-12</td></tr>
<tr><td>30</td><td>$CH_3CO_2 \rightarrow CH_3 + CO_2$</td><td colspan="2">2.2 E+10</td></tr>
<tr><td>31</td><td>$CH_3CO_3 + CH_3CHO \rightarrow CH_3COOOH + CH_3CO$</td><td>5.3 E-17</td><td>5.7 E-17</td></tr>
<tr><td>32</td><td>$CH_3CO_3 + CH_3O_2 \rightarrow CH_3CO_2 + CH_3O + O_2$</td><td>1.5 E-11</td><td>1.0 E-11</td></tr>
<tr><td>33</td><td>$HO_2 + CH_3CO_3 \rightarrow CH_3COOOH + O_2$</td><td colspan="2">4.5 E-12</td></tr>
<tr><td>34</td><td>$HO_2 + CH_3CO_2 \rightarrow CH_3COOH + O_3$</td><td colspan="2">1.5 E-12</td></tr>
<tr><td>35</td><td>$CH_3CO_3 + HAcHO_2 \rightarrow CH_3CO_2 + HAcHO + O_2$</td><td>--</td><td>1.0 E-11</td></tr>
<tr><td>36</td><td>$2\ CH_3O \rightarrow CH_3OH + CH_2O$</td><td colspan="2">2.5 E-11</td></tr>
<tr><td>37</td><td>$\rightarrow CH_3OOCH_3$</td><td colspan="2">6.6 E-12</td></tr>
<tr><td>38</td><td>$CH_3O + CH_3O_2 \rightarrow CH_3OOH + CH_2O$</td><td colspan="2">1.0 E-11</td></tr>
<tr><td>39</td><td>$CH_3O + CH_2O \rightarrow CH_3OH + HCO$</td><td colspan="2">1.5 E-14</td></tr>
<tr><td>40</td><td>$CH_3OOH + h\nu \rightarrow CH_3O + OH$</td><td colspan="2">1.7 E-05</td></tr>
<tr><td>41</td><td>$H_2O_2 + h\nu \rightarrow 2\ OH$</td><td colspan="2">1.9 E-05</td></tr>
<tr><td>42</td><td>$OH + CH_3CHO \rightarrow CH_3CO + H_2O$</td><td colspan="2">1.6 E-11</td></tr>
<tr><td>43</td><td>$OH + CO \rightarrow CO_2 + H$</td><td colspan="2">2.7 E-13</td></tr>
<tr><td>44</td><td>$OH + CH_2O \rightarrow HCO + H_2O$</td><td colspan="2">9.2 E-12</td></tr>
<tr><td>45</td><td>$OH + CH_3OH \rightarrow CH_2OH + H_2O$</td><td colspan="2">1.0 E-12</td></tr>
<tr><td>46</td><td>$OH + CH_3OOH \rightarrow CH_2O + OH + H_2O$</td><td colspan="2">4.4 E-12</td></tr>
<tr><td>47</td><td>$\rightarrow CH_3O_2 + H_2O$</td><td colspan="2">5.6 E-12</td></tr>
<tr><td>48</td><td>$OH + H_2O_2 \rightarrow H_2O + HO_2$</td><td colspan="2">1.7 E-12</td></tr>
<tr><td>49</td><td>$OH + HO_2 \rightarrow H_2O + O_2$</td><td colspan="2">1.1 E-10</td></tr>
<tr><td>50</td><td>$CH_2OH + O_2 \rightarrow CH_2O + HO_2$</td><td colspan="2">2.0 E-12</td></tr>
</table>

Units: Photolysis and 1st k: s^{-1}; 2nd k: $cm^3.molecule^{-1}.s^{-1}$

LABORATORY STUDIES OF THE NITRATE RADICAL

R.P. Wayne and S.J. Smith
Physical Chemistry Laboratory, Oxford, UK

and

R.A. Cox and I.W. Hall
A.E.R.E., Harwell, Oxon, UK

Summary

Time resolved absorptions of NO_3 radicals, generated by the reaction of Cl with $ClONO_2$, were studied in both static and flowing systems at several temperatures. The build-up and decay of radicals is interpreted in terms of production and loss mechanisms, and the results give a measure of the reactivity of Cl and ClO towards NO_3; they also indicate that wall loss processes are of minor importance. In the presence of H_2 and O_2, HO_2 radicals are produced: a greatly enhanced loss rate for NO_3 shows that there is interaction between the two radicals, and an estimate of the rate constant is adduced from the data. Additional evidence for inefficient wall loss of NO_3 comes from experiments in which NO_3 is generated from the reaction of F with HNO_3. The elimination from consideration of significant wall effects permits further analysis of experiments using the reaction of NO_2 with O_3 as a source of NO_3. A relatively slow reaction of NO_3 with O_3 seems possible, and such a process is consistent with absorbance measurements designed to determine the absorption cross section for NO_3 at λ=662nm.

1. Introduction

Reactions of the nitrate radical are important in several ways in atmospheric chemistry. For example, reaction with NO_2 gives N_2O_5, and, *via* heterogeneous steps, HNO_3. Furthermore, NO_3 has been shown to be a potentially important oxidant for organic species in the atmosphere. Assessment of the importance of such atmospheric processes depends on reliable and accurate laboratory investigations of kinetic and photochemical processes involving the radicals. The present paper describes experiments designed not only to provide information about processes occurring in the atmosphere, but also to facilitate interpretation of other relevant laboratory data.

Sources of NO_3 used in laboratory experiments have included the thermal dissociation of N_2O_5[1], and the reactions between NO_2 and O_3[2,3]:

$$NO_2 + O_3 \longrightarrow NO_3 + O_2 \quad (1),$$

between F and HNO_3[4,5]:

$$F + HNO_3 \longrightarrow NO_3 + HF \quad (2),$$

and between Cl and $ClONO_2$[6,7]:

$$Cl + ClONO_2 \longrightarrow NO_3 + Cl_2 \quad (3).$$
k_3= 1.2 x 10^{-11} cm^3 $molecule^{-1}$ s^{-1}[8]

This last reaction has the attraction that Cl atoms are readily generated photolytically, so that time resolved studies are facilitated. We have already described the application of the so-called molecular modulation variant of the technique to studies of NO_3 chemistry[9,10], in which the cross section, σ, for absorption by NO_3 at λ=662nm was estimated, and the secondary reactions

$$Cl + NO_3 \longrightarrow ClO + NO_2 \quad (4)$$
k_4=5.5 x 10^{-11} cm^3 $molecule^{-1}$ s^{-1}[10]

and

$$ClO + NO_3 \longrightarrow ClOO + NO_2 \quad (5)$$
k_5=4 x 10^{-13} cm^3 $molecule^{-1}$ s^{-1}[10]

were identified. The present paper describes further work on the reaction system, and shows how the method may be extended to studies of reactions with other radical partners, such as HO_2

$$HO_2 + NO_3 \longrightarrow products \quad (6).$$

Reaction (1) is the most important source of NO_3 in the atmosphere, and should also be a clean source of radicals in the laboratory. At high [O_3] and low [NO_2], it should be possible to ensure that reaction (1) dominates completely over the process

$$NO_2 + NO_3 + M \longrightarrow N_2O_5 + M \quad (7),$$
k_7=1.2 x 10^{-12} cm^3 $molecule^{-1}$ s^{-1} at 1 Atm [8]

so that NO_2 is stoicheiometrically converted to NO_3. However, using [NO_3] determined in this way to calculate the cross section, σ[2,3,11], leads to values of σ markedly smaller than those obtained using the alternative sources of NO_3[4,12,13,14]. It is apparent, therefore, that there are losses of NO_3 in the ozone system for which allowance has not been made. Experiments are described in this paper using reaction (2) as a third technique for generating NO_3 in order to yield independent information about the behaviour of NO_3. An interaction of NO_3 with ozone, such as

$$NO_3 + O_3 \longrightarrow NO_2 + 2O_2 \quad (8),$$

will be shown to be favoured over the thermal decomposition

$$NO_3 + M \longrightarrow NO + O_2 + M \quad (9),$$

that has been suggested by other workers[15].

2. Experimental

The apparatus used in experiments with $ClONO_2$ has been described in detail elsewhere[16], and only an outline will be given here. A quartz reaction cell, 120cm long and 30mm in diameter, is surrounded by six fluorescent striplamps, which may be illuminated in any combination. The lamps contain a phosphor emitting primarily at $\lambda=350 \pm 50$ nm. Reaction mixtures in the cell were subjected to modulated photolysis, and absorption was used to monitor for HO_2 at $\lambda=220$nm (where there is some contamination from $ClONO_2$ absorption) and for NO_3 at $\lambda=662$nm. The absorption signal was either stored digitally, or displayed on a conventional recorder. In experiments with Cl_2, $ClONO_2$ and N_2 alone admitted, rise and decay times were sufficiently slow, and NO_3 absorption intensities sufficiently intense, for experiments to be performed in a single shot mode, using a period of 20 to 60s and a duty cycle of 50%. For these experiments, the reaction mixture was made up in the cell for each run. When HO_2 was present, reaction time constants were greatly reduced, so that repetitive photolysis was employed, with a period of typically 2s. Signal averaging over many cycles was used to improve the signal-to-noise ratio of the weaker absorptions, and the reaction mixture was continually replenished by maintaining a slow flow of reactants through the cell. The corrections to the absorption-time profiles needed to allow for the flow have been discussed elsewhere[11].

In experiments using the reaction between F and HNO_3 as a source of NO_3, a conventional low pressure discharge-flow was employed, with flow velocities in the range 2.5 to 3 m s^{-1}, and total pressures in the range 0.6 to 8 mmHg. The main carrier gas was helium, to which was added a small flow of F_2 in He that had passed through a microwave discharge. A second flow of He passed through a bubbler containing anhydrous HNO_3 held at 255 K, and joined the other gases after the discharge. At two points downstream of the mixing region, NO could be injected. Optical absorption of NO_3 was measured at $\lambda=662$nm using a four-pass White cell giving a total optical pathlength of 35.2cm.

The closed circulating system used to study the reaction between NO_2 and O_3 has been described previously[2,3,11]. Ozone, at concentrations in the range 2 to 15 x 10^{17} molecule cm^{-3} were diluted by O_2 to total pressures of around 100 mmHg. To the circulating gases was added NO_2 just upstream of an optical absorption cell (diameter 1.4cm and length 24.7cm); product N_2O_5 was removed downstream in a slush-bath cooled trap. Optical absorption at $\lambda=662$nm was monitored using a chopped double-beam AC method[11] with a monochromator (Applied Photophysics) having a bandpass of 1.1nm FWHM.

3. Results and Discussion

Figure 1 presents some concentration-time profiles for NO_3 generated in the modulated photolysis of Cl_2 + $ClONO_2$ mixtures. Absorbances were converted to concentrations using a value of σ equal to 2.28 x 10^{-17} cm^2 $molecule^{-1}$, a cross section currently thought to be the most reliable [14].

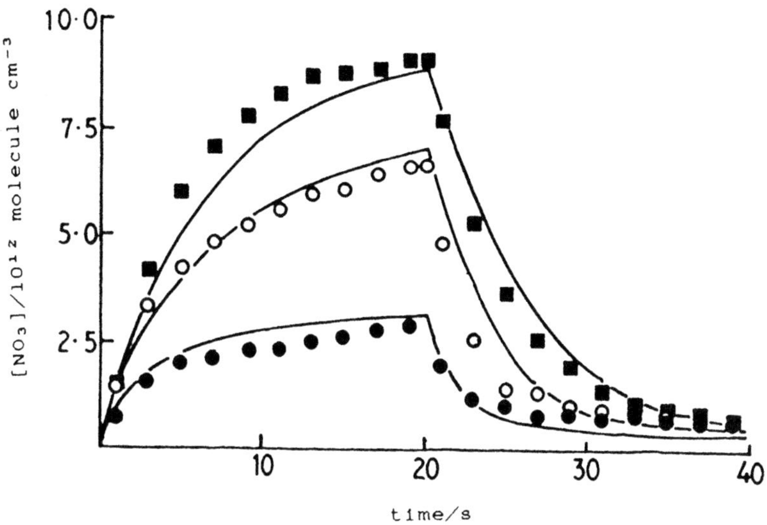

Figure 1. Concentration-time profiles for photolysis of $ClONO_2$ - Cl_2 - N_2 mixtures at 303K
■ $[ClONO_2] = 5.8 \times 10^{14}$ molecule cm^{-3}
○ $[ClONO_2] = 1.2 \times 10^{15}$ molecule cm^{-3}
● $[ClONO_2] = 2.4 \times 10^{15}$ molecule cm^{-3}

Solid lines are computer simulations.

Initial rates of NO_3 formation during the photolysis period should be equal to twice the rate of Cl_2 photolysis, since no loss processes will be significant during the early stages of the reaction. The mean initial rate of Cl_2 photolysis derived from the data in the figure is $3.3 \times 10^{-4}[Cl_2]$ molecule cm^{-3} s^{-1}. An independent check on the photolysis is available from measurements of the rate of loss of Cl_2, determined spectrophotometrically, without $ClONO_2$ added, but in the presence of a Cl atom scavenger. The rate measured using similar illumination to that employed in the $ClONO_2$ runs is $2.7 \times 10^{-4}[Cl_2]$ molecule cm^{-3} s^{-1}. Since the sets of experiments were performed separately, the agreement between the two methods of determining photolysis rate must be regarded as good. For further analysis, the rates actually determined in the NO_3 experiments are used.

As the reaction progresses, and NO_3 builds up, so reaction (4) begins to compete with (3) for Cl, and the secondary reactions (4), (5) and (7) limit the formation of further NO_3. However, a curious feature of the results, evident in figure 1, is that $[NO_3]$ towards the end of the photolysis period is higher for lower $[ClONO_2]$. The explanation is seen in terms of the thermal decomposition of $ClONO_2$

$$ClONO_2 + M \longrightarrow ClO + NO_2 + M \quad (10),$$

which is important because it yields two species that destroy NO_3. Indeed, it is impossible to interpret the exact shape of the build-up portion of the curve without the participation of this reaction. Figure 2 shows some computer simulations, to be discussed later, of the concentration-time curves using different values of k_4 and k_{10}. Although the limit to which

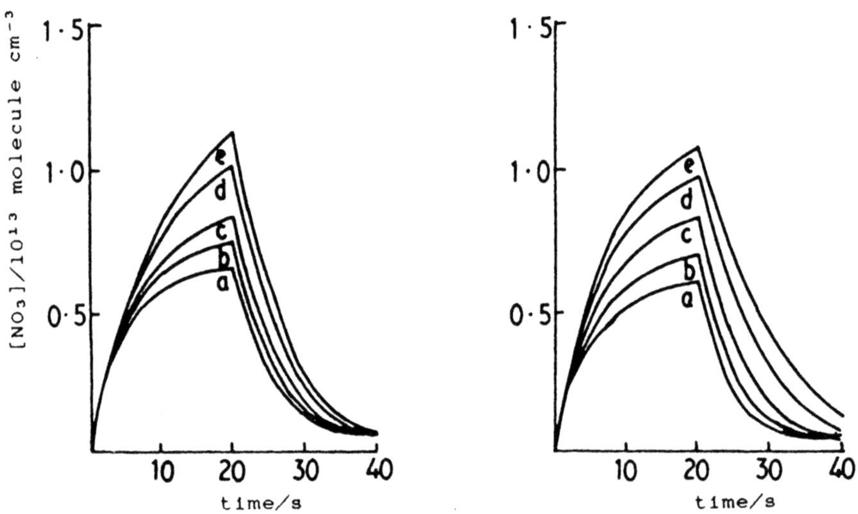

Figure 2. Effect of varying rate constants on simulated concentration-time profiles

Left side, k_4: a)10, b)7.5, c)5.5, d)2.5, e)1.0 $\times 10^{-11}$ cm^3 $molecule^{-1}$ s^{-1}
Right side, k_{10}: a)8, b)6, c)4, d)2, e)0.8 $\times 10^{-4}$ s^{-1}

$[NO_3]$ approaches can be brought near that observed towards the end of photolysis, the rise to the limit is always too steep. Rather, a loss of NO_3 is required that accelerates over the whole photolysis period . Reaction (10) provides that loss with a rate coefficient (6×10^{-4} s^{-1} at T=303K and P=1 Atm) in reasonable accord with other observations[17].

The computer simulations alluded to above use the reactions already discussed, together with the reverse reactions appropriate to processes (7) and (10). A fuller integration procedure that includes atomic oxygen and NO_x and ClO_x photochemistry shows that steps involving these species make little contribution to the total decay. The rate coefficients adopted are those shown under the reactions.

That decomposition of $ClONO_2$ in reaction (10) leads to loss of NO_3 restricts the rate of other loss processes for NO_3. Since reaction (10) continues to be a source of NO_2 and ClO in the dark, while [Cl] will fall rapidly after illumination ceases, it is appropriate to examine the decay portion of the curves for additional loss of NO_3. In fact, first order loss of NO_3, with products NO + O_2, makes the decay more rapid than that observed, and distorts the shape of the curves, for a rate coefficient greater than 0.08 s^{-1}. That value would then seem to be the upper limit for any homogeneous decomposition of NO_3. The solid lines in figure 1 show the how the simulated reaction profiles match the experimental data over light and dark periods.

Experiments performed at temperatures lower than 303K offer confirmation of the general proposition that most of the NO_3 loss after illumination ceases is a result of reaction with ClO and NO_2 produced from $ClONO_2$. Figure 3 shows some experimental concentration-time profiles for different temperatures. It is immediately apparent

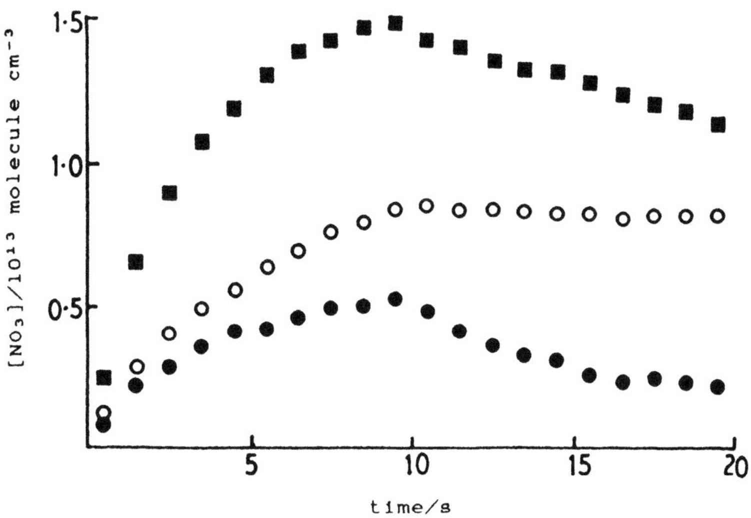

Figure 3. Concentration-time profiles for photolysis of $ClONO_2$ - Cl_2 - N_2 Mixtures at various temperatures.
O, 263K; ■, 283K; ●, 333K

that the post-irradiation decay of NO_3 is much slower at lower temperatures. Preliminary computer fits to these results gives values of k_{10} of 0.06, 1.5 and 800 x 10^{-4} s^{-1} at temperatures of 263K, 283K and 333K respectively. Arrhenius analysis suggests an activation energy of about 100 kJ mol^{-1}, a value close to that obtained in other studies[17], suggesting again that reaction (10) is the major cause of NO_3 decay. Data such as those presented in figure 3 should be capable of further, more detailed, analysis to yield information about the temperature dependences of other processes such as reaction (4), and that work is currently in progress.

We have undertaken[18] experiments using the F + HNO_3 source of NO_3 in order to validate the optical absorption system used in NO_2 + O_3 measurements[2,11]. The change in optical absorption at λ=662nm is measured as a function of [NO] added to the flow containing NO_3. Removal of NO_3 results from the rapid reaction

$$NO + NO_3 \longrightarrow 2NO_2 \qquad (13),$$

and, after making allowance for the secondary loss of NO_3 in reaction with NO_2, absolute changes in [NO_3] can be calculated. Our estimate of σ_{662} thus obtained is 2.1 ± 0.3 x 10^{-17} cm^2 $molecule^{-1}$, a value consistent with recent determinations[14]. We emphasize that we do not put forward our result as a new determination of σ, but rather use it to show that our optical system does not suffer from systematic error. However, the experiments described can also be used to provide an independent lower limit on the rate of heterogeneous loss of NO_3, since the titrant NO can be injected at two different jets so that NO_3 is in contact with the flow tube walls for different times before being removed. In fact, any difference in loss of NO_3 is smaller than the noise on the measurements (equivalent to 2.5 x 10^{11} molecule cm^{-3}). Taken in conjunction with the difference in contact times (around 200 ms), these data suggest an upper limit for wall loss of NO_3 of 0.03 s^{-1}. This result is comparable with that obtained in the molecular modulation system, with a similar size vessel fabricated of quartz rather than Pyrex. It should be observed that, although the pressure regimes are very different (<10 mmHg as against 1 Atm), diffusion effects will not be important for small wall efficiencies such as must be applicable here.

Studies of NO_3 formed in the NO_2 + O_3 system were earlier [2] interpreted to indicate a value of σ_{662} of 1.2 x 10^{-17} cm^2 $molecule^{-1}$. Given that σ_{662} is really around 2 x 10^{-17} cm^2 $molecule^{-1}$, and that the optical measurements themselves are shown not to be in error, it follows that there are losses of NO_3 for which no allowance had been made. The results described earlier mitigate strongly against such losses being heterogeneous ones. Possible homogeneous processes include interaction in reaction (8) with ozone, which is deliberately held at high concentration (say 10^{17} to 10^{18} molecule cm^{-3}) or in reaction (9) with the carrier gas (O_2) that is present at concentrations a few times those of O_3. The requisite pseudo-first order rate coefficient for loss, k_{loss}, is very approximately 5 s^{-1} if the product molecule is NO_2, so that the bimolecular rate coefficients for reactions (8) or (9) would have to be of the order of 10^{-17} or 10^{-18} cm^3 $molecule^{-1}$ s^{-1} respectively. Johnston et al.[15] have adduced evidence for a process very like reaction (9), but the suggested rate constant is orders of magnitude lower than we need (pseudo-first order rate constant of 3 ± 2 x 10^{-3} s^{-1} even at 1 Atm). We have tried to distinguish between the two possible reactions by computer fitting curves of [NO_3] against initial [NO_2], of the kind shown in figure 5. Only k_{loss} was used as a fitting parameter, all other rate constants being taken from the literature. The solid lines in figure 5 show

In the presence of added H_2 and O_2, the reaction takes a somewhat different course. The reactions

$$Cl + H_2 \longrightarrow HCl + H \quad (11)$$

$$H + O_2 + M \longrightarrow HO_2 + M \quad (12)$$

generate HO_2. Optical absorption measurements at λ=220nm show that the radicals are produced, and they manifest themselves by dramatically reducing the lifetime of NO_3, and thus the concentrations present in the system. At the same time, the lifetime of HO_2 itself can only be explained if there is an interaction with NO_3, as suggested in reaction (6). Figure 4 shows typical rise and fall curves for the two radicals, after the digitized signals have been treated to remove DC and flow

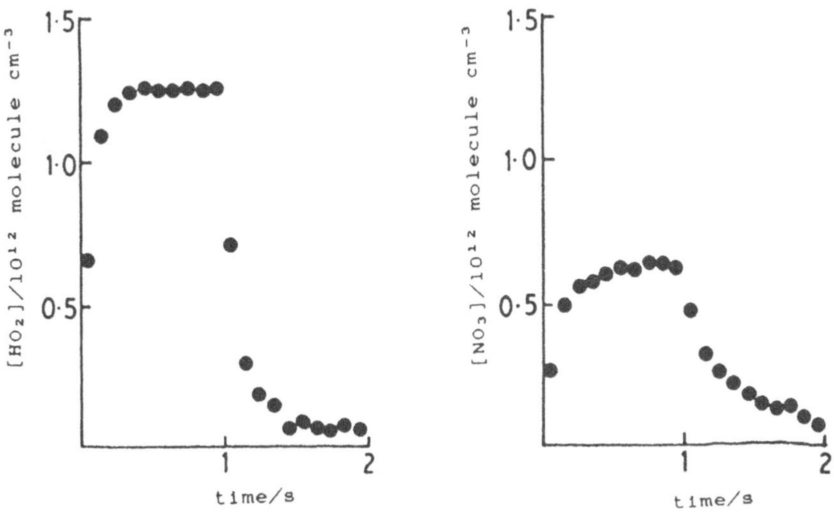

Figure 4. Concentration-time profiles for the photolysis of $ClONO_2$ - Cl_2 - H_2 - O_2 - N_2 Mixtures at 308K.

(a) HO_2 profile; (b) NO_3 profile

components. The coupled nature of the two radical concentrations makes virtually impossible any analysis other than a computer simulation. Such simulations mimic well the observed behaviour, and the influence of the rate constant for the reaction, k_6, is most keenly felt in the predicted modulation depth of the NO_3 absorption signal. First estimates of the rate constant at 308K put it within the range 3 to 8 x 10^{-12} cm^3 $molecule^{-1}$ s^{-1}.

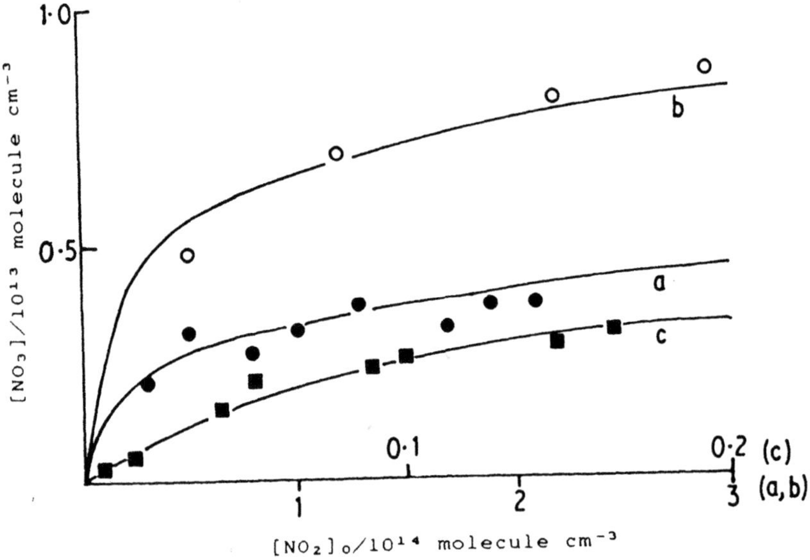

Figure 5. Plot of $[NO_3]$ against initial $[NO_2]$ in the reaction between NO_2 and O_3 at 298K

a) ●, $[O_3] = 1.5 \times 10^{17}$ molecule cm^{-3}
b) ○, $[O_3] = 3.9 \times 10^{17}$ molecule cm^{-3}
c) ■, $[O_3] = 6.3 \times 10^{17}$ molecule cm^{-3}

the computer-generated curves for the particular experiments. For the 27 experiments analysed, no discernible relationship between k_{loss} and [M] emerged. On the other hand, there was a strong positive correlation between k_{loss} and $[O_3]$. Forcing the data to a linear relationship passing through the origin yields a value of k_8 of about 1×10^{-17} cm^3 molecule^{-1} s^{-1}. As to the mechanism of this hypothetical reaction, two possibilities arise. Either the initial split of NO_3 may yield NO that reacts rapidly with O_3 to give NO_2, or the $O_2N—O\cdots\cdots O_3$ interaction may favour direct production of NO_2. Experiments with the object of investigating the reaction are in progress.

Acknowledgements

We thank the S.E.R.C. , the N.E.R.C, and the Department of the Environment for their partial support of this work.

References

[1] H.S Johnston, J. Amer. Chem. Soc., 73, 4542 (1951).

[2] D.N. Mitchell, R.P. Wayne, P.J. Allen, R.P. Harrison and R.J. Twin, J.Chem.Soc.Faraday II, 76, 785 (1980).

[3] M. Fowles, D.N. Mitchell, J.W.L. Morgan and R.P. Wayne, J.Chem.Soc.Faraday II, 78, 1239 (1982).

[4] A.R. Ravishankara and P.H. Wine, Chemical Physics Letts., 101,73 (1983)

[5] C.A. Smith, A.R. Ravishankara and P.H. Wine, J. Phys. Chem., 89, 1423 (1985)

[6] R.A. Cox, J.P. Burrows and G.B. Coker, Int. J. Chem. Kin., 16,445 (1984)

[7] J.J. Margitan, J. Chem. Phys., 87,674 (1983)

[8] NASA, Chemical kinetics and photochemical data for use in stratospheric modeling, Evaluation No. 7, JPL Pubn. 85-37, Pasadena, 1985

[9] R.A. Cox, R.A.Barton, E. Ljungstrom and D.W. Stocker, Chem. Phys. Letts., 108,228 (1984)

[10] R.A. Cox, M. Fowles, D.V. Moulton and R.P. Wayne, J. Phys. Chem., submitted (1986).

[11] M. Fowles, D.Phil thesis, Oxford, 1986

[12] R.A. Graham and H.S. Johnston, J. Phys. Chem., 82,254 (1978)

[13] W.J.Marinelli, D.W. Swanson and H.S. Johnston, J. Chem. Phys., 76,2864 (1982)

[14] S.P. Sander, J. Phys. Chem., 90,4135 (1986).

[15] H.S.Johnston, C.A. Cantrell and J.G. Calvert, J. Geophys. Res., 91, 5159, (1986).

[16] M.E. Jenkin and R.A. Cox, J. Phys. Chem., 89,192 (1985)

[17] G. Schönle, H.-D. Knauth and R.N. Schindler, J. Phys. Chem., 83,3297 (1979); W. Dasch, K.-H. Sternberg and R.N. Schindler, Ber. Bunsenges. Phys. Chem., 85, 611 (1981).

[18] C. Canosa-Mas, S.J. Smith and R.P. Wayne, in preparation

QUELQUES ASPECTS DE LA CHIMIE DES ALDEHYDES DANS LA BASSE TROPOSPHERE

P. CARLIER, P. FRESNET, H. HANNACHI, P. KALABOKAS, V. LESCOAT et G. MOUVIER.

Laboratoire de Physico-Chimie de l'Atmosphère, Associé au CNRS(UA717)
Université Paris VII, 2, Place Jussieu- 75251 PARIS CEDEX 05- FRANCE.

Résumé

L'étude de la chimie des aldéhydes dans la basse troposphère a été abordée par deux voies complémentaires : d'une part, la détermination de la réactivité des aldéhydes vis-à-vis des radicaux hydroxyles et nitrates sous conditions atmosphériques simulées en vue de l'estimation de leur temps de résidence troposphérique et d'autre part l'étude des facteurs sources et puits d'aldéhydes dans les masses d'air transitant au-dessus de la région parisienne.
La confrontation des résultats des deux études montre une importance prépondérante des aldéhydes secondaires dans la chimie des atmosphères urbaines et de leur panache sauf par temps couvert à proximité des sources intenses d'aldéhydes primaires.

1. INTRODUCTION

Les composés carbonylés jouent un rôle-clef dans la physico-chimie de la troposphère(1), car ils sont à la fois des contaminants primaires fréquents dans les zones urbaines et industrielles, des contaminants secondaires encore plus importants -la première étape de photooxydation des composés organiques dans l'atmosphère fournissant quasi systématiquement des composés carbonylés- et, de par leur photolyse, des promoteurs de radicaux libres dont l'importance devient prépondérante dans les atmosphères moyennement et fortement polluées. Il nous est apparu particulièrement inréressant de faire le bilan du problème de la chimie des aldéhydes dans la basse troposphère et plus particulièrement dans le panache des grandes agglomérations à la veille d'un changement technologique majeur (remplacement des supercarburants au plomb par des supercarburants à additifs oxygénés) susceptible de modifier profondément les données de base de cette chimie.

Dans ce but, deux études ont été entreprises par le laboratoire, l'une visant à déterminer la réactivité des principaux aldéhydes présents dans l'atmosphère vis-à-vis des radicaux $OH^{\cdot}$ et $NO_3^{\cdot}$ et de l'ozone (2), l'autre visant à estimer les différents facteurs sources et puits d'aldéhydes dans les masses d'air transitant au-dessus de la région parisienne prise comme site laboratoire (3). Dans le présent mémoire nous rappellerons les résultats les plus marquants obtenus à ce jour et confronterons les enseignements que l'on peut en déduire.

TABLEAU 1 : Constantes cinétiques des réactions aldéhydes + OH• (298 ± 2 K)

$k_{OH^{\bullet}} \cdot 10^{12}$ (cm^3 molécule^{-1} s^{-1})	Auteur	Année	Technique	Référence
$HCHO + OH^{\bullet} : k_1$				
>7	Herron et Penzhorn	1969	ED-SM	4
14 ∓ 3.5	Morris et Niki	1971	ED-SM	5
9.4 ∓ 1.0	Atkinson et Pitts	1978	PE-FR	6
14.4 ∓ 0.8	Niki et al	1978	MR	7
6.5 ∓ 1.5	Smith	1978	ED-SM	8
9.9 ∓ 1.1	Stief et al	1980	PE-FR	9
$k_1 = (10.43 \mp 0.92).10^{-12}$ cm^3 molécule^{-1} s^{-1} : ce travail				
$CH_3CHO + OH^{\bullet} : k_2$				
15 ∓ 3.8	Morris et al	1971	ED-SM	10
<20	Cox et al	1976	MR	11
16 ∓ 1.6	Atkinson et Pitts	1978	PE-FR	6
15.2 ∓ 1.6	Niki et al	1978	MR	7
12 ∓ 0.4	Kerr et Sheppard	1981	MR	12
12.2 ∓ 0.25	Semmes et al	1985	MR	13
$k_2 = (15.94 \mp 1.64).10^{-12}$ cm^3 molécule^{-1} s^{-1} : ce travail				
$CH_2{=}CHCHO + OH^{\bullet} : k_3$				
19 ∓ 2	Kerr et Sheppard	1981	MR	12
$k_3 = (28.22 \mp 0.78).10^{-12}$ cm^3 molécule^{-1} s^{-1} : ce travail				

TABLEAU 1 : suite

$C_6H_5CHO + OH^\bullet : k_4$				
12.8 ∓ 0.8	Niki et al	1978	MR	7
11 ∓ 2	Kerr et Sheppard	1981	MR	12
$k_4 = (6.70 \mp 0.22).10^{-12}$ cm^3 molécule^{-1} s^{-1} : ce travail				
$CH_3(CH_2)_4CHO + OH^\bullet : k_5$				
$k_5 = (19.60 \mp 0.79).10^{-12}$ cm^3 molécule^{-1} s^{-1} : ce travail				

ED = Réacteur à écoulement et à décharge
SM = Spectroscopie de masse FR = Fluorescence résonante
PE = Photolyse éclair MR = Mesure relative

TABLEAU 2 : Constantes cinétiques des réactions aldéhydes + $NO_3^\bullet$ (298 ± 2 K)

$k_{OH}.10^{15}$ (cm^3.molécule^{-1}.s^{-1})	Auteur	Année	Technique	Référence
$HCHO + NO_3^\bullet : k_1$				
0.323 ∓ 0.026	Atkinson et al	1984	MD	14
0.630 ∓ 0.035	Cantrell et al	1985	MD	15
$k_1 = (0.80 \mp 0.15).10^{-15}$ cm^3 molécule^{-1} s^{-1} : ce travail				

TABLEAU 2 : suite

$CH_3CHO + NO_3^{\cdot} : k_2$				
1.40 ∓ 0.35	Morris et Niki	1974	MD	16
1.34 ∓ 0.28	Atkinson et al	1984	MD	14
$k_2 = (2.65 \mp 0.36).10^{-15}$ cm^3 $molécule^{-1}$ s^{-1} : ce travail				
$CH_2{=}CHOCHO + NO_3^{\cdot} : k_3$				
$k_3 = (8.07 \mp 1.13).10^{-15}$ cm^3 $molécule^{-1}$ s^{-1} : ce travail				
$C_6H_5CHO + NO_3^{\cdot} : k_4$				
<8	Carter et al	1981	MR	17
1.13 ∓ 0.25	Atkinson et al	1984	MR	18
$k_4 = (3.72 \mp 0.72).10^{-15}$ cm^3 $molécule^{-1}$ s^{-1} : ce travail				
$CH_3(CH_2)_4CHO + NO_3^{\cdot}$				
$k_5 = (12.8 \mp 0.81).10^{-15}$ cm^3 $molécule^{-1}$ s^{-1} : ce travail				
MD = Méthode directe		MR = Méthode relative		

2. REACTIVITE DES ALDEHYDES VIS-A-VIS DE OH$^{\bullet}$, O_3 et $NO_3^{\bullet}$

La réactivité des précursseurs d'aldéhydes vis-à-vis de OH$^{\bullet}$, O_3 et $NO_3^{\bullet}$ est généralement pas trop mal connue. Par contre, la réactivité des aldéhydes eux-mêmes est très mal connue : ou les données manquent, ou elles sont très dispersées. Dans le but de préciser nos connaissances sur le temps de résidence atmosphérique de ces composés, nous avons mesuré ou remesuré, sous conditions atmosphériques simulées, les constantes de vitesse de cinq aldéhydes avec OH$^{\bullet}$, $NO_3^{\bullet}$ et O_3 à température ambiante par cinétique compétitive. Les valeurs trouvées et les valeurs de la littérature sont données dans le tableau 1 pour OH$^{\bullet}$ et le tableau 2 pour $NO_3^{\bullet}$. Pour l'ozone la constante cinétique de la réaction directe avec les aldéhydes en phase gazeuse homogène a été systématiquement trouvée inférieure à $0{,}75\ 10^{-20}$ cm^3 molécule $^{-1}$ s^{-1} (plus petite constante mesurable dans nos conditions opératoires).

En admettant, comme il est actuellement plausible, des concentrations diurnes en radicaux OH$^{\bullet}$ de 0,2 à 1×10^6 cm^{-3} dans les zones rurales (19-22) non polluées et de 10^6 à 10^7 cm^3 dans les zones polluées, et des concentrations nocturnes de 1 à 4 10^8 $NO_3^{\bullet}$ cm^{-3}, on trouve pour les cinq aldéhydes étudiés les fourchettes suivantes pour les temps de résidence dans la basse troposphère. (Tableau 3)

TABLEAU 3 : Temps de vie estimés des aldéhydes dans la basse troposphère

Composés	τ diurne (milieu rural non pollué)	τ diurne (milieu pollué)	τ nocturne
HCHO	1 j < τ < 5,6j	2,6h < τ < 24h	18 j < τ < 72 j
CH_3CHO	0,7j < τ < 3,6j	2 h < τ < 18h	5,5j < τ < 22 j
CH_2 = CHCHO	0,5j < τ < 2 j	1 h < τ < 10h	2,7j < τ < 7 j
C_6H_5 CHO	2 j < τ < 8 j	4 h < τ < 40h	4 j < τ < 15,5j
C_5H_{11} CHO	0,6j < τ < 3 j	1,5h < τ < 14h	1,2j < τ < 4,5j

3. ETUDE DES FACTEURS SOURCES ET PUITS D'ALDEHYDES DANS LES MASSES D'AIR TRANSITANT AU-DESSUS DE LA REGION PARISIENNE.

Cette étude est une partie d'une étude plus générale sur les facteurs sources et puits des photooxydants dans les masses d'air transitant au-dessus de la région parisienne, entreprise en collaboration avec une autre équipe de l'UA 717 (G. TOUPANCE et P. PERROS). Une première campagne a eu lieu pendant l'été 1985 (3). Pendant cet été 1986 nous effectuons une deuxième campagne avec des moyens plus importants. Rappelons les enseignements principaux de la campagne 1985.

1) Dans les zones rurales de l'Ile-de-France, hors de l'influence de l'agglomération parisienne, on observe des maxima journaliers de 10 à 15 ppbv en formaldéhyde et de 3 à 5 ppbv en acétaldéhyde par journée de beau temps chaud (t max 25 à 28 °C) et à l'opposé de 3 à 5 ppbv de formaldéhyde et de l'ordre de 1 ppbv ou moins en acétaldéhyde par temps frais couvert (t max 15 à 18°C).

2) Pour les données acquises au centre de la ville à l'altitude de 50 mètres il est encore possible de distinguer deux cas extrêmes. Par temps

frais couvert on observe des niveaux assez faibles, avec deux maxima journaliers (5-8 ppbv HCHO) correspondant aux heures de pointe de la circulation, ce qui indique une influence prépondérante des aldéhydes primaires émis par les véhicules automobiles. Par temps chaud ensoleillé on observe un seul maximum vers midi avec des concentrations nettement plus élevées, jusqu'à 30 ppbv HCHO et 8 ppbv CH_3CHO, ce qui indique que les aldéhydes secondaires l'emportent alors nettement sur les aldéhydes primaires.

3) Dans les zones rurales situées dans le panache de l'agglomération parisienne et à une distance de 40 km du centre ville, nous avons pu observer des teneurs en aldéhydes supérieures à celles observées simultanément à Paris, même par une journée de temps frais couvert. Compte-tenu que le temps de transport (2 à 3 h) du centre de la ville au point de prélèvement est du même ordre de grandeur que le temps de résidence estimé des aldéhydes, donc que les aldéhydes primaires émis dans l'agglomération ont déjà en majeure partie disparu , il est clair que ces niveaux élevés sont dus à la photooxydation des hydrocarbures pendant le transport, et ce malgrè la dispersion verticale de l'ensemble des polluants.

4. CONCLUSION

Les résultats montrent que, exceptés à l'approche immédiate des sources intenses de composés carbonylés, le problème des aldéhydes dans la basse troposphère est principalement un problème de polluants secondaires et que de ce fait, les problèmes majeurs peuvent se poser non dans le centre des grandes agglomérations mais souvent dans leur panache à une distance telle que les hydrocarbures émis (et sans doute principalement les oléfines et les alcoylebenzènes qui sont les plus réactifs) aient été déjà photooxydés et que d'autre part l'ensemble des contaminants n'ait pas été dispersé. Dans le cas particulier de notre étude (agglomération parisienne en été) il semble qu'une distance de 40 à 50 km,correspondant à des temps de transport de 2 à 3 heures (vitesse de vent : 15 à 20 km/h), satisfasse à ces deux conditions.

REMERCIEMENTS

Nous remercions le Centre National de la Recherche Scientifique (Programme interdisciplinaire de recherche sur l'environnement), le Ministère de l'Environnement, l'Union des Chambres Syndicales de l'Industrie du Pétrole et la fondation Alexandre Onassis pour les appuis financiers (contrats de recherche ou bourse de Thèse) qu'ils ont accordés à ce travail.

BIBLIOGRAPHIE

(1) P. CARLIER, H. HANNACHI et G. MOUVIER
Atm. Envir. sous presse.

(2) H. HANNACHI
Thèse de Doctorat. Université Paris VII (1986)

(3) P. KALABOKAS, P. CARLIER, P. FRESNET, G. MOUVIER, P. PERROS, M. SOEDOMO, K. TSALKANI et G. TOUPANCE.
in : Chemistry related to tropospheric Ozone. COST 611 (W.P.II) Cologne 12. 13 novembre 1986.

(4) J.T. HERRON et R.D. PENZHORN.
J. Phys. Chem., 73, 191-196 (1969)

(5) E.D. MORRIS Jr. et H. NIKI.
J. Phys; Chem., 55, 1991-1992, (1971).

(6) R. ATKINSON et J.N. PITTS Jr.
J. Chem. Phys., 68, 3581-3584, (1978).

(7) H. NIKI, P.D. MAKER, C.M. SAVAGE et L.P. BREITENBACH
J. Phys. Chem., 82, 132-134, (1978).

(8) R.H. SMITH
Int. J. Chem. kinet., 10, 519, (1978).

(9) L.J. STIEF, D.F. NAVA, W.A. PAYNE et J.V. MICHAEL
J. Chem. Phys., 73, 2254-2258, (1980)

(10) E.D. MORRIS Jr., D.H. STEDMAN et H. NIKI.
J. Amer. Chem. Soc., 93, 3570-3572, (1971).

(11) R.A. COX, R.G. DERWENT, P.M. HOLT et J.A. KERR
J. Chem. Soc. Fara., 1, 72, 2061-2075, (1976).

(12) J.A. KERR et D.W. SHEPPAD
Env. Sc. Technol., 15, 960-963, (1981).

(13) D.H. SEMMES, A.R. RAVISHANKARA, C.A. GUMP-PERKINS et P.H. WINE
Int. J. Chem. Kinet., 17, 303-313, (1985).

(14) R. ATKINSON, C.N. PLUM, W.P.L. CARTER, A.M. WINER et J.N. PITTS Jr.
J. Phys. Chem., 88, 1210, (1984).

(15) C.A. CANTRELL, W.R. STOCKWELL, L.G. ANDERSON, K.L. BUSARROW, D. PERNER, A. SCHMELTE KUPF, J.G. CALVERT et H. JOHNSTON.
J. Phys., Chem., 89-146, (1985).

(16) E.D. MORRIS et H. NIKI
J. Phys. Chem., 78, 1337, (1974).

(17) W.P.L. CARTER, A.M. WINER et J.N. PITTS Jr.
Env. Sci. Technol., 15, 829, (1981).

(18) R. ATKINSON, W.P.L. CARTER, C.N. PLUM, A.M. WINER et J.N. PITTS Jr.
Int. J. Chem. Kinet., 16, 887, (1984).

(19) J.A. LOGAN, M.J. PRATHER, S.C. WOPSY et M.B. Mc ELROY
J. Geophys. Res., 86, 7210, (1981).

(20) B. WEINSTOLK, H. NIKI et T.Y. CHANG
In : Adv. Environ. Sci. Technol. (J.N. PITTS Jr. et R.L. METCALF ed)
10, 221, (1980). John Wiley and Sons (New-York).

(21) U. PLATT, D. PERNER, A.M. WINER, G.W. HARRIS et J.N.
Geophys. Res. Lett., 7, 89, (1980).

(22) U. PLATT, J. SCHRODER, D. PERNER, C. KESSLER, A.J. TOENNISSEN
J. Geophys. Res., 86, 11965, (1981).

KINETICS OF THE FORMATION OF NITROUS ACID FROM THE THERMAL REACTION OF NITROGEN DIOXIDE AND WATER VAPOUR

R.A. COX and M.E. JENKIN
The Harwell Laboratory, Oxfordshire, England

Summary

The kinetics of the formation of nitrous acid (HONO) and removal of NO_2 during the heterogeneous reaction of NO_2 and H_2O were investigated in the temperature range 277-311K and pressure range 0.7-297 Torr, using infra-red diode laser spectroscopy and uv/visible spectroscopy. Experiments were performed in a 19.8ℓ pyrex reaction vessel (s/v = 13 m^{-1}) with stainless steel end plates and gold coated multi-reflection 'White' optics. The initial kinetics of HONO production and NO_2 removal were found to be 1st order in both $[NO_2]$ and $[H_2O]$, and the observed 2nd order rate constant, independent of pressure at 296K, was:-

$$k = \frac{d[HONO]/dt}{[NO_2]\,[H_2O]} = 3.2 \times 10^{-22}\ cm^3\ molecule^{-1}\ s^{-1}$$

The yield of HONO during the early stages of reaction was ~50% relative to NO_2 removed, but as the reaction proceeded the kinetics became more complicated, and the yield was reduced. HNO_3 is almost certainly produced, but could not be detected by infra-red measurement at 890 cm^{-1}. HONO formation by this reaction may be an important precursor to OH radicals in the atmospheric boundary layer. An estimate of the production rate of HONO in the night-time boundary layer has been made using the results of this work, and a comparison is made to those observed in the field.

1. INTRODUCTION

The importance of the OH radical in atmospheric oxidation mechanisms arises from its ability to react with a wide range of primary pollutants converting them to secondary pollutants such as aldehydes, acids and PAN with concurrent production of O_3 and regeneration of an OH radical.

At all altitudes, OH is produced from the reaction of $O(^1D)$ with H_2O, the $O(^1D)$ atoms arising from the photolysis of O_3 (λ<310nm). In the troposphere, the near uv photolysis of nitrous acid (HONO) might represent a significant additional source of OH (reaction (1)), provided HONO is produced by a mechanism which does not involve the removal of OH.

$$HONO + h\nu\ (330\ nm < \lambda < 380 nm) \longrightarrow OH + NO \qquad (1)$$

The importance of the heterogeneous thermal reaction of NO_2 with H_2O, believed to be reaction (2), for HONO generation has become apparent from smog chamber studies(1), where it was noted that OH radicals were formed from a precursor of heterogeneous origin.

$$2NO_2 + H_2O \longrightarrow HONO + HNO_3 \qquad (2)$$

More recent work(2,3) characterising this reaction indicates that it may be of atmospheric importance particularly in NO_x polluted air masses. Further support for this has been provided by direct observations of HONO in the boundary layer atmosphere using long path length absorption spectroscopy (4-6), which have indicated a strong relationship between ambient HONO and NO_2.

In this paper, the results of experiments performed to characterise the kinetics and mechanism of the reaction of NO_2 and H_2O are presented. An estimate of production rates of HONO in the night-time atmospheric boundary layer is also made, using the results of this work.

2. EXPERIMENTAL

The experiments were performed in a 19.8ℓ cylindrical pyrex reaction vessel (s/v = 13 m^{-1}) with stainless steel end plates and containing 'White' reflective optics consisting of 3 gold coated mirrors on stainless steel mounts. For some experiments, the pyrex surface was coated with P.T.F.E., and for one experiment the pyrex and stainless steel surfaces were coated with halocarbon wax. Two series of experiments were performed at ambient temperatures which were 296(±1)K and 292 (±1)K. For a limited number of experiments, the temperature of the pyrex surface was controlled at 277K and 311K. This was achieved by circulation of a thermostatically controlled 30% glycol/water mixture through P.V.C. tubing (10 mm i.d., 1 mm wall thickness) which was tightly coiled around the exterior of the vessel.

The majority of experiments were performed at pressures below 10 Torr, but a few experiments were carried out at pressures up to 297 Torr. The desired pressure of water vapour (Analar) in the range 0-10 Torr was measured into the reaction vessel from a vacuum manifold using an MKS Baratron capacitance manometer and was allowed to equilibrate for 30 minutes.

A concentration of purified NO_2 between 10^{14} and 10^{15} molecule cm^{-3} was injected and monitored by conventional absorption spectroscopy at 400 nm, and the subsequent time dependence of NO_2 was observed in the same manner.

The reaction mixture was also monitored in the infra-red using a tunable diode laser (Spectra Physics LS3) which was coupled to the multi-reflection White optics mounted internally in the reaction vessel as described before(7). The production of HONO was monitored using a vibration-rotation line of the band at 853 cm^{-1} (cis-HONO, υ_4). Infra-red measurements were also used to verify the time dependence of NO_2 (υ_2, 823 cm^{-1}) and to look for the formation of HNO_3 (υ_5, 891 cm^{-1}).

3. RESULTS

3.1 Kinetics of NO_2 removal in the vessel with PTFE coating

Early experiments performed with the pyrex surface of the reaction vessel coated with PTFE and at a temperature of 292K, concentrated on characterising the decay kinetics of NO_2 during its reaction with H_2O. In each experiment a steady decay of NO_2 was observed in the early stages but this was soon reduced in such a manner that the decay profiles did not obey simple kinetics. Variation of initial $[NO_2]$ at a constant $[H_2O]$ of 3.5 Torr resulted in a proportionate change in the initial decay rate, as shown in Figure 1, suggesting that NO_2 removal was first order, with an apparent constant, k_I, given by the slope of the line.

The variation of k_I with $[H_2O]$ is shown in Figure 2. The linear relation indicates that the reaction is also first order in water vapour concentration and the slope, 7.0×10^{-22} cm^3 $molecule^{-1}s^{-1}$, is the apparent second order decay constant characteristic of this vessel. The intercept represents the first order decay constant for NO_2 in the absence of H_2O which is presumably due to wall removal.

3.2 Kinetics Studies in the uncoated vessel

(i) Kinetics of NO_2 removal:

The results of a similar set of experiments performed in the uncoated vessel at 296K are shown in Figure 3. At water vapour pressures above approximately 2 Torr, the first order decay constant for NO_2, which was calculated once again from measurement of initial slopes, displayed a similar positive linear dependence on water vapour pressure. At pressures below 2 Torr, a deviation from linearity was observed which is attributed to a wall removal process for NO_2 which becomes more important at lower surface coverages of water. A least squares regression of points corresponding to $[H_2O] \geq 2$ Torr provided an apparent second order constant (slope) of 5.76×10^{-22} cm^3 $molecule^{-1}s^{-1}$ which is 18% lower than the value obtained in the last section.

(ii) Kinetics of HONO formation:

The production of HONO was also monitored during experiments performed in the uncoated vessel. Figure 4 shows a 0.8 cm^{-1} portion of the i.r. spectrum near 853 cm^{-1} during one experiment. The absorption path length was 60m. The intensity of one line corresponding to cis HONO was calibrated by generation of a sample of HONO ($\sim 10^{14}$ molecule cm^{-3}) from the reaction of 10% H_2SO_4 and 0.05M $NaNO_2$ as described previously(8), which was simultaneously monitored in the uv. at 230 nm. This line was used subsequently to determine the time dependence of HONO during the reaction of NO_2 and H_2O.

HONO concentration vs. time curves for 2 experiments from a series of 5 performed at 296K are shown in Figure 5. Initial $[NO_2]$ were similar in each case, but the pressures of water vapour were 0.68 and 2.01 Torr. The quantities of HONO produced in all these experiments increased with both NO_2 and H_2O, but clearly HONO was being removed since steady state is almost achieved after 100 minutes reaction time. The removal process for HONO was shown to be first order as indicated by plots of ln $([HONO]_{ss} - [HONO]_t)$ vs t. Production rates (B) of HONO were calculated for each experiment using the expression

$$B = k_r [HONO]_{ss}$$

where k_r is the first order removal constant for HONO.

A first order rate constant for HONO production was derived for each experiment and was defined as $k_I' = B/[NO_2]_i$, and Figure 6 shows the variation of k_I' with $[H_2O]$. A least squares regression of the 5 points gives a value of the apparent 2nd order rate constant for the $NO_2 \rightarrow HONO$ reaction (slope) of 3.16×10^{-22} cm^3 $molecule^{-1}$ s^{-1} at 296K, which is ~55% of the value calculated from NO_2 decay measurements, and suggests a reaction stoichiometry of 2:1, NO_2:HONO as would be predicted by equation (2).

The yields of HONO at 100 minutes reaction time expressed as a fraction of NO_2 removed are shown in Figure 7 as a function of H_2O. In the experiments performed at 2 Torr and above, the yield is consistently between 30% and 35%. These are somewhat lower than 50% owing to the subsequent removal of HONO in the system. The yield at low $[H_2O]$ is dramatically reduced and is further evidence for non-reactive adsorption of NO_2 in the low surface coverage regime as indicated earlier by the anomalous NO_2 removal data at low $[H_2O]$.

Attempts were made to observe gaseous HNO_3 by monitoring reaction mixtures at ~890 cm^{-1} (υ_5,HNO_3), but none was observed. A sample of HNO_3 was prepared by reaction of $NaNO_3$ with 100% H_2SO_4 at 0^oC(9), and it was found to be absorbed quite rapidly on the vessel walls, with a half life of ~5 minutes. It was also apparent that a concentration of 10^{13} molecule cm^{-3} would be detectable in the gas phase indicating that the yield of gaseous HNO_3 was < 3% during the NO_2/H_2O experiment performed whilst monitoring for HNO_3.

(iii) Temperature Variation Studies:

A further 4 experiments were performed with the temperature of the pyrex surface controlled at 277K and 311K (2 experiments at each temperature). The observed NO_2 removal constants, HONO production constants and yields are displayed in Figures 3, 6 and 7 respectively.

At 277K, HONO production rates and NO_2 removal rates did not differ significantly from the results at 296K. The yield of HONO, however, was reduced to ~17.5% indicative of more efficient HONO removal at the lower temperature.

At 311K, the lack of dependence of the NO_2 decay constant on $[H_2O]$ and the dependence of HONO yield on $[H_2O]$ suggest that measurements were being made in the low 'surface coverage' regime which would be expected to extend to higher $[H_2O]$ at the increased temperature. The dependence of the HONO production rate on $[H_2O]$ shown in Figure 6 is about 50% of that at 296K, suggesting a small negative temperature dependence.

(iv) Pressure Variation Studies

A study of the dependence of NO_2 removal rates on pressure was also made. Experiments were carried out at room temperature (296K) with water vapour pressures of 7.80 (±0.02) Torr which were high enough to ensure that measurements were not being made in the 'low surface coverage' regime. The results of 4 experiments performed in the pressure range 29-297 Torr displayed no dependence on pressure.

3.3 Coating of Vessel with Halocarbon Wax

A further experiment was carried out to observe how HONO production was affected by coating the vessel surface (pyrex body, stainless steel end plates and mirror mounts) with Halocarbon Wax. For T = 296K, and $[H_2O]$ = 4.5 Torr, the HONO production constant was lowered by a factor ~5 compared with the uncoated vessel.

4. Discussion

The results of the experiments described in the previous section indicate that NO_2 and H_2O react on the vessel surface with observed production of gaseous HONO, but not of gaseous HNO_3. The rate constants derived from observation of both removal of NO_2 and production of HONO displayed a linear dependence on NO_2 and H_2O. The failure to detect gaseous HNO_3 is a clear indication of heterogeneous reaction. HNO_3 is sufficiently long lived in the vessel to be observed if it was produced in the gas phase. HNO_3 has been easily detected as a product of the gaseous reaction of NO_2 and with both CH_3OH and C_2H_5OH(10).

In the present study, similar results were obtained whether or not the vessel was coated with PTFE, but coating with halocarbon wax reduced the rate of HONO production by a factor of 5. Variation in the temperature of the pyrex surface in the range 277-311K had a small effect on the rate of reaction. This may be indicative that the reaction is occurring on the stainless steel end plates and gold mirrors. Variation of the total pressure up to 297 Torr had no effect on NO_2 removal kinetics confirming that the reaction is too slow to be diffusion controlled.

The yield of HONO compared to NO_2 removed was approximately 50% in the early stages of reaction, suggesting that reaction (2) is a good representation of HONO producing reaction, with HNO_3 remaining at the wall

$$2\ NO_2 + H_2O \rightarrow HONO + HNO_3 \qquad (2)$$

Subsequent reaction of HONO occurred, probably resulting in regeneration of NO_2 since the decay of NO_2 was observed to decrease at longer reaction times. It is possible that the reaction of HONO with HNO_3 at the wall might be occurring (i.e. the reverse of reaction (2)) as observed previously by Kaiser and Wu(11). Under these circumstances an equlibrium would be established with H_2O, NO_2 and HONO present largely in the gas phase, but with HNO_3 in the adsorbed phase and both forward and reverse reactions would be occurring at the wall. The values of $\Delta H^o{}_2$ and $\Delta S^o{}_2$ at 298K are - 36.87 KJ mol^{-1} and -152.9 J $mol^{-1}K^{-1}$(12) if all species are gaseous. The derived equilibrium constant, K_p = 0.03 atm^{-1} would predict an equilibrium ratio of $[HONO]/[NO_2]$ of 0.014 for H_2O in excess at 5 Torr. After 100 minutes in the present experiments, a steady state had almost been achieved with $[HONO]/[NO_2] \approx 0.18$ when $[H_2O]$ = 5 Torr. An apparent shift in the equilibrium ratio of $[HONO]/[NO_2]$ to higher values could result from a major fraction of the HNO_3 being adsorbed at the wall. A reaction scheme consistent with all observations is as follows:-

$$H_2O\ (g) \rightleftarrows H_2O\ (ads) \qquad (a)$$

$$NO_2\ (g) \rightleftarrows NO_2\ (ads) \qquad (b)$$

$$NO_2\ (ads) + H_2O\ (ads) \rightarrow NO_2.H_2O\ (ads) \qquad (c)$$

$$NO_2\ (ads) + NO_2.H_2O\ (ads) \rightarrow HNO_3\ (ads) + HONO\ (g) \qquad (d)$$

$$HONO\ (g) + HNO_3\ (ads) \rightarrow 2NO_2\ (g) + H_2O\ (g) \qquad (e)$$

Reaction (c) is rate determining, resulting in a first order dependence of the reaction on both $[NO_2(g)]$ and $[H_2O(g)]$.

$$\frac{d\,[HONO]}{dt} = -0.5\ \frac{d\,[NO_2]}{dt} = K_a\ K_b\ k_c\ [NO_2(g)]\ [H_2O(g)]$$

5. ATMOSPHERIC PRODUCTION OF HONO

Using the kinetics results obtained in the laboratory it is possible to estimate production rates of HONO in the atmosphere arising from the heterogeneous reaction of NO_2 and H_2O, although this must be done with some caution since there are many obvious differences between the experimental conditions and those found in the atmosphere.

The surfaces available for reaction are provided by either the ground or by particulate matter, both of which are composed of a large variety of surfaces. The ground and associated surfaces may be comprised of metals, concrete soil and plants, etc., whereas the atmospheric aerosol, although largely made up of ammonium sulphate and nitrate, may have a significant proportion of organic carbon and minerals from suspended soil dust in the urban environment. The water soluble components will be in the form of droplets at higher relative humidities.

The surface area available for reaction on particles, as indiated by visibility measurements, is of the order of 0.01 m^{-1} for a reasonably polluted air mass (atmospheric visibility = 1 km). The surface area available at the ground is related to surface roughness and can therefore be significantly larger than the bulk surface area. However, s/v is usually quoted as bulk surface area/boundary layer height, which is 10^{-3} m^{-1} for a typical daytime boundary layer of 1km in depth and 0.02 m^{-1} for a typical night-time boundary layer of 50m depth.

If the kinetics results obtained in the laboratory are scaled to account for the following atmospheric conditions,

(i) s/v = 3×10^{-2} m^{-1} (night-time)

(ii) $[H_2O]$ = 1.36×10^{17} molecule cm^{-3} (T=278K, Rel.Humidity = 60%)

the corresponding $NO_2 \rightarrow$ HONO conversion rate is 0.036% hr^{-1}, which is a factor of 17 lower than the value of 0.6% hr^{-1} observed by Kessler and Platt(6). It should be noted that the majority of s/v is supplied by the ground, and provided transport to the ground occurs by processes other than molecular diffusion (diffusion distance ≈ 1m in 12 hours), then the value for the conversion rate calculated here is a lower limit since the ground was assumed to be smooth. It seems feasible, therefore, that the heterogeneous reaction of NO_2 and H_2O might explain the production rates of HONO observed in the atmosphere at night.

During the day the percentage conversion rate of NO_2 to HONO by heterogeneous processes would be expected to be smaller owing to the increased boundary layer height and decreased s/v.

ACKNOWLEDGEMENT

This work was supported by the U.K. Department of the Environment and by the C.E.C. Environmental Research Programme.

REFERENCES

(1) CARTER, W.P.L., ATKINSON, R., WINER, A.M. and PITTS, J.N. Jr., Int. J. Chem. Kinet. 14 1071 (1982)

(2) SAKAMAKI, F., HATEKEYAMA, S. and AKIMOTO, H., Int. J. Chem. Kinet. 15 1013 (1983)

(3) PITTS, J.N., Jr., SANHEUZA, E., ATKINSON. R., CARTER, W.P.L., WINER, A.M., HARRIS, G.W. and PLUM, C.N., Int. J. Chem. Kinet. 16 919 (1984)

(4) PLATT, U, PERNER, D., HARRIS, G.W., WINER, A.M. and PITTS, J.N.,Jr Nature 283 312 (1980)

(5) HARRIS, G.W., CARTER, W.P.L., WINER, A.M. and PITTS, J.N., Jr., Env. Sci. Technol. 16 414 (1982).

(6) KESSLER, C, and PLATT, U., Nitrous acid in polluted air masses - sources and formation pathways, CEC, Proc. of the 3rd European symposium on physico-chemical behaviour of atmospheric pollutants, Varese, Italy 10-12 April 1984

(7) COX, R.A. and COKER, G.B., J. Atm.Chem. 1 53 (1983)

(8) COX, R.A., DERWENT, R.G., KEARSEY, S.V., BATT, L. and PATRICK, K.G., J. Photochem 13 149 (1980)

(9) COTTON, F.W. and WILKINSON, G., Advanced inorganic chemistry, 3rd Ed; Witey Interscience: New York, 1972; p179

(10) NIKI, H., MAKER, P.D., SAVAGE, C.M. and BREITENBACH, L.P., Int. J. Chem. Kinet. 14 1199 (1982)

(11) KAISER, E.W. and WU, C.H., J. Phys Chem. 81 187 (1977)

(12) JANAF Thermodynamic Tables, 2nd Ed NSRDS - NBS 37 (1971)

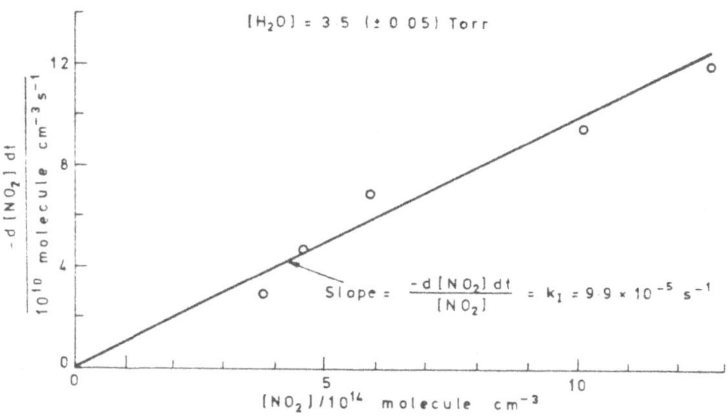

FIG. 1 VARIATION OF NO_2 DECAY RATE WITH INITIAL $[NO_2]$ IN TEFLON COATED CELL (T = 292 K)

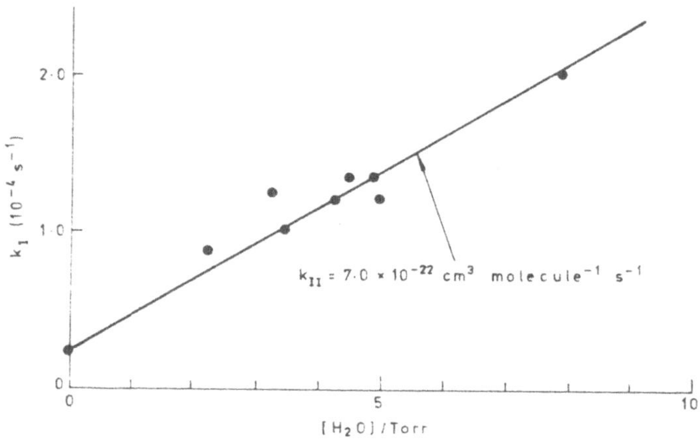

FIG 2 VARIATION OF 1st ORDER RATE CONSTANT WITH $[H_2O]$ IN TEFLON COATED CELL (T = 292 K)

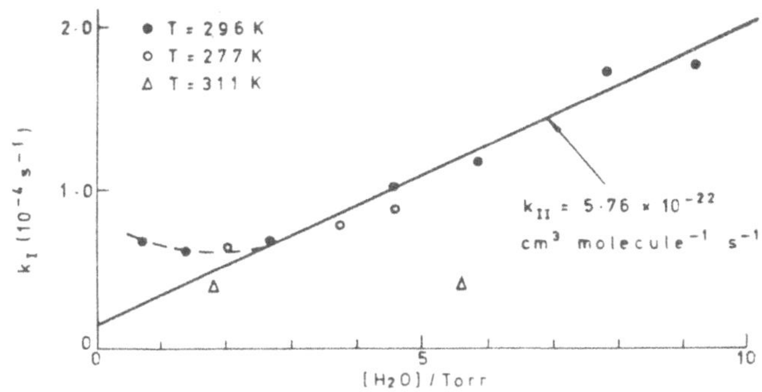

FIG 3 VARIATION OF 1st ORDER NO_2 REMOVAL CONSTANT WITH [H_2O] IN UNCOATED CELL

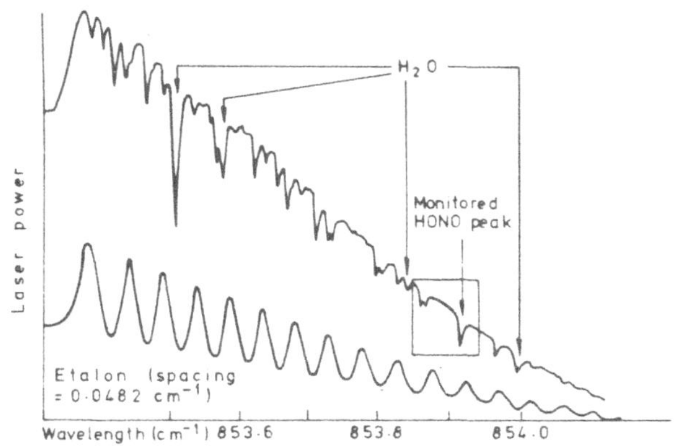

FIG. 4. INFRA RED DETECTION OF HONO

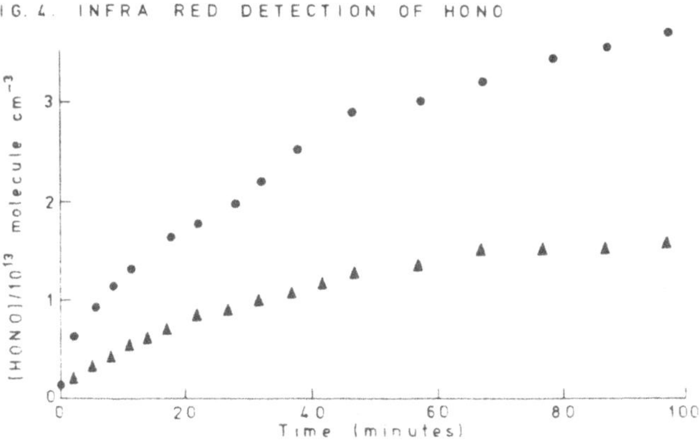

FIG. 5 TIME DEPENDENCE OF HONO PRODUCTION DURING EXPERIMENTS PERFORMED AT 0·68 TORR H_2O (▲) AND 2·01 TORR H_2O (●) AND 296 K

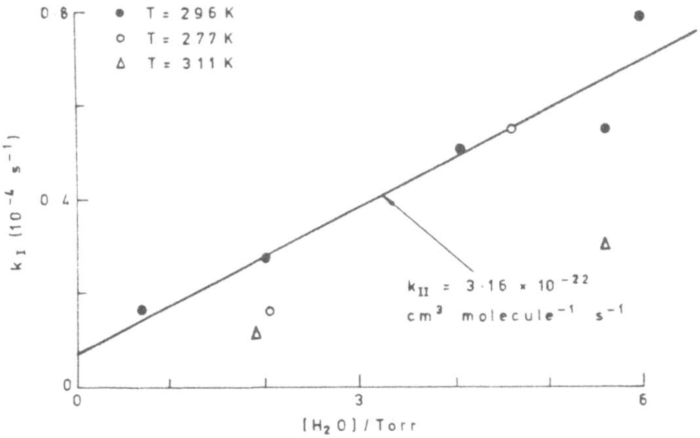

FIG. 6 VARIATION OF 1st ORDER HONO PRODUCTION CONSTANT WITH H_2O IN UNCOATED CELL

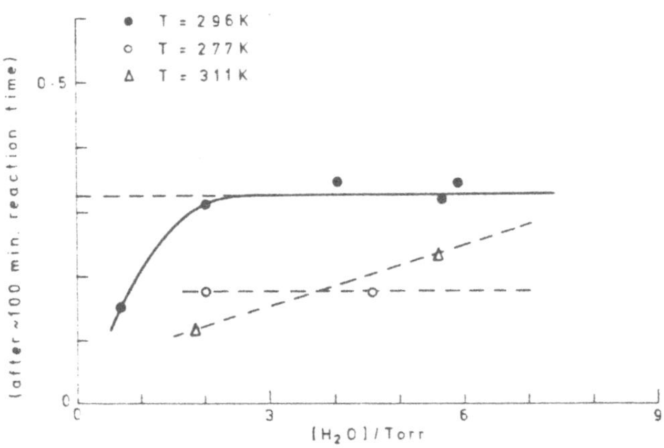

FIG 7 VARIATION OF HONO YIELDS WITH PRESSURE OF H_2O

FT-IR LABORATORY STUDIES OF KINETICS OF THE NO_x/O_3 SYSTEM OF TROPOSPHERIC INTEREST

J. HJORTH, G. OTTOBRINI, F. CAPPELLANI and G. RESTELLI
Commission of the European Communities
Joint Research Centre - Ispra Establishment
21020 Ispra (Va) - Italy

Summary

Two reactions of relevance to the nighttime equilibrium between the nitrogen oxides in the system NO_x/O_3 and to the formation of nitric acid at night have been studied by the use of FTIR-spectroscopy and large (1.5 m^3) Teflon bags as reaction chambers. The reaction, previously proposed by other authors, $NO_3 + CO \rightarrow NO_2 + CO_2$ was investigated by the use of isotopically labelled carbon monoxide. For the reaction rate constant an upper limit of 4×10^{-19} $cm^3 molec^{-1} s^{-1}$ at 296 K was found, which means that the reaction is of negligible atmospheric importance. The reaction $N_2O_5 + H_2O \rightarrow 2HNO_3$ was also investigated and the rate constant was found to be $(1.48 \pm 0.42) \times 10^{-21}$ $cm^3 molec^{-1} s^{-1}$ at 296 K. Due to the possible influence of the heterogeneous reaction between N_2O_5 and water, this must be considered also an upper limit for the homogeneous gas phase reaction. The results of these studies are discussed in view of the experimental procedures used and the difficulties encountered. Absolute band intensities of two bands of N_2O_5 and two bands of HNO_3, measured by curve of growth analysis in the context of these studies, are also presented.

1. INTRODUCTION

It has been recognized that the NO_3 radical and N_2O_5 formed mainly by the reaction between NO_2 and O_3 and built up in the nighttime troposphere may play an important role in the trasformation of nitrogen oxides to nitric acid (1), in the radical chemistry of the nighttime troposphere (2) and as a sink for some organic compounds. The reaction with organics, in turn, might be of interest because of the possible toxicity of the products formed (3).

Atmospheric observations of NO_3 have shown concentrations well below what one would expect from the rate of the formation reaction and the constant of the rapid equilibrium $NO_3 + NO_2 \overset{M}{\longleftrightarrow} N_2O_5$.

Although atmospheric observations of N_2O_5 are not available as for the case of NO_2 and NO_3, an erroneous value of the equilibrium constant does not seem to be a satisfactory explanation for the NO_3 deficit (4). The existence of a sink for the nitrogen oxides in the troposphere at night, via NO_3 and N_2O_5 reactions, might be of importance for the tropospheric budget of NO_y species. In this context it is then of interest to know whether the NO_3 reactions lead mainly to conversion of NO_3 to NO_2 or NO, which would be ineffective with respect to removal from the photochemical cycle or to HNO_3, a process which would really represent a sink for the nitrogen oxides.

The results of two studies are given in the following. They fit into this background and concern the evaluation of the kinetics at room temperature of the reactions:

$$NO_3 + CO \rightarrow CO_2 + NO_2 \qquad (1)$$

$$N_2O_5 + H_2O\ (gas) \rightarrow 2HNO_3 \qquad (2)$$

Reaction (1) has been discussed as a potential explanation of the observed deficit of NO_3 in the troposphere (4,5). The unknown scavenger as hypothesized by Noxon (4) should be ubiquitous and present at relatively high concentrations in air, two criteria that CO does fulfil. Very recently, the reaction between NO_3 and CO has been investigated by Burrows et al. (6) as part of a wider study of NO_3 chemistry using a flow system; an upper limit equal to 4×10^{-16} cm^{+3} $molec^{-1}$ s^{-1} has been found for the rate constant.

The importance of reaction (2) has been the subject of extended discussion. Observed NO_3 concentrations in air are significantly lower at higher values of relative humidity and it is also well known from laboratory experiments that N_2O_5 reacts with H_2O to form nitric acid. However, it has been found difficult to evaluate the rate of the homogeneous gas phase reaction without interference from the analogous heterogeneous reaction with liquid water. The first study performed by Morris and Niki in 1973 (7) using a 67 litre Pyrex glass cell as reaction chamber, gave an upper limit to the rate constant equal to 1.3×10^{-20} cm^{+3} $molec^{-1}$ s^{-1} at 298 K. Later observation (5,8) were in favour of a lower rate constant and this was recently confirmed by the laboratory study of Tuazon et al. (9). These authors using two large Teflon lined smog chambers found a rate constant, after allowing for the contribution of the heterogeneous reaction, equal to 1.3×10^{-21} cm^{+3} $molec^{-1}$ s^{-1} from an originally estimated upper limit equal to 2.4×10^{-21} at 298 ± 1K.

In these studies the reaction NO_3 + CO was evaluated using a static reactor and isotopic tracing to derive an upper limit for the rate constant equal to 4×10^{-19} cm^{+3} $molec^{-1}$ s^{-1} at 296 ± 2K. The reaction $N_2O_5 + H_2O$ (gas) was investigated taking advantage of the observed low reactivity of N_2O_5 in large FEP-Teflon bags (1.5 m^3) with respect to surface reactions. An upper limit was derived equal to $1.48 \pm 0.42\times10^{-21}$ cm^{+3} $molec^{-1}$ s^{-1} at 296 ± 2 K, together with evidence that the rate constant for the homogeneous reaction should most probably be significantly lower.

As part of this work, absolute band strengths were evaluated for the two bands of N_2O_5 at 745 and 1250 cm^{-1} and the two band systems of HNO_3 at 875 cm^{-1} (ν_5, $2\nu_9$ and $\nu_5 + \nu_9 - \nu_9$) and 1310 cm^{-1} (ν_3, ν_4).

2. EXPERIMENT

NO_2 (500 ppmv in air) was mixed in 1.5 m^3 FEP-Teflon (Dupont, type A) bags with surface to volume ration 6 m^{-1}, with purified air added with an excess of O_3 to convert NO_2 to NO_3 and N_2O_5; N_2O_5 was obtained at concentrations in the range 0.5-3 ppm with a half life in the bag in the range 3-7 hours depending on water content. O_3 was prepared by passing purified air from an AADCO generator through a silent discharge. Water vapour concentrations were kept in the range 300-2000 ppm. In the CO-NO_3 experiments, $C^{13}O^{18}$ and $C^{13}O^{16}$ were added to yield concentrations in the bags of 0.2-1.2 ppm. The bags filled at about 740 torr pressure were kept in the dark at 296 ± 2 K during the experiment.

O_3 concentrations were measured using a Dasibi instrument while CO, N_2O_5, HNO_3 and H_2O concentrations were determined by Fourier transform infrared spectroscopy (FTIR).

Infrared spectra were obtained by transferring by Teflon tubing samples from the bag to an evacuated 25 litre volume 70 m optical beam path Pyrex glass cell mounted in the sample compartment of a Bruker IFS 113 V FT spectrometer. Repeated samplings of bags with different concentrations of the reactants and the use of different conditions of transfer to the measuring gas cell did not show any evidence that the transfer procedure used had an influence on the measured spectroscopic data. Spectra were recorded at 2, 1 and 0.06 cm^{-1} instrumental resolution with Happ-Genzel apodization. For each sample three spectra were subsequently collected coadding 100 scans per spectrum. The sequence of spectra at regular time intervals allowed the correction of N_2O_5 absorbance values for the heterogeneous loss of N_2O_5 at the cell wall following injection of the sample. Half life values in the range 10-20 minutes were, in fact, measured for the experimental conditions used. Since water vapour slowly permeated through the Teflon film, its concentration inside the bag was checked frequently during each experiment. N_2O_5 and HNO_3 concentrations were measured from the integrated absorbances of their bands at 1125-1270 cm^{-1} and 1270-1350 cm^{-1}, respectively. While the N_2O_5 band was practically free from interferences by other absorptions, spectral subtraction was used previous to calculation of the HNO_3 band to correct for H_2O absorption in the region (Fig. 1).

The concentrations of $C^{13}O^{16}$ and $C^{13}O^{18}$ were calculated from the peak heights of interference free absorption lines in the regions 2145-2130 cm^{-1} and 2085-2070 cm^{-1}, respectively. Water concentrations were estimated from the line at 2136.144 cm^{-1}. $C^{13}O^{16}O^{18}$ was evaluated from the absorption lines at 2252.697 cm^{-1} and 2276.347 cm^{-1}.

Concentrations were derived from the spectral absorbance data using peak absorption coefficients in the case of isolated lines ($K_\nu = 1/cL \ln_{10}(I_o/I)$) or integrated band intensities ($IBI = 1/cL \int_{band} \lg_{10}(I_o/I)d\nu$) for unresolved band features.

In order to calibrate the instrument for the N_2O_5 an HNO_3 measurements, N_2O_5 was generated in a 450 litre evacuable Pyrex glass-Teflon gas cell with a 76 m multiple reflection system on line with a Bruker IFS 113 V FT spectrometer by mixing NO_2 and O_3. Pressure and instrumental conditions were similar to those applied in the study of the N_2O_5-H_2O reaction. Between one spectrum and the following one, a known volume of the gas phase (2-3 percent of the total) was taken from the centre of the cell using a Teflon tube, and bubbled through a 0.1 percent NaOH solution contained in two impingers in series. The pressure in the cell was kept constant by supplying pure air from a Teflon bag. The nitrate and nitrite ion content of the impingers was then measured by ion chromatography. N_2O_5 was found to form only nitrate ions in the NaOH solution where it was absorbed in the first impinger with approximately 100 percent efficiency. N_2O_5 was always accompanied in the gas phase by variable amounts of HNO_3 due to the reaction with water.

From a series of such measurements, the integrated band intensities (IBI) of HNO_3 and N_2O_5 could be found by a least-square fitting procedure. In a subsequent experiment, only HNO_3 vapour was introduced into the 400 litre cell and IBI values were found by a similar procedure. The latter values were in agreement with those previously derived within 5 percent.

The IBI depends on the effect of various parameters such as instrumental resolution, band structure, total gas pressure and optical depth (cL) used and does not necessarily coincide with the spectral parameter absolute band strengh $S = \int_{band} K_\nu d_\nu$ where K_ν is the absorption coefficient $K_\nu = 1/cL \ln I_o/I_\nu$ and the integration is performed over the band limits. Band strengths were then derived from a regression to the linear part of curves of growth (10) (equivalent width vs cL) constructed and calibrated as previously discussed for the determination of IBI values. The results of IBI and band strength determinations are shown in Tables I and II, respectively, where they are compared to results presented by other authors.

$C^{13}O^{16}$ and H_2O absorption coefficients were evaluated from the AFGL compilation of spectral data (11). Line strengths of $C^{13}O^{18}$ are not available; so they have been assumed equal to those of the corresponding transition of the $C^{13}O^{16}$ band. From a comparison of analogous transitions for $C^{12}O^{16}$, $C^{13}O^{16}$ and $C^{12}O^{18}$ we can expect to obtain with this approximation an accuracy better than 10%. The strengths of the two $O^{16}C^{13}O^{18}$ lines were also evaluated from AFGL data. An analysis by computer generated spectra show that the intense lines of $O^{16}C^{13}O^{18}$ all interfere with lines of other isotopic species of CO_2 and to a lesser extent of N_2O and H_2O vapour. However the two absorption lines selected interfere only with lines of CO_2 and N_2O, whose concentration remains essentially constant during the experiment; they permit evaluation of $O^{16}C^{13}O^{18}$ changes down to a few ppbv in ambient air.

RESULTS

1) $NO_3 + CO$

NO_3 concentrations were calculated from the concentrations of N_2O_5 and O_3 using the steady-state approximation for the reactions:

a) $N_2O_5 \overset{M}{\rightleftarrows} NO_2 + NO_3$

b) $2NO_3 \rightarrow 2NO_2 + O_2$

c) $NO_2 + O_3 \rightarrow NO_3 + O_2$

The steady-state concentration of the nitrate radical $[NO_3]_{ss}$ is then found from the equation:

$$[NO_3]_{ss} = \left(\frac{K\ h}{2g}\right)^{1/3} [N_2O_5]^{1/3} [O_3]^{1/3} \quad (1)$$

where K is the equilibrium constant for reaction (a) above and g and h are the rate constants for reactions (b) and (c), respectively. In this approximation some presumably minor reactions have been ignored. By using the iterative procedure of Graham and Johnston (11) based on a more elaborate reaction scheme, changes in $[NO_3]_{ss}$ of less than 20 percent were found.

Time integrated NO_3 concentrations were calculated from the simple steady-state approximation and plotted versus $\ln([CO]_o/[CO]_t)$ for the two isotopic species of CO (Fig. 2). By performing linear regression on these points, an upper limit for the rate constant could be estimated as

$$R_{max} = R' + ts$$

where R' is the regression coefficient, s is its standard deviation and t the appropriate t values. The upper limit, established as the 90% confidence limit for the rate constant at 296 ± 2 K, is equal to 4×10^{-19} cm^3 $molec^{-1}$ s^{-1}.

Unfortunately, few data were available to estimate the reaction rate from the formation of $O^{16}C^{13}O^{18}$, so no definitive conclusion can be reached from those. They did, however, confirm the impression of a rate equal to or lower than the value given above.

2) $\underline{N_2O_5 + H_2O}$

The results have been evaluated using a procedure of linear regression on the data of $\ln([N_2O_5]_o/[N_2O_5]_t)$ versus the product $[H_2O]$xt. The experimental data are shown in Fig. 3. From the analysis a rate constant at 296 ± 2 K equal to $(1.48 \pm 0.42) \times 10^{-21}$ cm^3 $molec^{-1}$ s^{-1} at a 90 percent significance level was derived with the error limits estimated by t-statistics. The intercept of the regression line with the y-axis was not significantly different from zero. Most of the scatter of the experimental data can be explained by a spread associated to the $[N_2O_5]$ values of ± 5 percent.

This rate constant must, obviously, be considered an upper limit for the homogeneous gas phase reaction because part of the N_2O_5 decay takes place at the bag wall. Because of the presence of O_3 in the gas mixture, NO_3 concentrations of typically about 100 ppb should be present, so wall reactions of this species could also contribute to the observed decay of N_2O_5.

The wall loss of N_2O_5 could be due to a second order heterogeneous reaction between N_2O_5 and H_2O as well as to a first order reaction. In order to eliminate the influence of the possible first order wall loss term we have done the analysis shown in Fig. 4, where the intercept with the vertical axis gives the rate constant for the second order reactions between N_2O_5 and H_2O (homogeneous and heterogeneous), which is found equal to $(0.90 \pm 0.21)x10^{-21}$ cm^3 $molec^{-1}$ s^{-1}. The rate constant of the first order wall loss of N_2O_5 is found to be $(1.8 \pm 0.7)x10^{-5}$ s^{-1}, which corresponds to a half life of N_2O_5 of about 11 hours with respect to this loss mechanism.

DISCUSSION

If the above mentioned upper limit for the rate constant of the reaction between NO_3 and CO is applied to tropospheric air conditions, it confirms the hypothesis that CO should play a negligible role in the fate of NO_3. The scavenging rate of NO_3 with respect to the reaction with CO, assuming a concentration of this species in air equal to 300 ppb would, in fact, be less than $3x10^{-6}$ s^{-1}, while measurements (4) indicate a scavening rate of about 10^{-3} s^{-1} in both clean and slightly polluted tropospheric air.

The upper limit of the N_2O_5+H_2O rate constant confirms the results of Tuazon et al. (9) and the analysis of the data suggests that the true rate constant might be lower. As pointed out by Tuazon et al., however, even a rate constant of the magnitude indicated here could imply a significant contribution to tropospheric nitric acid formation.

REFERENCES

(1) PLATT, U.F., WINER, A.M., BIERMANN, H.W., ATKINSON, R. and PITTS, J.N. Jr. (1984). Measurements of nitrate radical concentrations in continental air, Env. Sci. Technol., 18 (5), 365.

(2) STOCKWELL, W.R. and CALVERT, J.G. (1983). The mechanism of NO_3 and HONO formation in the nighttime chemistry of the urban atmosphere, J. Geophys. Res. 88 (C11), 6673.

(3) SHEPSON, P.B., EDNEY, E.O., KLEINDIENST, T.E., PITTMAN, J.H. and NAMIE, G.R. (1985). The production of organic nitrates from hydroxyl and nitrate radical reaction with propylene, Env. Sci. Technol., 19 (a), 849.

(4) NOXON, J.F. (1983). NO_3 and NO_2 in the Mid-Pacific troposphere, J. Geophys. Res. 88 (C15), 11,017.

(5) PLATT, U., PERNER, D., WINER, A.M., HARRIS, G.W. and PITTS J.N., Jr. (1980). Detection of NO_3 in the polluted troposphere by differential optical absorption, Geophys. Res. Lett., 7 (1), 89.

(6) BURROWS, J.P., TYNDALL, G.S. and MOORTGAT, G.K. (1985). Absorption spectrum of NO_3 and kinetics of the reactions of NO_3 with NO_2, Cl and several stable atmospheric species at 298 K, J. Phys. Chem., 89, 4848.

(7) MORRIS, E.D., Jr. and NIKI, H. (1973). Reaction of dinitrogen pentoxide with water, J. Phys. Chem., 77 (16), 1929.

(8) ATKINSON, R., LLYOD, A.C. and WINGES, L. (1982). An updated chemical mechanism for hydrocarbon NO_x/SO_2 photooxidations suitable for inclu-

sion in atmospheric simulation models, Atmos. Environ., 16 (6), 1341.
(9) TUAZON, E.C., ATKINSON, R., PLUM, C.N., WINER, A.M. and PITTS, J.N., Jr. (1983). The reaction of gas phase N_2O_5 with water vapor, Geoph. Res. Lett., 10 (10), 953.
(10) GOODY, R.M., "Atmospheric Radiation", Oxford University Press, London, 1964, Vol. 1.
(11) AFGL - atmospheric absorption line parameters compilation: 1982. Edition L.S. Rothman et al., Appl. Opt. 1983, 22, 2247.
(12) GRAHAM, R.A. and JOHNSTON, H.S. (1978). The potochemistry of NO_3 and the kinetics of the N_2O_5-O_3 system, J. Phys. Chem., 82 (3), 254.
(13) CANTRELL, C.A., STOCKWELL, W.R., ANDERSON, L.G., BUSAROW, K.L., PERNER, D., SCHMELTEKOPF, A. CALVERT, J.G. and JOHNSTON, H.S. (1985). Kinetic study of the NO_3-CH_2O reaction and its possible role in nighttime tropospheric chemistry, J. Phys. Chem., 89, 139.
(14) MASSIE, S.T., GOLDMAN, A., MURCRAY, D.G. and GILLE, J.C. (1985). Approximate absorption cross sections of F12, F11, $ClONO_2$, N_2O_5, HNO_3, CCl_4, CF_4, F21, F113, F114 and HNO_4, Appl. Opt., 24, 3426.
(15) LOVEJOY, R.W., CHACKERIAN, L., Jr. and BOESE, R.W. (1980). Integrated band intensities of gaseous N_2O_5, Appl. Opt. 19, 744.
(16) GOLDMAN, A., KYLE, G.T. and BONOMO, F.S. (1971). Statistical band model parameters and integrated intensities for the 5.9, 7.5 and 11.3 μ bands of HNO_3 vapour, Appl. Opt., 10, 65.
(17) GOLDMAN, A., BONOMO, F.S., VALERO, F.P.J., GOORVITCH, D. and BOESE, R.W. (1981). Temperature dependence of HNO_3 absorption in the 11.3 μm region, Appl. Opt., 20, 172.
(18) GIVER, L.P., VELERO, F.P.J., GOORVITCH, D. and BONOMO, F.S. (1984). Nitric-acid band intensities and band-model parameters from 610 to 1760 cm^{-1}, J. Opt. Soc. Am. B, 1, 715.

TABLE I - Integrated band intensities IBI = $1/cL \int lg(I_o/I)d\nu$ cm^{+1} $molec^{-1}$ used to convert spectral absorbances into molecular concentrations. The data are compared to other values in the literature. Because of the possible dependence of the IBI values upon band structure and experimental conditions used such as instrumental resolution, total pressure, optical density, these parameters are indicated.

	N_2O_5		HNO_3		
Ref.	725-760 cm^{-1}	1225-1270 cm^{-1}	840-930 cm^{-1}	1270-1350 cm^{-1}	840-930/1270-1350
This work	$1.88x10^{-17}$	$1.76x10^{-17}$	$1.01x10^{-17}$	$2.02x10^{-17}$	0.50
Cantrell et al. (13)	$2.01x10^{-17}$	$1.91x10^{-17}$	$0.59x10^{-17}$	$1.22x10^{-17}$	0.48
Massie et al. (14)	--	1.79	0.86	1.62	0.53

This work: 740 torr; 3-23 torr cm; 1 cm^{-1} instrumental resolution.
Ref. (13): 700 torr; 20-70 torr cm; 1 cm^{-1} instrumental resolution.
Ref. (14): 0.15 torr; 6.75-14.7 torr cm; 0.03 cm^{-1} instrumental resolution; for the H_2O_5 band, the integration limits are 1225-1265 cm^{-1}.

TABLE II - Absolute integrated band intensities $S = \int K\nu d\nu \ cm^{+1} molec^{-1}$, measured in this work by curve of growth analysis, compared to other data in the literature.

Ref.	Absolute band intensities S ($cm^{+1} molec^{-1}$)			
	N_2O_5		HNO_3	
	710-780 cm^{-1}	1225-1270 cm^{-1}	840-930 cm^{-1}	1270-1350 cm^{-1}
This work	$4.57 \pm 0.5 \times 10^{-17}$	$3.81 \pm 0.4 \times 10^{-17}$	$2.21 \pm 0.2 \times 10^{-17}$	$4.29 \pm 0.6 \times 10^{-17}$
Graham and Johnston (12)*	--	4.51×10^{-17}	--	--
Lovejoy et al. (15)	$2.78 \pm 0.13 \times 10^{-17}$	$2.36 \pm 0.09 \times 10^{-17}$	--	--
Goldman et al. (16)**	--	--	$2.60 \pm 0.13 \times 10^{-17}$	$5.06 \pm 0.25 \times 10^{-17}$
Goldman et al. (17)	--	--	$2.39 \pm 0.12 \times 10^{-17}$	--
Giver et al (18)	--	--	$2.57 + 0.11 \times 10^{-17}$	$5.14 + 0.26 \times 10^{-17}$

* The value of the band intensity derived from the measured curve "optical cross section vs wavelength" is quoted in Ref. (15).

** In Ref. (13) and Ref. (17) the integration limits for the bands are 850-920 cm^{-1} and 1275-1350 cm^{-1}, respectively.

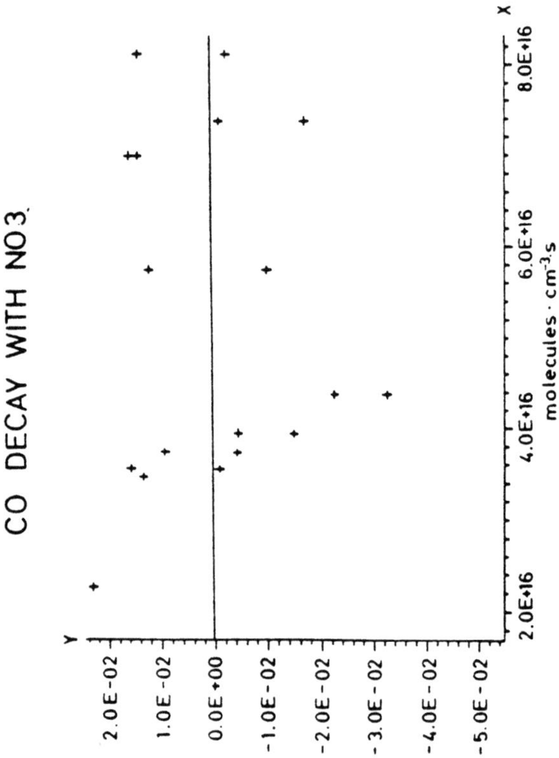

CO DECAY WITH NO3
molecules · cm⁻³·s
2.0E+16
4.0E+16
6.0E+16
8.0E+16
2.0E-02
1.0E-02
0.0E+00
-1.0E-02
-2.0E-02
-3.0E-02
-4.0E-02
-5.0E-02
X
Y

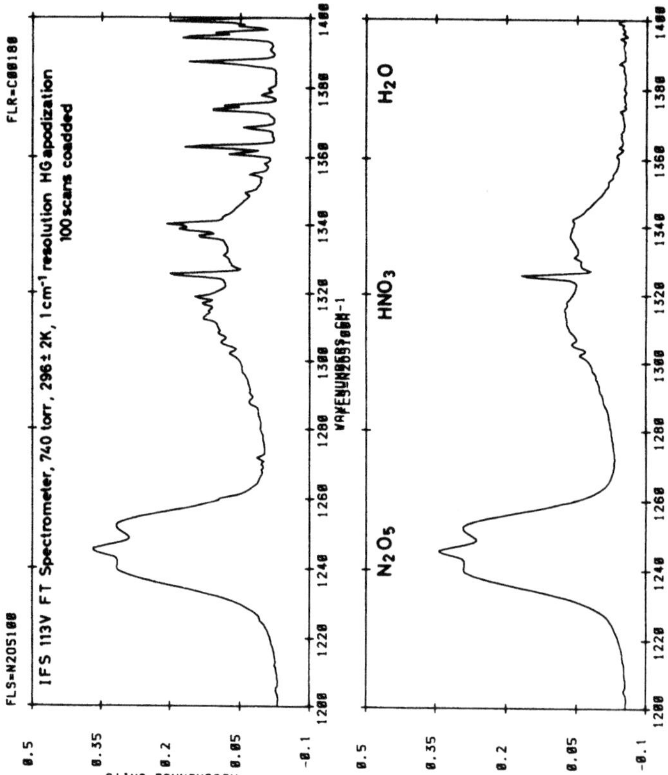

FLS=N205100
FLR=C00100
IFS 113V FT Spectrometer, 740 torr, 296 ± 2K, 1 cm⁻¹ resolution HG apodization
100 scans coadded
ABSORBANCE UNITS
WAVENUMBERS CM-1
N₂O₅
HNO₃
H₂O

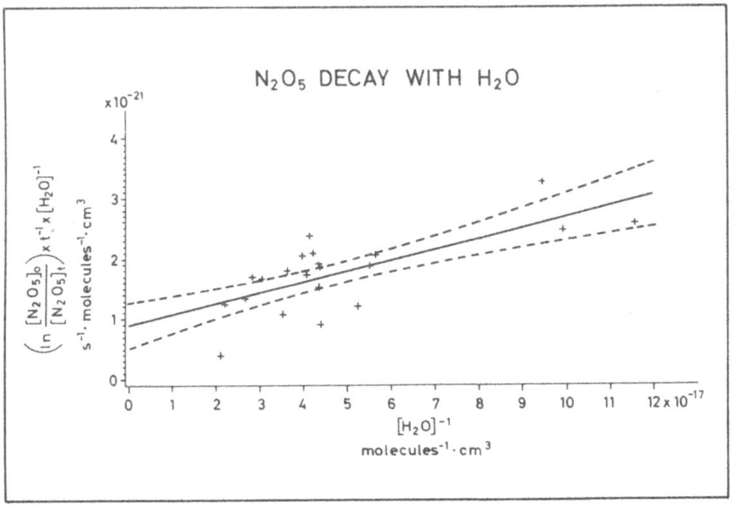

Fig. 3 - N_2O_5 decay ($\ln[N_2O_5]_0/[N_2O_5]_t$)) plotted versus the product of time and water concentration; linear least-squares fit to the data with 90 percent confidence limits.

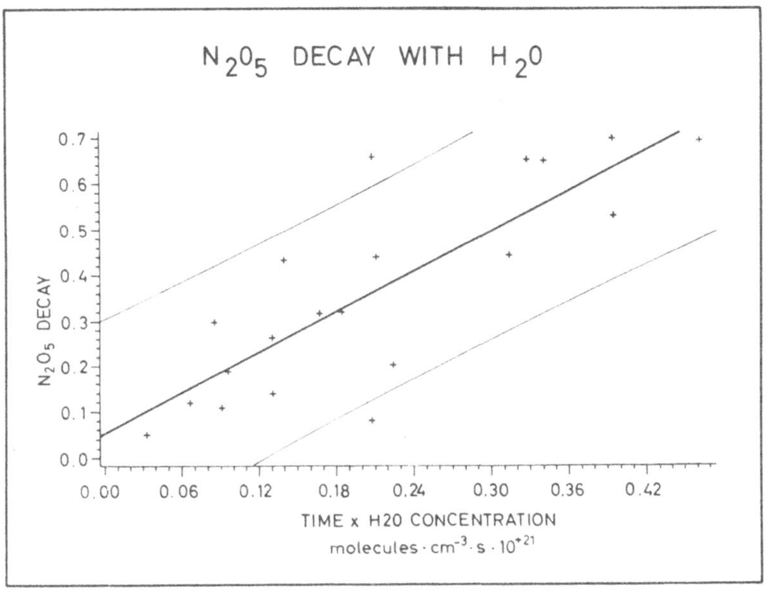

Fig. 4 - Observed rate constant for the homogeneous reaction $N_2O_5 + H_2O \rightarrow 2HNO_3$ versus $[H_2O]^{-1}$. $[H_2O]$ is the measured water concentration in the gas phase.

OH-INITIATED OXIDATION OF P-XYLENE UNDER ATMOSPHERIC CONDITIONS

K.H. Becker and Th. Klein
Physikalische Chemie, Fachbereich 9
Bergische Universität-Gesamthochschule Wuppertal, FRG

Summary

The OH-initiated oxidation of p-xylene has been studied under simulated atmospheric conditions in a 420 l reaction chamber using in situ long-path FT-IR absorption spectroscopy to determine product distributions. NO_x/p-xylene/hν/air and HO_2NO_2/NO were used as photolytic and 'dark' hydroxyl radical sources, respectively.
In the HO_2NO_2/NO-system convincing evidence was found for the formation of cis-3-hexene-2,5-dione with a yield of 12 % carbon. In addition in this system other ring-cleavage products, methylglyoxal and glyoxal were detected and account for 11 % and 9 % carbon, respectively. p-tolylaldehyde was formed with a yield of 10 % in the photolysis system. The further reactions of the dicarbonyl products have also been briefly studied under simulated atmospheric conditions.

1. INTRODUCTION

Aromatic hydrocarbons are important constituents of the polluted atmosphere. The concentrations range from ~30 pptv to > 40 ppbv in remote and urban areas, respectively. The xylenes account for 30-50 % of the total atmospheric aromatics (1,2).

Despite numerous experimental and computer modeling studies, the reaction pathways of the atmospheric photooxidation of aromatic hydrocarbons are still not fully understood (3). The major loss process for atmospheric aromatic compounds is initiated by the attack of hydroxyl radicals followed by reactions with O_2 and NO_x. These initiating OH radical reactions have been shown to proceed via two pathways:

1. H atom abstraction from the substituent alkyl groups

and

2. electrophilic (reversible) addition of the OH radical to the aromatic ring.

For the xylenes pathway 1 accounts for 4-15 % of the overall reaction and can lead to the formation of tolylaldehydes (3-6). Many uncertainties remain concerning the subsequent chemistry of the OH-aromatic-adduct under atmospheric conditions. The observation of ring-cleavage products with significant yields parallel to the formation of hydroxy and nitro aromatics shows that ring-cleavage is also an important overall reaction pathway (3).

In this work the OH-initiated oxidation of p-xylene and the behaviour of dicarbonyl products at atmospheric pressure in air and at 298 $\pm$ 2 K have been studied and the results obtained will be discussed.

2. EXPERIMENTAL

All experiments were carried out at atmospheric pressure and room-temperature in a 420 l cylindrical Duran-glass reaction chamber with Teflon-coated aluminium end flanges which has been described in detail elsewhere (7).

In situ long-path FT-IR absorption spectroscopy (NICOLET 7199, HgCdTe detector) was used to determine product distributions. A path length of 50.4 m was used with a spectral resolution of 1 cm^{-1}.

Absolute concentrations of the reactants and products were determined by computer-aided substraction using calibrated reference spectra of authentic samples where available. Depending on the substance the lower detection limit was of the order of 5×10^{12} molecules cm^{-3}.

NO_x/p-xylene/hν/air, MeONO/NO/hν/air and pernitric acid (HO_2NO_2)/NO were used as hydroxyl radical sources (7,8).

The light intensity corresponded to a NO_2 photolysis rate of 6.2×10^{-3} s^{-1} using 24 photolysis lamps (PHILIPS, type TL/05, 40 W).

All substances were used without further purifications. Initial concentrations of the reactants were in the ppmv range (1 ppmv(298 K) $\cong$ 2.5×10^{13} molecules cm^{-3}). Synthetic air was used as the diluent gas.

3. RESULTS AND DISCUSSION

A rate constant of $(1.4 \pm 0.4) \times 10^{-11}$ cm^3s^{-1} was determined for the reaction,

$$\text{p-Xylene} + OH \longrightarrow \text{products},\ k_1$$

in 1000 hPa air at 298 ± 2 K relative to cyclohexane ($k = 7.4 \times 10^{-12}$ cm^3s^{-1}, (8)) using the MeONO/NO/$h\nu$-air system as the OH radical source. Literature data for k_1 are in the range $(1.05\text{-}1.53) \times 10^{-11}$ cm^3s^{-1} (3).

Identified products of the OH-initiated oxidation of p-xylene using the photolysis and dark soures for OH generation are listed along with the known literature data in Table 1.

TABLE 1: Product distribution for the reaction of p-xylene + OH

product	% carbon				
	HO_2NO_2/NO	NO_x/$h\nu$		MeONO/NO/$h\nu$	
			(9)	(10)	(11)
dimethylphenol	+	+			
p-tolylaldehyde	+	10	8#		
cis-3-hexene-2,5-dione	12	4	6*		(<19)
methylglyoxal	11	7.5	4.5	4.2	4.0
glyoxal	9	2.5	6	3	5.6
peroxyacetylnitrate	2		+		
formic acid	1.5	1	+		
formaldehyde			2.1		
carbon dioxide	+	+			
carbon monoxide	2.5	3			
sum	38	28	26.6		

+ : qualitatively detected

: 4-15 % carbon (4-6)

* : as trans-3-hexene-2,5-dione (THD), estimated 18 % carbon for THD + cHD; 3-hexene-2,5-dione also was detected by Takagi et al. (12), though in low yields

After substraction of all known products unidentified absorption bands in the regions 1100-1400 cm^{-1} and 1700-1800 cm^{-1} are still present in the residual spectrum.

In the HO_2NO_2/NO-system convincing evidence was found for the formation of cis-3-hexene-2,5-dione (CHD) and perhaps for 2,5-dimethylphenol. CHD has been infered from bands at 1165 cm^{-1} and 1720 cm^{-1} which remain after spectral stripping of all known products and which correlate well with the authentic spectrum of CHD. The yields of identified products in the HO_2NO_2/NO-system have not been corrected for the further reactions of the products.

From a consideration of the product yields and with the assumption that ring-hydrogen abstraction is negligible a feasible initiating reaction mechanism is the following:

1) hydrogen abstraction from the substituent alkyl groups with the formation of p-tolylaldehyde with a yield of 10 %

and with

2) electrophilic addition of the hydroxyl radical to the aromatic ring with a yield of approximately 90 %.

Besides ring-cleavage products, which account for 65-85 % of detected carbon in our experiments, reaction 2 can lead to the formation of phenolic and nitro aromatic hydrocarbons with a fraction which is presently unknown. The total amount of detected carbon is about 40 %. The remaining carbon may be fixed in low volatile or highly polar substances which were deposited on the chamber wall.

Despite the detection of various products the operative mechanism of the ring cleavage of the OH-aromatic-O_2-adduct under atmospheric conditions still remains uncertain.

Preliminary studies on the photolysis and OH kinetics of glyoxal, methylglyoxal and trans-3-hexene-2,5-dione have been performed in 1 atm air at room temperature. The measured OH rate constant for glyoxal and the apparent photolysis rates for all the dicarbonyls are presented in Table 2 along with the available literature values.

TABLE 2: OH rate constants and apparent photolysis rates for dicarbonyls

	k_{OH} (10^{-11} $cm^3 s^{-1}$)	$k_{h\nu}$ (10^{-4} s^{-1})	$k_{h\nu}/k_{h\nu}$ (10^{-2})
glyoxal	1.4 ± 0.2 1.15 ± 0.04 (13)	1.4 ± 0.3	2.1 ± 0.6 0.8 ± 0.5 (13)
methylglyoxal	 1.73 ± 0.13 (13)	1.3	2.1 1.9 ± 0.5 (13) 15 (14) 4.5 (15)
trans-3-hexene-2,5-dione	 5.3 ± 0.5 (16)	7#	11 7.6 ± 1.9 (16)

k_{OH}(glyoxal) relative to cyclo-hexane

: photoisomerisation: $\left([trans]/[cis]\right)_{eq} = 0.45 \pm 0.1$

$= 0.55 \pm 0.05$ (16)

TABLE 3: Product distributions for the photolysis and OH initiated oxidation of dicarbonyls

	glyoxal	methylglyoxal		trans-3-hexene-2,5-dione	
products in % carbon	$+h\nu$	$+h\nu$	$NO_x/h\nu$	$+h\nu$	HO_2NO_2/NO
cis-3-hexene-2,5-dione				> 80	
methylglyoxal					32*
peroxyacetylnitrate			25		16
formic acid	11	< 2	< 0,5		3
formaldehyde	< 0.3 6.5#	10	21		
carbon dioxide		+	+		
carbon monoxide	84	25	34	≲ 3	5

+ : qualitatively detected

* : Tuazon et al. (16): not detectable

: Plum et al. (13)

The k_{OH} value of glyoxal is in fair agreement with a recent determination by Plum et al. (13). The ratio of the photolysis rate constants for the dicarbonyls with that for NO_2 allow a rough comparison with other studies. Products distributions of photolysis and OH-reaction experiments are shown in Table 3.

The reasons for the poor agreement of product yields in some cases are not yet understood.

4. CONCLUSIONS

The OH-initiated oxidation of p-xylene in the atmosphere is still incompletely understood. The initiating oxidation steps and a number of final products are fairly well known but the reaction mechanism of the ring-cleavage processes and the role of NO_x thereby is highly speculative. Also the product distributions of some dicarbonyl compounds under atmospheric conditions are not characterized in detail.

Further work on product distributions of the OH-initiated oxidation of aromatic hydrocarbons, with lower NO_x and aromatic hydrocarbon concentrations, and on the behaviour of the products under atmospheric conditions is necessary in order to establish a complete reaction mechanism for these processes.

REFERENCES

(1) NUTMAGUL, W. and CRONN, D.R. (1985). Determination of selected atmospheric aromatic hydrocarbons at Remote continental and oceanic locations using photoionization/flame-ionization detection. J. Atmos. Chem. 2, 415-433

(2) SINGH, H.B. et al. (1985). Distribution of aromatic hydrocarbons in the ambient air. Atmos. Environ. 19, 1911-1919

(3) ATKINSON, R. (1985). Kinetics and mechanisms of the gas-phase reactions of the hydroxyl radical with organic compounds under atmospheric conditions. Chem. Rev. 85, 69-201

(4) PERRY R.A., ATKINSON R. and PITTS Jr., J.N. (1980). Kinetics and mechanism of the gas phase reaction of OH radicals with aromatic hydrocarbons over the temperature range 296-473 K. J. Phys. Chem. 81, 296-304

(5) KENLEY R.A., DAVENPORT J.E. and HENDRY D.G. (1981). Gas-phase hydroxyl radical reactions. Products and pathways for the reaction of OH with aromatic hydrocarbons. J. Phys. Chem. 85, 2740-2746

(6) NICOVICH J.M., THOMPSON R.L. and RAVISHANKARA A.R. (1981). Kinetics of the reactions of hydroxyl radical with xylenes. J. Phys. Chem. 85, 2913-2916

(7) BARNES, I. et al. (1979). Proceedings of the "First European Symp. on the Physicochem. Behaviour of Atmospheric Pollutants", Ispra, Italy, Oct. 16-18, Versino B. and Ott H., Eds.

(8) BECKER K.H. et al., Ed. (1984). Methods of the Ecotoxicological Evaluation of Chemical Photodegradation in the Gas Phase. Volume 6: OH Reaction Rate Constants and Tropospheric Lifetimes of Selected Environmental Chemicals. Report 1980-1983. Jül-Spez-279, November 1984, KFA-Jülich

(9) BANDOW H. and WASHIDA N. (1985). Ring-cleavage reactions of aromatic hydrocarbons studied by FT-IR spectroscopy. II. Photooxidation of o-, m-, and p-xylenes in the NO_x-air system. Bull. Chem. Soc. Jpn. 58, 2541-2548

(10) TUAZON E.C. et al. (1984). Yields of glyoxal and methylglyoxal from the NO_x-air photooxidations of toluene and m- and p-xylene. Environ. Sci. Technol. 18, 981-984

(11) TUAZON E.C. et al. (1986). α-Dicarbonyl yields from the NO_x-air photooxidations of a series of aromatic hydrocarbons in air. Environ. Sci. Technol. 20, 383-387

(12) TAKAGI H. et al. (1982). Observation of 3-hexene-2,5-dione in the photooxidation of 1,2,4-trimethylbenzene in the NO-H_2O-air system. Spectrosc. Lett. 15, 145-152

(13) PLUM C.N. et al. (1983). OH radical rate constants and photolysis rates of α-dicarbonyls. Environ. Sci. Technol. 16, 479-484

(14) ATKINSON R. et al. (1980). A smog chamber and modeling study of the gas phase NO_x-air photooxidation of toluene and the cresols. Int. J. Chem. Kinet. 12, 779-836

(15) KILLUS J.P. and WHITTEN G.Z. (1982). A mechanism describing the photochemical oxidation of toluene in smog. Atmos. Environ. 16, 1973-1988

(16) TUAZON E.C., ATKINSON R. and CARTER W.P.L. (1985). Atmospheric chemistry of cis- and trans-3-hexene-2,5-dione. Environ. Sci. Technol. 19, 265-269

PRODUCTS AND KINETICS OF THE OH INITIATED OXIDATION OF SO_2, CH_3SH, DMS, DMDS AND DMSO

I.BARNES, V. BASTIAN AND K.H. BECKER
Physikalische Chemie/FB 9, Bergische Universität Gesamthochschule
Wuppertal, D 5600 Wuppertal 1, W. Germany

Summary

The kinetics of the reactions of OH radicals with SO_2, CH_3SH, DMS, DMDS and DMSO have been investigated in glass reaction chambers at atmospheric pressure using the relative rate technique with either the photolysis of CH_3ONO in the presence of NO or the photolysis of H_2O_2 as the OH source. The photolysis of CH_3ONO as OH source was found to be unsuitable for the study of the reaction of OH with CH_3SH and DMS due to complications caused by secondary chemistry. The photolysis of H_2O_2 in N_2 was used as the OH source for the investigation of these reactions. Preliminary product analyses of the reactions of OH with CH_3SH and DMS in air using the H_2O_2 photolyic OH source show SO_2 to be the major sulphur containing product of both reactions. The SO_2 yield is significantly higher than those obtained in other studies which were carried out in the presence of relatively large concentrations of nitric oxides.

1. INTRODUCTION

Natural sulphur compounds comprise approximately 50% of the global sulphur burden /1/. Although the anthropogenic source strengths are reasonably well known the natural sulphur source strengths are ill defined and are thought to arise mainly from the emissions of reduced sulphur compounds produced by biological activity in soils, water and vegetation. Reduced sulphur compounds detected in air include H_2S, CH_3SH, CH_3SCH_3 (DMS), CH_3SSCH_3 (DMDS), CS_2 and COS /2,3,4/. Of these compounds DMS represents the major source of reduced sulphur, at least to the marine atmosphere, and is believed to be responsible for the background levels of SO_2, excess sulphate and methane sulphonic acid (CH_3SO_3H) over oceanic areas /4/. Since natural sulphur compounds play a dominant role in the global atmospheric sulphur cycle the troposheric oxidation mechanisms of these compounds need to be thoroughly understood in order to reliably assess the contribution of anthropogenically produced sulphur to the formation of acid rain.

With the possible exception of DMS the principal recognized daytime sink for most of the reduced sulphur compounds is believed to be reaction with OH readicals /5/. Although the results from recent field measurements on DMS correlate well with model calculations involving the OH oxidation mechanism other measurements lead to very short atmospheric residence times for DMS which cannot be explained by the OH sink /7,8/. In the latter it would appear that another species in addition to OH is responsible for the DMS decay. It has been suggested that the reaction of IO radicals with DMS could be responsible for the observed short lifetime /9/. IO radicals can be formed in marine atmospheres containing CH_3I and O_3 via the reactions /10/,

$$CH_3I + h\nu \longrightarrow CH_3 + I,$$

and

$$I + O_3 \longrightarrow IO + O_2.$$

Two recent laboratory investigations of the reaction using different techniques are in good agreement and yield rate constants of $(3.0 \pm 1.5) \times 10^{-11}$ /11/ and $(1.5 \pm 0.5) \times 10^{-11}$ $cm^3\ s^{-1}$ /12/ for IO + CH_3I. The result implies that for regions were the IO radical concentration is in excess of 5×10^5 molecules cm^{-3} then the reaction for IO with DMS could be a more important sink for DMS than the corresponding reaction with OH.

The available kinetic data for the reaction of OH radicals with organic sulphur compounds has recently been reviewed by Atkinson /5/ and Heicklen /13/. For CH_3SH and DMS there are considerable discrepancies between the rate constants measured using absolute methods in inert diluent gases and those measured in air using relative rate techniques. Wine et al. /31/ have concluded that for CH_3SH the discrepancy cannot be due to an O_2 effect as has been reported for the reaction of OH with CS_2 /15/. In contrast, for the reaction of OH with DMS a small O_2 effect has been observed by Wine et al. /14/. Very recent investigations show that the relative techniques which rely on the photolysis of systems containing nitric oxides in air for OH production lead to erroneously high rate constants for the reactions of OH radicals with mercaptans /16/ and sulphides /17/. Radical chains possibly involving oxygenated CH_3S radicals are thought to be responsible for the observed effects /16/.

Presented here are the results on the reactions of OH radicals with SO_2, CH_3SH, DMS, DMDS and DMSO (CH_3SOCH_3). The competitive kinetic method was used to determine the rate constants using either the photolysis of CH_3ONO in the presence of NO or the photolysis of H_2O_2 (NO_x free) as the OH source. Although the rate constant for the reaction of OH with SO_2 has been determined many times /18-22/ it is still uncertain by a factor of 1.5 /22/. Since it is a major product of the reaction of OH with thiols and sulphides /5,13/ and is important for modeling the transformation of SO_2 to sulphate it has been included in the present investigations.

The concentration of DMSO in seawater is much greater than that of DMS /23/ but because it is not very volatile it has not been included in sulphur transport models. DMSO and $DMSO_2$ ($CH_3SO_2CH_3$) are possible oxidation products of DMS and both have recently been detected in marine rain and marine air masses /24/. To our knowledge we report here the first measurements on the gas phase reaction of OH with DMSO.

Exsisting product analyses to date of the reaction of OH with thiols and sulphides have been carried out in the presence of relatively large concentrations of nitric oxides which do not correspond to atmospheric conditions and could possibly influence the product distributions. Reported here are the preliminary results from product studies on the reaction of OH with CH_3SH, DMS and DMSO in an NO_x free system using the photolysis of H_2O_2 as the OH source.

2. EXPERIMENTAL

The kinetic measurements were carried out in glass reaction chambers of 20, 40 and 420 l volume at 760 Torr total pressure and room temperature using either N_2 or synthetic air as the diluent gas. The photolysis of either CH_3ONO in the presence of NO /25/ using Philips TLA 40 W/05 fluoresence lamps or H_2O_2 vapour at 254 nm /16,26/ using a germicidal lamp (Philips, TUV 40 W) was used as the OH source. Experimental details can be found elsewhere /16,27,28/.

The competitive kinetic technique /5/ was used to measure the rate constants for the reaction of OH radicals with the sulphur compounds relative to the reaction of OH with either n-butane ($k=1.68\times10^{-11}\exp(-559/t)$ $cm^3\ s^{-1}$, /22/), ethene ($k=1.66\times10^{-12}\exp(474/T)$ $cm^3\ s^{-1}$, /22/), propene ($k=4.1\times10^{-12}\exp(544/t)$ $cm^3\ s^{-1}$, /22/), cis-2-butene ($k=1.04\times10^{-11}\exp(488/T)$ cm^3

s^{-1}, /6/) or trans-2-butene ($k=1.22x10^{-11}exp(549/T)$ cm^3 s^{-1}, /22/). Reactant concentrations were in the range 2-15 ppmv and the concentrations of the OH precusors, CH_3ONO and H_2O_2, were between 20 and 40 ppmv. The reference hydrocarbons were monitored by gas chromatography with flame ionization detection (Hewlett Parkard, Model 5710 A) using a 2 m stainless steel Porasil C column. The organic sulphur compounds, with the exception of DMSO, were analysed using a Hall electrolytic conductivity detector (Tracor Inc., Model 700 A) after separation on a 1.5 m PTFE column packed with acetone washed Porapak QS mounted in a Tracor gas chromatograph (model 540). DMSO was monitored in the 420 l reactor using in situ long-path Fourier transform absorption spectroscopy (FTIR) /28/.

Product analyses of the reactions of OH with CH_3SH, DMS and DMSO were also performed in the 420 l reactor using the built-in FTIR facility and the photolysis of H_2O_2 as the OH source.

3. RESULTS AND DISCUSSION

3.1 Reaction OH + SO_2

The reaction of OH with SO_2 was investigated in a 40 l reactor relative to OH + n-butane in 760 Torr of air at 300 K using the photolysis of CH_3ONO as

(1) $OH + SO_2 \longrightarrow$ products, k_1

(2) $OH + C_4H_{10} \longrightarrow$ products, k_2

the OH source. From 15 experiments a rate constant ratio k_1/k_2 of 0.42 ± 0.07 was obtained which with $k_2=2.6x10^{-12}$ cm^3 s^{-1} yields a rate constant of $(1.1 \pm 0.2)x10^{-12}$ cm^3 s^{-1} for the reaction of OH with SO_2. The quoted 1σ error represents precision only. Further details of this investigation can be found elsewhere /16/.

The value determined here is in reasonable agreement with the various literature values /18,19,21/ and is also the same as the value of the rate constant currently recommended for atmospheric modeling studies /18/. However, the latest absolute study by Wine et al. /20/ using a FP-RF technique gives a value of $(8.0 \pm 0.8)x10^{-13}$ cm^3 s^{-1} for atmospheric conditions which is ~30% lower than the recommended value. Preliminary studies in this laboratory using a LP-LIF technique give a value of $(8.6\pm2.0)\times10^{-13}$ cm^3 s^{-1} for the reaction in 1 atm of air at 295 K. The value is in good agreement with the extrapolated value of Wine et al. /20/ and lies just outside the error limits of the value determined in this work. Considering the importance of this reaction for atmospheric modeling studies this small but significant discrepancy needs to be clarified by further studies.

3.2 Reaction OH + CH_3SH

Using the photolysis of CH_3ONO in 760 Torr air at 300 K as the OH source a rate constant ratio k_3/k_4 of 0.20 ± 0.05 was obtained for the reaction of

(3) $OH + CH_3SH \longrightarrow$ products, k_3

(4) $OH + C_3H_6 \longrightarrow$ products, k_4

OH with CH_3SH relative to that of OH with propene /16,29/. With $k_4=2.5\times10^{-11}$ cm^3 s^{-1} this gives a value of $(12.6\pm3.2)\times10^{-11}$ cm^3 s^{-1} for the reaction of OH with CH_3SH. The measured rate constant was also observed to decrease with decreasing O_2 partial pressure and a value of $(3.4\pm0.4)\times10^{-11}$ cm^3 s^{-1} was obtained for conditions of 760 torr N_2 with 2 torr O_2 /30/. Further the measured rate constant was found to be strongly dependent upon the concentration of NO present in the reaction system /16/. As mentioned in the introduction detailed analyses of such reaction systems /16,17/ has led to the conclusion that that kinetic studies by the relative method in air and in the presence of nitric oxides are subject to secondary chemistry and will not yield reliable OH rate constants for thiols in general.

Because of the above mentioned difficulties the reaction of OH with CH_3SH was investigated in a 20 l reactor in 760 Torr of N_2 at 300 K using the photolysis of H_2O_2 as the OH source. Under these conditions from a total of 14 experiments a rate constant ratio k_3/k_4 of $1.43\pm.14$ was obtained which yields a value of $(3.6\pm0.4)\times10^{-11}$ cm^3 s^{-1} for the reaction of OH with CH_3SH in 760 Torr N_2. Neglecting the value of Cox and Sheppard /30/ which was obtained using a relative method with the photolysis of HONO in air as the OH source the present value is in reasonable agreement with the other literature values /5,16/.

Replacing the N_2 diluent gas by synthetic air caused the measured rate constant to increase by 10-20%. The increase is probably due to secondary chemistry and not to an O_2 effect as has been reported for OH + CS_2 /15/ since Wine et al. /31/ have failed to detect an O_2 dependence for the rate constant using an absolute method.

Preliminary product analyses on the reaction in air have been carried out in the 420 l reactor using the photolysis of H_2O_2 as the OH source and FTIR to monitor reactant and reaction products. SO_2 was found to be a major product with a yield of $\gtrsim$80% and other products which have been qualitatively identified include HCOOH, HCHO and CO. After spectral stripping of all known products a peak remains at $\sim$1050 cm^{-1} which has been tentively assigned to hydroperoxy-hydroxy-methane (HO_2CH_2OH). Hydroperoxy-hydroxy-methane is known to be formed in the reaction between HO_2 radicals and HCHO /33/. In the present work the source of HO_2CH_2OH is probably also due to this reaction the HO_2 radicals and HCHO coming from the reaction of OH with H_2O_2 and DMS, respectively. The yield of the SO_2 observed here contrasts sharply with those obtained in studies performed in the presence of NO_x where the yield of SO_2 was typically 29%. Methane sulphonic acid (CH_3SO_3H) was also a major product ($\sim$40%) in the presence of NO_x /13,32/.

3.3 Reaction OH + DMS

Very high rate constants were obtained for the reaction of OH with DMS when the photolysis of CH_3ONO in air was used as the OH source /29/. The measured rate constant also showed dependence upon the O_2 partial pressue and the concentration of NO present in the reaction system. Similar observations have recently been made by Wallington et al. /17/ who studied the reaction as a function of the O_2 partial pressure using a CH_3ONO-NO-air photolysis system and the N_2H_4-O_3 dark reaction system to generate OH radicals. It would appear that as for the thiols relative methods performed in air and in the presence of nitric oxides are subject to secondary chemistry and are not suitable for the study of the reactions of OH with sulphides in general.

The reaction of OH with DMS was studied in a 420 l reactor at 760 Torr total pressure of N_2 at 300 K using the photolysis of H_2O_2 as the OH source. The measurements were made relative to the reaction of OH with ethene.

(5) $OH + DMS \longrightarrow$ products, k_5

(6) $OH + C_2H_4 \longrightarrow$ products, k_6

A rate constant ratio k_5/k_6 of 0.55 ± 0.05 from a total of 7 experiments was obtained which with $k_6 = 8 \times 10^{-12}$ cm^3 s^{-1} yields a value of $(4.4 \pm 0.4) \times 10^{-12}$ cm^3 s^{-1} for the reaction of OH with DMS. When the diluent gas was replaced by synthetic air the measured rate constant ratio rose to 1.00 ± 0.06 which corresponds to a rate constant for OH with DMS of $(8.0 \pm 0.5) \times 10^{-12}$ cm^3 s^{-1}.

The literature rate constant data for OH + DMS exhibit significant discrepancies /5/, however, the most recent values obtained using absolute techniques in the absence of O_2 are in reasonable agreement the rate constants lying in the range $(3.2-4.4) \times 10^{-12}$ cm^3 s^{-1}. Wine et al. /14/ have reported that the reaction of OH with DMS shows a small O_2 effect the rate constant increasing from 4.8×10^{-12} cm^3 s^{-1} in the absence of air to 5.18×10^{-12} and 5.8×10^{-12} cm^3 s^{-1} in the presence of 343 and 593 torr total pressure of air, respectively.

The rate constant for OH with DMS obtained in this study is in good agreement with the most recent values obtained using absolute methods in the absence of O_2. Although in this work an increase in the measured rate constant was observed on replacing the N_2 diluent gas by air, but because of the nature of the method it is not clear whether the increase is due to an oxygen effect or is being caused by secondary chemistry.

Product studies on the reaction of OH with DMS using the photolysis of H_2O_2 as the OH source are currently in progress in this laboratory. Preliminary results show that SO_2 is a major product the yield being $70 \pm 10\%$. Dimethyl sulphone ($DMSO_2$) is also formed, however, its origin is not fully certain at present. It is formed in a dark reaction between DMS and H_2O_2 but the concentration observed during the reaction of OH with DMS cannot be attributed solely to this source. DMSO was not observed to be formed in the reaction system. Other products which have been qualitatively identified include HCOOH, HCHO and CO. After spectral stripping of all known products a peak remains at 1050 cm^{-1} which as for the studies on OH with CH_3SH has been assigned to HO_2CH_2OH formed the reaction between HO_2 and HCHO. Again the yield of SO_2 of 70% obtained in this work differs significantly from those obtained from studies performed in the presence of relatively large concentrations of NO_x where the yield of SO_2 was typically 30% and methane sulphonic acid was also formed with a yield of 50% /29,32/.

3.3 Reaction OH + DMDS

The reaction of OH with DMDS was studied in the 420 l chamber in 760 Torr N_2 at 300 K relative to OH + trans-2-butene. The photolysis of H_2O_2 was used as the OH source. In N_2 the photolysis of DMDS by the 254 nm mercury line contributed 4% to the measured DMDS decay. A rate constant ratio k_7/k_8=

(7) OH + DMDS ⟶ products, k_7

(8) OH + butene ⟶ products, k_8

0.33±0.04 was obtained which with $k_8=6.98\times10^{-11}$ $cm^3\ s^{-1}$ yields a rate constant of $(2.1\pm0.2)\times10^{-10}$ $cm^3\ s^{-1}$ for OH with DMDS. The value is in excellent agreement with the available literature value /5/.

Changing the diluent gas from N_2 to synthetic air had no effect upon the measured rate constant. Using the photolysis of CH_3ONO in air as the OH source give a value of $(1.9\pm0.2)\times10^{-10} cm^3\ s^{-1}$ for the reaction. The rate constants obtained for OH + DMDS using relative methods which use the photolysis of HONO or CH_3ONO for OH generation are also in good agreement with values obtained using absolute techniques. As for the studies on OH with CH_3SH and DMS secondary chemistry is almost certainly occuring in the relative studies on OH + DMDS. However, since the measured rate is close to the collision number interference from side reactions is probably negligible.

3.4 Reaction OH + DMSO

The reaction of OH with DMSO was studied in a 420 l reactor relative to OH + cis-2-butene in 760 Torr air using the photolysis of CH_3ONO as the OH source. A rate constant ratio k_9/k_{10} of 0.90±0.36 was obtained which with $k_{10}=5.2\times10^{-11}$ $cm^3\ s^{-1}$ yields a rate constant of $(5.8\pm2.3)\times10^{-11}$ $cm^3\ s^{-1}$ for the reaction of OH with DMSO. The determined rate constant showed no dependence upon the O_2 partial pressure or the concentration of NO in the reaction system. To our knownledge it represents the first determination of this rate constant.

Preliminary product investigation indicate that $DMSO_2$ is the major reaction product.

4. CONCLUSIONS

Relative rate techniques employing OH photolysis sources in air containing nitric oxides will not yield reliable OH rate constants for thiols and sulphides. The rate constants for the reactions of OH radicals with SO_2, CH_3SH and DMDS seem to be reasonably well established. For DMS the most recent rate constant determinations in the absence of O_2 are in good agreement. In the presence of O_2 the rate constant shows an O_2 effect the magnitude of which is not fully certain. Considering the importance of this compound in the global sulphur cycle more studies on the effect of O_2 on the rate constant using more direct methods are needed.

The preliminary product analyses reported here show that in the absence of nitric oxides the main sulphur containing product from the reactions of OH with CH_3SH and DMS is SO_2. In the presence of nitric oxides the yield of SO_2 falls sharply and methane sulphonic acid is formed in large yields. The oxidation mechanism is presently not clear. It is probable that in the absence of NO, CH_3S radicals and its O_2 adducts determine the product spectrum. In the presence of NO, CH_3SO is probably being formed leading subsequently to to the formation of CH_3SO_3H.

ACKNOWLEDGEMENTS

This research was supported by the "Bundesminster für Forschung und Technologe (BMFT)" and the "Minster für Wissenschaft und Forschung des Landes NW".

REFERENCES

/1/ D. Möller
Atmos. Environ. 18(1984)19.

/2/ T.E. Graedel
Rev. Geophys. Space Phys. 15(1977)421.

/3/ S.M. Turner and P.S. Liss
J. Atmos. Chem. 2(1985)223.

/4/ M.O. Andreae, R.J. Ferek, F. Bermond, K.P. Byrd, R.T. Engstrom, S. Hardin, P.D. Houmere, F. LeMarrec and H. Raemonck
J. Geophys. Res. 90(1985)12,891.

/5/ R. Atkinson
Chem. Reviews 85(1986)69.

/6/ R. Atkinson, K.R. Darnall,A.C. Lloyd, A.M. Winer and J.N. Pitts, Jr.
Advances in Photochemistry 11(1979)375.

/7/ B.C. Nguyen, B. Bonsang and A. Gaudry
J. Geophys Res. 88(1983)10,903.

/8/ P. Carlier, C. Luce, R. Girard, J. Morelli, T. Marehal,
L. Girard-Reydet and G. Mouvier
submitted for publication (1986)

/9/ P. Carlier
Atmospheric Ozone (Eds. C.S. Zerefos and A. Ghazi)
D. Reidel Publ. Co., Dordrecht (1985), pp815-813.

/10/ W.L. Chameides and D.D. Davis
J. Geophys. Res. 85(1980)7383.

/11/ I. Barnes, K.H. Becker, P. Carlier and G. Mouvier
Int. J. Chem. Kinetics (in preparation 1986)

/12/ D. Martin, J.L. Jourdain, G. Laverdet and G. LeBras
(private communication 1986)

/13/ J. Heicklen
Rev. Chem. Intermediates 6(1985)175.

/14/ see reference /5/

/15/ I. Barnes, K.H. Becker, E.H. Fink, A. Reimer, F. Zabel and H. Niki
Int. J. Chem. Kinetics 15(1983)631.

/16/ I. Barnes, V. Bastain, K.H. Becker, E.H. Fink and W. Nelsen
J. Atmos. Chem. (1986, in press).

/17/ T.J. Wallington, R. Atkinson, E.C. Tuazon and S.M. Aschmann
Int. J. Chem. Kinetics (1986, in press).

/18/ D.L. Baulch, R.A. Cox, R.F. Hampson, Jr., J.A. Kerr, J. Troe and
R.T. Watson
J. Phys. Chem. Ref. Data 13(1984)1259.

/19/ J.J. Margitan
J. Phys. Chem. 88(1984)3314.

/20/ P.H. Wine, R.J. Thompson, A.R. Ravishankara, D.H. Semmes, C.A. Gump,
A. Torabi and J.M. Nicovich
J. Phys. Chem. 88(1984)2095).

/21/ K. Izumi, M. Mizuochi, M. Yoshioko, K. Murano and T. Fukuyama
Environ. Sci. Technol. 18(1984)116.

/22/ R. Atkinson and A.C. Lloyd
J. Phys. Chem. Ref. Data 13(1984)315.

/23/ M.O. Andreae
Oceanogr. 25(1980)1054.

/24/ G.R. Harvey and R.F. Lang
Geophys. Res. Letters 13(1986)49.

/25/ H. Niki, P.D. Maker, C.M. Savage and L.P. Breitenbach
Chem. Phys. Letters 80(1981)499.

/26/ T. Ohta
Int. J. Chem. Kinetics 16(1984)879.

/27/ I. Barnes, V. Bastian, K.H. Becker, E.H. Fink and F. Zabel
Atmos. Environ. 16(1982)545.

/28/ I. Barnes, V. Bastian, K.H. Becker, E.H. Fink, Th. Klein, V. Kriesche, W. Nelsen, A. Reimer and F. Zabel (1984)
Untersuchung der Reaktionssyteme $NO_x/ClO_x/HO_x$ unter troposphärischen und stratosphärischen Bedingungen.
Gesellschaft für Strahlen- und Umweltforschung mbH, ISSN 0176/0777
BPT-Bericht 11/84.

/29/ I. Barnes, V. Bastian, K.H. Becker and E.H. Fink
Physico-Chemical Behaviour of Atmospheric Pollutants, (B. Versino and G. Angeletti, editors), D. Reidel Publ. Co., Dordrecht (1984),pp149-157.

/30/ R.A. Cox and D. Sheppard
Nature(London) 284(1980)330.

/31/ P.H. Wine, R.J. Thompson and D.H. Semmes
Int. J. Chem. Kinetics 16(1984)1623.

/32/ S. Hatakeyama and H. Akimoto
J. Phys. Chem. 87(1983)2387.

/33/ F. Su, J.G. Calvert, J.H. Shaw, H. Niki, P.D. Maker, C.M. Savage and L.D. Breitenbach
Chem. Phys. Letters 65(1979)221

Absolute rate constant for the reaction of OH with SO_2 in the presence of water at atmospheric pressure.

O.J. Nielsen, P. Pagsberg, and A. Sillesen
Chemistry Department
Risø National Laboratory
DK-4000 Roskilde, Denmark.

Summary.

The gas phase reaction of OH with SO_2 in the presence of water was studied at atmospheric pressure. OH radicals were produced by pulse radiolysis of water/Ar mixtures. The formation and decay of OH were followed by monitoring the transient light absorbtion at 309 nm. $k(OH+SO_2)$ in the presence of water at 10 mbar was determined to 0.8×10^{-12} $cm^3 molec^{-1} s^{-1}$. A search for the absorption spectrum of the HSO_3 radical has not yet been succesful.

1. Introduction

There is great concern over the potentially harmful effect of "acid rain" both in Europe and in North America. It seems that both gas and liquid phase chemical processes are important sources of acidification. SO_2, organic sulfides, NO_2, and hydrocarbons are oxidized to acids in the gas phase through many different pathways. On the regional scale of Europe organic sulfides are only of minor importance as a source of sulphuric acid, and it is now generally recognized that the most important daytime gas phase oxidation reaction of SO_2 is:

(1) $$SO_2 + OH + M \rightarrow HOSO_2 + M$$

Rate parameters for this important reaction have been obtained by different experimental techniques and the reported values are presented in Table 1.

The fate of the $HOSO_2$ radical depends on the chemical environment in the troposphere and further oxidation of $HOSO_2$ is still not fully understood. Two product channels have been considered:

(2a) $HOSO_2 + O_2 + M \rightarrow HOSO_2O_2 + M$

(2b) $HOSO_2 + O_2 \rightarrow HO_2 + SO_3$

(3) $HOSO_2O_2 + NO \rightarrow HOSO_2O + NO_2$

(5) $SO_3 + H_2O + M \rightarrow H_2SO_4 + M$

(4) $HOSO_2O + HONO \rightarrow H_2SO_2 + NO_2$

Evidence for the formation of the peroxy radical $HOSO_2O_2$ has been obtained by FTIR spectroscopy of photochemical reaction products trapped in low temperature Ar/O_2 matrices (1). Structural parameters for $HOSO_2$ and $HOSO_2O_2$ have been derived by comparison of the observed vibrational frequencies with predicted normal modes and optimized geometries obtained by ab-initio model calculations. The ultraviolet absorption spectra of $HOSO_2$ and $HOSO_2O_2$ have not been observed so far. Indirect evidence for reaction (2b) has emerged from photochemical smog chamber studies of $HONO/NO/NO_2/H_2O/CO/SO_2$-mixtures (2). The observation that SO_2 did not affect the rate of CO oxidation via $CO + OH \rightarrow CO_2 + H$ implies that SO_2 is being

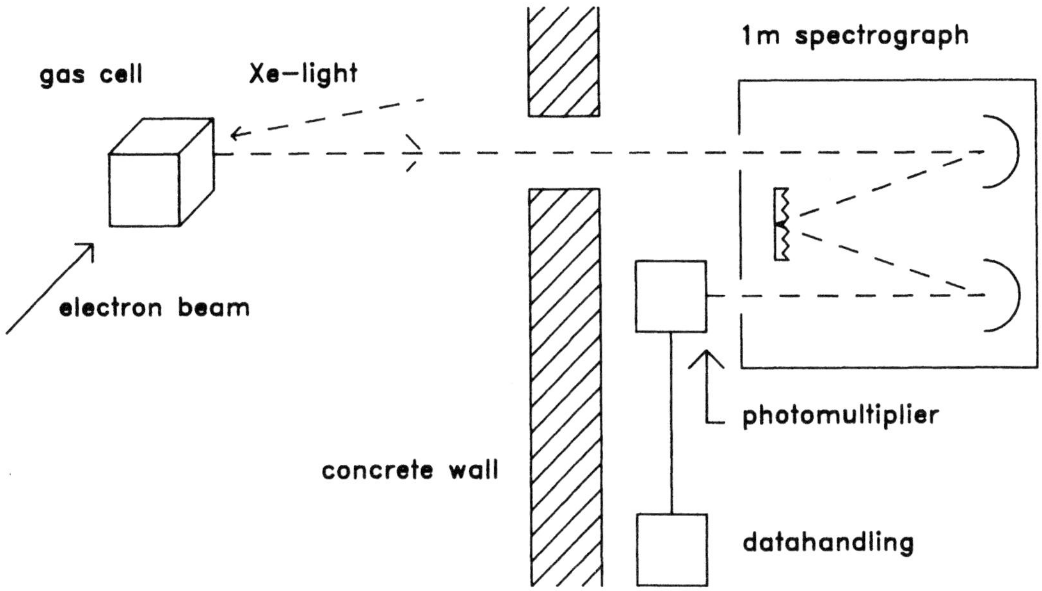

Figure 1. Schematic diagram of the experimental set-up.

oxidized without a net loss of OH. Reaction (2b) followed by $HO_2 + NO \rightarrow OH + NO_2$ was proposed as a plausible mechanism accounting for the experimental results.

This project was started in order to obtain more information about the mechanisms of gas phase SO_2 oxidation using a technique that has not yet been applied to this reaction system.

2. Experimental

The experimental set-up is shown schematically in Figure 1, and has been descibed in detail by Hansen et al. (3). The field emission accelerator (FEBETRON 705B) provides single pulses of 2-MeV electrons with a pulse duration of 30 nsec and a maximum current of 3000 amperes. The gas samples are contained in a 1-liter stainless steel irradiation cell mounted directly onto the accelerator. The optical arrangement involving an internal set of conjugate mirrors was first described by White (4). By this device the optical path length may be increased by multiple passages of the analyzing light beam through the gas sample. Most of the experiments were carried out using twelve traversals corresponding to an optical path length of 120 cm. Gas mixtures were prepared by admitting one component at a time and reading the corresponding partial pressure using a MKS Baratron model 170 absolute electronic membrane manometer with a resolution of 10^{-5} Bar. The analyzing light source is a pulsed 150-watt high-pressure xenon lamp with short axial arc and parabolic reflector. An optical train of suprasil lenses and front-surface aluminized mirrors carries the analyzing light through the sample cell to the entrance slit of the Hilger and Watts 1-meter monochromator. The xenon lamp pulser is most essential to the performance in the ultraviolet region where the light intensity is increased by a factor of 20-50 in light pulses stabilized to within a few percent and with a duration of several milliseconds. The application of a total of seven lenses implies a large loss of light intensity which makes the use of the lamp pulser indispensable. However, a major advantage of the system is that the lenses act as a pre-monochromator resulting in a very low level of stray light (2% at 2000 Å). The analyzing light is dispersed in the monochromator which has an aperture ratio of 1/10. For the ultraviolet region we apply a ruled grating with 1200 grooves/mm and blazed at 3000 Å. The reciprocal dispersion in the plane of the exit slit is 8 Å/mm. The light intensity over the wavelength band-pass selected by the exit slit is monitored using a Hamamatsu R955 photomultiplier coupled to a current input operational amplifier with adjustable offset. The transient signals are stored in a PDP8 minicomputer. Plots of transient absorption versus time, as well as simple first- and second-order kinetic plots are immediately available on a display screen and may be displayed on an X-Y recorder or stored on magnetic tape for further processing on a large central computer system.

We use a chemical simulation computer program developed by O. Lang Rasmussen (5). As input, this program accepts reaction schemes in the usual chemical notation, e.g., $H + OH \rightarrow H_2O$, $k=2x10^{10}$, etc. The program translates the chemical equations into the pertinent set of differential equations which is solved by numerical integration after specification of initial concentrations, irradiation dose, etc. Computer simulations are conducted much faster than real experiments and the effect of parameter variation on the model kinetics can be studied very quickly using a graphical computer terminal. In this way we have been able to find out in which cases it is safe to apply simple analytical procedures to calculate rate constants from the experimental decay curves.

3. Analysis of pulse radiolysis data

By pulse radiolysis of water vapour backed-up with Ar to 1 atm, Argon ions and excited Argon (Ar●) is initially formed. OH is formed mainly through the reaction:

(6) $$Ar\bullet + H_2O \rightarrow H + OH + Ar$$

The formation of OH is rapid compared to the timescale of the OH decay. In the absence of additives the fate of H and OH is governed mainly by the reactions:

(7) $$OH + OH + M \rightarrow H_2O_2 + M$$
(8) $$H + OH + M \rightarrow H_2O + M$$
(9) $$H + H + M \rightarrow H_2 + M$$

Consecutive reactions of H and OH with accumulating product molecules H_2 and H_2O_2, play only a minor role in singlepulse highdose rate experiments. Under these conditions the decay of OH and H is described to a good approximation described by the differential equations:

(I) $$-d[OH]/dt = 2k_7[OH]^2 + k_8[H][OH]$$

(II) $$-d[H]/dt = 2k_9[H]^2 + k_8[H][OH]$$

The decay of hydroxyl radicals can be studied by monitoring the transient absorbance of one of the characteristic rotational lines of the OH ($A^2\Sigma^+-X^2\Pi$) rovibronic transition in the region 3000-3200 Å.

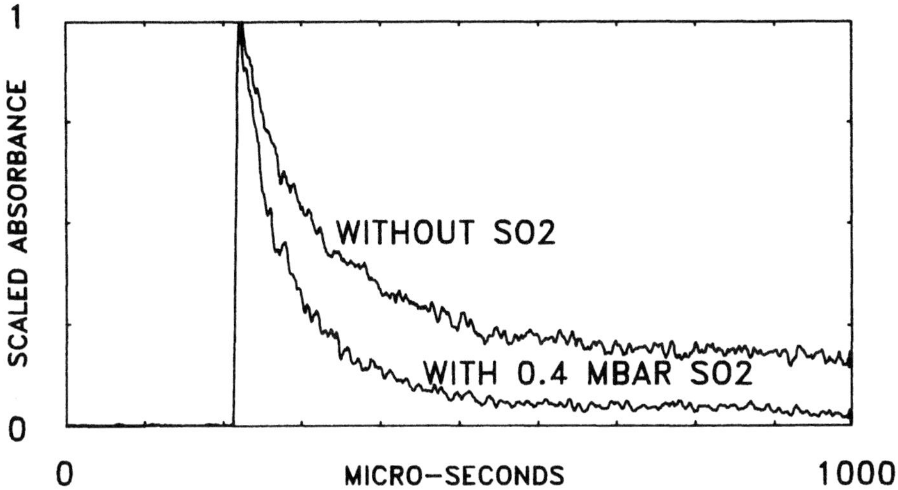

Figure 2. OH kinetics in the pulse radiolysis of a H_2O/Ar mixture, with and without SO_2 added.

A modified version, $A=(\varepsilon lc)^n$, of Beer´s law has been used (see references (6) and (7) for further details). The observed transient absorbance is a direct measure of the concentration of hydroxyl radicals at any given time during the decay. Figure 2 shows an example of the transient absorbance of OH monitored at 3089.6 Å in the absence of additives. The absorbance decreases to half of the maximum value in the course of 100 μsec and this "natural half-life" must be ascribed to the combined rates of reactions (7) and (8). Another characteristic feature is the long "second-order tail" following the steep initial part of the decay.

Addition of a "substrate", SO_2, reacting with OH has a characteristic effect on the observed decay of OH, as shown in Figure 2. Comparison of the two curves in Figure 2 reveals a substantial suppression of the "second--order tail" and also that the half-life is reduced relative to the "natural decay". In the presence of SO_2 an additional term must be added to (I),

$$\text{(III)} \quad -d[OH]/dt = 2k_7[OH]^2 + k_8[H][OH] + k_1[RH][SO^2]$$

This differential equation serves as the basis for our determinations of the rate constants k_1 for the elementary reaction between OH-radicals and SO_2. The simultaneous occurrence of (1),(7),(8) and (9) creates the problem of separating (1) from (7),(8) and (9) in the observed overall decay rate. In general, this can be be solved by studying the effect of reaction (1) over a sufficiently large range of substrate concentrations. In the limit of very high SO_2-concentrations, (i.e. when $k_1[SO_2] \gg 2\,k_7[OH]$ and $k_1[SO_2] \gg k_8[H]$) the differential equation (III) degenerates into:

$$\text{(III|)} \qquad -d[OH]/dt = k_1[SO_2][OH]$$

corresponding to a simple exponential decay:

$$[OH] = [OH]_o \exp(-k_1[OH]_o t)$$

with a characteristic half-life

$$\tau = \ln 2/k_1[SO_2]_o$$

This expression may be used to obtain a quick estimate of the value of k_1 from a single experiment in which the SO_2-concentration was high enough to push the observed halflife down to a small fraction of the "natural half-life". Problems arise in studies of slow reactions where it may be difficult to fulfill the requirement $k_1[SO_2]_o > 2k_7[OH]_o$. Based on computer simulations we find that it is still possible to calculate a reliable value of k_1 based on a relatively small variation in half-lives by applying the equation:

$$\text{(IV)} \quad 1/\tau = 1/\tau_o + \ln 2/k_1[SO_2]_o$$

According to this equation a plot of reciprocal half-lives versus the corresponding SO_2-concentrations is a linear function with a slope equal to $\ln(2)/k_1$. This is a simple method which allows a quick determination of the rate constant k_1 based solely on measured half-lives and the corresponding substrate concentrations. A more rigorous procedure which allows a clear distinction between first- and second-order kinetics is to analyze plots of ln(A) versus time, where A is the transient OH-absorbance. Such a plot becomes strictly linear when the decay degenerates into pseudo-first-order kinetics at high substrate concentrations in accordance with (III). Any significant contributions from reactions of OH with itself or other short-lived radicals will show up in terms of a more-or-less pronounced curvature. Thus, if a particular curve is found to be linear (within the experimental noise) it means that the decay is almost completely controlled by the last term in Equation (III). In more complicated cases a way of correcting for radical-radical reactions have been found (8).

The concentration of SO_2 was determined in two ways. After evacuating of the sample cell the SO_2 was introduced via a vacuum line; the partial pressure $p(SO_2)$ was measured with an electronic membrane manometer and the concentration was calculated as $c = p/RT$ where the gas temperature T was measured with a platinum resistance thermometer. The second method is based on measurement of the light absorption of SO_2 at 309nm which is a proportional to the SO_2 concentration. By application of the second method we discovered that the initial partial pressure of SO_2 decreased during the subsequent procedure where water vapor and Ar were introduced into the sample cell. This effect was caused by back flow of SO_2 into the liquid water reservoir. Therefore, the actual SO_2 concentrations were calculated from the UV-light absorption of the final $SO_2/H_2O/Ar$ mixtures immediately before pulse irradiation of the sample cell. The uncertainties in the measured SO_2 concentrations are estimated to be about ±10%.

4. Results and discussion

The plot of reciprocal OH half-life, τ^{-1}, versus SO_2-concentrations is shown in Figure 3. From a least-squares fit of the straight line, k_1 is determined to 0.8×10^{-12} $cm^3 molec^{-1} s^{-1}$.

In Table 1 our value of k_1 is compared to all reported values at room temperature and atmospheric pressure. It is seen that our result is slightly higher, within a factor of 1.5, than the average value of the four

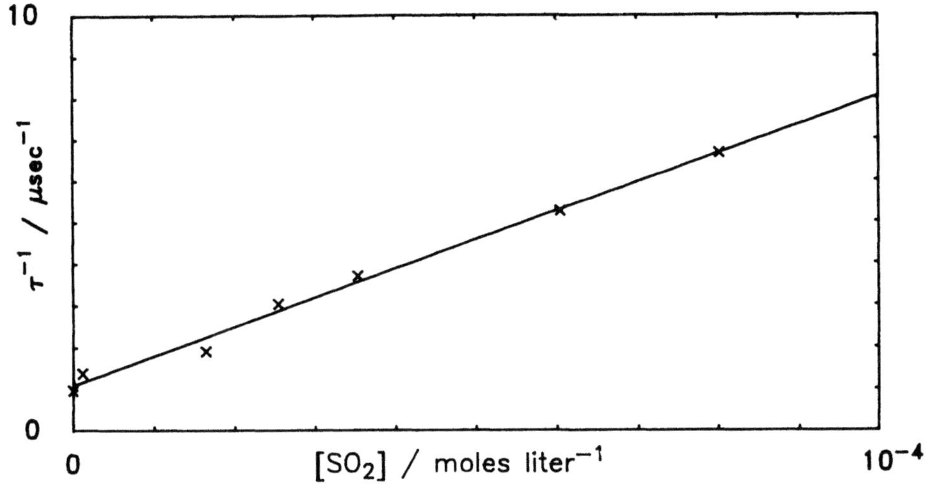

Figure 3. OH reciprocal half-life versus SO_2 concentration

previous measurements using Ar as a third body. The reason for this slightly higher value could be the 1% H_2O present. There is an indication that H_2O has a high third-body efficiency in this reaction (8). However, it is doubtful that the H_2O third-body efficiency will be high enough to explain our k_1 value. We intend to measure k_1 in 1 atmosphere H_2O at 373K. Another explanation for our higher k_1 value could be secondary reactions consuming OH. Even with OH reacting fast with the HSO_3 produced and $HSO_3 \cdot (H_2O)$ (hydrated HSO_3), concentrations of OH and HSO_3 are too small to influence the pseudo-first order decay.

If hydration of the $HOSO_2$ intermediate is important as suggested by Friend et al. (9) the effect of $[H_2O]$ on k_1 would provide a test of this mechanism.

5. Conclusion

A number of experimental problem have been solved. Our determination of k_1 was the first using pulse radiolysis combined with kinetic spectroscopy. The value of k_1 is in reasonable agreement with previous measurements using different techniques. Work on determining the effect of $[H_2O]$ is in progress as well as work on the temperature and pressure dependence.

Table 1: Summary of reported rate constants for the reaction $OH+SO_2+M \rightarrow HSO_3+M$ at room temperature and appr. pressure 1 atm.

M	k in $cm^3 molecule^{-1} s^{-1}$	ref
He	0.34×10^{-12}	(9)
He	0.27×10^{-12}	(10)
Ar	0.39×10^{-12}	(9)
Ar	0.65×10^{-12}	(11)
Ar	0.37×10^{-12}	(10)
Ar	0.67×10^{-12}	(12)
N_2	0.95×10^{-12}	(13)
N_2	0.60×10^{-12}	(14)
SF_6	1.00×10^{-12}	(9)
SF_6	1.61×10^{-12}	(11)
dry air	1.22×10^{-12}	(15)
air	1.1×10^{-12}	(16)
air	0.72×10^{-12}	(17)
Ar/H_2O	0.8×10^{-12}	This work

6. References

1) S. Hashimoto, G. Inoue, and H. Akimoto; Chem.Phys.Lett. 107, 198 (1984)
2) W.L. Stockwell and J.G. Calvert; Atm.Environ. 17, 2231 (1983)
3) K.B. Hansen, R. Wilbrandt, P. Pagsberg; Rev.Sci.Instrum. 50, 1532 (1979)
4) J.U. White; J.Opt.Soc.Am. 32, 283 (1942)
5) O.L. Rasmussen; Risø-R-395 (1984)
6) E. Bjarnov, J. Munk, O.J. Nielsen, P. Pagsberg, and A. Sillesen; Risø-M-2366 (1983)
7) O.J. Nielsen; Risø-R-480 (1984)
8) O.J. Nielsen, P. Pagsberg, and A. Sillesen; "Physico-Chemical Behaviour of Atmospheric Pollutants (3rd European Symp.)", B. Versiono and G. Angeletti, Eds., D. Reidel, Dordrecht, Holland, 1984, pp 283-292.
9) P.H. Wine, R.J. Thompson, A.R. Ravishankara, D.H. Semmes, C.A. Gump, A. Torabi, and J.M. Nicovich; J.Phys.Chem. 88, 2095 (1984)
10) D.D. Davis, A.R. Ravishankara, and S. Fischer; Geophys.Res.Lett. 6, 113 (1979)
11) G.W. Harris, R. Atkinson, and J.N. Pitts, Jr.; Chem.Phys.Lett. 69, 378 (1980)
12) R. Atkinson, R.A. Perry, and J.N. Pitts, Jr.; J.Chem.Phys. 65, 306 (1976)
13) G. Paraskevopoulos, D.L. Singleton, and R.S. Irwin; Chem.Phys.Lett. 100, 83 (1983)
14) A.W. Castleman, Jr.; J. Photochem. 6, 349 (1976)
15) K. Izumi, M. Mizuochi, M. Yoshioka, K. Murano, and T. Fukuyama; Environ.Sci.Technol. 18, 116 (1984)
16) I. Barnes, V. Bastian, K.H. Becker, E.H. Fink, and W Nelsen; J.Atm.Chem. (to be published 1986)
17) R.A. Cox and D. Sheppard; Nature 284, 330 (1980)
18) S. Gordon and W.A. Mulac; Int.J.Chem.Kinet.Symp. 1, 113 (1975)
19) J.P. Friend, R.A. Barnes, and R.M. Vasta; J.Phys.Chem. 84, 2423 (1980)

THE ROLE OF TRANSITION METAL IONS IN ATMOSPHERIC DROPLET CHEMISTRY

A.T. Cocks and W.J. McElroy
Central Electricity Generating Board, Central Electricity Research Laboratories, Leatherhead, KT22 7SE, UK

SUMMARY
A critical assessment of the factors determining the role of transition metals in cloud and rain chemistry has been made and their quantitative effects have been estimated using simple physical models. It is shown that recent work has overestimated the general importance of Fe^{3+} and Mn^{2+} as catalysts by using unrealistically high droplet concentrations. This is exacerbated in the case of Fe^{3+} by an incorrect consideration of Fe^{3+} solubility. For typical average atmospheric concentrations of transition metals reported in the USA, transition metals are predicted to play a minor role in droplet chemistry. However, using UK atmospheric concentrations based on a limited data set for iron and manganese levels in rain, with proper consideration of solubility effects, it is predicted that iron catalysed aerobic oxidation of SO_2 may be as important as oxidation by ozone in precipitation systems. A more thorough assessment of soluble iron concentrations in atmospheric droplets is needed before a realistic estimate of the influence of transition metals on SO_2 wet removal, and hence on the proportionality of sulphate deposition to SO_2 emissions can be made.

1. INTRODUCTION

Transition metals are ubiquitous in the troposphere being present in varying concentrations in all atmospheric aerosols, and it has been known for some time that the soluble fractions of these airborne metals play a role in the aqueous chemistry of cloud and rain droplets. Several recent reviews of the aqueous oxidation fo SO_2 in the atmosphere have critically assessed the available experimental evidence for each of the known oxidation processes (1,2). In general these reviewers concluded that H_2O_2 and O_3 are the major oxidants of SO_2 in cloud and rain systems. Metal-catalysed oxidation of SO_2, while not negligible, was only predicted to be important in urban fog where transition metal concentrations tend to be high (2) and this pathway of sulphate formation can dominate all others, particularly at night.

A recent paper by Graedel, Weschler and Mandlich (3) presented results of computer simulations using a detailed model of raindrop chemistry which incorporates the reactions of transition metal ions. Contrary to evidence presented hitherto, it was claimed that at pH 4, transition metal catalysed pathways account for 30-55% of the oxidation of SO_2 to sulphate in raindrops the remainder being attributed to oxidation by H_2O_2. If true, these results have far reaching implications in terms of our current understanding of aqueous tropospheric chemistry.

In this paper a critical assessment of the role of transition metals in cloud and rain chemistry is made. Particular attention is given to the concentrations of transition metal ions and the rate expressions for SO_2 oxidation employed in the model calculations.

2. THE OXIDATION OF SO_2 IN CLOUD AND RAIN DROPS

Rate Expressions $Fe(III)/O_2$, $MN(II)/O_2$ and H_2O_2

Whilst a number of transition metal ions catalyse the oxidation of SO_2 by O_2 in aqueous solution, only iron and manganese are important for typical environmental conditions (1). In the case of iron it is Fe(III) and not Fe(II) which promotes SO_2 oxidation, the former being the predominant oxidation state of iron in atmospheric aerosols. Graedel et al. (3) used the rate expression derived by Martin (1) for 25°C given in equation [1] where $[S(IV)] = [HSO_3^-] + [SO_3^{2-}] + [H_2SO_3]$.

$$-\frac{d[S(IV)]}{dt} = 0.82[S(IV)](Fe(III)]/[H^+]\ M\ s^{-1} \qquad \ldots [1]$$

While the form of this expression is consistent with other studies, the rate is about four times faster. This discrepancy has not been resolved but it has been suggested that chloride ion catalysis, arising from the use of HCl as a concentration buffer, may have been in part responsible (1). This rate expression thus represents an upper limit. Equation [1] is only valid over the pH range 0-4. Above pH 4, ferric ion is not stable and iron(III) is precipitated as hydrous iron(III) oxide.

The solubility of iron(III) in aqueous solution is highly dependent on pH due to the tendency of the ferric ion to hydrolyse and its ability to form complexes. The extent of hydrolysis is governed by the following equilibria:-

$$Fe(OH)_3 + 6H_2O \underset{}{\overset{K_{sp}}{\rightleftharpoons}} (Fe(H_2O)_6)^{3+} + 3OH^-$$

$$(Fe(H_2O)_6)^{3+} \overset{K_1}{\rightleftharpoons} (Fe(H_2O)_5OH)^{2+} + H^+$$

$$(Fe(H_2O)_5)^{2+} \overset{K_2}{\rightleftharpoons} (Fe(H_2O)_4(OH)_2)^{+} + H^+$$

$$2(Fe(H_2O)_6)^{3+} \overset{K_3}{\rightleftharpoons} (Fe_2(H_2O)_8(OH)_2)^{4+} + 2H^+.$$

In cloud and rainwater ferric-sulphito complexes may also be formed:-

$$(Fe(H_2O)_6)^{3+} + HSO_3^- \overset{K_4}{\rightleftharpoons} (Fe(H_2O)_5SO_3)^{+} + H^+.$$

Complexes with other species, such as sulphate, are much weaker and are present at negligible concentrations (2).

Values for K_{sp}, K_1, K_2 and K_3 of $1.58 \times 10^{-39}\ M^3$, 6.46×10^{-3} M, 3.09×10^{-4} M and 1.26×10^{-3} M respectively, have been recommended by Smith and Martell (4). While there is some uncertainty as to the exact values of the individual equilibrium constants, these data appear to form a set which is consistent with observations of total Fe(III) solubility (5). A detailed study of the aquation of the complex formed between sulphur(IV) and iron(III) has been carried out (6). The formation constant was estimated to be ≤0.4 and this maximum value is assigned to K_4. While Graedel et al. (3) use the recommended values of Smith and Martell (4) for K_1, K_2 and K_3, they employ a value of 7.9×10^{-38} M^3 for K_{sp}, based on the data of King (7). This results in estimates of iron(III) solubility which would appear to be inconsistent with the recommended values for the other hydrolysis equilibria.

Manganese(II) catalysed oxidation of dissolved SO_2 by molecular oxygen shows a zero-order dependence on S(IV) in the concentration range 10^{-3}-10^{-4} M (1). However at lower S(IV) concentrations (<10^{-6} M) and in the pH range 0-2, Martin (1) found that the rate was first-order in [S(IV)] and proposed the rate expression given in equation [2]. At higher pHs (3-7.5) Martin (1) recommended an expression [3] based on the data of Ibusucki and Barnes (8). The two expressions [2,3] match near pH 2. Despite this recommendation, Graedel et al. (3) employed equation [2] in their model simulations at pH 3 and pH 4 so that the rate of SO_2 oxidation by this route was probably overpredicted by factors of 5 and 50 respectively.

$$-\frac{d[S(IV)]}{dt} = 25[H^+]^{-1} [Mn(II)][S(IV)] \text{ M s}^{-1} \qquad \ldots [2]$$

$$-\frac{d[S(IV)]}{dt} = 5000[Mn(II)][HSO_3^-] \text{ M s}^{-1} \qquad \ldots [3]$$

There is evidence of catalytic synergism between iron and manganese in the aqueous oxidation of SO_2 (1,9). This rate enhancement in the mixed catalyst system is reasonably well characterised for S(IV) concentrations in the range 10^{-3} to 10^{-5} M but the complete functional dependence at lower S(IV) concentrations (<10^{-6} M) typically found in cloud and rainwater remains unknown. Preliminary data (1) suggest that in the low concentration regime, the rate of SO_2 oxidation in the mixed Mn(II)/Fe(III) system is about three times faster in the pH range 0-2 than the linear sum of the individual metal-catalysed rates (equations [1] and [2]). A hypothetical mechanism for Fe-Mn synergism has been reported [2]. The use of the higher rate constant in equation [1] probably encompasses the likely synergistic effect in the present work.

Available rate constants for the oxidation of S(IV) by hydrogen peroxide have been reviewed recently (10). Graedel et al. (3) employed the rate coefficient of Kunen et al. (11) determined over the pH range 4.0-5.8 (equation [4]).

$$-\frac{d[S(IV)]}{dt} = 8.03 \times 10^7 [H^+][HSO_3^-][H_2O_2] \text{ M s}^{-1} \qquad \ldots [4]$$

This study (11) was carried out at realistic atmospheric concentrations of S(IV) and H_2O_2 and confirmed earlier work (12,13) with regard to the functional form of the rate expression, although the rate coefficient is almost a factor of two higher. Martin and Damschen (12) and Cocks et al. (13) used sodium sulphite as a source of S(IV) in solution whereas Kunen et al. (11) used sodium metabisulphite. This may account for the difference in the observed rates. The rate expression obtained by Cocks et al. (13) is given in equation [5].

$$-\frac{d[S(IV)]}{dt} = 4.9 \times 10^{7} [H^{+}][HSO_3^{-}][H_2O_2] \text{ M s}^{-1} \quad \ldots [5]$$

<u>Transition-Metal Ion Concentrations in Rain</u>

Representative concentration ranges and mean values for Mn(II) and Fe(III) are given in Table 1 for rain collected at a number of urban and rural locations in the UK. These have been corrected for dry deposition using the data of Peirson et al. (14). Trace metal concentrations in rain falling over the UK are relatively high with mean total values of 3.6×10^{-6} M and 2.5×10^{-7} M for iron and manganese respectively (14). In most rainwater surveys to date, samples have not been filtered before analysis for trace metals. The analytical techniques most commonly used to determine metal concentrations are atomic absorption and inductively coupled plasma emission spectroscopy and both methods will include the contribution from insoluble fraction in determining the total concentration. Since 60-90% iron and 20-40% manganese present in atmospheric aerosols is insoluble (15), a significant correction to reported concentrations is required to obtain the concentration of the active soluble metal species. Assuming 70% Fe and 20% Mn to be insoluble, the typical UK values are reduced to 1.1×10^{-6} M and 2.0×10^{-7} M respectively. These concentrations are used in illustrative calculations of metal-catalysed aqueous oxidation rates in urban areas.

Calculations were also performed for a cloud system where the transition metal ion concentrations were taken to be ten times those in rain, which is a typical ratio obtained for cloud and rain measurements of ionic composition. For a typical rain system, mean concentrations lie between those of cloud and rain, and calculations were executed assuming twice the value in rain.

Graedel et al (3) assumed total soluble concentrations of iron and manganese in rain drops of 2×10^{-5} M and 2×10^{-6} M respectively. These values are more representative of fog water in the most severely polluted urban areas (2) and are unreasonably high as "typical" atmospheric concentrations. Average rainwater concentrations of transition metals in rural areas of the USA are much lower. Pratt, Coscia and Krupa (16) reported mean values of 1.4×10^{-7} M and 1.2×10^{-7} M for iron and manganese in rain falling in Minesota and Wisconsin.

<u>SO_2 Oxidation in Cloud and Rain Systems</u>

The relative contribution to SO_2 oxidation by the $Fe(III)/O_2$, $Mn(II)/O_2$ and H_2O_2 systems in passive cloud and in rain systems were calculated using simple models.

If a passive cloud is regarded as a batch reactor with initial boundary layer concentrations, equations [1], [3] and [5] can be integrated as a function of time and the resulting expressions solved using simple numerical procedures. For a given set of initial conditions the temporal variation in the SO_2 concentration and the oxidation rate can be obtained for each route. In a precipitating system, however, a closed reactor approximation is inapplicable as material is lost from the system. For these cases, a model involving the mixing of a 1 km boundary layer with 3 km of upper air and a first order loss process in precipitation. The differential equations were solved using a Gear algorithm (NAGLIB DO2QBF) on an interactive IBM VMSP system.

A boundary layer hydrogen peroxide concentration of 1 ppb, representing a maximum likely value from current modelling predictions, was assumed and a free tropospheric value equivalent to 1/3 ppb at NTP was also assumed. Liquid water contents for cloud and rain systems were taken as 0.5 g m^{-3} and 1.0 g m^{-3}. Detailed kinetic data for transition metal catalysed aerobic oxidation of SO_2 is only available at 25°C and all calculations refer to this temperature. The Henry's law and first dissociation constants of SO_2 in aqueous solution were taken to be 30.1 (dimensionless) and 1.7×10^{-2} M respectively (17). A value of 2.2×10^{6} was employed for the Henry's law constant in the case of H_2O_2 (18). The input conditions for the 12 scenarios considered are summarized in Table 2.

3. RESULTS AND DISCUSSION

Table 3 lists the initial rates of SO_2 oxidation (cases 1 and 4) expressed in terms of % hr^{-1} for the three oxidation pathways considered. For comparative purposes, the rate of aqueous SO_2 oxidation by O_3 is also given, based on the rate expression of Maahs (19) and a gas-phase ozone concentration of 50 ppb. At t = 0 hydrogen peroxide is potentially the major oxidant in agreement with the calculations of Graedel et al. (3), even if the concentration of this species is an order of magnitude lower. However, as H_2O_2 is consumed in the reaction, the rate of oxidation by this route eventually falls below that for trace-metal catalysed oxidation. At all times the rate of Fe(III) catalysed oxidation exceeds that due to Mn(II) in marked contrast to the predictions of Graedel et al. (3), even if the rate of the former process is reduced by a factor of 4 (see Section 2). This arises from the use of equation [2] by Graedel et al. (3) rather than equation [3] to predict the rate of Mn(II) catalysed SO_2 oxidation. As discussed above, equation [2] is invalid above pH 2.

Since the solubility of SO_2 in aqueous solution is inversely proportional to $[H^+]$ while the rate of oxidation by H_2O_2 is linearly dependent on this quantity, the production of sulphate by this route is independent of pH when the rate is expressed in terms of the gas-phase SO_2 concentration. In contrast, metal-catalysed oxidation is suppressed as the pH of the solution decreases and proceeds at a negligible rate at pH 3 (Table 3).

The extent of SO_2 conversion in passive cloud is illustrated in Figs. 1-3. While the maximum time scale for SO_2 oxidation in clouds is about 20 minutes, large scale rain events may last for periods of several hours. In both cases the available hydrogen peroxide is rapidly consumed ($t < 600$ s). For the conditions applying in Case 1 (Table 2), metal-catalysed oxidation accounts for only 15% of the sulphate produced in cloud after 20 minutes. When the concentration of metal ions in cloudwater is significantly higher than in rain, reactions (1) and (3) can account for a higher proportion of the sulphate produced (Case 2), although this is off-set by the rapid increase in acidity arising from the oxidation of SO_2 by any H_2O_2 present (Case 3).

In the rain system it is possible that a substantial proportion of the sulphate produced can be attributed to metal-catalysed oxidation over the long timescales involved. For all the conditions simulated (Cases 4-10; Figs. 4 and 5) the predicted contribution of the Fe(III) catalysed reaction exceeds that due to oxidation by H_2O_2 after 20,000 seconds. While the contribution of Mn(II) catalysed oxidation only approaches that of H_2O_2 in cases 5 and 8 the relative contribution of the former increases with respect to Fe(III) catalysed oxidation at longer times. This is due to the inverse $[H^+]$ dependence of the rate expression (equation (1)) in the case of Fe^{3+}.

For more realistic situations, where washout of metal and hydrogen ions is taken into account (Cases 7-9), the calculated SO_2 conversions show a significant increase over those for the "non-precipitating" rain system in the case of Fe(III) catalysed SO_2 oxidation (Fig. 5). The effect of the removal of Fe(III) from the rain system in precipitation is more than off-set by the accompanying increase in pH as the rate of removal of gas-phase SO_2 depends on the ratio $[Fe^{3+}]/[H^+]^2$. For Mn(II) a less marked enhancement in sulphate production is observed (Fig. 6) because the removal of SO_2 depends on $[Mn^{2+}]/[H^+]$.

The dilution of boundary-layer air by mixing with air from the free troposphere is considered in Case 11. While the Fe(III) catalysed oxidation of SO_2 is enhanced by this process, compared with Case 8, a decrease in the rate of sulphate production is predicted for Mn(II) catalysed oxidation (Fig. 5 and 6). This is again explained by the different $[H^+]$ dependences of the respective rate expressions (equations (1) and (3)).

Figure 7 shows the slight decrease in SO_2 oxidation predicted for Case 11 which takes into account the decrease in Fe^{3+} solubility for pHs above 4, which can be achieved in a diluted precipitating system. This effect would be smaller for non-diluted system where the pH is maintained at lower values. The initial soluble iron concentrations considered in the present work are close to the values calculated using the solubility equilibria discussed in Section 2.1 but for the higher concentrations considered by Graedel et al. (3) a large fraction of the Fe^{3+} would be predicted to be insoluble with respect to hydrous iron oxide, and hence, the catalytic effect would be overestimated.

SO_2 conversion by iron catalysis in Case 11 is compared with predictions for oxidation by 50 ppb ozone in Figure 18. It is predicted that the catalysed route is more effective under the conditions simulated.

The present work does indicate, however, that for measured values of transition metal concentrations in UK rain aerobic oxidation catalysed by Fe^{3+} may be an important process. In a real rain system, the acidity will vary throughout the system and the solubility effect might be more important than calculated for an initial pH of 4 used in this work, but this is unlikely to greatly diminish the contribution of the catalysed route. The metal concentrations derived from UK rain data are high compared to mean values obtained in the USA (16,20) and hence, it is important to establish the accuracy of the UK measurements. The CERL Flying Chemistry Programme measurements of cloud and aerosol composition (21) should provide indications of atmospheric iron and manganese concentrations and work has been initiated to produce these values.

4. ACKNOWLEDGEMENTS

This work was carried out at the Central Electricity Research Laboratories and is published by permission of the Central Electricity Generating Board.

5. REFERENCES

(1) Martin, L.R., 'SO_2, NO and NO_2 Oxidation Mechanisms, Atmospheric Considerations', edited by J.G. Calvert, Butterworth, London (1984)

(2) Hoffmann, M.R. and Jacob, D.E., as in (1) above

(3) Graedel, T.E., Weschler, C.J. and Mandlich, M.L., Nature, 317, 240-242 (1980)

(4) Smith, R.M. and Martell, A.E., 'Critical Stability Constants', Volume 4, Plenum, London (1976)
(5) Butler, J.N., 'Ionic Equilibria', Addison-Wesley, Reading, Massachussetts, USA (1964)
(6) Carlyle, D.W., Inorg. Chem., 10, 761-764 (1971)
(7) King, E.J., 'Qualitative Analysis and Electrolyte Solutions', Harcourt, Brace and World, New York (1959)
(8) Ibusuki, T. and Barnes, H.M., Atmos. Environment, 18, 145-151 (1984)
(9) Barrie, L.A. and Georgii, H.W., Atmos. Environment, 10, 743-749 (1976)
(10) Schwartz, S.E., as in (1) above
(11) Kunen, S.M., Lazrus, A.L., Kok, G.L. and Heikes, B.G., J. Geophys. Res., 88, 3671-3674 (1983)
(12) Martin, L.R. and Damschen, Atmos. Environment, 15, 1615-1621 (1981)
(13) Cocks, A.T., McElroy, W.J. and Wallis, P.G., Central Electricity Research Laboratories Report RD/L/2215N71 (1981)
(14) Pierson, D.H., Cawse, P.A. and Cambray, R.S., Nature, 251, 675-679 (1974)
(15) Rohbock, E., Georgii, H.W. and Perseke, J. Aerosol Sci., 14, 174-175 (1983)
(16) Pratt, G.C., Cascio, M.R. and Krupa, S.V., Atmos. Environment, 18, 173-182 (1984)
(17) Sillen, L.G., 'Stability Constants of Metal Ion Complexes. Part I; Inorganic Ligands', The Chemistry Society, London (1964)
(18) McElroy, W.J., Atmos. Environment, 20, 427-438 (1986)
(19) Maahs, H.G., J. Geophys. Res., 88, 10721-10732 (1983)
(20) Liljestrand, M.H. and Morgan, J.J., Envir. Sci. Technol., 15, 333-339 (1981)
(21) Clark, P.A., Fletcher, I.S., Kallend, A.S., McElroy, W.J., Marsh, A.R.W. and Webb, A.H., Atmos. Environment, 18, 1849-1858 (1984)

Table 1: Concentrations of Iron and Manganese in Rain

Corrected UK Values (Peirson, Cawse and Cambray, 1974)

Site / Metal	Leiston	Gresham	Patten
Iron	1.4×10^{-6} M	3.2×10^{-6} M	6.3×10^{-6} M
Manganese	1.8×10^{-8} M	1.9×10^{-7} M	3.0×10^{-7} M

Table 2: Input Conditions

Case	$[SO_2]_g$ ppb	$[H_2O_2]_g$ ppb	[Fe(III)] M	[Mn(II)] M	$[H^+]$ M	W_L $g\ m^{-3}$	
1	10	1	1.1×10^{-6}	2.0×10^{-7}	1.0×10^{-4}	0.5	PC
2	10	1	1.1×10^{-5}	2.0×10^{-6}	1.0×10^{-4}	0.5	PC
3	10	1	1.1×10^{-5}	2.0×10^{-6}	2.64×10^{-4}	0.5	PC
4	10	1	1.1×10^{-6}	2.0×10^{-7}	1.0×10^{-4}	1.0	NPR
5	10	1	2.2×10^{-6}	4.0×10^{-7}	1.0×10^{-4}	1.0	NPR
6	10	1	2.2×10^{-6}	4.0×10^{-7}	1.82×10^{-4}	1.0	NPR
7	10	1	1.1×10^{-6}	2.0×10^{-7}	1.0×10^{-4}	1.0	PR
8	10	1	2.2×10^{-6}	4.0×10^{-7}	1.0×10^{-4}	1.0	PR
9	10	1	2.2×10^{-6}	4.0×10^{-7}	1.82×10^{-4}	1.0	PR
10	10	1	2.2×10^{-6}	2.0×10^{-7}	1.82×10^{-4}	1.0	PR
11	10	1	2.2×10^{-6}	4.0×10^{-7}	1.0×10^{-4}	1.0	PRD
12	10	1	2.2×10^{-6}	2.0×10^{-7}	1.0×10^{-4}	1.0	PRD

PC = Passive cloud
NPR = Non-precipitating rain system
PR = Precipitating rain system
PRD = Precipitating rain system with dilution (by mixing with free tropospheric air)

Table 3: Calculated Initial Rates of SO_2 Oxidation

Oxidant	Initial SO_2 Oxidation Rate (% hr^{-1})			
	Case 1		Case 4	
	Cloud (W_L = 0.5 gm^{-3})		Rain (W_L = 1.0 gm^{-3})	
	pH = 3	pH = 4	pH = 3	pH = 4
H_2O_2	290	290	580	580
$Fe(III)/O_2$	0.06	6	0.12	12
$Mn(II)/O_2$	0.07	0.7	0.13	1.3
O_3	0.022	0.6	0.044	1.22

$[(SO_2)g)]_0$ = 10 ppb, $[(H_2O_2)g]_0$ = 1 ppb, [Fe(III)] = 1.1×10^{-6} M
[Mn(II)] = 2.0×10^{-7} M, $[O_3]$ = 50 ppb and T = 298 K.

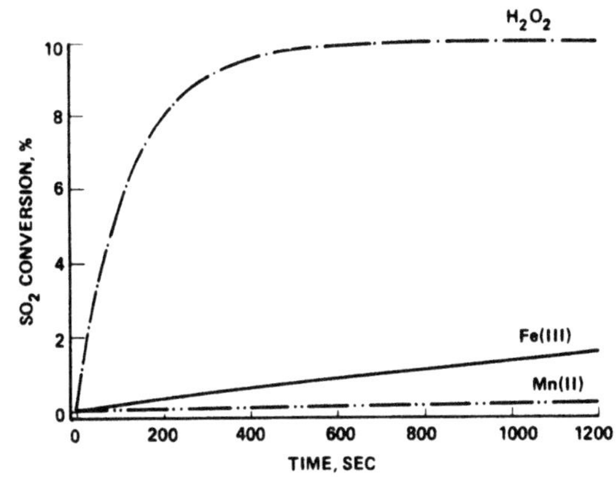

FIG. I THE EXTENT OF SO_2 CONVERSION TO SULPHATE (CASE 1)

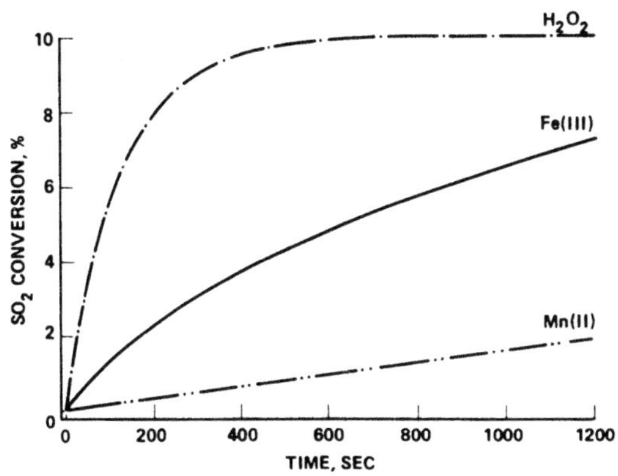

FIG. 2 THE EXTENT OF SO_2 CONVERSION TO SULPHATE (CASE 2)

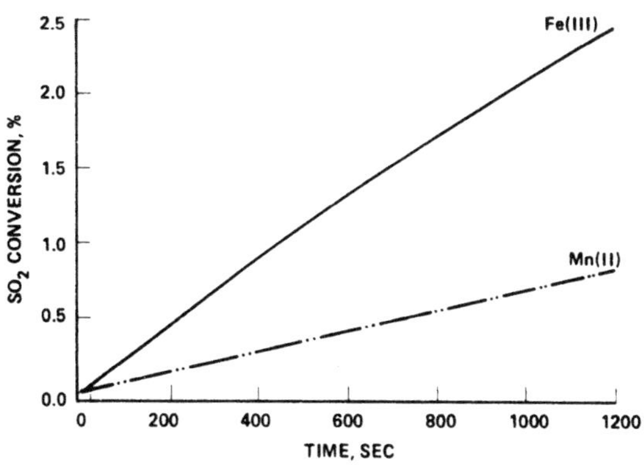

FIG. 3 THE EXTENT OF SO_2 CONVERSION TO SULPHATE (CASE 3)

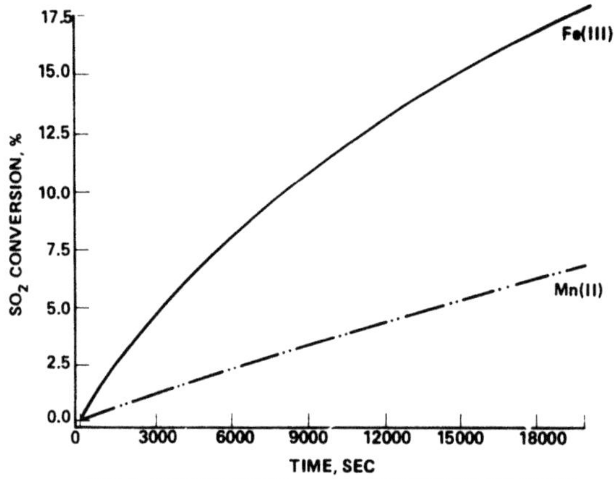

FIG. 4 THE EXTENT OF SO_2 CONVERSION TO SULPHATE (CASE 6)

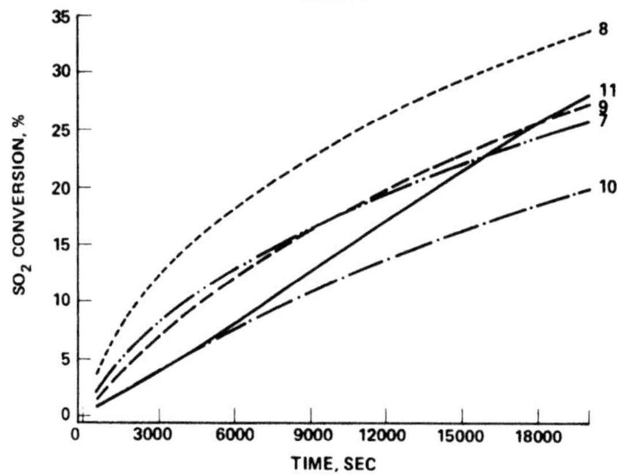

FIG. 5 THE EXTENT OF SO_2 CONVERSION TO SULPHATE DUE TO IRON CATALYSED OXIDATION (CASES 7-11)

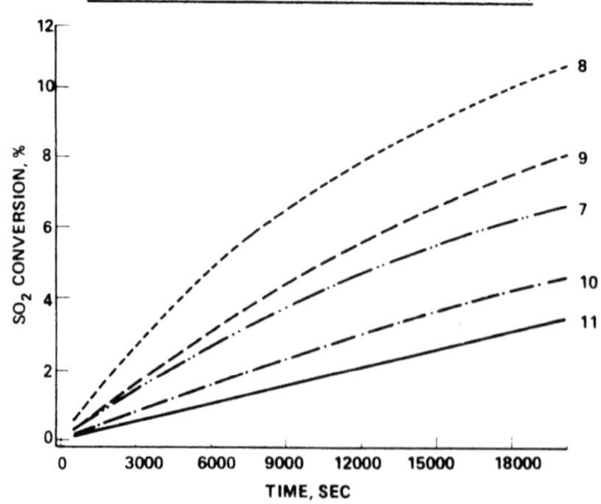

FIG. 6 THE EXTENT OF SO_2 CONVERSION TO SULPHATE DUE TO MANGANESE CATALYSED OXIDATION (CASES 7-11)

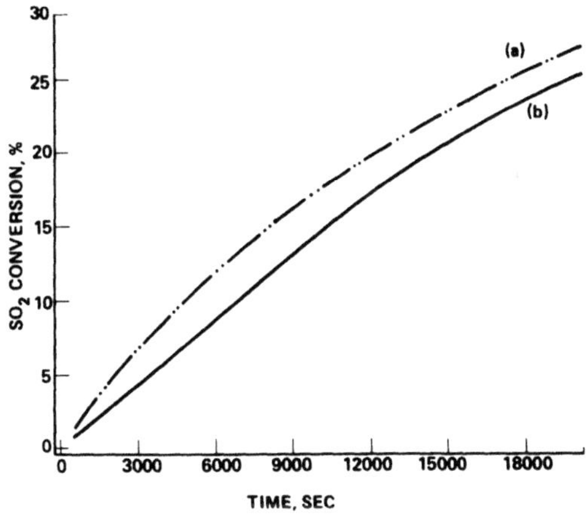

FIG. 7 THE EXTENT OF SO_2 CONVERSION TO SULPHATE DUE TO IRON CATALYSED OXIDATION (a) AS IN CASE 11 (b) TAKING Fe(III) SOLUBILITY INTO ACCOUNT (CASE 12)

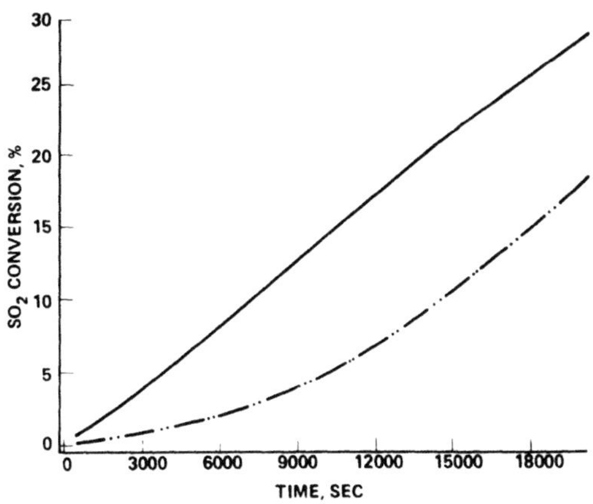

FIG. 8 THE EXTENT OF SO_2 CONVERSION TO SULPHATE (a) DUE TO IRON CATALYSED OXIDATION (CASE 11) (b) DUE TO OXIDATION BY O_3 (50 ppb).

WORKING PARTY 3
AEROSOL CHARACTERIZATION AND PARTICLE FORMATION

Chairman: J.A. SCOTT

Below and in cloud scavenging of particulates and gases

Mesure d'aérosols dans la fôret des vosges (France)

Anionic composition in size-fractionated atmospheric aerosol

Correlation analysis applied to a sulphate and nitrate aerosol measuring network in the Netherlands

Cendres volantes indutrielles et microspherules volcaniques: critères de distinction par microscopie analytique

Radiolytic aerosols: their production and inhibition

Chlorides in atmospheric aerosols

SUMMARY BY THE CHAIRMAN

J.A. SCOTT
University College Dublin, IRL

Research work carried out under the heading of Working Party 3 covers a wide spectrun of topics which can be conveniently classified under the title "Aerosols". However it would be too simplistic not to realise that this single title encompasses projects dealing with the formation of atmospheric aerosols their size, concentration and diurnal variation. Added to that this working party also concerns itself with the chemical characterisation of the aerosol and with measuring devices.

A feature of the projects in this working party is that they can be broadly grouped into two main categories (1) Laboratory Work and (2) Field Measurements. A further point to note is that research work carried out under Working Party 3 may be seen to overlap with similar type projects which may appear in some of the other four working parties. This of course is a good thing since most phenomena in the area of atmospheric pollution are highly complex and in order to achieve some understanding of the mechanisms multidisciplinary investigations are needed.

The symposia held under the aegis of the COST 611 programme serve the very useful role of making scientists from many different disciplines aware of current work in related fields. As such COST serves an important function.

Since the Third European Symposium, which was held in Varese in April 1984, research work under Working Party 3 has continued on a wide range of projects. In December 1985 a workshop on Aerosols and Acid Deposition was held at E.C.N. in Petten, The Netherlands. This proved to be a very interesting and worthwhile meeting enabling as it did project leaders to examine at first hand both current research work and measuring techniques.

The papers presented in the 4th Symposium at Stresa dealt largely with the chemical composition of aerosols and were in the main results of field measurements. These ranged from an in-depth analysis of samples of washout taken from beneath the plume of a 1000 MW coal fired station near the centre of the Netherlands to a comparison of the chemical composition of fly ash produced by coal fired stations with the composition of microspherules from the plume of Mount Etna. This latter work provided an interesting comparison between man-made and natural sources of pollution.

The results of work carried out in Madrid showed that the anionic composition of aerosols in that city are mainly sulphate, nitrate and chloride, with the nitrate and chloride displaying a bimodal size distribution while the nitrate exhibit a trimodal distribution. As part of the Acid Rain Programme in France the laboratory at Fontenay Aux Roses analysed, from the view point of size, concentration and chemical composition, aerosols sampled in the Vosges forest. They related their results to the meteorological conditions and the gaseous composition of the atmosophere.

An interesting paper on correlation analysis as applied to sulphate and nitrate aerosols showed that sulphate concentrations are mainly determined by large scale transport from eastern regions. This suggested a technique that could usefully be tested by other laboratories participating in the COST programme. Finally a paper on the production of aerosols by ionizing radiation showed that the production was dependent on both the concentration of sulphur dioxide and on the presence of organic vapours.

In summary then it is clear that the COST 611 programme provides an impetus to research laboratories to continue their work. However while present projects must continue and develop some effort must be made to enable intercomparisons between laboratories both as to sampling and measuring techniques and indeed into the complex question of aerosol formation. It is possible that this topic could usefully act as the theme for a future workshop meeting for Working Party 3.

BELOW AND IN CLOUD SCAVENGING OF PARTICULATES AND GASES

H.M. TEN BRINK, A.J. JANSSEN, P.I. VOORS and J. SLANINA
Netherlands Energy Research Foundation (ECN)
Petten, the Netherlands

Summary

The below-cloud scavenging of pollutants in the plume of a coal-fired power plant has been measured. The higher than expected scavenging ra te for particulates is attributed to growth of the particles by water accretion in the wet plume.
The two-fold increase in the acid deposition in the near vicinity of the installation is associated with and predominantly caused by the wash-out of hydrogen chloride from the plume. The almost negligible and erratic scavenging of sulfur dioxide in plumes is explained in a quantitative sense by detailed plume wash-out calculations.
As a first estimate it is calculated that below-cloud scavenging contributes about one third to the total wet deposition of acid related compounds in the Netherlands.

Highly efficient in-cloud scavenging of sulfate and submicron aerosol mass was recorded in aircraft flights.

1. INTRODUCTION

The reintroduction of coal as a fuel in power plants in the Netherlands may lead to increased deposition of potentially harmful compounds especially in the vicinity of the installations. In order to make an assessment of this extra deposition a combined experimental and theoretical study of the wash-out (scavenging) of plume constituents was performed. In the present study only the below-cloud scavenging process is considered since it is assumed that in the vicinity of the installation the plume will not reach cloud height. This assumption is corroborated by the fact that during the experiments the visible plume was clearly below cloud-base.

Apart from the plume wash-out study the below-cloud scavenging of ammonia, an important atmospheric pollutant in the Netherlands (1), was modeled in the present study. It is expected that below-cloud scavenging is a good first estimate of the local deposition of emissions, especially in the case of a surface source like ammonia. In this study use is made of the measured vertical concentration profiles of ammonia and associated compounds (1) and furthermore the influence of ammonia gas as a neutralizing agent
in the below-cloud scavenging of sulfur dioxide is also considered

in the model excercise.

It is discussed at length in literature (22, 3) that the below-cloud scavenging of submicron aerosol particles is negligible. It is only via in-cloud scavenging processes that these particles are incorporated in precipitation (2). The efficiency of the in-cloud scavenging process, however, seems to be rather efficient as discussed in detail elsewhere (4,5).

2. EXPERIMENTAL

Five successful plume wash-out experiments were performed in the vicinity of a 1000 MW coal-fired power plant near the center of the Netherlands. Standard rain collectors were placed downwind of the installation in a pattern as indicated in figure 1. For the specific analysis and conservation of the unstable species, (bi)- sulfite and hydrogen peroxide, each rain sample was divided into two, with monoethanolamine added to one fraction for stabilisation of the bisulfite.

Analysis of the rain samples was carried out with the usual wet-chemical techniques (ion-chromatography, atomic absorption spectrometry and the indophenol colouring method for ammonia). For the specific analysis of (bi)sulfite and hydrogen peroxide a calibrated amperometric method was used. Since Römer et al. (6) show that decomposition of hydrogen peroxide in rain samples is minor when analyzed within twenty-four hours after collection, care was taken here that the hydrogen peroxide analysis was finished within a few hours after collection.

Data on emission rates of pollutants observed in the rain under the plume were obtained from KEMA (7). Emission rates for the particular elements are based on coal composition which is a function of coal origin (8). Apart from the emission of sulfur dioxide and hydrogen chloride which is rather accurately known (uncertainty bounds of approximately 15%) the uncertainty in the emission of other components is likely to be of the order of 40% for bulk elements like aluminium (8) and a factor of two at best for trace elements like selenium (8). Because of very efficient particle filtration in the installation only the gaseous emissions of the semi-volatile elements like fluor, boron and selenium are of importance.

Meteorological data (windspeed at plume height, rainfall rate) were interpolated from data in the networks of the national research institutes (RIVM, KNMI).

Details of the in-cloud measurements of sulfate and submicron aerosol mass from an aircraft can be found in the proceedings of the previous conference (4) and in a forthcoming publication (5).

3. MODEL DESCRIPTION

In the calculations of the plume wash-out the so-called MPADD-model, developed by Hales et al. [9], was used. The model is a framework to be modified and extended by the user. Plume dispersion, including dry deposition, is combined with the calculation of the scavenging of pollutants by rain drops falling through the plume.

The rain drop trajectories depend on fall speed (which is a function of drop size) and wind velocity, which both are taken into account.

The scavenging (capture) of particulates by raindrops is described in full detail in literature (2, 3). The capture efficiency is a function of rain drop size and aerosol particle size. Uncertainties in the size of the emitted particles do not justify a detailed calculation of the scavenging of particles. In figure 1, it is shown that the emitted particulate mass is concentrated in particles for which a small uncertainty in size is associated with extremely large uncertainties in scavenging rate. Instead, an overall empirical scavenging rate for particulate wash-out was obtained from a comparison of measured and modeled wash-out of particles (par. 4.2).

The scavenging of the gaseous plume constituents (hydrogen chloride, fluoride and sulfur dioxide) was calculated with a coupled mass transport and chemical equilibrium submodel, which is analogous to the scheme developed by Adewuyi and Carmichael (10). The uptake of gases in falling rain drops is described in a two-film model for transport of gases and the dissolved species over diffusion layers at the drop boundary, with instantaneous protolysis of the dissolved gases. The increase of the liquid concentration during the fall of rain drops through the plume is calculated taking into account the gas concentration profile in the plume. The set of equations for mass-transfer are solved with the constraint of electroneutrality of the ionic species, both at the drop boundary and in the bulk solution. The appropriate values for the solubility, diffusion and protolysis equilibrium constants were obtained from handbooks.

In addition to the mentioned scheme the reaction of dissolved sulfur dioxide (as bisulfite) with hydrogen peroxide was incorporated in the model, using peroxide concentrations as measured in rain. The reaction was modeled in the so-called slow reaction limit with a reaction rate constant taken from literature (11).
Oxidation of bisulfite by dissolved ozone was neglected in view of the low concentration of ozone in the plume and the low solubility of ozone in rain. For proper modelling of the scavenging of sulfur dioxide from the plumes the presence of sulfur dioxide in the background air was taken into account.

The uptake of irreversible dissolving gases in rain as a function of rain drop size is relatively easy modeled. However as is shown in literature (12) the scavenging strongly depends on the size of the rain drops. Since in the experimental part of the present study the rain drop size spectrum is unknown, this spectrum is an adjustable parameter in the calculations.

For the calculation of the below-cloud scavenging of ammonia in the atmosphere the chemical submodel was extended. Since it is shown that the influence of carbon dioxide on rain chemistry is marginal (10) this compound was not incorporated in the present study.

4. RESULTS AND DISCUSSION

4.1. Plume wash-out experiments

Significantly increased concentrations of characteristic plume constituents were observed in rain collected below the plume as compared to the concentrations of these compounds in the rain outside the plume sector (tabel 1, figures 3 and 4). The wash-out of plume compounds was obtained after subtraction of the background deposition of these compounds. For this correction other components are used for which no plume wash-out is observed and which highly correlate with the plume constituents in the rain outside the plume sector (13). This correction procedure also serves as a means to identify and dismiss contaminated samples (figure 4).

The observed increase in acidity in the rain under the plume, cf. figure 3, is to a large extent caused by the scavenging of hydrogen chloride (figure 4). Whereas plume wash-out for certain gaseous pollutants, as indicated in figure 3 and 4 was obvious in all five experiments, it was only in two occasions that a measurable aluminium particulate wash-out was observed.

Wash-out of sulfur dioxide appeared to be minor in comparison to that of chloride and fluoride, while emitted in substantially larger quantities (if observed at all). This low-efficient and erratic scavenging of sulfur dioxide is in accordance with the results of others (14). In one experiment, with hydrogen peroxide in the background rain, extra sulfate was found in the rain below the plume, while the peroxide concentration below the plume showed an equivalent decrease. This is a certain indication of a reaction of peroxide and bisulfite in the rain falling through the plume.

4.2. Calculations of the plume wash-out

Using the appropriate meteorological data (15) the deposition at the position of the rain collectors was calculated and compared with the experimental results. Aluminium was used as the reference species for the particulate pollutants and chloride for the gaseous contaminants. Chloride was chosen since the combined uncertainty in emission rate and measured deposition was minimal.

For particulates a scavenging efficiency of approximately 0.25 was derived. With reference to figure 2 it is seen that this efficiency is higher than expected on the basis of the size of the emitted particles. The increased efficiency can be attributed to particle growth in the wet plume by water condensation or accretion, which results in a more efficient capture by the falling rain drops (by impaction). Water condensation could also be the explanation of the highly efficient scavenging of submicron aerosol in an other plume study (16) where the plume was directly emitted into the clouds.

It was found that chloride deposition was best predicted by the model calculations in case the standard Best distribution (3) for the rain droplet size spectrum was chosen. Since at least half of the precipitation in the Netherlands is delivered under meteorological conditions comparable to those during the experiments, the choice for Best's rain drop size distribution in the calculation of the (annual) deposition pattern around coal-fired installations seems justified. It is found that maximum acid deposition near the stack of a 600 MW coal-fired installation in the Netherlands could be comparable to the average acid deposition by precipitation (200 eq. $ha^{-1}.y^{-1}$). Because of advection of the rain drops with the wind and plume rise the maximum deposition is found at a distance of several hundred meters from the stack!

For the calculations of the sulfur dioxide wash-out, background concentrations of sulfur dioxide and rain acidity values were chosen which are representative of the Dutch situation. With respect to the results of these calculations the reader is referred to figure 5. This figure shows the predicted complex behaviour of the sulfur dioxide wash-out which is in accordance with the observations in the experimental study. The "negative" wash-out (figure 5) indicates that during plume passage of raindrops sulfur dioxide (bisulfite) is lost by desorption due to uptake of the strong acidic hydrogen chloride. With reference to figure 5 the absence of a measurable sulfur dioxide wash-out in the experimental study is understood.

According to the model calculations a significant wash-out of selenium, emitted as a gaseous compound, was expected. A possible explanation of the absence of a measurable plume wash-out is the assumption that after emission and cooling, "selenium" consenses with condensation preferentially on the smallest submicron particles (16) for which wash-out is negligible (figure 2). Such a condensation after further cooling is expected since the majority of the selenium already condenses on small particulates in the hot flue gas (8).

If it is assumed that the gaseous compounds are present in the plume as irreversibly dissolving gases the plume wash-out should be proportional to the emission. To validate this assumption the emission ratios of the relevant gaseous components were translated into wash-out ratiosat the position of the rain collectors (fig. 1) after correction for the following difference in wash-out rate derived from the model calculations

$$W \sim D^{0.75}$$

with W the wash-out rate and D the gas diffusion constant.

The calculated and measured wash-out ratios for boron, fluoride and chloride are compared in table 2. The higher measured wash-out for "boron" and fluoride is a strong indication that the emission of these compounds is higher, probably due to a volatility of these compounds that is higher than expected (see chapter 2).

4.3. Below-cloud scavenging of ammonia

Below-cloud scavenging of ammonia and related compounds in the Netherlands was calculated using the following assumptions: Cloud base at 800 m and a constant vertical concentration for particulate ammonium. For ammonia (NH_3) and nitric acid (HNO_3) gas the following average vertical profile in the concentrations, measured in a series of precipitation-free experiments (1), was used:
NH_3; ground level: 0.4 $\mu mol.m^{-3}$, linear decrease to 0.1 $\mu mol.m^{-3}$ (200 m) and a constant concentration upwards.
HNO_3; ground level: 0.02 $\mu mol.m^{-3}$, linear increase to 0.05 $\mu mol.m^{-3}$ (200 m) and a constant concentration upwards.
In the calculations background rain acidity (at cloud-base) was chosen in such a way that calculated ground-level rain concentrations were in accordance with measured acidity. Decrease in the concentration of the compounds due to wash-out during rain is accounted for.

It appeared that below-cloud scavenging of the gases (ammonia, sulfur dioxide, nitric acid) contributes about one third to the total wet depostition of acid related substances in the Netherlands. As expected for submicron particles, the scavenging of particulate ammonium is negligible. Due to the uptake of the alkaline ammonia an equivalent extra amount of sulfur dioxide, as bisulfite, is scavenged. In a sensitivity test it was shown that a constant vertical profile with a concentration equal to the ground level concentration would lead to a serious overestimate of the actual ammonia deposition.

4.4. In-cloud scavenging

Through analysis of a series of data from cloud flights in the eastern US, to be discussed in detail in a forthcoming paper (5), earlier preliminary findings of highly efficient (>90%) scavenging of sulfate and submicron aerosol mass in clouds were substantiated. In the flights additional parameters (e.g. ozone, temperature) were recorded for tracing the origin of the cloud air, while flight tracks were made in the clouds as well as above and below the cloud layer.

ACKNOWLEDGEMENT

The plume wash-out study was performed under contract with the Management Office for Energy Research in the Netherlands (PEO) and was financed by the Ministry of Economic Affairs. The plume experiments were carried out under technical supervision of J. Möls (ECN). The background corrections for the plume wash-out data were made by J. Baard (ECN). Mr. De Vos and Mr. Asselman (RIZA) are gratefully acknowledged for their assistance at the collection site.

The cloud data, obtained in flights with the Brookhaven aircraft, were analyzed at Brookhaven National Laboratory, Upton NY, USA (May 1985).

REFERENCES

1. A.W.M. Vermetten, W.A.H. Asman, E. Buijsman, W. Mulder, J. Slanina and A. Waijers, "Concentrations of NH_3 and NH_4^+ over the Netherlands", Proceedings VDI-Colloquium "Forest Decline", June 1985, Goslar, FRG, 241-251;

2. W.G.N. Slinn, in: "Air-Sea Exchange of Gases and Particles", P.S. Liss and W.G.N. Slinn eds. Reidel, Dordrecht (1983), 299;

3. H.R. Prupacher and J.D. Klett, "Microphysics of Clouds and Precipitation", Reidel, Dordrecht (1978), Ch. 14; pg. 464;

4. H.M. ten Brink, P.H. Daum and S.E. Schwartz, in: "Physico-Chemical Behaviour of Atmospheric Pollutants", Proceedings of the Third European Symposium held in Ispra, Italy, April 1984; Reidel, Dordrecht (1984), 356;

5. H.M. ten Brink, P.H. Daum and S.E. Schwartz, "Efficient Scavenging of Sulphate in Liquid Water Clouds ", paper submitted to Atmospheric Environment, September 1986;

6. F.G. Römer, A.A. Veldkamp and P. van Galen, see ref. [4], pg 74;

7. L.H.J.M. Jansen (KEMA, Arnhem, The Netherlands), personal communication, May 1985;
R. Meij (KEMA, Arnhem, The Netherlands), personal communication, May 1985;

8. R. Meij, J. van der Kooij, H.A. van der Sloot, J.M. Koppius-Oddink and L.J. Clement, "Emissions and Control of Particulates of Coal-Fired Power Plants", proceedings 2d US-Dutch International Symposium on Aerosols, Williamsburg VA, USA, May, 1985, in press;

9. J.M. Hales, T.W. Horst and M. Terry Dana, "MPADD-A Generalized Code for Reactive Scavenging and Deposition Calculations", report DW 930197-01-1, Pacific Northwest Laboratory, Richland, WA, USA (1983);

10. Y.G. Adewuyi and G.R. Carmichael, Atmos. Environ. 16 (1982), 719;

11. S.A. Penkett, B.M.R. Jones, K.A. Brice and A.E.J. Eggleton, Atmos. Environ 13 (1979), 123;

12. S.Z. Levine and S.E. Schwartz, Atmos. Environ. 16 (1982), 1725;

13. J. Slanina, F.G. Römer and W.A.H. Asman, in: "Precipitation Scavenging, Dry Deposition and Resuspension", Vol. 1, H.R. Pruppacher et al. eds., Elsevier, Amsterdam (1983), 349.

14. M. Terry Dana and A.A.N. Patrinos, idem pg 335;

15. A.J. Janssen and H.M. ten Brink, "Plume Wash-out: Modeling, Calculation and Validation", ECN-report 170 (July 1985), in Dutch;

16. M.M. Millan, S.C. Barton, N.D. Johnson, B. Weisman, M. Lusis and R. Vet, Atmos. Environ 16 (1982), 2709;

17. S. Friedlander, "Smoke, Dust and Haze, Fundamentals of Aerosol Behaviour", Wiley, London (1977), chapter 9, pg 251.

Table 1: Concentrations of some of the components in the rain downwind of the 1000 MW coal-fired Amer power station, 18 October 1984. See figure 1 for position of the rain-collectors.

Rain collector Nr.	H ppb	NH4 ppm	Na ppm	Ca ppm	SO4 ppm	NO3 ppm	Cl ppm	RAIN GRAM	IONBAL +/-
36	12	0.98	0.29	0.57	2.98	0.97	0.42	84.4	1.32
37	8	1.04	0.23	0.83	2.62	0.88	0.37	101.4	1.64
38	14	1.17	0.14	0.45	2.95	0.94	0.35	98.5	1.35
39	13	1.15	0.11	0.46	2.59	0.85	0.36	101.5	1.47
40	12	1.17	0.30	0.69	2.66	0.95	0.70	110.3	1.54
41	10	1.23	0.25	0.63	2.82	0.94	0.66	110.7	1.43
42	22	1.06	0.23	0.35	3.21	0.93	0.62	83.9	1.18
43	18	1.15	0.28	0.69	3.15	0.95	0.85	78.5	1.33
44	37	0.98	0.23	0.39	3.30	0.97	0.98	98.3	1.16
45	52	0.92	0.13	0.52	2.82	1.02	1.59	82.6	1.18
46	80	0.90	0.12	0.40	3.38	1.01	2.06	103.0	1.12
47	71	1.11	0.23	0.49	3.47	1.09	1.83	89.6	1.24
48	62	1.71	0.16	0.40	5.64	1.12	1.30	122.4	1.11
49	58	1.69	0.14	0.40	3.61	1.01	0.70	102.9	1.68
50	34	1.23	0.20	0.44	3.86	0.99	0.50	109.6	1.32
51	38	1.15	0.12	0.52	4.53	1.01	0.41	121.3	1.15

Raincollector Nr.	F ppb	B ppb	Al ppb	Se ppb	Mn ppb	Fe ppb
36	34	3.4	34.0	0.1	10.2	90.2
37	27	2.4	70.9	0.2	19.1	119.0
38	42	4.0	40.9	0.1	13.4	90.9
39	46	19.2	41.0	0.1	13.9	101.0
40	67	7.8	52.7	0.1	15.6	119.0
41	71	9.4	32.5	0.1	13.3	37.6
42	75	9.3	32.2	0.1	10.8	58.5
43	101	11.8	63.2	0.1	15.1	91.0
44	142	16.9	92.0	0.1	9.9	73.1
45	235	31.2	84.2	0.1	9.5	94.2
46	310	41.4	80.7	0.1	10.7	113.0
47	275	40.5	72.8	0.1	11.9	117.0
48	270	41.6	64.7	0.1	11.3	65.7
49	209	33.2	51.0	0.1	10.1	50.2
50	140	19.7	58.9	0.1	11.8	85.4
51	96	11.5	48.8	0.1	14.2	102.0

Table 2: Calculated and measured ratios of the wash-out of elements from the plume of the Amer power station; uncertainty in the measured wash-out ratios less than 20%.

Date	F/Cl		B/Cl	
	calculated	measured	calculated*	measured
4 Sept.	0.08	0.16	0.026	0.023
21 Sept.	0.08	0.50	0.042	0.111
9 Oct.	0.08	0.22	0.031	0.067
18 Oct.	0.08	0.39	0.052	0.100
21 Nov.	0.15	0.77	0.062	0.051

* Assuming a gas diffusion coefficient for the boron compound of 0.2 $cm^2.s^{-1}$. A more realistic diffusion coefficient (for H_3BO_3) of 0.1 $cm^2.s^{-1}$ would lead to a decrease by a factor of 1.7 in the tabulated values.

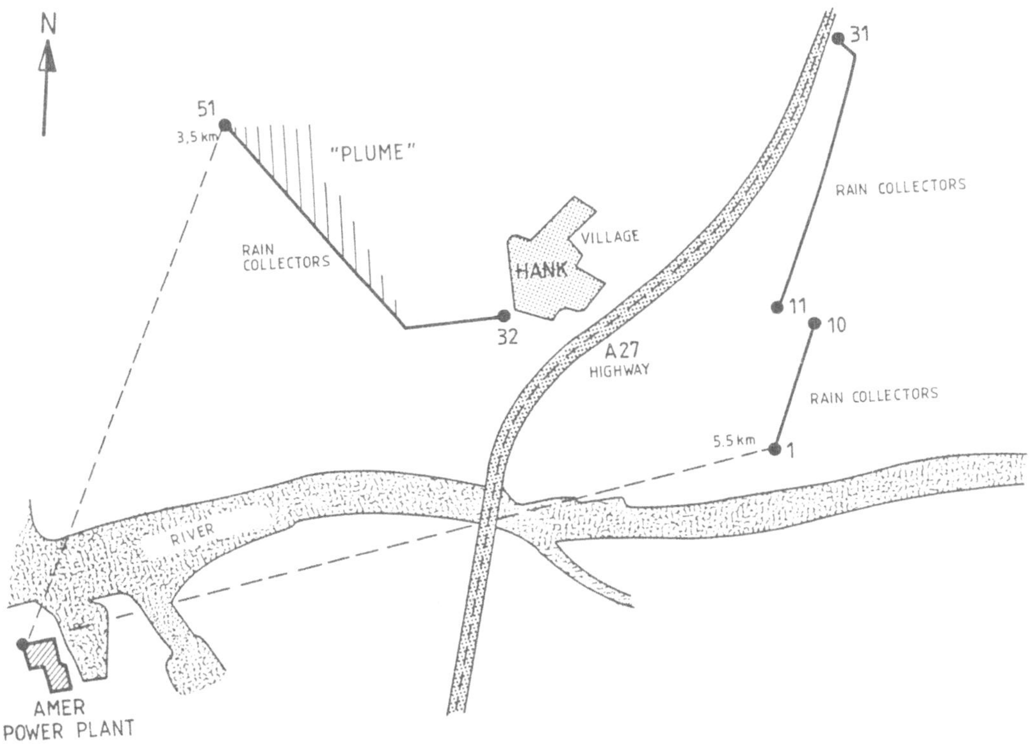

Fig. 1. Position of the rain collectors (1-51) near the Amer power station in the wash-out experiments, with a local situation sketch and an indication of the rained-through plume.

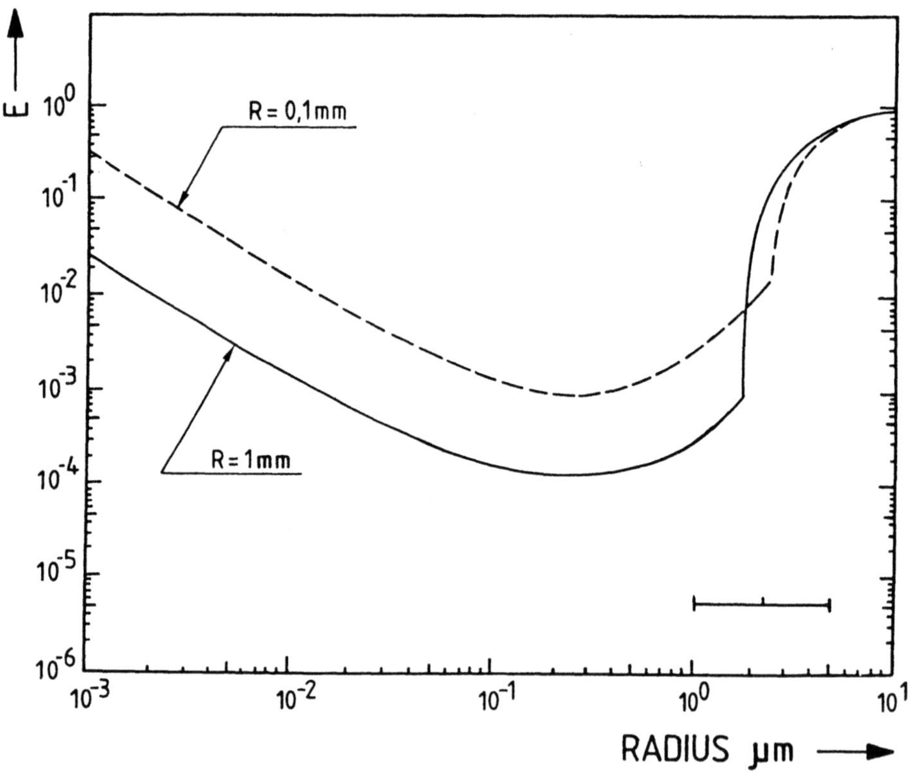

Fig. 2. Efficiency (E) of capture of particles by falling raindrops as a function of aerodynamic size of the particles at two raindrop sizes R (radius). Mass distribution of fly-ash particles after the ESP-filters [8] indicated in the figure (median and range). Figure taken from ref. [2].

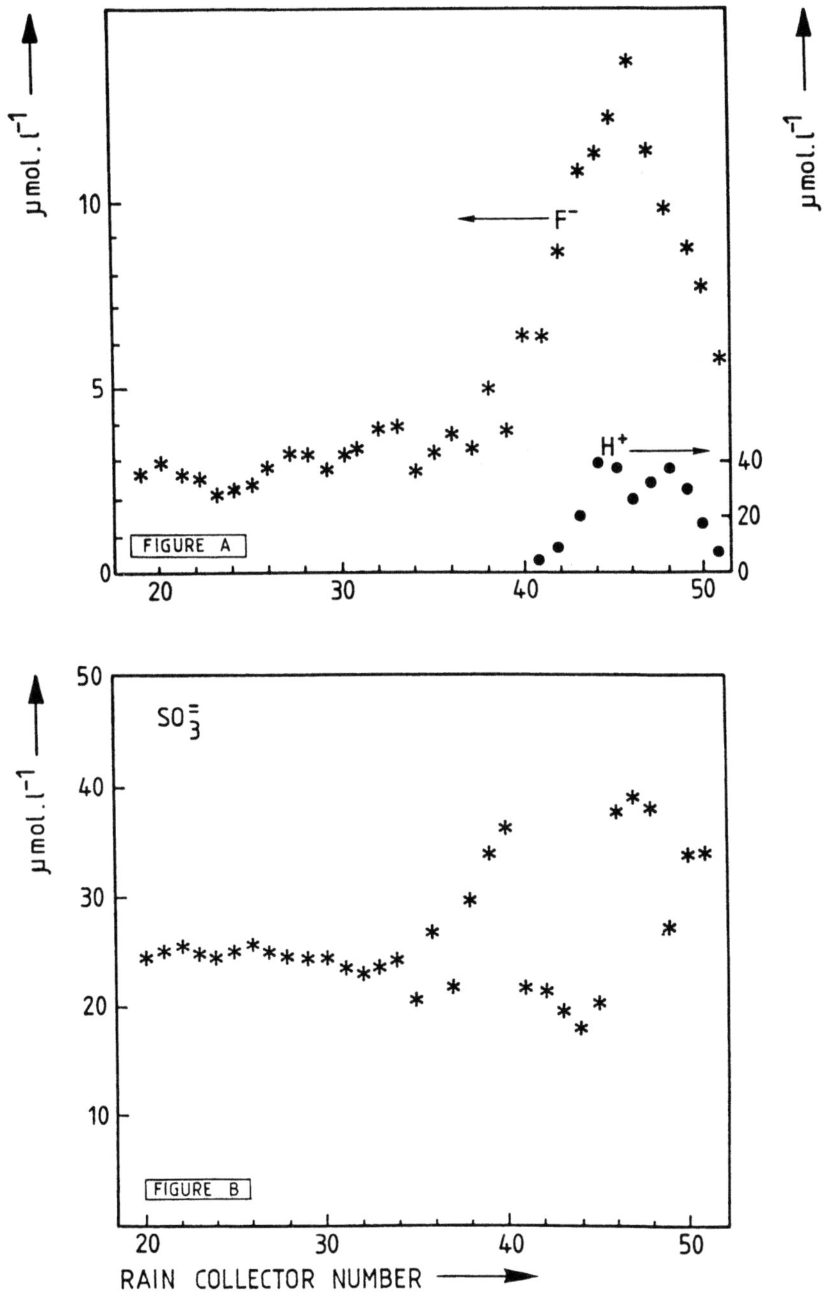

Fig. 3. Concentration of fluoride and acidity (A) and (bi)sulfite (B) in rain collected under and next to the plume of the coal-fired Amer power station on September 4, 1984. For position of rain collectors see Fig. 1.

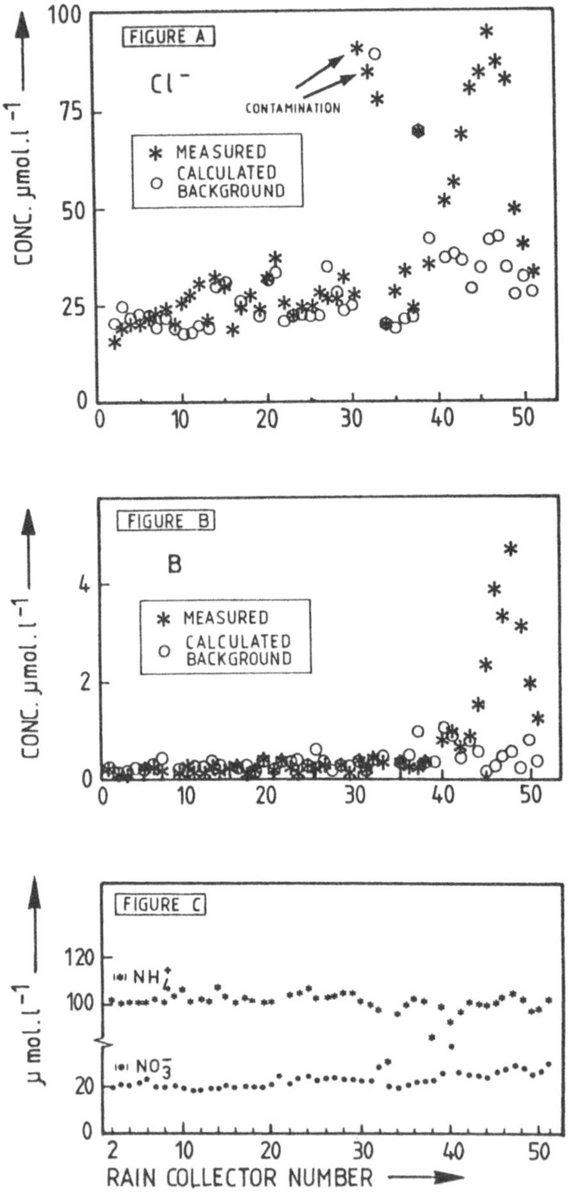

Fig. 4. Concentration of the indicated component in rain for which plume wash-out is observed (A,B) and for which plume wash-out is apparently absent (C). For position of rain collectors see Fig. 1.

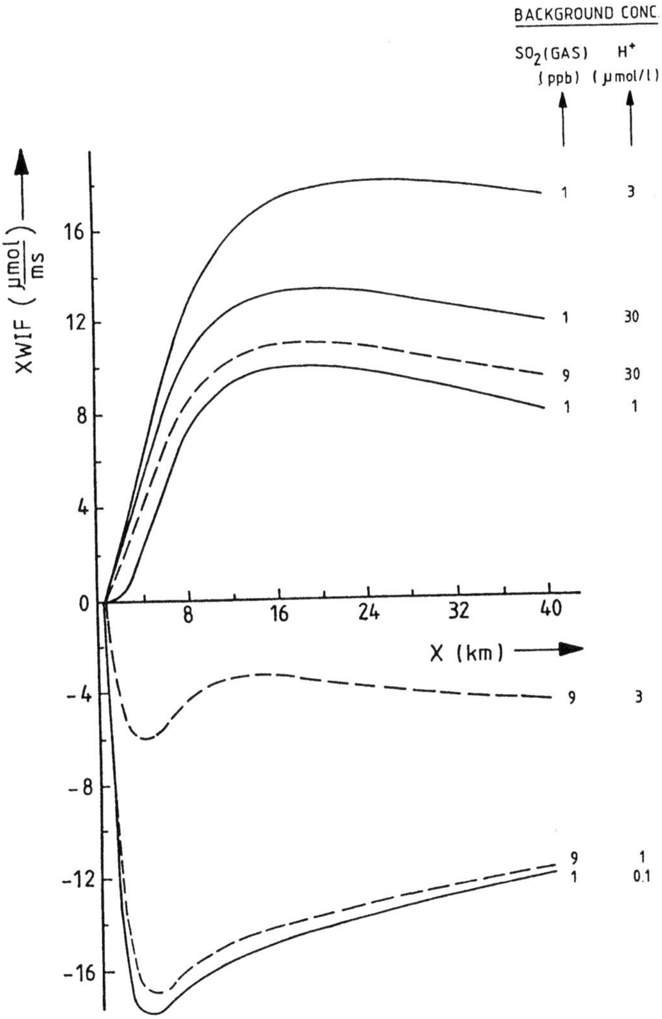

Fig. 5. Deposition of (bi)sulfite by plume wash-out of sulfur dioxide in a cross-section under the plume (XWIF, Cross-Wind Integrated Flux) as function of the distance from a coal-fired installation (1000 MW), at different background concentrations as indicated. Wind velocity 10 m/s, rainfall rate 1.5 mm/h. Emissions: SO_2 90 kmol/h; HCl 8.6 kmol/h; HF 1.8 kmol/h.

——— wash-out with SO_2-gas background concentration 1 ppb
------ wash-out with SO_2-gas background concentration 9 ppb.

MESURE D'AEROSOLS DANS LA FORET DES VOSGES (FRANCE)

M.L. PERRIN, C. FRAMBOURT, G. MADELAINE
Commissariat à l'Energie Atomique,
IPSN, DPT/SPIN, Laboratoire de Physique et de Métrologie des Aérosols
BP 6, 92265 FONTENAY AUX ROSES, CEDEX, France

Summary

On the French acid rains program DEFORPA (Dépérissement des Forêts attribué à la Pollution Atmosphérique), we have to characterize the atmospheric aerosol in the Vosges Forest, in the east of France. This paper concerns field measurements we have made in September 1985 and provides results on the particles concentration, size distribution (between 0.01 µm and 15 µm) and chemical composition. They are related to the regional meteorological conditions and the gaseous pollutants concentrations.

1. INTRODUCTION

Retenue comme hypothèse majeure du dépérissement des forêts, la pollution atmosphérique ne pourra être désignée de manière certaine comme responsable qu'après l'aboutissement d'une série de travaux tels que la mesure in situ du niveau des polluants gazeux et particulaires, l'identification des sources de ces polluants, la mise en évidence expérimentale de leurs responsabilités et de leurs mécanismes d'action. "La pollution atmosphérique est-elle responsable du dépérissement des forêts ?" est une des questions à laquelle le programme français DEFORPA (Dépérissement des Forêts attribué à la Pollution Atmosphérique) tente de répondre.

Dans le cadre de ce programme, nous sommes chargés de caractériser l'état particulaire, c'est-à-dire, de déterminer la concentration, la dimension et la composition chimique des aérosols présents, du site du Donon. Situé dans la forêt des Vosges, à l'est de la France, ce site a été retenu pour y installer une station de mesure de la pollution, certains peuplements forestiers autour du Col du Donon présentant des signes de dépérissement.

Deux campagnes de mesure des aérosols y sont prévues, pour des conditions météorologiques différentes.

Cet exposé traite des mesures que nous avons faites pendant une semaine, en septembre 1985.

2. MATERIELS ET METHODES DE MESURE

La figure 1 est une carte de la région des Vosges qui situe la station expérimentale du Donon.

Cette station est actuellement équipée d'analyseurs de gaz qui permettent la mesure en continu des concentrations en SO_2, NO_x, O_3.

Les données météorologiques, vitesse et direction du vent, nous sont fournies par la station de Phalsbourg (figure 1). Ces données ne tiennent pas compte de microméteorologie locale mais représentent cependant la tendance météorologique de la région.

Notre station "aérosols" était équipée du matériel suivant :

- un analyseur de mobilité électrique, basé sur les propriétés électriques des aérosols, qui mesure la granulométrie et la concentration des particules dont le diamètre est compris entre 0,01 µm et 1 µm ;
- un compteur optique, basé sur les propriétés optiques des aérosols, qui détecte les particules dont le diamètre est compris entre 0,5 µm et 10 µm ;

- un impacteur virtuel dichotomique qui collecte sur filtres les particules en deux classes, l'une supérieure à 2 µm et l'autre inférieure. L'analyse des filtres soit par absorption atomique, soit par chromatographie ionique, permet de connaître la composition chimique des particules collectées.

Les mesures physiques des aérosols se sont déroulées en continu du 18/09/1985 au 26/09/1985 ; pendant cette période 18 prélèvements sur impacteur ont été effectués, leur durée variant entre 2 heures et 12 heures.

3. RESULTATS

Pendant un premier épisode qui s'est prolongé sur presque toute la durée de la campagne (18/09/1985 au 25/09/1985 vers 12 heures LT), le vent a été orienté sud-ouest avec une vitesse variant entre 0 et 4 m/s. La journée du 21/09/1985 caractérise bien cette situation, comme l'indique la figure 2. Après le 25/09/1985, vers 12 heures LT intervient un deuxième épisode marqué par le vent qui tourne au secteur est.

Les mesures de polluants gazeux, sur le site, nous sont fournies par L'ASPA (Association pour la Surveillance de la Pollution Atmosphérique) de Strasbourg. Les concentrations en dioxyde de soufre sont quasiment nulles pendant toute la durée de la campagne, sauf une légère pointe à 3 ppb (9 $\mu g/m^3$) le 26/09/1985.

Pendant la période qui nous intéresse les données en monoxyde et dioxyde d'azote ne sont pas disponibles, du fait d'une panne des appareils. Les concentrations en ozone sont comprises entre 22 et 46 ppb (47 et 98 $\mu g/m^3$).

Les résultats des mesures physiques des aérosols montrent qu'au cours du premier épisode :

- la concentration en particules submicroniques est restée inférieure à 10 000 particules par cm^3, avec un diamètre numérique moyen de 0,06 µm, une surface de 400 $\mu m^2/cm^3$ et un volume de particules de 25 $\mu m^3/cm^3$. La figure 3 représente l'évolution de la concentration de ces particules mesurée le 21/09/1985. La figure 4 représente une granulométrie moyenne mesurée au cours de cet épisode.

- la concentration en particules supérieures au micron est de l'ordre de 10 par cm^3.

Lors du deuxième épisode lorsque le vent a tourné à l'est le 25/09/1985 vers 12 heures LT, on a pu constater une très nette augmentation du nombre de particules submicroniques et, en particulier, des particules très fines. Ceci est très net sur la figure 5, sur laquelle on a indiqué la concentration en particules de 0,01 µm. Cette augmentation est bien corrélée à une variation du dioxyde de soufre mesuré sur le site.

Du point de vue de la composition chimique des particules, on a relevé pendant le premier épisode entre 1,5 et 2,5 $\mu g/m^3$ d'ions sulfates sur la partie fine de l'aérosol, inférieure à 2 mg, et moins de 0,1 gm/m^3 sur la partie supérieure. Ces résultats sont portés sur la figure 6. Les faibles valeurs mesurées pour le prélèvement n° 13 sont dues à une chute du débit et ne sont donc pas à prendre en compte. Les concentrations en ions nitrates sont inférieures à 1 $\mu g/m^3$. Les teneurs en métaux tels que Hg, Mg, Mn et Br sont extrêmement faibles et bien souvent en dessous des limites détectables. Les teneurs en Pb sont de l'ordre de 10^{-3} à 10^{-2} $\mu g/m^3$, en Ni comprises entre 10^{-4} et 10^{-2} $\mu g/m^3$, et en Fe comprises entre 10^{-3} et 10^{-2} $\mu g/m^3$. L'aérosol apparait chimiquement comme des fines gouttelettes d'une solution contenant 60 à 100 $g.l^{-1}$ de SO_4^{--}, moins de 40 $g.l^{-1}$ de NO_3^-, 40 à 400 $mg.l^{-1}$ de Pb.

4. CONCLUSIONS

Dans le cas de masses d'air faiblement polluées comme celles que nous avons rencontrées dans cette première campagne du Donon nous pouvons bien caractériser l'état physique des particules : nous avons observé au cours du premier épisode une concentration en particules fines inférieure à 10 000 par cm^3, un diamètre moyen en nombre de 0,06 µm. Par contre en ce qui concerne l'état chimique, seules étaient au-dessus des limites de détection les teneurs en ions sulfates (teneur en sulfates inférieure à 2,5 µg/m^3), ainsi que les teneurs en Pb, Ni, Fe.

Les quelques mesures dont nous disposons par vent de secteur Est (deuxième épisode) montrent que le changement de l'origine des masses d'air, même faiblement polluées, s'observe immédiatement sur l'état physique des particules (dans notre cas, on observe par exemple l'augmentation du nombre de particules fines). Nous espérons pouvoir dans une prochaine campagne rencontrer des situations où les masses d'air auraient été chargées de polluants industriels.

A titre indicatif, le tableau suivant fournit les concentrations et diamètres moyens que nous avons mesurés lors de précédentes campagnes :

Site	concentration moyenne en particules par cm^3	diamètre moyen (µm)
Guissény / 1 /	3 000	0,02
Fontenay-aux-Roses / 2 /	100 000	0,05
Fos-sur-Mer / 3 /, / 4 /, / 5 /	300 000	0,02

Guissény est un site maritime situé sur la côte ouest du Finistère en Bretagne, et est intéressé essentiellement par des masses d'air d'origine atlantique.

Fontenay-aux-Roses, site urbain, fait partie de l'agglomération parisienne.

Sur le site industriel de Fos-sur-Mer, port pétrolier près de Marseille, les teneurs en ions sulfates de la partie fine du spectre granulométrique peuvent atteindre 20 µg/m^3.

REFERENCES

/ 1 / PERRIN M.L., MADELAINE G., RENOUX A. (1980)
Evolution of fine particles naturally generated in a marine atmosphere
Proceedings of the 8th GAeF Conference, Schmallenberg (R.F.A)

/ 2 / PERRIN M.L. (1981)
Etude la dynamique d'aérosols fins produits artificiellement. Application à l'atmosphère
Rapport CEA-R-5062

/ 3 / PERRIN M.L., MADELAINE G., NADAL R. (1982)
Fine particles measurements around Fos-Berre area
Proceedings of the 10th GAeF Conference, Bologne (Italie)

/ 4 / PERRIN M.L., BOURBIGOT Y., MADELAINE G. (1982)
Fine particles measurements around an industrial area
Aer. Sci. and Technology 2, 165

/ 5 / PERRIN M.L., MADELAINE G., FRAMBOURT C. (1984)
Mesures de particules fines dans une zone polluée
Proceedings of the third European Symposium on Physical-Chemical
Chemical behaviour of atmospheric pollutants, Varese (Italie).

REMERCIEMENTS

Les analyses chimiques ont été effectuées par le Service Applications de l'Electricité et Environnement de la Direction des Etudes et Recherches d'Electricité de France.

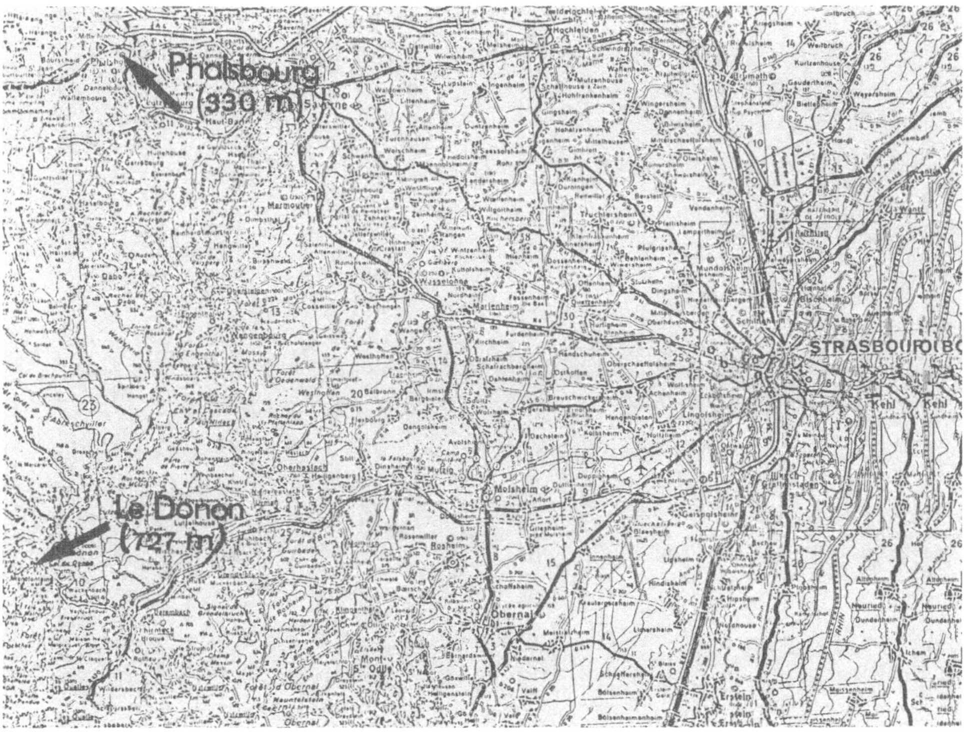

Figure 1 - Carte de la région des Vosges

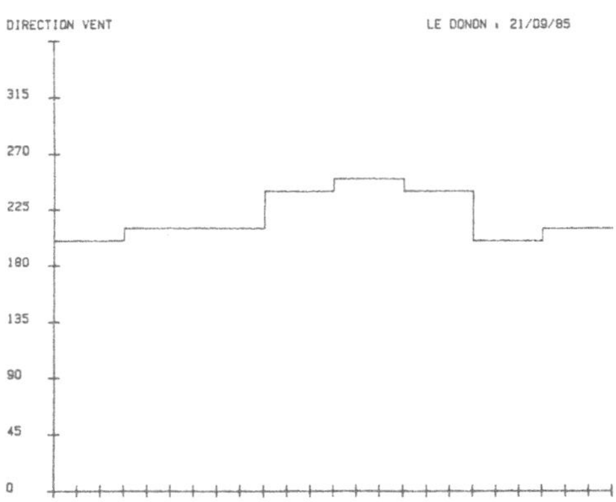

Figure 2 - Direction du vent enregistrée le 21/09/1985

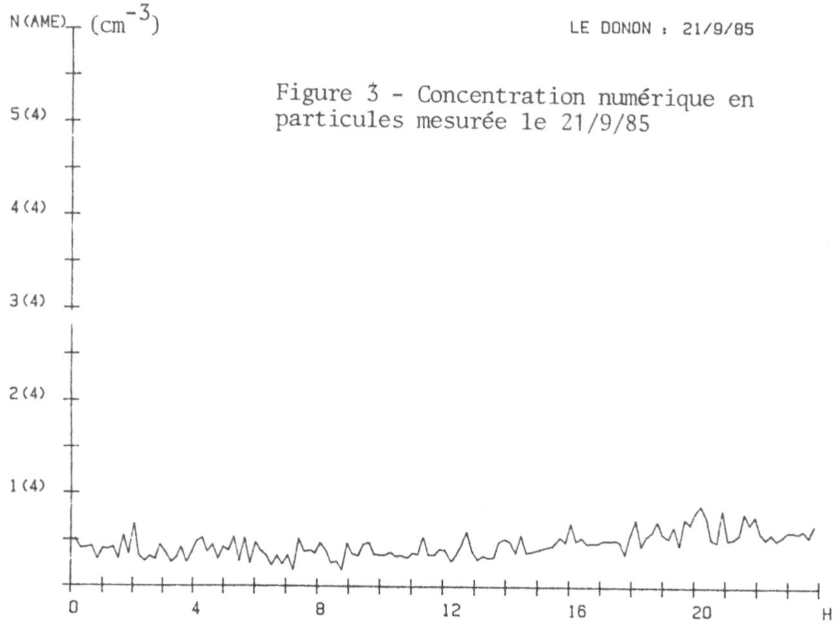

Figure 3 - Concentration numérique en particules mesurée le 21/9/85

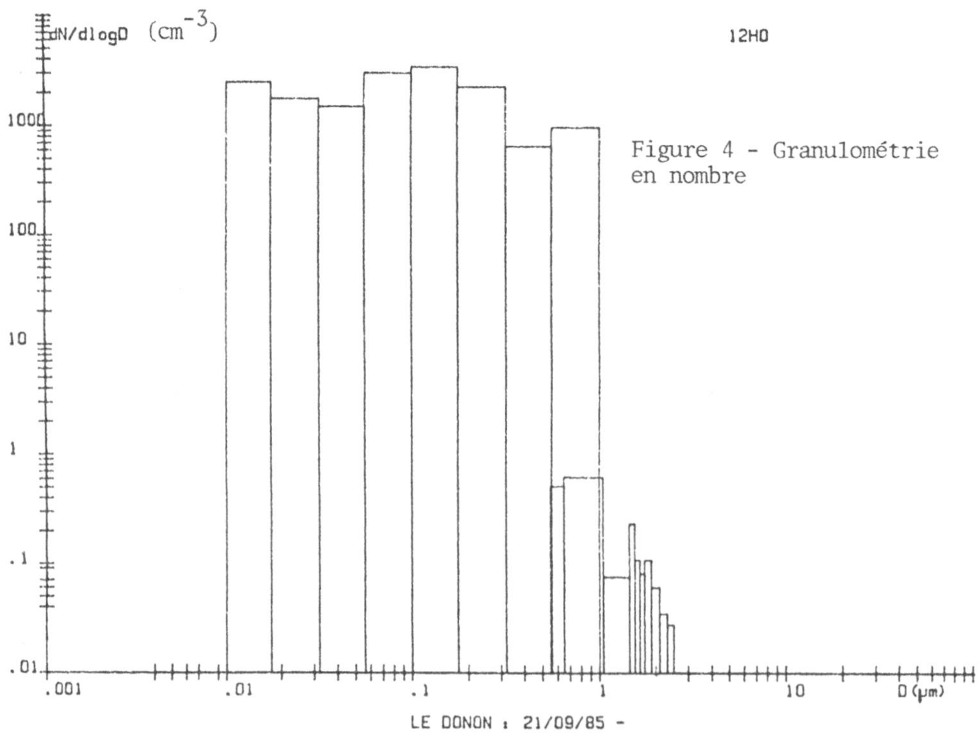

Figure 4 - Granulométrie en nombre

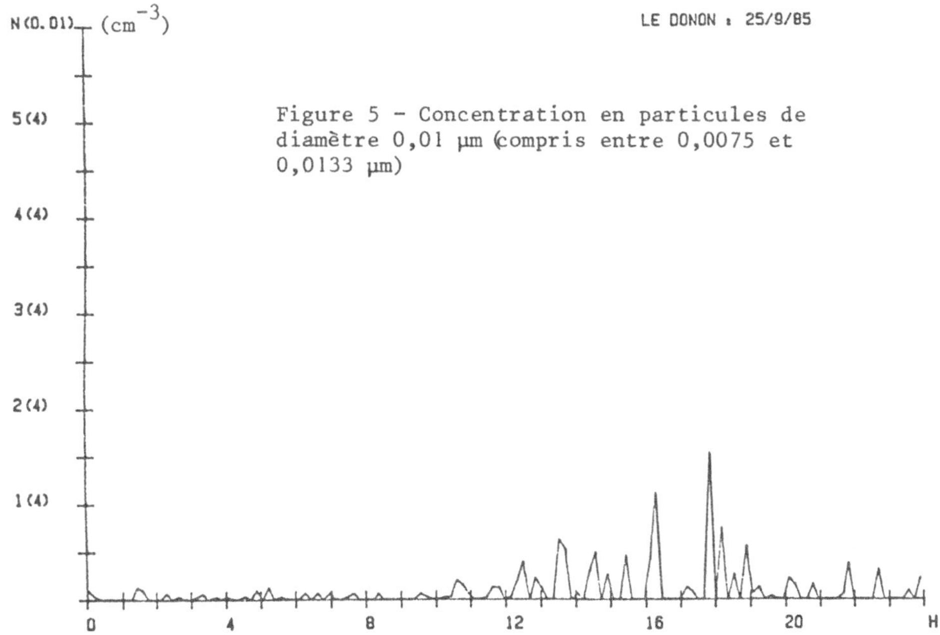

Figure 5 - Concentration en particules de diamètre 0,01 µm (compris entre 0,0075 et 0,0133 µm)

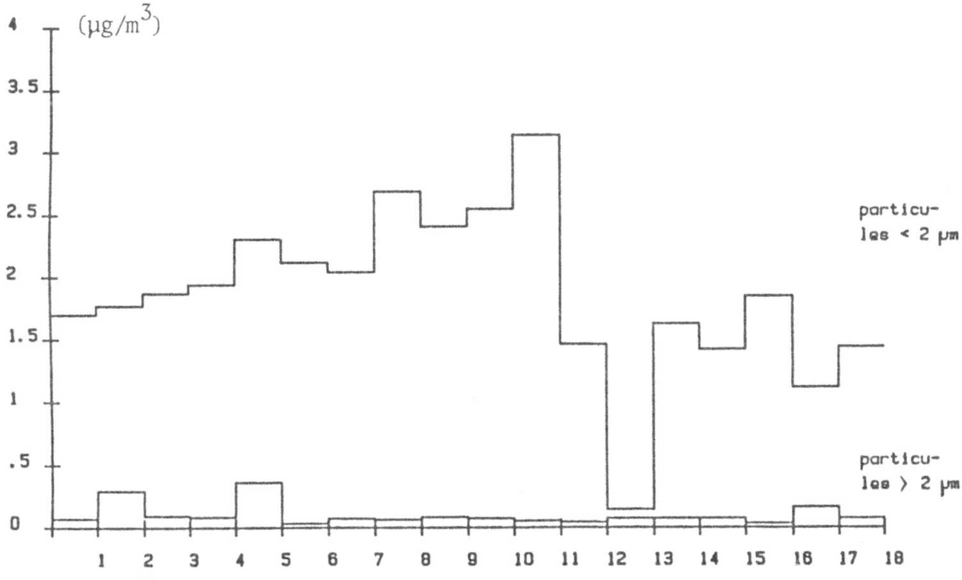

Figure 6 - Concentration en ions sulfates mesurée sur nos 18 prélèvements du 18 au 26/09/85

ANIONIC COMPOSITION IN SIZE-FRACTIONATED ATMOSPHERIC AEROSOL

J. DE LA SERNA, R. FERNANDEZ PATIER, F. PEREZ CARLES, J.GARCIA SANCHEZ and M. ESTEBAN LEFLER.
Departamento de Sanidad Ambiental, Escuela Nacional de Sanidad.
Ciudad Universitaria. 28040. Spain.

ABSTRACT

The physical-chemical characterization of atmospheric aerosol at University Campus in Madrid was performed.

Atmospheric aerosol was collected on glass fiber filter, at a 5-stages Hi-Vol cascade impactor. The mass fractions were determined gravimetricaly and, after subjected to a watery extraction by an ultrasonic bath. The anionic composition of each watery extract was analyzed by ion-chromatography.

In addition, it was determined the size distribution of the principal anions (Cl^-, NO_3^- and SO_4^{-2}) and the size distribution of TSP.

INTRODUCTION

In recent years, the acidification of the lower atmosphere, has led to increase the studies of atmospheric aerosols. One of the most important aspect of this study is the knowledge of the size distributions and the chemical composition of each size of the aerosols. Both aspects are related with mechanism of formation, transport and deposition of the particles and also with their effect on human health.

In many cases the size of the particles is realted with their sources, and with their chemical composition. For example, maritime air particles that are observed in the range 2-50 um. have chlorides as the predominant anion, with an average mode at 5.5 um. They are accompanied in lower proportions by sulphate ions, both associated with sodium cation in coarse particles. Fine particles of marine aerosol are mainly formed by ammonium chloride and sulphate, from secondary origin.

The predominant anions in polluted air masses are sulphate and nitrate, formed by the reaction between gases or between gases and particles from manmade origin. Sometimes, these anions have a bimodal distributions, with modes at around 3.5 um. and 1 um. in form of sodium and ammonium salts respectively. Chlorides in urban air, are present in fine particles as ammonium chlorides, coming from vehicle exhaust.

Following these topics, on this paper, we present a study, of the size distribution of the aerosols collected in Madrid, and the anionic chemical composition of each size of particles collected. This paper is a part of a sum of laboratory studies, directed to the physico-chemical analysis of atmospheric aerosols (1,2,3,4).

EXPERIMENTAL

Samples of atmospheric aerosols were collected weekly during a period of 6 months (May-December) at the University Campus in Madrid (40º-25'N, 3º-43'W).

This sampling station, outskirts of the city, is a calm place with universitary building among lawns and trees, with heavy traffic only in determined hours along the day.

Atmospheric particles were collected on glass-fiber filter at a 5 stages hi-vol cascade impactor (Sierra-mod) and at a flow rate of 60 $m^3 . h^{-1}$.

The impactor allowed us to separate six different size of particles 0.49 um. (back up) ; 0.95 -0.49 um. ; 1.5 - 0.95 um.; 3.0 -1.5 um. and 7.2 um.

The separate mass fractions, after weighted, were extracted in water by an ultrasonic bath. Water extracts were filtered through a Millipore GSTF filter of 0.22 um. pore and the anions analyed by an ion chromatograhp (Dionex 2010 i).

The anions analyzed were chlorides, nitrates and sulphates. Small amounts of other anions (fluoride, formiate, nitrite) without relevance for this study, were present in the chromatogramms.

RESULTS AND DISCUSSION

Seventeen samples were analyzed and the size distributions plotted on semi-logarithmic paper. Graph. 1 shows the relation $\Delta M/\Delta \lg D_p$ vs $\lg D_p$ where ΔM represents the mass of particles (as ug m^{-3}) in a given fraction, and Dp, the equivalent aerodinamic diameter of the particle. Dp. is calculated as the geometric mean of the maximum and minimum diameter in each stage.

Graphs 2,3, and 4, shows in the same way the size distribution of sulphate, nitrate and chloride in the particles samples.

It has been found in Graph 1, that the atmospheric aerosol studied shows a bimodal distribution, and present two modes: one at around 1.0 and other between 5 and 10 um. Withby (1975) asserts that aerosols sampled far from emission sources usualły have a bimodal size distribution, as it happens in this case.

In this work, sulphates have a bimodal distribution (Graph 2) with modes at around 0.1 and 1.0 um; accordingly with the results found by Harrison and Pio, (5) who also found another secondary mode at 3.5 um., that in our results do not appear. This discrepancy maybe due to the fact that the aerosol studied by those authors corresponds to a mixed marine and urban air in Lancaster. Instead of it, our values correspond to a inland place without connection with marine air.

With reference to the Graph 3, it shows the size distribution of nitrate with there modal peaks, at around 0.1, 1.0 and 5.0 um. corresponding to fine and coarse particles.

Chlorides in Graph 4, shows a mode at around 0.1 um. whose origin probably coming from automobile exhaust. Another mode at around 5.0 um. would correspond to primary particles.

CONCLUSIONS

The main anionic composition of aerosol in Madrid are sulphate, nitrate and chloride. Small amount of other anions are also found as fluoride, formiate and nitrite.

The size fractionated study of the aerosols has given a bimodal distributions for TSP, sulphate and chloride, and a trimodal ones for nitrates.

BIBLIOGRAPHY

(1) J. de la Serna; R. Fernández Patier and F. Perez Cárles (1984): " A study of the concentration of sulphates in the particulate matter". In the "Proceedings third European Symposium Physico-Chemical behaviour of atmospheric pollutants". Ed. B. Versino and G. Angeletti. Varese-

Italy 10-12 April. D. Reidel Pub. Co. pp. 322-330.

(2) J. de la Serna; R. Fernández Patier and F. Perez Cárles (1985): " Physico-chemical analysis of atmospheric deposition in the Centre of Spain". In "Advancementes in air pollution monitoring equipment and procedures". Freiburg. F.R.G. - 2-6 June. Ed. W. Grosch, Federal Ministry of the Interior. Bonn. pp. 131-137.

(3) J. de la Serna; R. Fernández Patier and F. Perez Cárles (1985): "Development of atmospheric aerosols and precipitation studies in Spain". In the "Proceedings of a workshop Aerosols and acid deposition". Petten (Nederland) 9-10 December. pp. 36-42.

(4) J. de la Serna; R. Fernández Patier; M. Fernández San Miguel and F. Perez Cárles (1986) :
" An air pollution episode in the Centre of Spain, in January 1985 (Contribution to the study of a severe air pollution episode in Europe). Presented at " International Symposium on prediction, monitoring and control of urban pollution episodes". Brussels (Belgium) pp. 46 June.

(5) K.T. Whitby (1975) "Modeling of the atmospheric aerosol particle size distributions. Sampling and analysis of atmospheric aerosols". University of Minnisota. Particle Technology Laboratory. Report No. 253.

(6) R,M. Harrison and C.A. Pio (1983): "Size - differentiated composition of inorganic atmospheric aerosols of both marine and polluted continental origin". Atmosph. Envir. 17 -pp. 1733-1738.

GRAPH 1.
Mass/size distribution of TSP

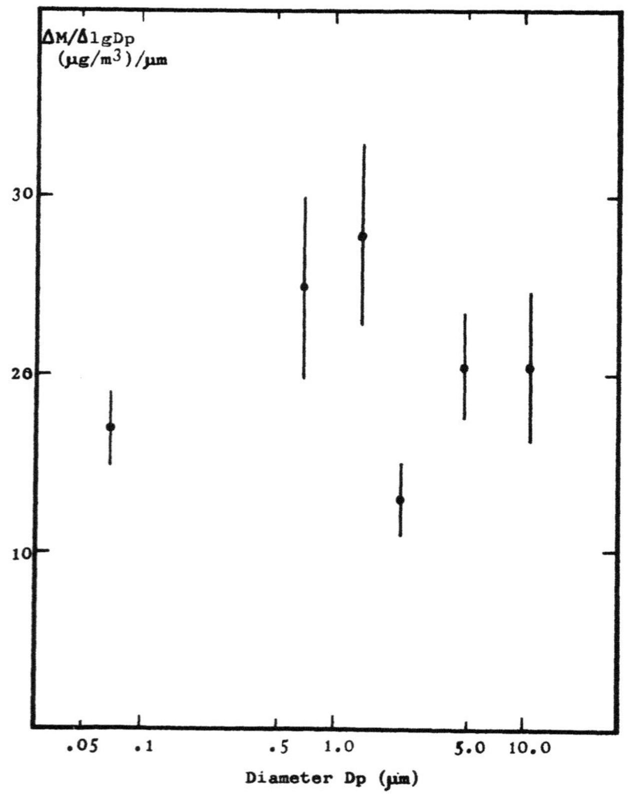

GRAPH 2.
Mass/size distribution of sulfate

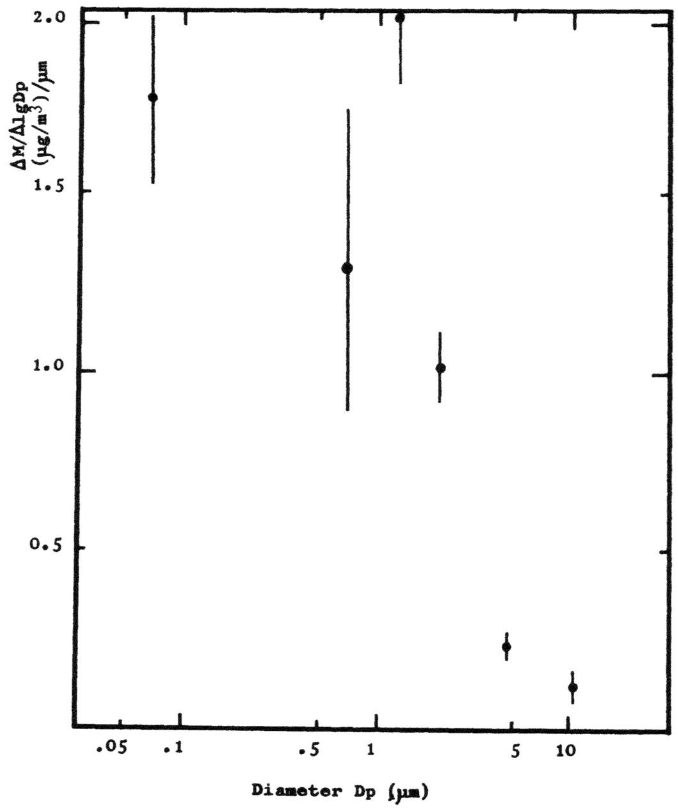

GRAPH 3.
Mass/size distribution of nitrate

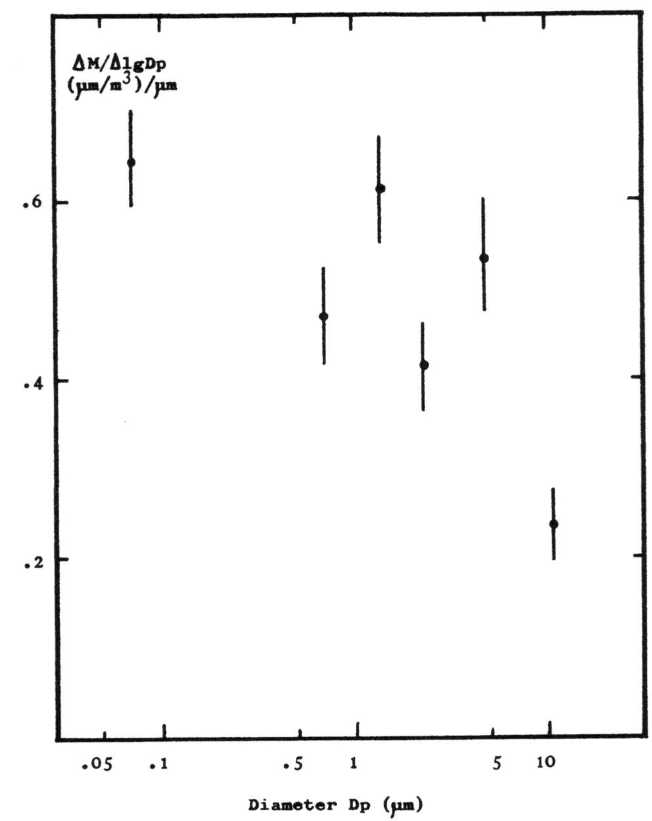

GRAPH 4.
Mass/size distribution of chloride

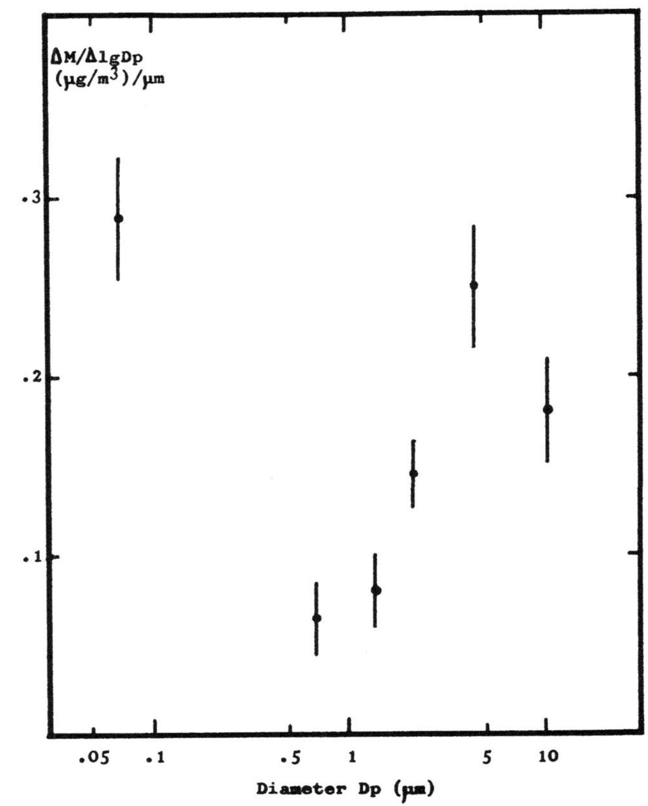

CORRELATION ANALYSIS APPLIED TO A SULPHATE AND NITRATE AEROSOL MEASURING NETWORK IN THE NETHERLANDS

L.H.J.M. Janssen, H. Visser, R.J. Vaessen, F.G. Römer and A.J. Elshout
Environmental Research Department, N.V. KEMA
P.O. Box 9035, 6800 ET ARNHEM, THE NETHERLANDS

Summary

Eigenvector analysis is applied to get insight into the spatial representativity of a set of sulphate and nitrate measuring stations in The Netherlands over the period 1979-1986. The variance explained by the first eigenvector of sulphate is about 80%. This means that the concentrations of sulphate are mainly determined by large-scale conditions and notably by large-scale transport from eastern directions. The same holds for nitrate in the winter period. Because their correlation distances of about 140 km have the same order of magnitude as the dimensions of The Netherlands, one measuring point would give a reasonable first approximation of sulphate and nitrate concentrations in The Netherlands. Nitrate concentrations in summer months however are mainly determined by local conditions reflected in the much shorter correlation distance of about 30 km. To obtain representative values of nitrate concentrations a denser measuring network would be required.

1. INTRODUCTION

In The Netherlands, part of the highly industrialized area of western Europe, concentrations of atmospheric pollutants are not only determined by local emissions, but also by large-scale transport. This holds for the gaseous components SO_2 and NO_2 but especially for secondary air pollutants such as sulphate and nitrate containing aerosols. To separate large-scale and regional contributions eigenvector analysis can be applied as shown by e.g. van Egmond and Onderdelinden (1). In the present paper eigenvector analysis has been applied to concentrations of total sulphate (sulphuric acid and sulphates) and particulate nitrates in aerosol and to the gaseous components SO_2, NO_2, NO and O_3. To measure concentrations of sulphate and nitrate aerosol over The Netherlands the Common Aerosol Measuring network of the Electricity utilities has been in operation during the period 1979-1986 (2). At first it consisted of nine (figure 3) and later of five measuring stations, over the years 1979-1982 (GAME) and 1982-1986 (TREND) respectively. In the first period the sulphate and nitrate concentrations were based on 6 hours of sampling using high volume samplers, on working days between 10h00 and 16h00, and during the second period on 24 hours of sampling using medium volume samplers, daily between 10h00 and 10h00. Data of SO_2, NO_2, NO and O_3 concentrations, together with information about wind direction and wind velocity, were obtained from nearby measuring stations of the Dutch air pollution monitoring network.

Analysis was performed to obtain insight into the spatial representativity of a set of fixed aerosol measuring stations and to distinguish different air pollution fields on the scale of The Netherlands. This has been done with a view to further use of the measured sulphate - and nitrate concentrations in aerosol, e.g. studying the physical and chemical behavior of different air pollutant components.

The measuring data were processed to be used by computer program packages such as SPSS. The following analysis will be restricted mainly to the first period, because the database of the second period 1982-1986 is still incomplete at this moment.

2. SULPHATE AND NITRATE CONCENTRATIONS

To give an impression of sulphate and nitrate concentrations, the mean and 50, 95 and 98 percentiles of four measuring stations, differentiated in summer and winter half year periods, are given in table I (period 1979-1982).

Table I Summer (April-September) and winter (October-March) mean values; 50, 95 and 98 percentiles ($\mu g/m^3$) of sulphate and nitrate concentrations at measuring stations 1, 4, 7 and 8. Period: April 1979 - April 1982

	MP 1		MP 4		MP 7*		MP 8	
	SO_4	NO_3	SO_4	NO_3	SO_4	NO_3	SO_4	NO_3
summer mean	11.1	5.0	10.8	4.6	11.4	5.9	12.4	4.1
50	8.3	3.4	8.0	2.8	9.5	3.6	9.6	2.5
95	28.4	14.9	29.3	16.0	27.0	20.6	31.6	13.8
98	39.7	24.7	38.2	20.8	38.5	31.0	37.4	24.4
winter mean	12.1	6.6	12.0	7.0	10.2	6.4	13.5	7.3
50	8.4	4.7	8.5	5.6	7.3	5.0	9.1	4.9
95	32.4	21.4	34.3	22.3	30.8	17.6	39.7	22.7
98	43.8	25.7	44.5	28.8	39.1	24.3	53.3	30.6

In general sulphate and nitrate concentrations in winter are higher than in summer, except at measuring point 7*. This was due to a failure of the high-volume sampler at this station in the summer of 1980 by which on relatively few days only (high) concentrations were measured.

Comparing the mean concentrations of the half-year periods of the different stations, it is clear that they differ little. So it can be expected that the half-year mean concentration of one station is a reasonable first approximation of the mean concentration in The Netherlands.

Monthly mean sulphate - and nitrate concentrations as measured at station 4, Utrecht, are shown in figures 1a and 1b.

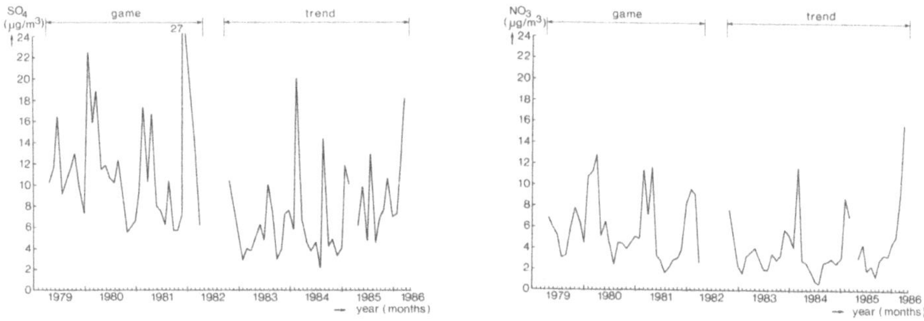

Figure 1a and 1b Monthly mean sulphate and nitrate concentrations as measured at site 4 (Utrecht)

These figures show that sulphate and nitrate concentrations on this location can vary a factor 3 or more from month to month. The next question to be answered is therefore: how representative are concentrations taken over a <u>short</u> period at one station, in relation to concentrations in a larger area? Therefore a so called correlation distance for the components sulphate, nitrate, SO_2, NO_2, NO and O_3 will be derived.

3. A CORRELATION DISTANCE FOR SULPHATE, NITRATE, SO_2, NO_2, NO AND O_3

Concentrations of air pollutants vary as a function of time and space. During the first period of three years sulphate, nitrate, SO_2, NO_2, NO and O_3 concentrations were measured using a network of nine stations with a grid length of about 60 km.
To investigate the variability of the concentrations over longer distances, the notion of <u>correlation distance</u> is used. The correlation distance is defined as a distance A_o between two measuring stations x and y for which 50 % of the variance of the concentrations at site x can be explained by the concentration fluctuations at site y. This corresponds with a correlation coefficient r_{xy} of 0.71.

Many statistical techniques use the assumption that processes follow a normal stochastic pattern. It has been verified that the logarithm of the concentrations satisfies this condition. For reasons of simplicity concentrations are standardized to have zero mean and variance 1.00. Both transformations lead to:

$$c_{ij} = \frac{c'_{ij} - m'_i}{S'_i} \qquad (1)$$

Index i denotes the measuring station (i = 1 - 9), and j denotes day-number (j = 1 - 784). Concentration c_{ij}', denotes the natural logarithm (ln) of the measured concentrations; m_i is the station mean of the set of logarithmic concentrations and S'_i its standard deviation.

The concentrations c_{ij} form the elements of a 9 x 784 matrix C. The correlation matrix R with the elements r_{xy} is calculated as:

$$R = \frac{1}{N} C \cdot C^T \qquad (2)$$

N denotes the number of measurements (784) and the matrix C^T the transposed of the matrix C. The 9 x 9 matrix R is a symmetrical matrix ($r_{xy} = r_{yx}$) which consists of the correlations between the measuring stations. The correlation coefficients as a function of interstation distance are presented in figure 2. Correlation distances are given in table II.

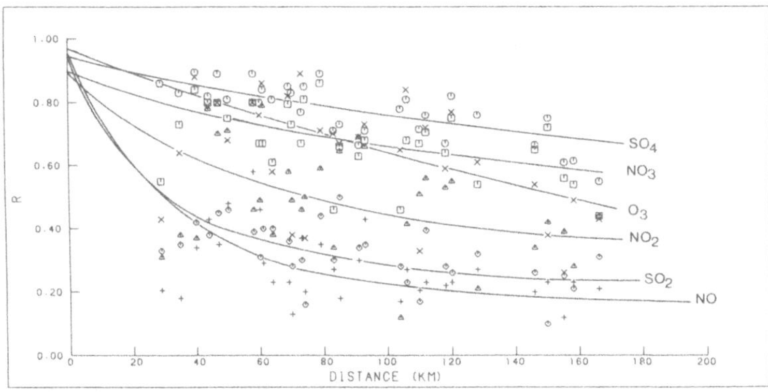

Figure 2 Spatial behaviour of the correlation coefficient of SO_4 ○, NO_3 □, O_3 x, NO_2 △, SO_2 ◇ and NO +

All correlation coefficents are supposed to be smaller than one at zero interstation distance as a consequence of a sampling error in the measured concentrations (1). Because of the rather large grid length of 60 km no reliable estimate of the sampling error can be inferred from figure 2 and the cut-off points are chosen in accordance with values of the gaseous components SO_2, NO_2, NO and O_3 as reported by Van Egmond and Onderdelinden (1).

Table II Calculated correlation distances based on a coherent measuring period of three years (1979 - 1982)

SO_4	A_o = 140 km	NO_2	A_o = 25 km
NO_3	A_o = 75 km	SO_2	A_o = 15 km
O_3	A_o = 75 km	NO	A_o = 15 km

Figure 2 shows that each component has its specific spatial behaviour. This behaviour is determined by:

- emission sources in the network
- non-linear dispersion and conversion processes
- the network configuration and the fact that the measured concentrations are put together for all wind directions
- large-scale background fluctuations.
- the measuring period (in this case a coherent period of three years).

Therefore, based on these issues, NO and SO_2 concentrations with their small correlation distance, can be considered as largely determined by local emission sources. NO_2 can be recognized as a secondary air pollutant which is formed when its primary component NO is converted, after dispersion and reactions with for instance O_3 (3). Ozone is a large-scale background component affected by local NO emission sources. Sulphate and to a lesser extent nitrate, (to be discussed later in more detail), behave like large-scale background components which means that concentrations measured at different stations at a rather large distance from each other, are strongly correlated.

The calculated correlation distances of NO, SO_2 and NO_2 agree well with those calculated for a period of two months in winter by Van Egmond and Onderdelinden (1). The correlation distance of ozone (75 km) differs from the value of 40 km based on calculations by Van Egmond and Onderdelinden for ozone concentrations during one month in summer. This will be discussed together with a seasonal effect on nitrate in section 5.

4. DESCRIPTION OF CONCENTRATION FIELDS USING EIGENVECTOR ANALYSIS

The correlation matrix R can also be used to produce different air pollution patterns for The Netherlands. To that purpose eigenvector analysis is applied, which method will be outlined shortly. For further references see Van Egmond and Onderdelinden (1), Stidd (4), Sellers (5) Peterson (6) and Harman (7). Eigenvector analysis is applied to get insight into the spatial behaviour of uncorrelated air pollution fields. To that end the correlated concentrations c_{ij} of matrix C are transformed into new uncorrelated concentrations q_{kj} of a 9 x 784 matrix Q. Matrix Q is calculated as:

$$Q = E\,C \tag{3}$$

E is a 9 x 9 matrix formed by the nine orthogonal eigenvectors $[e_k]$ of matrix R. They are placed in descending order according to their eigenvalues. The e_k are written as row vectors :

$$E = \begin{pmatrix} [e_1] \\ \vdots \\ [e_9] \end{pmatrix} \tag{4}$$

The largest eigenvalue λ_1, corresponds with eigenvector e_1, the next largest eigenvalue λ_2 corresponds with the eigenvector e_2 etc. Furthermore, the eigenvectors are scaled to have unit length. Matrix E is orthogonal, i.e. $E^T = E^{-1}$.

Its eigenvectors and eigenvalues can be interpreted in terms of explained variance because, when using standardized concentrations (c_{ij}), the fraction of the total variance explained by the ith eigenvector equals λ_i/n, and n the number of eigenvectors. The row vector $[q_{kj}]$, also called the kth main component, with the index k corresponding with the kth eigenvector is linear independent and can be written as:

$$q_{kj} = \sum_{i=1}^{9} e_{ki} \; c_{ij} \tag{5}$$

So q_{kj} is a linear combination of measured concentrations c_{ij} multiplied by weighing factors e_{ki}. When writing the concentration matrix C again as:

$$C = E^{-1} Q, \text{ or } C = E^{T} Q \tag{6}$$

and taking only the main components corresponding with the largest m eigenvalues into account, the concentrations c_{ij} can be approximated by:

$$\hat{c}_{ij} = \sum_{k=1}^{m} e_{ik} \, q_{kj} \quad \text{and } m < 9 \tag{7}$$

From equation (6) it can be derived that concentration matrix C, which contains space-time information, can be considered as a product of matrix Q, which row vectors vary only in time and matrix E^T, which column vectors vary only in space. The nine different orthogonal eigenvectors of E^T can be considered to represent nine uncorrelated air pollution patterns. The first pattern represents the mutual coherence (covariance) of the nine measuring stations, the measure of which is determined by the magnitude of the corresponding eigenvalue λ_1. Because each element e_{ik} of the ith eigenvector (e_i) uniquely belongs to a measuring station nine maps can be drawn based on the nine eigenvectors, each representing a characteristic air pollution pattern.
As an example the first three eigenvector patterns of sulphate and nitrate together with the variance explained by each vector are given in figures 3a and 3b.

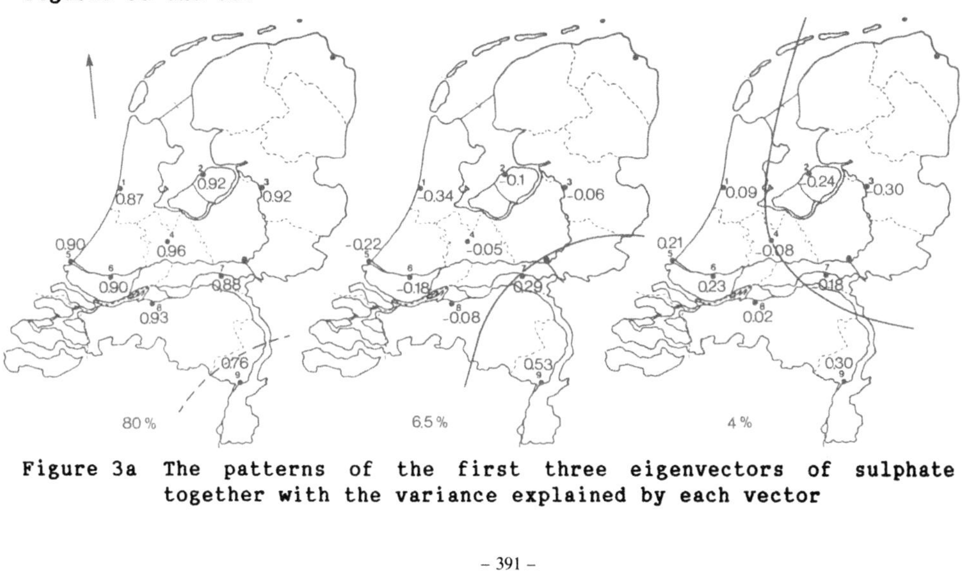

Figure 3a The patterns of the first three eigenvectors of sulphate together with the variance explained by each vector

The first eigenvector of SO_4 which explains about 80% of the total variance, shows that sulphate concentrations are mainly determined by a large-scale air pollution field covering the whole of The Netherlands. This means that the concentrations occurring at the measuring stations are strongly correlated, and that high concentrations at one station concur with high concentrations at other stations at the same time!

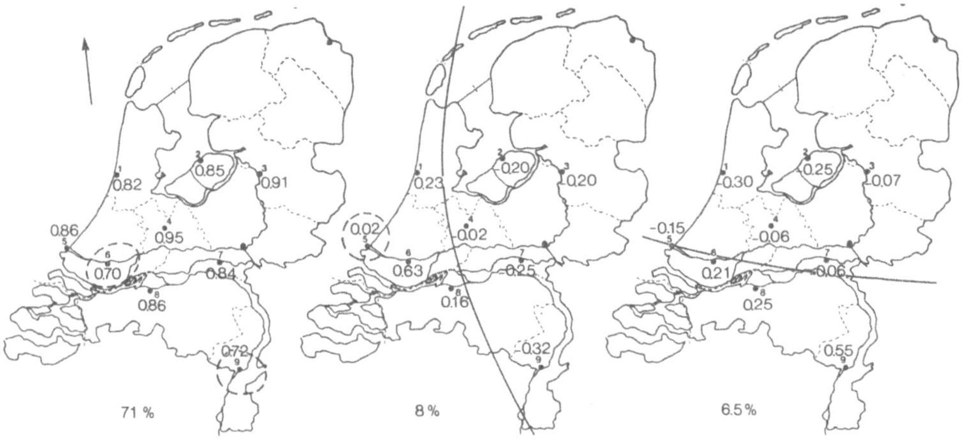

Figure 3b The patterns of the first three eigenvectors of nitrate together with the variance explained by each vector

As can be inferred from table I, the 95 and 98 percentiles of sulphate measured on different stations do not differ much and are related to the <u>same</u> periods of time. The 95 and 98 percentiles measured at one representative station are therefore also a reasonable first approximation of the high percentiles for the whole area of The Netherlands.

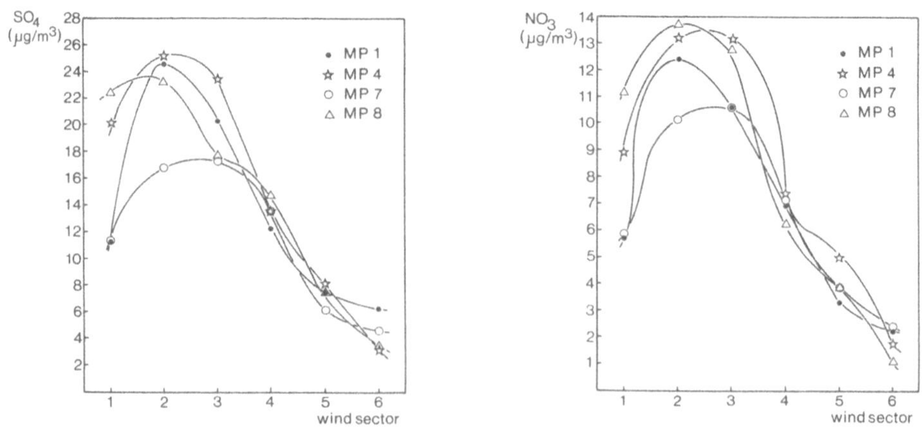

Figures 4a and 4b Mean concentrations of sulphate and nitrate differentiated to wind direction for three winter months (December-February) in the period 1979-1982

High sulphate concentrations are related to transport of air pollution from eastern direction. This is shown in figures 4a and 4b.The wind directions are divided in sectors as follows:
1: 346° - 45°, 2: 46° - 105°, 3: 106° - 165°, 4: 166° - 225°,
5: 226° - 285°, 6: 286° - 345°

Because of the high percentage of variance explained by the first eigenvector, it can be concluded, that if concentrations of sulphate and nitrate in winter at one measuring point increase because of transport of air pollution, other measuring stations obviously have the same wind field, and there concentrations will also increase.

The second and third eigenvectors explaining respectively 6.5 % and 4 % of the variance, show south-east and east gradients which are probably related to transport from foreign source areas. Although nitrate more or less shows the same behaviour as sulphate, there is a marked seasonal effect which will be discussed in the next section.

5. SEASONAL EFFECTS ON THE CORRELATION STRUCTURE

Because meteorological conditions affect transport and conversion processes of atmospheric pollutants, calculations of matrix E were extended. After determining matrix E for a coherent three year period for sulphate and nitrate, also calculations were carried out separately for the three summer months June-August and three winter months December-February. For sulphate no great difference could be seen between summer and winter periods. The variance explained by the first eigenvector remained very high; in the three summer months only a little less (79%) than in the three winter months (83%). However the nitrate concentration field showed a great difference between summer and winter months, see table III.

Table III Variances (%) of nitrate explained by the first six eigenvectors, for summer and winter periods of three months

periods	1	2	3	4	5	6
summer	46.6	14.2	9.8	7.5	6.5	5.1
winter	85.6	4.9	2.8	2.2	1.7	1.3

The smaller percentage of the variance explained by the first eigenvector in the summer period means that in summer months more than in winter months, nitrate concentrations are influenced by local conditions like emissions and meteorological factors.

Measurements of HNO_3 and particulate nitrate for example by Meixner et al. (8) also show an influence of season of the year and time of the day on the measured concentrations which indicates a meteorological influence.

The breaking up of the large-scale correlation structure of nitrate in the three summer months is also demonstrated in figure 5.

The correlation distance of nitrate in the three winter months of more than 140 km is reduced to about 30 km in the three summer months. This means that in these summer months nitrate concentrations are more influenced by local conditions and the network is probably not dense enough to give a representative image of nitrate concentrations in The Netherlands.

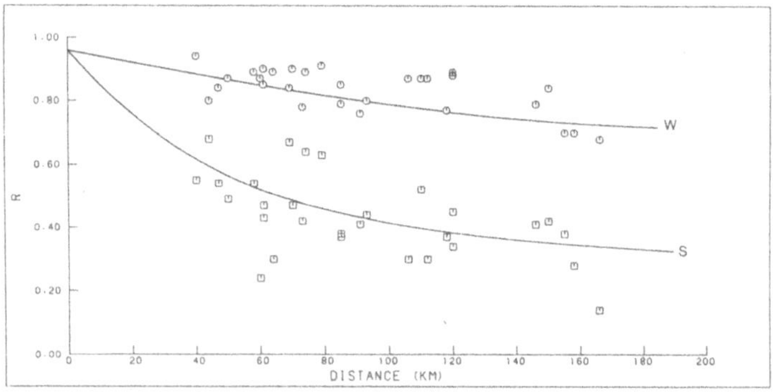

Figure 5 Spatial behaviour of the correlation coefficient of nitrate, differentiated for three summer and three winter months

The dependence of the correlation on season of the year, as demonstrated for nitrate, also holds for ozone This explains the difference between the calculated correlation distance of 40 km by Van Egmond and Onderdelinden (1) for one summer month (June) and the correlation distance in this paper of 75 km related to a coherent period of three years.

Because nitrate concentrations in summer are more determined by local conditions than in winter, the influence of large-scale transport of air pollutants from eastern directions is less clear for nitrate in summer. This is demonstrated in figures 4b and 6b. Figure 4b shows that in winter nitrate concentrations are indeed influenced by large-scale transport; the dependence of the nitrate concentrations on wind direction in summer months is less (figure 6b).

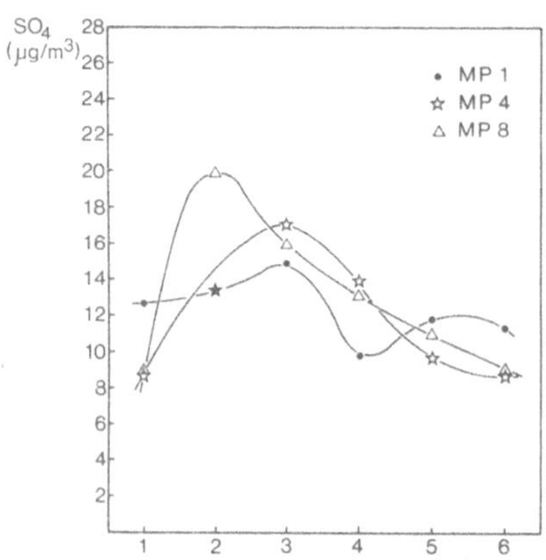

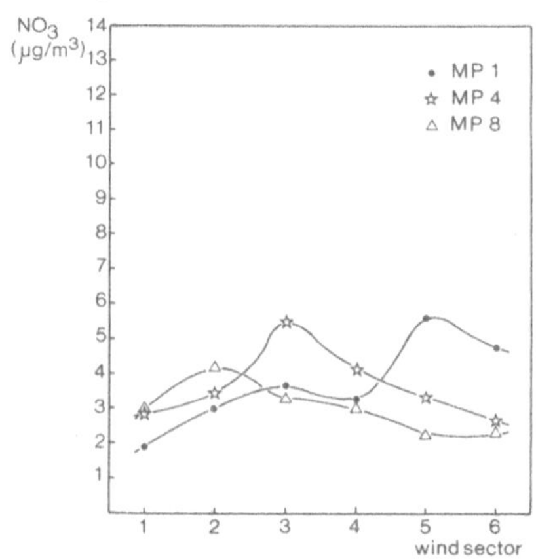

Figures 6a and 6b Mean concentrations of sulphate and nitrate differentiated to wind direction for three summer months (June-August) in the period 1979-1982

6. CONCLUSIONS

Eigenvector analysis is applied to analyse the spatial representativity of a set of sulphate and nitrate measuring stations in The Netherlands. Contrary to their primary components SO_2 and NO, which are mainly determined by local conditions, the concentrations of sulphate in summer and winter and of nitrate only in winter, are controlled by large-scale transport from eastern directions. If it is assumed that linear conversion and deposition processes play a minor role, concentrations measured at one representative station, give a reasonable first approximation of actual sulphate and nitrate concentrations in the whole area of The Netherlands, which dimensions are of the same order of magnitude as the correlation distance of sulphate and nitrate (only in winter). Because concentrations of nitrate in summer are determined by local conditions, a denser network of measuring stations is required to describe the concentration pattern. Further research will therefore be directed to investigate conversion processes under different meteorological conditions.

ACKNOWLEDGEMENTS

The National Institute of Public Health and Environmental Hygiene (RIVM) is acknowledged for supplying data of SO_2, NO, NO_2 and O_3 concentrations, dr. J.H.A. van Wakeren for his contribution in the discussion about the correlation distances, and mrs. J. M.Rietbergen for carrying out the chemical analyses.

REFERENCES

(1) VAN EGMOND, N.D. and ONDERDELINDEN, D. (1981). Objective analysis of air pollution monitoring network data; spatial interpolation and network density. Atmospheric Environment 15, 1035-1046.

(2) RÖMER, F.G., VAN DUUREN, H. and ELSHOUT, A.J. Nitrate and sulphate levels in The Netherlands: Dicussion of some recent measuring data. Symposium on physico-chemical behaviour of atmosperic pollutants, Ispra, Italy, October 1979.

(3) JANSSEN, L.H.J.M. (1986). Mixing of ambient air in a plume and its effects on the oxidation of NO. To be published in Atmospheric Environment.

(4) STIDD, L.K. (1967). The use of eigenvectors for climate estimates. Journal of applied meteorology 6, 255-264.

(5) SELLERS, W.D. (1968). Climatology of monthly precipitation patterns in the western United States, 1931-1966. Montly Weather Review, 96, 585-595.

(6) PETERSON, J.T. (1970). Distribution of sulfur dioxide over metropolitan St. Louis as described by empirical eigenvectors, and its relation to meteorological parameters. Atmospheric Environment 4, 501-518.

(7) HARMAN, H.H. (1960). Modern Factor Analysis. University of Chicago Press.

(8) MEIXNER, F.X., MüLLER, K.P., AHEIMER, G. and HöFKEN W.S. (1985). Measurements of nitric acid and particulate nitrate. Symposium paper Cost action 611, Bilthoven, September 1985.

CENDRES VOLANTES INDUSTRIELLES ET MICROSPHERULES VOLCANIQUES : CRITERES DE DISTINCTION PAR MICROSCOPIE ANALYTIQUE

R. LEFEVRE[1], A. GAUDICHET[1][2], J.F. MULLER[3] et J.P. SAINTIVES[4]

(1) Laboratoire de Microscopie Analytique Appliquée aux Sciences de la Terre - Université Paris XII - F. 94010 CRETEIL.
(2) Laboratoire d'Etude des Particules Inhalées - DASES - F. 75013 PARIS
(3) Service LAMMA de l'Université - F. 53000 METZ.
(4) Laboratoire de Mécanique Physique - F. 78210 SAINT-CYR L'ECOLE.

Summary - Analytical Microscopy of Atmospheric mineral microspherules: characterization of industrial fly-ash and similar volcanic particles.

Two methods of Analytical Microscopy have been performed to characterize, by their individual chemical composition, inorganic microspherules sampled by filtration of air:
1- Transmission or Scanning Electron Microscopy fitted with X-Ray microanalysis (Electron Microprobe), and 2-Secondary Ion Mass Spectrometry under Laser Excitation (LAMMA Microprobe). The first method allows quantitative analysis of major elements (Si,Al, Fe, K, Ca, Na, Mg, Ti, Mn) whereas the second allows qualitative detection of major and minor components and specialy light elements and alkaline/alkaline-earth elements. The individual compositions of microspherules intercepted in the permanent plume of Mount Etna have been compared with those of fly-ash produced by coal-fired power plants. The concentrations in major elements for two hundred of such microspherules have been plotted on the triangular diagram SiO_2-Al_2O_3 -other oxides. They are into two distinct areas according to their volcanic or industrial origin. In the same way, the minor elements detected by mass spectrometry are different from one source to another. The same microanalytical methodology has been applied to study microspherules intercepted in air filtered in the lower Mediterranean Atmosphere (southern cape of Sicily): their chemical compositions were in good agreement either with possible volcanic or industrial origins. These results were consistent with the trajectories of air masses sampled.

1. INTRODUCTION

Des auteurs ont décrit des particules minérales de forme sphérique et de taille micronique captées dans l'atmosphère, ou ayant eu une partie de leur parcours dans l'atmosphère, en leur attribuant des origines très variées : cosmique (1-11, 14), volcanique (2,3,4, 12-20), végétale (18), industrielle (21-24) voire même, récemment, nucléaire (25).

Il est ainsi difficile de remonter aux sources des microsphérules minérales atmosphériques à partir de la seule observation de leur forme et de leur taille. D'autres paramètres discriminatoires doivent être recherchés, parmi lesquels la composition chimique. Cependant, les méthodes globales d'analyse, en fournissant la seule composition moyenne des microparticules, ne peuvent aider de façon simple et décisive à effectuer une telle discrimination. Aussi, faut-il s'adresser à des méthodes de caractérisation chimique individuelle: parmi elles, la microscopie électronique analytique et la spectrométrie de masse ponctuelle offrent de telles possibilités. Nous les avons utilisées pour étudier les microsphérules émises par deux types de sources: cratère d'un volcan et foyers de centrales électriques à charbon. Ces sources émettent bien des microsphérules de morphologie et de taille comparables, mais de chimies différentes. Nous avons ensuite effectué la démarche inverse: tenter de reconnaître dans des lots de microsphérules recueillies dans une région sous influences variées, entre autres volcaniques et industrielles (le Bassin méditerranéen moyen, à la pointe méridionale de la Sicile), celles qui pouvaient avoir une origine caractérisable par leur chimie. Les résultats obtenus ont été corroborés par la reconstruction des trajectoires des masses d'air qui avaient transporté ces microsphérules.

2. ECHANTILLONNAGE DES MICROSPHERULES: FILTRATION DES PANACHES OU DE L'AIR AU TRAVERS DE MEMBRANES CALIBREES.

Les étapes successives de l'étude chimico-minéralogique des microsphérules ont été: la filtration de l'air, du panache volcanique ou encore des particules piégées dans les dépoussiéreurs des centrales électriques ; puis, leur observation et leur analyse, soit par microscopie et microanalyse électroniques, soit par spectrométrie de masse après pulvérisation individuelle sous l'impact d'un laser.

Les membranes et filtres utilisés pour la filtration furent de trois types:

a) membranes monocouches en polycarbonate (porosité: 0,4 μm). Ces membranes ont l'avantage de permettre l'observation et l'analyse directes des microparticules déposées à leur surface, dans un microscope électronique à balayage analytique, sans que la structure du filtre soit une gêne. Par ailleurs, ces membranes sont utilisables pour les études par microscopie électronique en transmission analytique car elles permettent le transfert direct des microparticules sur des grilles métalliques (26,27). Pour ce faire, la membrane a été recouverte d'un film de carbone avant la filtration. Après celle-ci une seconde couche de carbone est évaporée. L'ensemble est posé sur la grille de microscope et traversé par du chloroforme sous aspiration modérée. La membrane est ainsi dissoute et les microparticules demeurent comprises entre les deux couches de carbone qui les maintiennent.

b) membranes en esters de cellulose (porosité 0,4 μm) Les particules sont piégées en partie dans l'épaisseur de la membrane filtrante. Elles devront ensuite être déposées par filtration en phase liquide sur une membrane du type a/ après que la membrane de type b/ aura été brûlée à basse température dans un four à plasma d'oxygène.

c) filtres en fibres de verre. Ils sont insensibles aux variations d'humidité de l'air et sont ainsi utilisés pour la pesée précise des particules recueillies. Cependant, celles-ci sont irrécupérables pour la microscopie analytique. Il faut donc utiliser ces filtres c/ conjointement avec les membranes a/ et b/.

Les membranes a/ et b/ et les filtres c/ sont contenus dans des cassettes qui s'adaptent directement à l'extrémité d'un tube, sans manipulation directe de la membrane ou du filtre sur le site de prélèvement. Le tube est rélié à une pompe électrique Réciprotor, éventuellement alimentée par un groupe électrogène en l'absence locale de source électrique (en site volcanique, par exemple). L'air filtré sortant de la pompe traverse un compteur volumétrique Gallus. Le débit de la pompe est généralement réglé sur 15 l/mn (0,9 m^3/h).

Les microsphérules provenant des centrales électriques à charbon ont été captées par des dépoussiéreurs industriels à la sortie des chaudières. Elles ont été mises en suspension dans l'eau et dispersées par ultrasons avant d'être filtrées sur membranes de type a/.

3. MICROSCOPIE ANALYTIQUE DES MICROSPHERULES.

Les microsphérules ont d'abord été étudiées au moyen d'un Microscope Electronique à Transmission (JEOL 100 C) muni d'un détecteur solide sélectionnant les énergies des Rayons X émis sous l'impact des électrons (EDAX 711). Ce dispositif permet d'avoir simultanément des données sur:

a) la morphologie et la granulométrie individuelle des microsphérules.

b) leur proportion par rapport à l'ensemble des microparticules présentes dans le prélèvement, par un comptage systématique dans des champs choisis au hasard.

c) leur composition chimique exprimée en poids d'oxydes de Na, Mg, Al, Si, K, Ca, Fe, Ti, Mn comme le sont classiquement les analyses de minéraux et de roches. La faible épaisseur (généralement inférieure à 2 μm) des microsphérules permet d'en faire l'analyse par sonde électronique en prenant Si comme étalon interne et en appliquant la méthode des rapports des hauteurs de pics de Cliff et Lorimer (28). La microanalyse électronique a l'avantage d'être quantitative, mais les inconvénients de ne pas être appliquable aux éléments légers (Li, Be, B et F) ni aux éléments en trace (existence d'un fond continu dans l'émission X).

d) enfin, un microscope électronique à transmission permet d'étudier la structure des microsphérules par microdiffraction des électrons transmis. L'absence de cliché de microdiffraction, dès lors qu'on est certain de la perméabilité des microsphérules aux électrons (épaisseur inférieure à 0,2 μm), permet d'affirmer leur état amorphe. Nous n'avons jamais observé de diagramme de diffraction sur les plus fines microsphérules, ce qui ne nous permet pas d'affirmer que les plus grosses soient, elles-aussi, amorphes.

La spectrométrie de masse sous l'impact d'un laser à été effectuée au moyen d'une microsonde LAMMA 500 (Laser Microprobe Mass Analysis, Leybold-Heraeus) (29-31). Les microsphérules, contenues dans leur double film de carbone posé sur la grille de microscope électronique, sont soumises individuellement à l'impact d'un laser qui les pulvérise (cette étude est donc faite la dernière, car elle est destructice). Les produits ionisés de cette pulvérisation sont analysés dans un spectromètre de masse à temps de vol qui les caractérise par le rapport m/e de leur masse à leur charge électrique (fig. 1). Cette méthode a l'avantage d'être particulièrement sensible aux éléments légers, aux éléments en trace (pas de fond continu comme dans l'émission X) et de permettre la discrimination entre les différents isotopes d'un même élément chimique. Elle a cependant l'inconvénient de ne pas être quantitative du fait, entre autres de l'existence d'ions polyatomiques dans les produits de la pulvérisation des microsphérules par le laser excitateur. L'analyse quantitative se heurte encore au fait que la totalité des produits pulvérisés n'est pas sous forme ionisée.

4. LES MICROSPHERULES DU PANACHE PERMANENT DE L'ETNA. Ce volcan de 3300m d'altitude émet dans l'atmosphère de la Méditerranée moyenne deux types de panaches. Le premier, noir ou brun foncé, correspond à des bouffées intermittentes et courtes constituées de matériaux anciens et altérés tombés des parois internes du cratère à la surface du magma. Ils y forment un bouchon sous lequel s'accumulent des gaz dont la pression monte jusqu'à l'explosion. Il y a alors éjection violente des matériaux dont la taille peut être importante (plusieurs m^3). Ils retombent donc rapidement sur les flancs externes du volcan sur lesquels ils forment des couches successives

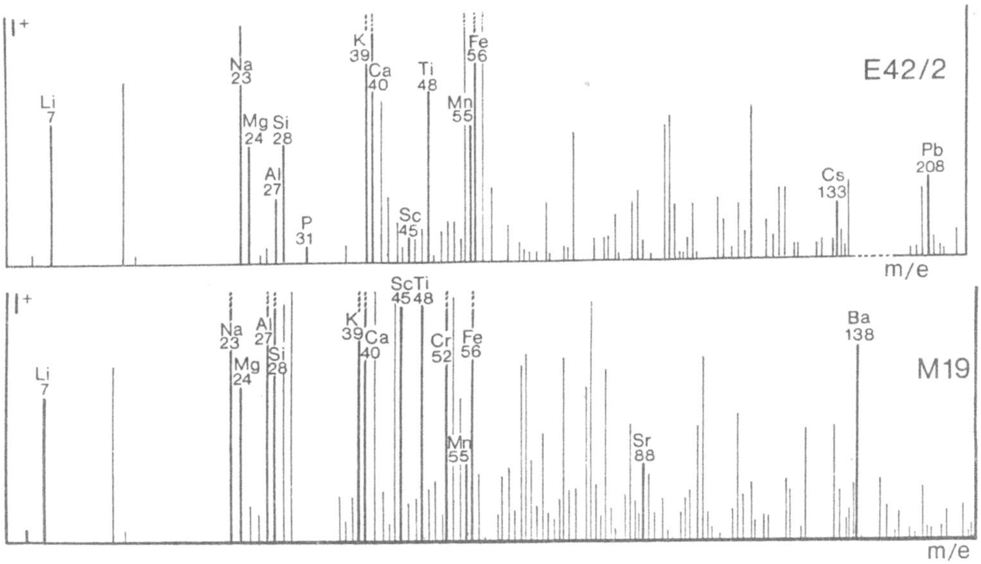

<u>ETNA</u> :
<u>Li</u>, - , <u>Na</u>, <u>Mg</u>, <u>Al</u>, <u>Si</u>, P, S, Cl, <u>K</u>, <u>Ca</u>, Sc, Ti, Cr, Mn, <u>Fe</u>, Co, Zn, <u>Rb</u>, -, <u>Cs</u>, - , Pb.

<u>CENTRALES ELECTRIQUES A CHARBON</u> :
<u>Li</u>, Be, <u>Na</u>, <u>Mg</u>, <u>Al</u>, <u>Si</u>, P, -, -, <u>K</u>, <u>Ca</u>, Sc, Ti, Cr, Mn, <u>Fe</u>, Co, Zn, Rb, Sr, Cs, <u>Ba</u>, -.

<u>Fig. 1</u>: <u>Exemples de spectrogrammes de masses des ions positifs</u> émis, sous l'impact d'un laser, dans une microsonde LAMMA-500 - <u>E42/2</u>: Microsphérule volcanique du panache de l'Etna - <u>M 19</u>: Fly-ash émise par la combustion du charbon dans une centrale électrique. Les éléments chimiques retrouvés dans l'ensemble des spectrogrammes réalisés ont été reportés sous ces deux exemples. Les éléments soulignés sont constamment présents, les autres n'apparaissent que sur certains spectogrammes.

(téphras). L'échantillonnage des produits de ce panache explosif ne peut évidemment se faire directement dans le cratère lui-même, en raison de la violence des explosions, mais à distance. On ne recueille alors que ce qui retombe sur le site de prélèvement. Le second type de panache s'élève pendant les périodes intermédiaires de calme et de dégazage libre du magma: il est blanchâtre et essentiellement formé de vapeurs de H_2O, SO_2,CO_2,HCl, HF... (32,33). L'échantillonnage de ce panache ne présente pas les risques du précédent. Nous l'avons fait au moyen d'une canne souple s'avançant dans le cratère d'environ 2 m et y plongeant d'environ 1 m. A son extrémité, la canne portait une cassette munie d'une membrane de type a/.

La présente étude porte sur le second type de panache. En effet, il comporte de fines particules minérales (19,20): oxydes, chlorures, sulfates..., silicates et verres volcaniques, ces derniers sous forme d'esquilles, de filaments étirés et de microsphérules. Les microsphérules qui ont été étudiées ici ont des <u>tailles</u> qui varient de 0,1 à 10 μm . Les tentatives de microdiffraction électronique se sont révélées négatives sur

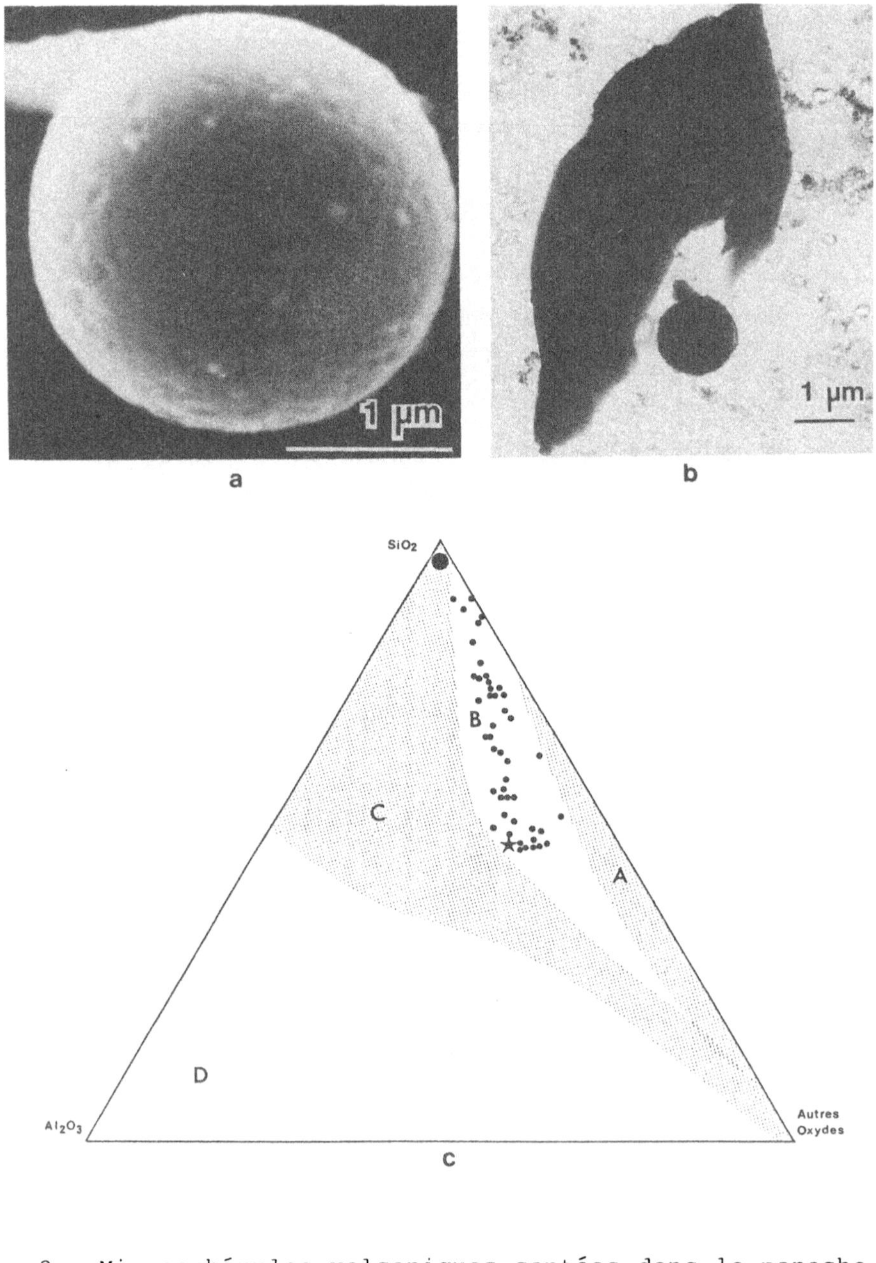

Fig. 2 - Microsphérules volcaniques captées dans le panache de l'Etna. a : Microscopie Electronique à Balayage d'une microsphérule à surface lisse discontinue. b : Microscopie Electronique en Transmission d'une microsphérule et d'une écharde de verre volcanique. c : compositions chimiques individuelles de 54 microsphérules (●) comparées à la composition moyenne des laves de l'Etna (★), reportées sur un diagramme triangulaire (24). Les aires A, B, C et D sont définies dans le texte.

les plus fines et impossibles sur les plus grosses. La morphologie externe, étudiée par microscopie électronique à balayage, après lavage à l'eau, montre des surfaces lisses mais discontinues, parfois recouvertes de petits cristaux, ou des surfaces granuleuses évoquant des agrégats de microparticules (fig.2). Ces microsphérules peuvent s'interprêter comme les produit d'une fine pulvérisation ("spray") due soit à des fontaines de lave, soit à des explosions phréatomagmatiques (par contact brutal d'eau météorique froide avec le magma dont la température avoisine 1100°) (16-18, 34-37).

La singularité des microsphérules du panache permanent de l'Etna tient à leurs compositions chimiques variées: elles vont de celle du basalte, tel qu'on peut l'analyser à la faveur de coulées latérales (38), jusqu'à celle de la silice pure. L'Etna est le seul volcan connu à ce jour, qui soit le siège d'un tel phénomène d'enrichissement considérable en silice de verres microsphériques. Toutes les microsphérules décrites sur des volcans basaltiques terrestres ou lunaires ont la composition du magma qui les a émises. Le mécanisme de l'enrichissement en silice, encore mal élucidé, pourrait faire intervenir l'eau météorique présente en quantités considérables sur le volcan ; le soufre, lui aussi en quantités très importantes ; enfin, une surchauffe de la surface du magma par combustion des gaz volcaniques au contact de l'oxygène atmosphérique. Cette dernière hypothèse rapproche les microsphérules du panache de l'Etna de celles qui sont émises lors de la combustion du charbon dans les chaudières des centrales électriques.

Les compositions chimiques des microsphérules, obtenues par Microscopie Electronique Analytique, sont reportées sur le diagramme triangulaire SiO_2 - Al_2O_3- Autres Oxydes de la fig. 2 (24). On voit que ces compositions forment une suite entre la composition moyenne des laves de l'Etna (38) et le pôle SiO_2. Un spectrogramme de masses, obtenu au moyen de la microsonde LAMMA, est reproduit fig. 1. L'ensemble des 19 microsphérules volcaniques que nous avons ainsi étudiées révèle la présence constante de Li, Na, Mg, Al, Si, K, Ca, Fe, Rb et Cs, et la présence seulement dans certaines microsphérules de P, S, Cl, Sc, Ti, Cr, Mn, Co, Zn et Pb. Les métaux alcalins (Rb et Cs) semblent donc caractériser, au niveau des éléments mineurs, ce type de microsphérules.

5. LES CENDRES VOLANTES DES CENTRALES ELECTRIQUES A CHARBON.

L'échantillonnage a été fait à partir des dépoussiéreurs situés en aval des fours dans lesquels brûle le charbon préalablement pulvérisé. Les échantillons étudiés sont constitués en partie de microsphérules entrant dans la catégorie des cendres volantes ou fly-ash. Leur taille varie de 0,2 à 20 μm. Leur morphologie externe, très variée, comporte les types décrits précédemment dans la littérature (21,24). Des compositions chimiques sont reportées sur le diagramme triangulaire SiO_2 - Al_2O_3 - Autres Oxydes de la fig. 3 (24). Elles s'y projettent dans une aire bien distincte de celle des microsphérules de l'Etna (fig.2). Cette aire a été précédemment définie par RAMSDEN et SHIBOAKA (24) dans une étude des fly-ash émises par des centrales électriques à charbon australiennes.

Un spectrogramme des masses de ces fly-ash, obtenu au moyen de la microsonde LAMMA, est reproduit fig.1. Une vingtaine de fly-ash ont été étudiées par cette méthode. Les éléments présents dans tous les spectogrammes réalisés sont: Li, Na, Mg, Al, Si, K, Ca, Fe et Ba. Certaines fly-ash révèlent, en plus des éléments cités ci-dessus: Be, P, Sc, Ti, Cr, Mn, Co, Zn, Rb, Sr et Cs. Elles ne montrent jamais la présence de S et Cl. On remarque donc que parmi les éléments mineurs ce sont les métaux alcalino-terreux (Be, Sr, Ba) que l'on retrouve le plus souvent dans ces

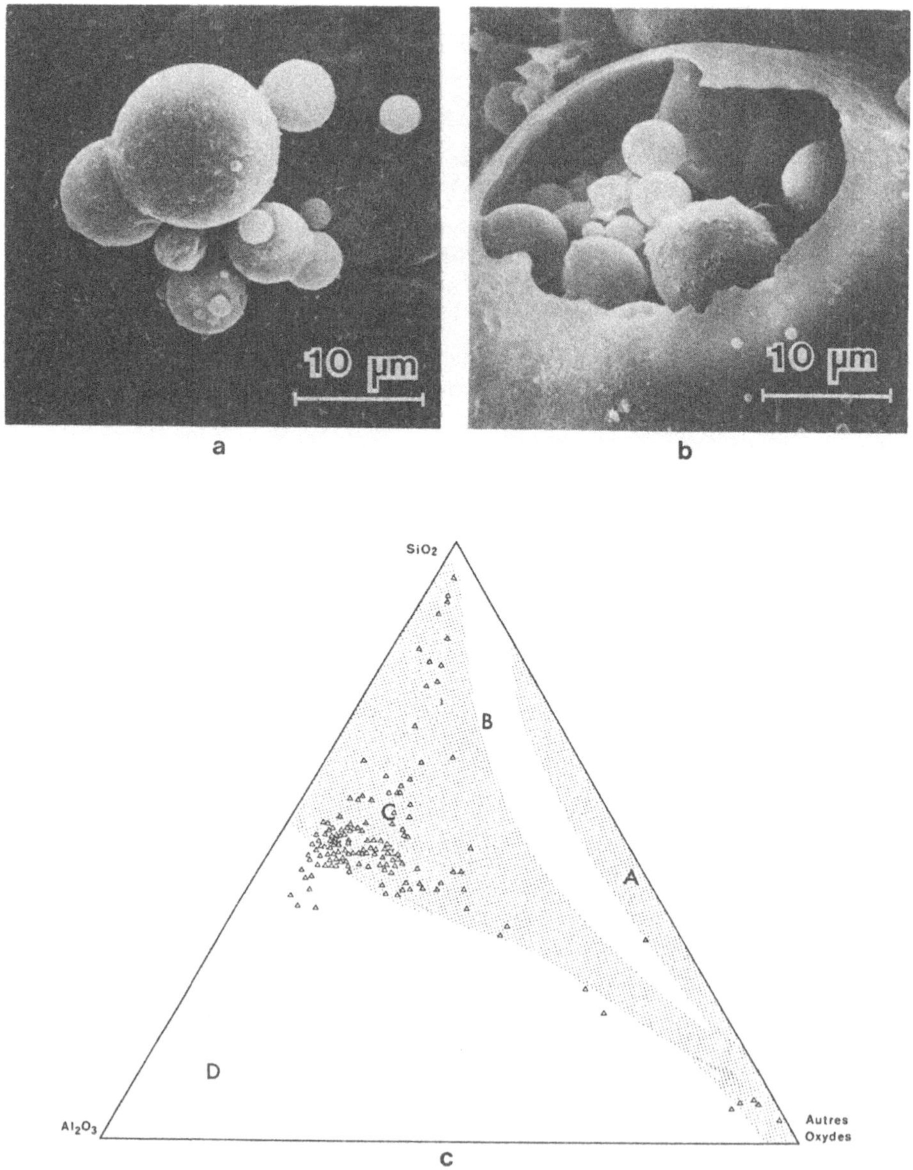

Fig. 3 - Cendres volantes de centrales électriques à charbon. a et b : Microscopie Electronique à Balayage. c : compositions chimiques individuelles de 150 cendres volantes (Δ), reportées sur un diagramme triangulaire (24). Les aires A, B, C et D sont définies dans le texte.

fly-ash. Ce caractère les oppose clairement aux microsphérules volcaniques. Par contre, d'autres alcalins et alcalino- terreux (Li, Na, Mg, K et Ca) se retrouvent dans les deux catégories de microsphérules, essentiellement au niveau des éléments majeurs (sauf Li).

6. LES MICROSPHERULES RECUEILLIES DANS LA BASSE ATMOSPHERE DE LA MEDITERRANEE MOYENNE.

Une campagne d'échantillonnage a été effectuée du 28 juillet au 13 août 1984 à la pointe méridionale de la Sicile, à proximité de la station météorologique de Cozzo Spadaro (village de Portopalo) (39). Parmi les filtres recueillis, deux ont été choisis pour être décrits ici, car ils correspondent à des périodes de prélèvement pendant lesquelles les situations météorologiques ont été bien contrastées: vent dominant de Nord pour le premier (E 20), vent dominant d'Ouest pour le second (E 32). Dans chacun des microsphérules ont été retrouvées mais en quantité beaucoup plus importante par vent de Nord. Par ailleurs, les compositions chimiques sont bien différentes d'un prélèvement à l'autre (fig. 4): par vent d'Ouest ces compositions se projettent exclusivement, sur le diagramme triangulaire SiO_2 Al_2O_3- Autres Oxydes, dans l'aire C des fly-ash industrielles définie par RAMSDEN et SHIBOAKA (24). Par vent de Nord, deux populations de microsphérules apparaissent: la première se projette encore dans l'aire C des fly-ash industrielles, ou près du pôle alumineux D ou enfin, au contraire dans une zone hypoalumineuse A ; la seconde population se projette dans l'aire B des microsphérules du panache de l'Etna. Une des microsphérules recueillies par vent d'Ouest se projette aussi dans cette aire B, mais l'examen détaillé de sa composition chimique (en particulier sa faible teneur en K_2O) fait écarter une origine volcanique.

La reconstitution des trajectoires des masses d'air ayant apporté des microsphérules à la pointe méridionale de la Sicile a été faite à partir du calcul du vent géostrophique à 3000 m d'altitude, déduit des données du Bulletin Météorologique Européen (fig. 4). Ces trajectoires confirment que les vents d'Ouest ont parcouru très peu de zones industrialisées, alors que les vents de Nord ont véhiculé des masses d'air ayant parcouru l'Europe industrielle et les Volcans d'Italie du Sud.

7. CONCLUSION.

Les méthodes de la Microscopie Analytique d'abord appliquées à deux sources choisies de microsphérules ont permis de montrer qu'on pouvait les caractériser par leur chimie (éléments majeurs et traces). Bien que de chimie variée, les microsphérules du panache de l'Etna se caractérisent par une lignée évolutive bien située sur un diagramme judicieux comme celui proposé par RAMSDEN et SHIBOAKA (24) et repris ici (fig. 2, 3, 4). Par contre, les cendres volantes des centrales électriques à charbon montrent des compositions monotones, reflets de celle des minéraux siliceux, silicatés ou carbonatés associés au charbon.

Des prélèvements effectués à distance de sources volcaniques ou

---

Fig. 4: Microsphérules captées à la Pointe Sud de la Sicile durant l'été 1984 a et b: Microscopies Electroniques à Balayage et à Transmission. c et d: Trajectoires possibles des masses d'air arrivant à la Pointe Sud de la Sicile les 28, 29 et 30 juillet et les 12 et 13 août 1984. e: Compositions chimiques de 71 microsphérules reportées sur un diagramme triangulaire (24). Les aires A, B, C et D sont définies dans le texte. (□): microsphérules captées par vent d'Ouest - (+): microsphérules captées par vent de Nord.

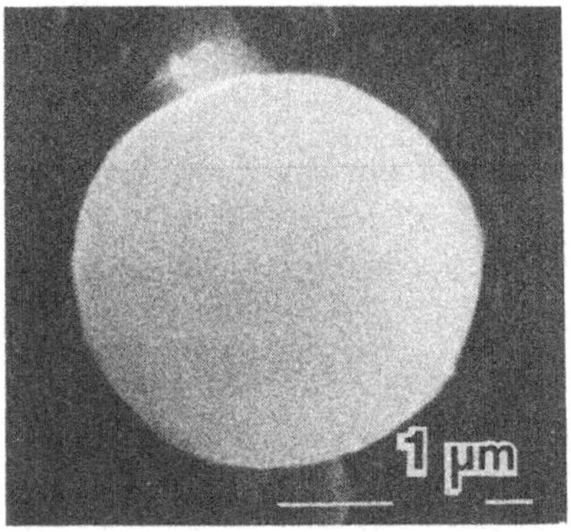
1 µm

a

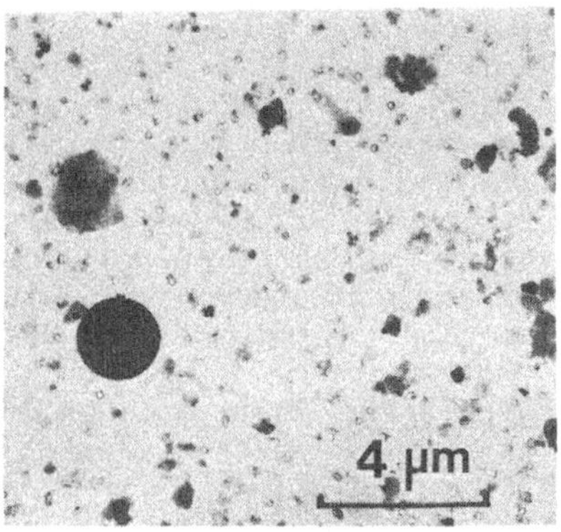
4 µm

b

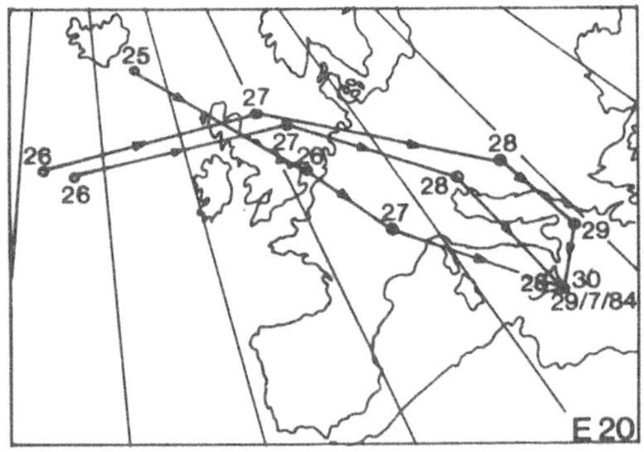
25
27
26
26
27
28
28
27
29
30
29/7/84
E 20

c

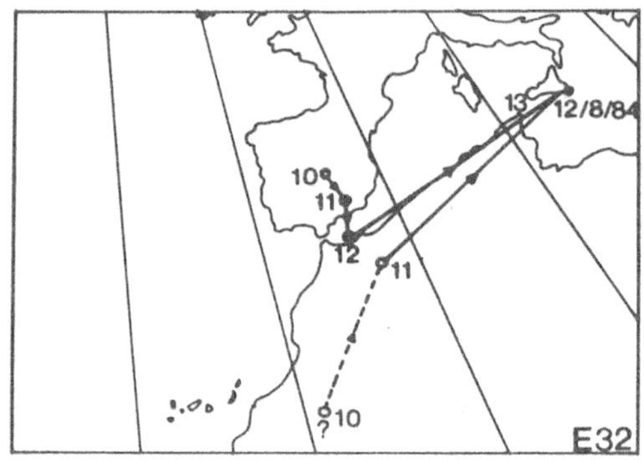
13
12/8/84
10
11
12
11
?10
E32

d

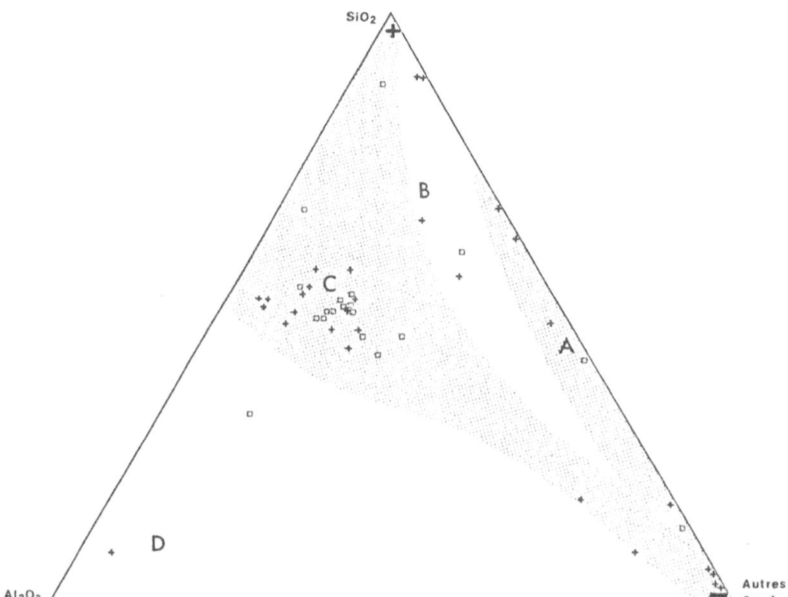
SiO_2
B
C
A
D
Al_2O_3
Autres
Oxydes

e

industrielles contiennent des microsphérules que leur chimie permet de rapporter à leurs sources, ce que confirment les trajectoires empruntées par les masses d'air qui les ont transportées.

Cette étude montre donc essentiellement que les panaches industriels ne sont pas les seuls émetteurs atmosphériques de microsphérules, et que la microscopie analytique est un outil précieux pour tenter de différencier par leur chimie les diverses sources possibles de microparticules présentes dans un prélèvement d'air effectué dans un site alimenté par des sources multiples inconnues. La complication introduite par la découverte de "cendres volantes" naturelles (volcaniques) est donc compensée par l'utilisation d'un outil analytique bien discriminant.

BIBLIOGRAPHIE

1-FIREMAN E.L. et KISTNER G.A.(1961). Geochim. Cosmochim. Acta,24,10-22.
2-WRIGHT F.W., HODGE P.W. et al.(1963). J. Geophys. Res.,68,19,5575-5587.
3-HODGE P.W. et WRIGHT F.W.(1964). J. Geophys. Res.,69,12,2449-2454.
4-WRIGHT F.W. et HODGE P.W.(1964). Ann. N. Y. Ac. Sc.,119,1,287-297.
5-SCHMIDT R.A. et KEIL K.(1966). Geochim. Cosmochim. Acta,30,471-478.
6-BROWNLEE D.E.(1978). Cosmic Dust, McDonnel ed.,295-335, Wiley, N.Y.
7-CLANTON U.S., GABEL E.M. et al.(1982). Cosmic Dust Catalog, NASA, Lyndon B. Johnson Space Center, Houston, Publ. 63, 3, 1, 95 p.
8-Mc KINNON I.D.R., Mc KAY D.S. et al.(1982). Proc. 13th Lunar Plan. Sc. Conf., 1, J. Geophys. Res.,97, supp.,A 413-421.
9-GLASS B.P. (1967). Nature,214,373-374.
10-GLASS B.P. (1974). Geol. Soc. Amer. Bull.,85,1305-1314.
11-GRJEBINE T., LALOU C. et ROS J.(1964). Ann. N. Y. Ac. Sc.,119,143-165.
12-FREDRIKSSON K. et MARTIN L.R.(1963). Geochim. Cosmochim. Acta,27,245-248.
13-HODGE P.W., WRIGHT F.W. et al.(1964). J. Geophys. Res.,69,14,2919-2931.
14-KING E.A. et WAGSTAFF J.(1980). Antarctic J. of the U.S.,15,5,78-79.
15-WRIGHT P.W. et HODGE F.W.(1965). J. Geophys. Res.,70,3889-3898.
16-HEIKEN G.H. et WOHLETZ K.(1985). Volcanic Ash,258p. Univ. of California Press Berkeley.
17-HEIKEN G.H. et LOFGREN G.(1971). Geol. Soc. Amer. Bull.,82,1045-1050.
18-BAKER G. (1967). Miner. Mag.,36,1012-1023.
19-LEFEVRE R., GAUDICHET A. et al.(1985). C. R. Ac. Sc.,Paris,301,II,20, 1433-1438.
20-LEFEVRE R., GAUDICHET A. et al.(1986). Nature, sous presse.
21-FISHER G.L., PRENTICE B.A. et al.(1978). Env. Sc. Tech.,12,4,447-451.
22-HULETT L.D.Jr., WEINBERGER A.J. et al.(1980). Science,210,1356-1358.
23-ROY W.R., THIERY R.G. et al.(1981). Env. Geology Notes,96,Illinois State Geological Survey, 1-69.
24-RAMSDEN A.R. et SHIBAOKA M.(1982). Atm. Env.,16,9,2191-2206.
25-DEVELL L., TOUVEDAL H. et al.(1986). Nature,321,192-193.
26-SEBASTIEN P., BILLON-GALLAND M.A. et al.(1978).Arch.Mal.Prof.,39,229-248.
27-GAUDICHET A.(1984). J. de Physique,45,C2,789-792.
28-CLIFF G. et LORIMER G.W.(1973). J. Microsc.,103,203-207.
29-WECHSUNG R., HILLENKAMP F. et al.(1978). Microsc.Acta,supp.2,281-296.
30-KAUFMAN R. et WIESER P.(1980). N.B.S., Spec. Publ.533,199-223.
31-BRUYNSEELS F. et VAN GRIEKEN R.(1985). Atm. Env.,19,11,1969-1970.
32-BUAT-MENARD P. et ARNOLD M.(1978). Geophys. Res. Letters,5,4,245-248.
33-FAIVRE-PIERRET R., MARTIN, D. et al.(1980). Bull. Volc.,43,3,473-485.
34-REID A.M. et al.(1973). EOS,54,606-607.
35-HEIKEN G.H. et al.(1974). Geochim. Cosmochim. Acta,38,1703-1718.
36-WOHLETZ K.H. et Mc QUEEN R.G.(1984). Geology,12,591-594.
37-WOHLETZ K.H.(1983). J. Volcan. Geotherm. Res.,17,31-63.
38-CRISTOFOLINI R. et ROMANO R.(1982). Mem. Soc. Geol. It.,23,99-115.
39-LEFEVRE R., GAUDICHET A. et al.(1986). C.R. Ac. Sc.,Paris,sous presse.

RADIOLYTIC AEROSOLS: THEIR PRODUCTION AND INHIBITION.

P. Goodman,*
and
J.A. Scott,
Physics Department,
University College, Dublin.

SUMMARY

The work described in this paper deals with the production of aerosols (condensation nuclei) formed when an air sample is irradiated with α- particles. In the experiments synthetic air was used to which various trace gases were added in measured amounts. The reactions took place in a cylindrical teflon chamber and the aerosol measurements were made using a condensation nucleus counter. It was found that the aerosol concentration produced in the reaction was dependent on the concentration of SO_2 present in the air sample. Organic vapours when present in the air sample prior to irradiation were found to act as an inhibiting agent to aerosol formation. Detailed studies were made under controlled conditions in order to try to quantify the effects of the organic vapours and to explain their action. A number of possible explanations are discussed.

INTRODUCTION

It is known that combustion processes are major sources of atmospheric aerosols (condensation nuclei). However it has also been shown that aerosols can be generated by the action of ionizing radiation. Megaw & Wiffen (1), Renzetti & Doyle (2), McGreevy (3), Billard et al (4), Burke & Scott (5) and Cox (6) were among those who have studied the production of aerosols in air by ionizing radiation. Megaw & Wiffen (7) reported a strong link between the number of aerosols produced and the levels of SO_2 present, prior to irradiation. Cox showed that most of the aerosols produced by this method are sulpheric acid aerosols. In view of the present high levels of acidic precipitation it was decided to investigate the processes of radiolytic aerosol formation.

Galloway (8) reported that in remote areas where the concentrations SO_2 and HNO_3 are low the contribution of HCOOH to the acidity of cloudwater may be significant, and that this may also effect the rates of sulphate production. Richards et al (9) showed that the liquid phase oxidation of S(iv) by H_2O_2 was inhibited by the presence of HCHO. Burke & Scott, and Coghlan (10) showed how some organic vapours could totally inhibit the radiolytic production of aerosols. It was decided to investigate further the effects of organic vapours on radiolytic aerosol production, and to see if any further insight could be obtained into the processes which give rise to the formation of aerosols by ionizing radiation.

* now with the Irish Meterological Service.

Experimental

The experimental arrangement used in this work (Figure I) is similar to that used by Burke & Scott. Synthetic air from a pressurised cylinder, [O_2=19.5%, 1.5%; CO 1ppmv; CO_2 300ppmv; NO_x 0.1ppmv; total Hydrocarbons 1ppmv, N_2 79%] was passed through teflon tubing and into a teflon experimental chamber. This chamber which was cylindrical in shape had a volume of 3.5 L. The air samples within the chamber were irradiated by alpha particles from an Americium source. The source, in the form of a metal foil of activity 0.35 C ($1.3x10^4$ Bq), was mounted in a stainless steel container equipped with a shutter mechanism which could be remotely operated. The air samples were irradiated for a period of 120 seconds after which they were transferred into a Nolan-Pollak Nucleus Counter, and a measurement was made of the aerosol concentration. When the sample consisted of solely synthetic air no aerosol production was found. However when some SO_2 was injected into the chamber prior to irradiation, after irradiation aerosols were detected.

In order to prepare samples of trace gases with a range of known concentrations a glass vacuum system was constructed. The method of obtaining such sample is basically one of a series of dilutions. It can be summarised as follows: An initial sample of the trace gas of known concentration is released into an evacuated flask. The total pressure in the flask is then brought up to atmospheric pressure by addition of synthetic air. By opening a vacuum tap this diluted sample of gas is allowed to fill a larger previously evacuated chamber. This too is brought up to atmospheric pressure by the addition of more synthetic air. This sample can then be further diluted by passing a known volume of it into another evacuated chamber and again bringing it up to atmospheric presssure by the addition of synthetic air. This process can be repeated as often as required, until a suitable concentration of trace gas/air mixture is obtained. This final mixture can be stored in a sealed flask. Samples of this mixture can be obtained by using a syringe via a vacuum tight septa.

A series of experiments were carried out in which the concentration of SO_2 was varied and the resulting concentration of aerosols recorded. These findings are shown in Figure 2. The findings were found to be reproducible with an accuracy of 10. Thus it was possible to know what mean aerosol concentration to expect from an SO_2 sample of given concentration.

In order to determine the effects of organic vapour on the rate of production of aerosols a further series of experiments were carried out . In all of these experiments the concentration of SO_2 was kept constant while the concentration of vapour added prior to irradiation was varied. Each experiment was carried out using only one organic vapour and SO_2 in the air mixture at one time. The results are shown in Figure 3.

Discussion of Results

Figure 2 shows the relationship between the concentration of SO_2 and resulting aerosol concentration after irradiation. The main points to note are (i) the graph is highly non-linear (ii) there appears to be a required SO_2 threshold concentration and (iii) at high SO_2 concentrations the graph forms a sort of plateau. This non-linearity was also observed by Megaw & Wiffen and Burke & Scott, but what is interesting is that for a small increase in SO_2 levels it is possible that a disproportionate amount of acidic aerosol may be produced after irradiation. For example, if the concentration of SO_2 is increased from 6 parts in 10^{10} to 7.6 parts in 10^{10} the resulting (average) aerosol concentrations goes from 14×10^3 per cm^3 to 26×10^3 per cm^3 which is almost a doubling of the aerosol concentration. It can be seen that even a small variation in SO_2 levels may give rise to considerably larger effects in aerosol production rates.

Megaw & Wiffen also observed that a threshold radiation dose is necessary before aerosol production can be detected. This was also observed by Burke & Scott and by Shaw (11). However concomitant with this concept of a threshold radiation dose the present investigation would seem to suggest that for a given radiation dose there is a threshold SO_2 concentration required before aerosol production can be observed.

Thus using the equipment described it was possible to detect SO_2 when it was present in air samples with concentrations less than 1 part in 10^{11}, which indicates the sensitivity of this technique for SO_2 detection. Van Luik and Rippere (12) also describe how the technique may be used for gas analysis.

Experiments were also conducted where NO_2 was injected into the chamber with the synthetic air (in other words in the absence of SO_2) Again aerosols were found to form. It must be noted however that the concentration of NO_2 required to observe this effect was of the order of one part in 10^7 and this yielded only approximately 100 aerosols per cm^3.

The plateau region in Figure 2 has also been observed by a number of workers, Raes et al (13) and Luria (14). Most seem to be in agreement that once the concentration of $SO_2 \approx$ 1ppm then a plateau should be observed. Raes and Janssens postulate that at high SO_2 concentrations all the available energy for aerosol production is used up. It is felt that this is a reasonable assumption as traces of SO_2 were found to remain in the chamber after experiments were carried out in their plateau region. When lower SO_2 concentrations were used it appeared that all the available SO_2 was used up. It should be noted that when the concentration of SO_2 was of the order of 1 part in 10^8 small numbers of aerosols were found to form in the absence of any applied radiation. It is interesting to note that this level of SO_2 concentration is close to that sometimes found in urban centres (15).

The Role of Organic Vapours

As previously mentioned organic vapours are known to play a significant role in Aerosol Physics, Brigg et al (16) and Garrett (17) discuss how organic vapours can inhibit droplet growth, but Burke & Scott showed how aerosol formation in the first place could be inhibited. It was felt that if certain vapours could inhibit aerosol formation, then these vapours must be in some way competing with the mechanisms for aerosol formation. Stockwell & Calvert (18) report that the reaction of the hydroxyl radical with SO_2 is generally recognised today as the major rate controlling step in the gas phase oxidation of SO_2 in the Troposphere. They give the following mechanism for the formation of sulpheric acid aerosols;

$$HO + SO_2\ (+M) \rightarrow HOSO_2\ (+M) \qquad [(1)]$$

$$HOSO_2 + O_2 \rightarrow HO_2 + SO_3 \qquad [(13)]$$

$$SO_3\ H_2O \rightarrow H_2SO_4 \qquad [(14)]$$

(The numbers in brackets refer to their equation numbers).

They conclude that equation (13) should be applied in preference to the over simplified $HO + SO_2 \rightarrow H_2SO_4$. It is felt that if an organic which has a high reaction rate with OH radicals is introduced into the experimental chamber then this vapour would react with the OH, leaving less OH available for reaction as in equation (1). This effect was observed and is shown in Figure 3. It can be seen that as the concentration of organic vapour is increased the aerosol production decreases. It was expected that those compounds with a high OH reaction rate would cause the most pronounced effects. Thus one would expect those curves in Figure 3 with the bigger gradients to represent such compounds. However this relationship was not found to exist. If for example one considers the reaction rate of OH radicals with cyclohexane ($k=6.2 \times 10^{-12}$ cm^3 per molecule-sec) and with trichoroethane, ($k=2.2 \times 10^{-14}$ cm^3 per molecule-sec) which has a considerably lower affinity for OH radicals,it is the latter which produces one of the most pronounced inhibiting effects. It should be noted that all of these experiments were conducted in dry air where the concentration of $H_2O < 1$ppmV. However an experiment was conducted where the concentration of H_2O was increased and this too was observed to have a mild inhibiting effect on aerosol formation. It is felt that the mechanisms of equation [(1)] are not as simple as is at present believed.

The inhibiting effects which were observed in this work seem to resemble the 'quenching' mechanisms used in Geiger counters. Trost (19),and Friedman (20) report that self quenched counters have been made using SO_2 and nitrous oxide. Efforts to link the present data to ionization potential values in relation to the inhibiting effect of the vapours failed to show any conclusive trends. It is felt that there may be a strong link between quenching and inhibiting effects. Seigneur and Saxena (21) state that the oxidation of SO_2 to sulphates involving Criegee intermediates is a possible important pathway when humidities are low. They note that in the gas phase and at high humidities the reaction of SO_2 + OH appears to be the major pathway to sulphate production. In the liquid phase oxidation of SO_2 it has been shown by Richards and by Adewuyi, et al (22) that liquid phase oxidation of H_2O_2 was inhibited by the presence of HCHO. This is particularly true at high concentrations of HCHO when the inhibition of sulphate production is well pronounced. They report that the rate of inhibition is controlled by the kinetics of formation of HMSA (Hydroxymethanesulphanate). They also found that the sulphate production was inhibited due to formaldehyde - free radical interaction in solution. These findings of Adewuyi et al, resemble closely those found in this laboratory, however they were dealing with aequous solutions while present work deals with dry gas phase conditions.

Conclusions

It is our opinion that the gas phase oxidation of SO_2 to sulphate is a major pathway for acidic aerosol production, but that not enough attention has been paid to the role of organic vapours in these processes. From our results it appears that the simple scavanging of the OH radicals by the organic vapours, thus leaving less OH available for suphate production is an over simplied mechanism. In fact the present experiments conducted in dry air more closely resembled findings which might be expected in aqueous state reactions. It was not possible to find any definite link between the abilities of organic vapours to inhibit sulphate production and their ionization potentials. However it is felt that there is a probable but yet unknown relationship between the inhibition mechanism observed by us, and the quenching mechanisms in Geiger counters. The same compounds that exhibit quenching are also found to inhibit aerosol production. This in our opinion is an indication of just how strong the connection between the mechanism may be. It is felt that further investigation of the inhibition of sulphate production by organic vapours should be carried out. The work outlined here is only part of ongoing experiments and it is hoped that with new equipment a complete analysis of the gases within our experimental chamber will be possible, in conjunction with the measuring of aerosol concentrations. We feel that further work in the area of inhibition will yield a key to the pathway for sulphate aerosol production from the irradiation of the gaseous phase.

Summary

It was shown that the production of sulphate aerosol from the irradiation of gas phase SO_2 in air is a highly non-linear process. It was also shown that this process can be inhibited by the presence of organic vapours. It was felt that the scavanging of OH radicals is an over simplification of the inhibiting mechanism. It is also believed that there is a strong connection between this inhibiting mechanism and the quenching in Geiger counters. Further investigations into this inhibiting of aerosol production are proposed.

References

(1) MEGAW, W.J. and WIFFEN, R.D. and CHAMBERLAIN. Geofisica Pura & Applicta 1957,36,p 233
(2) RANZETTI & DOYLE, "Photochemical aerosol formation in an SO_2 Hyrocarbon system. Int. J. Air Pollution 1960 2 p 327
(3) McGREEVY, G: "Towards the Production of reproducible aerosol by radiolytic method". Pure & Applied Geophysics, 71, p 190 1968
(4) BILLARD,F., BRICARD, J., MADELAINE, G. and CLERC. N. "Formation and evolution of Radiolysis particles"Proc. of 7th Int. Conference on Condensation and Ice Nuclei Prague Vienna, 1969 p 169
(5) BURKE, T.P., & SCOTT, J.A.: "Production of Condensation Nuclei by alpha radiation". Proc. R.I.A. 1973 73 Section A no 11.
(6) COX, R.A.: "Some experimental observations of aerosol formation in the photo-oxidation of SO_2". J.Aerosol Sci. 4 p 473, 1973.
(7) MEGAW, W.J. and WIFFEN, R.D.: "The generation of condensation nuclei by ionizing radiation". Geofisics Pura & Applicata, 1961, 50 p 118.
(8) GALLOWAY,J., LIKENS, G.E., KEENE, W.C. and MILLER, J.M., "The composition of precipitation in remote areas of the world". J. Geo-phys. Res. 87 p 8771 - 8786 1982
(9) RICHARDS, L., ANDERSON, J.A. BLUMENTHAL D.L. and McDONALD, T.A. "Hydrogen Peroxide and Sulpher (iv) in Los Angeles cloud water". Atmos. Envir. 17 no 4. p 911, 914. 1983.
(10) COGHLAN, M. and SCOTT, J.A., Proc in Atmospheric Electricity. p 40 1983.
(11) SHAW,G.E., "Nuclei Production in Alaska makes under the action of short wave (1850 A) U.V. irradiation".J. Aerosol Sci. 14 no 4 p 475
(12) VAN LUIK, R. and RIPPERE, R.,: "Condensation nuclei - a new technique for gas analysis". Analytical Chemistry, 34 no 12 1962.

(13) RAES, F., JANSSENS, A. and EGGERMONT G."A synergism between untraviolet and gamma radiation in producing aerosol particles from $SO_2 - H_2SO_4$ Laden Atmosphere".Atmos. Envir. 19 no 7 p 1069 - 1073, 1985
(14) LURIA, M., STOCKBURGER, L., OLSZYNE, K.J. and MEAGHER, J.F., "Dynamics of Sulphate Particle production and Growth in smog chamber experiments".
Atmos. Envir. 16 no. 4. p 697 1982.
(15) Dublin Corporation Atmospheric Air Pollution Report 1977 -1978
(16) BRIGG, E., BROWNSCOMBE, J. and THOMPSON, W."Fog Modification with long chain alcohols" J. Appl. Meteorology. 8 p 75 1969.
(17) GARRETT, W.D. "Retardation of water drop evaporation with monomolecular surface films". J. of Atmos. Science 28 p 816 1971
(18) STOCKWELL, W.R. and CALVERT,J.C."The mechanisms of the $HO-SO_2$ Reaction".Atmos. Envir. 17no 11 p 2231 - 2235, 1983
(19) TRUST,A:"Untersuchungen an Zahlrohen mit der Braunschen Rohne". Z. Physics 36 p 801 1936
(20) FREIDMAN H.:"Geiger Counter Tubes". Proc. of the I.R.E. 37 p 791 July 1949
(21) SEIGNEUR C. and SAXENA,P."A Study of Atmospheric Acid Formation in different environments". Atmos. Envir. 18 no 10 p 2109 - 2124 1984
(22) ADEWUYI, Y.G., CHO, S.Y., TSAY, R.P. and CARMICHAEL G.R. "Importance of Formaldehyde in cloud chemistry".Atmos. Envir. 18 p 2 413 - 2420, 1984

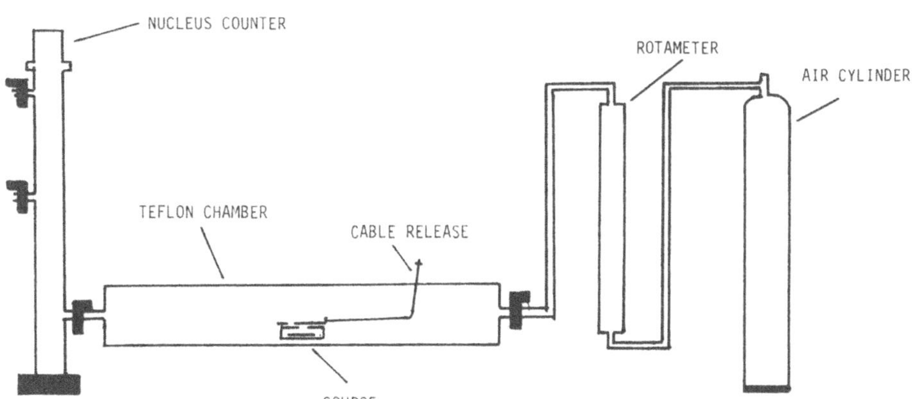

FIGURE 1: EXPERIMENTAL ARRANGEMENT

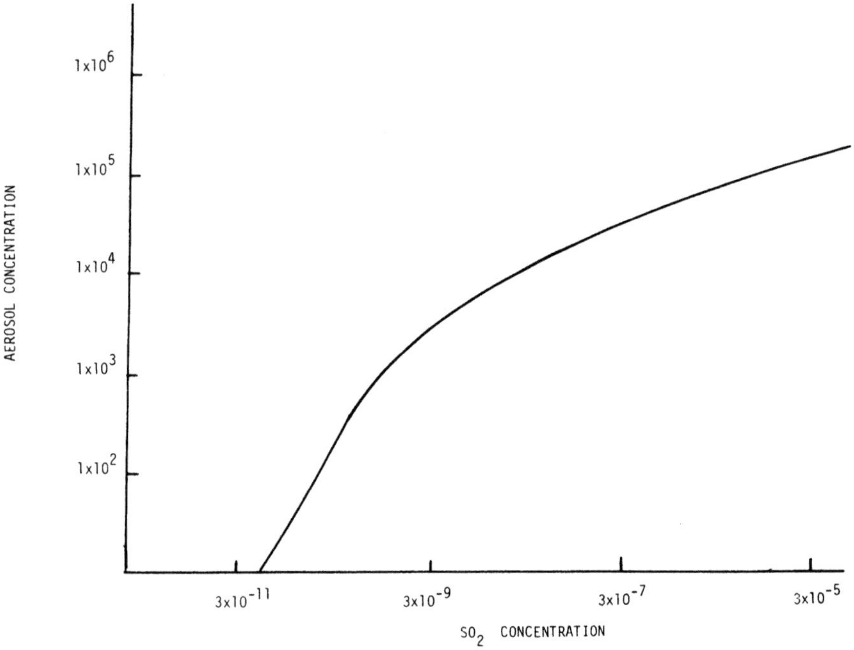

FIGURE 2: EFFECTS OF SO_2 ON AEROSOL PRODUCTION

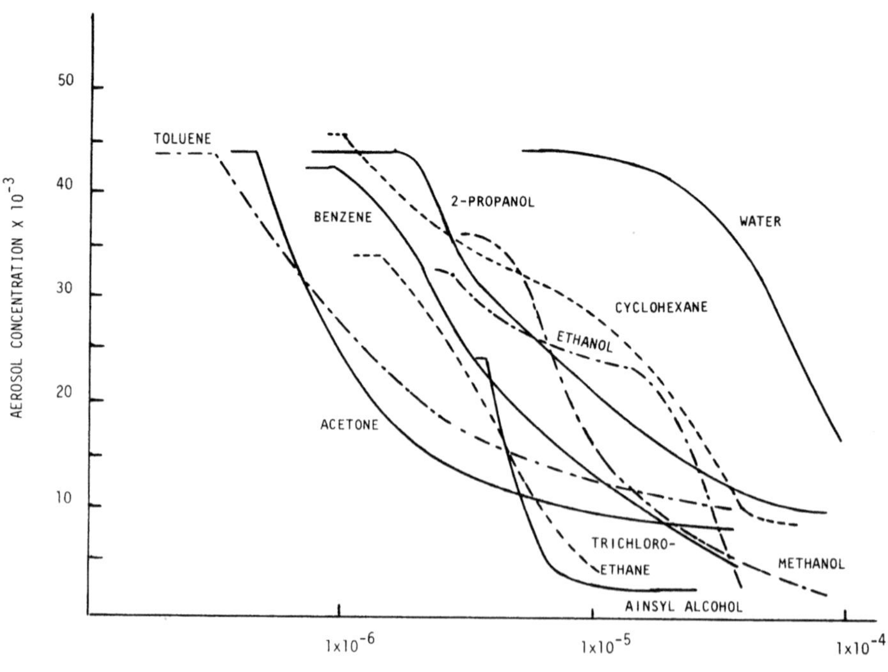

FIGURE 3: EFFECTS OF ORGANIC VAPOURS ON AEROSOL PRODUCTION

CHLORIDES IN ATMOSPHERIC AEROSOLS

M. MUÑOZ FERNANDEZ, J. de la SERNA, R. FERNANDEZ PATIER and F. PEREZ CARLES.

Departamento de Sanidad Ambiental. Escuela Nacional de Sanidad. Ciudad Universitaria. 28040. MADRID. Spain.

SUMMARY

During the period 1980-83, determinations of chlorides in TSP of the air in three different locations in Spain (Madrid, Bilbao and Puertollano) were carried out. The average level of atmospheric chlorides found in two of the sampling spots were similar. The third spot (Puertollano) gave half the average value (1.42 $ug.m^{-3}$) of the other locations. The statistical distributions of the chlorides concentrations, their correlation with TSP, as well as the bivariant distribution of these parameters have been studied.

INTRODUCTION

In connection with the studies on the physico-chemical characteristics of the aerosol carried out in our laboratory since 1976 (1) (2) a study of the concentration of chlorides in TSP, was made. The aim of this work was to determine the concentrations of chloride in the air in three different towns, in Spain, with different kinds of emission sources. It was done in order to determine if the variability of the chlorides levels in the micro-climates of these cities was dependent on those sources of emission, two of the cities experiences an intense automobile traffic, while the third has a moderate amount.

The characteristics of three cities are:

MADRID. Dry inland climate. Sampling point situated on the University Campus (40º N; 3º 43'W; 660m. of altitude). Residential area. Heavy traffic in certain hours.

BILBAO. Coastal city, situated in the North of Spain. Wet coastal climate. Sampling point situated down-town. Industrial and densily inhabited. Heavy traffic in certain hours (43º 18'N; 2º 55'N).

PUERTOLLANO. Medium sized city. Dry inland climate. (38º 40'N; 4º 5'W altitude 660m.) The town has an important petrochemical industry. Relatively little traffic.

EXPERIMENTAL.

As planned, air was sampled during 24 hours one day out of every seven, in such a way that week sampling corresponded to a different day in the week. Airborne particulates were collected on glass-fiber filters -WHATMAN GF/A - in hi-vol samplers, at a flow rate of 40 m^3 h^{-1}. (40 m^3 h^{-1}). A sufficient number of aliquots were taken from filters, placed in a 1M.HNO_3 solution and shaken in a beaker by mechanical stirring for 10 minutes at room temperature. The extracted solution was filtered though the same kind of filter. A second extraction was done in the same way and the two solutions were combined. We determined that extraction efficiency, using artificial samples, was 92 percent. The potentiometric titration was selected as the most appropiate method for this study. The intrument (Radiometer DTS 833) calculates the equivalence point by the second derivative method, instead of the Gran's theory. A prior study to test reliability of this method and to obtain the value of the average equivalence point was carried out. The results from 118 determinations of several standards gave the following value: Equivalence point = 261.1 $\pm$ 5.5. mv. Detection limit = 0.27 mg of Cl^-. Accuracy = 4%. Precission = 4%. The chloride determinations in TSP were performed at end-point (261.1 mv). Furthermore, a determination of the amount of chlorides in the blank filters was made. An average concentration of 0.43 mg of Cl^- was found in the GF/A filter of 15 cm. of diameter.

RESULTS.

In table 1 the average concentrations of chlorides and their standard deviations during the sampling period are shown.

TABLE 1. Concentrations of atmospheric chlorides.

Sampling Point	Period.	Number of data	Mean ug.m^{-3}	S.D ug.m^{-3}
Madrid	1981-83	69	2.98	1.63
BIlbao	1981-83	47	3.09	2.22
Puertollano	1980-83	118	1.42	0.71

A study of the bivariant distribution between TSP and chloride concentrations was carried out for each sampling point. Figure 1, shown data set for Madrid, Bilbao and Puertollano respectively.

Furthermore, it was considered of interest to study the correlation coefficients for the linear regression between the same parameters, the results of which are shown in Table 2.

TABLE 2. Regression parameters

Sampling point	Period.	Number data	Correlation Coefficient	Slope	Intercept.
Madrid	1981-83	66	0.51	0.02	1.80
Bilbao	1981-83	46	0.19	$7.08.10^{-3}$	2.45
Puertollano	1980-83	116	-0.04	$-4.12.10^{-4}$	1.45

DISCUSSION.

Comparing the average concentrations of chlorides found at the three different sampling point (Table 1) we observed a significant similarity between the chloride concentrations in the air of Madrid and Bilbao. The value for Puertollano is considerably lower. The comparison between chloride concentrations in Bilbao and Madrid led us to question the influence of marine aerosols in this coastal town. Certainly, Bilbao is situated at the end of an estuary, but not on the seaside (20 km. distant) and the influence of a marine breeze could be lesser than, for example, in the coastal county of Lancaster, where Harrison and Pio (3) found 4.16 $ug.m^{-3}$ in a spot situated 10 km. from the coast. This value is akin to those found by Chen and Duce (4) on the seaside in Bermuda (3.9 $ug.m^{-3}$). Puertollano has a concentration of 1.42. $ug.m^{-3}$, equivalent to half the average value of other locations. This result agrees with those obtained by Muller and Reuver (5) in an urban area (1.28 $ug.m^{-3}$) in W. Germany.

The study of the results of the bivariant distributions of chlorides in TSP in the theree towns sampled, does not permit a clear conclusion, as distributions are neither gaussian nor log-normal ones. This may be due to the low number of data used. For any interval of TSP, the concentrations of chlorides in Puertollano are lower (0-5 $ug.m^{-3}$) than those in Madrid and Bilbao (0-10 $ug.m^{-3}$).

The relations chlorides -TSP concentration shown in Table 2, give a good correlation for Madrid, and poor one for Bilbao and Puertollano. The good correlation in Madrid points out an influence of anthopogenic emissions (specially automobile exhausts) in the chloride concentrations; the low value of the correlation coefficient for Puertollano shows the chlorides concentrations are independent of TSP concentrations found. Moreover, the similarity of the average value of chlorides in Puertollano and the intercept of the regression line at the same spot compared with the background levels found in a close EMEP station (0.1 -0.9 $ug.m^{-3}$) lead us to believe that there may be a regional source of emission. With respect to the results obtained in Bilbao, it might be though that the rainout and washout processes that exist there, which remove chloride aerosols from the atmosphere, could give rise not only to the low correlations coefficient found, but also to the similarity between the levels existing in Bilbao and Madrid.

REFERENCES.

1. J. de la Serna, R. Fernandez Patier and F. Perez Carles, Physico-chemical analysis of atmospheric deposition in the center of Spain, in W. Grosch (Ed) Advarc. air poll. monit. equip. proc. Freiburg, F.R. Germany, June 2-6,85. 1985. pp 131-137.

2. J. de la Serna, R. Fernandez Patier and F. Perez Carles, Development of atmospheric aerosols and precipitation studies in Spain. Work-shop Aerosols and Acid. Deposition, Petten, Nederland, December 9-10,85 "in press"

3. R. M. Harrison and C. A. Pio, A. comparative study of the ionic composition of rainwater and atmospheric aerosols. Implications for the mechanism. of acidification of rainwater, Atmosph. Environ, 17 (1.983) 2.539 - 2.543.

4. L. Chen and R.S. Duce, The sources of sulfate, vanadium and mineral matter in aerosol particles over Bermuda, Atmosph. Environ, 17, (1.983) 2.055 - 2.064.

5. Muller, H. Reuver and D. Jost. Measurement of Cl^-, F^-, NO_3^- and $SO_4^=$ ions in rainwater and particulate matter by aid of ionic-chromatography, in B. Wersino and H. Ott (ED) Physico-Chemical behavoir of atm. poll. Varese, Italy, 29 Sep- 10 octu. 1981, D. Reidel Publ. Comp. 1982, pp 440-445.

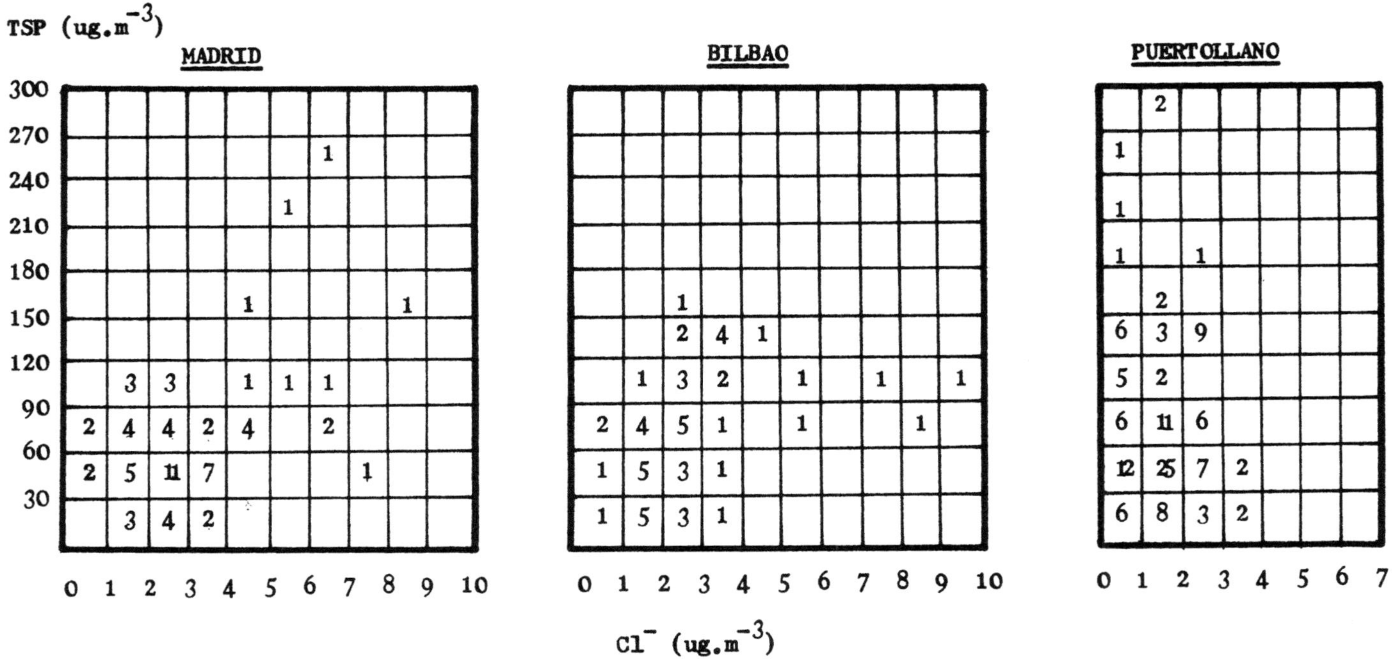

Fig. 1.- Bivariants distributions of chlorides in TSP.

WORKING PARTY 4
POLLUTANTS CYCLES

Chairman: S. BEILKE

Emissions of biogenic sulfur compounds from various soils

Dimethyl sulfide lifetime in the marine atmosphere

Natural and anthropogenic hydrocarbons and halocarbons over the atlantic ocean: latitudinal distributions of mixing ratios and sources

Characterisation of cloud and rain water chemical composition

Deposition measurements in Sweden

Measurements of the dry deposition rates of some trace gas species

Contribution of ammonia to the wet deposition of sulphur compounds

Impact of saharan dust on the rain acidity in the mediterranean atmosphere

Acid deposition in a snow field at 2500 m.a.s.l. in Switzerland

Acidity measurements of snowfalls collected over the VOSGES in relation to air mass trajectories

Nitrate and sulphate wet deposition in Europe

Seasonal and daily variations of ozone, sulfur dioxide and nitrous gases and the photochemical production and reactions with other trace gases of the ozone

A wet-only precipitation study in a mediterranean site, Patras, Greece

Daily trends of photochemical oxidants and their precursors in a suburban forested area. A useful approach for evaluating the relative contributions of natural and anthropogenic hydrocarbons to the photochemical smog formation in rural areas in Italy

Oxidant data collection in OECD-Europe 1985-87 (Oxidate): a co-operative project for rapid exchange of regional data on ozone, nitrogen dioxide and PAN

Observations of nitrous acid at roof and street level during air pollution episodes in Gothenburg

Speciation of ionic organolead compounds in the biosphere

Preliminary results on fog deposition

Determination, by PAH analysis, of the age of air mass contamination in a area remote from sources of pollution

Aircraft measurements of various sulfur compounds in a marine and continental environment

Flux d'azote inorganique dissous d'origine atmosphérique en système littoral d'europe occidentale

Meso-meteorological analysis of air pollution cycles in Spain

SUMMARY BY THE CHAIRMAN

S. BEILKE
Umweltbundesamt, Frankfurt, FRG

Working party 4 of COST 611 is concerned with the study of tropospheric pollutant cycles under special consideration of the processes involved in acid deposition and photochemical air pollution.
In September 1985 a discussion meeting of Working party 4 was held at the National Institute of Public Health and Environmental Hygiene(RIVM) in Bilthoven(The Netherlands) as a joint meeting together with Working party 5.
Most papers presented at this dicsussion meeting and at the 4th European Symposium were concerned with different aspects of the problems of acid deposition and photochemical oxidants on the regional scale of Europe.
On the following pages some aspects regarding the emission-transformation-deposition chain in acid deposition are discussed under special consideration of their treatment in regional and long-range transport models which try to establish a source-receptor relationship both on an episodic and long-term basis.

The principal precursors of acid deposition are SO_2,NO_x,NH_3 and NMHC which are emitted both by natural and man-made sources.
In general our present knowledge about the emissions of these substances is quite different for the various species concerned.
A broad data base exists for SO_2 for different European regions and countries.A major problem is probably the different quality of the emission data estimated for different regions and countries because definitions of emission sources and procedures for estimating emissions may differ from country to country in Europe(OECD,1985).In my opinion the uncertainties in national values are at best $\pm$ 20 % for SO_2 and much greater for other gases like NO_x,NH_3,HCl and NMHC.

There is growing concern about increasing emissions of NH_3 from agricultural activities in certain European regions and countries(OTT,1984;BUIJSMAN et al.,1985;ApSIMON et al.,1986).
As ammonia is of considerable importance for the acid deposition problem in Europe,more research is necessary to study its cycle on the regional scale of Europe including the development of long-range transport models for ammonia and ammonium.A first attempt to develop such a model was made by ASMAN and JANSSEN (1986).
As far as the emissions of non methane hydrocarbons(NMHC) are concerned some anthrpogenic emission inventories are available but these data seem to be even more uncertain than the corresponding data for SO_2,NO_x and NH_3.The uncertainty is increased by the largely unknown biogenic emissions of natural NMHC.

In order to reduce the uncertainties of the emission data it is recommended to measure fluxes of SO_2,NO_x and NMHC originating

from the main area sources in Europe using a combination of airborne and ground based techniques.
In some European regions or countries outside the industrial regions of central Europe the so-called "background" concentration and flux of acidic substances may contribute considerably to acidic deposition.Often the teım "background" is used to indicate the concentrations at the border of the area considered, for example Europe,a specific country or region.This "background" concentration may be caused by both anthropogenic and natural emissions and has clearly to be distinguished from the natural background which is the concentration level without any anthrpogenic influences(DELMAS and GRAVENHORST,1983;BUILTJES, 1986).The background concentrations and fluxes at the edges may be determined whereas the natural background concentrations can most likely not be determined exactly on the European continent.

As far as chemical transformation is concerned, a substantial knowledge exists mainly in the field of homogeneous gas phase reactions.Transformation processes are very important because deposition of the primary pollutants such as SO_2 and NO_x and that of their reaction products (sulfates,nitrates) are governed by quite different processes.
The treatment of chemical transformation processes in long-range transport models for acid deposition should be based on the present stage of knowledge which is very different for chemical processes occurring in the gas and liquid phase.
The most important gas phase reactions of SO_2 and NO_x seem to be relatively well known and should be incorporated into LRT models along with their thermodynamic and kinetic quantities.
A source of uncertainty seems to be the computation of the true atmospheric OH concentration and an exact quantification of the night time nitric acid generation due to NO_x oxidation.
As far as the situation in cloud droplets is concerned,the conversion of SO_2 and NO_x is more complicated than in the case of the corresponding gas phase reactions.
Laboratory experiments have shown that SO_2 oxidation by H_2O_2 proceeds very fast but the importance of this reaction is not really known under atmospheric conditions.
For NO_x conversion in atmospheric droplets into HNO_3 the real importance is even more uncertain.
From a chemical point of view,indications are strong that a significant non-linear conversion occurs for transformation of SO_2 to H_2SO_4 in cloud droplets(BEILKE,1986).

Considering the present stage of knowledge about acidity generation in clouds it is not yet possible to treat this type of transformation in long-range transport models in a similiar detailed manner as in the case of homogeneous gas phase reactions.At present the use of detailed chemical equations along with all their thermodynamic and kinetic quantities pretends an accuracy which is not given.
On the other hand it is also dangerous to use a parametrisation in long-range transport models which is too simple since,for example,SO_2 conversion alone can vary at least by two orders of magnitude depending on the physico-chemical conditions under which the oxidation proceeds.
In order to provide a more realistic chemical input to episodic

long-range transport models in a relatively short period of time it is suggested to measure SO_2 and NO_x conversion rates in clouds under different meteorological and chemical conditions as a function of distance and time from well defined area sources using aircrafts.

As far as deposition of acidic substances is concerned, there is considerable uncertainty about the relative contributions of acidic substances deposited by dry and wet deposition mainly due to a lack of reliable dry deposition measurements which makes estimates of this removal mechanism much less accurate than estimates for wet deposition.
In contrast to the difficult measurements of dry deposition, the integrated wet deposition is relatively simple to measure. As a result there exists a substantial data base of wet deposition from a series of European networks.
An important problem seems to be the existence of different national networks working to some extent with different methods.
For the years to come, an important task should be to harmonize and standardize the instrumentation for sampling, storage and chemical analyses of precipitation on a European scale.
Furthermore, the precipitation network should extend its programme in its geographical coverage particularly in countries where the station network is scarce as for example the Mediterranean area.
In general more research is necessary to study dry deposition of substances relevant to the acid deposition problem. Special attention should be paid on measuring dry deposition of nitrogen compounds such as NO_x and HNO_3.
For species which undergo a rapid chemical conversion like O_3, NO and NO_2 caution should be exercised when using the gradient method for determining deposition velocities because the observed gradients may be influenced by chemical reactions (KRAMM, 1985).

A serious problem associated with long-range transport modelling is that in some models dry deposition is treated parametrically by using a constant deposition velocity regardless of spatial coordinates and time. In order to obtain deposition fluxes these mean deposition velocities are weighted by concentrations. The major uncertainty arises from the wide variations of deposition velocities and concentrations with season, time of day, wheather conditions and surface properties.
On the basis of comparison of model outputs with measurements, parametrisation of dry deposition should be known within a factor of 2, but the range of measurements is much larger than that especially for gases other than SO_2 (NAS REPORT, 1983).
Instead of mean deposition velocities different classes of deposition velocities should be defined for various atmospheric transfer conditions and surface properties.

Another problem in modelling of LRT of substances relevant to the acid deposition arises from the uncertainty of particle dry deposition. A considerable controversy remains concerning the values of the deposition velocity for particles in the size range between 0.1-1 µm where most sulfate, nitrate and acidity is found. Some authors (see for example NAS REPORT, 1983; GRAVENHORST et al., 1983) advocate the use of values in the range of 1 cm/s whereas other authors (for example GARLAND, 1983) use values of

0.1 cm/s or less in this size range.The reasons for the considerable disagreement of particle deposition could be the influence of wind velocity and roughness height.
If the higher values of ca. 1 cm/s were representative for European regions with rough surfaces like areas covered with forests dry deposition of particles would be an important removal process for atmospheric acidity.

During the years to come,it is hoped that some of the specified problems can be investigated within project COST 611(1986-1990).

References:

ApSIMON,H.M.;BELL,N.and KRUSE,M.(1986)
Emissions of ammonia and their role in acid deposition.Paper presented at the Symposium on Interregional Air Pollution Transport,Budapest,April 22-24,1986.Inaugural Meeting of EURASAP.

ASMAN,W.A.H.and JANSSEN,A.J.(1986)
A long-range transport model for ammonia and ammonium for Europe.Paper presented at the Symposium on Interregional Air Pollution Transport,Budapest,April 22-24.EURASAP Meeting.

BEILKE,S.(1986)
Problems associated with long-range transport of air pollutants Lecture presented at Ispra course on Regional and Long-Range Transport of Air Pollution,Ispra(Italy),15-19 September 1986.

BUIJSMAN,E.;MAAS,J.F.M.and ASMAN,W.A.H.(1985)
Some remarks on the ammonia emission in Europe.Proceedings: Meeting of Working party 4,Project COST 611,Bilthoven(NL), 23-25 September 1985,pp.33-37.

BUILTJES,P.J.H.(1986)
Eurasap Newsletter Nr.1,May 1986

DELMAS,R.J.and GRAVENHORST,G.(1983)
Background precipitation acidity.Proceedings:CEC Workshop on Acid Deposition,Berlin,Reichstag,9 September 1982.Reidel Publ. Comp.,pp.82-108.

GARLAND,J.A.(1983)
Principles of dry deposition:Application to acidic species and ozone.VDI-Berichte 500,Acid Precipitation,Lindau,FRG,7-9 June 1983,pp.83-97.

GRAVENHORST,G.;HOEFKEN,K.D.and GEORGII,H.W.(1983)
Acid input to a beech and spruce forest.Proceedings:CEC Workshop on Acid Deposition,Berlin,Reichstag,9 September 1982,, pp. 155-172.

KRAMM,G.(1985)
Numerische Modellrechnungen zur Bestimmung der trockenen Deposition atmosphärischer Spurenstoffe.Proceedings:Bestimmung der Trockenen Deposition,Neuherberg,9-10 September 1985,BPT-Bericht 6/85.Editors:Hoefken,K.D.und Bauer,H.,pp.79-94.

NAS REPORT(1983)
Acid Deposition-Atmospheric processes in eastern North America, US National Academy of Sciences,Nat.Academic Press,Washington D.C.

OECD(1985)
OECD Environmental Data Compendium 1985,Paris 1985,297 pages.

OTT,H.(1985)
The acid deposition problem:scientific background,research priorities and abatement technologies.Paper presented at Symposium on Acid Deposition in Italy,Venice,24-25 October 1984.

EMISSIONS OF BIOGENIC SULFUR COMPOUNDS FROM VARIOUS SOILS

R. STAUBES, H.-W. GEORGII, G. OCKELMANN
Department of Meteorology and Geophysics
University of Frankfurt/Main, FRG

Summary

The emission of carbonylsulphide (COS), dimethylsulphide (DMS) and carbondisulphide (CS_2) from several soils in Germany has been examined by means of a dynamic gas emission chamber. The compounds were sampled by cryogenic technique and measured by subsequent GC/FPD-analysis. It was found that the sulfur emission rate strongly depends on a number of influencing factors: soil order, temperature and moisture appeared as the major factors controlling sulfur source strengths. Furthermore, diurnal and seasonal variations of sulfur emission rates were observed.

1. INTRODUCTION

The interest in the atmospheric sulfur budget has developed rapidly in the last years due to growing sulfur emission rates from anthropogenic sources. An important aspect is the influence of sulfur gases on air-chemical processes such as air pollution, climatic effects and precipitation chemistry. Most of the emitted reduced sulfur gases appear as intermediate products in the atmosphere and are rapidly converted to SO_2 and $SO_4^=$ in the troposphere (1). An exception is COS which reaches the stratosphere due to a long lifetime of more than one year (2,3) and seems to be a dominant source of the stratospheric sulfate aerosol layer.

While anthropogenic source strengths have been computed with reasonable accuracy, large uncertainties still exist about the chemical speciation and the magnitude of natural sulfur fluxes in the atmosphere. Previous calculations of the total natural sulfur emissions are equivalent to the current anthropogenic emission of about 100 Tg S/yr (4,5). A significant contribution to the natural sulfur cycle is ascribed to biogenic emissions. Biogenic sources include a diverse series of biological and chemical processes which lead to volatilaziation of gaseous sulfur compounds such as COS, DMS and CS_2. Microbiological activity in soils is considered to be one of the major biogenic sulfur sources beside the emission from oceans and volcanoes (6,7,8). The existing measurements of sulfur emissions from soils show great variabilities between the sulfur fluxes from different soil types. Adams et al. (9) and Bürgermeister (10) found a strong dependence of sulfur emission from temperature. Headspace analysis (11) showed that soil moisture can also influence the emission of several sulfur compounds. Some investigators (12,13) found seasonal variations of sulfur fluxes from soils. This paper presents sulfur emission measurements from a variety of soil types. Samples were collected in 1984 and 1985 to gain informations about the quantity and the range of emitted sulfur as well as to show how far various factors such as soil order, temperature, moisture, season and daytime influence the sulfur emission from soils.

2. EXPERIMENTAL METHODS

Sampling System: A dynamic emission chamber enclosing 0,07 m^2 of soil surface was used for measuring fluxes of gaseous sulfur compounds from soils. The glass chamber is swept with purified sulfur-free synthetic air at a rate of 3 l/min. The flux from the specific soil under the enclosure is determined by measuring the concentration of COS, DMS and CS_2 in the chamber's effluent. The sampling procedure is shown in Figure 1.

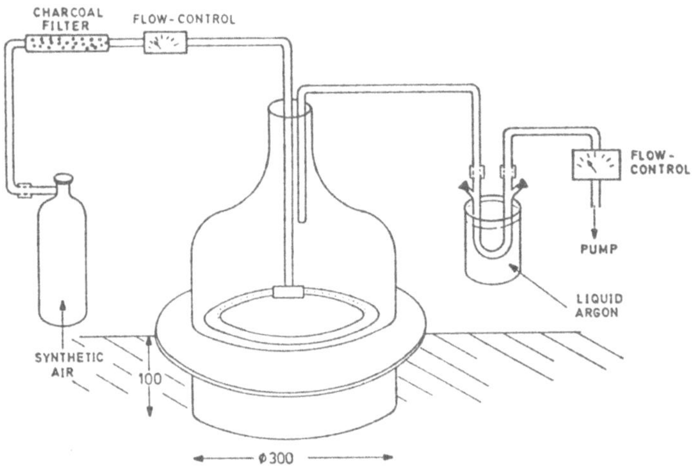

Figure 1: Emission chamber and sampling procedure

Preconcentration: A known volume of chamber air is drawn through a glass sample loop immersed in liquid argon to enrich the gaseous sulfur compounds.

Analysis: A gaschromatographic system (column: Carbopack BHT 100) equipped with a flame photometric detector was used for separation and analysis of the sulfur containing gases.

Detection limit: The sensitvity of the sampling and analyzing system is 0,1 ng of COS, DMS and CS_2 per sample corresponding to an emission rate of 1,37 ng/m^2 min for COS, DMS and CS_2 (sample volume: 3,0 l)

3. RESULTS

Average Emission Rates of different Soil Types

Table I shows the mean emission rates of COS, DMS and CS_2 during field studies at eight sampling sites from April to November 1985. All soils were examined in the same temperature range between 12°C and 23°C. Different soil orders are listed according to increasing humus content. Generally the humous soils appeared to be stronger sources for biogenic sulfur than those soils containing smaller amounts of organic matter. Fluxes ranged from under detection limit to 160 ng S/m^2 min.

TABLE I: Average Emission Rates

Soil Type	Mean Emission Rate S(COS)	S(DMS)	ng S/m² · min S(CS_2)	Total	Number of Measurements
Rendzina	3,13	4,22	1,69	9,04	42
Gleysol	3,00	1,19	2,60	6,79	63
Acrisol	3,26	4,10	2,54	9,90	41
Acrisol	4,54	4,67	2,79	12,00	16
Gleyic Acrisol	5,93	1,85	9,52	17,30	18
Hortisol	4,95	2,37	3,96	11,28	142
Fluvisol	5,66	7,84	5,88	19,38	83
Histosol	7,41	5,67	4,68	17,76	45

Influence of Temperature

Soil temperature is besides soil type the most important factor controlling the source strength as microorganisms enhance their activity with increasing temperature. Temperature was measured at soil surface, in 5 and 10 cm soil depth. While no dependence of emission rate upon temperatures in 5 and 10 cm soil depth could be found, surface temperature appeared as a determining factor for sulfur emission.

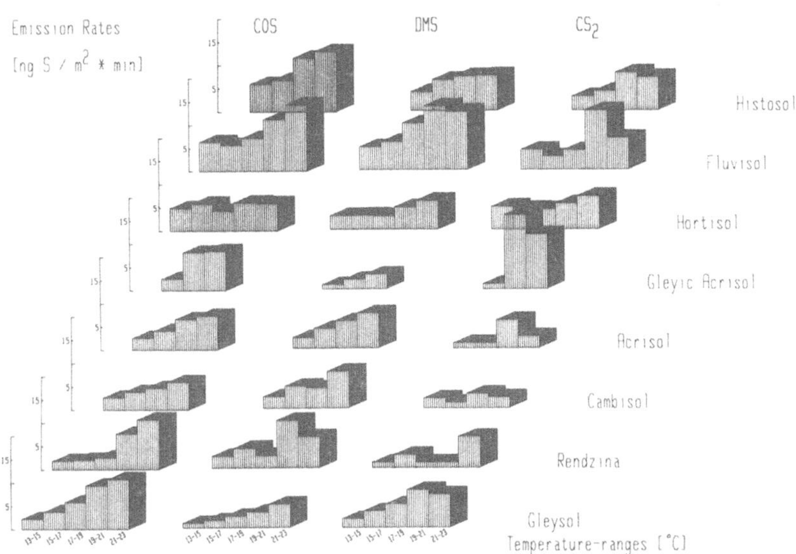

Figure 2: Temperature dependence of sulfur emission

In Figure 2 mean emission rates are presented within certain temperature intervals. In most cases it is found that the sulfur fluxes increased with soil temperature often by a factor of 2 when temperature rose by 5°C. Nevertheless, great standard deviation in emission rate occured in each of the temperature intervals. This may be a hint that other processes in addition to temperature influence the emission. This effect is most evident for CS_2, while DMS is the compound with the strongest dependence upon temperature.

Continuous measurements have shown that the emission rate is also depending on actual weather conditions and reacts strongly on sudden changes of temperature. For instance an abrupt warming leads to very high flux rates which exceed by far emissions examined at the same temperature, but after a long lasting warm period.

Influence of Soil Moisture

Soil moisture is another parameter influencing the sulfur emission by modifying source strength as well as the fraction of each compound on the total emission. It was found that the combination of low humus content and high moisture (gleysol, gleyic acrisol; Table I) reduces the DMS - and promotes the CS_2 - emission. The fraction of DMS emission from moist, low humus containing soils only amounts to 10-20% compared to 40% from dry soils (rendzina, acrisol, fluvisol); the fraction of CS_2 increases from 20% from dry soils to 40-50% from moist soils. An exception are the hortisol (low humus and moisture content) and the histosol (high humus and moisture content) where both influencing factors overlapped. The fraction of COS is about 35-40% of the total at all sampling sites.

During measurements at two histosols (4 sampling sites) in Bavaria it was found that the DMS emission from dry histosols increased 2-6 fold in opposition to strongly waterlogged histosols. There are only few indications that with growing water content the COS emission decreased and the CS_2 emission increased at these sampling sites.

Seasonal Variations

Monthly means of emission rates and soil surface temperature during measurements at a hortisol are presented in Figure 3.

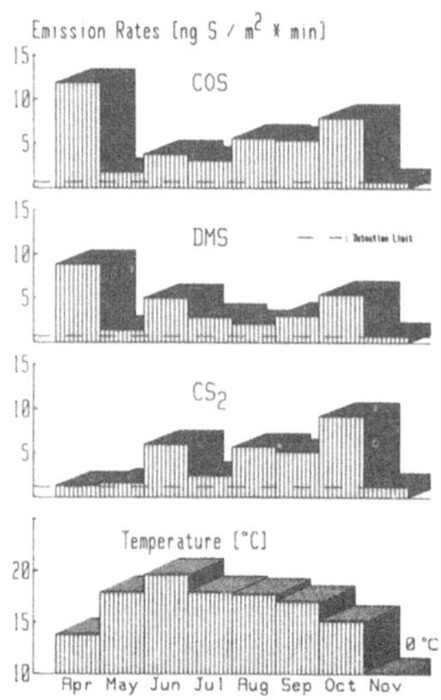

Figure 3: Monthly means of emission at a hortisol in 1985

Two remarkable results are found:

1. Druing late summer and early fall the emission of COS, DMS and CS_2 increased evidently although temperatures decreased. Especially the CS_2 emission showed this effect very clearly. A possible explanation is the higher yield of organic matter and the change of root respiration of plants in fall. In November the source strengths sank under detection limit according to very low temperatures.
2. In April the COS and DMS emissions were comparatively high even though temperatures were low. This may be due to a high activation of microorganisms caused by a sudden warming after the winter period. In contrast the CS_2 emission rates in April were still very low corresponding to low temperatures.

Diurnal Variations

Measurements were conducted during 24 hours at a hortisol and a fluvisol. Results are shown in Figure 4.

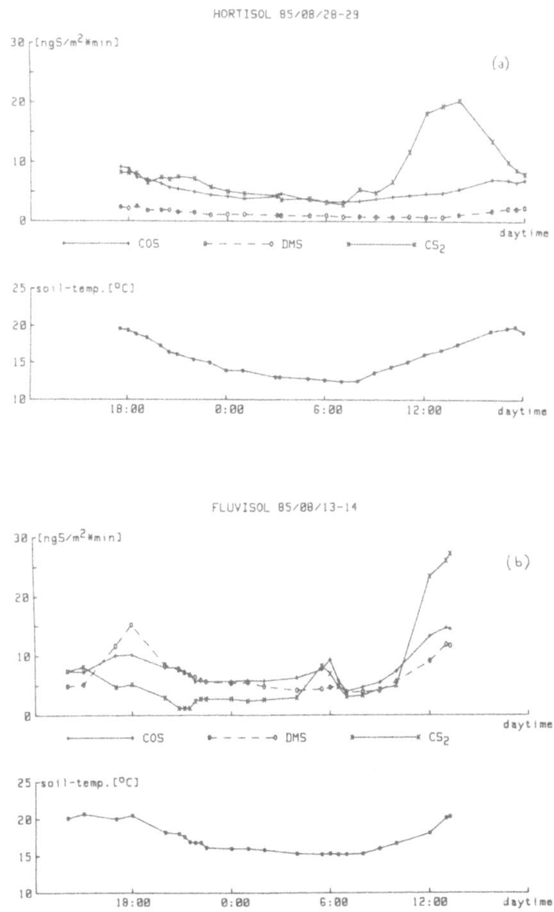

Figure 4: Diurnal variation of emission rate at a hortisol (a) and a fluvisol (b)

The diurnal variations of COS and DMS emissions followed the lapse of temperature: the highest COS and DMS emission occured at 18:00 (maximum temperature at 17:30), the lowest in the early morning hours (minimum temperature at 07:00) at both sampling sites. Quite different was the diurnal variation of CS_2 emission: an evident maximum appeared at midday, although the temperature was still rising and reached its maximum 4-6 hours later; from 14:00 - 18:00 the CS_2 emission rate decreased at the hortisol. It can therefore be assumed that the CS_2 emission is determined by other daytime effects apart from temperature variations.

During a strong thunderstorm the emission of all examined compounds increased (Figure 4b). Soil temperature and moisture remained constant during this period and were therefore not responsible for the change of emission rates.

Influence of other Parameters

Vegetation cover:

The vegetation cover can influence the surface conditions of soils e.g. by changing the humus cover and the moisture content. For this purpose the emission of three acrisols had been studied. Two of those were grown with deciduous trees, the other with conifers. The latter sampling site have a higher content of soil moisture, this causes a higher emission of CS_2 and a lower emission of DMS compared to the emission at the other acrisols.

Fertilizer:

Furthermore, an effect of soil fertilizing on sulfur emission has been determined. Our measurements showed that the sulfur emission from fertilized soils is 5-6 fold higher in contrast to unfertilized soils of the same type.

pH:

It appeared that the pH of soils has no influence on sulfur emissions. The occuring differences in sulfur source strengths between various soil types can not be explained with differences in soil-pH, but more likely with other soil attributes.

ACKNOWLEDGEMENT

The research in this paper has been supported by Deutsche Forschungsgemeinschaft through "Sonderforschungsbereich 73, Atmospheric Trace Substances" and by the EC-Commission.

REFERENCES

(1) GEORGII, H.-W. (1982): The Atmospheric sulfur-budget; in: H.-W. Georgii and W. Jaeschke, Chemistry of the unpolluted and polluted troposphere, 295-324. Reidel Publishing Company.

(2) COX, R.A.; SHEPPARD, D. (1980): Reactions of OH radicals with gaseous sulphur compounds. Nature Vol. 284, 330-331.

(3) KURYLO, M.J. (1978): Flashphotolysis resonance fluorescence investigation of the reaction of radicals with dimethylsulfide, OCS and CS_2. Chem. Phys. Letters, Vol. 58, 233-242.

(4) ANDREAE, M.O. (1985): The emission of sulfur to the remote atmosphere in: J.N. Galloway et al., The biogeochemical cycling of sulfur and nitrogen in the remote atmosphere, 5-25. D. Reidel Publ. Comp.

(5) JØRGENSEN, B.B.; OKHOLM-HANSEN, B. (1985): Emissions of biogenic sulfur gases from a Danish Estuary. Atm. Environment, Vol. 19,1737-1749.

(6) ANEJA, V.P.; ANEJA, A.P.; ADAMS, D.F. (1982): Biogenic sulfur compounds and the global sulfur cycle. Journal of the Air Poll. Control Ass., Vol. 32, 803- 807.

(7) ADAMS, D.F.; FARWELL, S.O.; PACK, M.R.; ROBINSON, E. (1981): Biogenic sulfur gas emissions from soils in Eastern and Southeastern United States. Journal of the Air Poll. Control Ass., Volume 31, 1083-1089.

(8) ADAMS, D.F.; FARWELL, S.O.; ROBINSON, E.; PACK, M.R.; BAMESBERGER, W.L. (1981): Biogenic sulfur source strength. Environmental Science and Technology, Vol. 15, 1493-1498.

(9) ADAMS, D.F.; FARWELL, S.O.; ROBINSON, E.; PACK, M.R. (1979): Assessment of biogenic sulfur emissions in the SURE Area. Final Report submitted to Electric Power Research Institute, Palo Alto,California.

(10) BÜRGERMEISTER, S. (1983): Gaschromatographische Messung von Bodenemissionen und atmosphärischen Konzentrationen des Dimethylsulfids. Diplomarbeit, Institut für Meteorologie und Geophysik, Universität Frankfurt/Main.

(11) BREMNER, J.M.; STEELE, C.G. (1978): Role of microorganisms in the atmospheric sulfur cycle; in: M. Alexander, Advances in Microbiol. Ecology, New York, 155- 201.

(12) STEUDLER, P.A.; PETERSON, B.J. (1984): Contribution of gaseous sulfur from salt marshes to the global sulfur cycle. Nature, Vol. 311, 455-457.

(13) STEUDLER, P.A.; PETERSON, B.J. (1985): Annual Cycle of gaseous sulfur emissions from a new England Spartina alterniflora marsh. Atm. Env., Vol. 19, 1411-1416.

DIMETHYL SULFIDE LIFETIME IN THE MARINE ATMOSPHERE

B.C. NGUYEN, S. BELVISO, B. BONSANG and G. LAMBERT
Centre des Faibles Radioactivités
Domaine du CNRS, B.P. 1
91190 Gif-sur-Yvette, France

Summary

The dimethyl sulfide (DMS) emitted by biological activity at the sea surface, is one of the predominant precursors of SO_2. Its lifetime in the atmosphere is still poorly known. During an oceanographic cruise in the South Pacific Ocean, we have made several vertical profiles of DMS in the low atmosphere, based on measurements at 1, 7 and 17 meters above the sea surface level. Very steep gradients of DMS were observed. These gradients correspond to a relatively short lifetime of DMS of the order of few minutes to few hours. Taking into account the kinetic constants of oxidation by OH radicals measured in laboratory by several authors, it appears that these short lifetimes we observed cannot be accounted for only by OH oxidation. Other oxidation mechanisms should exist. The different phenomenons which can interfere on the DMS lifetime in the marine atmosphere are discussed.

1. INTRODUCTION

The Dimethyl Sulfide (DMS) is produced by biological activity in surface sea water (1-4), volatile and little soluble, it is easily released into the atmosphere where it is one of the predominant precursors of SO_2. The lifetime of DMS in the atmosphere is still poorly known, but its knowledge is necessary to determine the contribution of this gas in the biogeochemical sulfur cycle and its role in the rainwater acidity.

2. EXPERIENCE AND MEASUREMENTS

During an oceanographic cruise in the South Pacific Ocean, from Panama to Hobart (figure 1), we have made on board "Lady Franklin" several vertical profiles of atmospheric DMS concentrations at the three levels 1m, 7m and 17m above the sea, simultaneously with DMS concentrations measurements in sea water.

The air was sampled during 20 minutes at 0.15 liter/minute through a Tenax trap at -100°C, on the front of the ship, winward. It is worthy to note that the "Lady Franklin's" chimney is close to the stern.

Water samples were collected from the bow by diving a 10 liter vessel directly in the sea.

The gases dissolved in the sea water were extracted by bubbling one hour 25 cm^3 of water with pure N_2 (100 cm^3 per minute), then trapped in Tenax at - 100°C, as precedingly. It was checked that this time is sufficient to degas completely the sample. The trap was then heated at +100°C and gases analysed by gas chromatography using a flame photometric detector and a 2 metre long FEP teflon column, 1/8" o.d., packed with chromosyl 310.

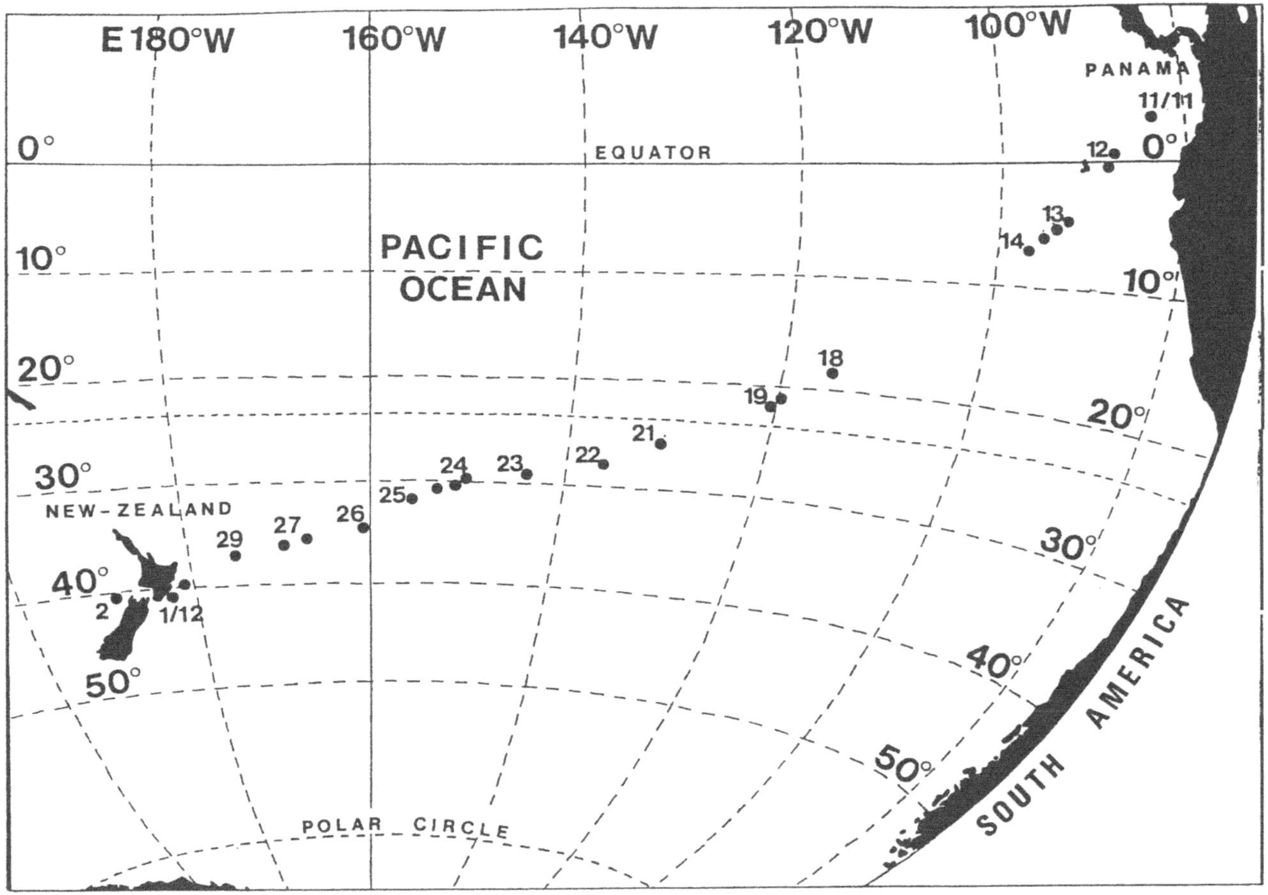

Figure 1 : Cruise track. The sampling positions are identified by the sampling date.

3. RESULTS AND DISCUSSION

DMS concentrations of sea water and air are shown in figure 2. With a Henry's constant of 0.3 (1), it may be observed that the sea water is supersaturated in DMS relatively to the atmospheric concentration, by a factor of about 300 times. Furthermore the atmospheric DMS concentrations decrease rapidly about a factor of 2 between the 1 m and 17 m levels. Assuming that :

- A steady state was established in the lower atmosphere from 1 m to 17 m during our measurements.
- The DMS oxidation is a first order reaction characterized by a constant λ which is the reversal of the lifetime.
- The vertical distribution of DMS in air is only controlled both by this constant λ and by the vertical eddy diffusion coefficient K_Z.

Therefore the DMS air concentration $C_{a,z}$ versus altitude Z is given by the exponential law

$$C_{a,z} = C_{a,o} \exp. \left(-\sqrt{\frac{\lambda}{K_z}}\; Z\right) \qquad (1)$$

in which we admit that K_Z has an uniform value in the layer considered, i.e. between 1 and 17 m above the sea.

The DMS oxidation constant can be calculated from air concentrations C_1 and C_2 measured at two levels Z_1 and Z_2, by using the equation (2) :

$$\lambda = \left(\frac{\mathrm{Ln}\frac{C_1}{C_2}}{Z_2 - Z_1}\right)^2 \times K_z \qquad (2)$$

K_z can be approached classically by the equation

$K_z = 0.41 \times U^* \times Z$ where 0.41 is the Von Karman's constant. U^* is related to the wind velocity U, measured at the 10 m level, by the Charnok's equation

$$U = \frac{U^*}{0.41} \;\mathrm{Ln}\; \frac{1000}{Z_o}$$

$$\frac{U^{2*}}{a\ g} \quad \text{where} \quad \begin{array}{l} a = 81 \\ g = 981 \end{array}$$

The wind velocities were measured at the 10 m level on board "Lady Franklin" and are indicated in table 1, simultaneously with the corresponding values of U and K_Z at the same 10 m level.

It can be observed in table 1 that the values of λ, as determined by this method, varied between 1.7×10^{-6} and 3.7×10^{-3} s^{-1}. However the values obtained for wind velocities greater than 5 m/s are between 1.7×10^{-3} s^{-1} and 3.7×10^{-3} s^{-1}, corresponding to lifetimes between 9.8 and 4.5 minutes.

These last figures were obtained in a condition of high turbulence and in presence of seaspray. On the opposite the DMS lifetime seems to be of the order of several hours in a very quiet atmosphere, i.e. in conditions similar to those which characterize laboratory experiments (5, 6).

4. CONCLUSION

It can be concluded from our experiments that the DMS lifetime is very variable, generally from 4 minutes to 16 hours.

DMS oxidation in atmosphere is assumed to be dependent on the OH$^\cdot$ radical during day time and on NO_3 radical during night time. Using reasonable typical OH concentration of 2×10^6 molecules/cm^3, the DMS lifetime should

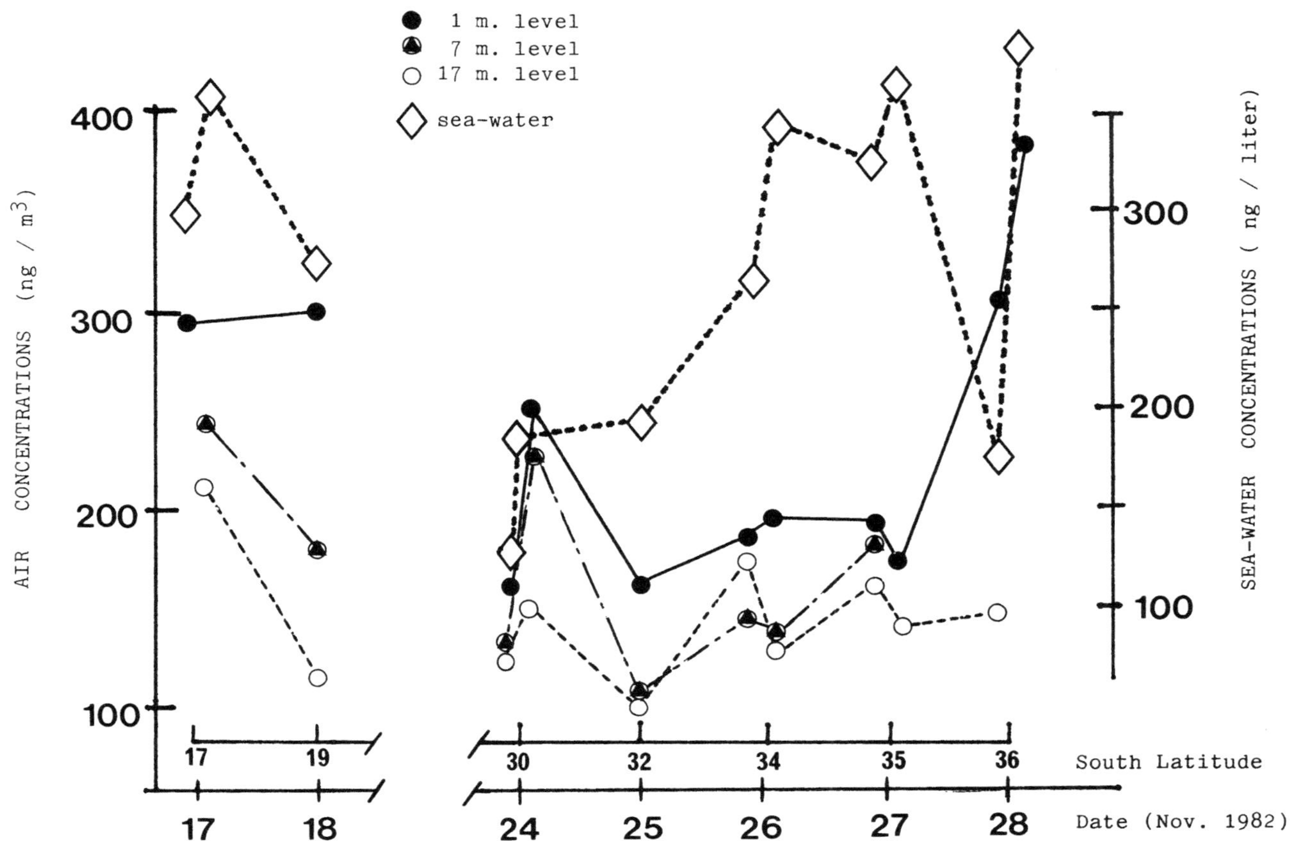

FIG.2 - DMS concentrations in air and in sea-water during "OCEAT 1" cruise between Panama and Hobart (Australia).

Table 1

Experimental results and calculations

Date, Hours	DMS water ng/1	DMS air (ng/1) 1 m	7 m	17 m	U m/s	λ s^{-1}	Lifetime
17 nov 14.00	361		0.245	0.213	2.2	4.7×10^{-5}	5h55'
18 nov 15.00	270	0.299	0.181	0.117	5.1	24.6×10^{-4}	7'
24 nov 9.00	127	0.161	0.118	0.118	3.4	5.2×10^{-4}	32'
24 nov 14.00	174	0.237	0.215	0.150	4.0	3.5×10^{-4}	48'
25 nov 14.00	193	0.162	0.108	0.108	1.0	2.1×10^{-4}	1h19'
26 nov 9.00	264	0.183	0.145	0.172	1.0	4.4×10^{-5}	6h19'
26 nov 14.00	345	0.194	0.140	0.140	1.0	1.4×10^{-4}	1h59'
27 nov 9.00	326	0.195	0.184	0.162	2.5	3.1×10^{-5}	8h58'
27 nov 14.00	363	0.174		0.152	2.5	1.7×10^{-5}	16h20'
29 nov 10.00	178	0.307		0.152	7.1	17.0×10^{-4}	10'
29 nov 14.00	395	0.386	0.276		8.2	37.6×10^{-4}	4'

vary from 15 hours (6) to 33 hours (7) during day time. At night NO_3 typical concentration of 2×10^8 molecules/cm^3 gives a DMS lifetime of 3 hours .

Other possible DMS oxidation mechanisms should exist, such as through Methyl Iodure CH_3I as proposed by Carlier et al. (8, 9). In a field experiment of coastal Britany (France), these authors found a DMS lifetime of a few minutes. Recently Martin et al. (10) found a DMS lifetime of about 0.18 hour by a laboratory experiment which consists of a DMS reaction with $IO°$ radical. These mechanisms seem to be in good agreement with experimental data shown in this paper.

5. ACKNOWLEDGMENTS

This work was supported by the administration of the Centre National de la Recherche Scientifique and Terres Australes et Antarctiques Françaises. The authors are grateful to Expéditions Françaises, to the crew of "Lady Franklin" for helpful assistance.

REFERENCES

(1) LOVELOCK, J.E., MAGGS, R.J. and RASMUSSEN, R.A. (1972). Atmospheric dimethyl sulfide and the natural sulfur cycle. Nature 237, 452-453.

(2) NGUYEN BA CUONG, GAUDRY A., BONSANG B. and LAMBERT G. (1978). Reevaluation of the role of dimethyl sulphide in the sulphur budget. Nature 275, 637-639.

(3) NGUYEN BA CUONG, BONSANG B. and GAUDRY A. (1982). The role of the ocean in the global atmospheric sulfur cyle. J. Geophys. Res., 88, 10 903 - 10 914.

(4) ANDREAE, M.O. (1986). The ocean as a source of atmospheric sulfur compounds. In The Role of Air-Sea Exchange in Geochemical Cycling, Ed. P. Buat-Ménard (D. Reidel Pub. Co.) 331-356.

(5) COX, R.A. (1975). Atmospheric photo-oxidation reactions, the gas phase reaction of OH radicals with some sulphur compounds. AERE-R 8132, 1-10.

(6) Mc LEOD, H., POULET, G. and LEBRAS G. (1983). J. Chem. Phys., 80, 287.

(7) WINE, P.H., KREUTTER, N.M., GUMP, C.A. and RAVISHANKARA, A.R. (1981). J. Phys. Chem., 85, 2660.

(8) CARLIER, P. (1984). Quadriennal Ozone Symposium IAMAP, Kalithea (Greece)

(9) CARLIER, P. and MOUVIER, G. (in press). Atmos. Chem.

(10) MARTIN, D., JOURDAIN, J.L., LAVERDET, G. and LE BRAS, G. (this issue). Kinetic studies of oxidation reactions of sulfur compounds : reaction SO_2 + OH in the presence of O_2 and reaction IO + DMS.

NATURAL AND ANTHROPOGENIC HYDROCARBONS AND HALOCARBONS OVER THE ATLANTIC OCEAN: LATITUDINAL DISTRIBUTIONS OF MIXING RATIOS AND SOURCES

K.J.W. Tille and K. Bächmann
Technische Hochschule Darmstadt
Fachbereich 8: Anorganische Chemie und Kernchemie
Hochschulstraße 4, D-6100 Darmstadt, FRG

Summary

During the first cruise of the German research vessel "DPFVS POLARSTERN" in spring 1983 whole air samples have been collected at Georg-von-Neumayer-Station in Antarctica and between Rio de Janeiro and Bremerhaven over the Atlantic Ocean. The samples were analysed in the laboratory by subambient temperature programmed capillary gas chromatography after cryogenic preconcentration. The separated compounds were detected by an ECD-FID tandem configuration. By this technique hydrocarbons and halocarbons could be analysed in a single run. The resulting concentration distributions of the various compounds show different trends due to different tropospheric lifetimes, sinks, and sources. To decide about natural or anthropogenic origins, we applied a simple one-dimensional four-box model to calculate the latitudinal distributions of source strengths.

1. INTRODUCTION

Nonmethane hydrocarbons (NMHC) and halogenated hydrocarbons (HalHC) play important roles in the unpolluted troposphere. They act as precursors for nitrogen containing oxidants (1, 2), as a source for tropospheric CO formation by reaction with OH (3, 4), as a sink for Cl atoms in the lower stratosphere (5, 6), and as a source for aggressive chlorine containing species such as phosgene and chloro-acetylchloride (7, 8). Furthermore, they are involved in photo chemical cycles leading to ozone production or destruction, depending on the respective NO_x concentrations (9).

Despite these significant impacts on tropospheric chemistry relatively few measurements of NMHC and HalHC outside urban areas are available. Global latitudinal distributions of ethane, methyl chloride, and some other organic volatiles have been reported by Singh et al. (10). Singh and Salas measured latitudinal distributions of selected hydrocarbons over the Pacific, Rudolph and Ehhalt measured C_2-C_5 NMHC over the North Atlantic (11, 12). The latitudinal distribution of methyl chloride has been investigated by Rasmussen et al. (13). Vertical profiles of light hydrocarbons at midlatitudes in the northern hemisphere (13, 14, 15) and in arctic air have been presented, as well as concentrations in background Brazil air (16, 17).

In this paper we will present additional measurements of NMHC and HalHC, carried out at Georg-von-Neumayer-Station in Antarctica and during the first cruise of the German research vessel "DPFVS POLARSTERN" from Rio de Janeiro to Bremerhaven in spring 1983. Calculations of the latitudinal distributions of emission rates, using our concentration data in a simple one-dimensional four-box model, will complete the paper.

2. EXPERIMENTAL

The samples were collected in evacuated (10^{-4} hPa) electropolished stainless steel containers (2 dm^3) with metal bellow valves. They were cleaned by purging with high purity He and by heating them up to 120 °C during evacuation for 12 h. Contamination and sample degradation are avoided by this pretreatment (12).

Gas chromatographic analysis was carried out in the laboratory (cf. Fig. 1). After connection to the inlet system, an aliquot of the air sample was passed from the container through an adsorbent packed precolumn (15 cm

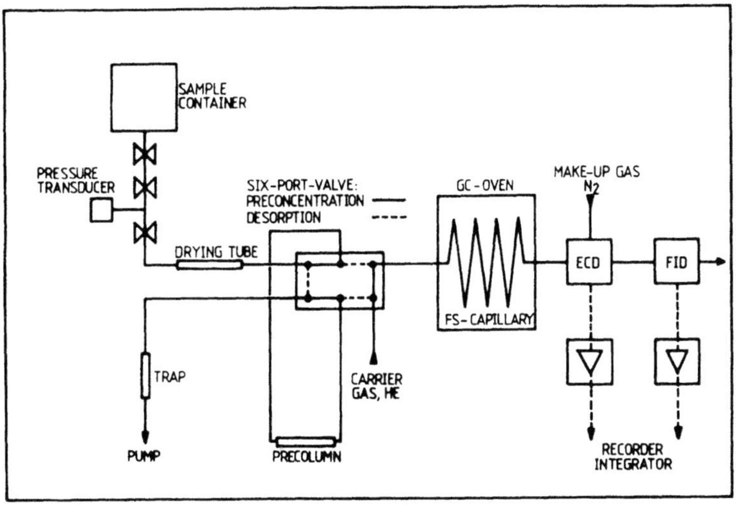

Fig. 1. Gas chromatographic system (schematic).

x 1/8 in. o.d. stainless steel, 10 cm Spherosil XOA 200, in LN_2) by means of a vacuum pump. The pressure in the sample container was measured before and after the withdrawal of the aliquot, respectively. From this pressure difference, container volume, and container temperature, the volume of the aliquot to be analysed was calculated and reduced to its value at standard conditions (STP: 0 °C, 1013.25 hPa). After thermal desorption by direct electrical heating to 250 °C, the compounds were transferred to the separation column (60 m x 0.32 mm, DB-1) by the carrier gas stream (He, 2 mL/min). Water was removed by passing the sample through a drying tube filled with magnesium perchlorate. No change of sample composition could be observed using this drying reagent.

During sample desorption (10 min) the fused silica capillary was cooled to -100 °C by LN_2. A temperature program (max. heating rate to -50 °C, 5 °C/min to 120 °C) was applied to seperate the compounds. They were detected by an ECD and FID in a tandem configuration. Chromatograms were plotted by a strip chart recorder and evaluated by an electronic integrator. Calibration was performed using a static dilution system. Detection limits were between 1 and 100 ppt with reproducibilities ranging from 1 to 50 %, depending on the respective compounds and concentration. These values were determined from subsequent analysis of five aliquots of representative air samples. Blanks were obtaind from containers which had been filled with precleaned high purity He. ECD and FID chromatograms of an Atlantic air sample are shown in Fig. 2.

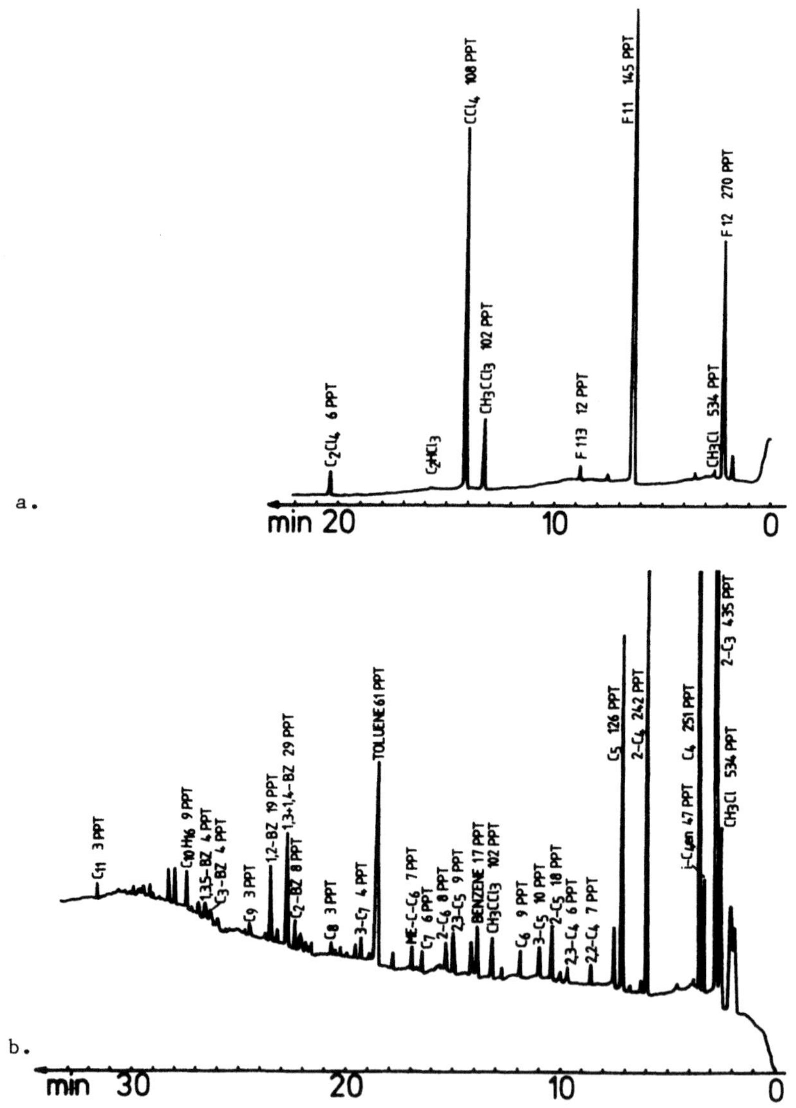

Fig. 2. ECD (a.) and FID (b.) chromatograms of an Atlantic air sample.

3. RESULTS AND DISCUSSION

3.1. Concentration distributions

Samples have been collected at Georg-von-Neumayer-Station in Antarctica (72 °S) and between there and Bremerhaven (53 °N) over the Atlantic Ocean. The route of "DPFVS POLARSTERN" is depicted in Fig. 3. Additional samples from a cruise betwen Capetown and Bremerhaven in spring 1982 have been analysed with respect to light NMHC and are included in this discussion (18).

Latitudinal concentration distributions of NMHC and HalHC exhibit very

Fig. 3. Route of "DPFVS POLARSTER from Antarctica to Bremerhaven in spring 1983.

different trends. Ethane, ethine, propane, the butanes, and tetrachloroethene show distinct north-south gradients, while tetrachloromethane, 1.1.1-trichloroethane and methyl chloride are rather equally distributed. Like the pentanes, benzene, toluene, and many of the minor constituents have their maximum concentrations at northern midlatitudes, as can be seen in Fig. 4. Other compounds, like ethene, propene, trichloromethane and i-butene exhibit high concentrations in the tropical regions.

Measurements of 222-Rn during the cruise (19) show significant influences of continental air between 9 °S and 53 °N, while the air masses at the other sampling sites were of marine origins.

3.2. Distribution of emission rates

Concentrations of volatile compounds in the troposphere do not necessarily reflect the emission rates in a certain region. Compounds such as tetrachloromethane and some of the freons are released predominantly in the industrialised countries at northern midlatitudes, but are distributed equally in the troposphere. This results from their long tropospheric residence times and their rather ineffective stratospheric sink, with respect to corresponding emission rates. So, for long lived compounds, transport processes have to be taken into account besides sink mechanisms to deduce emission rates.

For the calculation of the latitudinal distribution of emission rates from measured concentration distributions we used a simple one-dimensional four-box model (20):

$$\frac{dc_n}{dt} = P_n - a_n c_n - b_n(c_n - c_{nt})$$

$$\frac{dc_{nt}}{dt} = P_{nt} - a_{nt}c_{nt} - b_n(c_{nt} - c_n) - b(c_{nt} - c_{st})$$

$$\frac{dc_{st}}{dt} = P_{st} - a_{st}c_{st} - b_s(c_{st} - c_s) - b(c_{st} - c_{nt})$$

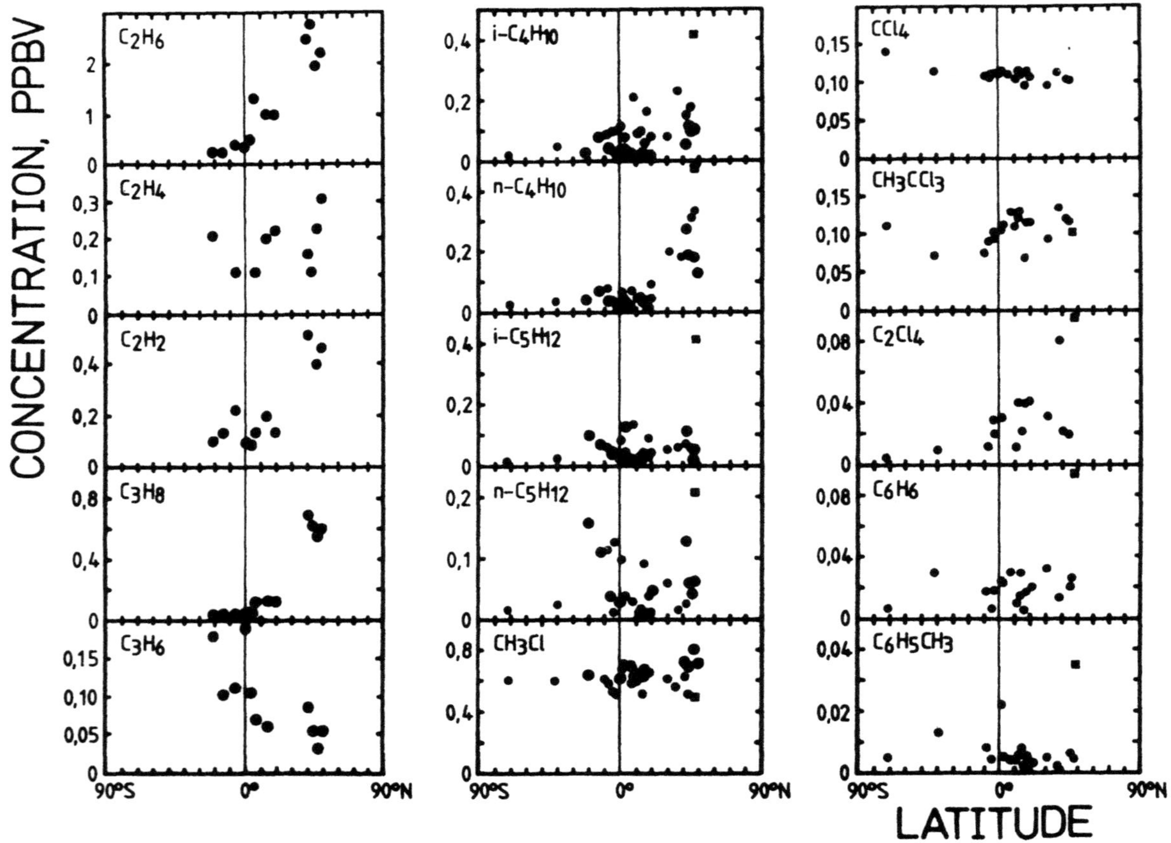

Fig. 4. Concentration distributions of NMHC and HalHC : large points: (18), small points: POLARSTERN cruise, squares: average values from measurements at Wank(FRG).

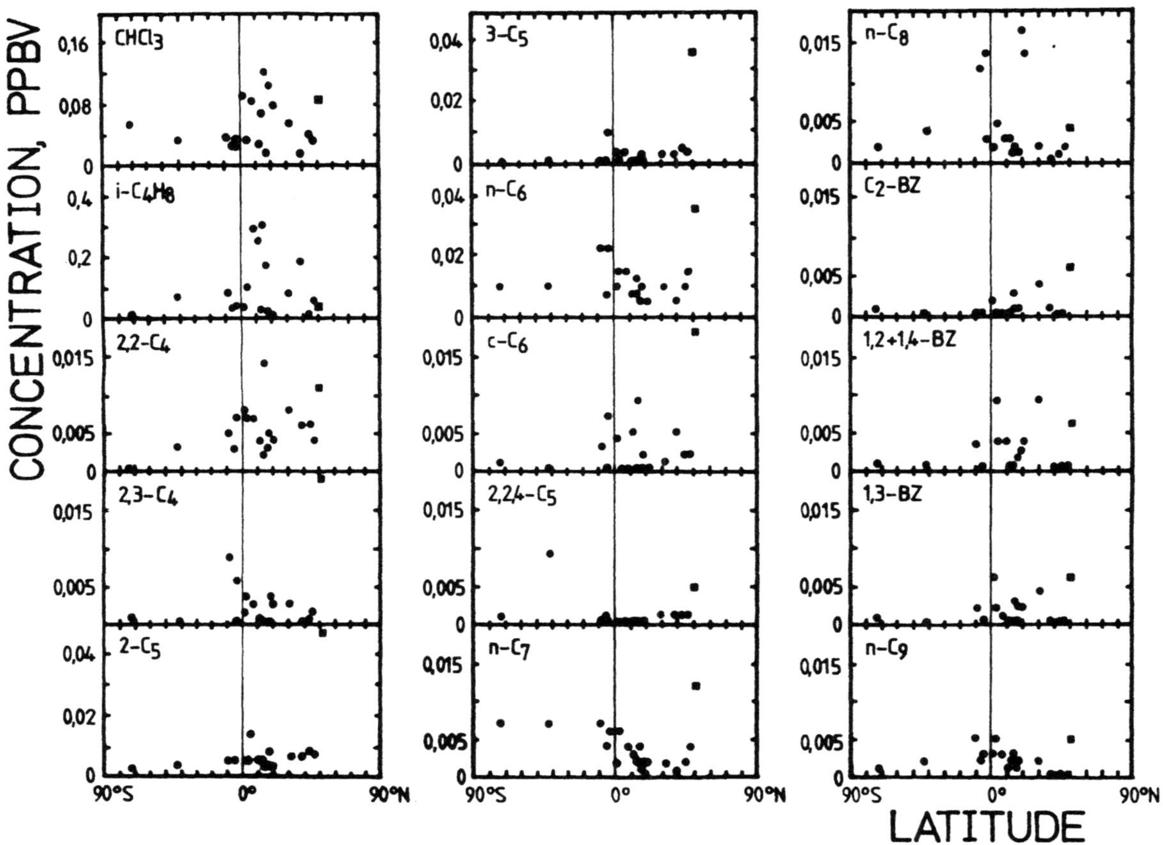
CONCENTRATION, PPBV
CHCl3
i-C4H8
2,2-C4
2,3-C4
2-C5
3-C5
n-C6
c-C6
2,2,4-C5
n-C7
n-C8
C2-BZ
1,2+1,4-BZ
1,3-BZ
n-C9
90°S
0°
90°N
LATITUDE

Fig. 4. cont'd.

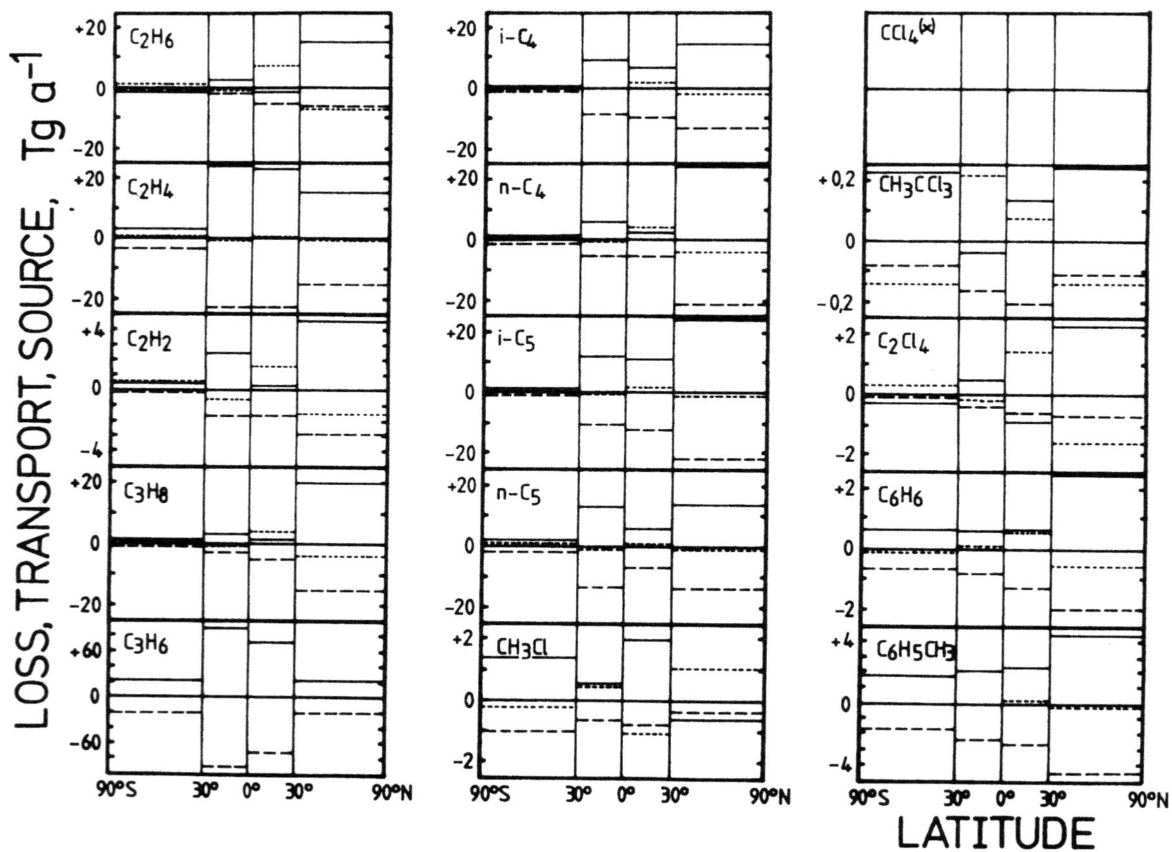

Fig. 5. Loss, transport, and emission rates: broken lines: tropospheric OH-sink, dashed lines: transport, full lines: required source. (x) = no emissions calculated.

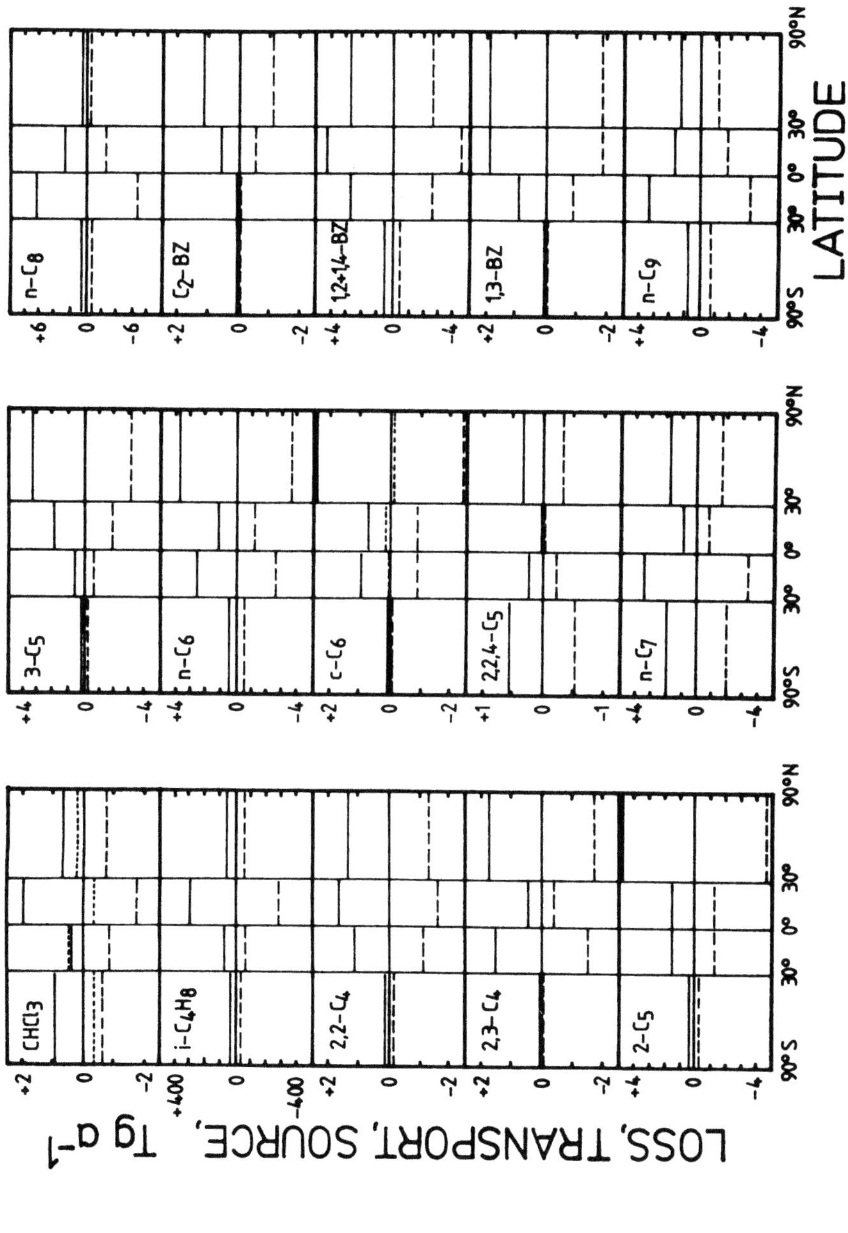

LOSS, TRANSPORT, SOURCE, Tg a^{-1}
LATITUDE
90°S
30°
0°
30°
90°N
CHCl3
i-C4H8
2,2-C4
2,3-C4
2-C5
3-C5
n-C6
c-C6
2,2,4-C5
n-C7
n-C8
C2-BZ
1,2+1,4-BZ
1,3-BZ
n-C9

Fig. 5. cont'd.

$$\frac{dc_s}{dt} = P_s - a_s c_s - b_s(c_s - c_{st})$$

The troposphere is divided into four compartments of equal volumes with the respective concentrations c_n (north), c_{nt} (north tropic), c_{st} (south tropic), and c_s (south). Intrahemispheric transport is characterized by the coefficients $b_i = 1/0.2$ a, and interhemispheric transport across the equator by $b = 1/1.2$ a (21). It is assumed that only ground level sources P_i exist, and that there are no effective sinks other than reactions of NMHC and HalHC with OH radicals. This tropospheric sink is determined by the factors a_i which stand for the inverse lifetimes of the compounds. Lifetimes were calculated from reaction rate constants (21, 22, 23, 24, 25, 26) and average OH concentrations in the individual boxes (9). For steady state conditions, which can be assumed for the compounds under study because of their relatively short lifetimes, the above equations can be solved for the P_i. Assuming a total tropospheric mass of 4.1×10^{21} g (h = 11 km) the P_i are converted to Tg/a.

The results of these mass balance calculations are shown in Fig. 5. For the short lived compounds transport between the boxes can be neglected, and the required source equals the OH sink. Within this class of compounds substances can be found with high emission rates in the tropics such as the unsaturated NMHC ethene, propene, and i-butene, but also methyl chloride, the dimethyl butanes, and the higher n-alkanes. This may indicate significant natural sources. Ethane, ethine, propane, the butanes and pentanes, tetrachloroethene, 1.1.1-trichloroethane, benzene, toluene, and some other minor constituents seem to be emitted predominantly by anthropogenic sources at northern midlatitudes, but natural contributions cannot be ruled out in some cases, e.g. i-pentane (15), 2.2-dimethyl butane, and the xylenes.

It must be stressed at this point, that the model calculations suffer from considerable uncertainties in concentrations, lifetimes, and transport coefficients. They may easily add up to a factor of 2. So the error is of the same magnitude as the one involved in the global methane budget (27), and the calculated emission rates can be considered as gross approximations at least. Total NMHC emission rates derived from the 29 measured compounds are given in Table I. 15 other identified and about 20

Table I. Total NMHC emission rates (Tg a^{-1})

90 - 30 °S	30 - 0 °S	0 - 30 °N	30 - 90 °N	in
83	264	436	263	box
347			700	hemisphere
	1047			troposphere

unidentified compounds in the lower ppt range have not been included in the calculations; so total emissions may be even higher. On the other hand we assumed constant mixing ratios for all compounds within the troposphere, which is certainly not true for short lived substances (15), which show decreasing mixing ratios with altitude. This effect will compensate for part of the contribution from the disregarded NMHC. Therefore, the calculated emission rates must be regarded as good estimations. They are higher than global emission rates derived from emission inventories (27), but compare rather well with values given by Penkett (24). This suggests the interpretation that emission inventories are a reliable basis to evaluate the anthropogenic release of NMHC and HalHC, but may underestimate natural sources. Another independent confirmation of our results can be deduced from model calculations (9) of the global CO budget that demand for a total source of 1500 Tg/a with 1100 Tg/a in the northern hemisphere and 400 Tg/a in the southern hemisphere showing a maximum in the northern tropics. These numbers correspond with the distribution of emission rates in Table I. Together with about 350 Tg/a of isoprene emitted by tree foliage in tropical forests the 1047 Tg/a of volatile NMHC may account for an effective CO source.

4. CONCLUSIONS

Using a very simple model for production, loss and transport mechanisms, respectively, it could be shown that there is evidence for significant contributions from natural sources to the global budgets of volatile NMHC and HalHC. Based on measurements during only one season of the year, these results can be considered as gross approximations only. To confirm the latitudinal distributions of mixing ratios and emission rates, measurements need to be performed over several years. Vertical distributions should be included in these investigations to minimize errors due to specific meteorological conditions and local emissions that are different from the average. Considerable effort is still required to improve our knowledge of the biogeochemical cycles of hydrocarbons.

Acknowledgements: We are grateful to the 'Deutsche Forschungsgemeinschaft' and the 'Alfred-Wegener-Institut für Polarforschung' for financial support. We thank Dr. Mart and Dr. Helas for their assistance.

REFERENCES

(1) Singh H. B. and Hanst P. C. (1981) Peroxyacetylnitrate (PAN) in the unpolluted atmosphere: an important reservoir for nitrogen oxides. Geophys. Res. Lett. 8, 941-944.

(2) Aikin A. C., Herman J. R., Maier E. J. and McQuillan C. J. (1982) Atmospheric chemistry of ethane and ethylene. J. geophys. Res. 87, 3105-3118.

(3) Chameides W. L. and Cicerone R. J. (1978) Effects of nonmethane hydrocarbons in the atmosphere. J. geophys. Res. 83, 947-952.

(4) Hanst P. L., Spence J. W. and Edney E. O. (1980) Carbon monoxide production in photooxidation of organic molecules in the air. Atmospheric Environment 14, 1077-1088.

(5) Rudolph J., Ehhalt D. H. and Tonnissen A. (1981) Vertical profiles of ethane and propane in the stratosphere. J. geophys. Res. 86, 7267-7272

(6) Hanst P. L. (1981) Nonmethane organic compounds in the atmosphere and their role in the scavenging of stratospheric chlorine atoms. Reprint, U.S. Environmental Protection Agency.

(7) Cicerone R. J. (1981) Halogens in the atmosphere. Rev. geophys. Space Phys. 19, 123-139.

(8) Gay B. W., Jr., Hanst P. L., Bufalini J. J. and Noonen R. C. (1979) Atmospheric oxidation of chlorinated ethylenes. Env. Sci. Technol. 10, 58-67.

(9) Crutzen P. J. and Gidel L. T. (1983) A two-dimensional photochemical model of the atmosphere 2: the tropospheric budgets of the anthropogenic chlorocarbons, CO, CH_4, CH_3Cl and the effect of various NO_x sources on tropospheric ozone. J. geophys. Res. 88, 6641-6661.

(10) Singh H. B., Salas L. J., Sigheishi H. and Scribner E. (1979) Atmospheric halocarbons, hydrocarbons and sulfur hexachloride: glo-

bal distributions, sources, and sinks. Science, Wash. 203, 899-903.
(11) Singh H. B. and Salas L. J. (1982) measurement of selected light hydrocarbons over the Pacific Ocean: latitudinal and seasonal variations. Geophys. Res. Lett. 9, 842-845.
(12) Rudolph J. and Ehhalt D. H. (1981) Measurements of C_2-C_5 hydrocarbons over the North Atlantic. J. geophys. Res. 86, 11959-11964.
(13) Rudolph J., Ehhalt D. H., Schmidt U. and Khedim A. (1982) Vertical distributions of some C_2-C_5 hydrocarbons in the nonurban troposphere. In Proceedings of the 2nd Symposium on the Composition of the Nonurban Troposphere, pp. 56-59. American Meteorological Society, Boston, MA.
(14) Cronn D. and Robinson E. (1979) Tropospheric and lower stratospheric vertical profiles of ethane and acetylene. Geophys. Res. Lett. 6, 641-644.
(15) Tille K. J. W., Savelsberg M. and Bächmann K. (1985) Airborne measurements of nonmethane hydrocarbons over Western Europe: vertical distributions, seasonal cycles of mixing ratios and source strengths. Atmospheric Environment 19, 1751-1760.
(16) Rasmussen R. A., Khalil M. A. K. and Fox R. J. (1983) Altitudinal and temporal variations of hydrocarbons and other gaseous tracers of arctic haze. Geophys. Res. Lett. 10, 144-147.
(17) Greenberg J. P. and Zimmerman P. R. (1984) Nonmethane hydrocarbons in remote tropical, continental and marine atmospheres. J. geophys. Res. 89, 4767-4778.
(18) Savelsberg M. (1983) Gas-chromatographische Untersuchung der globalen und regionalen Verteilung von leichtflüchtigen Kohlenwasserstoffen und Chlormethan in der Atmosphäre. Dissertation, Darmstadt.
(19) Metternich P. (1985) Untersuchungen zum Ferntransport partikelgebundener atmosphärischer Spurenstoffe mittels PIXE. Bericht Nr. 59 Institut f. Meteorologie und Geophysik der Universität Frankfurt/Main.
(20) Khalil M. A. K. and Rasmussen R. A. (1983) Sources, sinks, and seasonal cycles of atmopheric methane. J. geophys. Res. 88, 5131-5144.
(21) Wagner H. G. and Zellner R. (1984) Abbau von Kohlenwasserstoffen in der Atmosphäre. Erdöl und Kohle - Erdgas - Petrochemie vereinigt mit Brennstoff-Chemie, 37, 212-219.
(22) Howard C. J. and Evenson K. M. (1976) Rate constants for the reaction of OH with CH_4 and fluorine, chlorine and bromine substituted methanes at 296 K. J. Chem. Phys. 64, 197-202.
(23) Becker K. H., Biehl H. M., Bruckmann P., Fink E. H., Führ F., Klöpffer W., Zellner R. and Zetzsch C. (Editors) (1984) Methods of the ecotoxicological evaluation of chemicals: Photochemical degradation in the gas phase. Kernforschungsanlage Jülich GmbH, Jülich.
(24) Penkett S. A. (1982) Non-methane organics in the remote troposphere. In Atmospheric Chemistry (edited by Goldberg E. D.), pp. 329-355. Dahlem Konferenzen 1982. Springer, Berlin.
(25) Atkinson R. A., Darnall K. R., Lloyd A.C., Winer A. M. and Pitts J. N., Jr. (1979) Kinetics and mechanisms of the reactions of the hydroxyl radical with organic compounds in the gas phase. Adv. Photochem. 11, 375-488.
(26) Atkinson R., Carter W. P. L., Aschmann S. M., Winer A. M. and Pitts J. N., Jr. (1984) Kinetics of the reaction of OH radicals with a series of branched alkanes at 297+/-2 K. Int. J. Chem. Kinet. 16, 469-481.
(27) Ehhalt D. H. and Rudolph J. (1984) On the importance of light hydrocarbons in multiphase atmospheric systems. Kernforschungsanlage Jülich GmbH, Jülich.

CHARACTERIZATION OF CLOUD AND RAIN WATER CHEMICAL COMPOSITION

H.F.R. REIJNDERS, J. VAN ZIJL
National Institute of Public Health and Environmental Hygiene, Laboratory for Environmental Chemistry
P.O. Box 1, 3720 BA Bilthoven, The Netherlands
W.A.J. VAN PUL
Agricultural University,
Department of Physics and Meteorology
Duivendaal 1 and 2, 6701 AP Wageningen, The Netherlands
F.G. RØMER and H.J. SLANGEWAL
N.V. KEMA, Environmental Research Department
P.O. Box 9035, 6800 ET Arnhem, The Netherlands

Summary

In the period 1981 to 1985 flights were carried out above The Netherlands and the North Sea to investigate the chemical composition of clouds.

Based on the complete data set (primary gases, secondary aerosols, cloud and rainwater composition and meteorology) the occurrence of the most interesting components e.g. sulphate, nitrate, ammonium, hydronium and hydrogen peroxide is reviewed. Simple statistics are applied and relations between several components established.

Changes in chemical composition of cloud and rainwater due to transport over sea, land and industrial areas are studied. Meteorological conditions are also considered.

1. INTRODUCTION

In the period from 1981 up to 1985, 45 flights were carried out in order to make an inventory of the chemical composition of cloud and rainwater from precipitating clouds. The results were introduced in a computer, and a data base was structured. To facilitate the making of an ion balance for the samples it was attempted to determine at least the main components. As the sample volumes were often insufficient to allow a complete analysis, the determination of compounds which play a more important role in the acidifcation problem were given priority, i.e. hydronium, ammonium, sulphate. Moreover, hydrogen peroxide was measured because it is assumed to play an important role in the oxidation of dissolved sulphur dioxide. Other substances such as nitrite, sulphite, iron and manganese were measured occasionally but the data are not reported here. During most flights, measurements of primary gases (nitrogen oxides and sulphur dioxide) and ozone were performed at the same time. Measurements of secondary components (nitric acid, nitrate and sulphate) were only made in a few flights around industrial areas. For each flight the synoptic meteorological data were gathered and interpreted. Roughly the experiments were grouped according to the area above which the flights were carried out (above the North Sea or land). About 25 % of the flights were carried out upwind and downwind of industrial areas. The data were analysed in terms of frequency distributions for the concentrations of the several components in cloud water, rainwater as well as gases. In this paper only a part of the available data is presented.

2. INSTRUMENTATION AND CHEMICAL ANALYSIS

Aircraft equipment

The aircraft used were a Piper Navajo Chieftain and a Cessna 172 flown by Geosens B.V., Rotterdam. The Piper Navajo Chieftain was equipped with chemiluminescent monitors for NO, NO2, NOx (Teco 14D) and O3 (Bendix 8002) and a pulsed fluorescent SO2 monitor (Teco 43). Altitude, speed and temperature were measured continuously. The output signals of the instruments were recorded on a data logger (Monitor Labs 9400) and a cartridge recorder (Perex 6300). A full description of the aircraft and instruments is given elsewhere (1,2). The Cessna 172 was only operated for the collection of cloud or rainwater samples.

The sampling devices for cloud and rainwater were provided by Atmospheric Sciences Research Center, State University, New York (Albany). The samplers are passive collectors based upon the aero dynamic collection characteristics of one cylindrical rod for the collection of rainwater and seven cylindrical rods for the collection of cloud water (stagnation slot method) (2,3).

Aerosols were collected on quartz filters (Pallflex 2500 QAST) for the analysis of ammonium, chloride, nitrate and sulphate. The second filter in line was a nylon filter (Gelman) to collect nitric acid.

Chemical analysis

Before chemical analysis sample weights were determined. Hydronium ions were measured in a flow injection system with bromocresol green (4). Ammonium was analysed in a continuous flow analyser by the sodium salicylate method (5). Sulphate, nitrate and chloride were determined with an ion chromatograph (Dionex) (6). The method to determine hydrogen peroxide is based on the copper-catalysed chemiluminescent reaction with luminol (7). The determination was carried out in a flow injection system. In this system, before reaction with luminol, heavy metals are removed from the sample by a cation exchange resin. The pH was measured with a pH meter without any addition of sodium chloride solution. Hence, transition potentials are not taken into account.

Absolute volumes needed for the separate analysis are in the range of 50-400 uL. To run the analysis automatically, a total volume of 15 mL is needed for the single determination of anions, ammonium, hydronium ions and hydrogen peroxide.

Sodium and potassium were determined using atomic absorption spectrometry following the microsampling cup method (8). The measurement of calcium and magnesium was carried out with inductively coupled plasma atomic emission spectrometry. Graphite furnace atomic absorption spectrometry was used to determine iron and manganese.

3. RESULTS

Flights were carried out mainly from May until September and in November and December. The flight areas were situated above the Netherlands and a part of the adjacent North Sea. In figure 1 these areas are indicated by the codes, which are explained below. For all flights a more rough-cut Gross-Wetter-Lage (GWL) classification was used to group the weather pattern during these flights into twelve circulation types (9). The results are presented in table 1. From the twelve circulation types only eight occurred. Further meteorological characterization of the flights was done by classifying the air masses and cloud types. The results are given in tables 2 and 3.

Figure 1:
Map with flight areas

Table 1:
Grouping of Gross-Wetter-Lage classifications and relative occurrences of flights in these groups

	HNa, HNz, HB, Nz, Na	SEa, SEz	Sa, Sz, TB, TrW	SWa, SWz	Wa, Wz, Ws	NWa, NWz, TrM	Ww	TM	Ü
	N-circulation	SE	S	SW	W	NW	Ww	TM	
occurrence (%)	12	2	12	12	40	16	2	2	2

Table 2:
Classification of air masses

	N	%
Pm	31	72
Tm	9	21
Am	3	7

N: number

In general the atmospheric conditions in winter precluded flights, because of frequent occurrence of low icing levels which would cause technical problems.

Table 3:
Occurrence of cloud type

cloud type code	2	5	7	8	9
cloud type	Cu	Sc	St-fract., Cu-fract.	Cu, Sc	Cb
occurrences (%)	3	38	25	28	7

The flights were classified in those above the North Sea (NS) and those above land (L). 25 % of the flights above land were carried out around the industrial area of Rotterdam and are classified into flights upwind (U) and downwind (D) of this area. During some flights, tracks were flown above land (tL) as well as above the North Sea (tNS). The number of cloud water samples, grouped to the wind direction, which prevailed at the time of sample collection is given in figure 2.

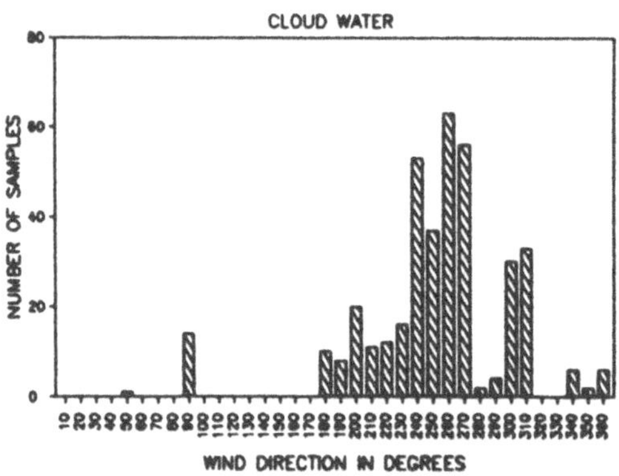

Figure 2:
Number of samples per wind sector of ten degrees

Frequency distributions were made for a first characterization of the data. The main part of the data was found in the low concentration range, so these distributions are skewed. Exceptions to this are the weight of sample and the pH. As an example the symmetrical distribution of the pH and the skewded distribution of ammonium in cloud water collected in air masses upwind of source areas (NS + tNS + U) are presented in figures 3a and 3b. In figures 3c and 3d the corresponding distributions are given of cloud water samples collected above land and partly downwind of source areas of sulphur dioxide, nitrogen oxides and ammonia (L + tL + D).

From the frequency distributions percentiles were calculated for all components. As the median is the most accurate value to characterize a distribution (10), as an example the median values for weight, pH, ammonium, sodium, chloride, nitrate and sulphate content for cloud and rainwater and the region above which the flights were carried out are given in table 4. In table 5 the median values of the components in cloud and rainwater as given in table 4 for the flight categories U and D are presented. The median values of mean gas concentrations for all flight categories are presented in table 6.

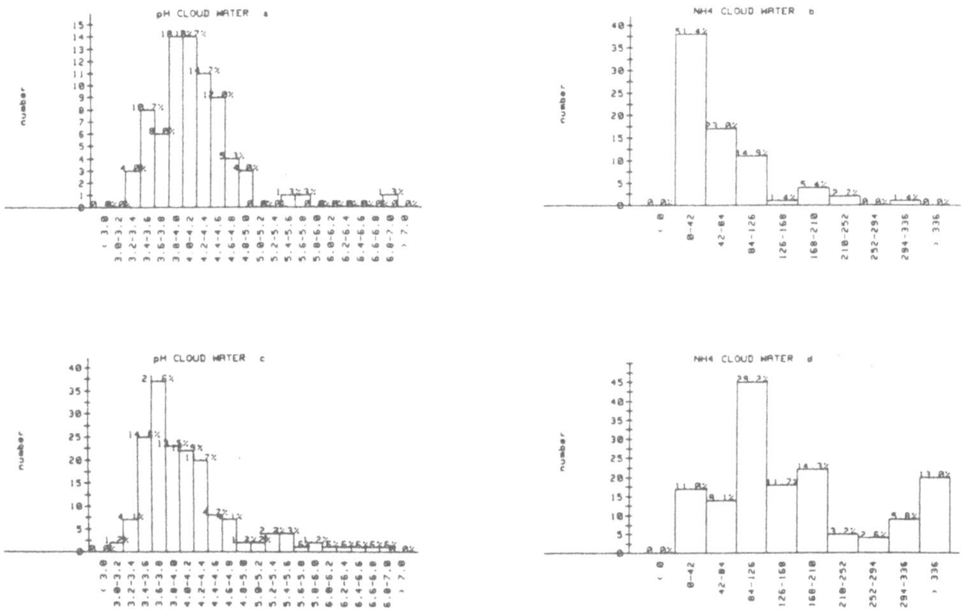

Figure 3:
Frequency distribution of pH (a) and ammonium (b) in cloud water collected in air masses upwind source areas (NS + tNS + U) and the pH (c) and ammonium (d) in cloud water collected in air masses above land (L + tL + D)

Table 4:
Median values for concentrations (umol/L), pH and weight (g) for cloud and rainwater collected above several regions

	pH				ammonium				sodium			
	c		r		c		r		c		r	
NS + tNS + U (99c, 72r)	4.05	(75)	4.40	(62)	41.5	(74)	17.5	(60)	25.1	(61)	8.92	(60)
L + tL + D (231c, 92r)	3.90	(170)	4.10	(62)	127	(154)	74.3	(77)	38.3	(131)	13.9	(65)
	chloride				nitrate				sulphate			
	c		r		c		r		c		r	
NS + tNS + U (99c, 72r)	36.5	(80)	18.3	(63)	36.9	(80)	21.1	(63)	47.2	(75)	19.0	(63)
L + tL + D (231c, 92r)	56.4	(196)	43.6	(82)	80.3	(193)	51.0	(82)	88.6	(193)	45.3	(82)
	weight				hydogen peroxide							
	c		r		c		r					
NS + tNS + U (99c, 72r)	10.8	(73)	17.4	(44)	6.06	(86)	6.03	(67)				
L + tL + D (231c, 92r)	15.0	(144)	16.8	(47)	3.97	(194)	3.64	(66)				

c: cloud water
r: rain water
(): number of results available

Table 5:
Median values for concentrations (umol/L), pH and weight (g) for cloud and rainwater collected above several regions

	pH				ammonium				sodium			
	c		r		c		r		c		r	
U (47c, 17r)	4.03	(38)	4.18	(11)	38.8	(38)	18.3	(15)	42.6	(35)	5.87	(12)
D (82c, 23r)	3.79	(74)	3.88	(16)	116	(75)	85.8	(20)	116	(75)	14.0	(19)

	chloride				nitrate				sulphate			
	c		r		c		r		c		r	
U (47c, 17r)	47.8	(42)	20.9	(15)	39.7	(42)	21.1	(15)	55.0	(42)	21.0	(15)
D (82c, 23r)	42.5	(78)	31.9	(21)	77.5	(78)	49.5	(21)	79.5	(78)	55.9	(21)

	weight				hydogen peroxide			
	c		r		c		r	
U (47c, 17r)	15.7	(40)	17.0	(8)	6.08	(37)	1.18	(13)
D (82c, 23r)	16.8	(75)	19.3	(15)	11.0	(71)	1.12	(12)

c: cloud water
r: rain water
(): number of results available

Table 6:
Median values for concentrations (umol/m3) of sulphur dioxide, nitrogen oxides and ozone

	sulphur dioxide	nitrogen oxides	ozone
NS + tNS + U (81)	0.067 (81)	0.36 (76)	1.96 (77)
L + tL + D (115)	0.18 (115)	0.80 (67)	1.38 (101)
U (24)	0.13 (24)	0.60 (22)	1.72 (23)
D (42)	0.18 (42)	0.89 (32)	1.72 (32)

(): number of results available

Differences in concentrations and masses for hydronium, ammonium, sulphate and nitrate were calculated for cloud and rainwater samples which were collected during the same time interval. These differences were tested for systematic differences with the sign test. The results of these calculations are given in table 7. In figure 4 the difference is plotted against the concentration level to show that the difference is systematic for sulphate.

To investigate the influence of emissions of a specific industrial area (mainly the Rijnmond area) on the cloud water composition, flights were carried out upwind (U) and downwind (D). In general one upwind and two downwind tracks were made. From nine of these flights per track the concentration (umol/L) ratio of sulphate/nitrate for cloud water was calculated. The results are given in table 8.

A relative increase of the concentration of sulphate and nitrate in cloud water between tracks was also established. Moreover the ratio of the relative increases in sulphate and nitrate between tracks and the sulphate to nitrate ratio of the tracks were calculated. The results of these calculations are given in table 9.

Table 7:
Results of the sign test applied to establish the significance of the difference between the concentrations in cloud and rainwater (Cc -Cr)

	N	n+	n-	n0	d	n	T5.0		T10.0	
hydronium	29	17	8	4	9	25	11	ns	11	ns
ammonium	29	20	3	6	17	23	11	s	9	s
sulphate	29	24	5	0	19	29	13	s	11	s
nitrate	29	11	14	4	-3	25	11	ns	11	ns

N: number of results, n+: number of positive differences, n-: number of negative differences, n0: number of zero differences, d: n+ - n-, n: N - n0, T5.0: critical value of two sided t-test at 5 % level, T10.0: critical value of two side t-test at 10 % level, ns: not significant, s: significant

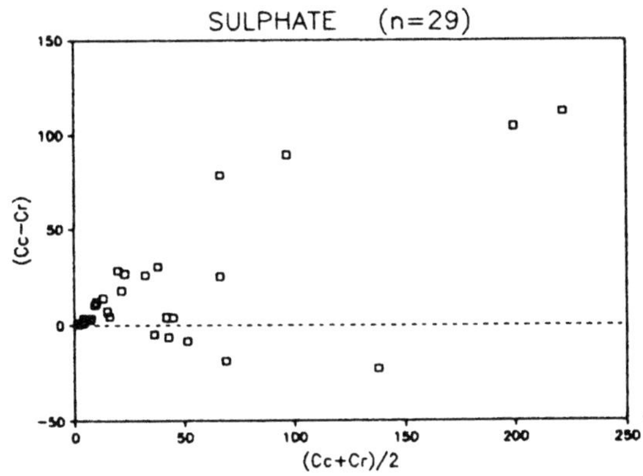

Figure 4:
Differences between the concentration (a) and mass (b) of sulphate in cloud and rainwater determined for samples collected in the time interval in relation to mean concentration and mass level

Table 8:
Molar ratio for the different tracks of the U (track 1 ~ t1) and D (t2 and t3) categories. Number of flights: nine

	t1	t2	t3
sulphate/nitrate molar ratio	1.28	1.42	1.63

Table 9:
Relative increases in sulphate and nitrate concentration between succesive tracks of the U (track 1 ~ t1) and D (t2 and t3) categories, and the ratio between these increases. Number of flights: nine

	t2/t1	t3/t1	t3/t2
nitrate	1.55	1.76	1.38
sulphate	3.29	4.88	1.62
sulphate/nitrate	2.12	2.77	1.17

4. DISCUSSION

From the rough GWL classification it follows that the major part (80 %) of the flights were carried out during north-westerly, westerly, south-westerly circulations. With these circulations mainly maritime polar air masses were advected (table 2). Nearly all flights were carried out during a frontal or trough passage accompanied by rain. The dominating lower cloud types in which flights were carried out were Stratocumulus and Cumulus.

Although a rather large number of flights were carried out during the programme , for several substances measured only a limited number of data are available to make frequency distributions (tables 4 and 5). Because of the small sample volumes it was often impossible to analyse all components of interest. Therefore for one type of flight and for almost all substances different numbers of data are available for the construction of frequency distributions. As a consequence medians have to be compared which were obtained from distributions that contain different numbers of data points.

The results in table 4 show that the concentrations in cloud and rainwater above sea are lower than the concentrations above land. This is not caused by evaporation because the ratio of the median weights of cloud and rainwater collected above land and sea is lower than the ratio of the concentrations measured above land and sea. From table 5 it can be deduced that in general the concentrations of the substances in cloud and rainwater collected downwind (D) industrial areas are higher than in cloud and rainwater collected upwind (U) industrial areas. However, for chloride and hydrogen peroxide in cloud water the opposite was found. Further study in this respect is necessary. In general the results in tables 4 and 5 show that the concentrations in cloud water are higher than in rainwater. This is not caused by evaporation, because the ratio of the concentrations in cloud and rainwater (Cc/Cr) is higher than the ratio between the volume of collected rain and cloud water samples (Vr/Vc). This may be expected because clouds are formed from condensation of water on particles and the scavenging efficiency of cloud drops is higher than of rain drops. On average the concentrations in cloud water are twice as high as in rainwater. This was investigated more accurately by sampling cloud and rainwater at the same time. Figure 4 shows clearly that the differences between the concentrations in cloud and rainwater depend on the concentration level. For sulphate and ammonium systematic differences are found when using the sign test (table 7). The sign test was chosen because it is a non-parametric test. However, with the sign test only great differences can be established. Therefore in further work the distribution of differences will first be tested for normality, and than the Student t-test will be applied.

For the regions mentioned in table 4 and 5 the medians of the concentrations of sulphur dioxide, nitrogen oxides and ozone were calculated (table 6). Roughly the differences in gas concentrations above sea and land and upwind and downwind industrial areas run parallel to the differences found for sulphate and nitrate in cloud and rainwater above the same areas.

From table 8 it is clear that in cloud water sulphate as well as nitrate concentrations increase with the distance from the source area. From both tables 8 and 9 it follows that, depending on the distance, the sulphate concentration becomes more important in relation to the nitrate concentration. It was verified that this increase is not caused by evaporation of cloud water by comparing the volume of the cloud water sample in relation to the sample period over the several U and D tracks. For the composition of table 8 and 9 a small set of data was used. Therefore the data should be considered as indicative.

Secondary components (sulphates, nitrate and nitric acid) were measured occasionally. No data are given here but these results wille be discussed in an other paper.

5. CONCLUSION

From the work the following conclusions can be drawn so far. The concentrations of the substances as given in table 1 are higher in cloud than in rainwater. Moreover, differences in concentrations are found between samples collected above the North Sea and also between land and samples collected upwind and downwind industrial areas. From the measurements upwind and downwind of industrial areas the increase in the sulphate and nitrate concentration is clearly demonstrated. The increase in sulphate concentration is larger than that of nitrate. This finding leads to the expectation that collecting data from cloud and rainwater analyses may provide a suitable basis for the calculation of mass fluxes.

ACKNOWLEDGEMENT

Mr. H. van Dop of the Royal Dutch Meteorological Institute provided us with several highly valuable meteorological data. We should like to thank him for his generous help. Mr. A.J. van den Eshof and Mrs. E. Smit are kindly acknowledged for their computer work. Thanks are due to Mr. G.T. Kempen (former co-worker of Geosens B.V.) for his meteorological assistance.

REFERENCES

(1) VILJEER, J.W. and DUUREN, H. van (1977). The Measurement by Aeroplane of Atmospheric Pollution, in Particular in Smoke Plumes (in Dutch). Elektrotechniek Vol. 55 540-547.

(2) RØMER, F.G. et al. (1979). Messungen aus Flugzeugen: Verbreitung und Umwandlung primärer luftfremder Komponenten in Rauchfahnen. VGB-Konferenz Kraftwerk und Umwelt 1979, Essen (FRG), 134-140.

(3) FALCONER, P.D. (1981). Cloud Chemistry and Meteorological Research at Whiteface Mountain: Summer 1980. A.S.R.C. Publ. 806, Atmos. Sci. Res. Center, State University of New York, Albany, N.Y..

(4) MOHNEN, V.A. 1980. Cloudwater Collection from Aircraft. Atmos. Technol. Vol. 12 20-25 (published by NCAR, Boulder, Colorado).

(5) SLANINA, J. (1982). Personal communication.

(6) TECHNICON INDUSTRIAL SYSTEMS (1977). Individual/Simultaneuos Determination of Nitrogen and/or Phosphorus in BD acid digests. Ind. Methods 229-74 W/b, Tarry Town, New York.

(6) SMALL, H., STEVENS, T.S. and BAUWMAN, W.C. (1975). Novel Ion Exchange Chromatographic Method using Conductometric Detection. Anal. Chem. Vol. 47 1801-1809.

(7) KOK, G.L. et al. (1980). Chemiluminescent Method for Determination of Hydrogen Peroxide in Ambient Atmosphere. Environm. Sci. Technol. Vol. 12 1072-1076.

(8) BERNDT, H. and JACKWERTH, B.H. (1979). Bestimmung von Li, Na, K, Mg und Ca mit einer mechanisierten Mikromethode der Flammenspektrometrie. J. Clin. Chem. Biochem. Vol. 17 71-76

(9) TONKELAAR, J.F. den (1985). Personal Communication.

(10) EEHRENS, H.C. and WIEL, H.J. van de (1986). Impact of the Measurement Deviation on Summary Statistics. To be published in the Proceedings of the Workshop on Data Quality, Freiburg.

DEPOSITION MEASUREMENTS IN SWEDEN

L. Granat
Department of Meteorology, University of Stockholm

Summary

In the National Swedish Air and Precipitation Chemistry Network some 90 individual collectors are presently in operation at 44 sites. Both wet-only and bulk collectors are used, sampling period is one month and the samples are analysed for nine major components. Some support for the sampling strategy is given. All major components of anthropogenic origin have similar concentration and deposition fields. High concentration values are found in the south and east of Sweden, low ones in the north-west near the Atlantic coast. High deposition values are found near the south-western coast where precipitation amount has a local maximum. Seasonal variation is similar for sulfate, ammonium, hydrogen ion and calcium with high values in late winter and early spring (February through May).

1. THE NETWORK AND ITS OPERATION

The sampling sites in the National Swedish Air and Precipitation Chemistry Network are shonw in Figure 1. Sampling period is one month per sample with continuous operation, both bulk and wet-only collectors are used and the samples are analysed for SO_4^{2-}, Cl^-, NO_3^-, NH_4^+, H^+, Na^+, K^+, Mg^{2+} och Ca^{2+}. The motives and the historical background for this programme are as follows.

Sampling sites

The National Swedish Air and Precipitation Chemistry Network is closely related to the National Swedish Environmental Programme (PMK in Swedish terminology). The main aim of this programme is (1)

- # to monitor long-term and large-scale changes in the environment
- # to collect data about environmental conditions in relatively unaffected tracts
- # to illuminate how pollutants are tranported.

To satisfy the need for this programme 14 new precipitation chemistry sites were established in or near the PMK study areas, most of them at the end of 1982. Eight new sites are in temporary operation (during 1986) to study the spatial trend over and adjacent to a large water body - "Kvarken" in the Gulf of Bothnia. At ten sites previous operation was continued in order to maintain historical records at sites which are considered adequate with regard to our present knowledge of sampling site representativeness. Most of these sites were put in operation in 1972, two started in 1965 (RK and Sn) and one (Fo) has been in operation at the same site since 1955. Thus, essentially all of the sites used in the 1950's and 1960's have been abandoned for various reasons - usually due to proven low spatial representativeness for one or several components.

In other parallel programmes trace metals are analysed from six of the above mentioned stations. There are also six sampling sites operating under the EMEP (European Monitoring and Evaluation Programme) scheme with daily sampling. Two of them are presently co-located with the national network (Br and Sn).

In this paper we will deal primarily with data from the new expanded network which has now been in operation for three full years (1983 through 1985).

The permanent network (not counting the nine temporary sites across the Gulf of Bothnia) covers all of Sweden with a fairly even distance between the sites, perhaps with exception for the mountainous area. The latter is above the timber line and the sampling conditions are too difficult to be handled with presently available resources. The network thus serves the dual purpose of giving an accurate information - without interpolation - about conditions right at the PMK areas for effect studies and watershed budget calculations and an estimate of the concentration/deposition fields for the measured components over Sweden.

Collection period

Selection of an optimum collection period with regard the purpose of the network and practical matters is essentially a choice between a long (i.e. month) or short (i.e. 24 h) sampling period or perhaps something in between (i.e. one week). Within given financial resources long period sampling will of course give room for many more sites - five to ten times more than short period sampling depending on sampling strategy. The following more principal arguments could be rised for either long or short sampling periods (below referred to as lp and sp, respectively).

1. Better spatial resolution with lp and less uncertainty in estimating concentration/deposition at a particular point - e.g. a PMK area.

2. Spatial trends can be revealed with lp which would otherwise not be found. The findings may have ecological impact and/or be of significance in atmospheric chemistry studies (cf. next chapter).

3. Determination of temporal trends will be more effective since some of the spatial variability in year to year variation is reduced with the larger number of sampling sites available with lp (compared to sp).

4. sp has the potential of showing spatial and/or temporal variation in the frequency distribution of concentration (in precipitation) or permit calculation of amount of precipitation with concentration above any given threshold level (with regard to ecological or other effects). Whether this is an important parameter or not still seems to be an open question.

5. sp has been much used for evaluation of transport and deposition models (i.e. within the EMEP project). It may be that models constructed to relate emission and deposition on a long-term basis also could be effectively evaluated with lp sampling for instance with regard to the seasonal and year to year variation and with regard to details in spatial trends.

6. Of some importance when selecting sampling period for the extended network was the tradition of precipitation chemistry measurements for almost three decades with monthly sampling periods.

In our opinion lp sampling was (and still is) well motivated with regard to the purpose of the network. However, the sampling period cannot be longer than that the integrity of the samples is maintained.

There is much evidence that precipitation collected at sites in Sweden will keep essentially unchanged for quite a long period of time. We shall here give one example. Some 70 rainwater samples were collected consequetively with 24 h sampling periods (during little more than one year). Each sample was split and aliqots were kept deepfrozen, in the cold (4-10°) and at room temperature with indoor illumination. Samples were analysed after a period ranging from 3 to 13 months. There was no significant difference between samples kept frozen and cool. Of the major components NH_4 seems to be the most susceptible for changes when stored at room temperature. NH_4 was on the average 10% lower after this long storing period at room temperature. It must also be mentioned that this included complete loss of NH_4 in 11 of the samples.

In view of the importance of this matter and the concern expressed in other countries about sample integrity during storage a new study is presently carried out within the network.

Sampling equipment

Two sampling systems are used, bulk and wet-only collectors. Bulk collectors for use in summer are made of a 200 mm diameter funnel, with a small net and a restriction to prevent evaporation, connected to a 5 l bottle, all in polyethylene. The bottle and part of the funnel is wrapped in aluminum foil (to keep collected water dark and prevent growth of algae - very important). The opening is at breast height. In winter a collector is used which can accomodate a full month of precipitation as snow and prevent evaporation from melted snow. A 1.3 m long polyethylene bag is fixed to a 200 mm diameter plastic ring mounted at about 2 m above the ground. A restriction near the bottom of the bag acts as an evaporation shield.

The wet-only collector is cylindrical in shape with a lid that swings away to the side during precipitation. Collected snow is slowly melting by thermostated heating. These collectors are designed and made at the Department of Meteorology, University of Stockholm.

The relative merits of bulk and wet-only collectors have been extensively discussed. Table 1 highlights this question with regard to sampling conditions in Sweden. During 1985 wet-only collectors were run in parallel with two bulk collectors at five sites from southern Sweden (Ap) to the northern inland (Br). The result further confirms previous findings that bulk and wet-only collectors give comparable results for SO_4, NO_3, NH_4 and H^+ indicating a low dry deposition flux (to the funnel!). Bulk samples have a lower acidity probably due to some dry deposition (insects, soil dust etc.). The difference (5%) may seem high and be of concern but one should also remember that the difference is only 0.02 pH units - close to the accuracy of the analytical method alone. It can also be noted that calcium has only a 10% higher concentration (0.7 μeq/l) in bulk collectors on the average. Larger dry deposition is inferred for seasalt components and for potassium.

With most of the PMK sites far away from access to electricity and with knowledge of the reasonably good agreement between wet-only and bulk collectors it was decided to make up the main part of the network with bulk collectors but with wet-only collectors continuously run in parallel to bulk collectors at several sites to provide a continuous record of the comparability. Bulk collectors are used in pairs with a distance of typically a few hundred metres. If possible a different surrounding is selected for each of the two collectors (for instance a bog versus a small clearing in a forest).

2. RESULTS

Data treatment

Data were inspected for gross contamination. For the period 1983-1985 a comparison of pairs of collectors and with data from surrounding sites revealed gross contamination quite easily. Such data were discarded - either for one or two elements or for all elements in a sample. About 5% of the data was removed in this way. During the period 1972-1982, when there was only one bulk collector at each site, the identification of contaminated samples could only be made by comparison to neighbouring quite distant sampling sites and in view of spatial variability only quite severe contamination could be detected. Data from the period before 1972 has been dealt with in other connections (2) and will not be discussed here.

Spatial distribution of concentration/deposition

In Figure 2 data from sampling sites are given for excess sulfate and for nitrate as volume weighted yearly mean concentration (excess sulfate: the seasalt sulfate has been removed from total sulfate). A concentration field is evaluated based on an objective analysis with subjective adjustments especially in coastal areas and at the border towards Norway and Finland to fit measurements in these countries.

During this evaluation spatial trends in rainfall amount has not been considered. The reason is that we have some indication from previous measurements in Sweden that the spatial variation in precipitation over distances of 20-200 km apparently does not affect the concentration field as far as we can judge.

Available records from very extensive precipitation measurements made by the Swedish Meteorological and Hydrological Institute (SMHI) show that the precipitation (amount) field has much larger small scale variability than the concentration field for major components (except for seasalt components). If there is only a week (spatial) relation between precipitation amount and concentration we can calculate the deposition field as the product of concentration (from the precipitation chemistry network) and precipitation amount (from the dense meteorological network), see Figure 3. For these calculations standard values of precipitation - from the period 1931-1960 - were used. The concentration field is from 1983-1985. Thus, Figure 3 can be considered as showing the contemporary wet deposition of respective components.

Seasonal trend

Data from four sites (Fo, Rk, Gn and Sn) which have comparable concentrations are used to show seasonal trends of major components. For each site and component a mean value was calculated for each month of the year during the period 1972-1984 and divided by the mean of all months. Finally the median value of the four sites for each month was calculated and given in Figure 4 for concentration and in Figure 5 for deposition. SO_4, NH_4, H^+ and Ca have about the same seasonal trend wich is different both from that of NO_3 and of seasalt components.

3. DISCUSSION AND CONCLUSIONS

The spatial trends in concentration of all major non-seasalt components are quite similar. High values are found in southern Sweden which gradually decrease towards south-central Sweden and further along the east coast. From the Gulf of Bothnia the concentration decreases rapidly in a westwardly direction. In southern Sweden the concentration always seems to be higher near the coast compared to locations further inland at the same latitude. Such features are not found in LRTAP model calculations for instance with the EMEP LRTAP model and might be related to more frequent start of rain in these areas.

Deposition rates of components (in contrast to concentrations) have pronounced maxima in an area of SW Sweden where precipitation amount also has a maximum. In NW Sweden high amount of precipitation does not give high deposition rates of non-seasalt components since most of precipitation in this region originates from the clean Atlantic air.

It can be instructive to compare the spatial distribution of several components in some more detail. Thus mean values of nitrate and excess sulfate at all sampling sites with data from 1983-1985 are compared in Figure 5. The relation can be expressed by the following equation

$$(\text{ex } SO_4) = 1.28\ (NO_3) + 18 \qquad (1)$$

where () denotes concentration expressed in μeq/l.

From similar plots we also obtain

$$(NH_4) = 1.00\ (NO_3) \qquad (2a)$$

except for seven southerly and coastal sites where the relation is

$$(NH_4) = 1.14\ (NO_3) \qquad (2b)$$

Calcium has a surprisingly good relation to nitrate (considering possible local influence) which is

$$(Ca) = 0.32\ (NO_3) \qquad (3)$$

while there will be an intercept when calcium is plotted versus excess sulfate. The intercept is about 20 μeq/l - quite comparable to what was found in (1) above. The relation betwen H^+ and excess sulfate is

$$(H^+) = 0.81\ (\text{ex } SO_4) \qquad (4)$$

except for sites in southern Sweden (Ap, Sä, Hl) where H^+ is about 10 μeq/l lower than predicted by the equation. Somewhat more of the sulfuric acid has here been neutralized - in very good agreement with higher NH_4/NO_3 ratio at these sites.

We find these relations to be quite useful as a quick reference when discussing the load to a particular ecosystem. It also tells us something about average transport from many different sources to points in Sweden - a matter which we are presently investigating

Deposition over large water bodies

Since the precipitation pattern over large water bodies is different from that over surrounding land areas it is difficult to estimate the deposition over such areas only from land-based measurements. The Baltic is one such area where atmospheric deposition of both nitrogen and trace components might be significant for certain elements or even dominating compared to other inputs. It is, however, far from self-evident how such deposition estimates should be made and the appropriate sampling sites should be selected. To obtain some experimental indication eight sites were established across the The Gulf of Bothnia and surrounding land areas (open circles in Figure 1). These sites provide a bridge between Swedish and Finnish national networks. In the area of Kvarken very small islands/light-houses in open waters are available and concentration gradients inland can be fairly well predicted.

ACKNOWLEDGEMENT

Financial support for the work described here has been obtained from the Swedish Environmental Monitoring Programme (PMK) of the

National Swedish Environment Protection Board and from the University of Stockholm.

REFERENCES

(1) MONITOR 1985, The National Swedish Environmental Monitoring Programme (PMK) ed.: Claes Bernes, Liber förlagsorder S-16289 Stockholm, Sweden.

(2) RODHE, H.and GRANAT, L. 1984. An evaluation of sulfate in European precipitation 1955-1982. Atmospheric Environment Vol. 18, No. 12, 2627-2619.

Table I. Comparison of wet-only and bulk collectors during 1985 at five sites (Ap, Sä, Fo, Rå and Br).

Compound	Volume weighted mean (wet-only) mm and µeq/l	Difference (bulk) - (wet-only) mm and µeq/ml	%
Precip.	663	53	8
SO_4	63	0.6	1
NO_3	30	0.0	0
NH_4	31	-0.8	-3
H_+	44	-2	-5
(pH)	(4.36)	(-0.02)	
Cl	18	2	11
Na	14	2	14
Mg	4.3	0.8	19
K	1.5	0.3	20
Ca	7.7	0.7	10

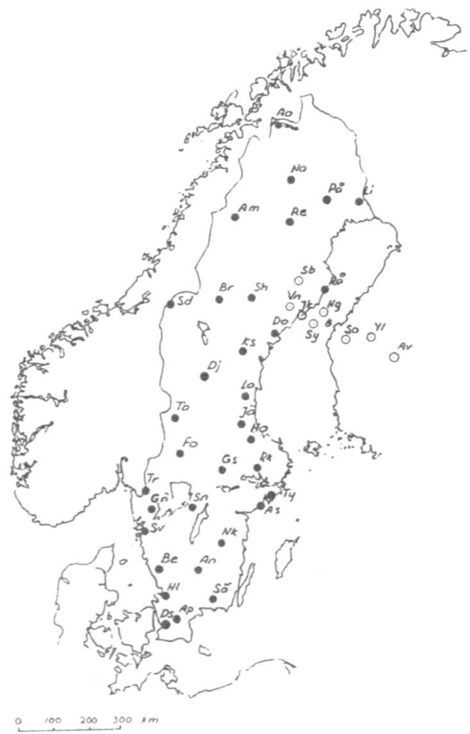

FIGURE 1: Sampling sites in the National Swedish Air and Precipitation Chemistry Network (The IMI/PMK network). Open circles: Sites in operation only during 1986- see further in the text.

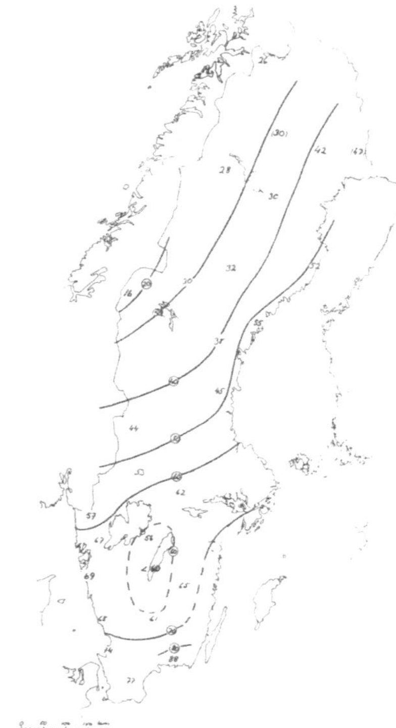

FIGURE 2a: Volume weighted mean concentration of excess sulfate for 1983-1985. Data from only two years are within parenthesis. Values in μeq/l.

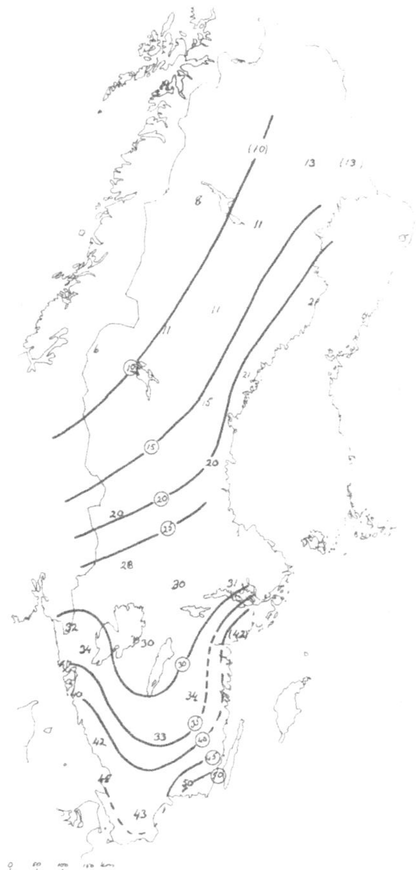

FIGURE 2b: Volume weighted mean concentration of nitrate for 1983-1985. Data from only two years are within parenthesis. Values in μeq/l.

FIGURE 3a: Deposition rate of excess sulfate. Based on concentration data from 1983-1985 - see further test. Values in meq/m^2, year.

FIGURE 3b: Deposition rate of nitrate. Based on concentration data from 1983-1985 - see further in text. Values in meq/m^2, year.

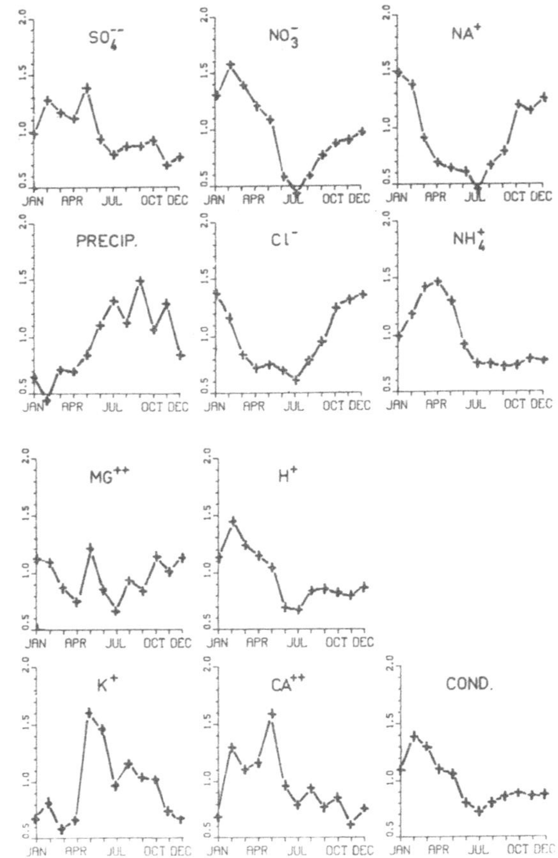

FIGURE 4: Relative seasonal trends in concentration of major components. Data from four sites (Fo,Rk,Gn and Sn) for the period 1972-1984.

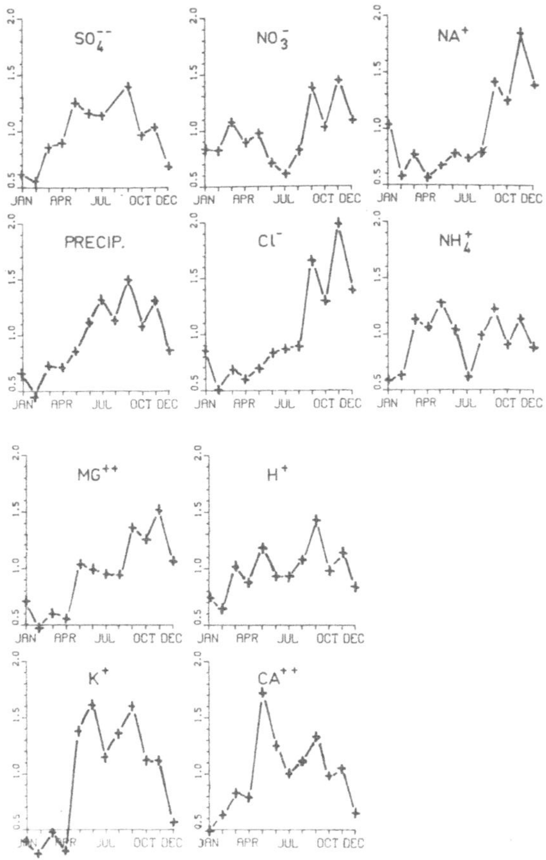

FIGURE 5: Relative seasonal trends in deposition rates of major components. Data from four sites (Fo, Rk, Gn and Sn) for the period 1972-1984.

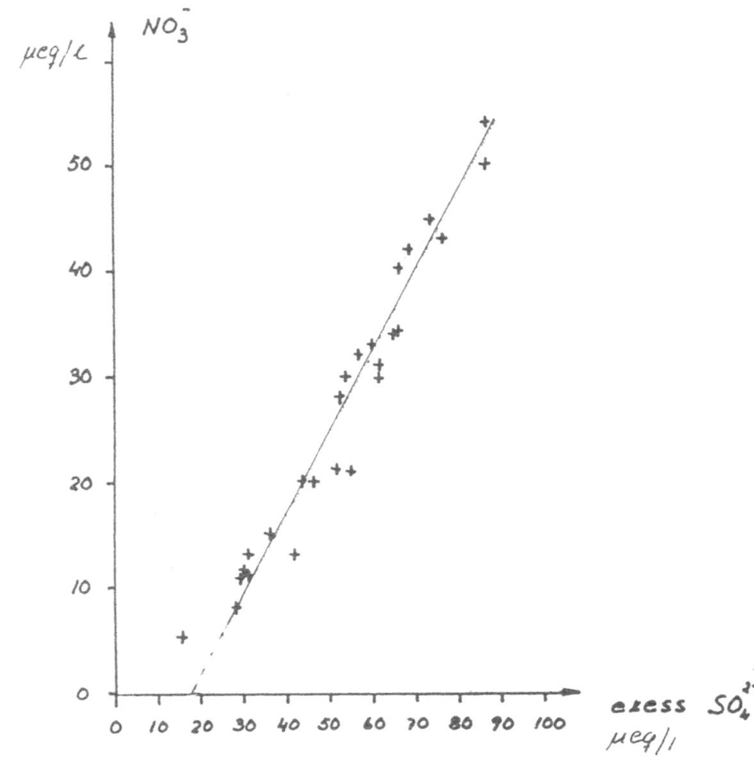

FIGURE 6: Relation between nitrate and excess sulfate. Based on mean values for 1983-1985 from sites in the National Swedish Network. Values in µeq/l.

MEASUREMENTS OF THE DRY DEPOSITION RATES OF SOME TRACE GAS SPECIES

G.J. Dollard, T.J. Davies and J.P.C. Lindstrom
Environmental and Medical Sciences Division,
The Harwell Laboratory, Oxon OX11 0RA, U.K.

Summary

Descriptions of the methods currently in use and some results obtained for the dry deposition fluxes of nitric acid, nitrogen oxides, ammonia and hydrogen peroxide and are described. In all cases a micrometeorological gradient technique was used.

Nitric acid was found to have strong downward fluxes with little surface resistance to uptake. The flux of NO_x was variable and on average there was little dry deposition to grass.

Analysis of filter samples for chloride suggested that hydrochloric acid behaved in a similar manner to nitric acid.

Measurements of ammonia made over a mature wheat crop revealed a strong emission of ammonia of the order 1 $\mu g\ m^{-2}s^{-1}$.

Application of the gradient technique has facilitated an estimate of about 1.0 cm s^{-1} for the deposition velocity of hydrogen peroxide to wheat.

The measurement of these species is continuing.

1. INTRODUCTION

Dry deposition, a process by which polluting gas species may be removed from the atmosphere without the intervention of rainfall is an important factor affecting the residence time of these species in the atmosphere. The rate at which a potentially acidifying gas or oxidant species is lost from the atmosphere has implications for the formation of acidic substances, their long range transport and for the ecosystems on which the materials are eventually deposited. It is important to quantify the dry deposition process in order to model realistically the effects of acid deposition and the likely results of any proposed emission control regulations.

This report presents some information collected to date on the fluxes of nitric acid, ammonia and hydrochloric acid to a cereal crop surface. Techniques and measurements relating to the deposition of nitrogen oxides and hydrogen peroxide are also given.

2. MATERIALS AND METHODS

For all the trace gas species a micrometeorological gradient technique was used (1)), based upon the measurement of gradients of concentration, windspeed and temperature with height.

The deposition flux densities, F, were calculated from the concentration gradient $\frac{d\chi}{dz}$ assuming the gases were transported by dz turbulence similarly to heat:

$$F = K_{H(z)}\, d\chi/dz$$

where the eddy diffusivity for heat at height z, $K_H(z)$ was obtained from the profiles of windspeed and temperature, with appropriate stability corrections. The flux was then calculated as

$$F = \frac{k^2\, du/d\ln(z-d)\; d\chi/d\ln(z-d)}{\phi_M \phi_H}$$

where k is the Von Karman constant (0.4), ϕ_M and ϕ_H are the dimensionless gradients for momentum and heat estimated after (2).

The deposition velocity, Vg, was then calculated as

$$Vg(z) = F/_{\chi(z)}$$

for a reference height z of 1m.

Some indication of the importance of the mechanisms controlling the deposition rate can be achieved by partitioning the total resistance $r_t = Vg(z)^{-1}$ between aerodynamic and surface terms (1). The aerodynamic resistance terms $r_a(z)$ and r_b represent the control exerted by the rate of transport through the atmosphere from height z to the surface. These terms may be evaluated from micrometeorological measurements (1). Any remaining resistance is attributed to the surface itself; very reactive gases would be expected to have little surface resistance.

The calculation of the deposition flux requires that the profiles are well adjusted to the underlying surface to allow development of a constant flux layer. Consequently a fetch of uniform surface extending upwind of the measurement site of at least one hundred times the measurement height is required.

The measurements reported here were made at two sites, one a grass field within the Harwell perimeter and the second a commercial wheat crop at a farm several miles from Harwell, both established on calcareous soil.

Temperature measurements were made at 5 heights above the surface using platinum resistance thermometers enclosed in polished metal radiation shields each ventilated by a small fan. Temperature differences could be measured to about 0.02°C by this method. Windspeed was measured using a vertical array of 5 cup amemometers. Profiles of concentration of HNO_3, HCl, NO_x, NH_3 and H_2O_2 were measured by the methods described below.

2.1 Nitric acid and hydrochloric acid

Concentrations of nitric acid were determined using a nylon filter technique (3). This method relies upon the quantitative removal of gaseous nitric acid by its reaction with the free amine groups of the hexamethylene diamine copolymer of the nylon filter. Nylon filters have been shown to be highly efficient at removing nitric acid from air (3), (4), penetration tests in this laboratory indicated a similar (>90%) high retention efficiency for HNO_3. Variable background nitrate levels on the nylon filters were reduced to acceptable levels by the routine pre-treatment of the filters through a repeated extraction with deionised water in a Soxhlet apparatus.

Concentrations of nitric acid were measured at 5 heights (0.25, 0.50, 1.0, 1.50 and 2.0 m) above the surface using 90 mm diameter filters (0.45 μm pore size), held in polypropylene open faced units. Particulate species were removed by a Teflon (1.0 μm pore size) pre filter. Measurements were made at a flow rate of 40 lit min^{-1} for periods of 30 to 120 minutes. The air flow to each filter was controlled by critical orifices and measured using dry gas meters.

Following exposure, the filters were extracted with 0.005 M NaOH in the laboratory and the nitrate determined using a continuous flow autoanalyser system.

Extractions of chloride from the filters were also analysed in order to derive information on gaseous HCl. The nylon filter technique has not generally been used for HCl measurements but the HCl/nylon reaction is expected to be similar to that for nitric acid and penetration tests carried out in the laboratory indicate similar high efficiency to that shown for nitric acid.

2.2 Ammonia

Measurements of the concentration of ammonia were made at 4 heights (0.25, 0.50, 1.0 and 2.0 m) above a mature wheat crop using denuder tubes coated with oxalic acid (5). Three tubes were exposed at each height and operated at just under 3 litres per minute using critical orifices driven by a single pump. Exposed denuder tubes were extracted into 5 ml of pure water and analysed for NH_4^+ on a continuous autoanalyser system.

2.3 NO_x measurements

Profiles of NO_x ($NO+NO_2$) were made using the equipment illustrated in Figure 1. Air was drawn continuously through four 12 mm ID teflon lines at a flowrate of 30 lit min^{-1}. These flow lines were of equal length and positioned at 4 heights above the surface (0.25, 0.50, 1.0 and 2.0 m). From each of the flow lines air samples were taken through 3 mm ID teflon tubing via a system of switched solenoid valves into a chemiluminescent NO_x analyser. By means of a mechanical timer sample air from each flow line was taken to the analyser on a 4 minute cycle allowing a one minute sampling period at each height. Profiles of NO_x concentration could be accumulated in this manner over extended periods. The output from the NO_x analyser was taken directly to a chart reader. Zero checks and calibrations were performed in the laboratory using standard gas sources prior to use in the field. Initial measurements using this system were made over a short grass surface.

2.4 Hydrogen peroxide

Hydrogen peroxide in the gas phase was made using a combination of the chemiluminescence technique (6) and a gas stripping coil system similar to that described by Heikes et al (7). The hydrogen peroxide analyser measures the chemiluminescence produced when hydrogen peroxide reacts with luminol in the presence of the catalyst microperoxidase. This instrument is superior to the original copper catalysed technique described by Kok et al (8) and compares very well to the fluorimetric technique of Lazrus et al (9), under field conditions (10). Ames (6) has thoroughly investigated interferences in the luminol method and points out that the more likely interference (metal catalysed oxygen effect), lies well below the detection limit of 50 nM lit^{-1} under normal circumstances.

Interferences generated in the stripping coil have been studied by Heikes et al (7) who also examined the production of H_2O_2 in the coil from ozone. They found that ozone concentrations of 40 and 100 ppbv generate H_2O_2 concentrations equivalent to 0.010 to 0.030 ppbv; considerably less than the values recorded during the measurements made to date.

In order to measure concentration profiles the apparatus illustrated in Figure 2 has been constructed. Air is sampled at 4 heights (0.25, 0.5, 1.0 and 2.0 m) above the surface, through 5 mm ID Teflon tubing at a flow rate of 3 lit min^{-1}. The air is taken through a stripping coil with a concurrent flow of pure water pumped into the coil at a rate of 0.8 ml minute. The H_2O_2 is stripped from the gas phase into the water, the air and water flows are then separated and the water sample collected directly into clean containers.

Water is collected from each of the 4 sample lines and stored for analysis on return to the laboratory. From the measured H_2O_2 concentration in the sample water, the ratio of the flow rates of water and air the gas phase concentration of H_2O_2 was derived.

The initial evaluations and measurements using this system were made over a mature wheat surface.

3. RESULTS

3.1 Nitric acid

Deposition velocities derived for a mature wheat crop, grass and a young wheat crop (15 cm length) are given in Table 1. The values obtained for nitric acid are subtantially larger than the range of 0.5 to 1 cm s^{-1} commonly quoted for SO_2 and other reactive species (11). The deposition velocities measured over the short grass and young cereal are similar to those reported from studies in the USA (12), (13). The higher values derived for the mature wheat are of a similar magnitude to values presented by Meixner et al (14) for deposition to a 12-18 m tall pine forest in West Germany. The results of the present study and others are consistent with little surface resistance to deposition of nitric acid. The dry deposition of this species would appear to be largely controlled by aerodynamic factors. This would account for the higher deposition rates to the taller wheat than the shorter surfaces. The similarities between the deposition velocity values to a forest quoted in (14) and the values derived for the mature wheat in the present study may be explained in part by the similar friction velocities encountered in both studies.

Deposition velocities for HCl derived from the analysis of the filters for chloride are given in Table 2. The values indicate a behaviour close to nitric acid.

3.2 Ammonia

Measurements have been confined so far to a mature wheat canopy. Measurements made at this stage of the crop growth indicate a strong emmission of NH_3. Figure 3 is an example of one of the experimental runs. The upward flux of ammonia from the surface in this case was 1.1 μg NH_3 $m^{-2}s^{-1}$.

This upward flux could be attributed to a source in the soil resulting from the volatilisation of fertiliser material or biogenic production of ammonia. In addition, the crop itself could act as a source of ammonia during the process of senescence as a consequence of the breakdown of organic material. Continuing measurements at the site following harvesting will clarify the source of the ammonia and give information on any seasonal changes in the ammonia flux.

3.3 Nitrogen oxides

The NO_x gradient system has been used in the field on several occasions. The direction and magnitude of the NO and NO_2 have been variable and often gradients could not be detected for either species.

Several factors can influence observed gradients of NO and NO_2 near to the ground. These include fast chemical reactions involving ozone (15), (16). Additionally, soil biogenic sources of NO and possibly NO_2 can be of importance (17), (18), (19).

Figure 4 summarises NO_x data as 20 minute average blocks collected during a 48 hour period of continuous measurements over a short grass surface in autumn. It is evident that both emission and deposition occurred during this period with apparent deposition values ranging from -1.5 to +0.9 cms^{-1} for No_x. There was an indication of diurnal pattern in the data with lower emission and deposition during late evening and nightime periods.

If averaged for the whole period of the measurements the deposition flux of NO_x to the grass surface is very limited.

3.3. Hydrogen Peroxide

Preliminary attempts to measure concentration gradients of H_2O_2 have been made over a mature wheat crop using the system described above. The first series of measurements made during August 1986 suggested the presence of very high concentrations of gaseous H_2O_2 (>50 ppbv). Controlled tests on the sampling system revealed an artefact production of H_2O_2. Initial runs were carried out using polycarbonate pots to collect the water from the stripping coil, it was found that there was a strong interaction between these pots and sunlight capable of generating H_2O_2 in solution (in strong sunlight to concentrations in excess of 100 μM lit^{-1}). The magnitude of the effect was found to be related to the volume of solution in the container and is thought to be a surface effect. Glass containers exhibited only a small effect and polycarbonate containers exposed inside glass containers still exhibited H_2O_2 production without the UV component of light filtered by the glass. Tests using catalase on the peroxide produced indicated that all of the peroxide was H_2O_2 (9).

The effect, though somewhat reduced, also occurs in diffuse sunlight and could account for some of the drift noted during the routine use of the hydrogen peroxide analyser in the laboratory.

The gradient system is now operated inside a lightproof envelope and all samples collected are immediately stored in glass containers in darkness. Recent experiments carried out over mature wheat have yielded H_2O_2 concentrations between 0.5 and 1.2 ppbv in air. A preliminary analysis of concentration profiles measured over mature wheat indicates a dry deposition velocity of about 1.0 cm s^{-1}.

4. CONCLUSIONS

The very high deposition velocities derived for nitric acid relative to the apparent low removal of NO_x by dry deposition point to the importance of nitric acid as a removal mechanism for atmospheric nitrogen species. The fact that there is little or no surface resistance for nitric acid deposition indicates that in conditions where atmospheric turbulence is enhanced e.g. over forests and towns the deposition flux of nitric acid could be very large.

Values derived for HCl deposition velocities suggest a similar behaviour to nitric acid and where concentrations of HCl are influenced by coal burning or the incineration of plastics this could be an important source of dry deposited acidity.

The release of ammonia by the ripening crop/soil system requires further study in order to evaluate any seasonal effects on the release. Emission of ammonia at the surface will be important in terms of the ammonia/nitric acid/ammonium nitrate equilibrium.

The preliminary application of the gradient method to investigate the dry deposition of H_2O_2 shows promise provided problems associated with artefact production of H_2O_2 are overcome. The observation that direct and diffuse sunlight can generated hydrogen peroxide in solution is interesting and could be of importance in the area of plant damage; it is possible that droplets of dew or rain on foliage could be a site of H_2O_2 generation. However, this requires investigation to ascertain its true significance.

ACKNOWLEDGEMENT

We are grateful to the UK Department of the Environment for the funding of this work.

REFERENCES

(1) GARLAND, J., 1977
The dry deposition of sulphur dioxide to land and water surfaces.
Proc.R.Soc.Lond. A. 354: 245-268.

(2) DYER, A.J. and HICKS, B.B. 1976
Flux gradient relationships in the constant flux layer
Q.Jl.R.Met.Soc. 96, 715-721.

(3) GOLDAN, P.D., KUSTER, W.C., ALBRITTON, D.L., FEHSENFELD, F.C., CONNELL, P.S., NORTON, R.B. and HUEBERT, B.J. 1983
Calibration and tests of the filter collection method for measuring clean air ambient levels of nitric acid.
Atmos. Environ. 17(7);1355-1364.

(4) GROSJEAN, D., 1982
Quantitative collection of total inorganic atmospheric nitrate on nylon filters.
Anal. Lett. 15(A9); 785-796.

(5) FERM, M., 1979
Method for determination of atmospheric ammonia.
Atmos. Environ.13, 1385-1393.

(6) AMES, D.L., 1983
A method of determining hydrogen peroxide in cloud and rainwater.
Central Electricity Generating Board report No TPRD/L/2552/N83

(7) HEIKES, B.G., LAZRUS, A.L. and KOK, G.L. 1985
Measurements of H_2O_2 in the lower troposphere.
Presented at 17th International symposium on free radicals. Snow Mountain, Granby CO, USA August 1985.

(8) KOK, G.L., HOLLER, T.P., LOPEZ M.B., NACHTRIELS, H.A. and YUAN M., 1978
Chemiluminescent Method for Hydrogen Peroxide in the Ambient Atmosphere.

(9) LAZRUS, A.L., KOK, G.L., GITLIN, S.N., LIND, J.A. and MCLAREN, S.E. 1985
Automated fluorimetric method for hydrogen peroxide in atmospheric precipitation
Anal. Chem. 57 p917-922.

(10) DOLLARD, G.J., JONES, B.M.R., CHANDLER, A.S. and GAY, M.J. 1986
Measurements of ambient SO_2 and H_2O_2 at Great Dun Fell and evidence of their reaction in cloud.
To be published in the proceedings of the NATO advanced research workshop on Acid Deposition processes at high elevation sites. Edinburgh September 1986.
(11) SEHMEL, G.A. 1980
Particle and gas dry deposition - a review
Atmos. Environ. 14: 983-1011.
(12) HUEBERT, B.J.
Measurement of the dry deposition flux of nitric acid vapour to grasslands and forest.
In: Precipitation Scavenging, dry deposition and resuspension Vol. 2. Coordinators Pruppacher, H.R. Semonin, R.G. and Slinn, W.G.N.
pp 785-793. Elsevier 1983.
(13) HUEBERT, B.J. and ROBERT, C.H.
The dry deposition of nitric acid to grass
J. Geophys. Res. 90(D1); 2085-2090 (1985).
(14) MEIXNER, F.X. 1986 To be published in Proceedings of 2nd meeting on Atmosphere Biosphere
Interactions Mainz, FRG 1986.
(15) LENSCHOW, D.H. 1982
Reactive trace species in the boundary layer from a micro-meteorological perspective.
J. Meteorol.Soc. Japan 60(1) 472-480.
(16) DUIJZER, J.H., MEYER, G.M., and VAN AALST, R.M. 1983
Measurement of dry deposition velocities of NO, NO_2 and O_3 and the influence of chemical reactions.
Atmos. Environ. 17, 2117-2120.
(17) GALBALLY, I.E. and ROY. C.R. 1978
Loss of fixed nitrogen from soils by NO exhalation.
Nature Lond. 275, 734-5.
(18) FILLARY, I.R.P, 1983
Biological denitrification. In Gaseous loss of nitrogen from plant soil systems
ed. by J.R. Freney and J.R. Simpson, 33-61, The Hague, Martinus Nijhoff/Dr Junk.
(19) COLBOURN, P., RYDEN, J.C. and DOLLARD, G.J. 1986
Emission of NOx from urine treated pasture
To be published in Environmental Pollution.

Table 1 Dry deposition velocities for nitric acid

Surface	Friction velocity U_* cm s^{-1}	Deposition velocity Vg cm s^{-1}
Short grass	0.30	3.6
h=15 cm, z_o=10 cm	0.26	1.7
Mature wheat	0.53	11.0
h=80 cm	0.29	11.0
z_o=10-22 cm	0.37	7.5
	0.56	26.0
	0.62	16.4
	0.73	19.1
	0.61	18.1
	0.67	4.9
	0.71	8.1
	0.70	13.3
Young wheat	0.27	2.9
h-10 cm, z_o=15 cm		

z_o = roughness length

Table 2 Dry deposition velocities for hydrochloric acid

Surface	Friction velocity U_* cm s^{-1}	Deposition velocity Vg cm s^{-1}
Short grass	0.30	4.7
h=15 cm, z_o=10 cm	0.26	1.2
Mature wheat	0.54	8.4
h=80 cm	0.53	9.8
z_o=10-19 cm	0.37	17.0
	0.67	17.0
Young wheat		
h-10 cm	0.27	3.9
z_o=15 cm		

z_o = roughness length

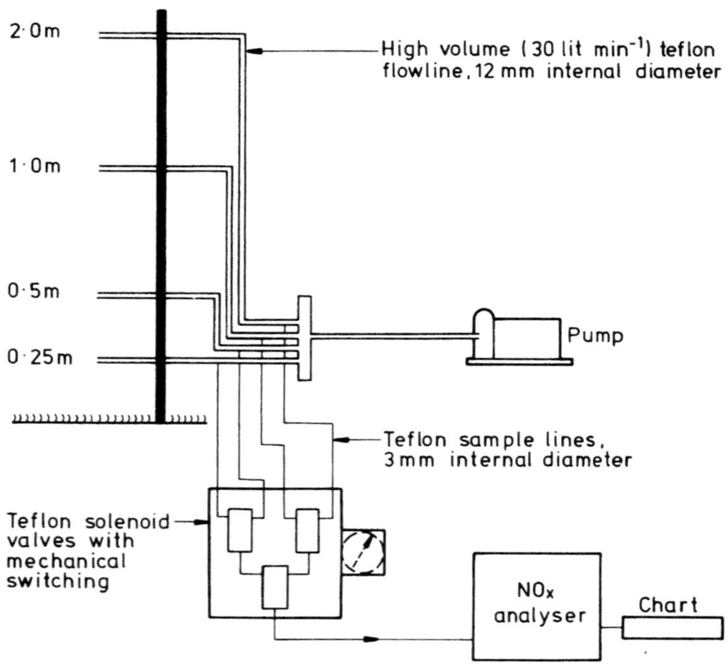

FIG. 1. SCHEMATIC REPRESENTATION OF NO_x GRADIENT SYSTEM

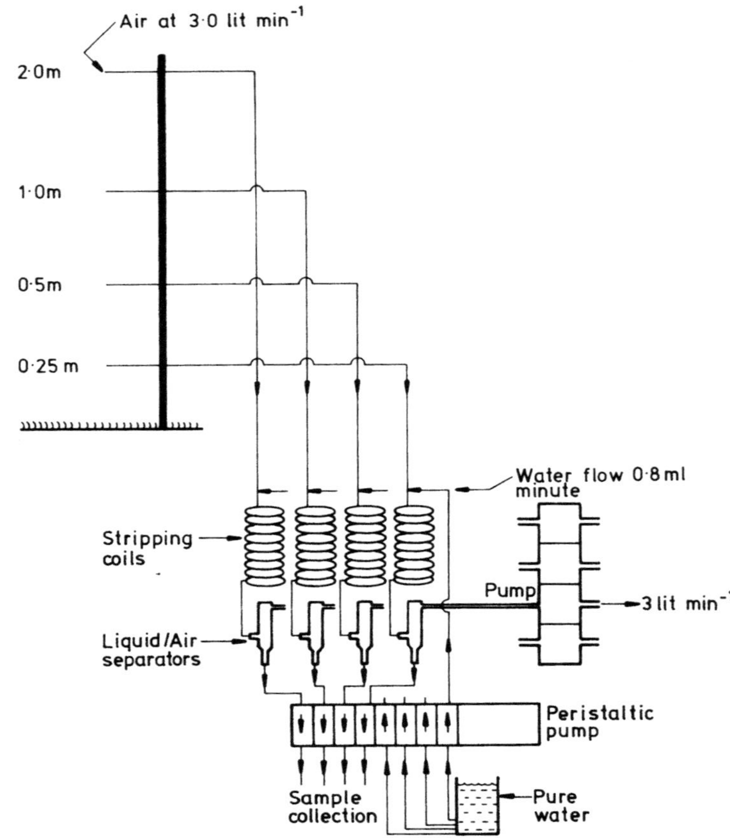

FIG. 2. SCHEMATIC REPRESENTATION OF PRELIMINARY H_2O_2 GRADIENT SYSTEM

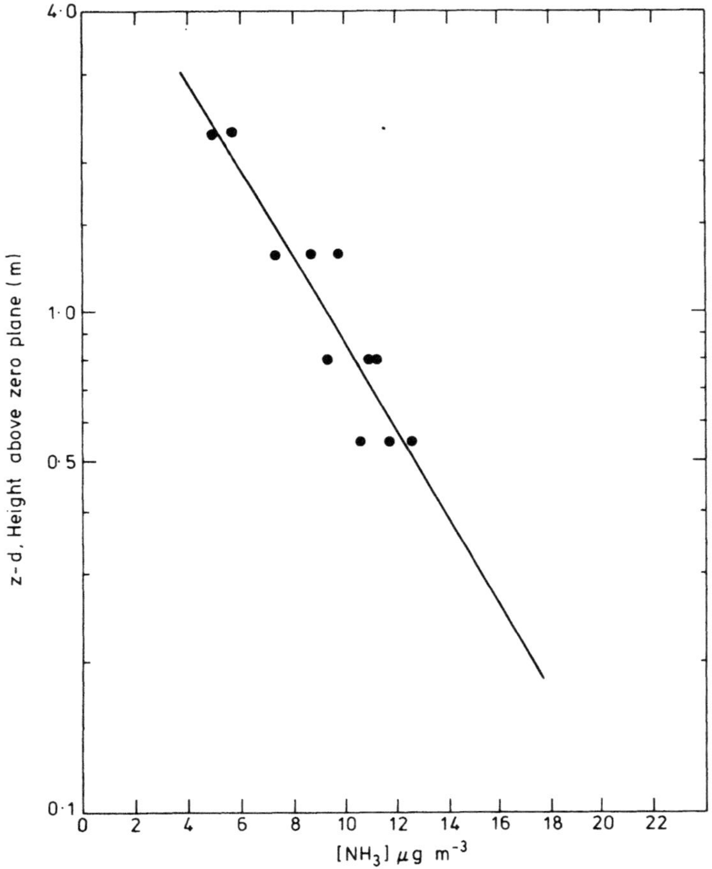

FIG. 3. AMMONIA CONCENTRATIONS WITH HEIGHT (z-d) ABOVE A WHEAT CROP

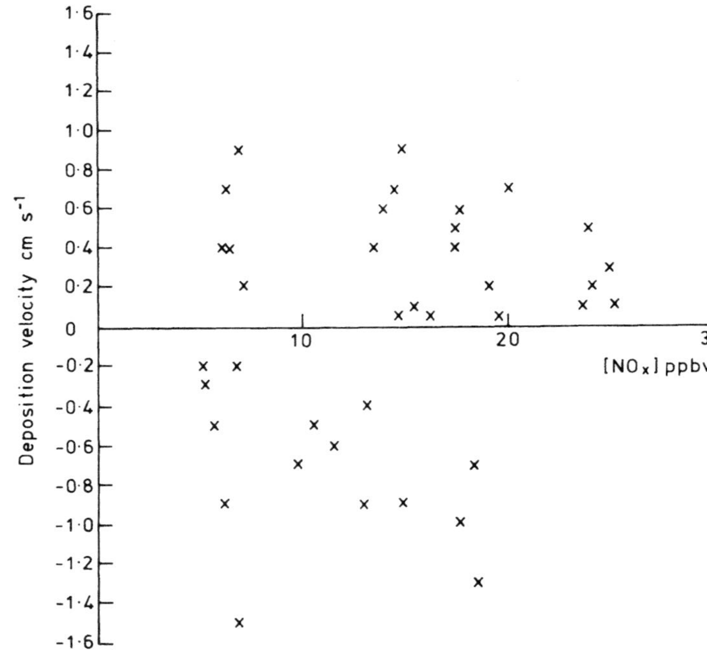

FIG. 4. NO DRY DEPOSITION VELOCITIES TO GRASS

CONTRIBUTION OF AMMONIA TO THE WET DEPOSITION OF SULFUR COMPOUNDS

F. Zürcher and B. Gisler
EAWAG, Swiss Federal Institute for Water Resources and Water Pollution Control, CH-8600 Dübendorf, Switzerland

Summary

The wet deposition of sulfur compounds from the atmosphere was investigated. Neutralizing agents such as ammonia and CaCO3 particulates alter the pH and therefore the absorption of SO2 into the liquid phase. The contribution of ammonia to the wet deposition of sulfuric acid is estimated. Additionally ammonium sulfate aerosols may be scavenged by rain, fog and snow. Relative contributions of both pathways to wet sulfur deposition will be discussed. A method for the determination of trace aqueous ammonia in low volume samples, utilizing the technique of gas diffusion flow injection will be described.

1. EINFÜHRUNG

Die anthropogene SO2- Emission ist die bedeutendste Quelle für die Belastung von Biota und Gewässer mit Schwefelkomponenten aus der Atmosphäre [Leutert 1985, Lindberg et. al. 1986] . Die atmosphärische Deposition von SO2, HSO3 /SO3 ,H2SO4, SO4 u.a. erfolgt auf verschiedenen Wegen. Bei der gasförmigen und trockenen Deposition sind zusätzlich zur Belastung auch die Eigenschaften der Rezeptoren ausschlaggebend. Die tatsächliche Belastung durch trockene oder gasförmige Deposition kann nur durch aufwendige Untersuchungen abgeschätzt werden[Lindberg et al 1986]. Dagegen erlaubt die Untersuchung der nassen Deposition Prognosen über die wirksame atmosphärische Belastung mit Regen und Schnee .
Im Vergleich mit der gesamten atmosphärischen S-Deposition stellt die nasse Deposition von Schwefelverbindungen (SO4 ,H2SO4, HSO3/SO3 einen quantitativ wichtigen Beitrag dar [Lindberg et. al. 1986, Oppenheimer 1983, Hertz 1986] .

Unabhängig von der Art der S-Deposition ist der Transfer zwischen Gas- und Wasserphase ein wichtiger Schritt. Die physikalisch / chemischen Prozesse bei der SO2 Aufnahme in ein Wolkentröpfchen (Regen, Nebel), die Gewässer oder das lebende Gewebe (Blattoberfläche, Lungengewebe) sind vergleichbar. Einige wichtige Prozesse lassen sich folgendermassen schematisch darstellen:

1. Der Transport von SO2 Gas zur Grenzfläche ist durch Diffusion und Konvektion bestimmt.

2. Das Absorption-Gleichgewicht für SO2 zwischen Gas- und Wasser - Phase wird durch den Henry-Koeffizienten kontrolliert . Die Absorptions-Geschwindigkeit hängt davon ab, wie rasch die Diffusionsschicht erneuert wird. Die physikalischen Randbedingungen, wie Tröpfchengrösse und der Eintrag mechanischer Energie (Wind, fallender Tropfen) spielen für die Absorptionskinetik eine wichtige Rolle [Carmichael and Peters 1979].

3. Die partielle Hydrolyse von SO2(w) wird durch die Dissoziationskonstante k1 bestimmt (pK1 = 1.9). Die Dissoziation von SO2*H20 ist vom pH der wässrigen Lösung abhängig. Die 2. Dissoziationskonstante k2 (HSO3 <-> SO3, pK2 = 7.2) ist für die SO2 Aufnahme unter pH 5.5 vernachlässigbar.

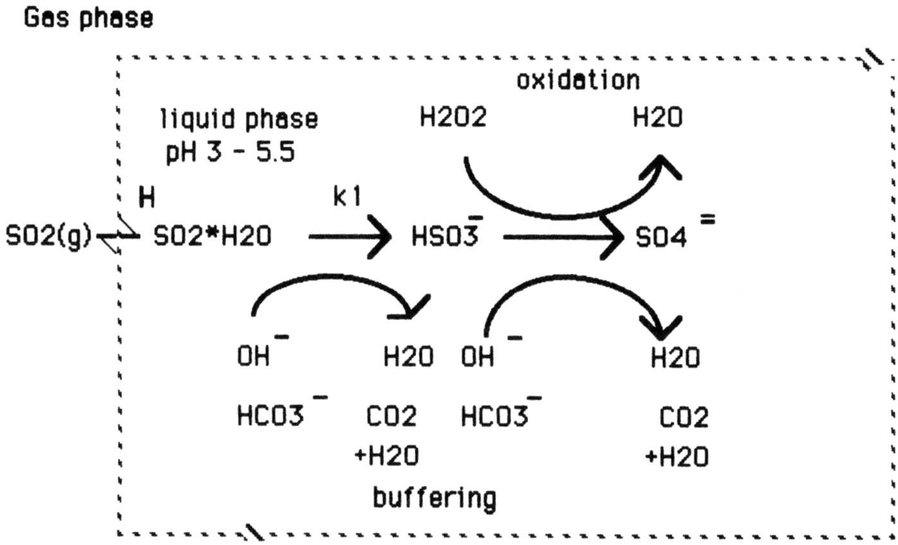

Verglichen mit der Absorption verläuft die Dissoziation etwa 100 mal schneller [Carmichael and Peters 1979]. Da die Dissoziation quasi gleichzeitig mit der Absorption abläuft,werden die beiden Prozesse zusammen betrachtet und unter dem Begriff SO_2-Aufnahme beschrieben [Schwartz 1984]. Die Absorption von SO_2 in ein Wolkentröpfchen (50 µm) ist bereits in weniger als 100 sec abgeschlossen [Carmichael and Peters 1979]. Gestützt auf eine mittlere Lebensdauer der atmosphärischen Wasserphase (Minuten bis Stunden) wird die Aufnahme von S-Spezies bei der Bildung nasser Depositionen durch das Dissoziationsgleichgewicht und nicht durch die Transportkinetik bestimmt.

Die pH abhängige SO_2-Aufnahme wird durch die sekundär ablaufenden Oxidationsreaktionen zusätzlich negativ beeinflusst (Reaktionsschema). Durch die Oxidation S(IV) -> S(VI) wird der pH der flüssigen Phase abgesenkt, was einerseits die $SO_2 * H_2O$ Dissoziation blockiert und anderseits die vom pH weniger beeinflusste Oxidation mit H_2O_2 begünstigt [Bøhler und Isaksen 1984]. H_2O_2 reagiert direkt mit HSO_3 über ein Zwischenprodukt zu SO_4 [Graedel 1981]. Das Ausmass der S(IV) Oxidation mit H_2O_2 ist von der photochemischen Produktionsrate abhängig [Bøhler und Isaksen 1984]. Grössere tageszeitliche und saisonale Schwankungen sind demzufolge bei der Oxidationsleistung zu erwarten. Die S(IV)-Oxidation in der flüssigen Phase ist verglichen mit der Gasphasenoxidation der dominierende Mechanismus für die S(VI)- Bildung [Calvert et. al. 1985].

Die S(IV)-Oxidation ist für die Aufnahme von S nur dann von quantitativer Bedeutung, wenn die wässrige Lösung genügend gepuffert ist. Ammoniak ist wegen seines natürlichen Vorkommens und der guten Wasserlöslichkeit einer der wichtigsten Puffer im atmosphärischen Wasser. Die Pufferwirkung von Ammoniak ist gerade im Bereich der sauren Depositionen mit pH-Werten unter 5.5 ausgeprägt. Besonders bei den Nassdepositionen mit niedrigem Flüssigwassergehalt (LWC) ist die NH_3-Pufferung gegenüber der Karbonatpufferung entscheidend [Lilijestrand 1986]. Ammoniak-Emmissionen sind von der biologischen Aktivität abhängig, jahreszeitlich bedingte Änderungen sind darum zu erwarten. Die Ammoniak- Gaskonzentrationen sind während der warmen Jahreszeit, Frühling-Sommer, höher und führen wegen der guten Wasserlöslichkeit von NH_3 auch zu erhöhten Konzentrationen in den Nassdepositionen [Hertz 1986, Calvert et. al. 1985].

Im Hinblick auf die wichtige Rolle von Ammoniak bei der Aufnahme von Schwefel in die atmosphärische Wasserphase haben wir versucht, Messdaten von unterschiedlichen Nass-Depositionen auf den Zusammenhang zwischen Ammonium- und Schwefel-Gehalt zu vergleichen. Konkret wollten wir auf folgende Fragen Antworten finden:

1. Wird die nasse Deposition von Schwefel-Spezies durch Ammoniak beeinflusst?
2. Welche quantitative Bedeutung hat der Einbau von gasförmigem SO_2 gegenüber dem Einbau von Sulfat-Aerosolen bei der nassen Deposition?
3. Welche quantitative Bedeutung hat die Ammoniak-Pufferung im Vergleich zur Karbonat-Pufferung bei der Absorption von SO_2 in Hydrometeors .

2. EXPERIMENTELLER TEIL

Probenahme:

Niederschläge (wet only) wurden zwischen November 1985 und Juli 1986 mit einem Depositionssammelgerät [Rohbock und Georgii 1982] in Polyethylentrichtern (25 cm ø) und Polyethylenflaschen gesammelt. Das Sammelgerät war auf dem Dach unseres Institutes in Dübendorf, ca. 1 m über der höchsten Kante aufgestellt. Die Proben wurden im Rhythmus von 14 Tagen entnommen.

Nebelproben sind mit einem aktiven Screen-Collector [Jacob et. al. 1985] gesammelt worden. Das Gerät wurde in der Nähe unseres Labors in Dübendorf, 1.5 m über dem Boden auf einer Wiese betrieben. Die Proben wurden in Intervallen von etwa 2 Stunden in Polyethylenflaschen gesammelt und anschliessend sofort auf den Gehalt an S(IV) untersucht. Die übrigen Parameter sind innerhalb von 24 h bestimmt worden.

Schneeprofile von einem Versuchsfeld 2540 m ü.M. (Eidg. Schnee und Lawinen-Forschungs Institutes Weissfluhjoch-Davos) wurden in 2-3 cm dicke Schichten aufgetrennt und beprobt. Die Proben wurden bis zur Analyse in gefrorenem Zustand gelagert und erst unmittelbar vor der Analyse unter Vermeidung des Kontaktes mit Laborluft geschmolzen [Neftel et al 1986]. Vor der Analyse sind die Proben durch ein Membranfilter (.45 µm) gepresst worden. Sämtliches Material das mit den Proben in Kontakt kam, wurde auf Kontaminationsfreiheit geprüft.

Analytik:

Die *Anionen* wurden mittels Ionenchromatographie (IC) bestimmt. Die Verwendete Anlage bestand aus einer Anionentrennsäule AS-3 (Dionex) kombiniert mit einem Fiber-Suppressor (Dionex) in Verbindung mit einem Leitfähigkeitsdetektor (SYKAM). Ein pneumatisch betriebenes Injektionsventil wurde verwendet, um die 500 µl Schnee-Proben direkt aus der Überdruckkabine in die IC Anlage einzuspritzen. Die Mehrzahl der IC Trennungen wurden mit einem Eluenten aus 1.4 mM $NaHCO_3$ und 1.1 mM Na_2CO_3 (pH 9.6) vorgenommen. Der totale S-Gehalt (S tot) wurde aufgrund des SO_4-Gehaltes, der nach der oxidativen Behandlung mit H_2O_2 gemessen wurde, berechnet. Die Bestimmungsgrenze lag für alle Anionen < .5 ueq/l.

Die wichtigsten *Kationen* Ca, Na und K wurden mit der Luft-Acetylen Flamme im AAS bestimmt. Für die Ca-Bestimmung wurde ein Fliessinjektions-Probenaufgabesystem (FIA) mit Lanthan-Zudosierung verwendet.

Die sichere Bestimmung von Ammonium-Spuren in Schnee- und Regenproben machte anfänglich Schwierigkeiten. Die Spuren- Bestimmung von Ammonium (NH_4) in sauren Proben ist wegen der Belastung der Laborluft (Mensch, nichtionogene Detergentien) erschwert. Die Probenvorbereitung für Schnee und Nasse Deposition wurde in einer Überdruckkabine vorgenommen um den Kontakt mit der Umgebungsluft zu vermeiden. Die Proben wurden nur in vorgereinigten Polyethylen-Flaschen aufbewahrt.

Die Ammoniumbestimmung wurde mit der Gaspermeations FIA- Technik durchgeführt . Die Messeinrichtung bestand aus einer komerziell erhältlichen Gasdiffusionszelle (Tecator). Die Diffusionsfläche war 75 mm lang und 2mm breit. Als Membran wurde ein Stück Teflon Dichtungsband verwendet. Die Verbindungsleitungen bestanden aus Teflonschläuchen mit einem i. D. von .5 mm. Reagentien und Probe wurden in einem FIA Analyzer (Tecator Modell 5020) mit den angegebenen Flussraten gefördert (Abb1). Prinzip : 20-200 µl Probe werden in einen alkalischen Trägerstrom eingespritzt und dabei NH4+ zu NH3 umgewandelt. NH3-Gas dringt durch die makroporöse Teflonmembran in die schwach saure Rezeptor-Lösung mit dem pH-Indikator Bromthymolblau. Die messbare Farbveränderung ist proportional zur NH4-Konzentration der Probe und wird photometrisch erfasst.

Die Spurenbestimmung von NH4+ mit der Teflon-Membran wird durch HCO3/CO2 Verunreinigungen in der Indikatorlösung gestört.Um diese Störung zu beseitigen wurde das deionisierte Wasser für die Bromthymolblau-Lösung CO2 befreit (belüften mit N2 unter Vakuum). Nach der Einstellung der Indikatorlösung auf pH 6.5 mit 0.01n Ba(OH)2 wurde die Lösung in einem dickwandigen zusammenfaltbaren Polyethylen-Vorratsbehälter aufbewahrt. Für empfindliche Messungen wurde eine 10 µM Bromthymollösung pH 6.5 verwendet. Mit einem maximalen Injektionsvolumen von 200 µl haben wir im Bereich zwischen 0-300 µg NH4-N/l (0.1-20 µmol/l) eine lineare Eichkuve erhalten, mit einer Standardabweichung von+- .02 µmol/l bei 0.35 µmol/l. Die Methode wird durch Trübstoffe (< .1 mm) nicht gestört und verlangt deshalb auch keine Probenvorbereitung (Filtration). Allfällige flüchtige Amine würden miterfasst, diese sind aber in unseren Proben im Vergleich zur NH4-Konzentration vernachlässigbar.

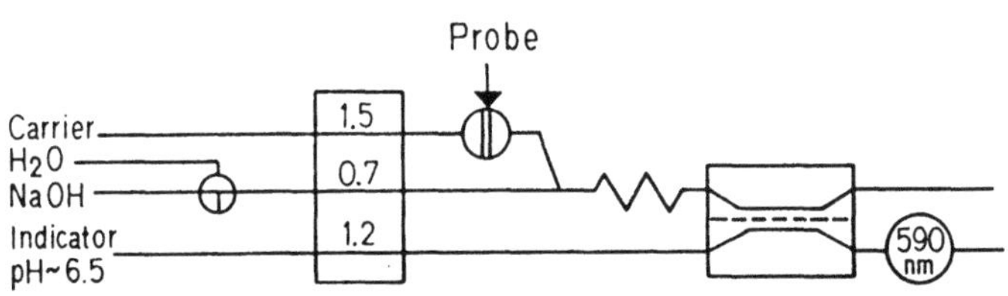

Abb 1 Ammonium determination by gas diffusion FIA, using macroporous Teflon membrane. Flow diagram; Carrier: pure Water, NaOH: 0.1 n, indicator: 10 µmol bromthymol blue. Injected sample volume: 200 ul.

3. SAISONALE VARIATIONEN VON S-tot UND NH4+ IM NIEDERSCHLAG

Die mittleren Konzentrationen von S tot , HSO3 und NH4 aus den 14 tägigen Sammelproben zwischen Nov. 85 und Junli 86 sind in Abb. 2 dargestellt. Die 440 m ü.M. liegende Probenahmestelle in Dübendorf grenzt im Sektor der Hauptwindrichtung (W, NW) an die Stadt Zürich. Der Konzentrationsverlauf von Stot und NH4 ist sehr ähnlich und stimmt auch bei den Extremwerten überein. Stot besteht fast ausschliesslich aus S(VI). In den Wintermonaten wurden geringe S(IV) Konzentrationen gemessen.Gestützt auf die IC-Analyse und den pH-Wert der Proben handelt es sich dabei um HSO3-Spezies. Während der Lagerung der Sammelproben wird S(IV) relativ rasch zu SO4 oxidiert. Anhand der gemessenen Werte kann die tatsächliche Verteilung zwischen S(IV) und S(VI) somit nicht zuverlässig abgeschätzt werden. Ein erhöhter S(IV) Gehalt in Winter- gegenüber Sommerniederschlägen konnte jedoch bei der sofortigen Analyse von Einzelereignissen gemessen werden. Diese Ergebnisse deuten auf eine SO2-Gasabsorption hin, die in den Wintermonaten mit 5-10 fach höheren SO2-Gaskonzentrationen deutlicher ausfällt.

Die SO2-Gasabsorption in einer sauren Lösung (pH 3.5-5) setzt eine entsprechende Pufferung voraus. Die Korrelation zwischen NH4 und Stot (0.90) bestätigt die Rolle von Ammoniak als Puffergas für die SO2-Aufnahme. Ammoniak übernimmt eine Pförtnerfunktion, da NH3(g) im sauren ,anthropogen beeinflussten Hydrometeor gut löslich ist. Die saisonalen Variationen der nassen Deposition geben Anhaltspunkte für die Bedeutung der verschiedenen S-Einbauprozesse (Gas-Absorption gegenüber Aerosol-Einbau). Trotz deutlich höheren SO2-Gaskonzentrationen im Winter folgt die S-Deposition der NH4-Deposition während des ganzen Jahres. Eine Unterscheidung zwischen SO2(g)-Absorption und der Auswaschung von SO4-Aerosolen unterhalb der Wolke kann gestützt auf die NH4- und Ca-Deposition vorgenommen werden (TAB1). Die Ca-Deposition im Niederschlag (wet only) kann als Tracer für die Auswaschung von Partikeln, Aerosole oder Staubpartikel, unterhalb der Wolke (below cloud scavenging) betrachtet werden. Bei der untersuchten nassen S-Deposition sind sowohl der Einbau von SO2-Gas als auch die Auswaschung von Sulfataerosolen beteiligt. Wegen der relativ langen Probesammeldauer von 14 Tagen sind die gemessenen NH4-Konzentrationen z.T. kleiner als die tatsächlichen Werte (Nitrifikation). Bei der sofortigen Analyse von Proben aus Einzelereignissen war die Übereinstimmung zwischen S tot- und NH4- Konzentrationen besser und das Aequivalenten-Verhältnis nahezu 1:1.

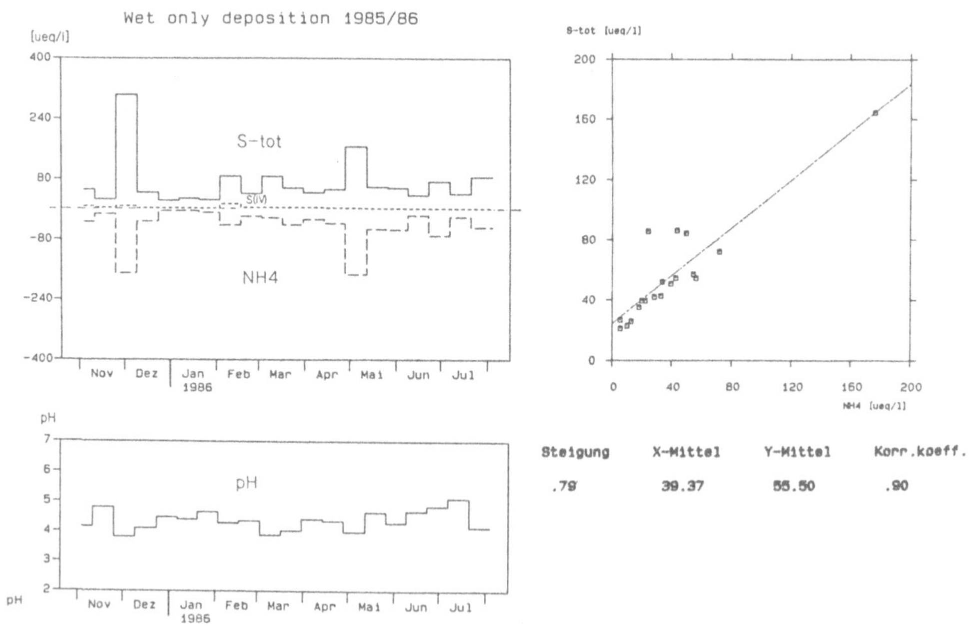

Abb. 2 Wet only deposition: Seasonal variations of S-tot and S(IV)-conc. in comparison with NH4-conc. Sampling period was 14 days. S-tot conc. was highly correlated with NH4 conc. Measured pH values varied between pH 4 and pH 5.

4. SCHWEFELDEPOSITION MIT SCHNEE

Die Deposition von S-Spezies mit Schnee wurde an den Proben vom Weissfluhjoch Davos (2540 m ü.M.) untersucht. Aus 3 vergleichbaren Schneeprofilen (Neftel et al dieses Symposium) wurde ein Abschnitt mit grosser Akkumulation, bei dem die trockene Deposition vernachlässigbar ist, ausgewählt. In Abb. 3 sind die Konzentrationsganglinien von S tot, NH_4 und Ca im Schnee vom 15. - 25. Jan. 86 entsprechend einer Wasserakkumulation von 60 mm aufgezeichnet. S tot wurde ausschliesslich in oxidierter Form als $(NH_4)_2SO_4$, H_2SO_4 oder $CaSO_4$ deponiert. S(IV)-Spezies waren bei einer Bestimmungsgrenze von <.1 ueq/l nicht nachweisbar. Die S tot Konzentrationsganglinie ist charakterisiert durch ein Maximum bei 384 mm Wasseräquivalent. Ganz ähnlich verläuft die NH_4 Ganglinie, sichtbar an der Summenkurve NH_4+Ca. Die S-Aufnahme in die Wasserphase vor der Schnee-Bildung wurde durch die NH_3-Pufferung beeinflusst, aber nicht ausschliesslich dominiert. Verglichen mit der S-tot -Konzentration ist die NH_4 Konzentration etwa halb so gross (TAB1). Die Korrelation zwischen den beiden Konzentrationen ist jedoch hoch signifikant. Trotz der S tot-Übersättigung zeigt die Regressionsgerade eine Steigung von nahezu 1. Daraus geht hervor, dass ein Teil der Schwefelsäure als $(NH_4)_2SO_4$ deponiert wurde. Aus den Messungen kann geschlossen werden, dass die S-Deposition im Schnee durch Scavenging von SO_4 Aerosolen in den Wolken erfolgte.

Die nicht als Neutralsalz deponierte S(VI)-Konzentration ist während dem Beobachtungszeitraum unverändert geblieben. Eine solche Hintergrundkonzentration von 2-3 ueq SO_4/l würde in einem offenen System im Gleichgewicht mit 2-3 ppb SO_2(g) erreicht. Die Schätzung für die S(IV)-Anreicherung in einem ungenügend gepufferten Wassertropfen von pH 5 stützt sich auf einen pseudo Henry-Koeffizienten von 10^3 M /atm [Schwarz 1983]. Weil der im Schnee gefundene Schwefel als S(VI) vorlag, ist ein weiträumiger Transport von gelöstem S(IV) anzunehmen. Der Einbau von SO_2(g) in den Schnee und eine mögliche Oxidation ist wegen der behinderten Diffusion auszuschliessen. Die fehlende Korrelation zwischen S-tot- und HNO_3-Deposition (Tab. 1: Korrelation S-tot/Σ-) weist auf unterschiedliche Einbauprozesse für H_2SO_4 und HNO_3 hin.

Die Frage der Karbonat-Pufferung durch Aerosol- oder Staubpartikel bleibt noch unbeantwortet. Die Ca-Konzentration blieb während der Beobachtungsperiode konstant. Bei Aerosol und Staub Auswaschungen oder Windverfrachtungen aus der näheren Umgebung wären grössere Schwankungen zu erwarten gewesen.

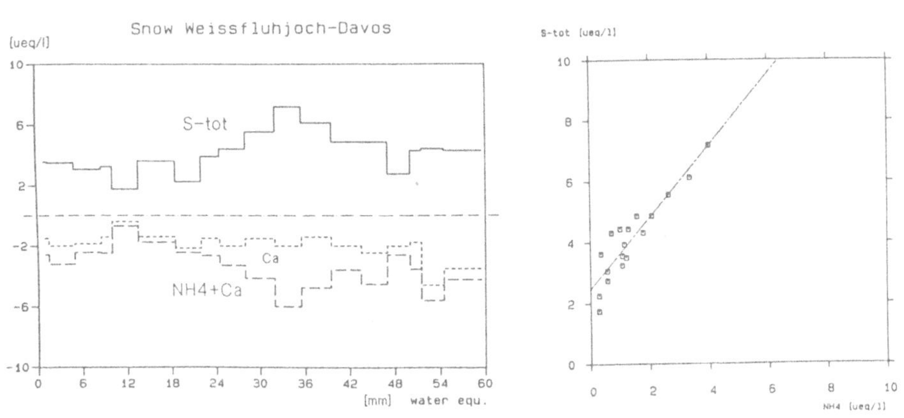

Steigung	X-Mittel	Y-Mittel	Korr.koeff.
1.17.	1.38	4.10	.91

Abb.3 SO_4-Deposition with snow at Weissflujoch-Davos(2500 m a.s.l.).
Comparison of SO_4-concentration with the Σ(NH_4 + Ca)- and Ca-concentration between 15th and 25th january 1986. Correlation of S-tot (SO_4) with NH_4 is high.

5. AUFNAHME VON SO2 IM NEBEL

Physikalisch gesehen sind Wolken und Nebel sehr ähnlich, folglich sind auch die chemischen Prozesse die in Wolkentröpfchen oder Aerosolen ablaufen mit dem was im Nebeltröpfchen geschieht vergleichbar.

Abb.4 zeigt den Konzentrationsverlauf von S tot und NH4 während eines Nebelereignisses vom Dez. 1985. S-tot setzt sich aus SO4, HSO3, SO2*H2O und Hydroxymethansulfonsäure (HMSA) zusammen. Das Nebelereignis ist durch zwei Phasen gekennzeichnet. Die Messungen begannen bei dichtem Nebel, der sich vorübergehend nahezu auflöste. Anschliessend hat sich der Nebel wieder verdichtet und erst nach Mitternacht vollständig aufgelöst. Eine ausführlichere Diskussion über dieses Ereignis wird andernorts vorgenommen [Johnson et al 1986].

In der Umwandlungsphase des Nebels (LWC nimmt zu) wurde zuerst NH3 in die Tröpfchen aufgenommen. Die SO2 Löslichkeit im Nebeltropfen wird vom pH-Wert bestimmt. Während der Tageszeit wurde der aufgenommene S(IV) relativ rasch zu S(VI) oxidiert. In der Nacht, am Ende der Nebelphase wurde die Oxidation wahrscheinlich mangels Oxidantien (photochemisch erzeugtes H2O2) blockiert. Die Verschiebung im Verhältnis S(IV)/S(VI) hatte keinen Einfluss auf die S-Aufnahme. Die während der Oxidation freigesetzten Protonen hemmten eine weitere Absorption von SO2 aus der Gasphase.

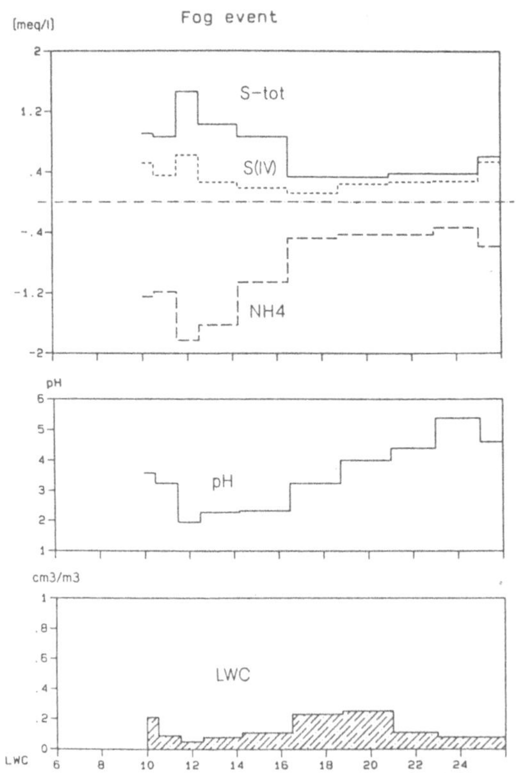

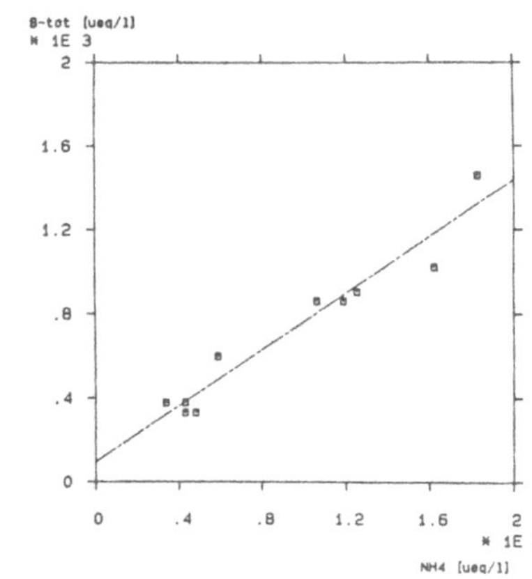

Steigung	X-Mittel	Y-Mittel	Korr.koeff.
.67	919.99	713.67	.97

Abb. 4 Concentration profiles of S-tot and S(IV) compared with NH4 during fog event at Dübendorf (1./2. december 1985). The pH of the collected fog water has changed over a wide range with time. The liquid water content (LWC) shows the variation of fog densities with time. The correlation of S-tot- and NH4-conc. was significant.

Während des ganzen Nebelereignises folgte die S tot Konzentration der NH4 Konzentration. Die Korrelation zwischen S tot und NH4 ist hoch signifikant. Im Bereich der hohen Konzentrationen weicht die Regressionsgerade stark von der 1:1 Sättigungsgeraden ab. Erst gegen Ende der Nebelperiode erreichte die S tot Konzentration den von der NH4-Konzentration abhängigen Sättigungswert. Im Vergleich zu NH3 spielte die CaCO3-Pufferung bei der S-Aufnahme im Boden-Nebel eine geringe Rolle (TAB 1).

TAB 1 Correlation of Total Sulfur-Spezies with Ammonium, Calcium and total Acid Anions for different Wet Atmospheric Depositions

DEPOSITION	Height of m.a.s.l	Water-Equiv. (mm)	Numb. of Sampl.	MEAN CONCENTRATIONS (geom. mean: ueq/l) Stot	S04	NH4	Ca	Σ -	KORRELATION Stot with NH4	Ca	Ca+NH4	S04	Σ -
Wet Deposition Dübendorf Nov 85-Jul 86 14 days interv.	440	560	20	53	50	31	20	112	.90	.86	.96	.99	.97
Fog Dübendorf 1./2. Dez 85	440	-	12	630	320	780	34	1540	.97	.79	.97	.87	.86
Snow Weissfluhjoch 15.-30.Jan 86	2540	150	18	3.9	3.9	1.1	1.8	14	.92	.21	.86	.99	.66

6. SCHLUSSFOLGERUNGEN

Die bei der S-Aufnahme in Wolkentröpfchen oder Aerosolen ablaufenden Prozesse lassen sich sehr gut im Nebel verfolgen. Bei der Übertragung auf die Nasse Deposition mit Regen oder Schnee ist zu berücksichtigen, dass die säurepuffernde NH3-Gaskonzentration mit zunehmender geographischer Höhe und Distanz zur Biota abnimmt. Eine empfindliche Methode für die Bestimmung von NH4-Spuren in kleinen Probevolumen ist für die Abklärung der S-Aufnahme-Prozesse in Hydrometeore notwendig.

Bei den untersuchten nassen Deposition wurde der S-Gehalt unabhängig von der Jahreszeit durch die absorbierten NH3-Mengen geprägt. Für den Nebel und die nasse Deposition in einer städtischen Agglomeration war der quantitative Beitrag der SO2-Gasabsorption wichtiger als die Aerosol-Auswaschung. Abhängig von lokalen Einflüssen wurde die nasse S-Deposition zeitweise auch durch CaCO3-Pufferung mitbestimmt. Für die S-Deposition im hochalpinen Schnee war der Einbau von H2SO4- und (NH4)2SO4-Aerosolen entscheidend.

VERDANKUNG

Die Arbeit wurde durch W. Stumm und L. Sigg unterstützt. Die Probenahme für die Nassdeposition und den Nebel wurde durch J. Zobrist und C. Jaques besorgt. Für die wertvolle Unterstützung bei der Durchführung der chemischen Analysen wird C. Mäder, U. Michel und D. Kistler gedankt.

REFERENZEN

Børhler T. and I. S. A. Isaksen: The Atmospheric Significance of Liquid Phase Oxidation of S02 to Sulphate by 03 and H202. Proc. of 3rd Europ. Symp. on Physicochemical behaviour of atmospheric pollutants (B. Versino & Angeletti, G. Eds.), 439-449, Dordrecht, D. Reidel Publ. Comp. 1984.

Calvert, J. G., A. Lazrus, G. L. Kok, B. G. Heikes, J. G. Walega, J. Lind and C. A. Cantrell: Chemical mechanisms of Acid Generation of the Troposphere. Nature, 317 (1985) 27-35.

Carmichael G. R. and L. K. Peters: Some Aspects of S02 Absorption by Water-Generalyzed Treatment. Atmospheric Environment, 13 (1979) 1505-1513.

Dasgupta P. K. and S. Dong: Solubility of Ammonia in Liquid Water and Generation of Trace Levels of Standard Gaseous Ammonia. Atmospheric Environment 20,3(1986) 565-570.

Fuhrer, J.: Study of Acid Deposition in Switzerland: Temporal Variation in the Ionic Composition of wet Precipitation at Rural Sites During 1983-1984. Environmental Pollution (Ser. B), in press (1986).

Graedel T. E. and C. J. Weschler: Chemistry within Aequeous Atmospheric Aerosols and Raindrops. Reviews of Geophysics and Space Physics, Vol. 19, 4 (1981) 505-539.

Hales J. M. and Drewes D. R.: Solubility of Ammonia in Water at Low Concentrations. Atmospheric Environment 13 (1979) 1133-1147.

Hertz, J.: Depositionsmessungen im Zürcher Waldprogramm. Persönliche Mitteilung 1986.

Jacob D. J., J. M. Waldman, m. Haghi, M. R. Hoffmann and R.C. Flagan: An Instrument to Collect Fogwater for Chemical Analysis. Review of Scientific Instruments 56 (1985) 1291-1293.

Johnson C. A., L. Sigg, J. Zobrist and W. Stumm: Chemical Composition of Fogwater During Two Fog Events; The Influence of Gaseous Acidic Components. in preparation

Liljestrand, H. M.: Average Rainwater pH, Concepts of Atmospheric Acidity, and Buffering in Open Systems. Atmospheric Environment (1986) in press

Leutert, G.: Luftverschmutzung in der Schweiz: Wo stehen wir? BUS-Bulletin 3/1985.

Lindberg, S. E., G. M. Lovett, D. D. Richter, D. W. Johnson: Atmospheric Deposition and Canopy Interaction of Major Ions in a Forest. Science, 231 (1986) 241-245.

Neftel A., A. Sigg, F. Zürcher: Acid Deposition in a Snow Field at 2500 m a. s. l. in Switzerland. This conference 1986

Oppenheimer, M.: The Relationship of Sulfur Emissions to Sulfate in Precipitationn-II. Gas Phase Processes. Atmospheric Environment 17,8 (1983) 1489-1495.

Rohbock E. und H. Georgii: Ein Depositionssammelgerät zur getrennten Erfassung der trockenen und feuchten Deposition atmosphärischer Schadstoffe. Forschungsbericht Proj. 104 02 712 Umweltbundesamt, Berlin 1982.

Schwartz, S. E.: Gas Aqueous Reactions of Sulfur and Nitrogen Oxides in Liquid-Water Clouds. In S02, NO and NO2 Oxidation Mechanisms: Atmospheric Considerations, Edited by J. G. Calvert Ann Arbor Publ. (1984), pp 173-208.

IMPACT DES POUSSIERES SAHARIENNES SUR L'ACIDITE DES PLUIES DANS L'ATMOSPHERE MEDITERRANEENNE/IMPACT OF SAHARAN DUST ON THE RAIN ACIDITY IN THE MEDITERRANEAN ATMOSPHERE

par/by M.D. LOYE-PILOT, J. MORELLI* J.M. MARTIN**
*J.M. GROS**and B. STRAUSS***

**Institut de Biogéochimie Marine - Ecole Normale Supérieure*
Unité Associée au CNRS n° 386, 1, rue Maurice Arnoux
92120 Montrouge, France.

***Météorologie Nationale, SCEM/D/ES, 2 avenue Rapp*
75230 Paris Cédex 7, France.

ABSTRACT

A geochemical study of atmospheric deposition was performed since 1980 in a pristine area of South Corsica (Bavella Pass). The sampling times ranged from several hours to one week. Samples represent bulk deposition (wet + dry) or fit with rainfalls. Many episodes of red rains were observed. The water samples were filtered indoors close to the field, using 0,4 um Nuclepore filters, in order to evaluate their content of solid particulate matter. The pH was measured immediately after sampling using a potentiometric method, major ions by atomic absorption spectrometry and colorimetric techniques.

Significant fluctuations of the rain acidity were pointed out and related to antagonistic influences of various sources of aerosols, on the basis of air masses trajectories arriving on the sampling zone. Acidic episodes are observed when incoming air masses passed over various remote european industrial regions (lowest pH values close to 4) and can be mainly attributed to a pollution by sulfuric acid. On the contrary, situations with air masses originating from arid regions of North Africa fit with the highest pH values : 6 to 8. They correspond to red rain episodes, high content in solid particulate matter including up to 30 % of calcite and high enrichment in dissolved calcium due to the partial dissolution of this component.

In situations with mixing of Saharan and European air masses, the intermediate pH values observed can be attributed to a neutralizing effect of the incoming desertic aerosol on the atmospheric acidity. It appears then that Saharan dust may counterbalance the effects of acidic rains in Mediterranean regions. Such interaction in the atmosphere and also at ground and sea level after atmospheric deposition may have implications for marine and terrestrial ecosystems of these regions.

Des recherches sur la géochimie des précipitations ont été effectuées depuis 1980 au Col de Bavella, à 1 200 m d'altitude, au Sud de la Corse (1). Cette région, pratiquement inhabitée et exempte d'activités industrielles, a été prise comme observatoire pour étudier la variabilité des caractéristiques du dépôt atmosphérique au Sud de l'Europe, en se plaçant relativement loin des grandes sources de pollution. En raison de la configuration de mer fermée de la Méditerranée, la qualité de l'air y est dans une large mesure soumise à l'influence de l'environnement continental. Selon la direction des vents, notre station de Corse peut notamment recevoir des apports atmosphériques en provenance de l'Europe industrielle et des poussières résultant de l'érosion éolienne des déserts de l'Afrique du Nord (1). Si on ne dispose encore que de peu de données sur la dispersion de la pollution atmosphérique dans le bassin méditerranéen, on sait, par contre, qu'il existe un transport très actif de poussières sahariennes au-dessus de cette région (1, 2, 3, 4). Ce phénomène est responsable des épisodes de pluies rouges ou de neiges rouges souvent constatés au Sud de l'Europe (5, 6, 7).

A partir des informations acquises au cours de 6 années de collection de la retombée atmosphérique, nous avons sélectionné quelques résultats significatifs. Ils s'appuient sur des analyses de la composition chimique des précipitations recueillies au moyen d'un pluviomètre en plastique de 400 cm^2 d'ouverture. Les durées d'échantillonnage utilisées sont variables et comprises entre quelques heures et au maximum une semaine. On a souvent cherché à les adapter sensiblement à la durée d'évènements pluvieux bien individualisés. Lorsque cela n'a pas été possible, on a alors prélevé la retombée totale (apport par l'eau de pluie + dépôt sec). Les échantillons ont été filtrés sur des filtres Nuclepore de 0,4 μm de porosité, dans un local proche du site de prélèvement, en vue de la détermination de leur contenu en matière particulaire. Le pH a été mesuré par la méthode potentiométrique. Les solutions filtrées ont été analysées par spectrométrie d'absorption atomique pour la mesure des ions Na^+, K^+, Ca^{++} et Mg^{++} et par colorimétrie pour celle des ions Cl^-, $SO4^{--}$, $NO3^-$ et $NH4^+$.

On soulignera d'abord ici l'ampleur du phénomène d'apport d'aérosols sahariens en milieu méditerranéen. Les résultats présentés apporteront aussi des indications sur l'impact de la pollution industrielle de l'Europe. L'accent sera mis sur l'antagonisme entre ces deux types d'influence vis-à-vis de la composition chimique et de l'acidité des précipitations.

Nous avons extrait d'une récente publication des données relatives aux pluies collectées pendant l'année 1984 au Col de Bavella (1). Sur un total de 60 précipitations, on a observé 19 "pluies rouges", ce qui donne une idée de la fréquence relativement élevée de ces évènements en milieu méditerranéen. Une étude des trajectoires des masses d'air arrivant sur la zone d'échantillonnage a permis de confirmer que les poussières rouges incorporées à la pluie provenaient des régions arides de l'Afrique du Nord. La figure 1 montre les variations simultanées du contenu en poussières rouges des échantillons d'eau de pluie, de leur enrichissement en calcium* et de leur pH. Il apparaît que certains épisodes sporadiques contribuent très largement au dépôt annuel en poussières rouges : 60 % pour les seuls évènements de mai 1984. Des examens minéralogiques ont indiqué que ces poussières rouges contiennent jusqu'à 30 % de calcite et quelquefois du gypse, en moindre quantité. La dissolution partielle de la calcite explique la très nette augmentation du pH : celui-ci, souvent voisin de 5 pour des pluies non alimentées par des apports sahariens, atteint alors des valeurs comprises entre 6 et 7. Elle est aussi dans une large mesure à l'origine de la forte augmentation de l'enrichissement en calcium observée, à laquelle contribue par ailleurs la dissolution du gypse.

Nous avons d'autre part sélectionné quelques évènements significatifs des années 1981, 1984 et 1985 en vue d'un examen des variations de la composition ionique de l'eau de pluie en relation avec l'origine des masses d'air qui ont alimenté les précipitations. Les résultats rassemblés dans le tableau 1 permettent de distinguer des épisodes avec influence de masses d'air provenant du continent européen, des apports de poussières sahariennes et des épisodes mixtes où il y a interaction entre ces deux types d'apports. Les trajectoires des masses d'air correspondant à chaque situation sont représentées sur les figures 2 à 12. Elles ont été établies par la Météorologie Nationale Française à 2 niveaux (A : 700 hPa, B : 850 hPa), en prenant des origines de temps de 4 jours avant l'arrivée des masses d'air sur le site de prélèvements, les segments représentés sur les trajectoires équivalant à des durées de 6 heures.

On a enfin déterminé le bilan ionique de nos échantillons en considérant la somme des concentrations des ions dissous (anions + cations) exprimées en microéquivalents par litre. Ceci nous a amené à évaluer la contribution des diverses espèces analysées exprimée en % des concentrations ioniques totales. La somme des contributions des anions et des cations sont chacunes le plus souvent comprises dans des limites de 50 + 1 %. A titre d'exemple, nous avons représenté sur les figures 13, 14 et 15 les résultats obtenus pour trois types de situations météorologiques.

Il apparaît que lorsque les masses d'air ont balayé différentes régions industrialisées de l'Europe (figures 2 à 5), le pH de l'eau de pluie est relativement acide et voisin de 4. Les échantillons présentent alors de fortes concentrations en sulfates, en nitrates et en ammonium, ces dernières étant toutefois moins élevées (tableau 1 et figure 13). Bien qu'il existe des mécanismes naturels d'acidification des précipitations (8, 9), ce résultat s'explique essentiellement par l'influence prépondérante d'un aérosol de pollution contenant de l'acide sulfurique et en moindre quantité de l'acide nitrique.

* l'enrichissement en calcium, qui est un indicateur de l'influence des apports en calcium d'origine non marine est défini par la relation
E Ca = Ca/Na eau de pluie / Ca/Na eau de mer

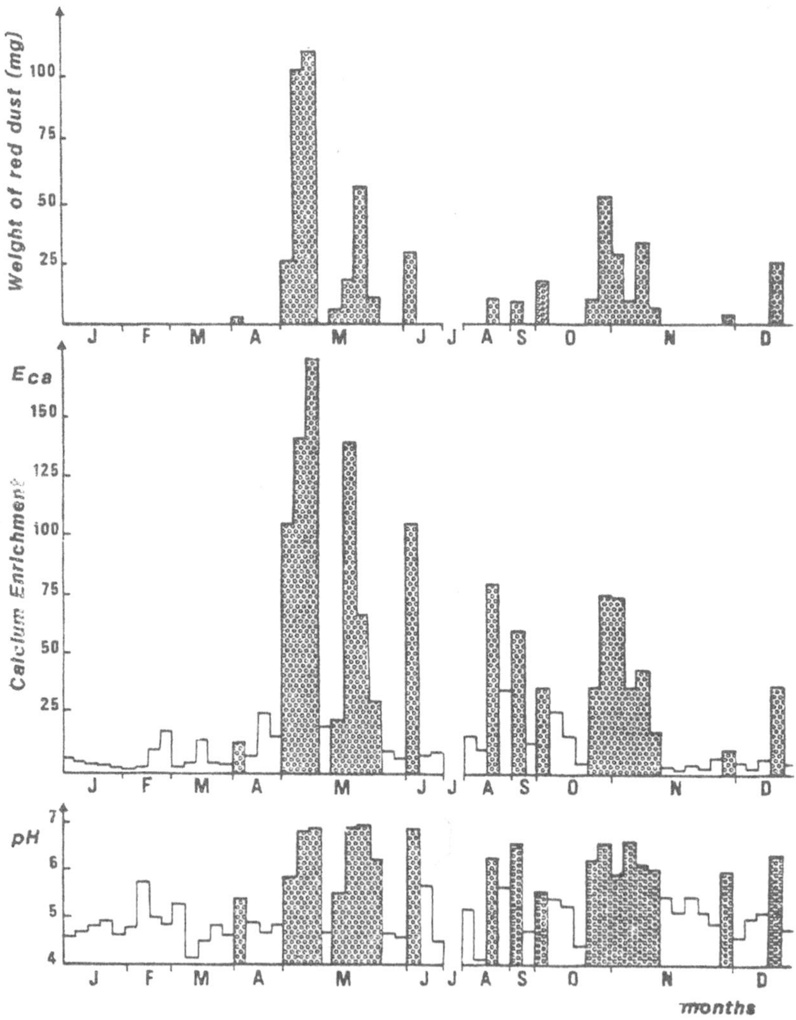

Figure 1. Variations simultanées du contenu en poussières rouges, de l'enrichissement en calcium et du pH des échantillons d'eau de pluie collectés pendant l'année 1984.

Simultaneous variations of the content of rain water samples in red dust, their calcium enrichment and their pH during the year 1984.

N° de l'échantillon	Date de l'épisode pluvieux prélevé	Origine des masses d'air	Hauteur d'eau (mm)	pH	Concentrations (en $mg.l^{-1}$)				Concentrations (en $mg.l^{-1}$)				
					Cl^-	SO_4^{--}	HCO_3^-	NO_3^-	Na^+	K^+	Ca^{++}	Mg^{++}	NH_4^+
EPISODES AVEC INFLUENCE DU CONTINENT EUROPEEN SITUATIONS WITH ATMOSPHERIC INPUT FROM THE EUROPEAN CONTINENT													
1	20 Fev 1981	voir fig. 2	7,0 (neige)	4,10	1,60	7,55	0,00	0,63	0,74	0,23	0,30	0,16	1,11
2	7-8 Nov 1981	voir fig. 3	11,8	4,07	2,00	4,50	0,00	3,10	0,87	0,21	0,32	0,10	0,90
3	8 Mars 1984	voir fig. 4	28,7	4,16	5,50	5,53	0,00	2,89	2,67	1,29	0,38	0,36	1,12
4	20-25 Nov 1985	voir fig. 5	3,1	3,89	18,60	10,90	0,00	8,10	9,12	4,52	0,75	1,05	1,15
EPISODES TYPIQUEMENT SAHARIENS SITUATIONS WITH INFLUENCE OF SAHARAN DUST													
5	28-29 Juin 1981	voir fig. 6	6,0	7,18	3,22	7,50	22,20	n.d	1,60	0,43	8,48	0,10	n.d
6	4-5 Mai 1984	voir fig. 7	42,0	6,81	3,14	3,95	11,10	1,28	1,04	0,16	5,72	0,26	0,15
7	12-13 Mai 1985	voir fig. 8	15,0	7,14	1,50	4,40	22,30	1,38	0,96	0,32	7,90	0,23	0,20
8	18-20 Mai 1985	voir fig. 9	8,0	6,82	1,10	4,10	10,40	2,10	0,50	0,25	3,32	0,22	0,37
EPISODES MIXTES MIXED SITUATIONS													
9	15-16 Nov 1985	voir fig.10	27,7	5,84	0,48	1,74	0,50	0,74	0,10	0,22	0,62	0,05	0,32
10	16-18 Nov 1985	voir fig.11	18,9	5,80	0,58	2,31	0,40	0,86	0,15	0,30	0,60	0,04	0,54
11	18-20 Nov 1985	voir fig.12	18,0	4,46	1,04	2,30	0,00	1,16	0,37	0,45	0,11	0,06	0,31

Tableau 1 : Composition chimique de précipitations collectées au Col de Bavella (Corse du Sud) n.d. = non déterminé
Chemical composition of precipitations collected at the Bavella Pass (South Corsica)

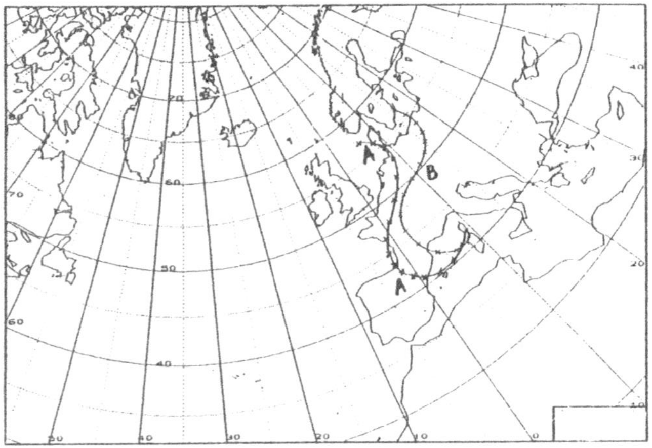

Figure 2. Situation du 20 février 1981 à 6 h T.U.
Air masses backtrajectories - A : 700 hPa, B : 850 hPa

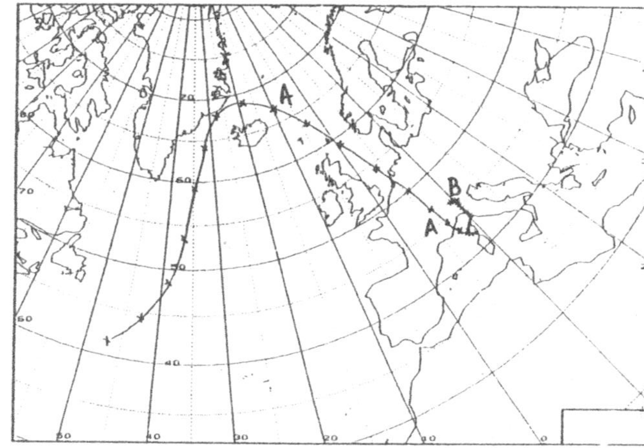

Figure 3. Situation du 7 novembre 1981 à 18 h T.U.
Air masses backtrajectories - A : 700 hPa, B : 850 hPa

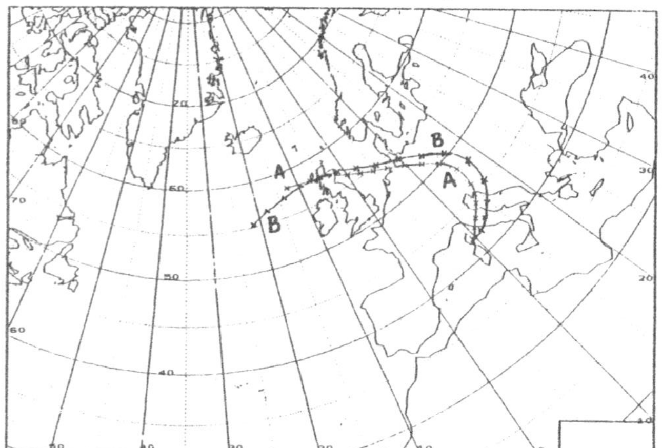

Figure 4. Situation du 8 mars 1984 à 18 h T.U.
Air masses backtrajectories - A : 700 hPa, B : 850 hPa

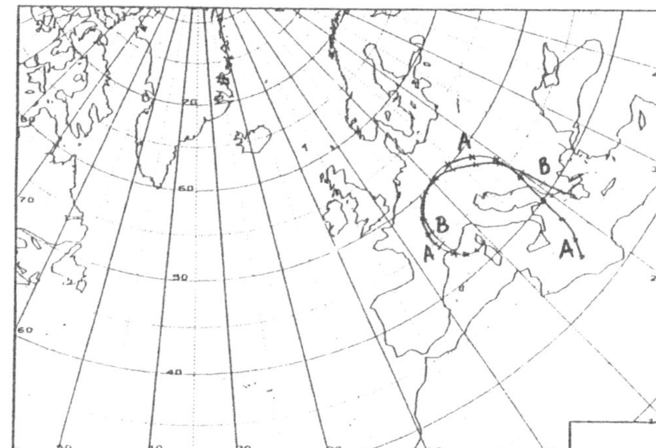

Figure 5. Situation du 23 novembre 1985 à 6 h T.U.
Air masses backtrajectories - A : 700 hPa, B : 850 hPa

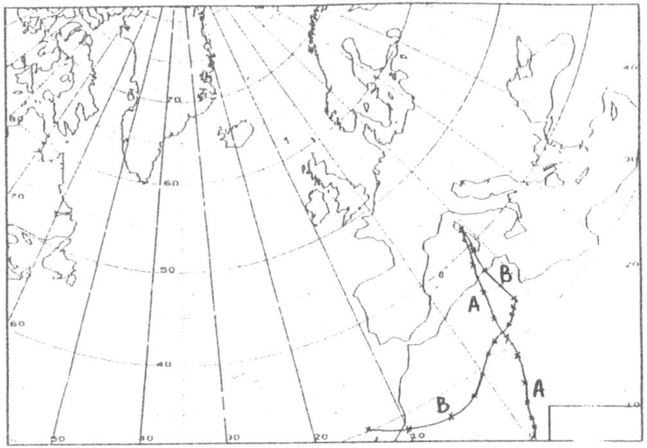

Figure 6. Situation du 28 juin 1981 à 12 h T.U.
Air masses backtrajectories - A : 700 hPa, B : 850 hPa.

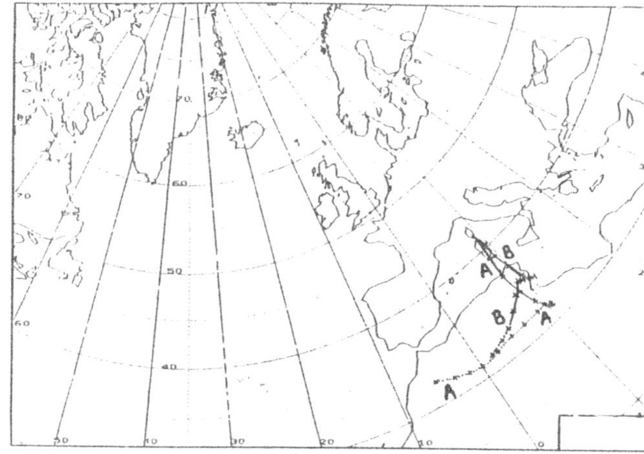

Figure 7. Situation du 4 mai 1984 à 0 h T.U.
Air masses backtrajectories - A : 700 hPa, B : 850 hPa.

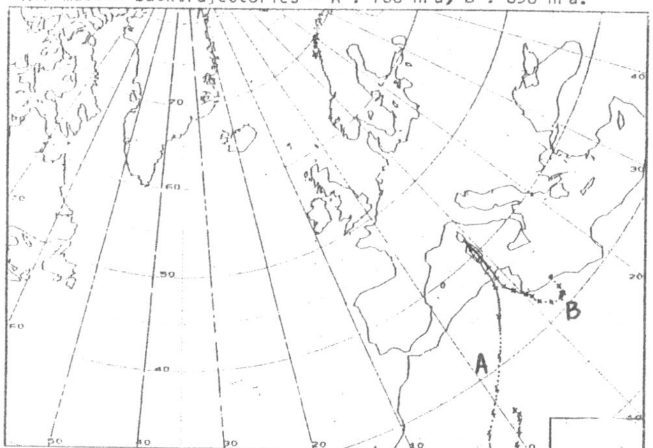

Figure 8. Situation du 13 mai 1985 à 12 h T.U.
Air masses backtrajectories - A : 700 hPa, B : 850 hPa

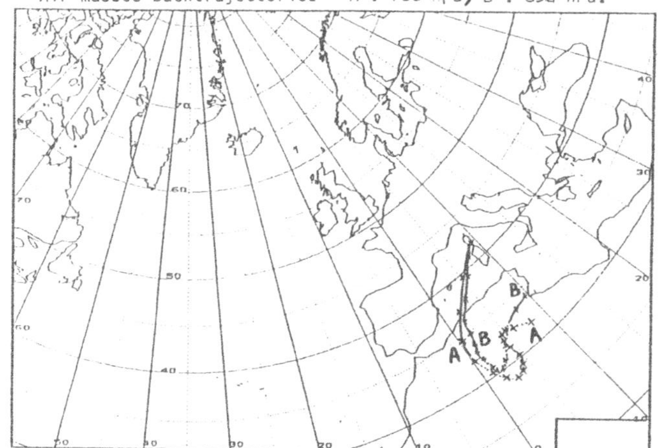

Figure 9. Situation du 19 mai 1985 à 0 h T.U.
Air masses backtrajectories - A : 700 hPa, B : 850 hPa.

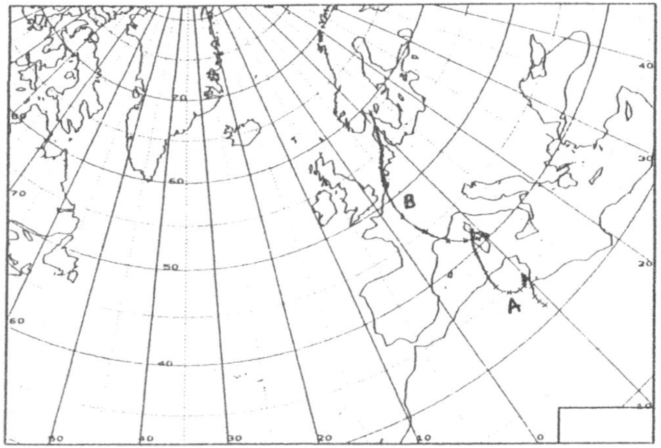

Figure 10. Situation du 16 novembre 1985 à 0 h T.U.
Air masses backtrajectories - A : 700 hPa, B : 850 hPa

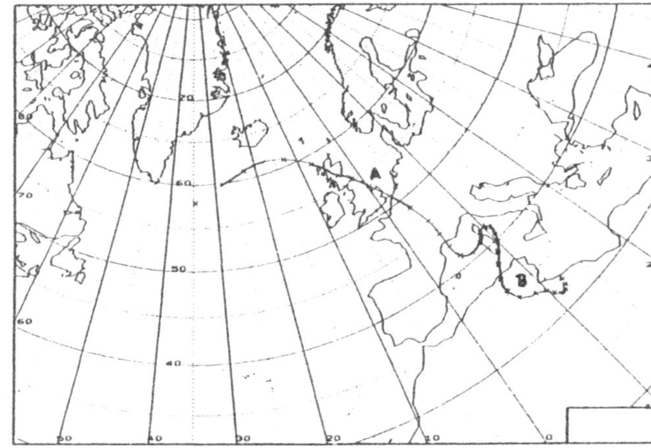

Figure 11. Situation du 18 novembre 1985 à 12 h T.U.
Air masses backtrajectories - A : 700 hPa, B : 850 hPa.

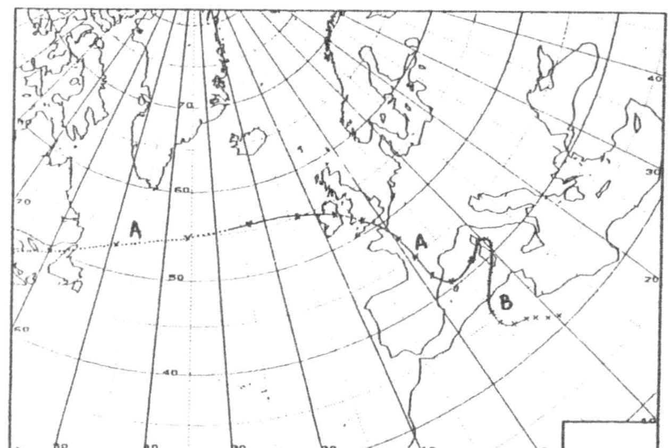

Figure 12. Situation du 19 novembre 1985 à 0 h T.U.
Air masses backtrajectories - A : 700 hPa, B : 850 hPa

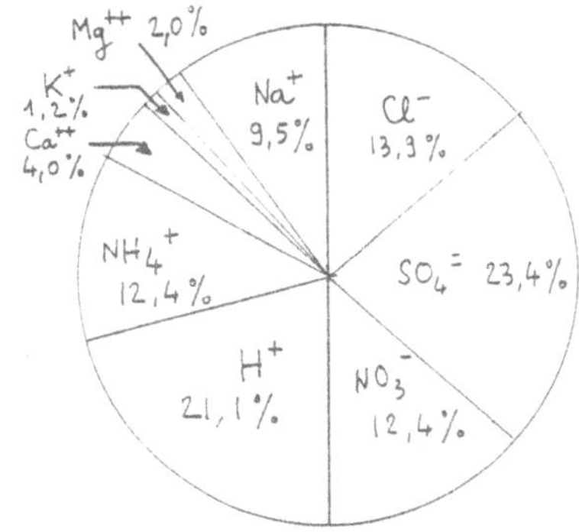

<u>Figure 13.</u> Bilan ionique de la pluie dans une situation avec influence d'apports atmosphériques venant du continent européen (7-8 novembre 1981)
Ionic balance for a situation with atmospheric input from the European continent (novembre 7-8, 1981).

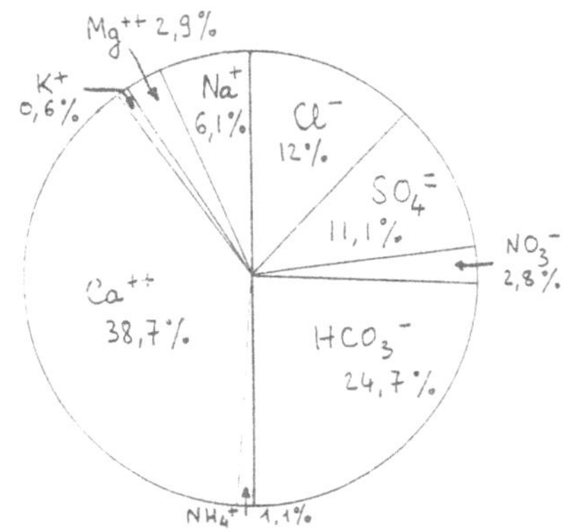

<u>Figure 14.</u> Bilan ionique de la pluie pour un épisode d'apport de poussières sahariennes (4-5 mai 1984)
Ionic balance for a situation with Saharan dust influence (May 4-5, 1984).

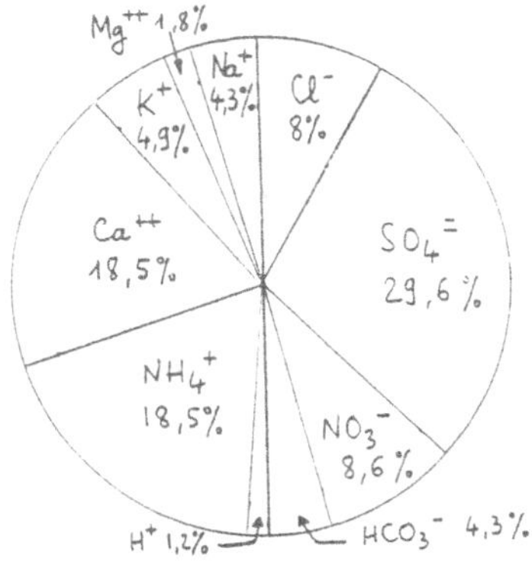

<u>Figure 15.</u> Bilan ionique de la pluie pour une situation mixte (16-18 novembre 1985)
Ionic balance for a mixed situation (November 16-18, 1985).

Dans le cas d'épisodes typiquement sahariens (figures 6 à 9) la composition ionique est complètement modifiée. Du fait de la dissolution de la calcite, les concentrations en bicarbonate et en calcium deviennent très fortes et contribuent à l'essentiel des concentrations ioniques totales (tableau 1 et figure 14). La dissolution de particules de gypse pourrait rendre compte en partie des concentrations en calcium et en sulfates observées, ces dernières restant relativement élevées (tableau 1).

Dans le cas d'interactions entre des masses d'air venues du Sahara et de l'Europe (figures 10 à 12), les résultats présentés correspondent à des pluies assez peu minéralisées. On constate que les sulfates constituent l'espèce ionique de loin la plus abondante (tableau 1 et figure 15). Ce résultat peut être attribué en partie à l'addition de sulfates résultant de la dissolution de particules de gypse d'origine désertique à un aérosol soufré de pollution. L'apport désertique contribue à l'essentiel du calcium dissous tandis que l'influence de la pollution européenne pourrait expliquer aussi les concentrations en nitrates et ammonium mesurées.

On remarque enfin que quelle que soit l'origine des masses d'air arrivant en Corse, leur passage au-dessus de la Méditerranée fait que l'influence de la source marine sur la composition de l'eau de pluie est loin d'être négligeable. Ainsi, si l'on admet que tout le sodium collecté provient d'un aérosol marin de pétillement reflètant les rapports d'abondances des espèces ioniques présentes dans l'eau de mer, il apparaît que selon les situations prises en compte, l'apport en sels marins recyclés intervient pour 10 à 60 % dans le bilan ionique. Il explique l'essentiel des chlorures et parfois du magnésium et contribue à une partie des sulfates marins, une autre partie de ceux-ci résultant de la transformation de gaz biogéniques organosoufrés émis à la surface de la mer (10, 11).

En conclusion, on retiendra que la dispersion de poussières sahariennes au-dessus du bassin méditerranéen est susceptible d'avoir un impact important pour cette région. Dans des situations de mélange entre masses d'air sahariennes et européennes, ces poussières tendent à neutraliser l'acidité atmosphérique. Elles constituent donc un allié inattendu contre les effets de la pollution atmosphérique au Sud de l'Europe. De plus, il existe des interactions entre ces poussières et les polluants de l'air après qu'ils aient été déposés sur les sols et à la surface de la mer. Ceci concerne notamment les carbonates des aérosols sahariens qui, n'étant que partiellement dissous dans l'eau des précipitations, pourront neutraliser au niveau du sol l'acidité des pluies ultérieures. Ces processus devraient avoir une incidence sur les écosystèmes terrestres et marins de l'environnement méditerranéen. On rappellera enfin qu'en 1984, le flux de poussières sahariennes déposées au Sud de la Corse à été évalué à 14 Tonnes/Km^2.an (1). A l'échelle de la moitié Nord du bassin méditerranéen occidental, un tel flux représente un apport équivalent à la quantité de matière solide que déverse le Rhône, soit environ 4.10^6 Tonnes.an.$^{-1}$. Même si cette extrapolation appelle des réserves, il s'avère donc que du point de vue de la géochimie marine l'apport atmosphérique dans cette région est un phénomène majeur.

Rermerciements

Ce travail a été réalisé avec le soutien du laboratoire de Biologie-Géologie de l'ENSJF et celui du CNRS - GRECO ICO (Interactions-Continent-Océan) et ATP Aerosols désertiques -. Nous remercions Messieurs N. Kielbasa et J.A. Lefevre pour leur collaboration technique.

REFERENCES

(1) LOYE-PILOT M.D., MARTIN J.M. & MORELLI J., Nature 321, N° 6068, 427-428 (1986).

(2) GANOR E. & MAMANE V., Atmos. Envir. 10, 1079-1084 1979).

(3) TOMADIN L., LENAZ R., LANDUZZI V., MAZZUCOTELLI A. & VANNUCCI R., Oceanol. Acta 7, 13-23 (1984).

(4) CHESTER R., SHARPLES E.J., SANDERS G.S. & SAYDAM A.C., Atmos. Envir. 19,929-935 (1984).

(5) CLEMENT R., RICQ de BOUARD M. & THOMAS A., La Météorologie (Paris) 24, 65-83 (1972).

(6) PRODI F. & FEA G., J. Geophys. Res. 84, C11, 6951-6960 (1979).

(7) BÜCHER A. & LUCAS C., La Météorologie (Paris) 33, 53-69 (1975).

(8) CHARLSON R.J. & RODHE H., Nature 295, 683-685 (1982).

(9) LEGRAND M., DELMAS R.J. & ZANOLINI F., in Proceedings of the 3rd European Symposium on Physico-chemical behaviour of Atmospheric Pollutants (Varèse, Italy), Reidel Publishing Company, Dordrecht,(1984).

(10) BONSANG B., NGUYEN B.C., GAUDRY A. & LAMBERT G., J. Geophys. Res., 85, 7410-7416, (1980).

(11) CARLIER P., LUCE C., GIRARD R., MOUVIER G., MORELLI J., GIRARD-REYDET L., MARCHAL T. & CADENE S., in Proceedings of the 3rd European Symposium on Physico-chemical behaviour of Atmospheric Pollutants (Varèse, Italy), Reidel Publishing Company Dordrecht, (1984).

Acid deposition in a snow field at 2500 m.a.s.l. in Switzerland.

Albrecht Neftel and Andreas Sigg
Physikalisches Institut, Universität Bern CH - 3012 Bern

Fritz Zürcher
EAWAG, Dübendorf CH - 8600 Dübendorf

Abstract

In Alpine regions the precipitation is stored as cold snow in the winter half year. Meltwater flowing through the snow cover in springtime carries the impurities along and leads to a strong input of pollution into lakes and underground water systems. During winter 1985/86 three profiles through a snow cover situated 2540 meter above sea level in Switzerland have been analysed for the main chemical trace species. The measurements show that the acidity is dominated by nitric acid, while sulfate is deposited mostly as neutral salt. Hydrochloric acid and formic acid are only minor contributors to the total acidity. As long as no meltwater or rain flows through the snow cover, the main chemical impurities are conservative tracers.

1.) Introduction

In alpine regions precipitation falls as snow during the winter half year and is conserved until spring. As long as the temperature in the snow cover is below 0°C, no melt water is flowing through the snow cover. Fractionation during the spring melt leads to spikes of acidity in the runoff water (Tranter et al, 1986).
We investigated the behaviour of chemical impurities in the experimental snow cover of the Swiss Snow and Avalanche Research Institute, Weissfluhjoch Davos. The field is located at an elevation of 2540 m.a.s.l.. Concentrations of H^+, Ca^{2+}, Na^+, K^+, NH_4^+, Cl^-, NO_3^-, SO_4^{2-} and $HCOO^-$ were determined in three profiles sampled at the 1/30, 2/28 and 4/15. Parallel to this analysis H_2O_2 was measured. We tried to give an answer to following questions:

- Are the main impurities conservative tracers?

- Can the acidity be specified?

- What are the mean concentrations at this altitude and how compare they to other european measurements from alpine stations?

H_2O_2 behaves differently from all other species. It is discussed in an other paper presented at this conference (Neftel et al, this conference).

2.) Experimental

The snow studies were done at the experimental snow field of the Swiss Snow and Avalanche Research Institute, Davos-Weissfluhjoch. Profiles through the snow cover are routinely taken each second week. The surface is marked with a thread, so that corresponding layers can be identified later. The samples were taken with a drawer made out of polycarbonate fixed on a mast with meter measure, in order to provide equally spaced samples. The snow samples were collected in 150ml polypropylene beakers and stored at -20°C until treatment.
The preparation of the samples for the subsequent analysis was performed in an overpressure box flushed with pure nitrogen in order to exclude contact with laboratory air. As a first step the samples were transferred into polyethylene bags and hermetically sealed. The snow in the bags was then rapidly melted by means of an IR spot light. 2ml of the raw samples were used for ammonia determination by gas diffusion injection (Zuercher and Gisler, 1986). The rest was filtered through .45μm membrane filter (Gelman Polysulfone ACRODISC φ25mm). The same prewashed filter assembly was used for one series (20-40 samples). An aliquot, adjusted to 0.05M KCl, was subsequently taken for pH determination (Johnsen and Sigg, 1985). Major cations such as Ca^{2+}, Na^+, K^+ were determined on a Perkin Elmer absorption spectrophotometer using an air-acetylene flame. For Ca^{2+} determination a flow injection system with lanthanum addition was used (Olsen et al 1983). Anions were determined by ion chromatography (IC). The IC apparatus was equipped with anion separator AS-3 from Dionex combined with a Dionex standard fiber suppressor and a Wescan conductivity detector. A pneumatic driven injection valve with .5ml sample loop was used. The injection port was separatly mounted in an overpressure box. As eluent a sodiumbicarbonate/carbonate mixture was used. Samples were spiked prior injection to have the same bicarbonate concentration as the eluent. Ion exclusion post column-separation was used to separate formiate from interferring carbonate, acetate and fluoride (Pimminger et al, 1985).

3.) Results

Fig. 1 shows the evolution of the accumulation as function of time starting at 1st of november. During the winter the snow completely recristallises by dry metamorphism enhanced by temperature gradients. Fig. 2 shows the temperature recorded in the three profiles. The bottom is slightly below 0°C, while the surface reflects the air temperature. The strongest temperature gradients are found near the surface, due to the daily air temperature variations. There also the fastest recristallisation occurs, leading to a substantial loss of

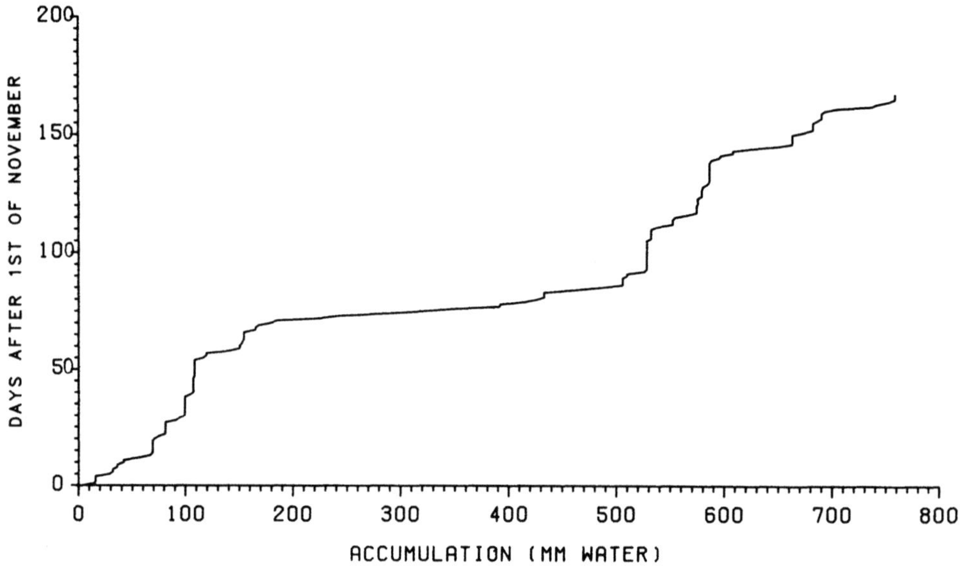

Fig. 1 Development of accumulation with time.

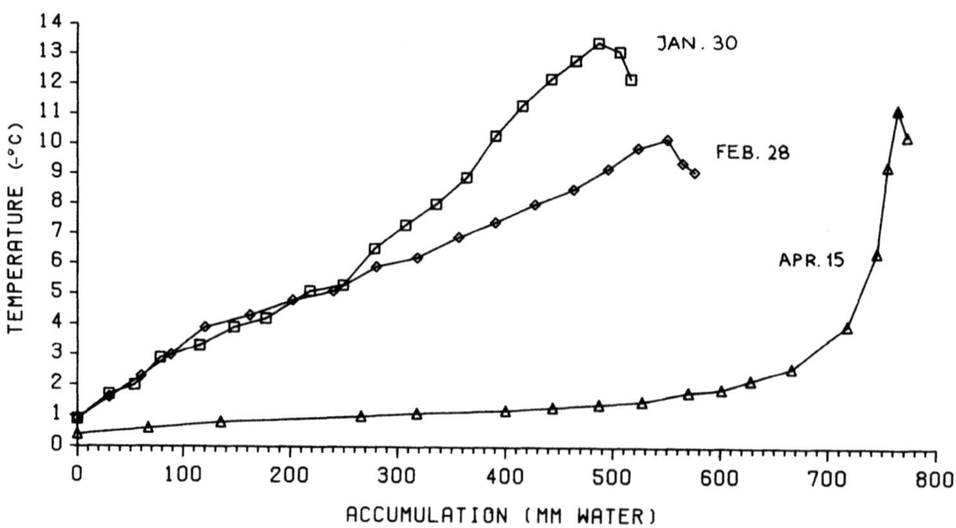

Fig. 2 Temperature distribution in the three profiles

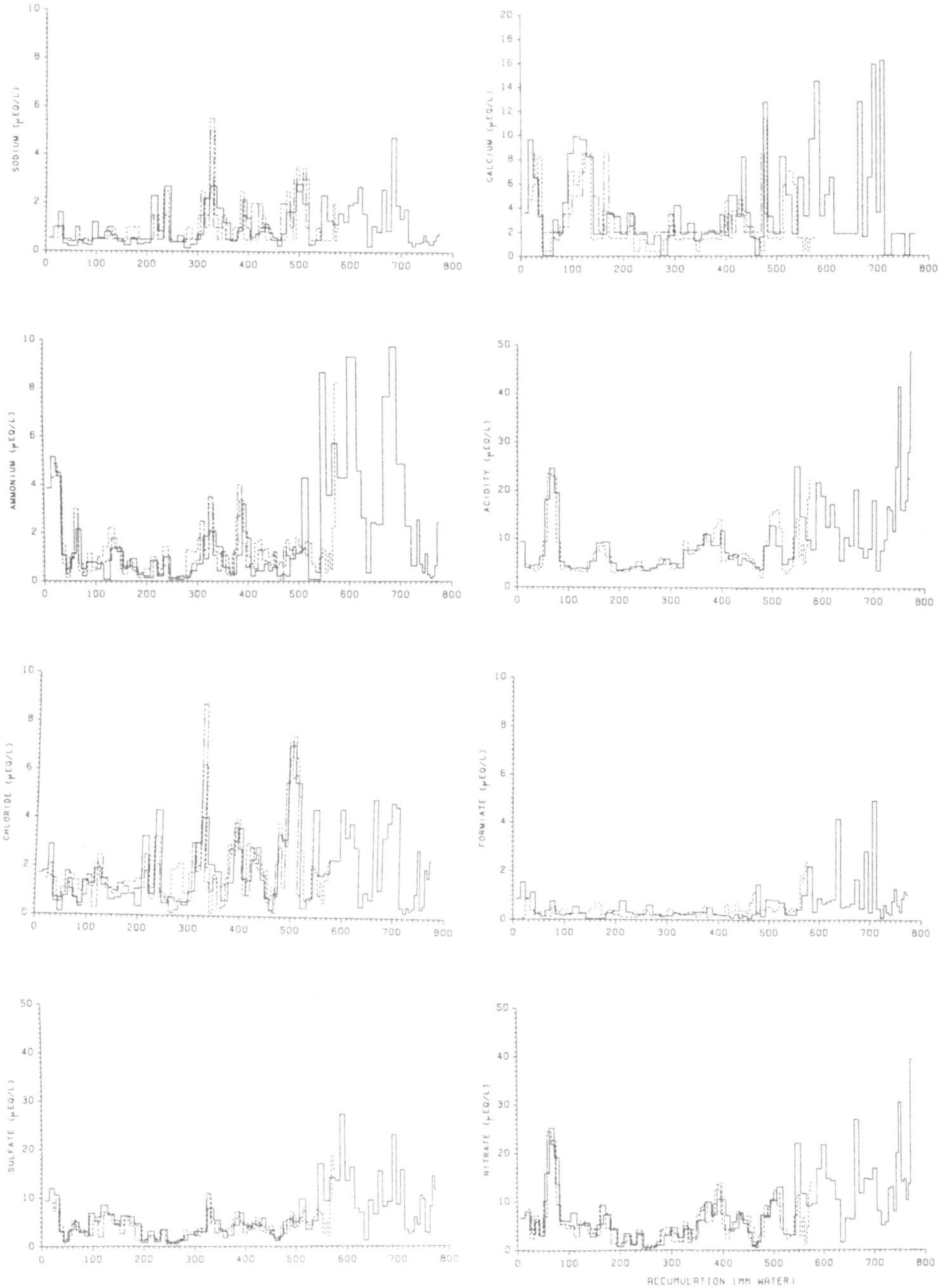

Fig. 3 Concentrations measured in the three profiles. H+ and HCOO- were not properly measured in the first profile and are not shown.

specific surface. Fig. 3 shows the measured concentrations in the three profiles as function of the accumulation in mm water equivalent above ground. The sample size corresponds to ca. 10mm water equivalent precipitation. The ion balance, plotted in Fig. 4 for the third profile is not always satisfied, due to not measured ions, such as Mg^{2+}, CH_3COO^-, HCO_3^- and oxalate, and cumulated analytical errors. The agreement between the three profiles is surprisingly good. An exeption is the top of the second profile (575mm w.equ.), where higher concentrations were measured in the third profile. The occurence of melt layers and contribution of dry deposition are reasons for this shift (see below). For comparison of the mean levels we subdivided the profiles in three equal accumulation intervals of 250mm w.equ. each. As it can be seen from Fig.1 the first and the third intervals cover a time span of 75 resp. 80 days, while the second interval covers only a time period of 10 days. The mean concentrations are listed in table 1. There is a slight indication of a concentration increase in the bottom part for SO_4^{2-} and NO_3^- which is barely significant, taking the error of the mean value, thus $\sigma_n/\surd n$ as representative for the uncertainty. In the second interval no concentration change is visible. From these considerations we conclude that the main impurities in the examined cold snow cover are conservative tracers. Also the spatial variabilty over a few meters must be small.
It can also be excluded, that the ventilated snow cover adsorbs pollutants such as NO_x or SO_2 which are later oxidized and absorbed by the snow crystals as acids.
This allows to determine the potential "frozen in acidity" in a snow cover with one profile, preferentially taken at the end of the winter, just before the melt water runs through.

Impurities in the snow originate from wet and dry deposition. There is no direct way to distinguish both contributions by only measuring the concentration profile in the snow cover. But the concentrations of the main impurities are not significantly higher after long periods without precipitation, thus dry deposition can only be a minor contribution to the total deposition. This is in agreement with reported deposition velocties in the literature (Georgii et al 1986, Cadle 1986). For SO_2 they lie in the range of 0.01 - 0.1 cm/sec. Assuming a mean SO_2 concentration of 2 $\mu g/m^3$ in the air at the Weissfluhjoch (NABEL, 1983) the contribution of dry deposition would amount to 1.6 to 16 mgS/m^2 for a period of 180 days, which is small compared to the total deposition of 180mgS/m^2 in the whole snow pack. This is supported by the fact, that we never found S(4) in the snow samples (detection limit: 1 ppbS).
The occurence of melt layers, i.e. melting of snow at the surface due to intensive radiation, leads to a redistribution of impurities along the path of the water, in our case only a few centimeters. Soluble impurities originally concentrated at the surface of the snow flakes or in grain boundaries will be carried with the melt water and concentrated in the refrozen layer. This is visible in the upper part of the third profiles, where six melt layers were recorded. The described

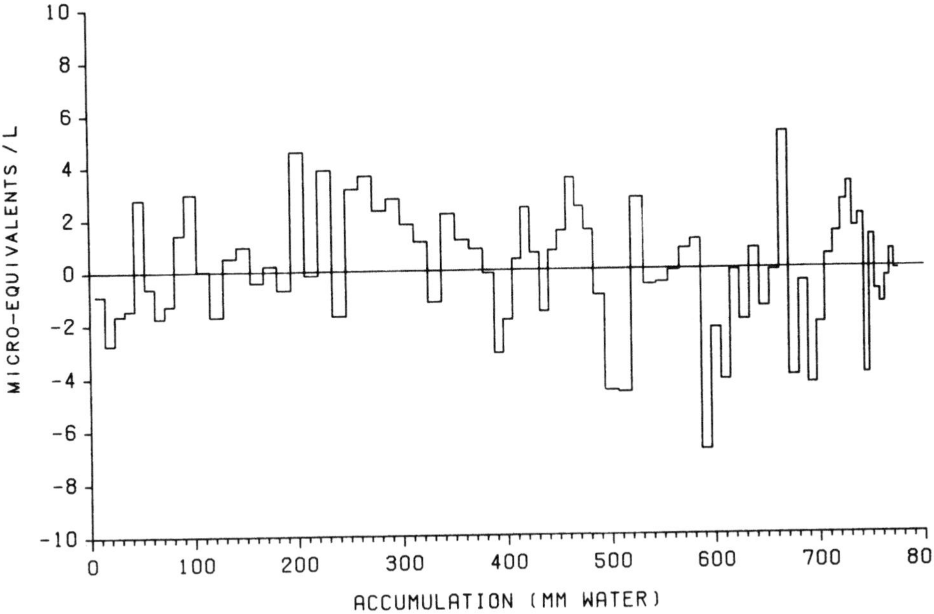

Fig. 4 Ionic balance of the second and third profile.

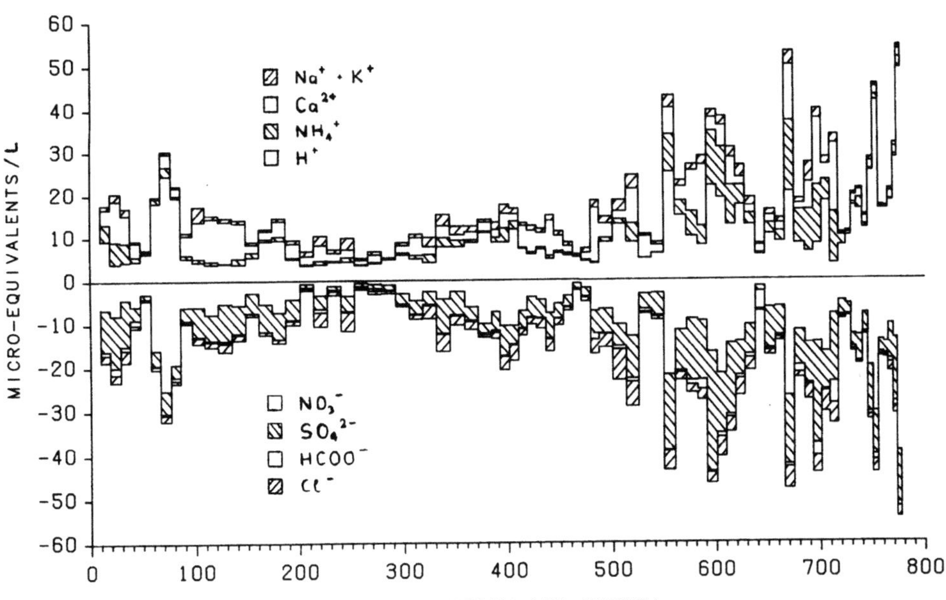

Fig. 5 Compilation of chemical species measured in the third profile.

TABLE 1. Mean values (μequivalent/liter) (H_2O_2: μmol/liter)

Species	Cl^-	NO_3^-	SO_4^{2-}	$HCOO^-$	H^+
Accumulation interval: 0 - 250 mm water					
Profile 1	1.2 ± 0.1	6.1 ± 1.2	3.6 ± 0.5	-	-
Profile 2	1.4 ± 0.1	6.3 ± 1.2	4.1 ± 0.4	0.3 ± 0.04	6.4 ± 1.1
Profile 3	1.3 ± 0.2	6.7 ± 1.2	5.0 ± 0.6	0.4 ± 0.08	7.3 ± 1.1
Accumulation interval: 250 - 500 mm water					
Profile 1	2.2 ± 0.4	5.5 ± 0.6	3.9 ± 0.4	-	-
Profile 2	2.2 ± 0.3	5.5 ± 0.7	4.0 ± 0.4	0.3 ± 0.04	7.3 ± 0.6
Profile 3	1.8 ± 0.4	5.1 ± 0.6	3.8 ± 0.4	0.3 ± 0.06	6.7 ± 0.4
Accumulation interval: 500 - 750 mm water					
Profile 3	2.3 ± 0.4	13.3 ± 1.6	9.6 ± 1.2	1.0 ± 0.2	15.3 ± 1.9

Table 1 (continued)

Species	Ca^{2+}	K^+	NH_4^+	Na^+
Accumulation interval: 0 - 250 mm water				
Profile 1	3.8 ± 1.2	0.2 ± 0.04	1.0 ± 0.2	0.6 ± 0.05
Profile 2	3.1 ± 0.5	0.3 ± 0.03	1.2 ± 0.2	0.6 ± 0.08
Profile 3	4.0 ± 0.6	0.4 ± 0.1	1.3 ± 0.3	0.6 ± 0.1
Accumulation interval: 250 - 500 mm water				
Profile 1	2.4 ± 0.2	0.3 ± 0.02	1.0 ± 0.2	1.2 ± 0.2
Profile 2	2.2 ± 0.4	0.3 ± 0.03	1.3 ± 0.2	1.1 ± 0.2
Profile 3	3.1 ± 0.6	0.3 ± 0.04	0.8 ± 0.2	0.9 ± 0.1
Accumulation interval: 500 - 750 mm water				
Profile 3	4.3 ± 0.9	0.4 ± 0.05	4.4 ± 0.9	1.1 ± 0.2

CORRELATION COEFFICIENTS (N = 78)

	Cl^-	NO_3^-	SO_4^{2-}	$HCOO^-$	H^+	Ca^{2+}	NH_4^+	Na^+
Cl^-	1	.36	.50	.24	.18	.44	.54	.82
NO_3^-	.36	1	.53	.13	.90	.12	.46	.20
SO_4^{2-}	.50	.53	1	.37	.39	.57	.81	.42
$HCOO^-$	.24	.13	.37	1	.03	.49	.39	.12
H^+	.18	.90	.39	.03	1	-.18	.19	.01
Ca^{2+}	.44	.12	.57	.49	-.18	1	.55	.37
NH_4^+	.54	.46	.81	.39	.19	.55	1	.56
Na^+	.82	.20	.42	.12	.01	.37	.56	1

concentration mechanism is not seen for $HCOO^-$. We believe that HCOOH is homogeneously incorporated in the snow and therefore not much affected by the melt layer formation.
We can restrict the discussion of the specification of the acidity on the last profile, where the longest time period is covered. Fig. 5 shows all concentrations measured in the third profile. Table 2 shows the correlation matrix of the measured species. Significant correlations on the 95% level are found between most of the species, with exeption of H_2O_2 and formiate. This reflect the somewhat trivial fact, that the general level of impurities is modulated by the meteorological conditions. The high correlation between H^+ and NO_3^- (>0.8) and the similar concentration levels show that the acidity is dominated by nitric acid. Also Cl^- and Na^+ as well as SO_4^{2-} and NH_4^+ show correlation coefficients >.8, but the anion concentrations exceed the cation concentrations. We conclude that NH_4^+ is entirely deposited as ammoniumsulfat and sodium is coming entirely from sea salt. The contribution of hydrochloric acid is then calculated to be between 0.6 and 1µequ/l, thus ca. 10% of the total acidity. The contribution of formic acid from November to February is 0.4µequ/l and rises to 1µequ/l in spring time.
It can principally not be discriminated whether calciumcarbonate is already neutralized in cloud droplets or after the samples were melted. Calciumcarbonate already incorporated into cloud droplets will play a similar role as NH_3 by enhancing the uptake of SO_2 with subsequent oxidation (Zürcher and Gisler, this conference).
Part of the sulfate in the snow cover is of local origin. Davos, a large ski resort with up to 40'000 inhabitants during the winter, acts as a large SO_2 source. Concentrations up to 100 $\mu g/m^3$ (daily average) have been measured in the winter half year (Moser, 1985). Part of this SO_2 reaches the Weissfluhjoch region where it will be oxidized and deposited as sulfuric acid in the snow. Because we never detected sulfite in the snow samples, the sulfur has to be oxidized before it is incorporated in the snow.

How compare our results with other measurements of precipitation from the alpine region? Continuous records from the last years exist from Zugspitze (2964 m.a.s.l.) located in the bavarian alps and from the Jungfraujoch (3540 m.a.s.l) in the bernese oberland. A principle difference exists of course in the way of sampling between our study and the other two stations, where precipitation was directly collected. Fresh snow has to be collected mostly with a wet-only collector during the precipitation event, which is often accompanied by strong winds. There is a tendency to collect with higher efficiency light precipitation which has often higher impurity concentrations. The mean levels recorded at these stations are effectively higher. Cl^-, SO_4^{2-} and NH_4^+ are 3 to 5 times higher in the precipitation from the Zugspitze (Reiter et al, 1985) than from the Weissfluhjoch, while the nitrate levels are twice as high. Nitrate concentrations in precipitation collected at the Jungfraujoch has the same order

of magnitude, while NH_4^+ and SO_4^{2-} are roughly twice as high (Fuhrer, 1986). At the Jungfraujoch the acidity is low compared to the sulfate and nitrate level. Most of the acidity seems therefore to be neutralized either by calciumcarbonate or ammonia. At the Zugspitze similar equivalents of ammonium and sulfate were reported.
Both Zugspitze and Jungfraujoch are located at the edge of the alps, while the Weissfluhjoch is situated in the interior. this might be an other explanation for the different levels found at these stations. Polluted air risen to high elevations could start to precipitate at the edge, leading to relatively high impurity concentrations in the snow. Precipitating air masses could well reach the Weissfluhjoch later, so that a large fraction of the aerosol load in the cloud is already washed out.

The general increase of the NH_4^+, SO_4^{2-}, NO_3^- and Ca^{2+} going versus spring time is visible in all three stations. Several reasons are responsible for this trend:

- Increasing solar radiation leads to a stronger vertical air mass exchange. Polluted air is more easily transported to higher elevations.
- Increasing solar radiation leads indirectly to a faster oxidation of the precursors of the strong acids.
- The snow cover is declining so that more dust, especially limestone particles are deposited.
- Increasing biological emissions from soil (NH_3 and NO_x).

Conclusions

The analysis showed that a cold snow cover can be regarded as a total deposition sampler which allows to investigate the acidic deposition in the alpine region during winter. The strong acidity in the snow cover is largely dominated by nitric acid, while sulfate is predominantly deposited as neutral salt. All investigated chemical species with the exeption of H_2O_2 are conservative tracers. A profile taken at the end of the winter allows to trace the acidic deposition pattern of the winter half year. The recorded concentrations are low compared to precipitation analysis from other alpine research stations (Zugspitze and Jungfraujoch).

Acknowledgments

We thank Prof C. Jaccard and H. Beck from the Swiss Snow and Avalanche Research Institute for the excellent support, Profs. H. Oeschger and W. Stumm for the opportunity to do this work and their continuous interest. Claudia Mäder, Ursula Michel, Bruno Gisler and David Kistler, all at the EAWAG, helped with the measurements. The project was supported by COST action 611.

References:

Cadle S.H. and Dasch J.M.
The contribution of dry deposition to the acidity of snow in northern Michigan. Proceedings of the NATO ASI summer school "Chemical dynamics of seasonal snow covers" held july 1986 in Lesarc, France

Fuhrer J. (1986)
Study of acid deposition in Switzerland: Temporal variation in the ionic composition of wet precipitation at rural sites during 1983 - 1984.
Environmental Pollution (Ser. B), in press

Georgii H.W., Perseke C. and Rohbock E. (1986)
Investigation of the wet and dry deposition of acidic and heavy-metal components in Chemistry of multiphase atmospheric systems,
edited by W. Jaeschke.
NATO ASI Series Vol G6 227-248

Johnson C.A. and Sigg L. (1985)
Acidity of rain and fog: Conceptual definitions and practical measurements of acidity.
Chimia 39, 59-61

Moser H. (1985)
Die lufthygienischen Verhältnisse im Hochtal von Davos.
Schlussberichte zum schweizerischen MAB-Programm Nr.16
Bundesamt f. Umweltschutz CH 3003 Bern.

NABEL (1983)
Messresultate des Nationalen Beobachtungnetztes für Luftfremdstoffe.
Bundesamt f. Umweltschutz CH 3003 Bern.

Neftel A., Sigg A. and Jakob P. (1986)
H_2O_2 in solid precipitation. This conference.

Olsen S., Pessenda L.R.C., Ruzicka J. and Hansen E.H. (1983)
Combination of flow injection analysis with flame atomic absorption spectrophotometry.
The Analyst 108, 905-917

Pimminger M., Puxbaum H., Kossina I. and Weber M. (1985)
Simultane Bestimmung von einigen anorganischen und organischen Anionen durch ein ionenchromatographisches Drei-Säulen-System.
Fres. Z. Anal. Chemie 320, 445-450

Tranter M., Brimblecombe P., Davies T.D., Vincent C.E., Abrahams P.W. and Blackwood I. (1986)
The composition of snowfall, snowpack and meltwater in the scottish highlands-evidence for preferential elution
Atmospheric Environment 20 517-525

Zürcher F. and Gisler B. (1986)
Contribution of ammonia to the wet deposition of sulphur compounds. (This conference).

ACIDITY MEASUREMENTS OF SNOWFALLS COLLECTED OVER THE VOSGES IN RELATION TO AIR MASS TRAJECTORIES

J.L. COLIN - J.L. JAFFREZO - V. LESCOAT
D. RENARD - J. PINART

Laboratoire de Physico-Chimie de l'Atmosphère
Université PARIS VII - 2, place JUSSIEU
75251 PARIS cedex 05 FRANCE

Summary

The pluridisciplinary programme DEFORPA was recently launched on the VOSGES mountain range. One of its aspects consists in systematic monitoring of the wet deposition. Our short-term contribution concerns, in particular, the acidity of snowfalls, observed by means of an original technique for sequential sampling of each precipitation.
A sampling campaign was carried out between March and April 1985, at the "Col du Donon". 29 events related to various types of meteorological situation were studied.
By means of the air mass trajectories calculated at 925 hPa and supplied by the National Weather Center (Météorologie Nationale), we were able to classify these events according to 4 geographical sectors.
For this sampling period it appears that the East and North sectors show much lower average pH values (3.19 and 3.71) than the West and South (4.33 et 4.80). Moreover , chemical analysis indicates that nitrates are distinctly more important than sulphates in the East and North sectors (NO_3^- / SO_4^{2-} = 1.77 and 1.99 respectively).
Finally, but taking narrow fractions of each event (0.1 mm by 0.1 mm) it becomes possible to follow the evolution of the concentration of the most abundant salts during precipitation and to determine the relative contribution of the inorganic acids present.

1. INTRODUCTION

Recently the pluridisciplinary programme DEFORPA was set up in the East of France to try and more closely define the different theories proposed about the decline of the VOSGES forest. One of the main axes of this programme concerns the physical chemistry of the atmosphere because of, in particular, the disturbing increase of the acidity in the precipitations over the last few years (1). In order to characterise this effect, the programme was designed to follow the wet deposit continuously by means of collectors which open automatically when precipitation occurs.
However, the collectors are insuitable in the case of snow

precipitation which, nevertheless contributes significantly to the flux of fall out observed at this mountainous site.

Our laboratory has therefore undertaken to define a sampling system adapted to this problem and by short-term campaigns to characterise the acidity and the ionic composition of this type of deposit.

We present in this paper the main results obtained during a first winter campaign in March and April 1985. The original application of the sequential sampling mode devised in this laboratory will be discussed and illustrated through a particular event

2. EXPERIMENTAL SECTION

The wet deposit collector has been designed so as to work either in the cumulative or the sequential mode. In the first mode, each event is taken completely in a single fraction. In the second, on the other hand, the event is cut into fractions of known volume. This option determine the size of the collector because it must take fractions of the rain small enough to describe the precipitation finely (2), while remaining compatible with the amounts of water needed for the various analytical techniques employed. A good compromise is achieved with a collector area of $1 m^2$ which corresponds thus to fractions of 100 ml for 0,1 mm of water sampled. The collector consists of a square based funnel obtained by fusing 4 sheets of polyane together at an angle of 60°C. It rests on a collapsible metal frame, and this is supported by 4 telescopic legs which set the opening 1.5 m from the ground. The funnel has polyane flanges which completely cover the metal part, thus avoiding all danger of contamination by the armature (3). The system is covered with a polyane lid which closes it hermetically between two events and thus prevents dry deposit (4). The collector is uncovered manually at the beginning of precipitation and closed at the end, which means that there have to be operators in the field. The snow collected in the $1 m^2$ funnel in then melted in a second funnel made of stainless steel in a thermostatted heating chamber. This fonction is disconnected in the case of rain.

Sampling depends also on the mode choosen. In cumulative mode, the water is collected in a 5 l can. Only the times at the beginning and the end of the precipitation are recorded. In the sequential mode, the water is stored in preweighted 150 ml polyethylene bottles. The time is noted when each fraction is taken so that the variation of the intensity during the event can be measured afterwards. The samples are filtered immediately after collection to avoid as far as possible variations in the concentrations of the chemical species, due mainly to absorption/desorption mechanisms on the suspended particles (5).

The analytical part has previously been described in detail elsewhere (6). We shall briefly recapitulated here the main elements to be analysed. One measures in the soluble phase :

- basic physico-chemical data : pH conductivity ;
- the major species : Cl^-, NO_3^-, SO_4^{2-} by ion chromatography ;

Na,K , Mg, Ca by atomic absorption ; HCO_3^- if necessary by titration ; NH_4^+ by colorimetry ;
- certain metals followed as atmospheric tracers : Al, Fe, Zn, by atomic absorption.
The insoluble phase deposited on the 0.45 µm millipore filter membrane is analysed by x-ray fluorescence by means of well established procedures (7) for Cl, S, K, Ca, Zn, Al, Si, Fe.
In what follows we shall discuss only those species related to the acidity.

3. RESULTS AND DISCUSSION

The sampling campaign was carried out at the Col du Donon in the VOSGES massif at the site selected for the DEFORPA programme. It continued without interruption from March 07 to April 09. The precipitations collected during this campaign at the end of the winter period are divided into two groups, snows (20 events) or rains (9 events).
The events were collected alternately in the sequential and cumulative modes. In fact, sequential collection considerably increases the number of samples to be analysed and it cannot therefore be used systematically. In the first part of the discussion the overall concentrations of sequential events have been recalculated by summing the fractions, so as to have 29 events in all.
Two mechanisms are responsible for the enrichment of wet deposits. The first concerns scavenging of the air column situated under the cloud by falling snowflakes (or by raindrops). This is a local process. The second is related to the incorporation of the aerosol and the gases within the cloud and attests to a medium or long range enrichment. The receptor site has an important effect upon the proportion of two contributions. An urban or thickly industrialised zone will, for example, be associated with intense local scavenging which will increase with the atmospheric load. On the opposite, at a clean site remote from all sources of immediate pollution this scavenging process will be of secondary importance.

The high VOSGES site in a mountain range enters into the latter category. The characterization of the effects of the emissions of source zones at medium or long distances should be the more marked. This is what we have tried to do by classifying the events by sector, using the air mass trajectories. They have been calculated by the National Meteorological Center for a pressure of 925 hPa and were carried at 4 days previous at 6 hours intervals. The result obtained by this classification is indicated in fig. 1.

During the campaign we enjoyed particularly favourable circumstances in that all the geographical sectors were covered, with 4 events in the East (14%), 6 in the North (21%), 11 in the West (35%) and 8 in the South (27%). We have given (see map (✫)) the mean pH and conductivity valours for the entire campaign. The pH value (4.18) is lower than that observed at the Phaslsbourg station (4.48) of the BAPMON network for the period 1984-1985.

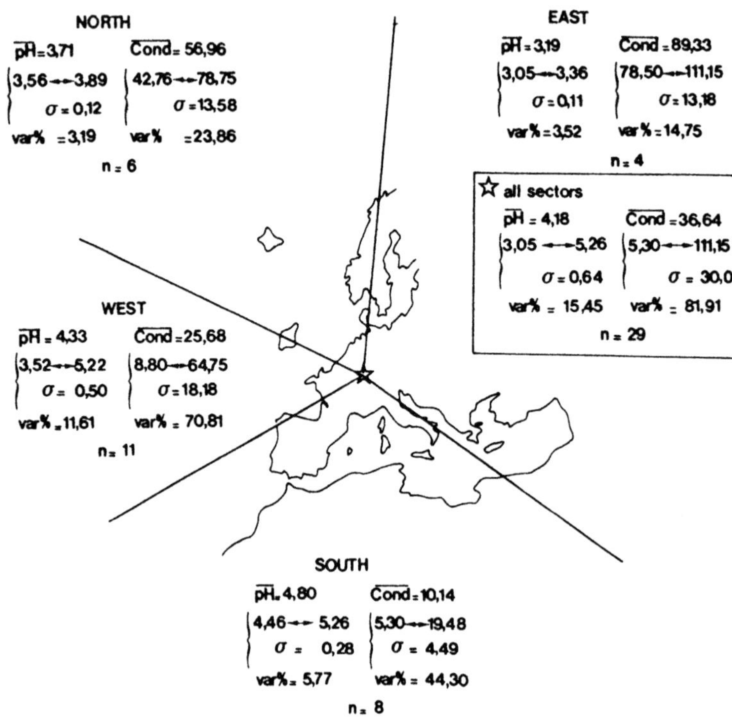

FIGURE 1 : pH and conductivity (μs/cm) for the entire campaign (✰) and for the 4 sectors.
Arithmetic mean, standard deviation and variability.

However the value which we have indicated should be considered whitin the context of the sampling campaign. We sampled during 4 weeks only and the frequency observed for each sector does not necessary correspond to the annual distribution. In particular the East sector is strongly represented here ; this shifts pH towards most acid values. In the same way, the high proportion of snow (69% of the events) could induce an increase in the average acidity attributable to this type of hydrometeor, too.

However, the object of this study was rather to characterise the source zones and in this respect the results obtained are highly significant. The East sector, as could be expected, gives the lowest average pH value (pH = 3.19) with an extreme value of 3.05 for a particular event (11 march 1985). This acidity is to be related to the fact that the zones traversed are very industrialised. This is also observed for the North sector (pH = 3.71) but it is nevertheless less pronounced. The pH then decreases from West to the

South. The conductivity, which is very affected by the activity of the H^+ ions, follows this general pattern with the highest values at the lowest pH's.

In addition, one can see a marked reduction in the variability (about 50%) of both the pH and the conductivity on going from the totality of the campaign to the different sectors. The amplitude of this reduction underlines the effect of the "source zone" parameter in the general process of enrichment of wet deposits. It will be remarked that the greatest reduction occurs for the East and North sectors. These two populations are in fact very homogeneous and this characteristic appears to express a certain constancy in the emissions from these two source regions during the sampling period. It would be interesting to check this hypothesis during longer campaigns at different times of the year. In comparaison, the West and South sectors are much more heterogeneous and include events which deviate considerably from the average population. For the West, certain acid pH values can be imputed to the presence of large industrial countries but the trajectories are not sufficiently precise to allow detailed discussion of this question here. As for the South, certain very neutral events of saharian origin (06 April 1985) are accompained by others more acid ; this very probably implies the industrialised source zones of Northern Spain, Italy and the Rhône Valley.

The figure 2 shows the results of the classification by sector both on the nitrate and sulphate concentrations.

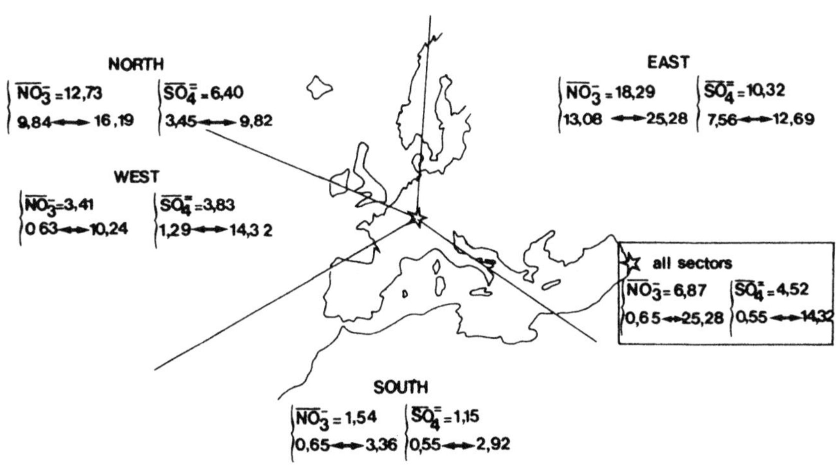

FIGURE 2 : Nitrate and sulphate concentrations in 10^{-5} M for the entire campaign (✰) and for the sectors. Arithmetic means and extreme values.

The average concentrations for each sector are seem to vary in much the same way as the pH and conductivity values. The nitrate and sulphate contents decrease by a factor of 10 on going in turn from the East to the North, West and South sectors. It can be seen that the nitrate concentrations are always higher than sulphate, particulary for the North and East (the concentration are expressed in molarity). In table I we have listed the ionic ratios with respect to sodium, of Mg^{2+}, Cl^-, SO_4^{2-} and NO_3^-, for each sector, as well as the values of these ratios in the seawater model (8).

sectors	WEST	NORTH	EAST	SOUTH
Mg/Na(0,11)	0,10	0,10	0,17	0,14
Cl / Na(1,16)	1,11	3,49	2,01	1,21
SO_4/Na(0,06)	2,31	5,66	9,87	2,61
NO_3/Na(–)	2,14	11,41	16,11	3,17

TABLE I : Ionic ratios with respect to sodium for the 4 sectors.

This way of presenting the results ratios makes it possible to evaluate the enrichment of these species relative to the reference value and thus to estimate the contributions of non marine origin. The first line of the Table concerns a typically marine element, magnesium. One observes that its ionic ratio stays very close to the reference value (0.11) whatever the sector. The chloride of the West and South sectors is also entirely marine in origin. However, there is a slight anthropogenic and continental enrichment for the East and especially the North. On the other hand, the sulphates appear to be highly enriched in all cases with a particularly high value for the East (with a factor about 160). Since there are no nitrates of marine origin, it is so not possible to compare them with a reference value. Nevertheless, it is noticeable that their ionic ratios vary in parallel with those of the sulphates.

However, the calculation of the acidity and the identification of the principal acids involved : HCl, HNO_3 and H_2SO_4, necessitate a further step. To obtain this information one has to subtract from the total concentrations of the ions the concentrations of the neutral salts due to the marine, terrigenic and anthropogenic contributions ; this latter fraction corresponds to the anions initially associated with strong acids which have been neutralised during transport by carbonates. In a previous article (6) we developed an original model based on sequential sampling for obtaining this result. We have applied this type of calculation to a particular event in this campaign. It concerns snow collected the 14 march 1985, from 01 h 03 pm to 08 h 46 pm, in 17 consecutive fractions. The associated trajectory classifies this episode in the East sector, in agreement with the very low pH's measured.

In figure 3, we have plotted the total Cl^-, NO_3^- and SO_4^{2-} concentrations and the pH against the duration of the event.

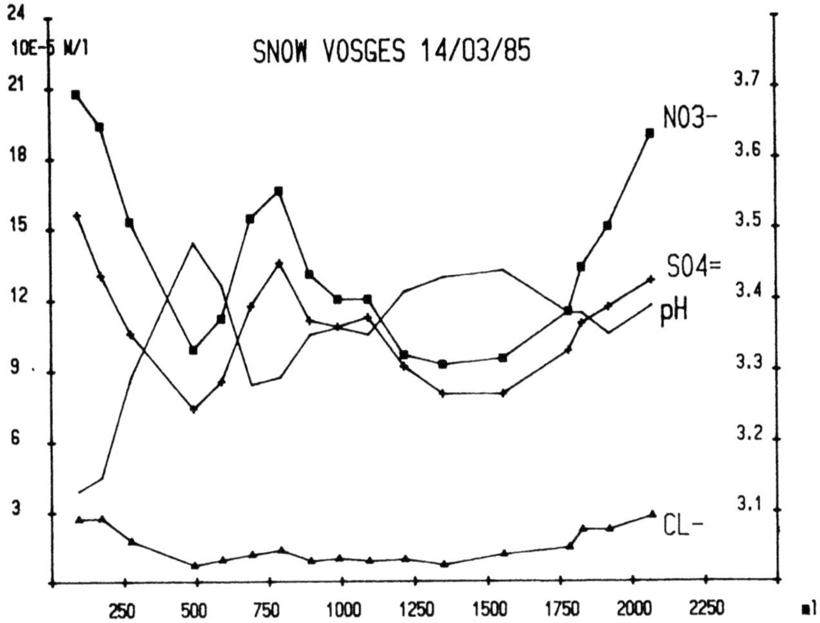

FIGURE 3 : Variation of the concentrations of pH, Cl^-, NO_3^-, SO_4^{2-} in each fraction during precipitation.

One can see an initial decrease typical of the beginning of a precipitation and corresponding to the scavenging of the local aerosol. This part is not very marked since as we have previously indicated, the Col du Donon can be considered as a clean site with a very thightly loaded local atmosphere. Quite large variations of the concentrations during the event can be observed too ; these are to be related to those of the intensity. The good correlation between the sulphates and nitrates expresses a certain identity in the behaviour of these two species with respect to enrichment. Figure 4 illustrates the result obtained by the acidity contribution model. One can demonstrate that sulphuric acid is more important than nitric acid during this episode in the East sector.

A previous observation (6) showed that nitric acid was more important in snow collected in Champagne (FRANCE) from the North sector. Other snow épisodes should be considered before attempting to discuss a tendency as regards the sectors. Table II shows the good agreement between the measured pH values and these calculated from the model.

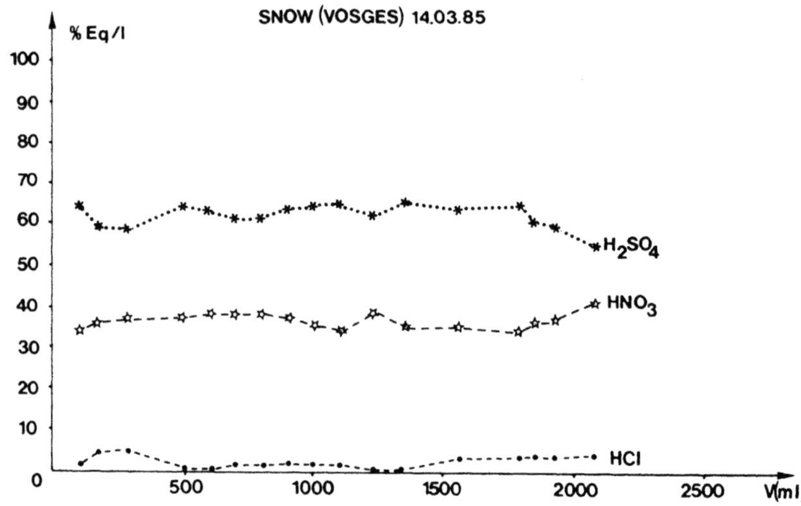

FIGURE 4 : Relative contributions of nitric, sulphuric and hydrochloric acids to the total acidity during precipitation

N°	1	2	3	4	5	6	7	8	9	10	11	12	13	14	15	16	17
pH calculated	3,46	3,54	3,56	3,72	3,62	3,46	3,39	3,53	3,50	3,49	3,59	3,64	3,63	3,55	3,48	3,45	3,37
pH measured	3,13	3,24	3,19	3,29	3,31	3,31	3,31	3,31	3,32	3,32	3,33	3,34	3,35	3,35	3,35	3,35	3,36

TABLE II : Measured and calculated acidities of different fractions taken during precipitation.

4. CONCLUSION

In this paper we have presented the first results concerning the acidity of snow collected during a sampling campaign in the VOSGES.

Sectorisation of the events demonstrates that there are clear differences between the source zones. The East and North sectors seem to be in this respect the most outstanding as regards acidity. This observation could moreover have implications concerning the sampling strategy. Since meteorological conditions change very rapidly, collection over too long a period is likely to mix amounts arriving from different sectors. The average values then observed might mask critical episodes which could carrie stress phenomene harmfull to the forest ecosystem. We insist therefore on the importance of observing discret events.

This study has also shown that snow is sometimes very acid and that, in the flux of fallout, this type of precipitation must be taken into account. Moreover in the context of this programme, this observation is highly significant in view of the time of contact between the snow flakes and the vegetation.

ACKNOWLEDGEMENTS

This work was supported by the CEE DG XII. We wish to thank Mrs STRAUSS and GROS from "Météorologie Nationale" for assistance on air mass trajectories discussion.

REFERENCES

(1) LIKENS, G., WRIGHT, R., GALLOWAY, J. and BUTLER, T. (1979). Acid rain. Sci. American. 43, 59-57.
(2) SEYMOUR, M.D. and STOUT, T. (1983). Observations of the chemical composition of rain using short sampling times during a single event. Atmospheric Environment. 17, 1483-1487.
(3) ASMAN, W.H. (1980). Draft, construction and operation of a sequential rain sampler. Wat. Air. Soil Pollut. 13, 235-245.
(4) GASCOYNE, M. and PATRICK, C.K. (1981). Variations in rainwater chemistry and its relation to synoptic conditions, at a site in North-West England. Intern. J. Environmental studies. 17, 209-214.
(5) PEDEN, M.E. and SKOWRON, L.M. (1978). Ionic stability of precipitation samples. Atmospheric Environment. 12, 2343-2349.
(6) COLIN, J.L., JAFFREZO, J.L., PINART, J. and ROULETTE-CADENE, S. Sequential sampling of snow in a rural area. Experimentation and identification of the acidifying agents. Atmospheric Environment, à paraître.
(7) ELICHEGARAY, C., DUTOT A.L., GRUBIS, B. and VIE LE SAGE, R. (1981). Dosage par fluorescence X des aérosols atmospheriques ; détermination des facteurs de correction. Analusis. 9, 492-497.
(8) HOLLAND, H.D. (1978). The chemistry of the atmosphere and oceans. Wiley Interscience, New York.

NITRATE AND SULPHATE WET DEPOSITION IN EUROPE

K. NODOP
Institute of Meteorology and Geophysics,
University of Frankfurt, FED. REP. OF GERMANY
present address:
Norwegian Institute for Air Research
P.B.130, N - 2001 Lillestrøm, NORWAY

Summary

The chemical composition of precipitation and its temporal and spatial pattern were evaluated for Europe. The results base upon data collected under the ECE's EMEP programme covering the period 1980 to 1984. The average annual nitrate and sulphate concentration in precipitation varies within Europe from 0.1 to 1.0 mgN/l and 0.2 to 2 mgS/l, respectively. The spatial pattern for both pollutants is very much the same. The main feature is the increase of the values towards central Europe. The annual cycle of nitrate and sulphate in precipitation is less pronounced than the annual cycle of their precursers nitrogen dioxide and sulphur dioxide in ambient air, and the highest concentrations occur later in the year. The nitrate to sulphate equivalent ratio demonstrates the relative contribution of nitrates and sulphates to acid deposition. On yearly average nitrate contributes up to 40% to the acidity in wet deposition. Generally the nitrate to non-marine sulphate ratio is higher in wintertime than in summertime, except at some stations in central Europe. The spatial pattern show higher values at stations in central and northern Europe and lower values in eastern Europe. The wet deposition of nitrate in Europe is governed by the precipitation pattern. Long-range transport of nitrate in precipitation is shown by sector analysed data at several sites. Generally, transport directions associated with high concentrations are directed towards major emission areas.

1. INTRODUCTION

Long-range transport of air pollutants was recognized as a problem several years ago. In the case of sulphur compounds large amounts of data have been collected to quantify the long-range transport. In recent years more emphasis has been placed to oxidized nitrogen compounds because of increasing nitrogen oxide emissions. There is a strong evidence of a growing contribution of nitrate to acid deposition. Among nitrogen compounds nitrate in precipitation is the far best documented. This paper tries to establish an overview on available data for nitrate in precipiation in Europe and the relation to sulphate.

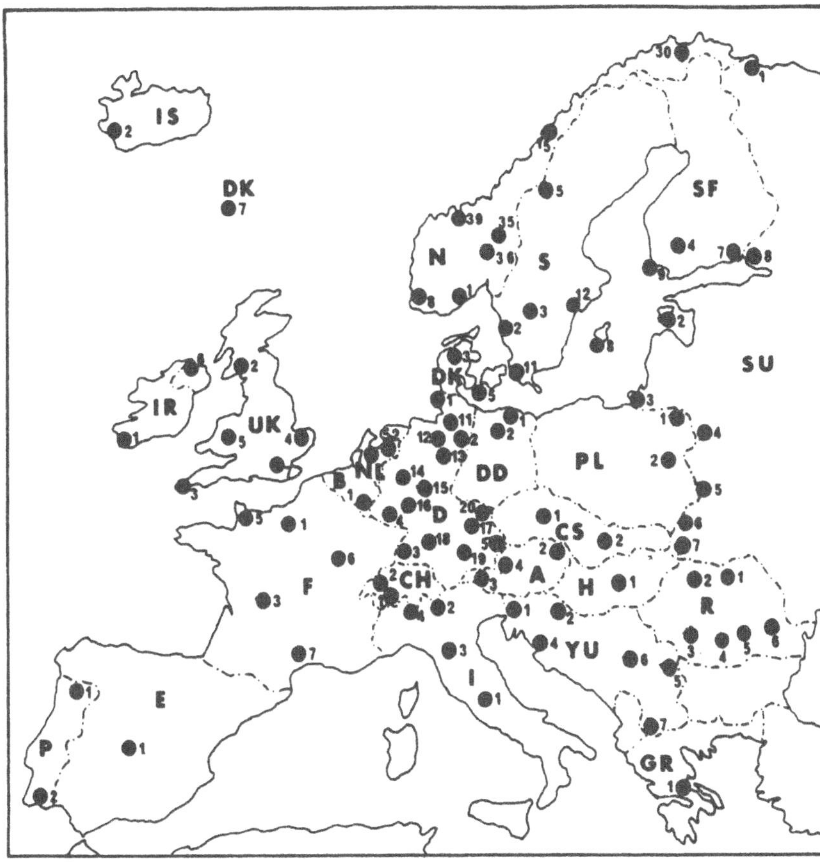

Fig. 1: The EMEP station network

2. <u>DATA SOURCES</u>

Most of the data used in this paper have been collected under the ECE convention´s "Co-operative programme on monitoring and evaluation of long-range transmission of air pollution in Europe" (EMEP). The EMEP network comprises only stations in rural areas not influenced by local sources. It is shown in figure 1. Calculations are made from available daily measurements from 1980 to 1984. In addition data from national and regional networks are used. The number of stations where nitrate in precipitation is measured is still limited compared to the number where sulphate is measured. Nevertheless, sufficient data is now available and a well established overview on concentrations and depositions could be evaluated.

3. NITRATE AND SULPHATE IN PRECIPITATION

The average annual nitrate and sulphate concentration in precipitation varies within Europe from 0.1 to 1.0 mgN/l and 0.2 to 2.5 mgS/l, respectively. The annual volume-weighted mean concentrations of nitrate and sulphate in 1982 are presented in figure 2a and b. The pattern of the isolines for nitrate and sulphate concentrations is quite similar, but the gradients are more pronounced for nitrate. The main feature is the increase of the values towards central Europe. The highest concentrations of both pollutants are found in central and south-east Europe. In Scandinavia and in the United Kingdom typical values are found to be between 0.1 and 0.5 mgN/l for nitrate and 0.2 to 1.0 mgS/l for sulphate. In central Europe the nitrate and sulphate concentrations vary around 0.7 mgN/l and 1.5 mgS/l, respectively.

The available data do not reveal any clear year-to-year variation of nitrate and sulphate in precipitation. At least from the data record 1980 to 1984 no uniform trend can be seen, however, for trend analysis the time period is too short.

4. NITRATE TO NON-MARINE SULPHATE EQUIVALENT RATIO

The main ions contributing to the acidity of wet deposition are sulphate and nitrate. The nitrate to sulphate ratio on an ion equivalent basis demonstrates the relative contribution of nitrates and sulphates to acid deposition. This ratio has become of great interest, because in most European countries nitrogen emissions were and still are increasing, whereas sulphur emissions may have slowed down. Therefore, the ratio of nitrate to non-marine sulphate is likely to increase as well. The ratios presented here have been calculated from averaged concentrations of the time period in question.

The annual mean ratios were calculated from 1980 to 1984. A few sites in Norway, Sweden, the United Kingdom and Hungary show an increase in the ratio over this period (fig. 3). At other sites in Europe, increases were not very pronounced. The year-by-year increase in the nitrate to sulphate ratio is an indication for the growing contribution of nitrogen oxides in acidification.

The spatial distribution of the annual mean ratios in Europe in 1982 is presented in figure 4. As it can be seen, the highest values are found in the area stretching from northern France to southern Finland. The typical value of the nitrate to non-marine ratio in this area is 0.6. Towards the north-west the ratio is decreasing to 0.4 (United Kingdom and Norway). The gradient to the south-east is much more pronounced. In south-eastern Europe the ratios vary around 0.3. Assuming that only nitate and sulphate contribute to total acidity, nitrate contributes up to 40 per cent to the acidity in wet deposition on a yearly average in wide areas in central and northern Europe. In eastern Europe the relative contribution of nitrate is in general 25 per cent.

To show the seasonal variations in the nitrate to non-marine ratio, they were calculated from summer (April to September) and winter (October to March) volume-weighted concentration means. Generally, the ratio is higher in wintertime

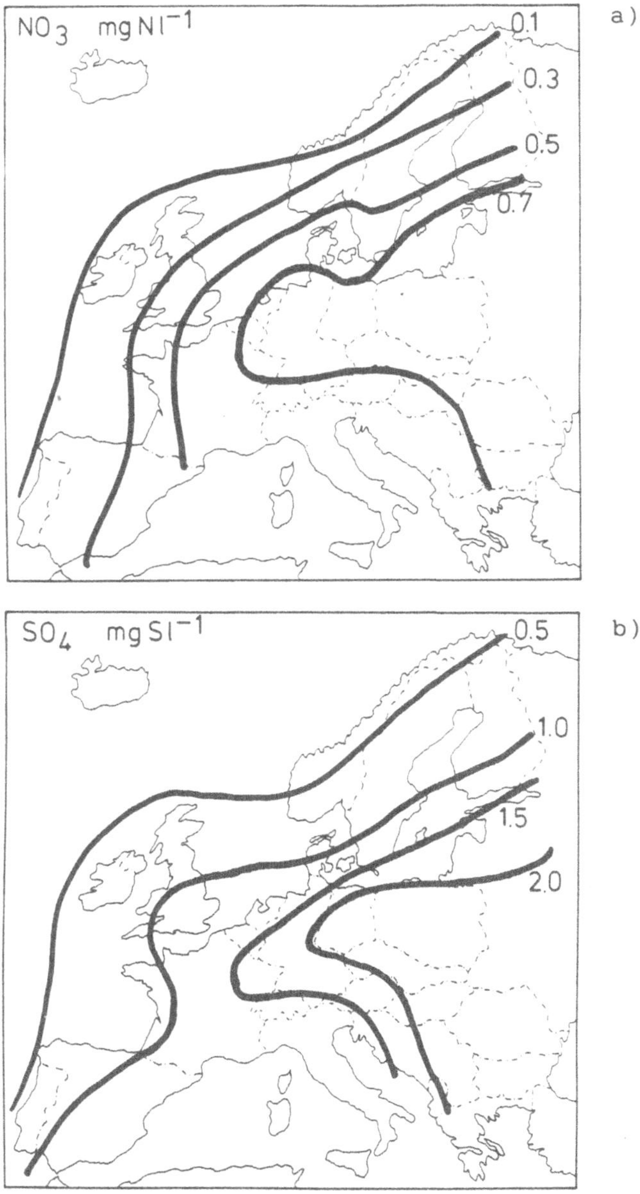

Fig. 2: Volume-weighted annual mean nitrate (a) and non-marine sulphate (b) concentration in precipitation in Europe in 1982.

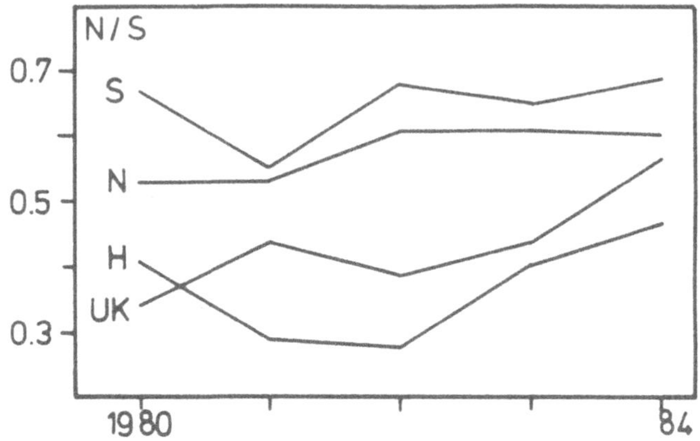

Fig. 3: Annual mean nitrate to non-marine sulphate equivalent ratio in precipitation at Rörvik (S), Birkenes (N), K-puszta (H) and Eskedalemuir (UK).

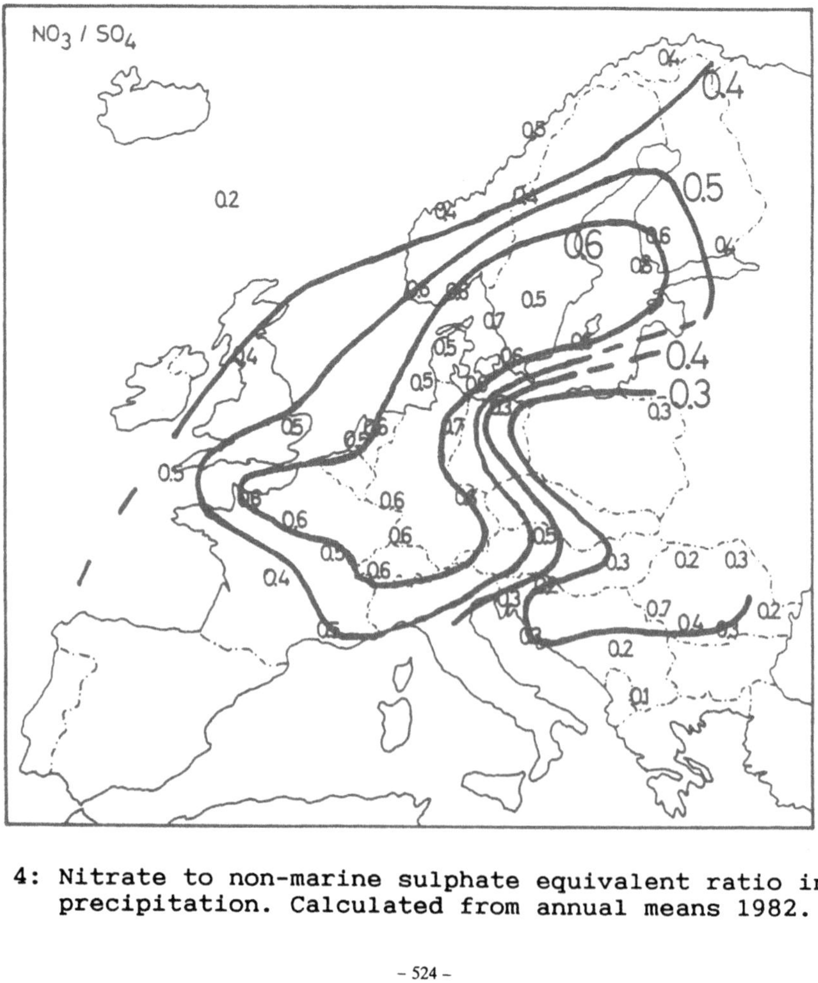

Fig. 4: Nitrate to non-marine sulphate equivalent ratio in precipitation. Calculated from annual means 1982.

than in summertime, except at some stations in the Netherlands and the Federal Republic of Germany (fig. 5). This greater relative contribution of nitrate during winter, especially at those stations for from major emissions areas, is evidence of a longer residence time for nitrate. For that reason, the long-range transport of nitrate in precipitation contributes strongly especially in wintertime to acid deposition in remote areas.

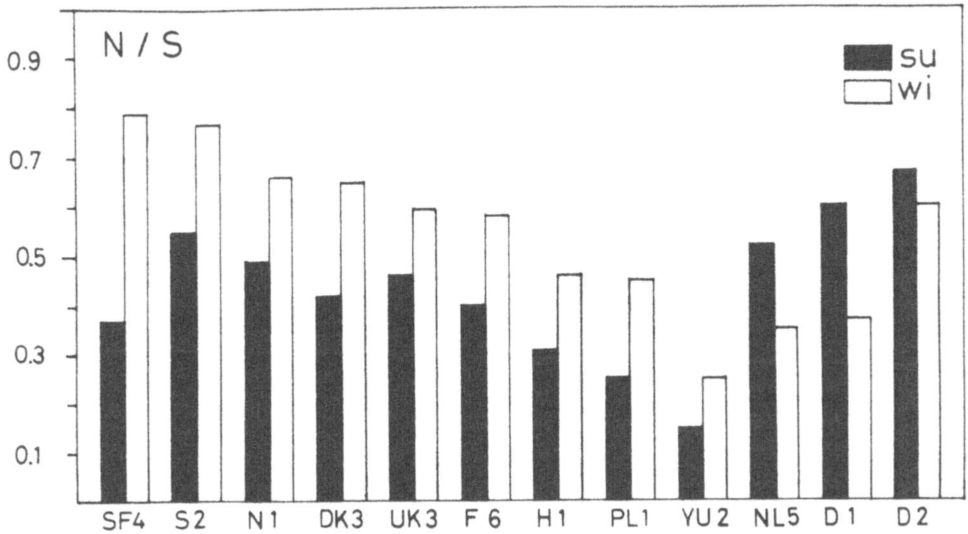

Fig. 5: Seasonal variation of the nitrate to non-marine sulphate equivalent ratio in precipitation at several sites in Europe (compare Fig. 1).

5. NITRATE WET DEPOSITION

The wet deposition is the product of concentration and precipitation. The spatial and temporal variability in sulphate and nitrate concentration is very small compared to the varability in precipitation as such. The amount of wet deposition is therefore governed by the precipitation rate at a given site.

The nitrate concentrations of the year 1982 were used in combination with the average precipitation pattern in establishing a wet deposition map of Europe (fig. 6). The estimates of wet deposition show a value of 0.5 gNm^{-2} in central and south-eastern Europe, and between 0.2 and 0.4 gNm^{-2} in the remaining regions. In coastal and mountainous areas, which have a higher precipitation rate, the wet deposition of nitrate was found to be between 0.9 and 1.4 gNm^{-2}, while in areas with very little precipitation the wet deposition is less than 0.1 gNm^{-2}. Quite large amounts of nitrate have been observed in remote areas of Europe owing to long-range transport of nitrates.

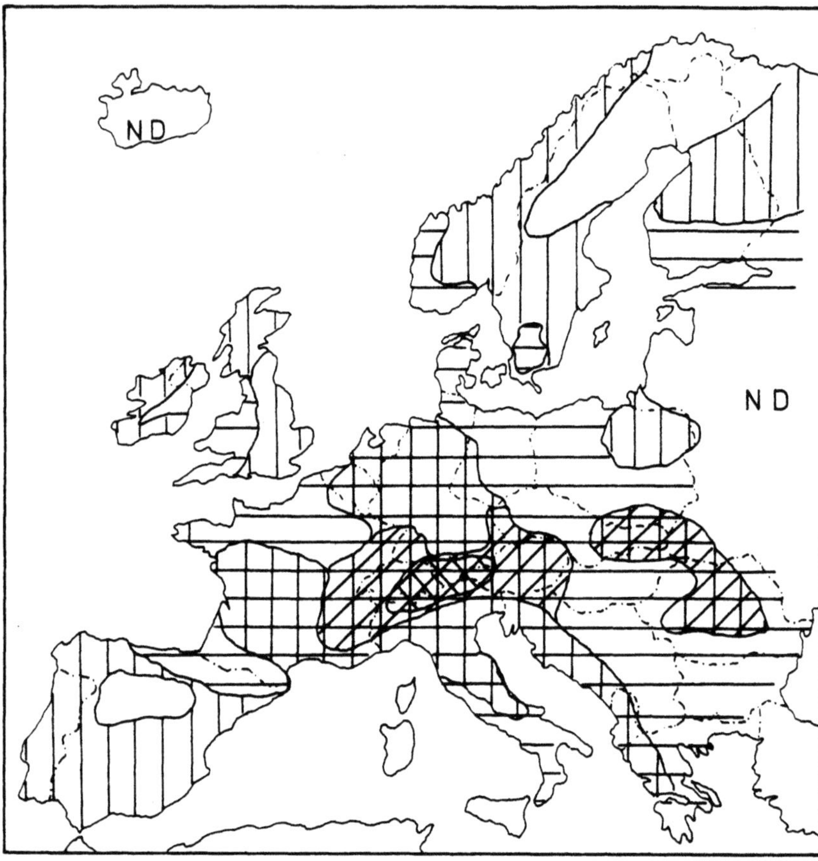

Fig. 6: Estimation of the annual wet deposition of nitrate in 1982 (gNm^{-2}).

(blank)	< 0.1 or NO Data		0.6 - 0.9
	0.1 - 0.3		0.9 - 1.4
	0.3 - 0.6		> 1.4

The seasonal variation of the wet deposition of nitrate is parallel to the temporal variation of precipitation. Wet deposition is often episodic, i.e. a large fraction of the deposition occurs on only a few days. At Goonhilly (United Kingdom), for example, 30 per cent of the total nitrate and sulphate in 1980 were deposited in a total of 7 and 5 days, respectively (1).

6. LONG-RANGE TRANSPORT OF NITRATE

Air mass trajectories calculated by the Meteorological Synthesizing Centre - West (2) have been used to perform a sector analysis of nitrate in precipitation at some selected EMEP sites. The origin of an air mass at a given location is determined by 850 mb isobaric trajectories traced back over

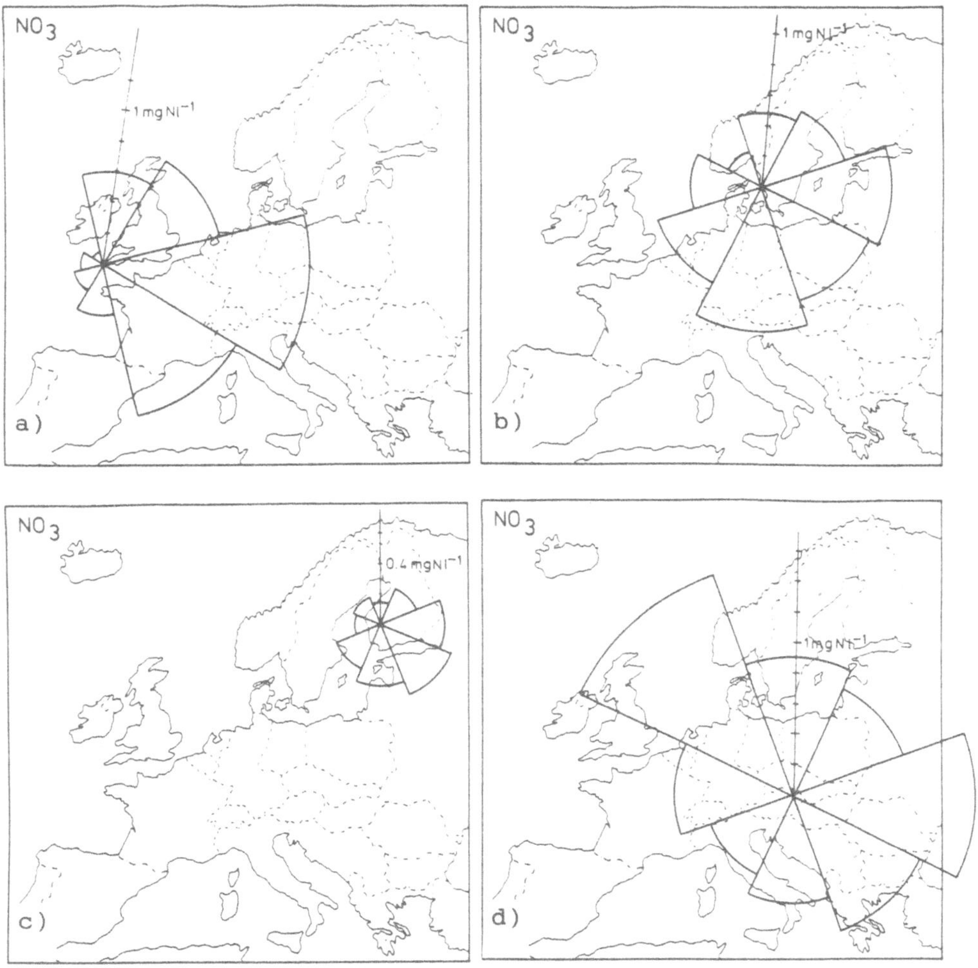

Fig. 7: Mean nitrate concentration in precipitation per sector at Goonhilly (a), Rörvik (b), Athäri (c) and Illmitz (d). Calculated from available data 1980 to 1984.

the last 72 hours. The transport directions are grouped into eight equal size sectors (45^0). Daily measurements of concentration together with daily sector numbers were used to calculate mean concentrations for each of the eight transport directions. Generally, sectors associated with high concentrations are directed towards major emission areas.

Long-range transport phenomena can be studied very well at stations located at the edge of Europe. Gonnhilly, located at the south-western tip of England, shows an extreme distribution of nitrate concentrations with respect to transport directions (fig. 7a). With the air masses originating from the Atlantic Ocean, concentrations of nitrate are low, while from the eastern and south-eastern sectors which directly point to England and continental Europe the concentrations of nitrate are very high. These configurations can only be explained by long-range transport.

The sector analysis was also made at Rörvik (Sweden), Athäri (Finland) and Illmitz (Austria). The fact that sectors showing the highest concentrations of pollutants are those directed to major emission sources is also valid at these stations. At Rörvik (fig. 7b) the highest nitrate concentration occurs in the south sector. This is in accordance with other oxidized nitrogen compounds (3). The station in Finland (fig. 7c) shows the highest concentration in the south and south-east direction. The main feature at Illmitz (fig. 7d) is the occurance of two main directions from where heavy loaded air masses arrive. Countries in the east and south-east direction contribute with the same large amounts to the wet deposition of nitrate as the north-western neighbours.

REFERENCES

(1) United Kingdom Review Group on Acid Rain (1983) 'Acid Deposition in the United Kingdom', Warren Spring Laboratory

(2) Eliassen, A. and Saltbones, J. (1983). Modelling of long-range transport of sulphur over Europe: a two-year model run and some model experiments. Atmos. Envir. 17 457

(3) Ferm, M., Samuelsson, U., Sjödin, Å. and Grennfelt, P. (1984) Long-range transport of gaseous and particulate oxidized nitrogen compounds. Atmos. Envir. 18 1731

SEASONAL AND DAILY VARIATIONS OF OZONE, SULFUR DIOXIDE AND NITROUS GASES AND THE PHOTOCHEMICAL PRODUCTION AND REACTIONS WITH OTHER TRACE GASES OF THE OZONE

R. REITER+), R. SLADKIVIC and H.-J. KANTER
Fraunhofer Institute for Atmospheric Environmental Reseach
Garmisch - Partenkirchen, West Germany

Summary

Since 1978 the concentrations of O_3, SO_2 and NO_x and since 1981 of Hydrocarbons are recorded at our mountain stations in 740, 1780 (BAPMoN STATION) and 2964 m asl, their horizontal distances being rather small. The typical seasonal and daily variations and long term behavior are shown. The amount of photochemical production of O_3 as function of height and solar radiation as well as the reaction of O_3 with other trace gases was investigated. The relatively week importance of stratospheric O_3 intrusions is mentioned.
By this way the knowledge of the typical time behavior, the sources and sinks of trace gases in the lower troposphere can be deducted. The results can be used to plan laboratory experiments under realistic conditions. Same is true for theoretical modelling.

1. INTRODUCTION

Since 1978 continuous trace gas recordings by applying careful calibration methods are carried out at our mountain stations (Zugspitze, 2964 m asl: O_3 since 1978; BAPMoN station Wank Peak, 1780 m: O_3 since 1978, SO_2 and NO_x since 1982; valley station Garmisch: O_3, SO_2 and NO/NO_2 since 1978 in 740 m asl). The data have been intensely evaluated and analyzes by using different aspects. All weather parameters measured at the stations, the vertical mixing intensity and regional meteorological data have been considered. However, in this paper we cannot go into details (for those see 1, 2). Nevertheless the most important variations with time and altitude of the trace gas conc. shall be demonstrated. Concerning O_3 the recent results on the photochemically production even in a region of relatively pure air are shown - including the influence of the amount of the recorded relative sunshine duration (in%) and the altitude above ground (valley floor). Because there are many measurements in or near polluted areas, data from remote sites and higher altitudes seem to be valuable, also with respect to the intensively discussed question of damages on trees which are also observed here.

2. OZONE

2.1. TIME VARIATIONS SINCE 1978

Figs. 1 a-c show the time variations at the 3 stations by 10-day mean values and annual means (horiz.bars). The small ver-

+) Adress: Fritz Müllerstr. 54, D-8100 Garmisch-Partenkirchen

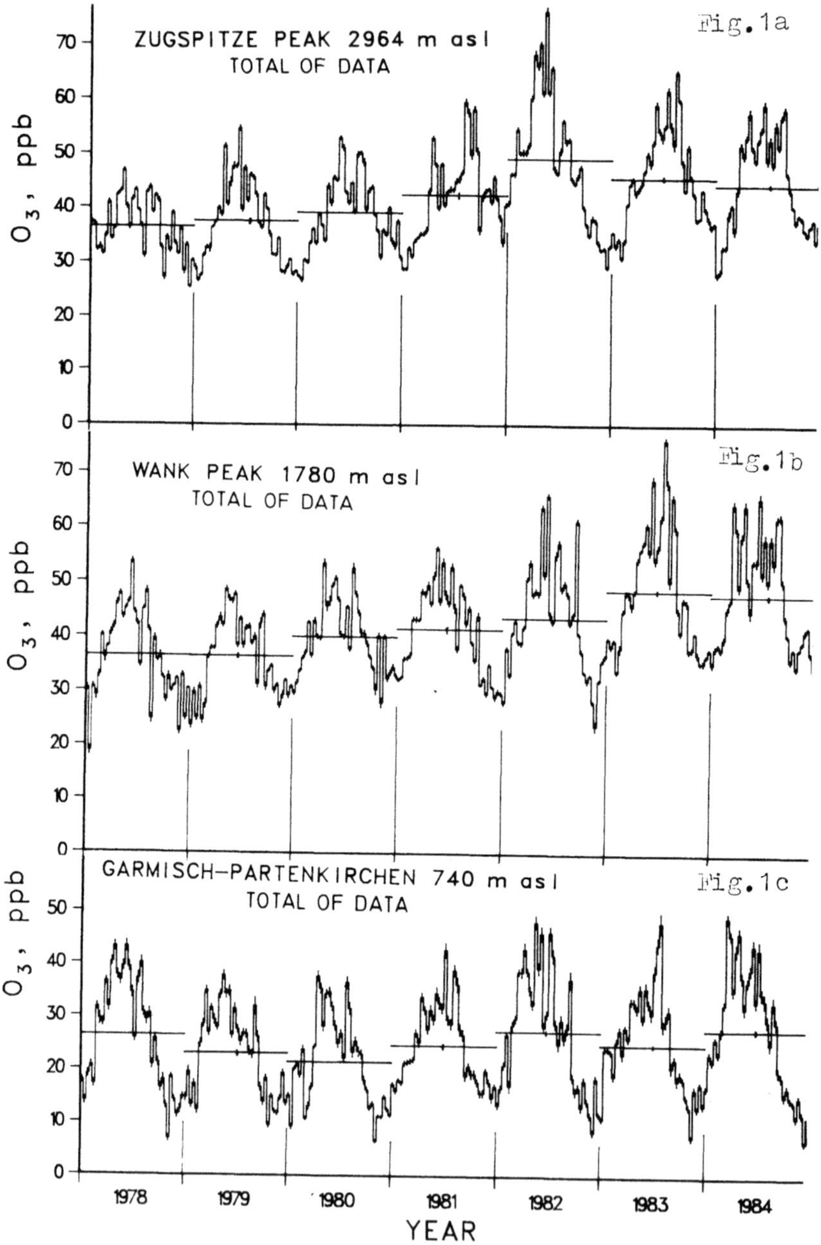

tical bars indicate the statistical significance per 10-day mean. At all 3 Stations the annual variation is clearely indicated: maximum in summer, minimum in winter. The absolute difference between both is strong in 740 m, weaker at higher altitudes. There is also no trend 1978 - 84. In 2964 and 1780m there is an increase until 1982, resp. 1983 which does not continue: the values are decreasing up to date. The concern about a possible steady increase of the tropospheric O_3 (3) obviously was premature. Even there exists an annual rhythm one can find many irregularities in the curves depending mainly on short term variations in weather and climate. This "noise" must be known on the basis of long term records in order to establish significant trends, may be caused by nature or anthropogenic activities.

2.2. MEAN ANNUAL AND DAILY VARIATIONS

Figs. 2 a-c show 3-dimensional plots of the O_3 conc. as a function of hour of the day and month of the year averaged over

2964 m asl

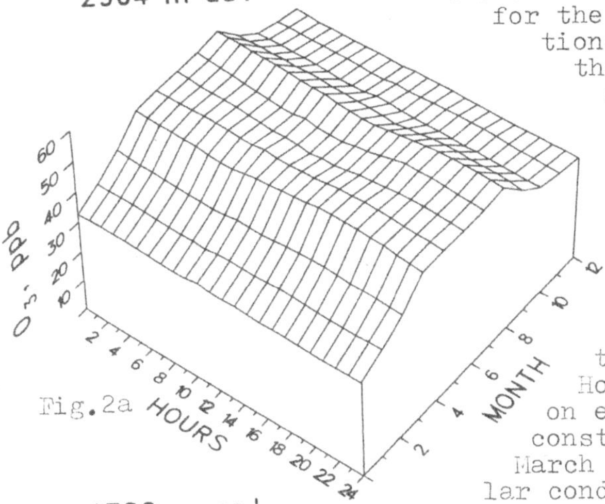

Fig.2a

1780 m asl

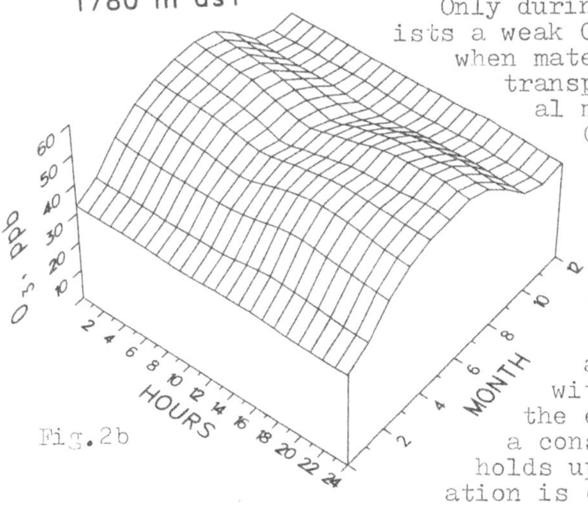

Fig.2b

740 m asl

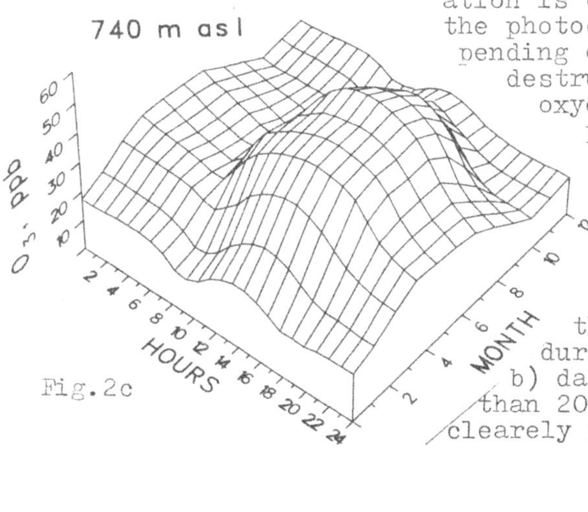

Fig.2c

the entire data from 1978 to 1984 for the 3 altitudes of the stations. At 2964 m (Fig 2a) there obviously does not exist any daily variation and this independently of the month of the year. This indicates that at 3 km altitude in the free atmosphere there are existing neither decay processes for O_3 nor a photochemical production correlated with the alteration of day and night. However, an annual variation exists with a plateau of constand O_3 conc. between March and August. We find similar conditions at 1780 m (Fig.2b). Only during summer time there exists a weak O_3 minimum around noon when materials destroying O_3 are transported upwards by vertical mixing.

Completely different from the O_3 behavior in altitudes higher than 1500 m asl is its variation in the valley (Fig.2c). A certain daily variation can be observed at all months of the year: an increase beginning at sunrise with a strong maximum in the early afternoon and then a consistent decrease which holds up to midnight. This variation is caused by 2 processes: the photochemically production depending on insolation and the O_3 destruction durching dark by oxydation of material suspended in the lowermost layer above ground.

2.3. THE PHOTOCHEMICAL O_3 PRODUCTION

In Fig. 3 we find 2 groups of O_3 data:

a) measured during more than 80% relative sunshine duration (RSD) = solid line.

b) data obtained during less than 20% RSD (broken line). We clearely see that during long RSD

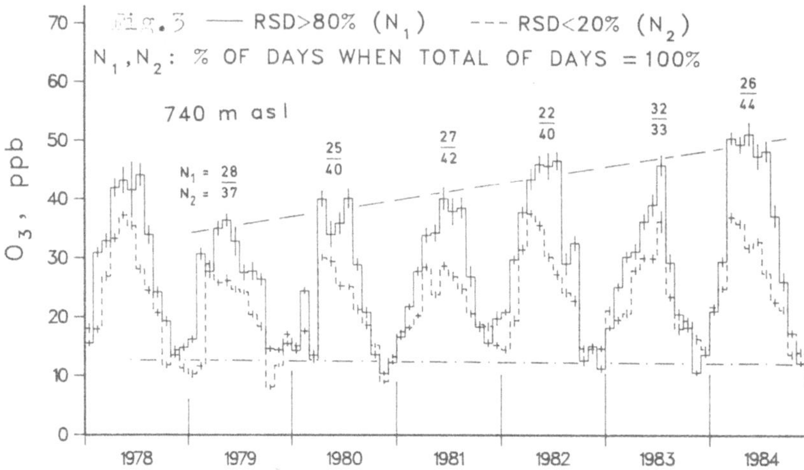

the O_3 conc. is much higher than in the opposite case. Furthermore we find that during high RSD there is an increase of the annual maximum of the O_3 conc. since 1979 until 1984. The reason for this may be that during the same period the concentration of those materials has increased which are necessary for any photochemical O_3 production.

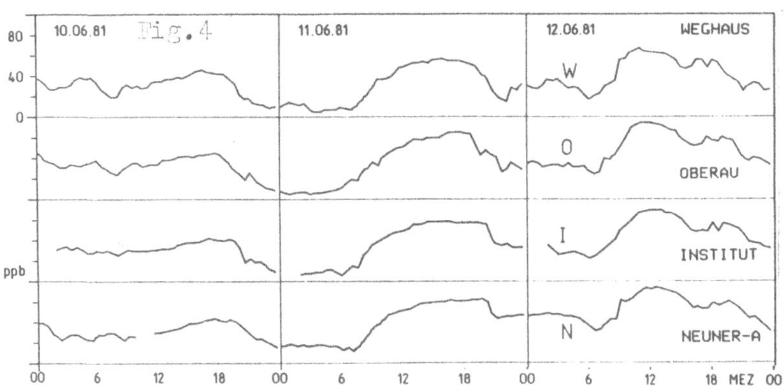

To become shure that the O_3 maximum near the ground after noon was not only induced by an O_3 transport from polluted areas north of us in the pre-alpine region into our valley by the ordinary valley wind (entering the valley from north) we installed a chain of O_3 recording stations (W,O,I and N - from north to south - in Fig 4) along the valley. W was situated at the northern outlet of the valley to the pre-alpine region. N was 20 km south of W inside the valley. In the case there would exist a O_3 transport by the valley wind on day (mean 10 km/hr), a time shift in the maxima and minima between W and N must have been observed. This obviously is not the case as Fig.4 shows. The photochemical O_3 production sets in at the same time at the different places.

2.4. THE VERTICAL OZONE PROFILES IN THE LAYER 1000 - 3000 m asl

A cable car operating between 1000 and 3000 m asl (up to Zugspitze) was used to on board recordings of the O_3 profile once per hour from morning to the late afternoon. Fig. 5 shows the isolines of O_3 conc. as normally observed during undisturbed weather and medium insolation. In a relatively thin layer of max. 500 m thickness the O_3 conc. increases from morning (25 ppb) to afternoon (45 ppb) as shown in Fig.5 whereas in higher altitudes of more than 1800 m the O_3 conc. remains constant

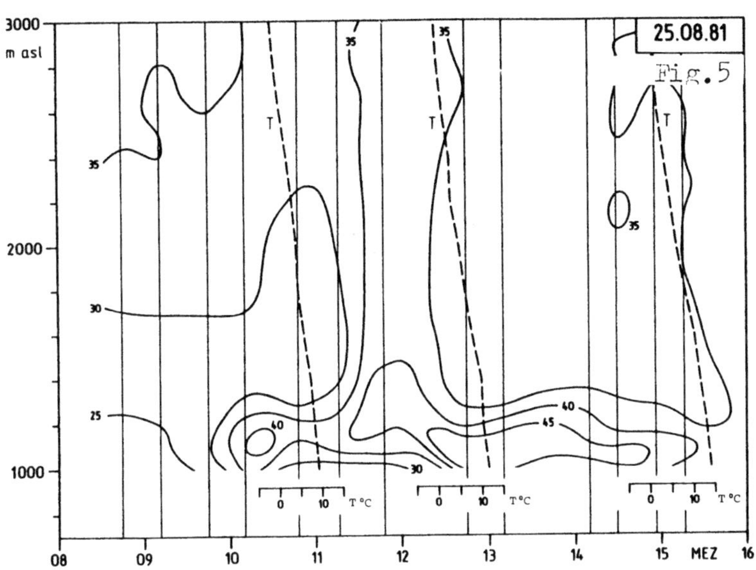

Fig.5

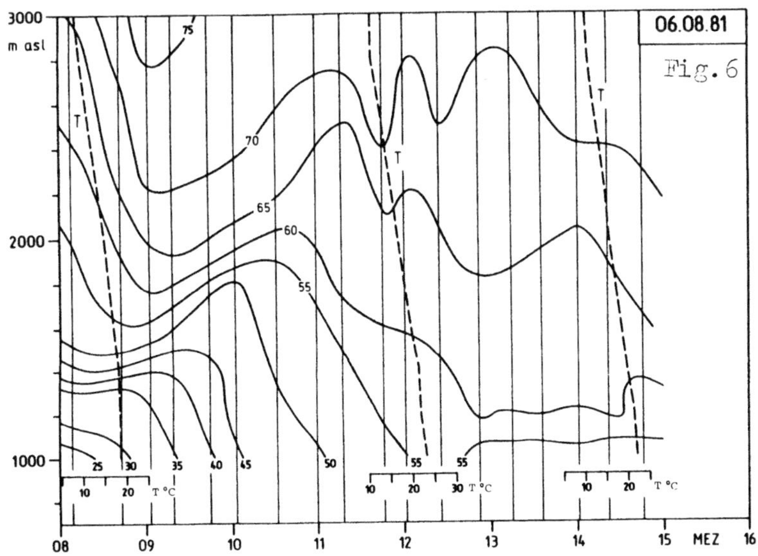

Fig.6

throughout the day. Only in very rare cases we found evidence for stratospheric O_3 intrusions. Fig 6 shows on the one hand the same daily O_3 variation up to 1400 m like in Fig.5, on the other it is obvious that in the early morning there happened a steep increase of O_3 with altitude up to 75 ppb in 3 km. By simultaneous records of stratospheric cosmogenic radionuclides (Be7, P32) it could be verified that peaks of O_3 at 3 km are caused by stratospheric intrusions. However, this stratospheric O_3 does not penetrate deeper than 1500 m asl into the troposphere and remains insignificant to maintain parts of the O_3 conc. in the lowest part of the troposphere. In general, the stratospheric O_3 input remains relatively low (4).

2.5. <u>CORRELATION OF O_3 WITH OTHER TRACE GASES</u>

As a first result Fig.7 shows (now confirmed by recent data) that in summer the O_3 conc. increases with the NO_x conc. However, in winter O_3 decreases with increasing NO_x conc. This correlation is trong in 740m (but also good in 1780 m). Fig.8 shows that there is absolutely no correlation between O_3 and CH_4 in 740 m asl. However there exists a strong relationship between O_3 conc. and the concentration of reactive hydrocarbons CnHm (see Fig.9). This relationship shows a homogeneous transition from summer to winter. On the basis of the result in

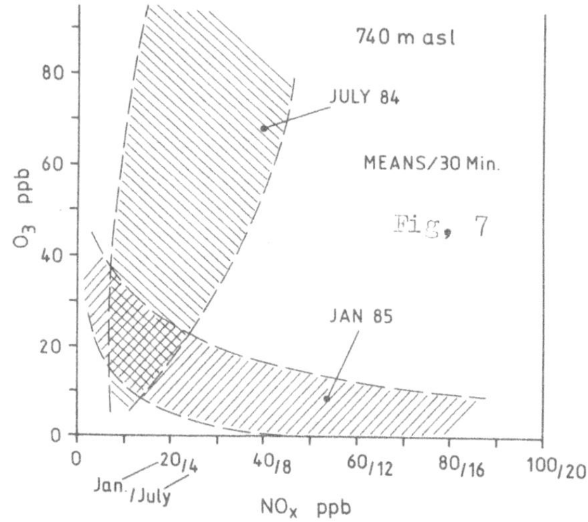

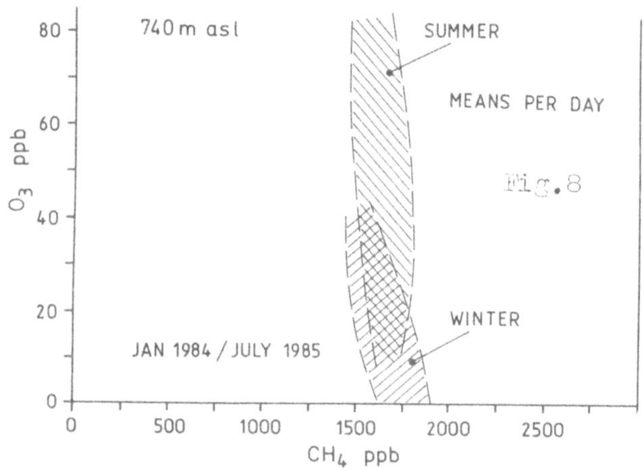

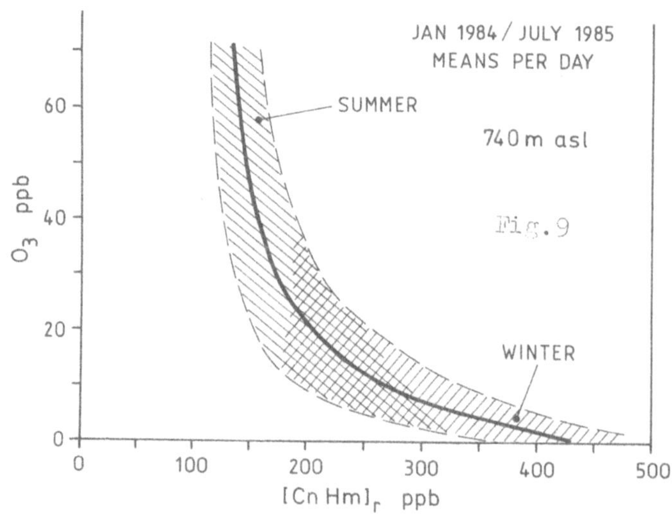

Fig 9 we can deduce that high concentrations of reactive hydrocarbons in winter reduce the O_3 conc. drastically. During summer, when the CnHm conc. is lower than 180 ppb the O_3 conc. is influenced by CnHm only very week.

Recordings of the reactice hydrocarbones are only executed in 740 m asl.

Very recent recordings and evaluations have shown that the relationship as shown in Fig. 9 is correct for all seasons of the year.

One remark: the hatched fields in the Figs. 7-9 cover a minimum of 90% of all data. "Exotic" ones, mainly cases of extremely conc. or anusual weather conditions are ignored.

3. SULFUR DIOXIDE

3.1. TIME VARIATIONS

In 1780 m asl (Fig.10 a) the SO_2 conc. shows strong irregular variations with time (plotted are again 10 day mean val.) but no annual rhythm. A weak decrease from 1982 to 1984 is indicated. This behavior suggests that even the vertical exchange intensity strongly depends on the season (and there are SO_2 sources in low altitudes), in about 2 km asl long range transportation of SO_2 superimposes the influence of regional sources. In contrast Fig. 10b shows that in the valley (740 m asl) the expected annual variation of the SO_2 conc. is well indicated with the maximum in winter and the minimum in summer. The reason for this is: enhanced heating activity in winter during a very weak vertical mixing intensity on the one hand and strong vertical mixing in summer which reduces the SO_2 conc. near the ground.

The extremely high peaks in Fig. 10b are - as investigated - caused during periods of strong inversions in the cold seasons.

The systematic decrease of the annual mean values in Fig. 10b suggests the success of reduced anthropogenic emission rates ordered by law. However, in 1985 the annual mean was again ingreased. This may be explained by the extremely deep air temperatures in Nov. and Dec. 1985 (more heating and stronger inversions.

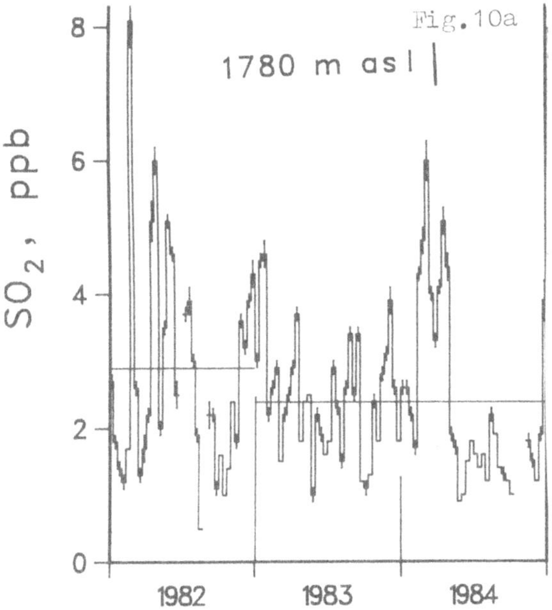

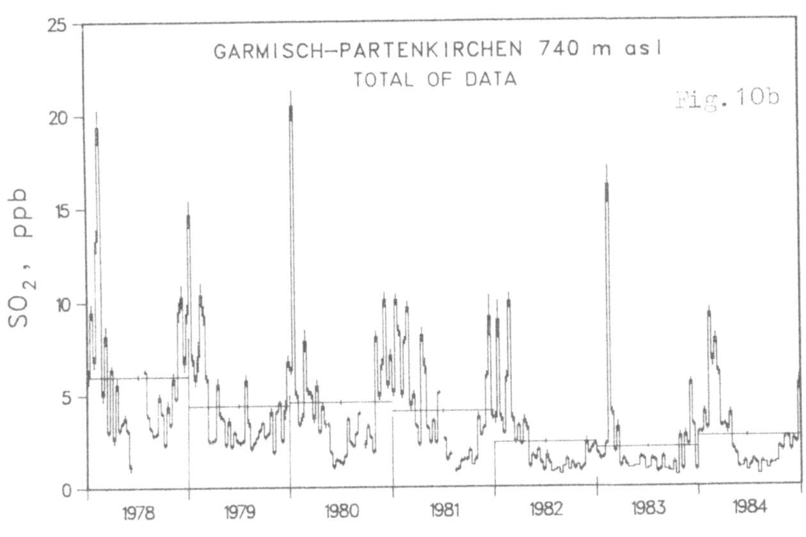

3.2. MEAN ANNUAL AND DAILY VARIATIONS

The Figs. 11 and 12 show 3-dimensional plots of the SO_2 conc. as function of hour of the day and month of the year again averaged over the entire data available. Whereas in 740 m asl the annual variation is well indicated for all hours of the day (a daily variation, however, mainly exists in the cold season) we cannot find a significant annual variation of the SO_2 concentration in 1780 m asl at all. This confirms by this kind of a summarizing evaluation what has been stated above: the importance of long range SO_2 transports in this altitude, mainly being independent of the local vertical mixing intensity. The behavior of the SO_2 conc. is strongly different from that of the O_3 (compare Figs. 2 a-c).

4. NITROUS GASES

4.1. TIME VARIATIONS

Because of the limeted space we restrict our presentation by reporting only on the measured NO_x conc. in 1780 and 740 m asl (also NO and NO_2 have been recorded separately).

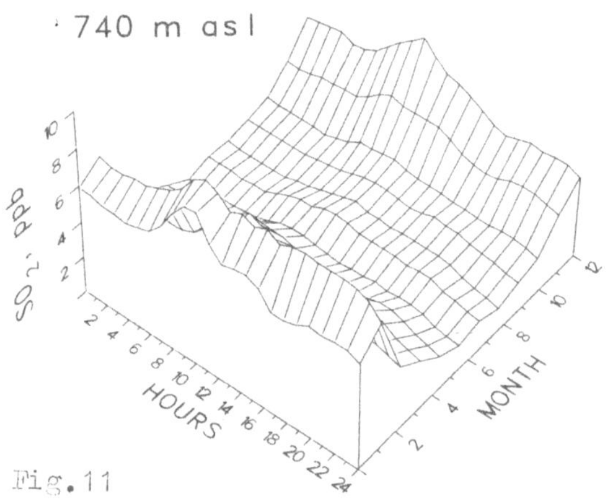

Fig.11

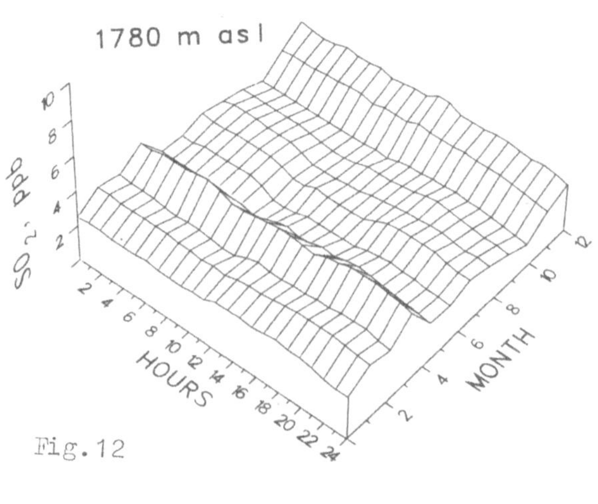

Fig.12

Fig. 13a demonstrates the NO_x conc. recorded at 1760 m asl (the plot shows again 10 days mean values and annual means as horizontal bars, same is true for Fig. 13b). Lika in the case of the SO_2 conc. we cannot find any annual variation at this altitude despite of the strong changes of the vertical mixing intensity throughout the year. Consequently we have to assume that not only SO_2 gas but also NO_x is transported over long range distances to the remote station in 1780 m asl. The annual mean value is - in contrast to the SO_2 conc. - increasing from 1982 until 84 (also to 85).
In 740 m asl, however, we find a strong annual variation of NO_x (see Fig. 13b) with high isolated peaks in winter. This again indicates the effect of heating and inversions as barring layers. The difference between minima in summer and maxima in winter are greater in the case of NO_x compared with those of SO_2 (Fig.10b), however, the NO_x recordings dont show such extreme isolated peaks as SO_2 in Fig. 10b. It has to be pointed on the fact that in the altitude of 740 m there is not any trend of the NO_x conc. indicated despite the reduction of the emission from anthropogenic sources by law.

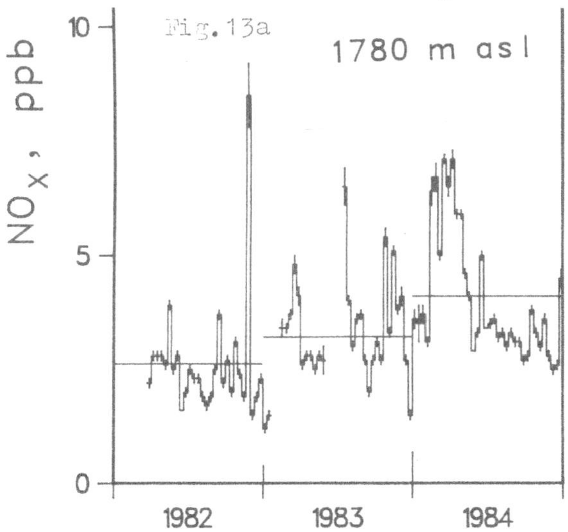

Fig. 13a

4.2. MEAN ANNUAL AND DAILY VARIATIONS

By the Figs. 14 and 15 we show the 3 dimensional plots of the NO_x conc. as function of hour of the day and month of the year using the available total of data. In 740 m asl we find a significant daily variation in winter, early spring and fall. During

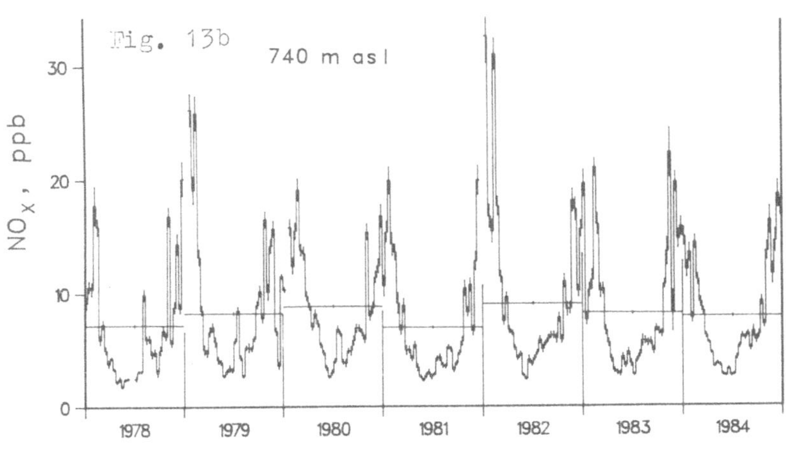

Fig. 13b

summer the daily variation is unimportant. The annual variation is well indicated during all hours of the day, however trongly pronounced in winter bevor sunrise (end of the accumulation of NO_x from combustion processes during night). Fig. 15 is totally different from Fig.14. In 1780 m asl no significant daily and annual variations exist. Regional source, their influence in 1780 m asl being strongly determined by the vertical mixing intensity are obviously out of any importance.

Fig.14

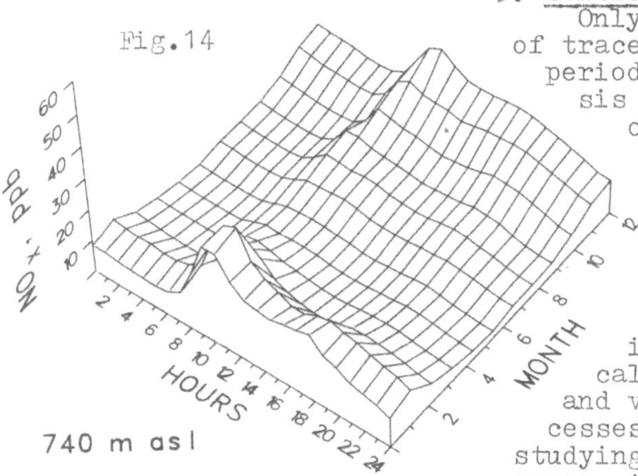

Fig.15

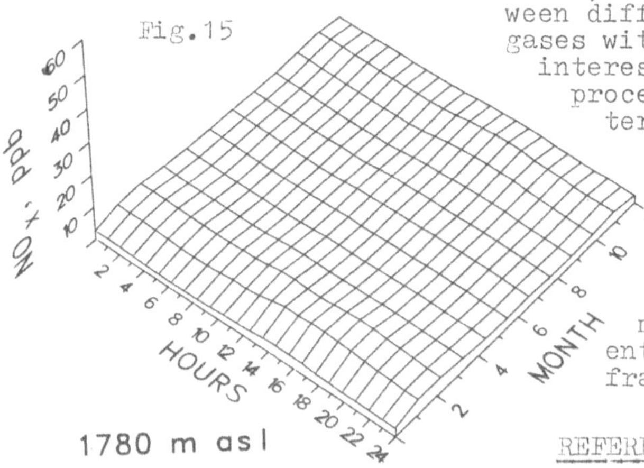

5. CONCLUSIONS

Only continuous recordings of trace gases over a long time period guarantee a sound basis for an understanding of the behaviour of concentrations and the atmospheric processes which are controlling those. Furthermore it is necessary to have simultaneous recordings in different altitudes to study vertical gradients, horizontal and vertical transport processes. The same is true for studying anthropogenic influences, inter-reactions between different trace gases and gases with aerosols. Of high interest are photochemical processes which mainly determine the O_3 conc. in the biosphere.

ACKNOWLEDGEMENTS

The research work has been mainly founded by the Bayerisches Staatsministerium für Landesentwicklung und Umweltfragen.

REFERENCES

(1) REITER, R. et al., Concentration of trace gases, Part II: Ozone. Meteorology and Physics of the Atmosphere, Springer-Verlag Wien, in press (1986)

(2) REITER, R. et al., same as (1), Part III: SO_2, NO_2 and NO same issue, Springer-Verlag Wien, in press (1986)

(3) ATTMANNSPACHER, W. et al., Langzeittendenzen des Ozons Meteorol. Rundschau, 37, 193 - 199 (1984)

(4) REITER, R. et al., Cosmogenic Radionuclides and Ozone Arch. Met. Geophys. Bioklim. Ser. B., 30, 131-160 (1983)

A WET-ONLY PRECIPITATION STUDY IN A MEDITERRANEAN SITE, PATRAS, GREECE

SOTIRIOS GLAVAS
University of Patras, Department of Chemistry
GR-26110 Patras, Greece

Summary

A wet-only precipitation study was conducted in Patras, Greece from October 1985 to April 1986. Very high pH variations were recorded ranging from pH = 4.12 to pH = 8.76, indicating high variability in the acid sources and in neutralization. Ghemical analysis was performed for H^+, SO_4^{2-}, NO_3^-, Cl^-, Ca^{2+}, Mg^{2+} and NH_4^+ ions to evaluate the contribution of each ion to the acidity and alkalinity of the precipitation. Air masses originating in the area of France and North-Eastern Spain correlate very well with the acidic episodes, whereas those from Northern Africa with neutral to basic precipitation.

1. INTRODUCTION

The association of acid rain with major ecological deterioration is quite well documented (1,2), thus a great number of studies (3) have been conducted on the precipition of the industrialized countries of Central Northern Europe as well as in the USA and Canada. However very scarce data are available for moderately industrialized Mediterranean sites.

Earlier pH measurements of the wet precipitation in Patras (4), conducted from January 1980 to May 1981 had indicated considerable variability in the measured pH values, extending from as low as 4.06 in March 1980 to as high as 7.95 pH units in February 1981. On an event basis 23 % of the occuring rain had a pH < 5.6 . Since the contribution of local emmission sources to wet precipitation is limited to cases of convective precipitation and in any way these sources for our region are rather small, combined with the fact that no seasonal variation was observed and acidic rain was randomly observed, indicated long range transport of acid rain.

We undertook this more systematic study, conducting chemical analysis of almost all of the major ions involved in the wet-precipitation, in order to evaluate the relative significance or the measured ions to the acidity and alkalinity of the occuring rain. To assist us with the trace of the air masses responsible for the observed rain, back trajectories were manually constructed.

2. EXPERIMENTAL

A wet-only precipitation collector of our own construction was used, whereby a polyethylene lid was automatically opened and closed at the beginning and end of the rain event, as controlled by a rain sensor, to exclude dry precipitation. The lid when opened is displased 40 cm off the funnel and lower than it, to avoid splashing of rain drops from the lid back to the funnel. The rain was also collected in a polyethylene bottle. The collecting funnel was situated 2.00 m above ground. The whole structure was supported-enclosed in a PVC column 250 mm diameter which prevents exposure of the collecting funnel and storage bottles to sun light.

Samples were collected on a 24 h basis and were analyzed immediately for pH, SO_4^{2-} , NO_3^- , Cl^- and NH_4^+ . Free acidity was determined from pH measurements using a Radiometer pH meter 84. Calibration was always carried out before the pH measurement using Orion's low ionic strength buffers. Sulfate anions were measured nephelometrically with a daily calibration curve. Nitrate and chloride anions were determined with Radiometer Specific electrodes properly calibrated. Ammonium ions were determined with an Orion ammonia gas electrode. Ca^{2+} and Mg^{2+} were measured every three weeks by flame atomic absorption spectrophotometry. The volume collected was determined by weight. The samples were stored at 4° C until analysis.

3. RESULTS AND DISCUSSION

A. Chemical Analysis

Every wet-precipitation event from 13 October 1985 until 20 April 1986 was collected. The amount measured was linearly correlated to the rain height as measured on a nearby situated rain gauge. 18 samples out of a total of 46 24 h samples were completely analyzed for the above ions. For the other samples the collected volume was too small for analysis for which a minimum volume of 50 ml was necessary for a complete analysis. The 18 samples analyzed constituted 58% of the total volume of the rain collected. As the rain of the months May-September is usually less than 5 % of the yearly precipitation amount the samples analyzed may be considered to represent very well the yearly wet precipitation.

pH was measured however for all samples and as in the years 1980-81 a great variability was observed. The lowest value recorded was pH=4.12 on 10 February 1986 and the highest value pH = 8.75 on 2 Feb. 1986. The following Figure 1 depicts the % relative frequency distribution of pH, volume weighted, for all sampled rains.

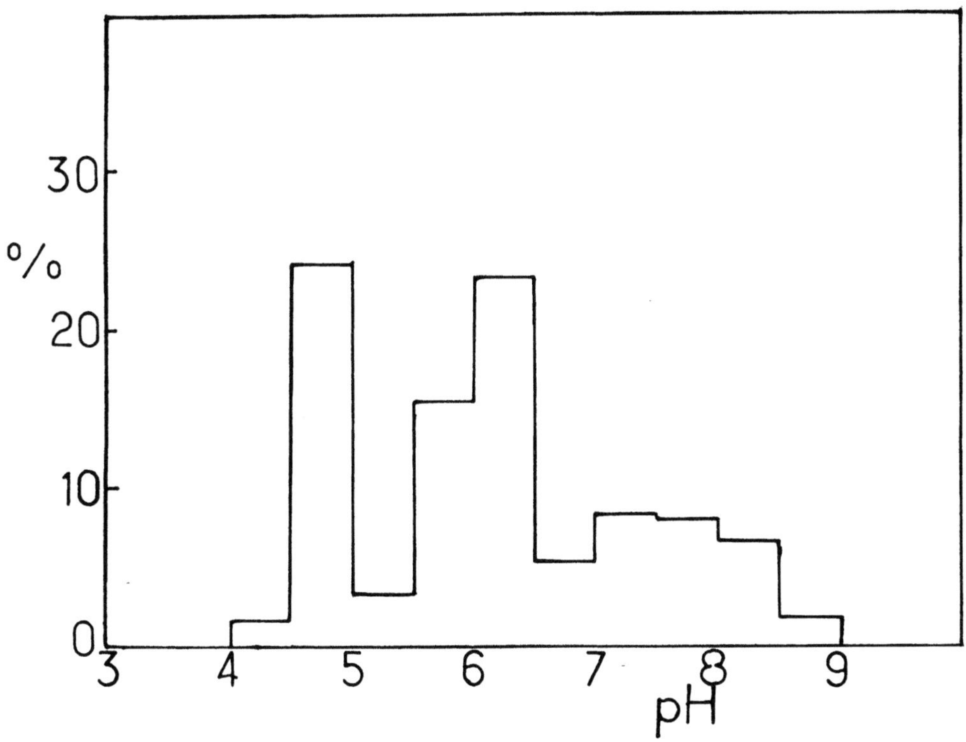

Fig. 1. Relative frequency distribution of pH, volume weighted.

The above istogram can be divided into three sections: a quite acidic section pH < 5, a moderately basic section 5.5 < pH < 6.5 and a basic section pH > 7, each representing 26 % , 39 % and 25 % respectively of the total amount of rain collected.

Apparently non uniform acid and neutralizing sources must operate in order to obtain such diverse pH values. The chemical analysis may assist to distinguish these sources. The volume weighted average concentrations of major ions, expressed in $\mu eq\ L^{-1}$, are given in Table I along with their concentration ranges :

Table I : Volume-weighted average concentrations of major ions and concentration ranges, in $\mu eq\ L^{-1}$.

	Non sea - salt corrected	Sea - salt corrected
Free acidity	1.7 (0.07 - 67.9)	
Ca^{2+}	113.3 (10 - 362)	Ca^{2+} = 108.6
Mg^{2+}	31.2 (0 - 235)	Mg^{2+} = 6.2
NH_4^+	15.7 (1 - 57)	
SO_4^{2-}	69.3 (21 - 138)	SO_4^{2-} = 56.8
NO_3^-	66.7 (30 - 250)	
Cl^-	118.1 (15 - 365)	
Sum of cations	161.9	
Sum of anions	254.1	
Sum of cations	sea-salt corrected	132.2 (without sodium)
Sum of anions	sea-salt corrected	123.5 (without chloride)

If one considers that Patras is a coastal city and assumes a ratio Cl^+/Na^+ of 1.16, the accepted sea water value which dominates the concentration of Cl^- and Na^+ for coastal sites (5), one obtains from the Cl^- concentration a Na^+ concetration of 101.8 $\mu eq\ L^{-1}$ assuming that no excess Cl^- concentration exists. Similar Cl^+/Na^+ ratio equal to 1.24 was also reported for Bermuda (6). This sodium ion is used to obtain the sea-salt corrected values of Ca^{2+} , Mg^{2+} and SO_4^{2-} (7). If one neglects the sodium and chloride values a good mass balance is obtained. The analytical methods employed for the determination of the anions should be mainly responsible for the deviation.

The main feature of the above Table I is the high Ca^{2+} ion concentration. If one accepts that the acidity is due to SO_4^{2-} and NO_3^- only, then the relative importance of Ca^{2+} as a neutralizing agent is 88 % and that of

NH_3 only 13 %. This is the reverse of what is observed in northern-central Europe (8,9), where NH_3 is the dominating neutralizing agent.

The other interesting feature of the above Table I is that sulfate contributes 46 % to the observed acidity to 54 % of the nitrate assuming that Cl^- does not contribute significantly. This is in contrast to the sulfate and nitrate values reported for northern Europe (10) . Possibly local NO_x emissions affect the observed nitrate values.

Mg^{2+} does not contribute significantly to the neutralization of acidity.

B. Air mass trajectories

To explain the data of the chemical analysis, 48 h back air mass trajectories based on 700 h Pa wheather charts were graphically constructed. 74 % of the volume of the occuring precipitation can be traced back mainly to northern Africa (Algeria, Marocco, Tunisia and somewhat Libya) and somewhat to southern Spain. 23 % can be traced to north-eastern Spain, France and Italy and only 3 % is assigned to air masses originating in the Soviet Union, in the periphery of the Black Sea.

Thus result three sectors according to the origin of the air masses as shown in Figure 2.

The pH differences associated with each one of the above sectors are remarkable. Thus masses originating in sector 1 have a volume weighted pH = = 7.01 , sector 2 has pH = 4.72 and sector 3 pH = 5.87.

The alkalinity of sector 1 can be attributed to the lack of acidic emission sources in this region and to the neutralizing calcareous soils of northern Sahara, in agreement with the high Ca^{2+} concetration observed in our rain, 215 μeq L^{-1} , sea salt corrected.

Sector 2 on the contrary is associated with air masses originating in industriallized north-western Europe and with a lot lower neutralizing calcium carbonate soils. The volume weighted Ca^{2+} concentration observed in those rains was 64 μeq L^{-1}. Thus the occuring rains are acidic and they could presumably have a lower pH if the Ca^{2+} concentrations were in the levels observed in Central Europe (9). Every single precipitation sample with pH < 5 was traced in this sector 2.

As is the case with the total data given in Table I, the nitrate anions, in the rain associated with sector 2, contribute approximately 54 % to the total acidity, to 36 % of the sulfate anions, contrary to the observations in north and central Europe (2,9). As stated earlier possibly local NO_x emissions increase the nitrate concentration.

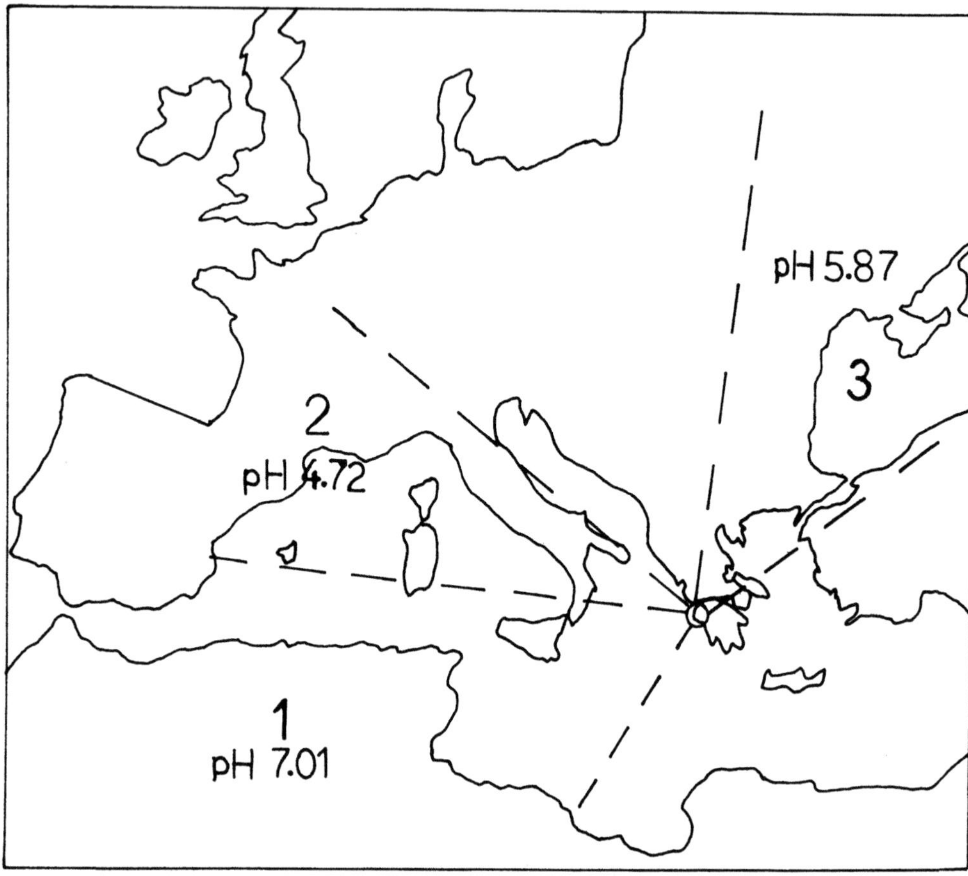

Fig. 2. The three sectors indicating the origin of air masses responsible for Patras precipitation, as obtained from back trajectories.

The 5.87 of sector 3 is very close to that of clean air 5.6 as expected, as neither acidifying nor neutralizing sources can be associated with these air parcels.

The fact that Ca^{2+} ion concentration is always high, even for the acidic sector, make us consider that local Ca^{2+} emissions from the calcareous soils of our area could also play a role, not very significant however.

4. CONCLUSIONS

Acidity transported from the north-western Europe is main acid source observed in our area. The soil dust transported from northern Africa, very rich in neutralizing calcareous soil is mainly responsible for the observed alkalinity. Under appropriate meteorological conditions these two contrasting sources are separated from one another and thus the occuring precipitation is acidic.

Local NO_x emissions, as well as Ca^{2+} rich soils of our area also affect, to a lesser degree, the occuring precipitation.

5. ACKNOWLEDGEMENTS

This work was supported by the Greek Ministry of Environment. I wish to thank Mrs. O. Kallitsi for carring out the most of the chemical analysis, Mr. Kakouros, of the Greek Meteorological Service for the weather charts and Dr. P. Winkler for tis suggestions.

REFERENCES

(1) Glass, N.R., et al. (1982). Effects of acid precipitation. Environ. Sci. Technol., 16 , 162A - 169A.

(2) Oden, S. (1976). The acidity problem, an outline of concepts Water, Air and Soil Polln., 6, (2,3,4) 137-166.

(3) Acid Deposition, Atmospheric Processes in Eastern North America, National Academy Press, Washington, D.C. 1983.

(4) Glavas, S. and J. Pneumaticos (1982). Measurements of the pH of precipitation in Patras, Bull. Hellenic Meteo. Soc., 7 (1,2,3) 7-19.

(5) Lebowitz, L.G. and R.G. de Pena (1985). Chloride and Sodium content in northeastern United States precipitation, J. Geophys. Res., 90 , D5, 8149-8154.

(6) Jichells, T., et al. (1982). Acid rain on Bermuda, Nature, 297 , 55-57.

(7) Granat, L. (1972) On the relation between pH and the chemical composition in atmospheric precipitation, Tellus 24, 550-560.

(8) Asman, W.A.H., P.J. Jonker, J. Slanina and J.H. Baard (1981). Neutralization of acid in precipitation and some results of sequential rain sampling, in Deposition of Atmospheric Pollutants, Ed. H.W. Georgii and J. Pankrath D. Reidel, Dordrecht, pp. 115-123.

(9) Fuhrer, J. (1984). Study of the chemical characteristics of wet and dry deposition in Switzerland, in Proc. Third. European Symp. on Physico-Chemical Behaniour of Atmospheric Pollutants, Varese, pp. 423-432.

(10) Rodhe, H., P. Crutzen and A. Vanderpol, (1981). Formation of sulfuric and nitric acid in the atmosphere during long range transport, Tellus, 33, 132-141, and references there in.

OXIDANT DATA COLLECTION IN OECD-EUROPE 1985-87 (OXIDATE): A CO-OPERATIVE PROJECT FOR RAPID EXCHANGE OF REGIONAL DATA ON OZONE, NITROGEN DIOXIDE AND PAN

P. Grennfelt[1] and J. Schjoldager[2]
[1] Swedish Environmental Research Institute (IVL), Gøteborg, Sweden
[2] Norwegian Institute for Air Research (NILU), Lillestrøm, Norway

Summary

The countries in OECD-Europe are invited to participate in a project with the purpose of exchanging photochemical oxidant data on the regional scale. In 1985, nine countries and 25 measurement stations were participating; 24 of the stations measured ozone, seven measured nitrogen dioxide, and one station measured PAN.

1. OBJECTIVE

The objective of the project is to stimulate research and abatement efforts of photochemical oxidants, by collecting and redistributing data from a number of regional measurement stations in Europe. The data of interest are hourly ambient concentrations of ozone, nitrogen dioxide and peroxyacetyl nitrate.

2. BACKGROUND

The project was first discussed in the fall of 1984 between the Swedish Environmental Research Institute (IVL), National Swedish Environment Protection Board (SNV), Norwegian Institute for Air Research (NILU), and Norwegian State Pollution Control Authority (SFT).

In October 1984, the plans were discussed at the OECD/UBA "Workshop on ambient monitoring data on photochemical oxidants - Availability and requirements", and received positive response from participants from a number of countries in Europe.

In January 1985, representatives of IVL, NILU and OECD prepared a "Proposal by Norway and Sweden for Central Oxidant Data Collection in Europe, 1985 through 1987". The proposal was presented at the meeting of OECD's Air Management Policy Group (AMPG) in March 1985 and received general support. It was agreed that the project should be a part of the OECD project "Control of Major Air Pollutants" (MAP).

The European OECD countries were contacted between March and May 1985 and officially invited to participate.

3. ORGANISATION

Each country nominates an institution responsible for collecting national data and submitting them to NILU. For measurements from the summer half year (April - September), the deadline for data submission is the end of December. For measurements from the winter half year (October - March), the deadline is the end of June. All data will be redistributed to the participating institutions at the end of March and September, respectively. The data will thus be redistributed six months after the end the measurement periods.

At the end of March each year, IVL and NILU will prepare a short project report with information on status of stations and measurements, and a summary of the most important oxidant episodes during the preceding year.

The costs of national data handling are covered by the individual countries. The costs of central data collection, redistribution and project management have been covered by the National Swedish Environment Protection Board and the Norwegian State Pollution Control Authority.

4. PROJECT STATUS, AUGUST 1986

Data from April-September 1985 have been received from these nine countries:

AUSTRIA, BELGIUM, DENMARK,
FEDERAL REPUBLIC OF GERMANY, FINLAND, NETHERLANDS,
NORWAY, SWEDEN, UNITED KINGDOM.

Total number of stations are 25. Ozone measurements are carried out at 24 stations, NO_2 measurements at 7 stations, and PAN measurements at 1 station. The locations of the stations are indicated in Figure 1. The contact institutions and persons, and the measurement stations and compounds are given in Tables 1 and 2.

The data were not always sent according to the tape/diskette specifications which the contact institutions had received in advance. There has thus been more work than expected to bring the data in order and generate the tapes/diskettes for redistribution.

We now plan redistribution of the 1985 data in September or October 1986. The presentation of the most significant oxidant episodes will be ready at the same time.

We assume that the project will be delayed only the first year. After that, the routines for generation and handling of data are probably well established

TABLE 1
OXIDANT DATA COLLECTION IN OECD-EUROPE 1985-87 (OXIDATE)

LIST OF CONTACT PERSONS AND INSTITUTIONS, 1985

COUNTRY	NAME, INSTITUTION	TELEPHONE
AUSTRIA	Dr. Ruth Baumann Umweltbundesamt Wien Abteilung für Lufthygiene Währinger Strasse 25a A-1096 WIEN	43-222-43 2504
BELGIUM	Dr. J. Beyloos Institut d'Hygiène et d'Epidémiologie 14, Rue Juliette Wytsman B-1050 BRUXELLES	32-2-642 5111
DENMARK	Dr. Finn Palmgren Jensen Miljøstyrelsen Luftforureningslaboratoriet Forsøgsanlæg Risø DK-4000 ROSKILDE	45-2-37 1137
FEDERAL REPUBLIC OF GERMANY	Dr. Rolf Sartorius Umweltbundesamt Bismarckplatz 1 D-1000 BERLIN	49-30-8903 511
	Dr. Wolfgang Grosch Umweltbundesamt Pilotstation Frankfurt Frankfurter Str. 135 D-6050 OFFENBACH	49-69-88 80 38
FINLAND	Mr. Heikki Lättilä Finnish Meteorological Institute (FMI) P.O.Box 50 SF-00810 HELSINKI	358-0-119 22
NORWAY	Mr. Jørgen Schjoldager Norwegian Institute for Air Research (NILU) P.O. Box 130 N-2001 LILLESTRØM	47-2-71 41 70
NETHERLANDS	Dr. W.F. Blom Air Research Laboratory Rijksinstituut vor Volksgezondheid en Milieuhygiene Postbus 1 NL-3720 BA BILTHOVEN	31-30-74 91 11
SWEDEN	Mr. Peringe Grennfelt Swedish Environmental Research Institute (IVL) P.O. Box 5207 S-402 24 GOTEBORG	46-31-81 02 80
UNITED KINGDOM	Dr. B. Sweeney Air Pollution Division Warren Spring Laboratory Gunnels Wood Road, Stevenage Herts SG1 2BX, ENGLAND	44-438-31 33 88

TABLE 2
OXIDANT DATA COLLECTION IN OECD-EUROPE 1985-87 (OXIDATE)

LIST OF COUNTRIES, STATIONS AND COMPOUNDS, 1985

COUNTRY/STATION	COMPOUNDS O_3	NO_2	PAN	PERIOD
AUSTRIA				
Illmitz	X			1 April-30 September
BELGIUM				
Gent, St.Kruiswinkel	X	X		1 April-30 September
DENMARK				
Risø	X			8 May-30 September
FED.REP. OF GERMANY				
Westerland	X			1 April-30 September
Waldhof (Langenbrügge)	X			" "
Deuselbach	X			" "
Brotjacklriegel	X			" "
Schauinsland	X			" "
FINLAND				
Utö	X			4 July-30 September
NETHERLANDS				
Witteveen	X	X		1 April-30 September
Eibergen	X	X		" "
Delft			X	" "
NORWAY				
Birkenes	X			1 July-30 September
Langesund	X			1 April-31 July
Jeløya	X			1 April-30 September
SWEDEN				
Wavihill	X			2 May-30 September
Rörvik	X			1 April-30 September
Aspvreten	X			" "
Ringamåla	X			" "
Norrakvill	X			23 June-30 September
Vindeln	X			3 July-30 September
UNITED KINGDOM				
Lancaster	X	X		1 April-30 September
Bottesford	X	X		8 May-30 September
Sibton	X	X		1 April-30 September
Harwell	X	X		" "

FIGURE 1
MEASUREMENT STATIONS IN PROJECT OXIDATE 1985

DAILY TRENDS OF PHOTOCHEMICAL OXIDANTS AND THEIR PRECURSORS IN A SUBURBAN FORESTED AREA. A USEFUL APPROACH FOR EVALUATING THE RELATIVE CONTRIBUTIONS OF NATURAL AND ANTHROPOGENIC HYDROCARBONS TO THE PHOTOCHEMICAL SMOG FORMATION IN RURAL AREAS IN ITALY.

P. Ciccioli, E. Brancaleoni, C. Di Palo, A. Brachetti and A. Cecinato
Istituto sull'Inquinamento Atmosferico - CNR - Via Salaria Km 29,300, CP 10, I-00016 - Monterotondo Scalo.

Summary

This paper presents the results obtained during two photochemical smog episodes occurring in the last two years in a suburban forested site located downwind of Rome.
The content of ozone, NO_2, PAN, PPN, methyl nitrate and selected aldehydes has been measured together with hourly variations of the most reactive natural (terpenes) and anthropogenic (alkylbenzenes) hydrocarbons. Daily trends of the various components, including some photochemical degradation products not yet identified, have been plotted together with the values of the meteorological parameters (solar radiation intensity, relative humidity, wind speed and direction) measured over the site.
During one episode, measurements were carried out above and below the canopy in order to evaluate the vertical profile of natural and anthropogenic hydrocarbons. A detailed analysis of the results, collected under different meteorological conditions, seems to suggest that naturally emitted components do not contribute significantly to increase the content of photochemical oxydants in the sampling site. Transport phenomena, well evidentiated by the trends of PAN and its homologs, seem instead to account for the relatively high levels of photochemical pollutants and free acidity observed in our site.
The results of aerometric measurements of natural components are in agreement with the emission rates estimated by means of the bag enclosure technique.

1 - INTRODUCTION

The main tropospheric sink for hydrocarbons (HC) in the air is linked to a sequence of photochemical processes leading to the formation of atmospheric pollutants potentially hazardous to men and his environment. They include ozone, aldehydes, peroxyacetylnitrate (PAN), free acids and particulate matter. Therefore, there is a continuous interest in the measure of these species in air when photochemical smog events are taking place. In the past, such measurements have been carried out almost exclusively in urban and industrialized areas where high emission of HC and NO_x occurs. Recently, the attention has been focused on the determination of HC and photochemical pollutants in forested areas in order to find out to what extent naturally emitted HC contribute to photochemical pollution in rural areas not directly exposed to strong anthropogenic emission (1-3). For this reason, we have developed, in the past years, sensitive techniques for the monitoring of natural and anthropogenic HC in air and monitoring campaignes have been undertaken to measure daily trends of HC and other pollutants relevant to photochemical smog formation. Such measurements, carried out within the canopy of a small forest located 30 Km downwind of Rome (2), have shown that the mixing ratios of natural HC are comparable to those of anthropogenic HC transported into the forest. The measurements have shown that natural HC react rapidly with OH radicals around noon, but it was difficult to evaluate the relevance of their contribution to the ozone formation. In this work we provide additional data which may be used to estimate the source strength of natural HC and help to assess their role in the formation of photochemical oxidants and some acidic species.

2 - EXPERIMENTAL

HC and aldehydes were collected on adsorption traps filled with Carbopack B adsorbent. The individual species, thermally desorbed by the trap, were identified and quantitated by GC-MS used in the selected-ion mode (2). Ozone was monitored by UV fluorescence. PAN, PPN and CH_3NO_3 were monitored by using a laboratory-made GC analyzer equipped with an ECD (4). NO_2 was monitored by chemiluminescence detection whereas HNO_3 was collected by using high-efficiency annular denuders and measured as nitrate by ion-chromatography. The small forest of pine trees was sorrounded by farmland and located 30 Km NE of Rome, not far from the mountains of central Italy (2).

3 - RESULTS AND DISCUSSION

3.1 RELEVANCE OF PHOTOCHEMICAL SMOG IN ITALY AND SELECTION OF THE SAMPLING SITE.

In order to give an idea of the relevance that photochemical pollutants, especially the phytotoxic ones, such as H_2SO_4, HNO_3, PAN and O_3, may have on the Italian environment it is useful to look carefully at Figure 1. In this Figure is reported a map of the Italian peninsula, the distribution of forested areas (dashed area) and the location of the most densely populated sites where higher is the anthropogenic emission (industrial and autoveicular). By considering the orographic situation, the large surface exposed to sea and land breezes, the high temperatures that can be reached during summer and the population density, it is not difficult to predict that: a) the frequency and intensity of photochemical smog events occurring during summer can be very high; b) the occurrence of such episodes can be widely distributed overall the country; c) photochemical oxidants generated into the cities can be easily transported into the forested areas since the traveling distance is, on the average, not longer than 50-100 Km at most.

Because of this situation, it is very important for us to investigate whether organic HC emitted from trees or plants could contribute signifi-

Fig. 1 - Map of the Italian Peninsula showing the distribution of forested areas and the location of the most intense anthropogenic emission sites (black spots).

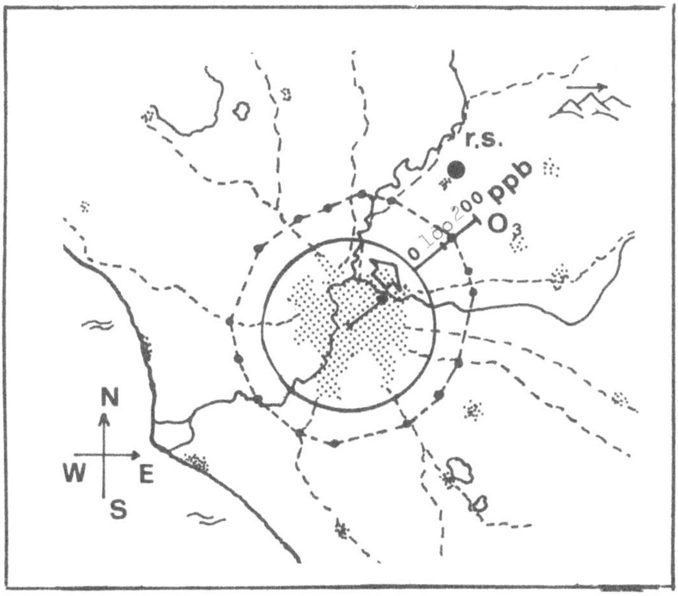

Fig. 2 - Values of O_3 (ppb) around Rome measured at noon during a photochemical smog episode observed in 1978 and relative position of our sampling site (r.s.).

cantly to increase the levels of photochemical oxidants, which, in the Mediterranean countries intensely exposed to solar radiation, can be a major source of forests decline.

In order to evaluate the influence of natural and anthropogenic HC to photochemical smog generation, we have selected a site located between a large city and the mountains acrossing Central Italy. This position is representative of the Italian situation and permits to see transport fenomena occurring from both the urban area and the forests. The site was sorrounded by a wide agricoltural area and sampling was carried out in a small forest containing pine trees together with different other species. The presence of trees permitted the evaluation of natural emission. In Figure 2 is shown the position of the site with respect to the urban area of Rome and the distribution of ozone measured around the city during a photochemical smog event observed in July 1978 (5) . The measurements were carried out by a 20 minutes flyght to have an almost instantaneous situation of the O_3 plume. At that time, an high pressure was established over the city with an inversion layer at ca. 1,000 m and light winds were blowing from the SW sector. The frequency of high pressure conditions, the prevalent wind direction and the frequency distribution of the height of the inversion layer, measured at midnight and noon at the nearest airport, are reported in Figure 3.

By combining the information of the last two Figures we can see that, when photochemical smog develops in Rome, it is likely that pollutants are acrossing the site and transported toward the mountains located ca. 60 Km W from the city. Previous measurements (4) have shown that the maximum O_3 concentration across the site is reached in the early afternoon.

By following diurnal trends of photochemical oxydants and their precursors on this site it would be possible, therefore, to estimate to what extent the rural and forestal areas sorrounding the site contribute to photochemical smog levels.

3.2 HOURLY VARIATIONS OF HC AND PHOTOCHEMICAL OXIDANT DURING PHOTOCHEMICAL SMOG EPISODES.

In Figure 4 and 5 are shown, respectively, daily trends of HC, O_3, PAN and NO_2 measured during two-days photochemical smog episodes observed in July 1984 and 1985.
In the former set of measurements, determinations of HC were carried out below (1 m from the ground) and above the canopy (12 m from the ground). The formers are indicated by solid lines whereas the latters are represented by dotted lines. In spite of the similarity of the metereological situation and season, the levels of photochemical pollution measured during these episodes were quite different. In July 1984 the content of O_3 and PAN was, actually, almost double than that measured in July 1985. Anthropogenic HC reached in 1984 higher levels than those measured in July 1985. A comparison between the mixing ratios measured above and below the canopy indicated that anthropogenic HC were well mixed within the inversion layer and consequently they were transported into the site by vertical an horizontal diffusion. A ratio between xylenes and ethylbenzene lower than 3.8 indicated

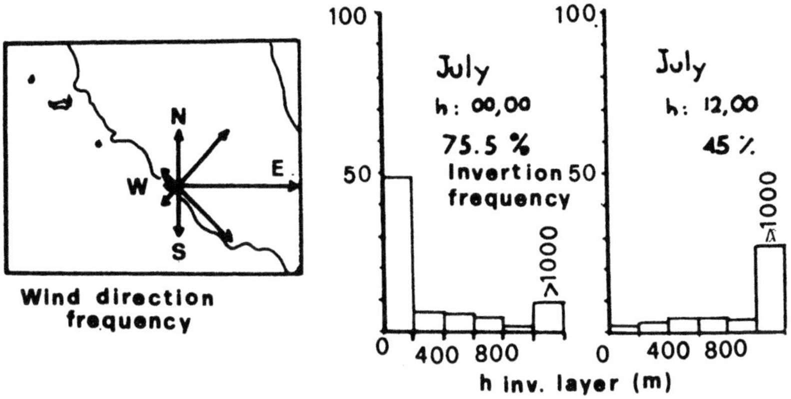

Fig. 3) Prevalent winds observed over the site in July and frequency of anticyclonic whether. Bars are used to indicate the frequency and height of the inversion layer at midnight and noon.

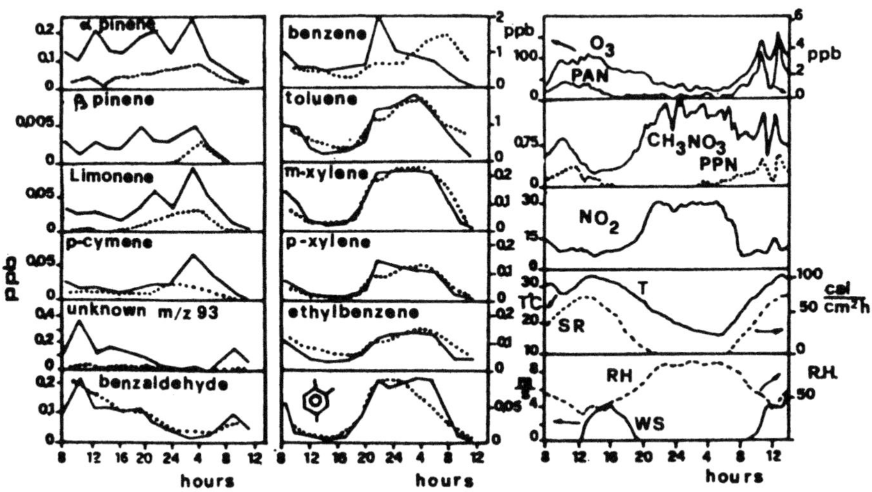

Fig. 4) Hourly variations of HC,photochemical oxidants and meteorological parameters occurring during a photochemical smog episode observed in July 1984 in our site. T:temperature,SR:solar radiation,RH:relative humidity WS:wind speed.————Measurements below the canopy --------- Measurements above the canopy.

also that the air mass was quite "aged" and local "fresh" emission was negligible. By contrast, the content of natural HC measured during the same episode (1984) was rather low if compared with that of anthropogenic HC, especially if we look at the values measured above the canopy. The vertical gradient is rather steep suggesting that the imput from grassland or trees, placed along the way between the city and the site, was neglegible. In both cases it is evident that, after sunrise, photochemical decomposition of HC occurs. While the convertion of anthropogenic HC is well seen by the drop in mixing ratios observed after sunrise, that of natural HC can be better detected by the trend of the unknown product (very likely an aldehyde) which has the same diurnal profile of benzaldehyde but is more abundant below the canopy. Daily trends of natural HC shown in Figure 4 are rather complex and this is due, in our opinion, to the fact that natural emission was, a that time, rather low (ca. 5 ug/gh) due to soil dryness and, therefore, increases in the HC emission rates, occurring during the day, were fully or partly compensated by the increase in reactivity.

The measurements carried out in July 1985 show, instead, that emission of natural HC was much higher than that measured the year before (20 ug/gh) and it is possible to see clearly the photochemical decomposition of α -pinene and other natural HC through their decrease in the mixing ratio occurring after sunrise. The trend of the unkown product seems to be, also in this case, consistent with the hypothesis of photochemical reaction. During this monitoring campaign we observed, however, that the content of anthropogenica HC, though always higher than that of natural HC, was 50% less than that measured in 1984. Since the ratio between xylenes and ethylbenzene was, also in this case, lower than 3.8, this decrease was certainly linked to a lower imput coming from the city.

A comparison between the values of PAN, O_3 and HC seems to suggest that the dominant effect in determining the photochemical pollution levels in our site is due to anthropogenic pollution generated in the urban area whereas the contribution of natural HC can be, as a first approximation, completely neglected. This conclusion is supported by other observations made when different meteorological conditions were established over the site. An example is shown in Figure 6 where daily trends of O_3, PAN and HNO_3 are reported. During this measure, carried out in July 1985, it was possible to distinguish clearly the background levels, generated around the site by natural emission and low anthropogenic activity, from the urban imput. As can be seen "natural" levels are much lower than the urban imput although some anthropogenic activity is always going-on in the country side and the natural HC are not the only components responsible for these levels. Actually, when breezes or northern winds are carrying the pollutants from the mountains to the site, the maximum values of O_3 and PAN never exceeded 30-40 ppb and 0.3-0.5 ppb respectively.

3.3 EVALUATION OF NATURAL AND ANTHROPOGENIC HC EMISSION AND ITS POSSIBLE IMPACT IN THE ITALIAN SITUATION.

As far as the emission rate of natural HC is concerned, measurements

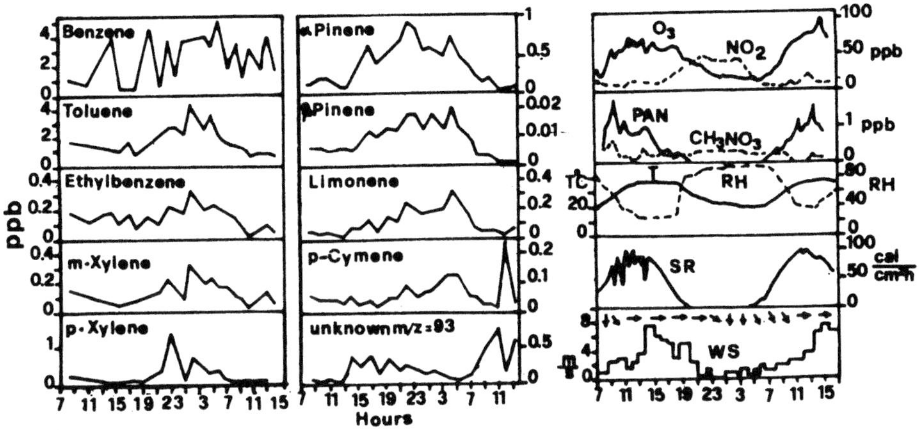

Fig. 5) Hourly variations of the same pollutants and meteorological parameters shown in Fig.4. The measurements were carried out during a photochemical smog episode observed in July 1985.

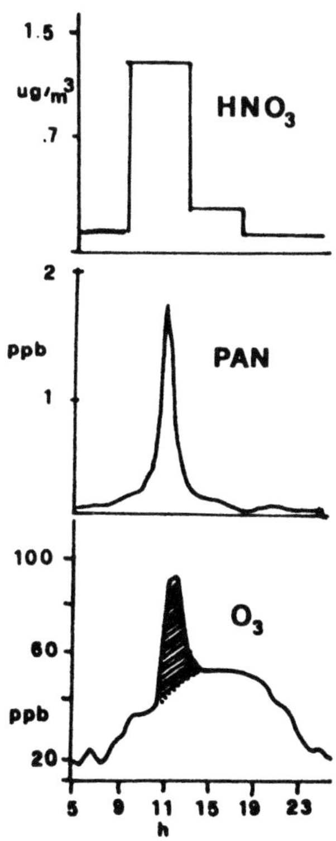

Fig. 6) Trends of O_3,PAN and HNO_3 measured during a day of July when it was possible to distinguish clearly the imput of O_3 coming from the city (shadowed area) from that originated from the natural and anthropogenic activity sorrounding the site. PAN and HNO_3 have been used as marker to detect photochemical transport from the urban area.

carried out with the bag enclosure technique gave values of total HC ranging between 2 and 20 ug/gh depending upon the physiological situation of the plant and the season. Although this technique has been questioned, the results are reasonable and can be used to estimate the natural emission in Italy. By combining the measurements carried out in 1982, 1983, 1984 and 1985 with the emission data we obtained, we can estimate that, in hot summer days (30-40^oC max. temperature), the emission rate is ca. 2,000 ug/m^2 h whereas during the winter season the value is more close to 200-300 ug/m^2h. We can assume, therefore, that the average emission rate over the year is ca. 800 ug/m^2h. By considering that pine trees cover approximately only 30% of the total forested area (50,000 Km^2) and the emission rate of deciduous forests is 50% less than coniferous forests and grassland (6), we can estima an average emission strenght of ca. 500 ug/m^2h. From these approximate figures we obtain that natural emission is ranging between 2 and 3 10^5 tons/y whereas the anthropogenic emission is estimated to be 1.76 10^6 tons/y. Since, in Italy, the forests are concentrated near the mountains where little NO_2 is produced, it is unlikely that they can influence detectably (few percents) the level of photochemical oxidants.

Instead, the effect of anthropogenic emission from the city toward the forests can be a source of great concern. The meteorological situation is such that photochemical pollutants accumulated in the city can be easily transported into the forests and because of the altitude, easily deposited into the tree leaves by several media (clouds, rain, snow and impaction). We believe that the synergistic effect of dry and wet deposition could cause severe damages to many Italian forests if action is not taken to reduce anthropogenic emission of HC and NO_x in densely populated areas. Among the various sources, special attention should be paid to reduce autovehicular emission which can reach extremely high values because of the ancient structure of many Italian cities. An average speed of 20 to 40 Km/h is not uncommon in many large cities as Rome, Naples or Florence. This, combined with a high traffic density, can give rise, easily, to emission strenght for HC close to 20-30 mg/m^2h generating mixing ratios of O_3 as higher as 200 ppb (7). Although, within the city, O_3 can be partly reduced by reaction with NO, 100-150 ppb have been recorded easily in our site and this amount was certainly transported, together with PAN, HNO_3 and H_2SO_4 into the near forests.

4 - REFERENCES

1) Y. Yokoucki, M. Okaniwa, Y. Ambe and K. Fuwa. Atm. Environ., 17, (1983), 743.
2) P. Ciccioli, E. Brancaleoni, M. Possanzini, A. Brachetti, C. Di Palo. Proceedings of the 3th European Symposium on the "Physico-Chemical Behaviour of Atmospheric Pollutants". 10-12 April, Varese, Italy. D. Reidel Publ. Co., pp. 62-73.
3) V.A. Isidorov, I.G. Zenkevich, B.V. Ioffe. Atm. Environ., 19. (1985), 1.
4) P. Ciccioli, E. Brancaleoni, V. DiPalo, M. Liberti and C. DiPalo

"Misura delle alterazioni della qualità dell'Aria da fenomeni di smog fotochimico. Sviluppo di una apparecchiatura semiautomatica per la misura del PAN,PPN e nitrato di Metile" ACQUA e ARIA; (1986) in press;

5) A. Anav and G. Dall'Oglio. Misure dell'O_3 atmosferico in quota in area urbana: misure preliminari sulla citta' di Roma. Seminario su "Apporto della meteorologia nel Controllo e nell'Abbattimento dello Inquinamento Atmosferico", Roma, 8-10 Nov. 1978, pubbl. CNR AC/3/111, p. 505-516, Roma, 1979.
6) A.T. Altshuller. Atmospheric Environment, 17, 2131, (1983).
7) K. Zellner and N. Moussiopoulos. "Simulation of the Ozone Formation Caused by Motorcar Emissions in Urban Areas", Workshop on "Pollutant Cycles, Transport Modelling and Field Experiments", RIVM, Bilthoven, NL, 23-25 Sept. 1985, (Proceedings), p. 138-142.

OBSERVATIONS OF NITROUS ACID AT ROOF AND STREET LEVEL DURING AIR POLLUTION EPISODES IN GOTHENBURG

A. Sjödin
Swedish Environmental Research Institute
P.O. Box 5207, S-402 24 Gothenburg, Sweden

Summary

The diurnal variation of nitrous acid at roof and street level during air pollution episodes in the early summer of 1986 was studied at a site heavily influenced by local traffic emissions. Measurements were made in parallel at the two sites by means of a denuder technique. Simultaneously chemiluminescent measurements of NO and NO_2 were made. Nighttime HNO_2-concentrations of up to 6 ppb as 1 hr averages have been observed simultaneously at ground and roof level. The observed diurnal variation of HNO_2/NO_x-ratios indicates that HNO_2 is formed in a thermal reaction in ambient air with an average conversion rate corresponding to 0.8% NO_2 hr^{-1}. An average effective OH-radical production rate of $7 \cdot 10^6$ molecules cm^{-3} s^{-1} was calculated for the HNO_2-photolysis at sunrise.

1. INTRODUCTION

Nitrous acid, HNO_2, is an important nitrogen species in the polluted troposphere, primarily due to its rapid single-step photolysis in daylight to the hydroxylradical:

$$HNO_2 + h\nu \rightarrow OH + NO \qquad (1)$$

Since OH is a key intermediate in photochemical smog formation, it is of great importance to be able to quantify the absolute and relative strengths of this OH-source under varying atmospheric conditions. Therefore, a lot of research work has been performed in laboratories as well as in field during the past decade, with the aim of elucidating sources and sinks of nitrous acid in the atmosphere (1-21). Thermal reactions of NO_2 in ambient air and direct emissions with vehicle exhaust have recently been identified as important HNO_2-sources in urban areas (14-18, 20, 21). Effective rates of radical production from reaction (1) of $1\text{-}3 \cdot 10^7$ molecules cm^{-3} s^{-1}, estimated from the decay of HNO_2 during morning hours, have been reported from field measurements in Cologne, Los Angeles and Gothenburg (11, 12,20).

The present study further illustrates the nighttime and daytime behaviour of nitrous acid in urban atmospheres. HNO_2, NO and NO_2 were simultaneously monitored close to ground level in a street canyon with dense traffic in the central part of

Gothenburg. In parallel the same species were monitored at roof level 20 m above the ground measuring point. Measurements were made during the early summer of 86', preferably on occasions with elevated levels of primary air pollutants. HNO_2 was measured with a previously described denuder technique, while NO- and NO_2 were measured by chemiluminescense. From the observed diurnal patterns of the ratios between HNO_2- and NO_x-concentrations, rates of the formation and the photolysis of HNO_2 were estimated.

2. EXPERIMENTAL

The technique for determining ambient air concentrations of nitrous acid by means of sodium carbonate coated denuders has been described in detail earlier (19, 20, 22). Concentrations of HNO_2 were determined as 1-3 hrs mean values, while NO and NO_2 were continuously monitored with chemiluminescent instruments (Monitor Labs dual channel nitrogen oxides analyzer 8840). Both instruments were carefully calibrated prior to and after each measurement campaign. The three nitrogen species were monitored simultaneously and in parallell at ground and roof level in connections to the facilities of the Air Department of the Swedish Environmental Research Institute in the central part of Gothenburg. This city is located on the west coast of Sweden and has approximately half a million inhabitants.

The ground level measuring point was located in a street canyon about 1 m above the street surface, close to the passing traffic. The roof level measuring point was almost streight above the ground point at a height of about 20 m.

The measurements reported in this study were carried out in May and June 1986. A total of nine measurement campaigns were performed. The weather on all occasions was characterized by clear or almost clear skies, high daytime temperatures (20-25 ^{o}C), and a weak or moderate seabreeze during the day, that ceased to calm after sunset. As a consequence, primary pollutants mainly arising from city traffic, were accumulated in the air of the Gothenburg basin during the subsequent nighttime hours. During morning hours, the air was diluted by the increasing turbulence and the initial seabreeze.

Each measurement campaign extended from 14-20 hrs LMT on the first day to 5-14 hrs LMT on the following day.

3. RESULTS

3.1 Observations of HNO_2

The maximal nighttime concentrations of HNO_2 at roof and street level observed on each of the nine measurement occasions are presented in Table 1.

Date	LMT for observed HNO_2-max	ppb HNO_2 roof level	ppb HNO_2 street level
860502	23-00	5.7	6.8
860503	23-00	2.8	3.8
860530	01-02, 23-00	2.6	5.2
860612	00-01, 01-02	1.2	2.0
860616	01-02	1.3	1.9
860617	23-00	4.8	4.7
860623	03-04, 00-01	2.6	3.4
860624	02-03	4.3	3.3
860626	23-00, 22-23	1.4	1.4

Table 1. Observed max nighttime concentration of nitrous acid.

The highest values reached 5.7 and 6.8 ppb, respectively. The average of all observations was 1.6 ppb at roof level and 1.8 ppb at street level. The total number of observations was 85 for each measuring point. From statistical treatment (linear regression analysis) the relation between the HNO_2-concentration in ppb observed at roof and street level could be expressed as:

$$[HNO_2]_{roof} = 0.68 \cdot [HNO_2]_{street} + 0.42 \ (p=0.77)$$

Typical concentration profiles for HNO_2 during a measurement campaign are given by Figure 1.

3.2 Observations of NO and NO_2

The NO_x-concentration varied from the order of a few ppb to several hundred ppb. The highest 1 hour mean value reached 450 ppb at street level. At such high concentrations most of the NO_x was in the form of NO. In general the NO_x-concentration was lower at roof level compared to street level. The average of 104 observations was 61 ppb at roof level and 98 ppb NO_x at street level. The following linear regression relationship was achieved:

$$[NO_x]_{roof} = 0.45 \cdot [NO_x]_{street} + 17 \ (p=0.81)$$

The difference in NO_2-concentrations was less compared to the difference in NO_x-concentration. The average of 72 observations was 31 ppb at roof level and 36 ppb NO_2 at street level, with the following statistical relationship:

$$[NO_2]_{roof} = 0.63 \cdot [NO_2]_{street} + 8 \ (p=0.76)$$

3.3 Estimations of HNO_2 formation rates from HNO_2/NO_x- and HNO_2/NO_2-ratios

Significant fractions of HNO_2 have been observed recently in dilute auto exhaust (16). Therefore, it was considered that the ratio between HNO_2- and NO_x-concentration was a better parameter than absolute HNO_2-levels in trying to quantify rates of HNO_2 formation and decomposition from the field observations. Since NO is a primary and the predominant oxidized nitrogen form during the observed air pollution episodes at roof as well as street level, the parameter HNO_2/NO_x compensates for the accumulation or the dilution of pollutants that occur due to changes of windspeed over Gothenburg. Furthermore it is more adequate to make comparisons between different occasions if this parameter is used, since air pollutant levels varied strongly from one occasion to another.

Figure 2 demonstrates the observed diurnal profile of the HNO_2/NO_x-ratio calculated as averages from the nine roof level observations. Despite the relatively large scatter in individual values at a certain timepoint of the day, a marked increase in HNO_2/NO_x-ratio is observed from early afternoon until early dawn (3 or 4 am): from less than 1% to approximately 5%. This indicates that HNO_2 is formed in the ambient air from atmospheric reactions. A similar pattern is obtained for the street level observations, cf. Figure 3. In this case, however, the average ratios generally are lower compared to the roof level observations. As can be seen from the figures, the increase in HNO_2/NO_x-ratio is followed by a rapid decrease after sunrise. This indicates that HNO_2 is decomposed under the influence of sunlight irradiation. If linear regression analysis is applied on the HNO_2/NO_x vs timeplot during the build-up period (15-04 hrs LMT), the following expression is obtained for the roof level observations:

$$[HNO_2]/[NO_x](\%) = 0.45 \cdot t\ (hrs) + 0.12 \quad (p=0.97,\ N=8)$$

and for the street level observations:

$$[HNO_2]/[NO_x](\%) = 0.28 \cdot t\ (hrs) + 0.30 \quad (p=0.91,\ N=10)$$

Between times t_1 and t_2 these two expressions can be written as:

$$\Delta\left(\frac{[HNO_2]}{[NO_x]}\right)_{roof} = 0.45 \cdot \Delta t\ (\%\ hr^{-1})$$

$$\Delta\left(\frac{[HNO_2]}{[NO_x]}\right)_{street} = 0.28 \cdot \Delta t\ (\%\ hr^{-1})$$

If the NO_x-concentration as a rough estimate is assumed to be independent of time t during build-up period, these expressions can be written as:

$$\frac{d[HNO_2]}{dt}_{roof} = 0.0045 \cdot [NO_x]_{roof}\ (ppb\ hr^{-1})$$

$$\frac{d[HNO_2]}{dt}_{street} = 0.0028 \cdot [NO_x]_{street}\ (ppb\ hr^{-1})$$

From these expressions two independent values of an average formation rate for HNO_2, r_{HNO_2}, in the ambient air during the build-up periods can be calculated. We obtain for the roof level observations, for which the average NO_x-concentration during build-up period is 65 ppb:

$$r_{HNO_2} = 0.0045 \cdot 65 \text{ ppb hr}^{-1} = 0.29 \text{ ppb hr}^{-1}$$

and for the street level observations (average NO_x 106 ppb):

$$r_{HNO_2} = 0.0028 \cdot 106 \text{ ppb hr}^{-1} = 0.30 \text{ ppb hr}^{-1}.$$

If HNO_2 is assumed to be formed from atmospheric reactions of NO_2, then an average rate of the conversion of NO_2 to HNO_2, $r_{NO_2 \rightarrow HNO_2}$, can be estimated:

$$r_{NO_2 \rightarrow HNO_2} = \frac{0.29}{31} = 0.94\% \text{ hr}^{-1}$$

for the roof level observations and:

$$r_{NO_2 \rightarrow HNO_2} = \frac{0.30}{36} = 0.83\% \text{ hr}^{-1}$$

for the street level observations.

A similar conversion rate was estimated from the HNO_2/NO_2 vs time plots shown by Figures 4 and 5. The linear regression plots for the build-up period were calculated to:

$$\left(\frac{[HNO_2]}{[NO_2]}\right)_{roof} (\%) = 0.88 \cdot t(hrs) - 0.32 \quad (p=0.96, N=8)$$

$$\left(\frac{[HNO_2]}{[NO_2]}\right)_{street} (\%) = 0.67 \cdot t\ (hrs) - 0.26 \quad (p=0.98, N=7)$$

i.e. according to the assumptions above

$$r_{NO_2 \rightarrow HNO_2} = 0.88\% \text{ hr}^{-1}$$

for the roof level observations and:

$$r_{NO_2 \rightarrow HNO_2} = 0.67\% \text{ hr}^{-1}$$

for the street level observations.

3.4 Estimations of HNO_2 photolysis rates from HNO_2/NO_x-ratios

As can be seen by Figure 3 and 4, the HNO_2/NO_x vs time-plot exhibits an exponential decrease after sunrise until noon. This indicates that HNO_2 is photolysed by sunlight irradiation to OH-radicals according to the first order reaction:

$$HNO_2 + h\nu \rightarrow OH + NO \qquad (1)$$

The integrated rate equation for this reaction can be expressed as:

$$-\ln[HNO_2] = k_1 \cdot t + \ln[HNO_2]_{t=0}$$

If the parameter $[HNO_2]/[NO_x]$ is assumed to describe the decrease in HNO_2-concentration during morning hours exclusively due to the photolytic decay more adequately than the absolute HNO_2-levels, then an average rate for the HNO_2-photolysis can be estimated from the plots of $\ln([HNO_2]/[NO_x])$ vs time after sunrise. Plots for roof level as well as street level observations are shown by Figure 6. From a linear regression analysis of these plots we achieve:

$$-\ln\left(\frac{[HNO_2]}{[NO_x]}\right)_{roof} = 0.24 \cdot t + 2.84 \; (p=0.98, N=6)$$

$$-\ln\left(\frac{[HNO_2]}{[NO_x]}\right)_{street} = 0.22 \cdot t + 3.25 \; (p=0.91, N=5)$$

The average value of the first order rate constant for reaction (1) from our observations becomes

$$k_1 = 0.23 \text{ hr}^{-1}$$

For a typical predawn HNO_2-concentration during a high pressure situation of 4 ppb the rate of OH-radical production from HNO_2-photolysis at sunrise then can be calculated to

$$\frac{d[OH]}{dt} = -\frac{d[HNO_2]}{dt} = k_1 \cdot [HNO_2] = 0.92 \text{ ppb hr}^{-1}$$

$$\frac{d[OH]}{dt} = 6.8 \cdot 10^6 \text{ molecules cm}^{-3}\text{s}^{-1}$$

4. CONCLUSIONS

During high pressure situations in the summer significant amounts, up to 6 ppb, of nitrous acid occurs during the night in the airshed of the Gothenburg basin. The increase in HNO_2-concentration around sunset is associated with a general accumulation of air pollutants over Gothenburg due to the stagnation of the seabreeze. However, the increase in HNO_2-concentration is much more rapid than the increase in NO- and NO_2-concentration, which indicates that HNO_2 is not only emitted directly with vehicle exhaust, but also is formed in the air from thermochemical reactions. Such build-up is observed at roof as well as street level. Furthermore, there is a strong correlation between individual HNO_2-values measured in parallel at the two measurement sites. The average HNO_2-level is 13 % higher, while the average NO_x-level is 61% higher for street level observations compared to roof level observations. Consequently, there is a continuous transformation of NO_x to HNO_2 during the transport of air from ground level to higher altitudes. The estimated average rate for the conversion of NO_2 to HNO_2, ≈0.8% per hour. agrees fairly well with values reported from earlier field studies. In Germany, conversion rates between 0.2 and 1.5% per hour were found in the boundary layer, with an average rate of 0.6% per hour (17). In a recent field study in Gothenburg we found the conversion factor 1.1% per hr^{-1} (20). From laboratory kinetic experiments a conversion factor of the same order of magnitude has been reported (14, 15, 18).

The calculated initial rate of the OH-radical production from HNO_2-photolysis at 4 ppb predawn HNO_2-concentration, $7 \cdot 10^6$ molecules $cm^{-3} s^{-1}$, can be compared with values reported from field measurements in Cologne and Los Angeles reaching $2-3 \cdot 10^7$ molecules $cm^{-3} s^{-1}$ as upper limits (11, 12).

ACKNOWLEDGEMENT

This work was financially supported by the Swedish National Environmental Protection Board.

REFERENCES

(1) NASH, T. (1974) Nitrous acid in the atmosphere and laboratory experiments on its photolysis. Tellus, 26, 175-179.

(2) COX, R.A. (1974) The photolysis of gaseous nitrous acid. J. Photochem. 3, 175-188.

(3) COX, R.A. (1975) The photolysis of gaseous nitrous acid - a technique for obtaining kinetic data on atmospheric photooxidation reactions. In: Proceedings of the symposium on chemical kinetics data for the upper and lower atmosphere. Int. J. Chem. Kinet. Symposium No. 1, 1975.

(4) ENGLAND, E. and CORCORAN, W.H. (1974) Kinetics and mechanisms of the gasphase reaction of water vapor and nitrogen dioxide. Ind. Engng. Chem. Fundam. 13, 373-384.

(5) ENGLAND, E. and CORCORAN, W.H. (1975) The rate and mechanism of the air oxidation of parts-per million concentrations of nitric oxide in the presence of water vapor. Ind. Engng. Chem. Fundam. 14, 55-63.

(6) CHAN, W.H., NORDSTRÖM, R.J., CALVERT, J.G. and SHAW, J.H. (1976) Kinetic study of HONO-formation and decay reactions in gaseous mixtures of HONO, NO, NO_2, H_2O and N_2. Env. Sc. Techn. 10, 674-682.

(7) KAISER, E.W. and WU, C.H. (1977) A kinetic study of the gas phase formation and decomposition reactions of nitrous acid. J. Phys. Chem. 81, 1701-7106.

(8) PERNER, D. and PLATT, U. (1979) Detection of nitrous acid in the atmosphere by differential optical absorption.

(9) PLATT, U. and PERNER, D. (1980) Direct measurements of atmospheric CH_2O, HNO_2, O_3, NO_2 and SO_2 by differential optical absorption in the near u.v. J. Geophys. Res. 85, 7453-7458.

(10) PLATT, U., PERNER, D., HARRIS, G.W., WINER, A.M. and PITTS, J.N. (1980) Observation of nitrous acid in an urban atmosphere by differential optical absorption. Nature. London, 285, 312-314.

(11) KESSLER, C., PERNER, D. and PLATT, U. (1981) Spectroscopic measurements of nitrous acid and formaldehyde-implications for urban photochemistry. Commission of the European communities - Physico-chemical behaviour of atmospheric pollutants. Proc. 2nd Eur. Symp., Varese, 29 September-1 October, 1981.
(12) HARRIS, G.W., WINER, A.M., PITTS, J.N., PLATT, U. and PERNER, D. (1982) Observations of nitrous acid in the Los Angeles atmosphere and implications for predictions of ozone-precursor relationships. Env. Sci. Techn. 16, 414-419.
(13) HARRIS, G.W., WINER, A.M., PITTS, J.N., PLATT, U. and PERNER, D: (1983) Measurements of HONO, NO_3 and NO_2 by long-path differential optical absorption spectroscopy in the Los Angeles basin. Optical Laser Remote Sensing, 39, 106-113.
(14) SAKAMAKI, F., HATAKEYAMA, S. and AKIMOTO, H. (1983) Formation of nitrous acid and nitric oxide in the heterogeneous dark reaction of nitrogen dioxide and water vapor in a smog chamber. Int. J. Chem. Kinet. 15, 1013-1029.
(15) PITTS, J.N., SANHUEZA, E., ATKINSON, R., CARTER, W.P.L., WINER, A.M., HARRIS, G.W. and PLUM, C.N. (1984) An investigation of the dark formation of nitrous acid in environmental chambers. Int. J. Chem. Kinet. 15, 919-939.
(16) PITTS, J.N , BIERMANN, H.W., WINER, A.M., and THAZON, E.C. (1984) Spectroscopic identification and measurement of gaseous nitrous acid in dilute auto exhaust. Atm. Env. 18, 847-854.
(17) KESSLER, C. and PLATT, U. (1984) Nitrous acid in polluted air masses - sources and formation pathways. In: Proc. Commission of the European communities. Phys, Chem. Behav. Atm. Pollut. 3rd Eur. Symp., Varese, Italy, April 10-12, 1984, 412-422.
(18) PITTS, J.N., WALLINGTON, T.J., BIERMANN, H.W. and WINER, A.M. Identification and measurement of nitrous acid in an indoor environment. Atm. Env. 19, 763-767.
(19) SJÖDIN, Å. and FERM. M. (1985) Measurements of nitrous acid in an urban area. Atm. Env. 19, 985-992.
(20) SJÖDIN, Å. (1986) Diurnal variation of nitrous acid in urban air in Scandinavia. Submitted to Env. Sci. Techn.
(21) SVENSSON, R., LJUNGSTRÖM, E., LINDQVIST, O. Kinetics of the dark reaction between nitrogen dioxide and water vapour. Report No. A8502 January 1985. Chalmers Univ. of Techn. and Univ. of Göteborg, S-412 96 Göteborg, Sweden.
(22) FERM, M. and SJÖDIN, Å. (1985) A sodium carbonate coated denuder for determination of nitrous acid in the atmosphere. Atm. Env., 19, 979-983.

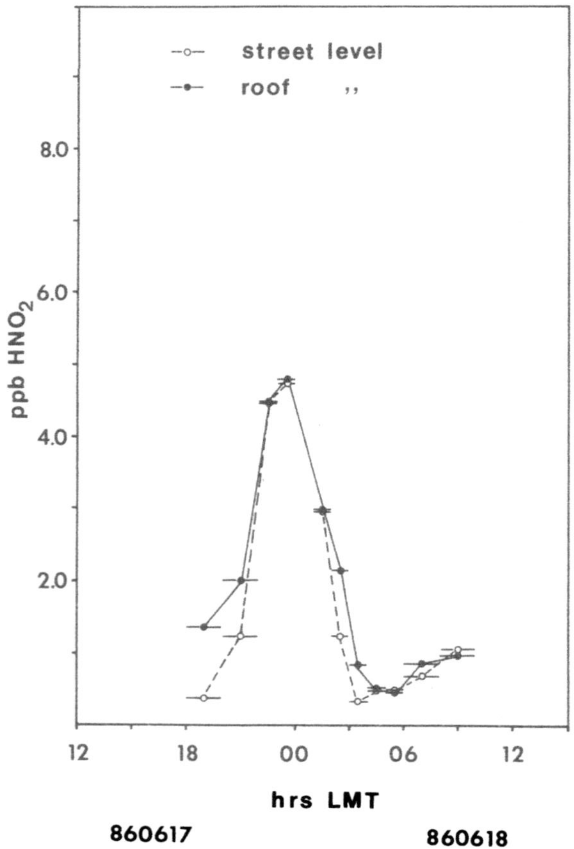

FIGURE 1: Typical diurnal profile of HNO_2 at street and roof level

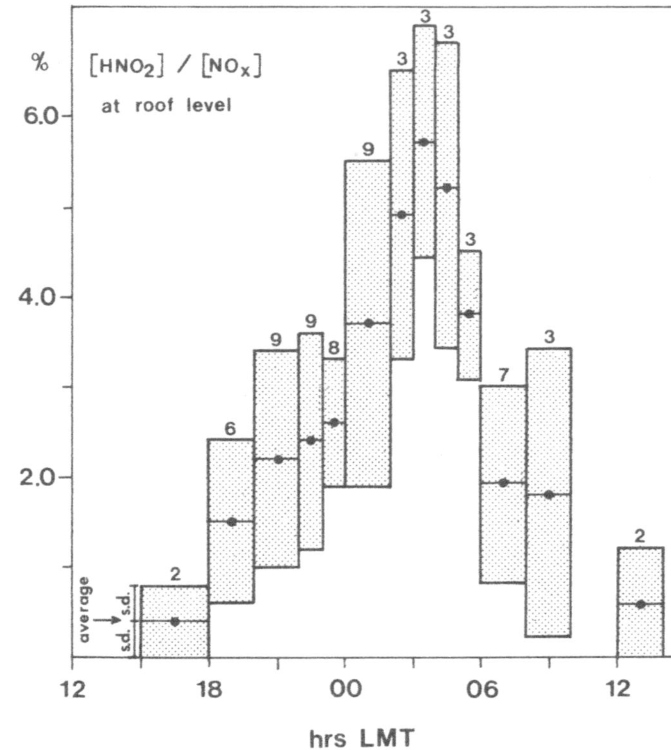

FIGURE 2: Average diurnal profile of the HNO_2/NO_x-ratio for the nine observations at roof level. The figure over each bar represents the number of observations averaged for each time period.

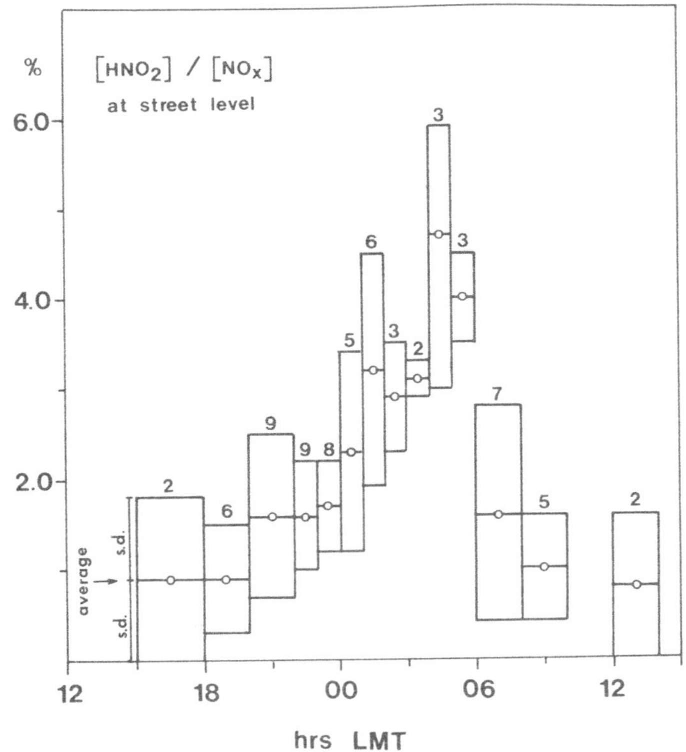

FIGURE 3: Average diurnal profile of the HNO_2/NO_x-ratio at roof level

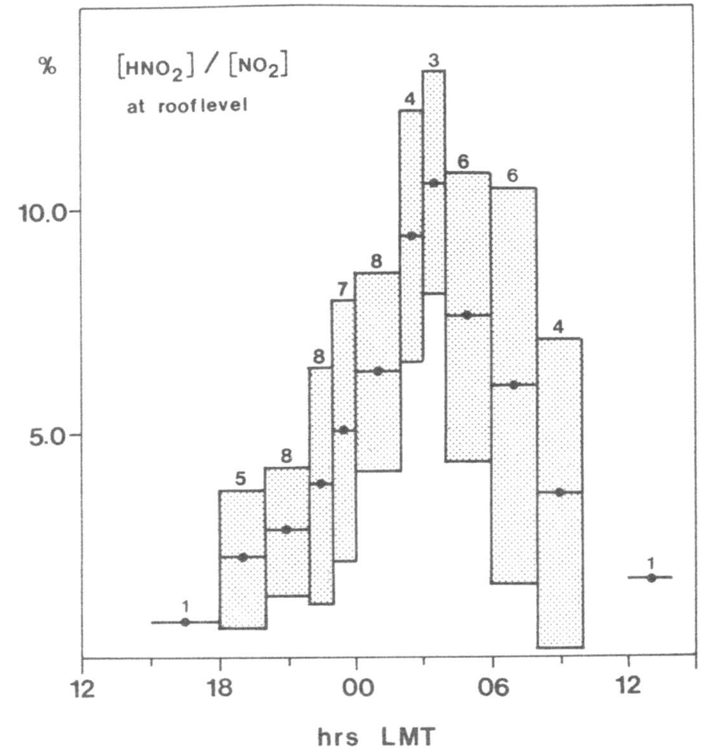

FIGURE 4: Average diurnal profile of the HNO_2/NO_2-ratio at roof level

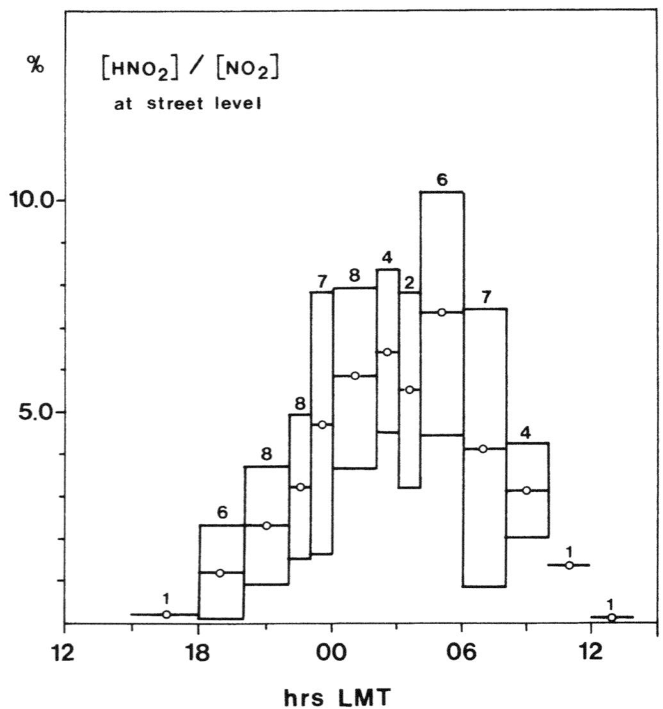

FIGURE 5: Average diurnal profile of the HNO_2/NO_2-ratio at street level.

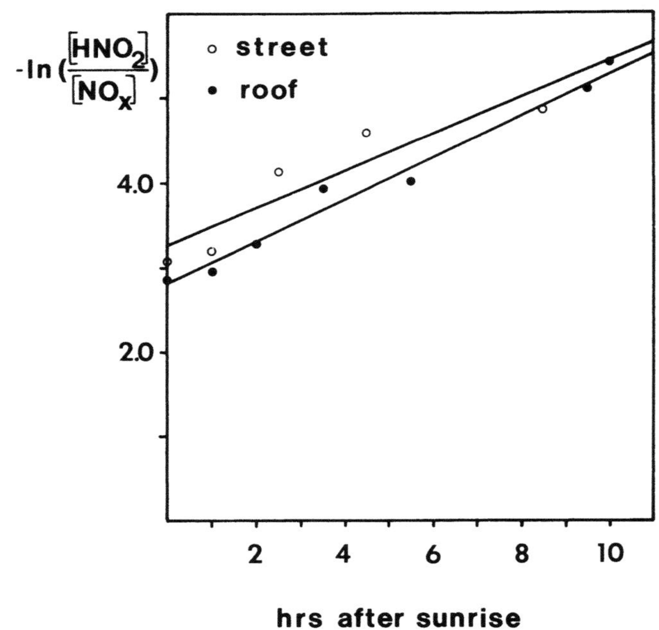

FIGURE 6: **Plot of ln([HNO_2]/[NO_x]) vs time after sunrise.**

SPECIATION OF IONIC ORGANOLEAD COMPOUNDS IN THE BIOSPHERE

R. VAN CLEUVENBERGEN, D. CHAKRABORTI*, and F. ADAMS
Department of Chemistry, University of Antwerp, B-2610 Wilrijk, Belgium
*on leave from Jadavpur University, Calcutta-32, India

Summary

Mounting concern about the occurrence and fate of organic lead compounds in the environment and living systems, has led us to optimize a sensitive speciation methodology for ionic alkyllead determination. The progress in this study is reviewed, the environmental applications summarized, and some apparent correlations between the individual results are indicated and discussed.

1. INTRODUCTION

Since the introduction of catalytic convertors on the automobile exhaust, implying the use of lead-free gasoline, one of the most impressive sources of lead contamination in the biosphere is gradually dealt with (1). An extensive methodology reflects that historically the concern has been centered on the environmental occurrence, fate and effects of "total" lead, regardless of the chemical form. More recently, it has been suggested that the impact of organic lead species (i.e. tetraalkyllead compounds and their ionic degradation products, the tri- and dialkyllead salts) on the ecosystem cannot yet be considered negligible. This is well illustrated by recent work claiming that the forest damage problem may be connected with the uptake of organolead compounds from rainwater (2).

Organolead species are considered to enter the environment mainly through anthropogenic sources, like vehicle emissions, effluents from the manufacturing process of the gasoline additive, and accidental spillage during handling and transport (3). In addition, although probably an inefficient process, biomethylation of organic lead may contribute to the organolead burden of the biosphere (4). A compilation of the sources and transformations in a biogeochemical cycle of lead in the environment, as repeatedly suggested (5-7), is at present still speculative in many of its chains. This has prompted us to undertake an overall investigation of organolead in the biosphere, particularly focusing our attention on the occurrence and fate of the ionic alkyllead intermediates. In this report, the progress in this study will be reviewed, and some preliminary correlations between the individual results will be indicated.

2. SPECIATION METHODOLOGY

In the analytical pathway which we follow for our survey of ionic alkyllead species in the biosphere, the following steps can be discerned :

- sampling
- preextraction of tetraalkyllead compounds
- leaching of the ionic alkyllead species from the matrix
- solvent extraction as diethyldithiocarbamate (DDTC) complex

- derivatization with butyl Grignard reagent
- gas chromatographic separation of the species, followed by quantization through atomic absorption detection

The techniques and instrumentation developed are described in detail elsewhere (8-10). In the present paper we will review the main features and include some minor changes made for optimization since our first report on this subject (8).

a) sampling

Rainwater collection is carried out with a conventional funnel and polyethylene bottle arrangement, to protect the analytes as effectively as possible from photochemical degradation. Although we could not detect so far any tetraalkyllead (TAL) compounds in rain, we assume that it cannot be excluded that TAL species - if present - could generate ionic alkyllead in the container, which is kept in the field usually for 1-2 weeks. Degradation studies did indicate that in situ breakdown of the analytes in the dark is a slow process, thus excluding a significant contribution of this process in routine determinations. Runoff water and dust are sampled directly from the street ; other water samples are obtained by filling manually polyethylene bottles. Aerosols are collected on a 24 h base on GF/A filter papers (1.1 μm cut off) using a high volume sampler with an average flow rate of ca. 400 L min^{-1}.

b) preextraction of TAL

Since in our analytical procedure the enrichment step after extraction, which provides the excellent detection limits, prevents a simultaneous determination of the tetraalkyllead compounds by GC-AAS, the obvious way is to remove the TAL species prior to extraction, particularly if a time-consuming leaching is involved. Thus any further degradation into ionic alkyllead species can be avoided. Hexane preextraction appears to be convenient for this purpose ; furthermore, it allows a separate measurement of the TAL compounds by GC-AAS, if desired. For solid samples like dust or aerosol, care should be taken to minimalize losses of particularly the less polar triethyllead species ; for this reason, an at least tenfold excess amount of deionized water is added to the hexane-solids mixture prior to filtration and phase separation (10).

c) leaching of the analytes from the matrix

Although the ionic alkyllead compounds are well soluble in water, the extraction from a solid matrix is not a straightforward operation. Through subjecting dust or aerosol samples repeatedly to a leaching treatment with water, it was demonstrated that part of the species resides with the particulate fraction after a single treatment. A solution containing citric acid and EDTA, on the other hand, leaches out quantitatively the analytes from the matrix. As these reagents are anyhow involved in the solvent extraction step, no further complication of the analysis arises.

d) solvent extraction

After filtration the aqueous sample containing 8 g L^{-1} of citric acid is buffered at pH 9.0 by adding ammonia. The analytes are then extracted into pentane as diethyldithiocarbamate complexes. Depending on the sample type, 2-10 g L-1 of EDTA is added before extraction to mask the inorganic lead ; if its coextraction is not largely prevented in this way, inorganic lead becomes partly butylated, giving rise to a high $PbBu_4$ peak in the chromatogram. Apart from implying tedious chromatographic clean-up procedures, this could eventually reduce the sensitivity of detection of the species of interest, and also decrease their recovery in the analytical procedure (8). It should be noted, however, that these problems do not manifest themselves unless excessive lead concentrations are involved. For samples in which a high inorganic lead content is expected, such as the road dusts or road runoff waters, its coextraction is more efficiently prevented by washing the pentane layer after extraction with a buffered EDTA solution (11). To avoid losses of the analytes, the same DDTC containing buffer solution as employed in the original extraction is applied.

The combined pentane layers from twicefold extraction are thereupon dried by adding a minimum amount of sodium sulfate. Particularly when an emulsion forms during extraction, this has proved to be essential to eliminate interference during the subsequent derivatization. The residue is rinsed several times with pentane to assure complete recovery. After vacuum rotoevaporation of pentane at 20°C, the complexes are dissolved in a microvolume of nonane (250 µL). This does result in a further enrichment of at least 100 times, which is reflected in the excellent detection limits of the analytical procedure.

e) Grignard derivatization

To enable an easy gas chromatographic separation, the tri- and dialkyllead complexes are derivatized with butyl Grignard reagent to tetraalkyllead species (PbR_3Bu, PbR_2Bu_2, R = methyl or ethyl), thus preserving their identity. The concentration of the reagent during butylation is a critical factor. Employing too high amounts may result in a rearrangement process, in which part of the organolead species is transformed into tetrabutyllead. In a water-free medium, the most suitable compromise for complete derivatization is employing 1 mL of 0.5 M Grignard reagent, which is a four times smaller amount than we did adopt originally (8).

f) measurement with GC-AAS

Coupling of an atomic absorption spectrometer as element-specific detector with a gas chromatograph results in a relatively simple but sensitive technique for measuring the organolead derivates in the nonane extract. Currently a short heated nickel transfer line (8) interfaces both instruments, leading the individual species after separation on a conventional glass column directly into a quartz furnace mounted on top of an air acetylene burner. The optimal parameters appear to be highly dependent on characteristics like the length of column and transfer line, and are furthermore related to the desired resolution between the signals of $PbMe_2Bu_2$ and $PbEt_3Bu$. At present we prefer a GC temperature program of 110-260°C at 15°C min^{-1} over the earlier reported one, as this shortens the analysis time markedly without any significant decrease in performance for tri- and dialkyllead determination. Peak height measurement is applied to quantify the species through comparison with available standards.

The overall procedure allows an interference-free determination of ionic alkyllead species with a detection limit of 1-2 ng L^{-1} (0.5 L sample volume) and a precision of about 10% RSD (at 250 ng L^{-1} level). Since higher sample volumes can routinely be processed, available detection limits extend to the pg L^{-1} range for liquid and pg g^{-1} range for solid samples. During the evaluation of the procedure, extraction recoveries are controlled by spiking aqueous solutions of tri- and dialkyllead compounds to the samples. For some different sample types, average recoveries are summarized in Table I.

Table I recovery and precision of the determination of ionic alkyllead compounds in spiked environmental samples (at least 3 replicates ; spiking level 100-500 ng Pb)

	recovery (%) and standard deviation			
	$PbMe_3^+$	$PbMe_2^{++}$	$PbEt_3^+$	$PbEt_2^{++}$
deionized water	105 ± 10	90 ± 8	102 ± 7	108 ± 8
road runoff	98 ± 4	96 ± 1	100 ± 6	103 ± 3
dust	93 ± 2	87 ± 5	89 ± 2	105 ± 5
aerosol	101 ± 5	109 ± 3	91 ± 1	98 ± 9

3. ENVIRONMENTAL APPLICATIONS

The optimized methodology for the speciation of ionic alkyllead compounds has been applied to assess the levels of these highly toxic pollutants in various parts of the biosphere. Individual results on rain, snow, surface water (9,12), road runoff and road dust (10) have been reported in detail elsewhere. In what follows we will compile some results obtained so far and discuss the trends which tend to become apparent. It should be emphasized that the data set should be increased before any definite correlations can be outlined. Particularly a simultaneous determination of alkyllead species in the different matrices is a prerequisite, and we are further focusing on this topic.

In Table II, the distribution of the total ionic alkyllead content over the individual species is represented for different environmental sample types. All these samples were collected in the same period of the year (i.e. late winter - early spring); results of single sampling campaigns during a three years period are combined. The location chosen, the vicinity of our university campus, is a typical residential one, situated about 7 km south of the downtown area of Antwerp, Belgium. It is characterized by low traffic density roads and car parking areas, two entrance roads with somewhat higher traffic density, at the crossing of which is situated a gasoline station, and the main highway between Antwerp and Brussels at about 1 km to the east. Analysis of the premium brand gasoline distributed in this station in spring 1986 showed a tetraalkyllead content of 0.35 g Pb L^{-1}, 88% of which was added as tetraethyllead and the remaining as tetramethyllead. The concentrations found in water and dust collected immediately adjacent to the station clearly reflect this composition by a high relative abundance of the ethyl degradation products, triethyllead and diethyllead.

It is still unclear why the former is predominant in road runoff and the latter in street dust ; this is a trend which also holds for the residential concentrations, though less explicitly. An earlier study on

Table II average concentration levels of ionic alkyllead in different environmental matrices, and the distribution (%) of the individual species (nd = not detected)

SAMPLE TYPE	NUMBER OF SAMPLES	PERIOD month,year	TOTAL CONCENTRATION OF IONIC ALKYLLEAD		DISTRIBUTION (%) $PbMe_3^+$	$PbMe_2Et^+$	$PbMeEt_2^+$	$PbMe_2^{++}$	$PbEt_3^+$	$PbMeEt^{++}$	$PbEt_2^{++}$
RESIDENTIAL											
aerosol	8	2,86	196.	$pg\ m^{-3}$	2	2	n.d	2	17	1	77
road dust	3	12,85-1,86	12.8	$ng\ g^{-1}$	9	2	1	4	14	2	70
road runoff	3	11,85	703.	$ng\ L^{-1}$	24	10	4	n.d	54	n.d	8
rain	14	2-5,84	63.	$ng\ L^{-1}$	49	6	n.d	16	11	n.d	17
	11	1-4,86	50.	$ng\ L^{-1}$	40	3	1	11	18	2	25
lake water	2	3,85	5.2	$ng\ L^{-1}$	73	n.d	n.d	6	12	n.d	10
GASOLINE STATION											
road dust	1	2,86	347.	$ng\ g^{-1}$	2	<1	<1	1	9	1	87
road runoff	1	2,86	4290.	$ng\ L^{-1}$	16	6	3	n.d	72	n.d	4

the determination of tetraalkyllead in simulated dust (13) indicates that the particulates may serve as substrates for conversion of sorbed tetraalkyllead into species which are not readily extracted with hexane or removed by moderate heating ; these species were, unfortunately, not further investigated. A quite rapid decomposition of particularly dialkyllead compounds in contact with the dust has been suggested as a probable explanation for their low recovery in a total organic lead determination (14). The high diethyllead levels present in our dust samples, however, are in contradiction with Harrison's conclusion. Our preliminary study on the degradation of ionic alkyllead species in dust does not indicate a significantly higher stability of diethyllead compared to triethyllead, whereas trimethyllead, contributing less to the ionic alkyllead content of dust, is clearly more stable. Consequently, the diethyllead stability seems not to offer the explanation ; on the other hand, it should be noted that these samples were kept in daylight in a beaker sealed with a perforated Parafilm membrane, a condition which may be quite different from a typical situation in a ventilated outdoors environment.

At present we tend to consider a hypothesis of De Jonghe and Adams (15) the most probable explanation for the high diethyllead levels in street dust as well as aerosol. Based on the sublimation point of the chloride salts, they suggested that the dominant proportion of trialkyllead will probably remain gaseous, with a minor fraction condensing and remaining onto particulate matter. Recently, the direct detection of gas-phase trimethyllead has been reported (16) and the species was shown to contribute significantly to the total alkyllead level of the atmosphere, its concentration in aerosol being claimed to be only about 1% of that in the vapour phase (7). The higher relative abundance of triethyllead in road runoff water seems in agreement with the degradation studies of tetraethyllead in water performed by our group as well as by several other research teams (17-19). In all these cases, a more or less rapid breakdown to triethyllead was demonstrated ; diethyllead was detected but in trace quantities, inorganic lead being the final degradation product.

As the tri- and dialkyllead compounds, present in the atmosphere, are well soluble in water, washout and rainout can be expected to be major sink processes and to constitute a source of these pollutants in the hydrosphere. Our pioneering work on this topic (9,12) points out that the distribution of the species in wet fall-out is intermediate between that found on particulate matter and that expected from the gaseous fraction of the air (16), both parts being subjected to the removal process and continuous degradation. As apparent from Table II, we did not notice any marked difference between the average distribution of the species in rain in the spring of 1984 and 1986. This is consistent with the regulations on gasoline lead in Belgium, which did not change during that time. Assuming, on this base, that extrapolation to 1980 is not too speculative, it is interesting to take into account the earlier measurements of tetraalkyllead species in the air by our group in the same residential area (20). At that time, the tetramethyllead contribution to the TAL content was about 75%, confirming a possible high influence of trimethyllead in the gaseous fraction of the atmosphere on the rain concentration level, as suggested above.

Considering that wet atmospheric fall-out processes transport ionic alkyllead species into the aquatic environment, we have investigated the concentration and distribution of these possible contaminants in different waters, e.g. lake and river water. This reveals that in the lake situated on the university campus premises, trimethyllead is clearly the predominant species. Since it also represents a significant part of the alkyllead level in rain and moreover appears to be the most stable of the tri- and dialkyllead compounds in natural water, as demonstrated by a preliminary degradation study, its high relative importance can easily be interpreted.

In addition to a correlation of the individual distributions, a comparison between the absolute concentration levels reveals information on the occurrence and fate of organolead in the biosphere. In the residential area which we have investigated, the average concentration of ionic alkyllead species in road runoff water sampled directly from the street is about an order of magnitude higher than in rainwater. The amounts in rain and lake water differ again by about an order of magnitude. The latter finding can reasonably well be explained by assuming that subsequent dilution and decomposition prevent accumulation of high concentrations. It is tempting to assume that washing of the road surface is responsible for the high amounts present in road runoff, the water partly scavenging additional local emissions by the automotive traffic. This process, on the other hand, has been suggested to have minor influence on the alkyllead composition of highway drainage water (21), a contradiction which emphasizes probably the dependence of organolead emissions on the traffic behaviour.

On the basis of the concentration levels found so far, it seems that ionic alkyllead pollution is likely to be non-hazardous for man and his environment, with the possible exception of particular cases such as gasoline manufacture and handling places. Indeed, close to distributing stations, much higher concentrations are encountered, which is in agreement with the elevated local levels of organolead in the air (22). As the effect of long-range exposure to significant alkyllead concentrations is still unclear, we cannot exclude that gasoline stations are a potential risk for alkyllead intoxication.

ACKNOWLEDGEMENT

This research was funded through EC Grant ENV-768-B/RS and FKFO grant 2.0091.85. R. Van Cleuvenbergen thanks the National Foundation for Scientific Research, Belgium, for a research assistant grant.

REFERENCES

(1) LAVESKOG, A., (1984). Gasoline additives : past, present and future. in : Biological effects of organolead compounds, Grandjean P. (Ed.), CRC Press Boca Raton (Florida), 5-12.

(2) FAULSTICH, H., STOURNARAS, C., (1985). Potentially toxic concentrations of triethyl lead in Black Forest rainwater samples. Nature, vol 317, 714-715.

(3) NIELSEN, T., (1984). Atmospheric occurrence of organolead compounds. in : Biological effects of organolead compounds, Grandjean P. (Ed.), CRC Press Boca Raton (Florida), 43-62.

(4) BEIJER, K., JERNELOV, A., (1984). Microbial methylation of lead. in: Biological effeets of organolead compounds, Grandjean P. (Ed.), CRC Press Boca Raton (Florida), 13-20.

(5) DE JONGHE, W., JIANG, S., ADAMS, F., (1981). Alkyllead compounds in the environment. Proc. Int. Conf. on Environmental Pollution, Thessaloniki, 183-189.

(6) RODERER, G., (1981). Fate and toxicity of tetraalkyllead and its derivatives in aquatic environments. Proc. Int. Conf. Heavy Metals in the Environment, Amsterdam, 250-253.
(7) HARRISON, R.M., HEWITT, C.N., RADOJEVIC, M. (1985). Environmental pathways of alkyllead compounds. Proc. Int. Conf. Heavy Metals in the Environment, Athens, vol. 1, 82-84.
(8) CHAKRABORTI, D., DE JONGHE, W.R.A., VAN MOL, W.E., VAN CLEUVENBERGEN, R.J.A., ADAMS, F.C., (1984). Determination of ionic alkyllead compounds in water by gas chromatography atomic absorption spectrometry. Analytical chemistry, vol. 56, 2692-2697.
(9) VAN CLEUVENBERGEN, R.J.A., CHAKRABORTI, D., ADAMS, F., (1986). Occurrence of tri- and dialkyllead species in environmental waters. Environmental Science and Technology, in press, scheduled for June 1986.
(10) CHAKRABORTI, D., VAN CLEUVENBERGEN, R., ADAMS, F., (1986). Determination of ionic alkyllead compounds in road runoff water and road dust by butylation and gas chromatography - atomic absorption spectrometry. Proc. Int. Conf. Chemicals in the Environment, Lisbon, in press.
(11) CHAKRABORTI, D., VAN CLEUVENBERGEN, R.J.A., ADAMS, F.C., (1986). Determination of total ionic alkyllead by electrothermal atomisation atomic absorption spectrometry. Journal of Analytical Atomic Spectrometry, in press, scheduled for August 1986.
(12) VAN CLEUVENBERGEN, R., CHAKRABORTI, D., VAN MOL, W., ADAMS, F., (1985). Ionic alkyllead species in atmospheric fall-out and surface water. Proc. Int. Conf. Heavy Metals in the Environment, Athens, vol. 1, 153-155.
(13) EDWARDS, H.W., ROSENVOLD, R.J., (1974). Uptake of tetraethyl lead vapor by atmospheric dust components. Proc. 2nd. Ann. NSF-RANN Trace Contaminants Conf., California, 59-63.
(14) HARRISON, R.M., (1976). Organic lead in street dust. Journal of Environmental Science and Health, vol. A11, 417-423.
(15) DE JONGHE, W.R.A., ADAMS, F.C., (1986). Biogeochemical cycling of organic lead compounds. in : Toxic metals in the atmosphere, Nriagu J.O. and Davidson C.I. (Eds.), John Wiley and Sons Inc. New York, 561-594.
(16) HEWITT, C.N., HARRISON, R.M., (1985). Total speciation of gas-phase alkyllead in the atmosphere. Proc. Int. Conf. Heavy Metals in the Environment, Athens, 171-173.
(17) GROVE, J.R. (1980). Investigations into the formation and behaviour of aqueous solutions of lead alkyls. in : Lead in the marine environment, Branica M. and Konrad Z. (Eds.), Pergamon Oxford, 45-52.
(18) JARVIE, A.W.P., MARKALL, R.N., POTTER, H.R., (1981). Decomposition of organolead compounds in aqueous systems. Environmental Research, vol. 25, 241-249.
(19) CHARLOU, J.L., CAPRAIS, M.P., BLANCHARD, G., MARTIN, G., (1982). Degradation of TEL in seawater. Environmental Technology Letters, vol. 3, 415-424.
(20) DE JONGHE, W.R.A., CHAKRABORTI, D., ADAMS, F.C., (1981). Identification and determination of individual tetraalkyllead species in air. Environmental Science and Technology, vol. 15, 1217-1222.
(21) HARRISON, R.M., RADOJEVIC, M., WILSON, S.J., (1986). The chemical composition of highway drainage waters IV. Alkyllead compounds in runoff waters. The Science of the Total Environment, vol. 50, 129-137.
(22) DE JONGHE, W.R.A., ADAMS, F.C., (1982). Measurements of organic lead in air - a review. Talanta, vol. 29, 1057-1067.

PRELIMINARY RESULTS ON FOG DEPOSITION

P. Winkler, G. Kroll

Deutscher Wetterdienst, Meteorologisches Observatorium Hamburg
Frahmredder 95, D-2000 Hamburg 65

ABSTRACT

A programm has been started in order to assess the humid deposition by fog interception on vegetation. A resistance modell will be used to calculate the deposited amount of water. In order to obtain realistic data, the model input will be taken from routine meteorological observations like wind speed and direction, visibility and so on. To get information on the chemical composition a fog water collector has been developed. Measurements on fog water resp. cloud water composition will be conducted at several places (in the Black forest, the Bavarian forest, the Wasserkuppe, the Kahler Asten, and Solling). At present some investigations have been made on fog frequency as function of height a.s.l. and month. It has also been found out that the liquid water content is a function of the time of year.

1. INTRODUCTION

Acidity is deposited not only by precipitation but also by fog interception. A large body of literature exists which demonstrates the importance of the deposition of fog resp. clouds in contact with mountains for the hydrological cycle (e.g. Grunow 1960, Bauer 1962, Baumgartner 1958, 1959). The waterdeposition induces a deposition of trace substances which, in contrast to wet deposition by precipitation, as denoted as humid or fog deposition. Fog deposition is not as uniform as wet deposition but depends in a complex manner on fog frequency, drop size distribution, geometry of vegetation, wind velocity, and on the trace substance content which can be expected to depend on the wind direction.

In the here presented program (which started in autumm 1985) we try to asses the humid deposition for several ions and acidity by means of a deposition model. The model works as resistance model (e.g. Shuttleworth 1977, Lovett 1984) and the input data will be taken from routine meteorological observations. In order to get a realistic assessment on the fog deposition it seems to be of importance that contemporarily observed data like fog and the observed visibility and wind speed and so on are used. At selected places fog samplers will be installed in order to get information on trace substance content as a function of season, liquid water content and wind direction.

In this paper the current state of work is described with respect to various aspects of the deposition process. In principle we can describe fog deposition by

$$\text{deposition} = \begin{matrix}\text{frequency}\\ \text{of events}\end{matrix} \times \begin{matrix}\text{deposited amount}\\ \text{per event}\end{matrix} \times \begin{matrix}\text{trace substance content}\\ \text{per event}\end{matrix}$$

In the following these factors are considered in more detail.

2. FREQUENCY OF EVENTS

There are several ways to approach this point. In a first step we evaluated the number of foggy days as a function of height a.s.l. for a number of stations of the climatological observation network. We obtained an increase of fog frequency with height but there was a large scatter of data points. It was therefore necessary to correct the height a.s.l. for the average height of the country from which the mountains raise. For example Munic with a height of 500 m is not in clouds if we observe clouds at 500 m height in Hamburg and assume northerly winds. Thus we corrected the height of the bavarian forest by -500 m. In doing so for all stations we obtained the height distribution depicted in fig. 1, with a largely reduced scatter. In the flat country or in valleys we find 50 days with fog, at greater heights this value increases to 250-280 days. Above 700 m we find no further increase.

In a second step we investigated the observed height of the cloud hase as a function of month for selected stations assuming that the cloud tops are beyond the height of the tallest mountains nearby. The result is shown in fig. 2. We have calculated the probability for every month for a given height above surface to be in clouds. It can be seen that clouds are more frequent in winter and that also the cloud bases are higher in summer. We have calculated the fog probability from cloud base observations for a mountain of given height above a flat land station and compared it with direct observations at a mountain nearby. Good agreement was found which means that in the region between a low and greater height missing observations of fog frequency can be supplied by cloud base observations.

3. DEPOSITED AMOUNT OF WATER

The deposited amount of water depends on the drop size distribution of fog droplets, the liquid water content, the geometry of plants, wind speed and the turbulent parameters which can be assessed from temperature and wind profiles.

The first aspect which we investigated in more detail concerned the liquid water (LWC) content and the drop size distribution which we intend to derive from the observed visibility using Best's drop size distribution for fog and cloud drops:

$$1 - F = \exp\left[-\left(\frac{2r}{\omega}\right)^n\right] \qquad (1)$$

with F = fraction of liquid water accumulated at drops with radii less than r, r = radius, n = exponent = 3.3, $\omega = (909 \times LWC)^{0.559}$, LWC = liquid water content.

We have calculated the visibility in a fog assuming eq. (1) for various LWC. It turned out that the parameter n was of minor influence. Thus we let n constant. The parameter ω had a large influence and during a detailed study we tried to compare our calculated relation between visibility and LWC with observed data (Eldridge 1966, Garland 1971, Chylek 1978, Pinnick et. al. 1979). Because the agreement was only poor we changed the relation given by Best for ω to $\omega = (3 \cdot 10^4\ LWC)^{0,33}$. With this new ω we achieved good agreement between observation and calculation over a wide range of liquid water content resp. visibility. This improvement seems to be justified because Best derived his relation from a relatively small data base. From this investigation it is possible to derive the LWC and the drop size distribution from visibility observation without large error

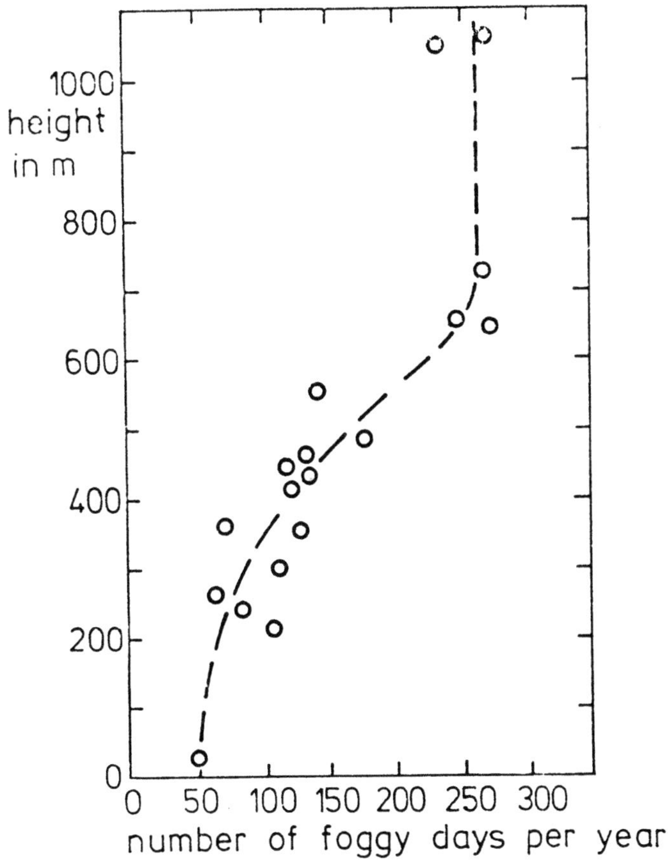

Fig. 1 Number of foggy days as a function of height.

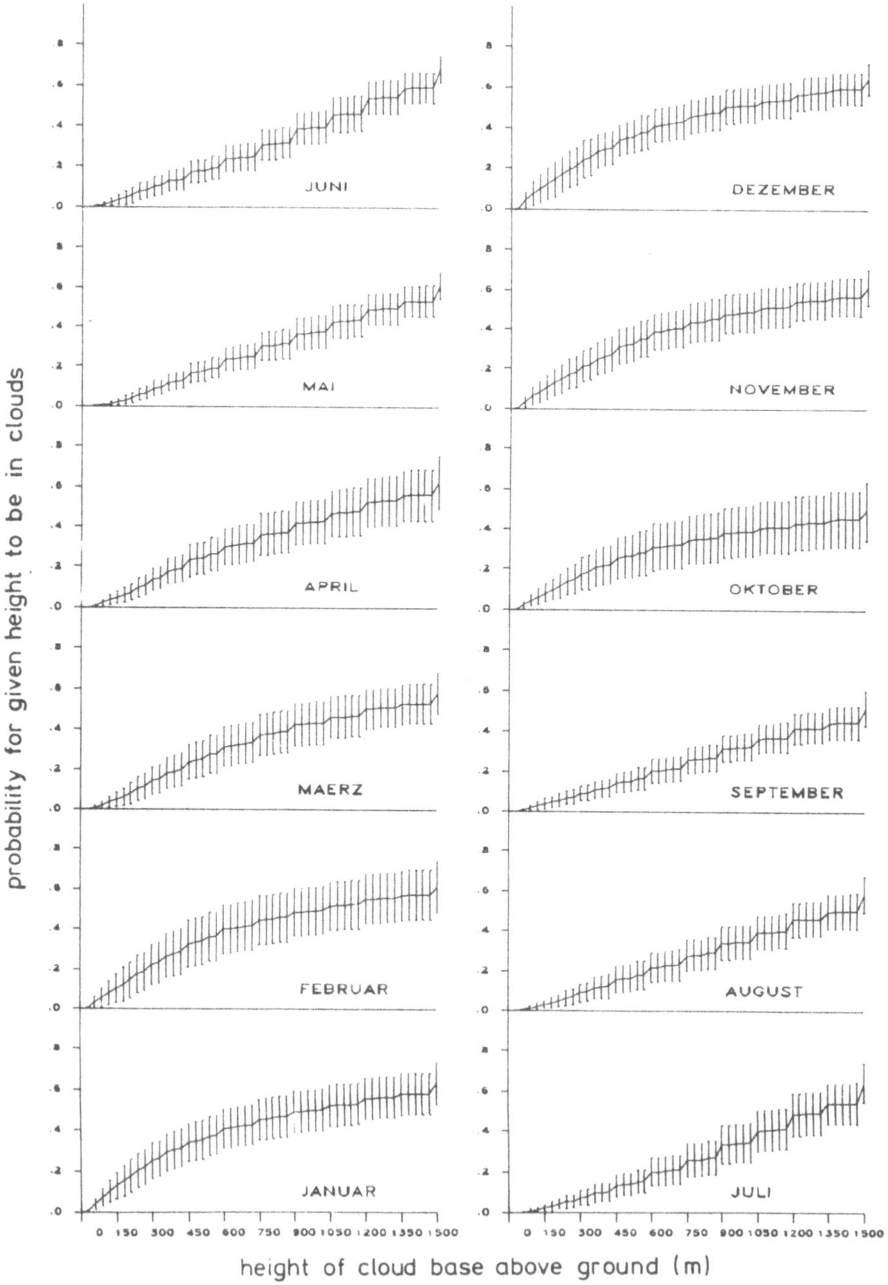

Fig. 2 Average probability for a given height above ground to be in clouds for each month. Data from Erding (near Munic), 1976-1980.

because the exponent n turned out to be only of minor importance.

It is now of interest to investigate the climatology of visibility for various stations. In fig. 3 we depict nonthly averages of the frequency of cases with visibility below 1000 m for flat country stations and a mountain site. a yearly cycle is recognized with highest frequency during the winter month. The yearly cycle is less pronounced for the mountain station. Fogs resp. clouds are not very effective in depositing water if visibility is around 1000 m because the LWC is small. We therefore also calculated monthly averages of the visibility less than 400 m. It can be seen that the frequency of fogs which are rich in liquid water follows essentially the same pattern as the frequency of all fogs, however, if we look into more detail we recognize for the flat country (resp. valley) stations that if we observe fog in the month of May or September or October the fogs are preferentially very dence i.e. they have a high liquid water content. For mountain stations fogs with high LWC occur in about 80% of all events throughout the year. In order to get deposited amounts of water the visibility values have to be combined with the contemporarily observed wind speed and the geometry of the receiver (tree, forest, grass ...) which will be investigated in our future program.

4. TRACE SUBSTANCE CONTENT

To collect and analyze fogwater a fog collector has been developed which works as an impactor (see fig. 5). The fog impactor has a twin nozzle through which air is sucked at a rate of 95 m^3/hr. Behind the nozzles specially designed deposition bodies are placed. In the center of the body is a vertically oriented hole which is connected to the impaction surface by numerous capillary holes. The center hole is connected to the pump exit of the instrument by a tube so that a small underpressure is applied sustaining a slight air flow through the capillary holes. The deposited water is sucked into the center hole and drains into the collection bottle. The deposition body has a small rim at either side in order to prevent deposited water from being riped off and carried away with the fast air stream. With this design a rapid separation of the droplets from the air is reached and problems like evaparation or ongoing reactions are minimized. Behind the nozzles the air is guided by semicircular surfaces to the exit in order to avoid turbulence.

At present some data on fog water composition have been collected only at Hamburg (Winkler 1986). It is planned to collect fogwater also at several places (Schauinsland, Black Forest; Große Ohe, Bavarian Forest; Wasserkuppe, Röhn; Kahler Asten, Sauerland; Solling, Kl. Feldberg, Taunus). If we study the literature data e.g. of SO_4^{--} an appreciable variability of observed concentrations can be seen (table 1).

Unfortunately only in rare cases concentration data are published along with data on the LWC. This would be necessary in order to be able to intercompare data from different sampling sites since Fuzzi et.al. have shown that the trace substance content of fog water varies inversely to the LWC. We recommend the following form for data publications on fog or cloud water composition:

(1) measured original data

(2) normalized concentrations for an LWC of e.g. 0.1 g/m^3 along with the range of observed LWC

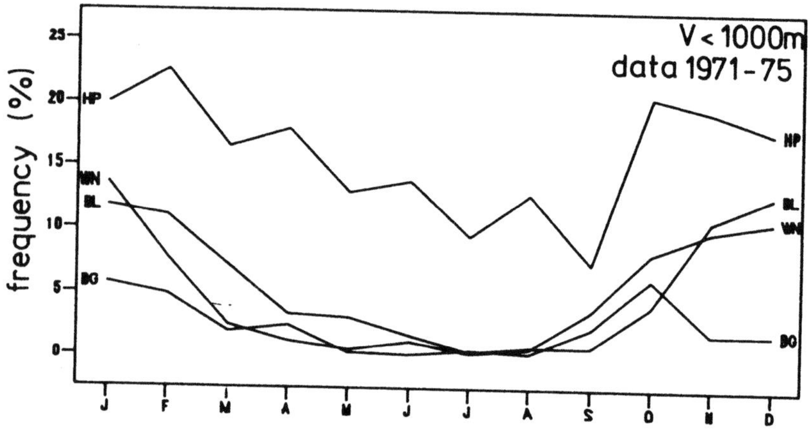

Fig. 3 Frequency of hourly observations with visibility below 1000 m. HP - Hohenpeißenberg, WN - Weihenstephan, BL - Braunlage, BG - Braunschweig.

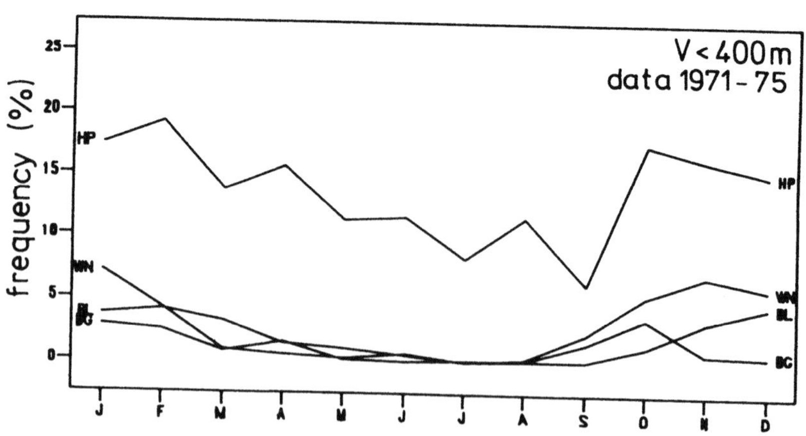

Fig. 4 Frequency of hourly observations with visibility below 400 m. These events characterize cases with fogs which have a high liquid water content. Abbreviations as in fig. 3.

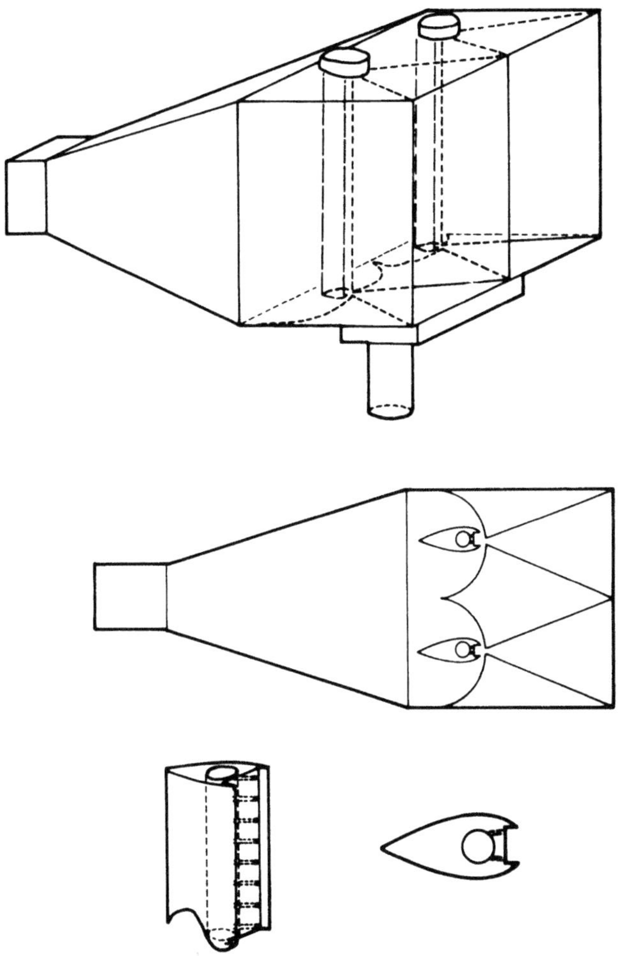

Fig. 5 Fog water collector

(3) averages of the normalized concentrations with standard deviations and the frequency distribution of LWC.

This is of high importance because the physico-chemical and chemical processes which govern the incorporation of trace substances into fog droplets are the same for different places but the liquid water content can change on a rather short term time scale for example due to adiabatic cooling of air flowing up hill.

TABLE 1: Sulfate concentrations in fog resp. cloud water

Polluted areas:		range of values (μeq/l)
Dresden (Germany)	Mrose 1966	3300
Bologna (Italy)	Fuzzi 1983	400-1000
Hamburg (Germany)	Winkler 1986	200-9500
Bakersfield (Calif.)	Jacob 1984	200-9500
Dübendorf (Switzerland)		200-7000
Mountains		
Feldberg (Taunus)	Georgii	40- 800
Schauinsland (Black forest)	UBA	150
Brocken (Harz)	Mrose 1966	800
Whiteface(Adirondacks)	Falconer et al. 1981	33- 900
	Castillo et al. 1976	60- 150
Mt Noribura (Japan)	Okita 1968	250-1300
Mt Tsukaba		350-2300
Clouds		
Kiew (USSR)	Petrenchuk 1964	400-2200
USA	Hegg a. Hobbs 1981	12- 400
England	Oddie 1962	40
Netherlands	Römer et al. 1985	10- 200

5. SUMMARY

(1) The frequency of fog resp. clouds in contact with ground increases with height. Fogs are more frequent in winter than in summer. Fog density is approximately constant over the year for mountain stations. Flat country stations have the most dence fogs during May, September and October.

(2) The liquid water content and the drop size distribution can be estimated from the visibility. The LWC is a function of the time of the year.

(3) In order to make measurements on chemical composition from various places comparable it is absolutely indispensable to determine the liquid water content during collection of fog water. Thus the composition for a standardized liquid water content (e.g. 0.1 g/m^3) can be calculated.

(4) A fog-water collector is described which uses the impactor principle and which rapidly separates the collected water from the sampled air.

REFERENCES

Bauer, E.(1962): Ergebnisse der mehrjährigen Nebelniederschlagsmessungen in Hunsrück und Eiffel. Dt. Gewässerkundl. Mitt. 6, 118-119.

Baumgartner, A. (1958): Nebel und Nebelniederschlag als Standortsfaktoren am Großen Falkenstein (Bayrischer Wald). Forstwiss. Centralblatt 77, 257-272.

Baumgartner, A. (1959): Das Wasserangebot aus Regen und Nebel sowie die Schneeverteilung in den Wäldern am Großen Falkenstein (Bayrischer Wald). Wald und Wasser 3, 45-54.

Best, A.C. (1951): Drop-size distribution in cloud and fog. Quart. J.Royal Met. Soc. 77, 418-426.

Castillo, R.A., J.E. Jiusto (1984): A preliminary study of the chemical modification of cloud condensation nuclei in stratiform clouds at Whiteface Mountain, New York. Atm. Environment 18, 1933-1937.

Chylek, P. (1978):Extinction and liquid water content in fogs and clouds. J. Atm. Sciences 35, 296-300.

Eldridge, G.R. (1966): Haze and fog aerosol distributions. J. Atm. Sciences 23, 605-613.

Falconer, R.E., P.D. Falconer (1980): Determination of cloud water acidity at a Mountain Observatory in the Adirondack Mountains of New York State. J. Geophys. Res. 85, 7465-7470.

Fuzzi, S., G. Orsi, M. Mariotti (1983): Radiation fog liquid water acidity at a field station in the Po valley. J. Aerosol Sci. 14, 135-139.

Garland, J.A. (1971): Some fog droplet size distribution obtained by an impaction method. Quart. J. Royal Met. Soc. 97, 483-494.

Grunow, J. (1960): The productiveness of fog precipitation in relation to the cloud droplet spectrum. Monograph No.5, American Geophys. Union, 110-117.

Jacob, D.J., J.M. Waldman, J.W. Munger, M.R. Hoffmann (1984): A field investigation of physical and chemical mechanisms affecting pollutant concentrations in fog droplets. Tellus 36 B, 272-285.

Lovett, G.M. (1984): Rates and mechanisms of cloud water deposition to a subalpine balsam fir forest. Atm. Environment 18, 361-371.

Mrose, H. (1966): Measurement of pH and chemical analysis of rain, snow and fog water. Tellus 18, 266-270.

Oddie, B. (1962): The chemical composition of precipitation of cloud levels. Quart. J. Royal Met. Soc. 88, 535-538.

Okita, T. (1968): Concentration of sulfate and other inorganic materials in fog and cloud water and in aerosol. J. Met. Soc. Japan 46, 120-127.

Pinnick, G.R., S.G. Jennings, P. Chylek, H.J. Auvermann (1979): Verification of a linear relation between IR extinction, absorption and liquid water content of fogs. J. Atm. Sciences 36, 1577-1586.

Römer, F.G., J.W. Viljeer, L. van den Beld, H.J. Slangeval, A.A. Veldkamp, H.F.R. Reijnders (1985): The chemical composition of cloud and rainwater. Results of preliminary measurements from the aircraft. Atm. Environment 19, 1847-1858.

Shuttleworth, W.J. (1977): The exchange of wind-driven fog and mist between vegetation and the atmosphere. Boundary-layer Meteorology 12, 463-489.

UBA (1984): Monatsberichte aus dem Meßnetz 5, 51-53.

Winkler, P.(1986): Observation on fog water composition in Hamburg. Atmospheric Pollutants in Forest Area - Their Deposition and Interception. Ed. H.-W. Georgii, Reidel Publ. Comp. Dordrecht/Holland.

DETERMINATION, BY PAH ANALYSIS, OF THE AGE OF AIR MASS CONTAMINATION IN A AREA REMOTE FROM SOURCES OF POLLUTION

MASCLET P., PISTIKOPOULOS P., BERGAMETTI G., LOSNO R. and MOUVIER G.
Laboratoire de Physico-Chimie de l'Atmosphère, associé au CNRS, Université Paris VII, 2 Place Jussieu
75251 Paris Cedex 05 FRANCE

The various particulate PAH present in the atmosphere are anthropogenic. In a typical urban atmosphere the concentration of each PAH is the result of mixing several sources. In pratice the average PAH composition, depends little on the sampling site in a very polluted urban atmosphere, but rather on the temperature conditions. In a lightly polluted area, several hundred kilometers far from any source, the average PAH composition can differ considerably; in fact PAH travel over average and long distances, but some very reactive PAH, such as BaP or Anthracene, are degraded by chemical and photochemical reactions, while the concentrations of others, non reactive compounds, are not modified during transport. Consequently the PAH histogram, at a sampling site remote from pollution sources depends mainly on the history of the air-mass and the time of travel between the production area - in the broad sense - and the sampling site. It is therefore possible to evaluate the transport time of the air-mass by measuring the characteristic ratios between reactive and non reactive PAH's. Morever concentration ratio of two non reactive compounds can be modified by a special sources (fuel, or vehicles emissions) during transport. The concentration ratios of minor pollutants can be used to determine the temporal origin of air masses, providing an incontestable physico-chemical parameter for modelling pollutant transport phenomena and for dating contaminations.

The PAH which are present in the atmosphere are all anthropogenous. Their origin is combustion of fossil combustibles or combustion of wood. The average concentration in the urban atmosphere or close to pollution sources is the result of the mixing of the different sources. In real terms, for the different cities of Europe the profile or the relative composition of PAH differs slightly from one city to another under the condition that air temperature are similar. In fact, the profile of PAH in particle form depends more on this factor - on condition that there is an equilibrum between the gas-particle conversions and the possible revolatilisation - than the sources themselves which in Europe are principally: domestic heating, vehicles using gasoline, diesel motors, power plants and rubbish incinerators[4,6,9,12,14]. On the other hand the average profile is different in USA[11], essentially because of the absence of lead in gasoline and the use of catalytic pots.

The mean profile of the characteristic histogram of the urban european air of a city of big importance is presented on fig.1. Nevertheless the average profile strongly depends on temperature. This, on figure 1, corresponds in a mean temperature of 19°C. The comparison with the measurements taken over a long distance is possible only if the temperatures between source point and receptor site are close to each other. Otherwise a correction should be made.
Within a short distance of the pollution sources the profile is very similar, even if the absolute concentrations are weaker, because of the dilution of pollutants in the atmosphere.
But the profile can differ for reactive compounds after a long range transport.

Studies over a medium, or a long range transport have been made by differents authors[3,1,7,8]. The experimental techniques used at that time perhaps were not as developped as now. The authors concluded that PAH were transported over a long distance but they do not observe different concentrations ratios between the sources and the receptors sites. If this result was confirmed no variation of the PAH profile can be constated.
In fact the concentrations of PAH decrease differently ; the reason is that their life time is more or less reduced[5,10,16].
This being greatly different from one PAH to another, concentrations will vary little or much according to the long or short life time of the PAH[2,13,15,17]. These life times are relative to expected non-reactive compounds like BePyrene or Bb/Bk Fluoranthene.
The BaP and the Fluoranthene also have concentrations close to each other at the origin of the air mass, but that of BaP will decrease much faster than that of the Fluoranthene because of the strong reactivity of the first and the very weak of the second.
In order to verify the evolution of PAH concentrations in medium scale, a study has been carried out in october 1985 in Corsica (at Cap Cavallo). The site is remote from any ambiant pollution and the measurements which have been taken represent the background pollution, coming from far sources.
The histogram of every sampling has to give us the temporal origine or the age of the contamination and eventually to show that the air mass has been polluted by a pollution in particle form during its passing over a site if the contribution of a PAH tracer of that particular pollution is found to be significantly increased.
On the other side, we use the trajectographies of the air

masses at 925 mbar so that to be able to follow a trajectory of the air mass and verify also the eventual passing over a very polluted area. The combination of the two informations give us a good index about the age as well as the contamination area. Nevertheless we should enter a reservation in respect to that, temperatures between sources areas and receptor sites must not be too different.

Among the different informations provided by the histograms, it is shown that some PAH are particulary more interesting ; we are going to consider the concentration ratio between the reactive compounds and the non reactive ones, in order to determine the age of the contamination. The ratio will be as much weaker as the age of the contamination will be stronger since a more rapid decrease of the reactive compound is observed.

Two ratios are, in this respect, significative :
Ba Pyrene / Be Pyrene and Anthracene / Fluoranthene.

The average value of these ratios at the source area is given on the figure 1. The reference compound BeP always represents almost 15% of the total mass of PAH.

The concentration ratio between a non reactive PAH tracer of a particular source and a non reactive-non tracer PAH are considered, two ratios more have been hold :
Fluoranthene / Bb Fluoranthene , characteristic of an eventual pollution by fuel if this ratio is superior to 5,2 (source area)and Bghi Perylene / Be Pyrene characteristic of the pollution by vehicles using gasoline (0,71 at the source).

The figures 2 to 5 show four cases differenciated by the characteristic ratios then by the age of the contamination. At the opposite side we present the trajectographies in order to verify the hypothesis and to determine the possible source area.

The altitude of air mass -not represented here- constitutes a confirmation of the result.

RESULTS

For the first day the ratios BaP/BeP and Ant/Fla are 1.45 and 0.95 respectively. The BaP is slighly decreased , indicating a contamination age of the order of 18-24 hours. The increase of the Ant/Fla ratio shows a gas-particle conversion for Anthracene which is essentially gaseous at the point sources. This conversion rate is faster than the decreasing concentration rate of Anthracene by photooxydation. The Fla/BbF and BghiP/BeP ratios have not varied perceptibly. Nevertheless they indicate that Fluoranthene is slightly more reactive than B b Fluoranthene confirming a previous hypothesis[15].

For the second day, the ratio BaP/BeP is strongly decreased and the Ant/ Fla ratio is pratically equal to that of a urban source. The contamination is about 2 days older and the source area can be the North Italy. It is remarked that the stronger reactivity of Anthracene compensates the

gas-particle conversion.
The Fla/BeP ratio is found to be strongly increased showing so the passing over an area where fuel is widely used; in return the important part of the air mass trajectory over the sea explains the decrease of the ratio BghiP/BeP = 0.20 which is characteristic of the remoteness of the motor vehicles sources.

The third day, the total charge was weaker than the previous ones because of the dilution and the characteristic ratios of a contamination age further decreased : BaP/BeP = 0.30 and Ant/Fla =0.24.

The contamination age can be estimated up to three days. The source area can be situated in the South of Germany. The same remarks, which preceed, can be make in respect to the BghiP/BeP and Fla/BbF ratios.

The last day, the charge was much stronger (3,6 ng/m3) and the ratio BaP/BeP very weak (0,07) indicating a pollution older than 4 days. The ratio Ant/Fla is not modified because of the concentration decrease of Fluoranthene which is also found in the ratio Fla/BbF (0,5). The air mass is not enriched any more by fuel characteristic PAH. On the other hand, the ratio BghiP/BeP is equal to that of a very urbanished area. The source area is West Germany which is very industrialized. Since the trajectography has a spiral form, the air is doubtlessly very contamined by staying for many days over that area.

In conclusion, the study of the concentrations of the trace organic compounds like PAH, helped by the contribution of trajectographies, can estimate the age and the contamination area by anthropogenous pollutants. It can also indicate the eventual contribution of the particular sources. The concentration ratios of these minor pollutants can provide a useful parameter in the modelisation of the air masses transport phenomena.

ACKNOWLEDGMENTS

We want to thank the Metereological staff of Cap Cavallo for his comprehensive collaboration.

BIBLIOGRAPHY

1) ALFHEIM I. and MOLLER N.
Sci. of Total Environ. 1979,13, 275.

2) BEHYMER T.D. and HITES R.A.
E.S.T., 1985,19, 1004.

3) BJORSETH A., LUNDE G. and LINDSKOG A.
Atm. Env. 1979,13, 45.

4) BRODDIN G., CAUTREELS W., VAN CAUVENBERGHE K.V.
Atm. Env. 1980,14, (8), 895.

5) BUTLER J.D. and CROSSLEY P.
Atm. Env. 1981,15, 9.

6) CRETNEY J.R., LEE H.K., WRIGHT G.J. and SMALLOW W.
E.S.T., 1985,19, 397.

7) DAISEY J.M.
Environ. Int., 85,11, 285.

8) FISHER B.E.A.
Atm. Env., 1983,17, 1865.

9) FUNCKE W. and KONIG J.
Atm. Env., 1981, 807.

10) GIBSON T.L., RICCI A.I. and WILLIAMS R.L.
Rapport GMR, 3478, ENV. 95 (1983).

11) GORDON R.J.
E.S.T., 1976,10, 370.

12) GRIMMER G., NAUJACK K.W. and SCHNEIDER D.
Fresenius 7., Anal. Chem., 1982,311, 475.

13) KATZ M., CHAN E., TOSINE M. and SUKUMA T.
PAH ed. by JONES P.W. and LEBER P., 1979, p. 171.
Ann Arbor Publ. U.K.

14) MASCLET P., MOUVIER G. and NIKOLAOU K.
Atm. Env., 1986,20, 439.

15) NIKOLAOU K., MASCLET P. and MOUVIER G.
Sci. Total Environ., 1984,36, 383.

16) PITTS J.N. Jr.
Env. Health Perps., 1983,47, 115.

17) PITTS J.N. Jr., ZIELINSKA B., SWEETWAN J.A. and ATKINSON R.
Atm. Env. 1985,19, 911.

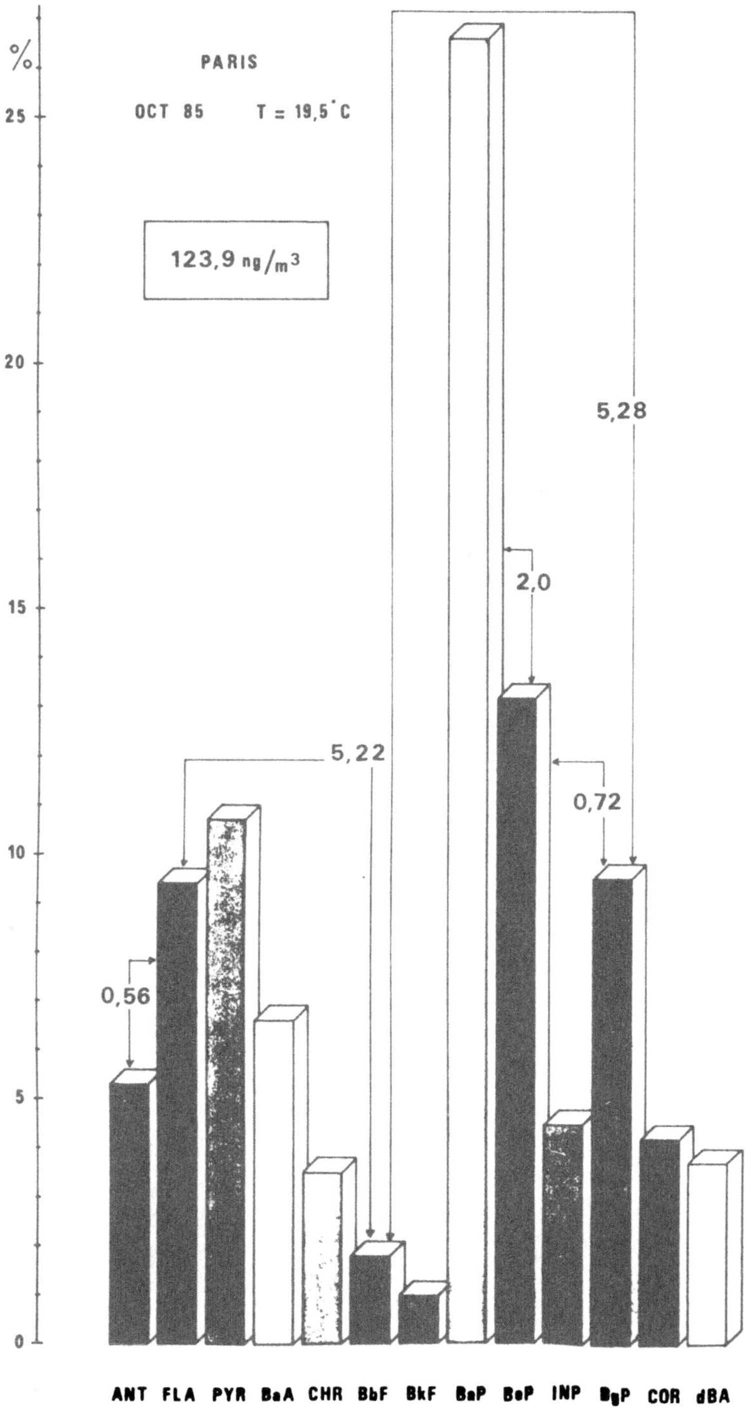
%
PARIS
OCT 85 T = 19,5°C
123,9 ng/m³
25
20
15
10
5
0
5,28
2,0
5,22
0,72
0,56
ANT FLA PYR BaA CHR BbF BkF BaP BeP INP BgP COR dBA

FIGURE 1

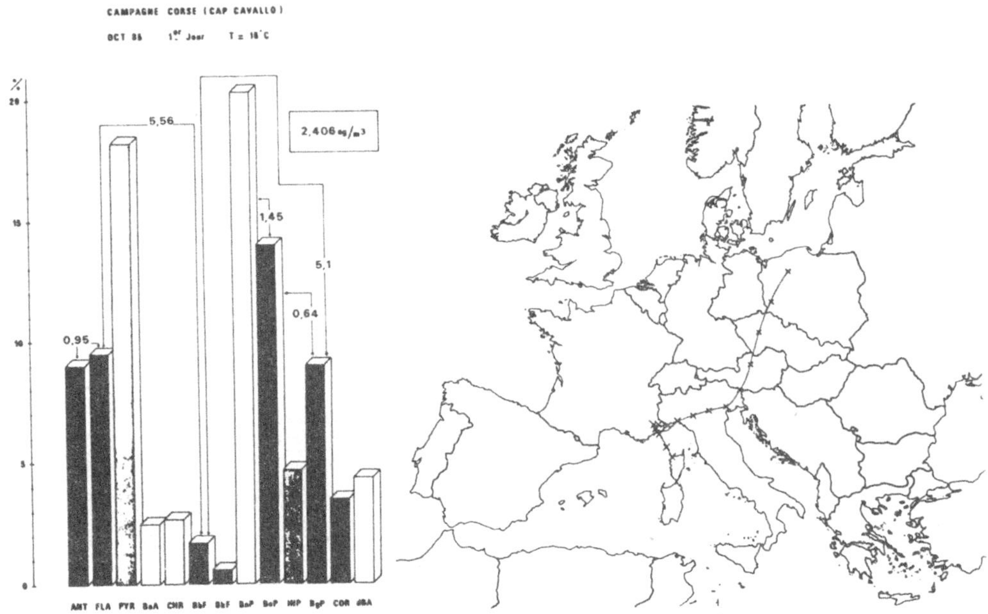

FIGURE 2

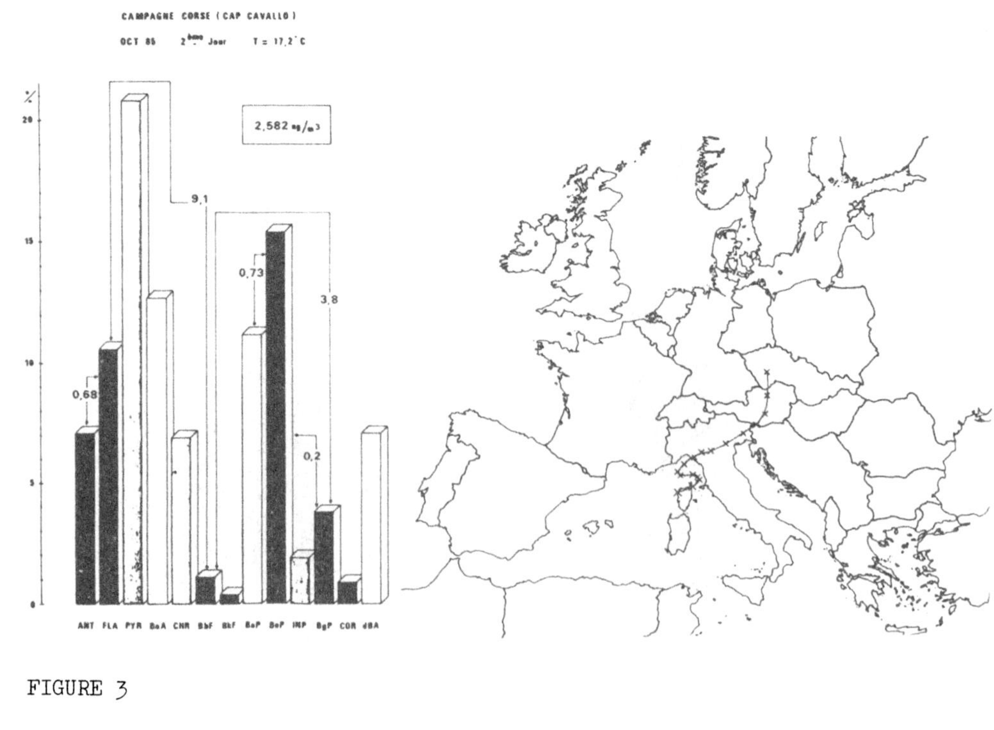

FIGURE 3

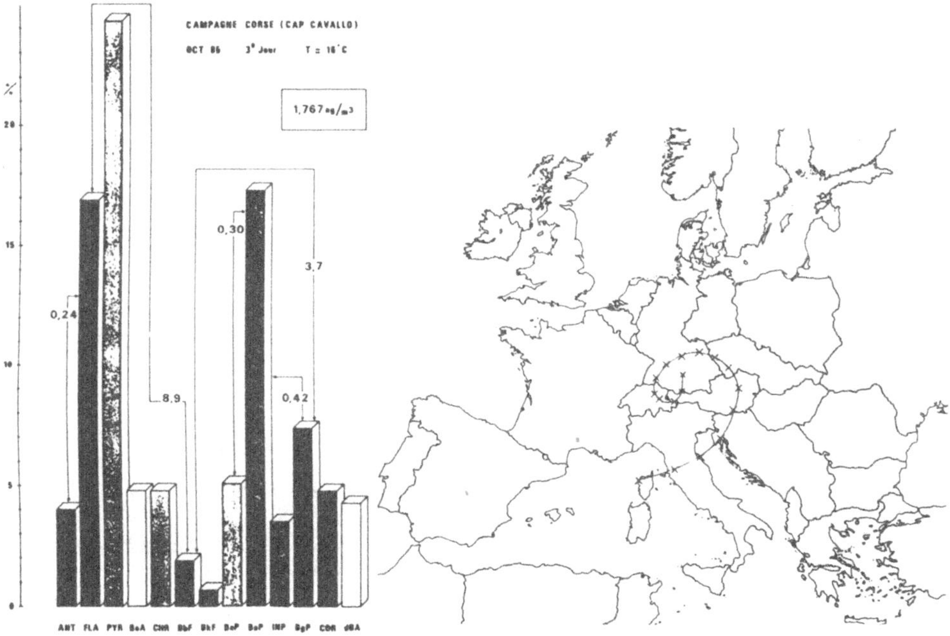

FIGURE 4

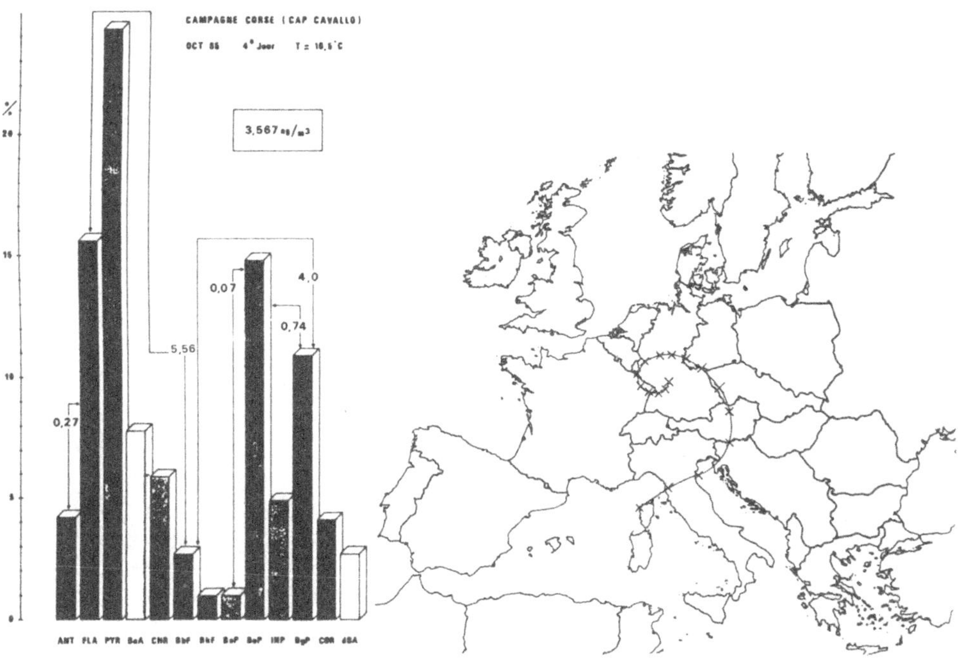

FIGURE 5

AIRCRAFT MEASUREMENTS OF VARIOUS SULFUR COMPOUNDS IN A MARINE AND CONTINENTAL ENVIRONMENT

G. Ockelmann, S. Bürgermeister, R. Ciompa and H.W. Georgii
Department of Meteorology and Geophysics
University of Frankfurt/Main, FRG

Summary

Measurements of gaseous sulfur species were performed in the boundary layer and in the free troposphere around the Azores (Portugal) and over Southern Bavaria (FRG). GC/FPD-, fluorescence and chemiluminescence techniques were applied for the detection of COS, CS_2, DMS, H_2S and SO_2. At the Azores CS_2 was detectable only in the boundary layer with values between 3 and 6 pptv. SO_2 decreased sharply from 300 pptv in the boundary layer to 20-100 pptv in the free troposphere wheras the concentration of COS showed only little variations. Over Bavaria SO_2-values lower than 1 ppbv indicated the advection of clean airmasses. With the exception of COS the mixing ratio of the other compounds decreased considerably with height (most pronounced for DMS). In the free troposphere DMS concentrations were generally below the detection limit (0.2 pptv) while CS_2 was measured in the range of 2-4 pptv and H_2S below 20 pptv. Rawinsonde data and trajectories show that the vertical distribution of the sulfur compounds is influenced by meteorological conditions as well as by their sources and sinks.

1. INTRODUCTION

In the research of the atmospheric sulfur cycle a growing interest is directed to the biogenic sulfur compounds. They contribute considerably to the atmospheric sulfur burden and are a dominant precursor of atmospheric SO_2 in remote areas. The most important compounds are dimethylsulfide (DMS), hydrogen sulfide (H_2S), carbon disulfide (CS_2) and carbonylsulfide (COS).

DMS is produced mainly by microbiological activity in ocean water and soils (1-4). Andreae and Raemdonck (5) estimated that the global sea to air flux is about 38.5 Tg DMS-S/y. From flux measurements in the USA Adams et al. (3) extrapolated that the global DMS emission by soils amounts to 13.4 Tg DMS-S/y. This means that the total flux of DMS to the atmosphere equals about half of the anthropogenic SO_2-emission (6). The atmospheric residence time of DMS is limited

to 8 hrs - 2 days (7) due to the rapid oxidation by OH- and No_3- radicals (8,9).
H_2S is mainly produced under anaerobic conditions e.g. in marshes, wetlands and swamps (10,11). Also some parts of the ocean are assumed to be sources of H_2S (12,13). The reactivity of H_2S is comparable to DMS.
In contrast COS is a stable component in the troposphere. Its lifetime due to reaction with OH is of the order of several years (14). Consequently COS is homogenously distributed in the troposphere. An important sink is the diffusion into the stratosphere and subsequent photooxidation (15). Carbonyl-sulfide has natural and anthropogenic sources, estimates of global emissions range from 2-5 Tg COS/y (16).
Much less information is available on the source strengths and atmospheric concentrations of CS_2. Estimations of the atmospheric lifetime range from 2 to 14 days (7, 16).From the chemistry CS_2 is a precursor of COS and SO_2 as well (17).

The measurements were carried out with a Do 28 aircraft in altitudes up to 3600 m. The results are discussed in the following chapters.

2. EXPERIMENTAL METHODS

A gaschromatographic system equipped with a flame-photometric detector was used for the measurement of DMS, COS and CS_2. DMS was preconcentrated by chemisorption on goldwool (18,19). For the sampling of COS and CS_2 a cryogenic technique similar to Maroulis (20) was applied. The detection limit is 0.2 pptv for DMS and 2 pptv for COS and CS_2.
The measurements of H_2S and SO_2 were accomplished by sampling on $AgNO_3$ respectively TCM impregnated filters with subsequent fluorescence analysis for H_2S (21) and a chemiluminescence technique for SO_2 (22). The detection limit of both methods is 10 pptv.

3. SAMPLING SITES

The measurements in the marine environment were performed in September 1984 at the Azores, about 1600 kilometers west of Portugal. Due to the large distant from the continent this area is favoured for the study of the distribution of trace gases in a clean marine atmosphere. Nevertheless sources of sulfur compounds are present on the islands, mainly of anthropogenic or volcanic activity. Typical for the meteorological situation was the stationary subtropical anticyclone with center close to the islands. Winds were usually weak and mostly from NE in all heights. For the measurements three flight levels were chosen (Figure 1):
0.3 - 0.7 km (below the cloud base)
1.8 - 2.2 km (just above the cloud cover)
3.6 km (free troposphere)
For the campain in a continental environment the area of Southern Bavaria (FRG) was chosen. The aircraft flights which took place in fall 1985 were carried out at the northern slope of the Alps (Figure 2). This region is mainly agricultural with widespread marshes which are known sources of biogenic sulfur compounds. The meteorological situation was

characterized by a strong westerly flow over Central Europe. The air samples were taken in four altitudes as listed below (reference height: mean sea level, the field elevation was 600 meters):

0.8 km (planetary boundary layer)
1.2 km (planetary boundary layer)
1.8 km (adequate to 850 hPa- level)
3.0 km (adequate to 700 hPa- level)

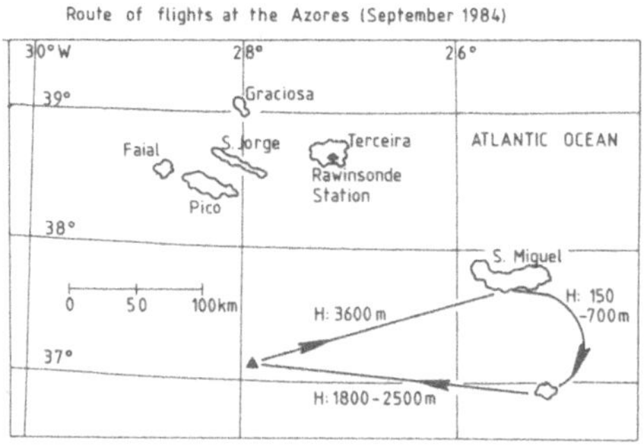

Figure 1

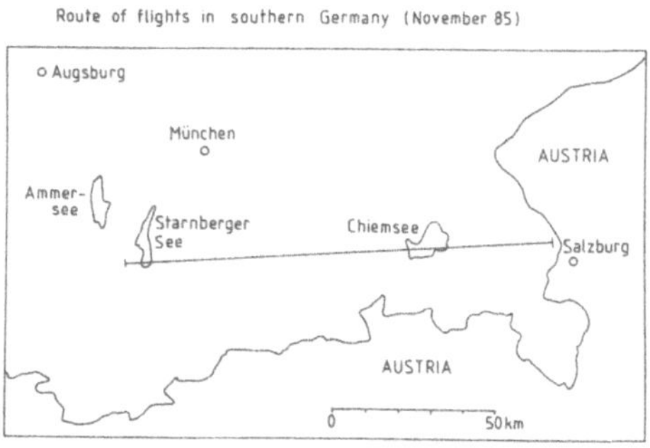

Figure 2

4. RESULTS AND DISCUSSION

Results of the measurements of COS, CS_2 and SO_2 at the Azores are shown in Figures 3-5. Additionally the rawinsonde data of temperature, dew point and wind are plotted.
The concentrations of CS_2 were always small with values of less than 10 pptv in the boundary layer decreasing to undetectable amounts of less than 2 pptv in the free troposphere. This confirms that there are only minor sources of CS_2 in this area. The SO_2 mixing ratios were much higher. During flight 1 and 2 the concentrations and type of distributions were similar with values of 300 pptv in 500 meters dropping rapidly to 40 pptv in 2000 m and increasing again to 90 pptv in 3600 meters. During the third flight the vertical distribution was distinctly different with lower values of about 100 pptv in the PBL decreasing to 60 pptv in 2000 m and 40 pptv in 3600 meters height. From earlier SO_2 measurements over the Atlantic (23) which showed much smaller concentrations of 40 pptv in the PBL and from the fact that during the measuring period components of anthropogenic origin were clearly enhanced (W. Seiler, pers. comm.), we argue that the rather high mixing ratios of SO_2 in the PBL were propably caused by local anthropogenic sources although some care was taken to prevent their influence. From rawinsonde data and cloud observations it is concluded that the SO_2- profile is dependant on the stratification of air: During flight 1 and 2 there was a strong inversion in about 1700 meters and the convection was small as seen from flat cumulus clouds. In case of the third flight the inversion was less pronounced and

Results of flights at the Azores

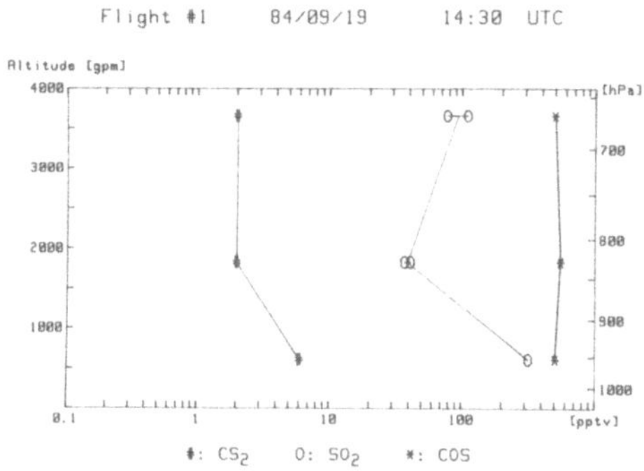

Figure 3a

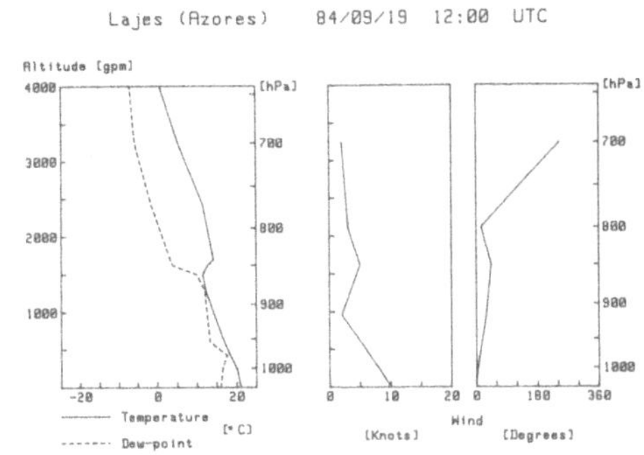

Figure 3b

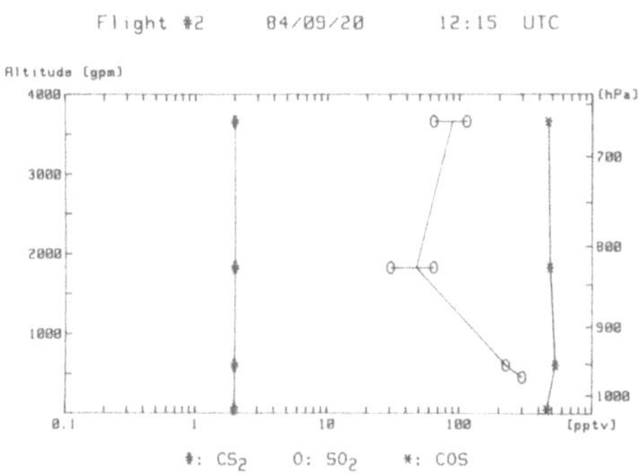

Figure 4a

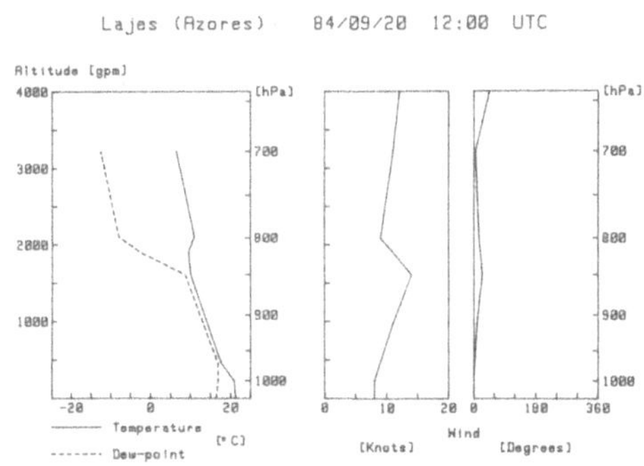

Figure 4b

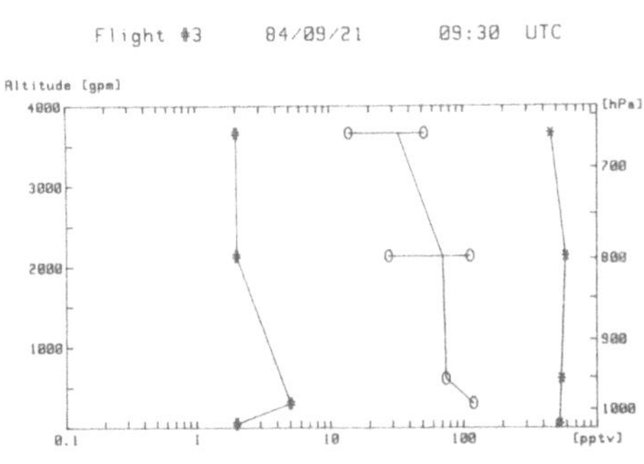

Figure 5a

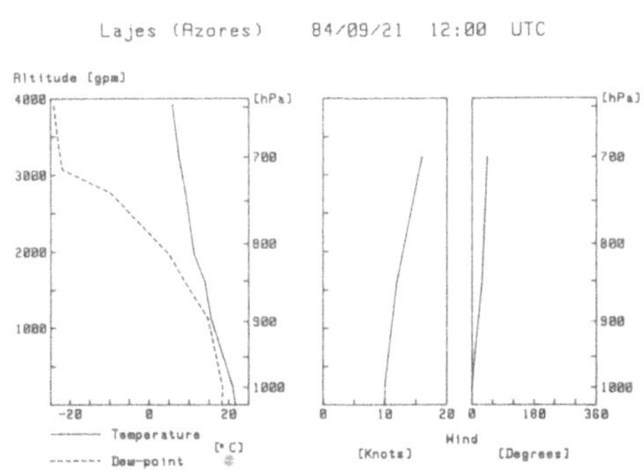

Figure 5b

Results of flights over Southern Germany

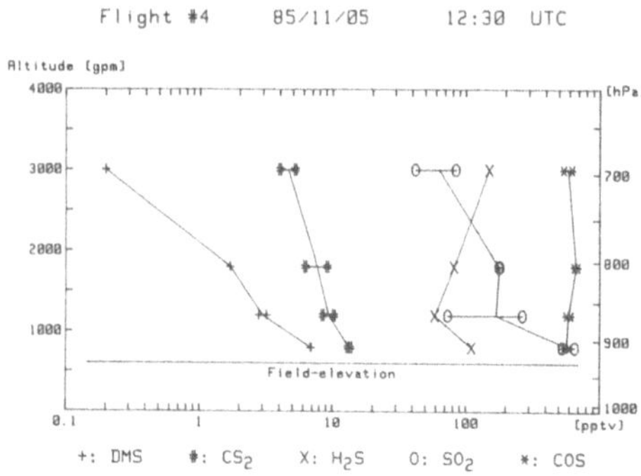

Figure 6a

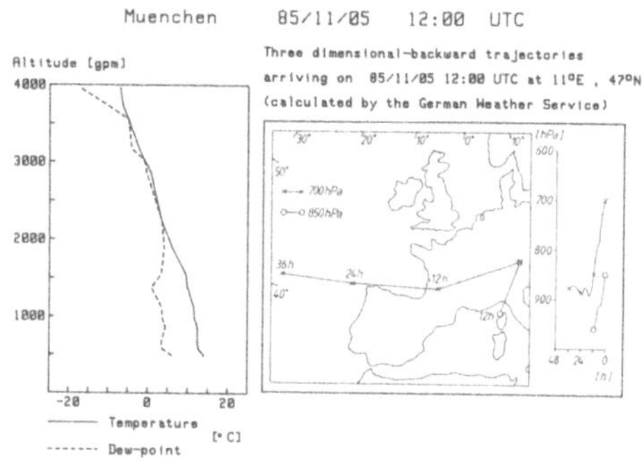

Figure 6b

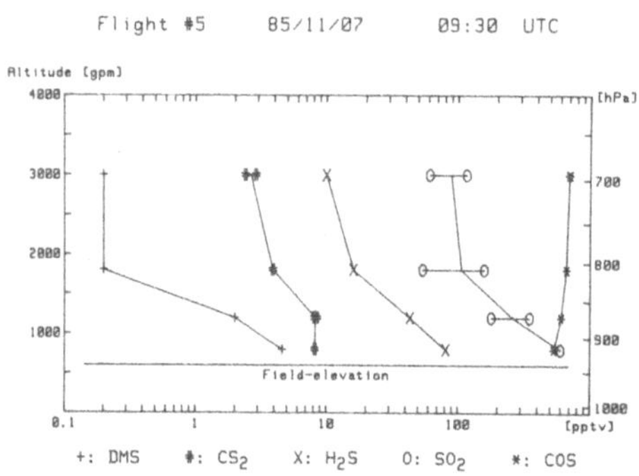

Figure 7a

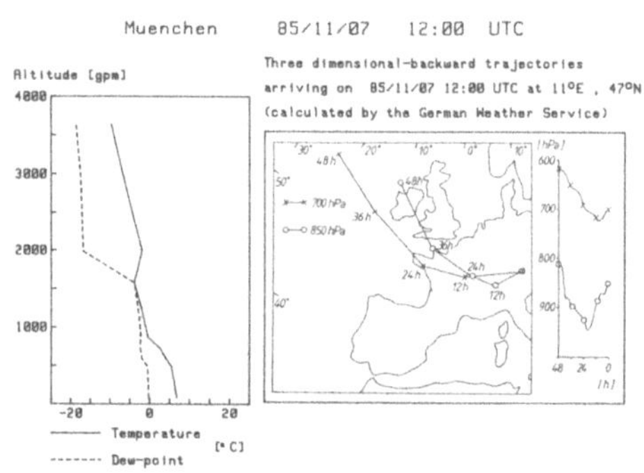

Figure 7b

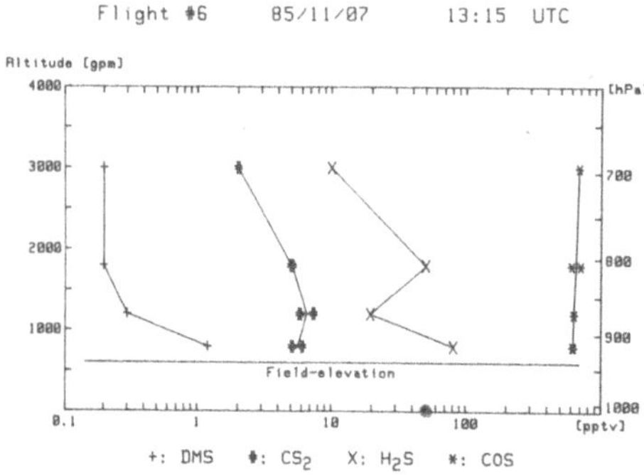

Figure 8a

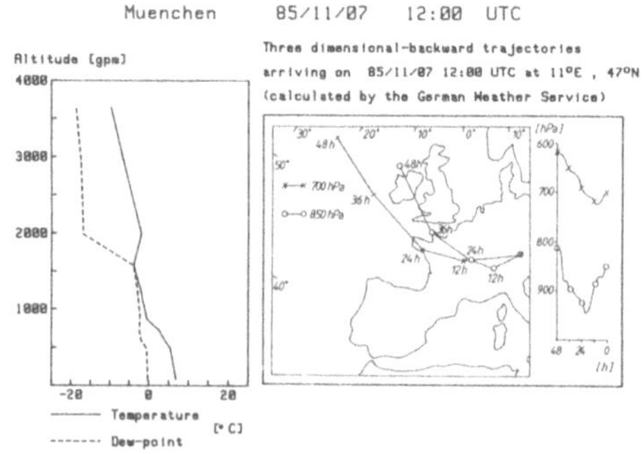

Figure 8b

cumulus development was much more intensive with cloud tops rising to 3500 m. This caused a much better vertical mixing and consequently a weaker concentration gradient.
The mixing ratios of COS always showed small variations between 450 and 550 pptv. However a small concentration maximum was visible in all of the flights and its height increased considerably with stronger convection.
The results of the measurements in continental regions are presented together with meteorological data in Figures 6-8. As can be seen by the trajectories the air was transported in only 12-36 hours from the Atlantic or Mediterranean coast to Southern Germany. This means that the airmasses were originally marine but influenced by advection over the continent for about 1 day.
The SO_2 mixing ratios of less than 1 ppbv measured over Bavaria were very low for the area of Central Europe and the values in 3 km altitude were similar to those in the marine atmosphere. The COS concentration again showed the most homogenous vertical distribution and the same slight increase with height as over the Atlantic. The mixing ratio of this compound exceeded with values of 500-700 pptv those measured under marine conditions by more than 100 pptv. The CS_2 concentrations were also higher than over the Atlantic. They decreased from 5-15 pptv in the PBL to 2-4 pptv in the free troposphere. Earlier CS_2 measurements are reported from Bandy et al. (24). In the USA concentrations of about 30 pptv near the ground were found and values lower than 3 pptv in the free troposphere are quite consistant with our results. Compared to the other sulfur compounds DMS had the lowest concentrations and the strongest decrease with height. In the free troposphere the mixing ratios were always below the detection limit of 0.2 pptv, while in the boundary layer 1-5 pptv were measured. The reduction of the DMS concentration on November 7 can be attributed to the oxidation of DMS at daytime by OH- radicals. The measured DMS values are in the same order of magnitude as those found by Valin et al. (25) in continental air at the south coast of the USA. There 7 ± 3 pptv in the boundary layer and no detectable quantities of DMS ($<$ 5 pptv) in heights above 1000 m were measured. Aircraft measurements performed by Ferek et al. (6) in the Carribean regions showed substantial higher DMS concentrations in the lower and mid troposphere which can be attributed to highly productive sources at the ground and convective transport. The H_2S mixing ratio over Bavaria decreased from 70-100 pptv in 200 m over the ground to 20-60 pptv in the second level and to values below 20 pptv above the cloud tops. During flight 4 the H_2S mixing ratio increased distinctly with height between 1.2 and 3 km. An interpretation is offered by the trajectory analysis: In the 700 hPa level boundary layer air from coastal regions of France with a possibly higher H_2S- concentration was advected within the last 12 hours.

5. CONCLUSION

Among the measured compounds COS showed the most

homogenous vertical distribution due to its long atmospheric residence time. The slight increase of the COS concentration with height in the lowest two kilometers during most of the flights can be taken as an indication that the ocean as well as soils and vegetation in the regions studied represent a net sink for COS. The low concentrations of CS_2 in the area of the Azores coupled with only small amounts of DMS measured by Bingemer in the same area (19) demonstrate that this part of the Atlantic is of minor importance as source of biogenic sulfur compounds. Therfore we conclude that the high SO_2 values measured in the PBL can not be attributed to the oxidation of reduced sulfur species. The measurements over Bavaria demonstrate that DMS is the most reactive compound as shown by its rapid decrease with height. From our results we estimate that DMS is oxidized three times faster than CS_2. The mixing ratios of H_2S over Bavaria exceeded those of CS_2 and DMS by far. This can be explained by H_2S emissions from swamps and agricultural activity. Flux measurements which were performed with a chamber technique over wetlands during the campain showed that emissions of H_2S were about ten times higher than those of DMS and CS_2.

ACKNOWLEDGEMENT

The financial support by the "Deutsche Forschungsgemeinschaft" and the Commission of the European Community is gratefully acknowledged.

REFERENCES

(1) BREMNER, J.M. and STEELE, C.G. (1978). Role of microorganisms in the atmospheric sulfur cycle. Adv. Microbial Ecology (M. Alexander, ed.), 2, 155-201, Plenum Press, New York.

(2) BARNARD, W.R. et al., BINGEMER, H. et al. (1982). The flux of dimethylsulfide from the oceans to the atmosphere. J. Geophys. Res., 87, 8787-8793.

(3) ADAMS, D.F. et al. (1981). Biogenic sulfur source strengths. Environ. Science & Technology, 15, 1493-1498.

(4) BÜRGERMEISTER, S. (1983). Gaschromatographische Messung von Bodenemissionen und atmosphärischen Konzentrationen des Dimethylsulfids. Diplomarbeit, Inst. f. Meteorologie u. Geophys., J.W. Goethe Univ., Frankfurt/Main.

(5) ANDREAE,M.O. and RAEMDONCK, H. (1983). Dimethylsulfide in the surface ocean and in the marine atmosphere: A global view. Science, 221, 744-747.

(6) FEREK, R.J. et al. (1986). Vertical distribution of dimethylsulfide in the marine atmosphere. Nature, 320, 514-516.

(7) CHATFIELD, R.B. and CRUTZEN, P.J. (1984). Sulfur dioxide in remote oceanic air: Cloud transport of reactive precursors. J. Geophys. Res., 89, 7111-7132.

(8) COX, R.A. and SHEPPARD, D. (1980). Reactions of OH-radicals with gaseous sulfur compounds. Nature, 284, 330-331.

(9) WINER, A.M. et al. (1984). Gaseous nitrate radical: Possible nighttime atmospheric sink for biogenic organic compounds. Science, 224, 156-159.
(10) JAESCHKE, W. et al. (1979). Messung und Verteilung von Schwefelwasserstoff in reiner und verunreinigter Atmosphäre. Staub- Reinhalt. Luft, 39, 174-177.
(11) JORGENSEN, B.B. and HANSEN, B.O. (1985). Emissions of biogenic sulfur gases from a Danish estuary. Atm. Env., 19, 1737-1749.
(12) DELMAS, R. and SERVANT, J. (1982). The origins of sulfur compounds in the atmosphere of a zone of high productivity (Gulf of Guinea). J. Geophys. Res., 87, 11019-11026.
(13) HERRMANN, J. and JAESCHKE, W. (1984). Measurements of H_2S and SO_2 over the Atlantic Ocean. J. Atm. Chem., 1, 111-123.
(14) RAVISHANKARA, A.R. et al. (1980). Rate of reaction of OH with COS. Geophys. Res. Lett., 7, 861-864.
(15) CRUTZEN, P.J. (1976). The possible importance of CSO for the sulfate layer of the stratosphere. Geophys. Res. Lett., 3, 73-76.
(16) KHALIL, M.A.K. and RASMUSSEN, R.A. (1984). Global sources, lifetimes and mass balances of carbonylsulfide (COS) and Carbondisulfide (CS_2) in the earth´s atmosphere. Atm. Environ., 18, 1805-1813.
(17) JONES, B.M.R. et al. (1983). Atmospheric chemistry of carbondisulfide. J. Atm. Chem., 1, 87-105.
(18) AMMONS, J.M. (1980). Preconcentration methods for the detection of gaseous sulfur compounds in air. Ph. D. dissertation, Univ. of South Florida, Miami.
(19) BINGEMER, H. (1984). Dimethylsulfid in Ozean und mariner Atmosphäre- Experimentelle Untersuchung einer natürlichen Schwefelquelle für die Atmosphäre. Ber. Inst. f. Meteorologie u. Geophys., Univ. Frankfurt/Main Nr. 58.
(20) MAROULIS, P.J. et al. (1977). Atmospheric concentrations of carbonylsulfide in the Southwestern and Eastern United States. Geophys. Res. Lett., 4, 510-512.
(21) JAESCHKE, W. (1978). New methods for the analysis of SO_2 and H_2S in remote areas and their application to the atmosphere. Atm. Environ., 12, 715-721.
(22) MEIXNER, F.X. and JAESCHKE, W. (1981). The detection of low atmospheric SO_2 concentrations with a chemiluminescence technique. Int. J. Environ. Analyt. Chem., 10, 51-67.
(23) OCKELMANN, G. (1982). Untersuchung der Verteilung des atmosphärischen Schwefeldioxids über dem Atlantik und der Arktis. Diplomarbeit, Inst. f. Meteorologie u. Geophys., J.W. Goethe Univ., Frankfurt/Main.
(24) BANDY, A.R. et al. (1981). Evidence for a short tropospheric residence time for carbondisulfide. Geophys. Res. Lett., 8, 1180-1183.
(25) VAN VALIN, C.C. et al. (1986). Natural sulfur flux from the Gulf of Mexico: Dimethylsulfide, carbonylsulfide and sulfurdioxide. Atm. Environ., submitted 1986.

FLUX D'AZOTE INORGANIQUE DISSOUS D'ORIGINE ATMOSPHERIQUE EN SYSTEME LITTORAL D'EUROPE OCCIDENTALE.

P. SOUCHU et P. TREGUER

Laboratoire de Chimie des Ecosystèmes Marins - J.E. CNRS 914692
Faculté des Sciences et Techniques, 6, avenue le Gorgeu, 29287
BREST - CEDEX.

RESUME

Les flux d'azote inorganique dissous (NID), ammonium, nitrates + nitrites, d'origine atmosphérique (précipitations) sont suivis de juin 1985 à août 1986 pour trois sites d'Europe Occidentale (Bretagne, France). Les fortes teneurs en ammonium mesurées sur le site du centre Bretagne sont attribuées à l'activité rurale de l'arrière pays. Les concentrations élevées, lorsque les vents sont du secteur Est à Nord-Est, s'expliquent par l'arrivée de masses d'air polluées d'origine continentale.
En comparant ces flux aux apports de NID d'origine fluviale et à ceux engendrés par le recyclage à l'interface Eau-Sédiment, pour un écosystème littoral type, leur impact éventuel sur l'activité du phytoplancton est étudié pour les quatre scénarios saisonniers.

ABSTRACT

In order to study the deposition of Dissolved Inorganic Nitrogen (DIN) by precipitation, ammonia and nitrate + nitrite have been measured in rainwater from june 1985 to august 1986. High concentrations of ammonia found on the Brittany center site are ascribed to rural activities. To explain strong DIN inputs, a long range transport of polluted air from eastern areas was also suggested. This inputs from precipitations have been compared to waste natural inputs and water-sediment exchanges in a coastal marine ecosystem, the impact on primary production is studied.

I. CONTEXTE DES RECHERCHES ET OBJECTIFS

Depuis quelques années, les chimistes marins s'intéressent aux apports d'origine atmosphérique sur le cycle des nutriments (1)(2)(3). Si le cycle de l'azote a été étudié depuis plusieurs années dans les écosystèmes littoraux d'Europe Occidentale (4)(5), aucun travail n'a été fait sur l'interface eau-atmosphère. Les flux d'azote inorganique dissous (NID) : ammonium et nitrates + nitrites sont suivis sur trois sites de l'extrême pointe de l'Europe Occidentale et présentent des caractères différents (influences marines, anthropiques, etc...)

Les objectifs de ce travail sont les suivants :

- Suivre l'évolution saisonnière des concentrations en NID dans les précipitations,
- Identifier les principales sources responsables des apports en NID d'origine atmosphérique,
- Evaluer l'impact éventuel de ces apports sur la production phytoplanctonique en écosystème macrotidal.

II. ZONE ETUDIEE ET TECHNIQUES MISES EN OEUVRE

. Les sites

Face à l'Atlantique, la Bretagne est une région peu industrialisée dont l'économie est essentiellement tournée vers les activités agricoles. Trois sites ont été retenus pour la collecte des précipitations (Fig. 1) : Brest (200 000 habitants) est une ville située au bord d'une rade de 180km^2 alimentée en eaux douces par deux fleuves : l'Aulne et l'Elorn ; Rostrenen se situe au coeur d'un secteur essentiellement rural ; Le Pouliguen est un site marin à l'abri des pollutions locales.

. Méthodologie

Les précipitations sont recueillies à l'aide de collecteurs ouverts entièrement en PYREX (diamètre : 20 cm). L'échantillonnage est effectué toutes les 24 heures. Les collecteurs sont rincés à l'eau déionisée après prélèvement. Les échantillons sont stockés au congélateur dans des flacons en verre. La méthode automatique de Tréguer et Le Corre (6) permet la mesure des concentrations en ammonium et nitrates + nitrites avec une précision de 1 %. La forme nitrite étant mineure, on exprime par "nitrates" la quantité "nitrates + nitrites". Les hauteurs des précipitations et la direction des vents sont fournies par la météorologie nationale aux stations météorologiques les plus proches. Le pH des eaux de pluies est contrôlé sur le seul site de Brest.(Il est en général de 4.5 à 5.5 sauf en période de vents d'Est où il descend à 3).

III. RESULTATS ET INTERPRETATION

. Etude comparative des trois sites et origines de l'azote inorganique dissous dans les précipitations

De façon générale, les concentrations mesurées restent faibles par rapport à celles mises en évidence en Europe Septentrionale (7). Elles se caractérisent, toutefois, par un niveau élevé en ammonium. Si pour les trois sites, les concentrations en ammonium et en nitrates mesurées sont très variables (0.6 à 400 µM-N), les teneurs les plus élevées correspondent généralement aux précipitations les plus faibles (Fig. 2) (8)(9). Une pluie fortement concentrée n'est donc pas toujours synonyme de dépot important.

Sur la figure 3, sont représentés les flux mensuels en ammonium et en nitrates de décembre 1985 à avril 1986. En moyenne, les dépots les plus importants sont mesurés à Brest, et les plus faibles au Pouliguen. Pour ces deux sites, ammonium et nitrates suivent une évolution parallèle, sur le site de Rostrenen, un excès systématique en ammonium par rapport aux nitrates est décelé au printemps avec un maximum au mois de mai. La figure 4 représente une étude comparative des flux pour les trois sites, en fonction de l'orientation des vents. Pour les sites de Brest et du Pouliguen, les flux orientés à l'Ouest sont faibles. En revanche, pour les trois sites, les vents d'Est et en particulier de Nord-Est sont à l'origine d'apports beaucoup plus importants (situation observée en février).

Cet ensemble de résultats suscite quatre remarques :

- Les masses d'air océaniques contribuent faiblement aux apports en NID. L'océan n'est pas une source de NID importante pour l'atmosphère (10).
- La Bretagne, dans son ensemble, subit l'influence de pollutions plus ou moins lointaines à partir de masses d'air s'étant préalablement déplacées au-dessus de l'Europe du Nord-Est (11).
- La ville de Brest est à l'origine d'une pollution locale. La combustion du fuel et autres activités urbaines participent à la charge en NID dans l'atmosphère (12).
- A Rostrenen, les fortes teneurs en ammonium observées sont à relier aux activités agricoles du pays. L'épandage de lisiers et l'application d'engrais chimiques au printemps sont sans doute les causes de cette pollution locale. Cette situation n'est pas sans analogie avec celle observée aux Pays-Bas (Asman, Communication personnelle).

. Impact éventuel des apports en NID d'origine atmosphérique sur l'activité phytoplanctonique d'un écosystème macrotidal

L'azote est un élément essentiel du contrôle de l'activité biologique des milieux littoraux. En rade de Brest, il constitue le principal facteur limitant d'une production primaire analogue à celle des autres écosystèmes des zones tempérées (13). La Rade de Brest reçoit des apports en NID multiples qui assurent sa fertilisation (Tableau 1). Ce sont les fleuves qui apportent la contribution annuelle la plus importante au stock de NID à l'écosystème (14). En période estivale, les échanges eau-sédiment dominent dans les apports (15). Si dans certains écosystèmes marins, les flux d'origine atmosphérique jouent un rôle important dans la fertilisation, il n'en est pas de même pour la Rade de Brest (Tableau 2). Cependant, en période automnale et dans des conditions maximales de stabilité des eaux (Mortes-eaux, basse-mer) un flux de 350 $\mu m.m^{-2}.j^{-1}$ NID (Brest, 3 novembre 1985), peut constituer une source importante d'azote immédiatement disponible dans des eaux de surface très appauvries en NID et susceptible d'alimenter un développement d'espèces phytoplanctoniques opportunistes.

IV. CONCLUSION

Située à l'écart des grands centres industriels, la Bretagne est souvent considérée comme une région à l'abri des pollutions atmosphériques. Ces résultats montrent que des masses d'air d'origine continentale (Europe du Nord-Est) contribue à la composante anthropogénique en NID. Les importantes activités agricoles qui caractérisent cette région de l'Europe constituent une source importante d'azote ammoniacal pour l'atmosphère.

Si les précipitations jouent un rôle moteur dans le transfert de composés d'origine anthropique aux écosystèmes littoraux d'Europe Occidentale, la voie la plus "efficace" reste celle du lessivage et de l'enrichissement des fleuves. Cependant, les apports directs par des précipitations particulièrement riches en ammonium, dont l'importance reste marginale au regard du bilan annuel, sont susceptibles en scénério automnal d'accroître l'activité phytoplanctonique.

REMERCIEMENTS

Nous exprimons nos vifs remerciements

- à Madame et Monsieur SOUCHU pour les prélèvements effectués au Pouliguen ; au personnel de la Météorologie Nationale pour ceux de Rostrenen,
- à l'Etablissement Public Régional de Bretagne pour son aide financière.

REFERENCES

(1) DUCE, R.A. (1985). The impact of atmospheric nitrogen, phosphorus and iron species on marine biological productivity. NATO advanced study institute on the role of air-sea exchange in geochemical cycling. Bombannes (France).
(2) PAERL, H.W. (1985). Enhancement of marine primary production by nitrogen-enriched acid rain. Nature. Vol. 315, 747-749.
(3) CORELL, D.L. and FORD, D. (1982). Comparison of precipitation and Land runoff as sources of estuarine nitrogen. Est. Coast. Shelf Sci. Vol. 15, 45-56.
(4) DELMAS, R. et TREGUER, P. (1983). Evolution saisonnière des nutriments dans un écosystème eutrophe d'Europe Occidentale (La Rade de Brest). Interactions marines et terrestres. Oceanol. Acta. Vol. 6, 345-356.
(5) LE JEHAN, S. et TREGUER, P. (1984). Evolution saisonnière de composés organiques dissous dans un écosystème eutrophe d'Europe Occidentale (Rade de Brest). Oceanol. Acta. Vol. 7, 181-190.
(6) TREGUER, P. et LE CORRE, P. (1975). Manuel d'analyses des sels nutritifs dans l'eau de mer (utilisation de l'auto-analyser II Technicon) LOC-UBO 2ème édition, pp.110.
(7) GRENNFELT, P. et HULTBERG, H. (1985). Effects of nitrogen deposition on the acidification of terrestrial and aquatic ecosystems. International symposium on acidic precipitation. MUSKOKA, September 15-20.
(8) DOKIYA, Y. et KATSUGARI, Y. (1985). Chemical components in the precipitation at Komoro and Tsukuba, Japan in 1983. J. Environ. Sci. Health, A20 (3), 269-291.
(9) ZEPHORIS, M. (1984).Pluies acides en France.Pollut.Atmos. 103, 159-291.
(10) TSUNOGAI, T. (1971). Ammonia in the oceanic atmosphere and the cycle of nitrogen compounds through the atmosphere and the hydrosphere. Geochem. J. Vol. 5, 57-67.
(11) MORELLI, J., GIRART-REYDET, L. MARCHAL, T., MASNIERE, P., FEDOROFF, M., ROUCHAUD, J.C., DEBOVE, L. et CARLIER. P. (1984). Fluctuations temporelles fines de la composition chimique de l'aérosol côtier. Third European symposium on Physico-Chemical behaviour of atmospheric pollutants Varèse (Italy).
(12) FREYER, H.D. (1978). Seasonal trends of NH_4^+ and NO_3^- nitrogen isotope composition in rain collected at Jülich, Germany, Tellus, Vol. 30, 83-92.
(13) HAFSAOUI, M., QUEGUINER, B. et TREGUER, P. (1985). Production primaire et facteurs limitant la croissance du phytoplancton en Rade de Brest (1981-1983). Oceanis, Vol. 11, 181-183.
(14) SOUCHU, P. (1986). Contribution à l'étude du cycle de l'azote dans un écosystème eutrophe d'Europe Occidentale (La Rade de Brest). Thèse Spéc. Univ. Bretagne Occidentale (en préparation).
(15) SOUCHU, P. et TREGUER, P. (1986). Transferts biogéochimiques d'azote à l'interface eau-sédiment en écosystème littoral (en préparation).

(16)Ministère de l'Environnement (1981). Etude acceptabilité du milieu. Rapport d'avancement, pp.120.

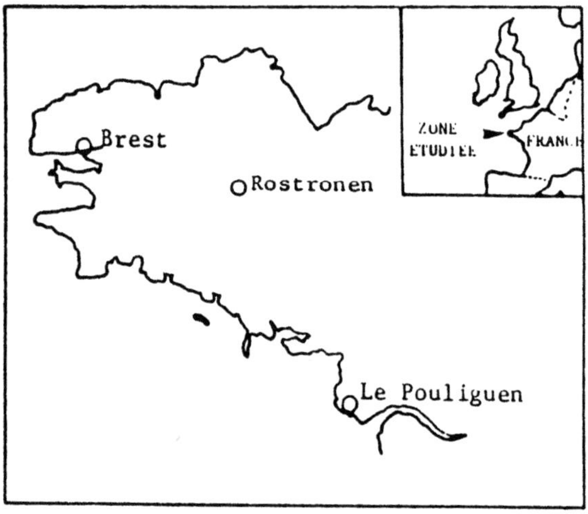

Figure 1 Localisation des sites

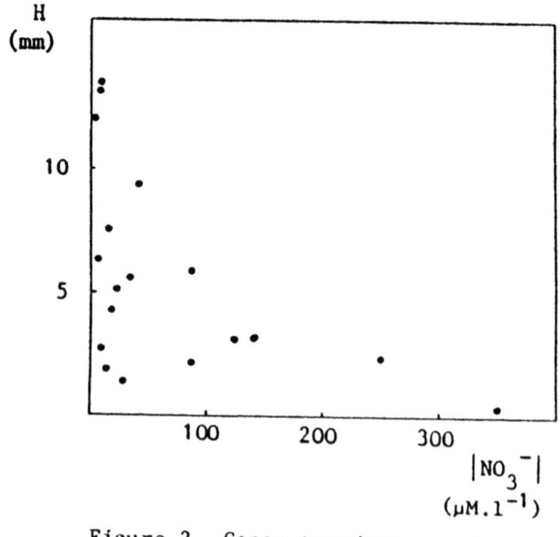

Figure 2 Concentrations en nitrates dans les pluies en fonction des hauteurs de précipitations (avril 1986, Brest).

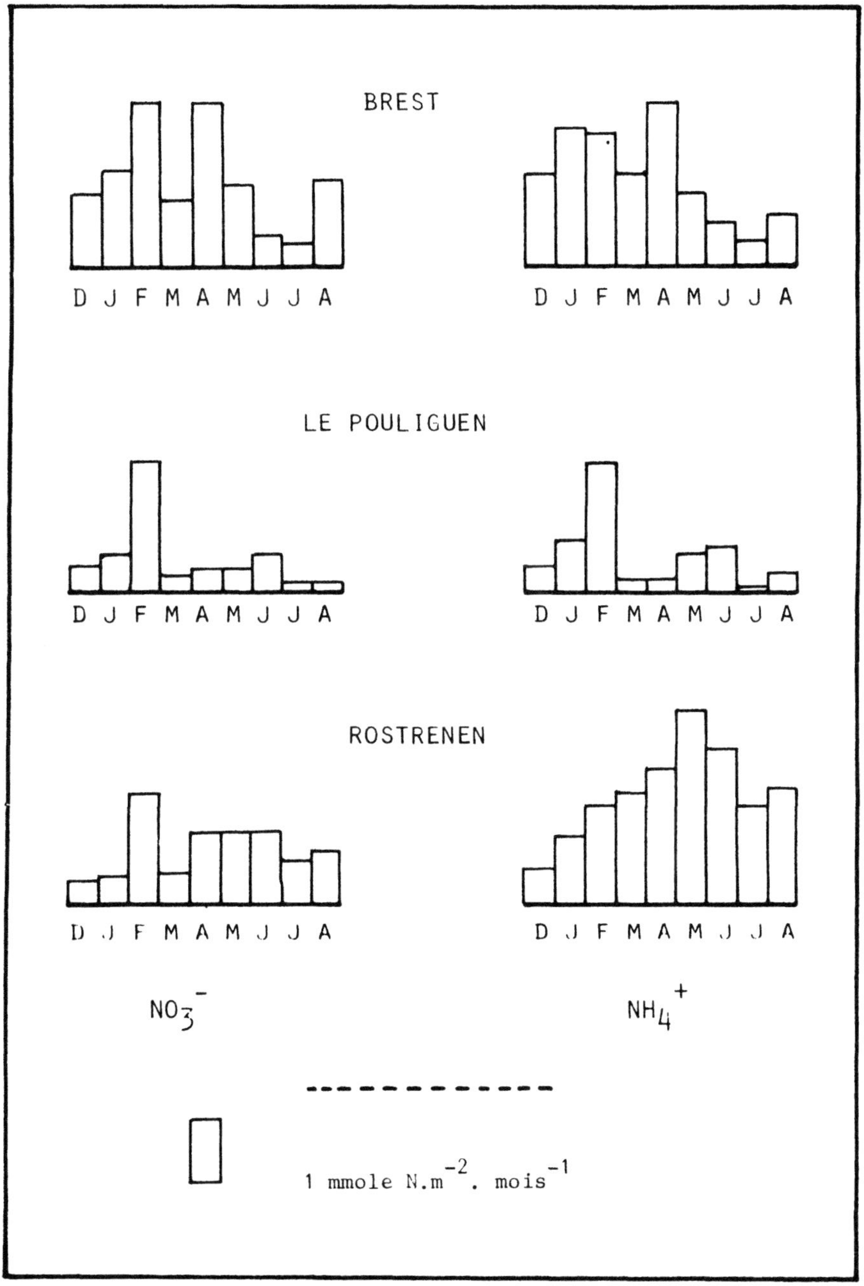

Figure 3 : Flux d'azote inorganique dissous apporté par les précipitations sur les trois sites de Décembre 1985 à Août 1986.

4a) Direction des vents et flux du NID à Brest

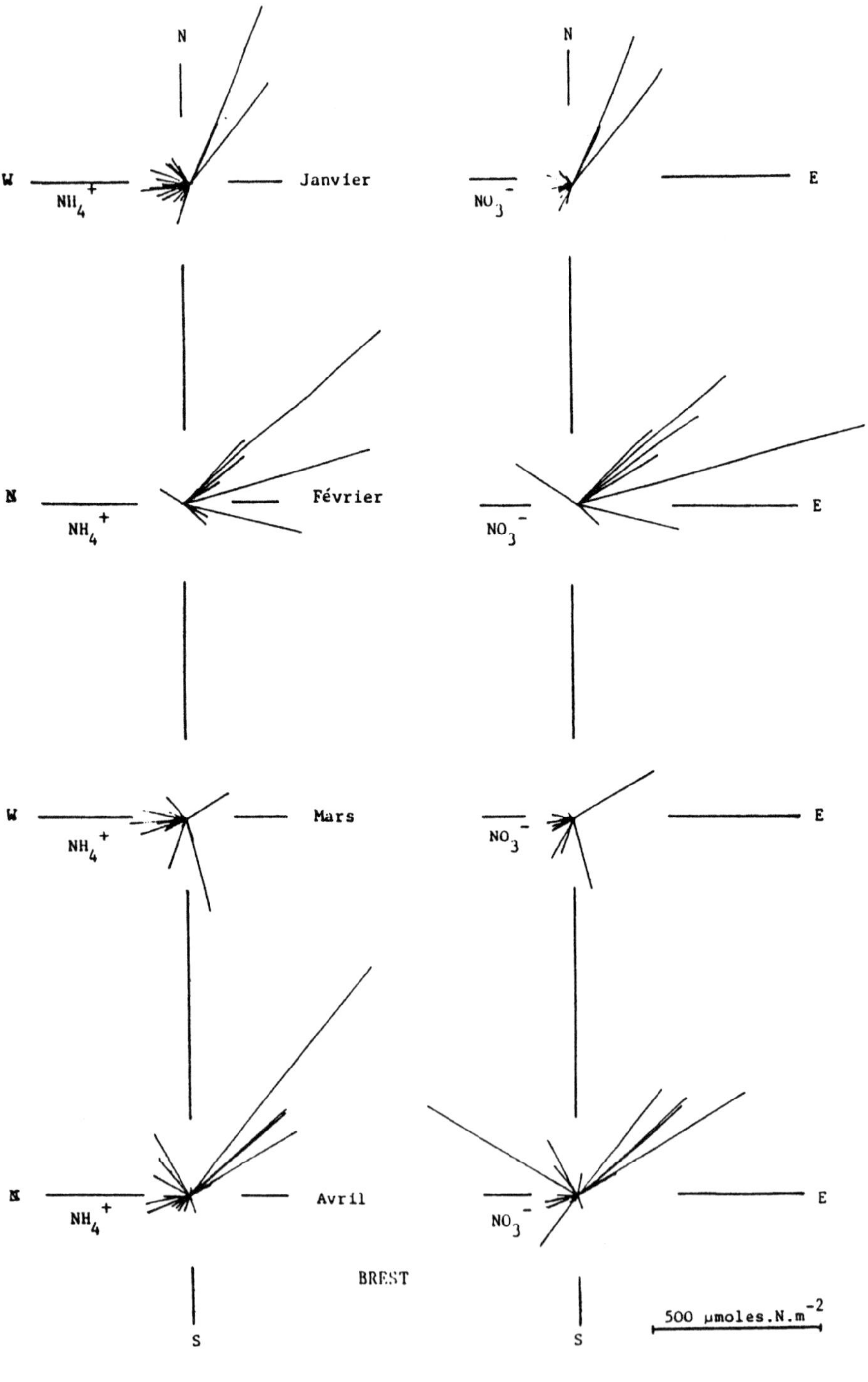

4b) Direction des vents et flux de NID au Pouliguen

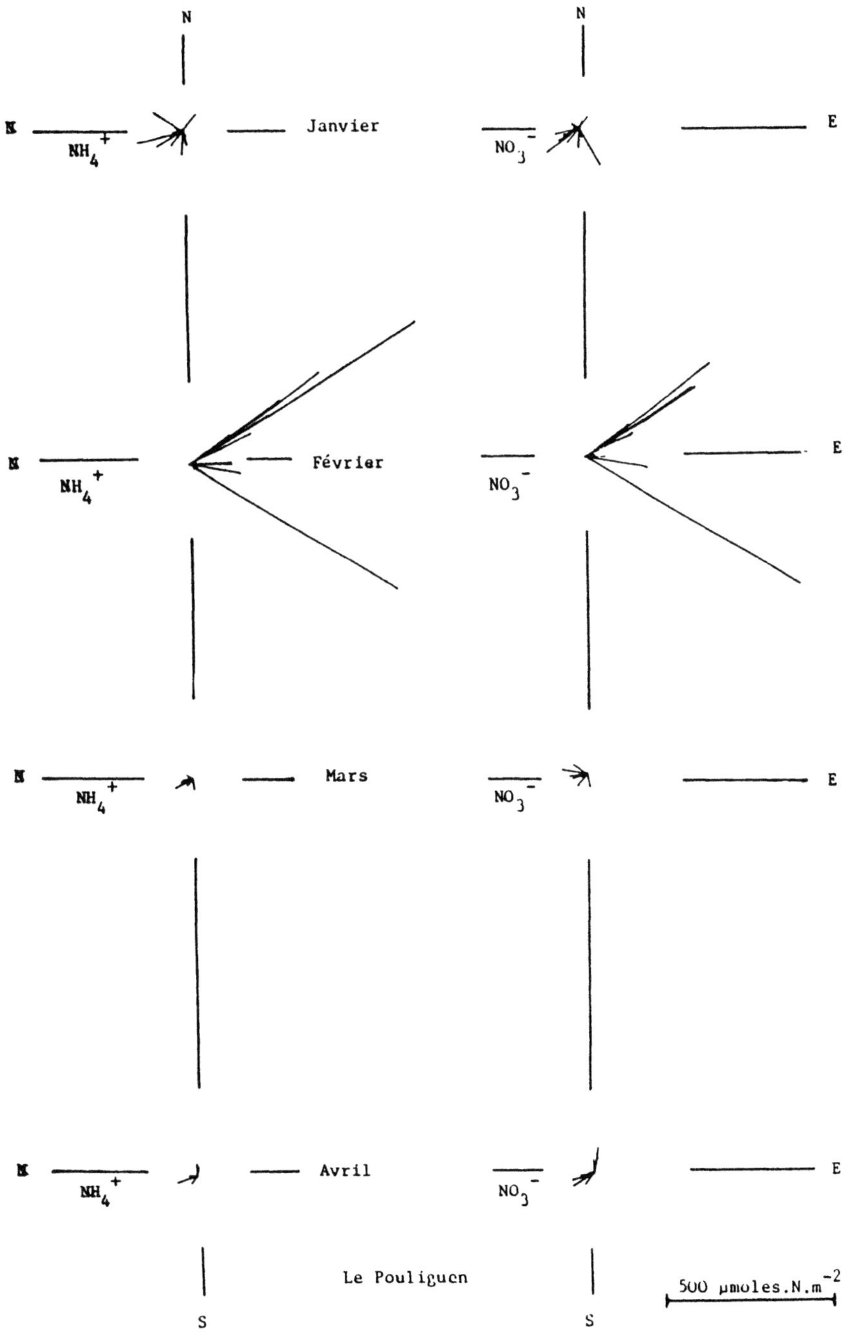

4c) Direction des vents et flux de NID à Rostronen

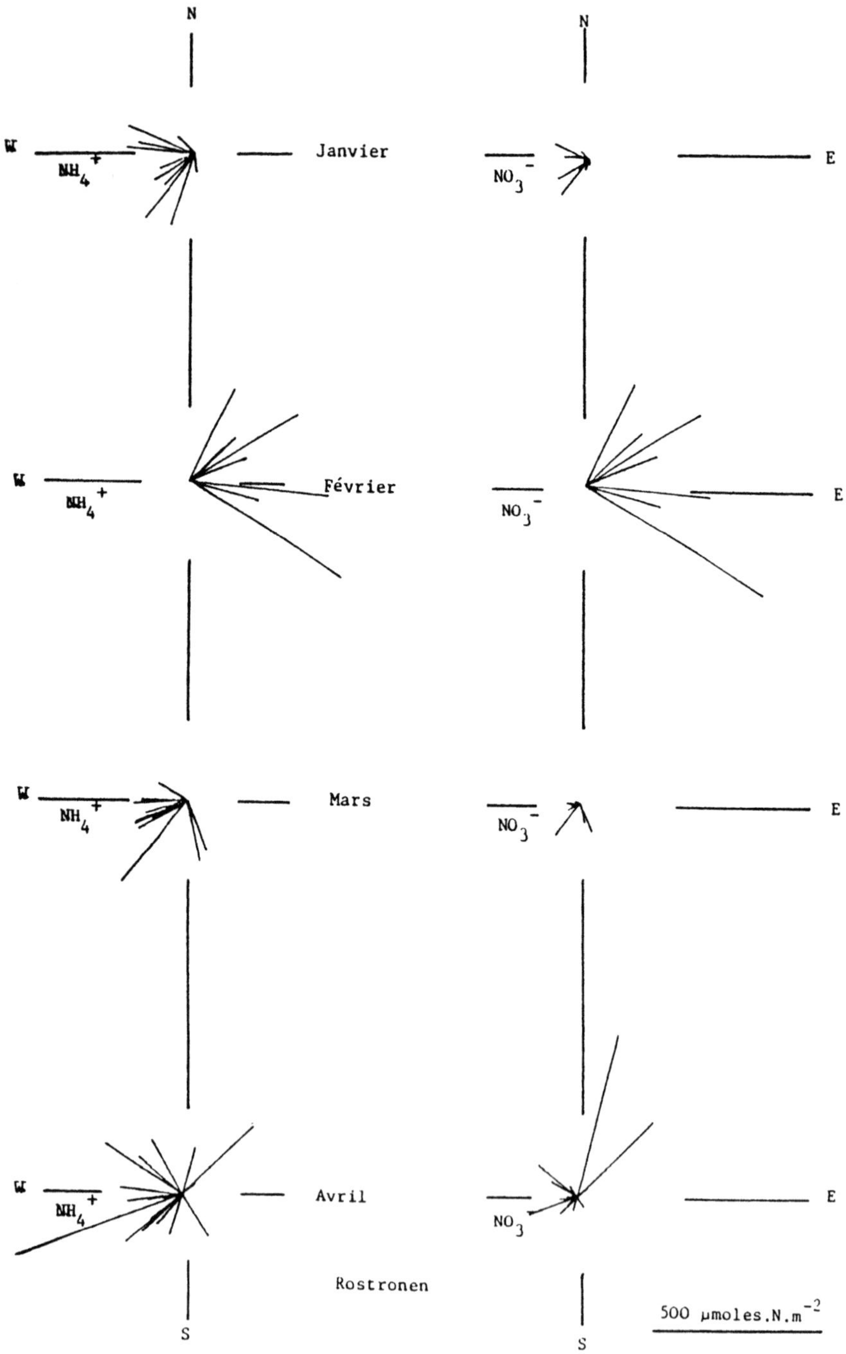

	HIVER	PRINTEMPS	ETE	AUTOMNE	
Apports fluviaux	48.0	29.1	2.3	1.6	(14)
Echanges Eau-Sédiment	0.7	-2.0	13.0	2.5	(15)
Rejets urbains	1.0	1.0	1.0	1.0	(16)
Apports par les précipitations	0.3	0.3	0.1	0.2	

Tableau 1 : Apports et consommations en Azote Inorganique Dissous dans la Rade de Brest pour les quatre scénarios saisonniers. Les résultats sont exprimés en Tonnes d'azote par jour. Le signe - correspond à une consommation.

Ecosystèmes	Flux du NID par les précipitations $\mu m.N.m^{-2}.j^{-1}$	Contribution aux apports nets en azote	Références
Mer des Sargasses (Oligotrophe) Atlantique Nord	40	8 à 70 %	(1)
Rhod River (Côtier, USA)	144	12 %	(2)
Rade de Brest (Europe Occidentale)	88	1 %	-

Tableau 2 : Contribution des flux moyens de NID d'origine atmosphérique sur un an aux apports totaux en azote pour différents écosystèmes.

MESO-METEOROLOGICAL ANALYSIS OF AIR POLLUTION CYCLES IN SPAIN

Millán M. Millán, Marino Navazo and Agustin Ezcurra
CIEMAT-JEN, Av. Complutense 22, E-28040, Madrid

Summary

Three years of field measurements in the coastal area of Bilbao in Northern Spain show that most episodic conditions in this area occur as a result of: (a) anticyclonic stagnation over Central Europe and (b) pollutants being trapped in local recirculating flows. In general, the dispersion behaviour in this area of complex terrain can be classified into several well defined patterns, which correspond to certain meteorological conditions. Results to date show that complex re-entry cycles occur and that synoptically different flows may be undistinguishable by wind sensors at ground level (affected by channelling), and yet result in totally different observed pollution levels by a fixed monitoring network (affected by topographical effects). A hitherto undocumented phenomenon occurs during the summer as the result of the diurnal formation of the Iberian thermal low. It can form every day over the central plateau and forces an inward flow of coastal emissions which can spiral inland over successive days. This inflow can originate episodic conditions in Madrid at the end of the summer when local emissions are at their lowest.

1. INTRODUCTION

Diffusion and transport in complex topography is still one of the outstanding problem areas in air pollution (1-3). This is the case of the city of Bilbao and its industrial area, located in the Basque country in North Central Spain.

The area is located along an estuary which runs nearly 16 km from the centre of the city to the sea and is aligned in a general SE-NW direction (Fig. 6). Two mountain ranges run parallel to the waterway; the one SW of the city reaches over 700 metres and the other in the NE reaches 300 m. Four main airsheds can be identified:

1.- The upper Nervion and Ibaizabal valleys which merge SE of the city; these are dotted with small and medium-sized industrial plants located along the river banks.
2.- The Cadagua valley drains from the SW, cutting through the higher mountain range to join the estuary just NW of the city.
3.- The broader and shallower Asua valley is located NE of Bilbao at the other side of the lower mountain range. It is connected directly to the sea through two shallow passes, which are easily overcome by the sea breeze. With respect to very stable flow, however, it drains into the estuary via a short and narrow gorge.
4.- The lower Nervion valley proper, or estuary, starts NW of the city where the heavy industries are located. It drains directly into the sea through the port area. This could be considered a fifth airshed on its

own due to the important effects caused by its body of water. It is roughly trapezoidal in shape, enclosed by high land (>70 m) on three sides and is nearly 17 Km2 in area.

Because of its location, air flow within the whole area is strongly influenced by the local topography under all meteorological conditions. This is particularly important under some anticyclonic conditions (i.e. High NE of the Iberian Peninsula) with clear skies, when atmospheric motions in the area are dominated by meso-meteorological phenomena of local origen.

Bilbao has long been known as an area with serious pollution problems derived from its unfavorable topography, large total emissions, too close a housing-industry arrangement (mostly at fence length) and old industrial plants. An automatic monitoring network was installed by the UAM/IBM Scientific Center of Madrid in 1977. It consists of eight SO_2 (Philips) monitors and seven other stations (four of them co-located with the SO_2 sensors) which measure some meteorological parameters. Data is telephone linked to a central computer and can be obtained either in hourly or five minute averaged values. Early attempts to use Gaussian plume and/or statistical models met with little success in understanding or forecasting episodic conditions and in 1981, after a serious pollution episode (4), the Basque Government decided to sponsor a different approach to the problem.

2. EXPERIMENTAL PROGRAM

As its guiding concept, this program emphasized a detailed experimental characterization of the dispersion dynamics in the area as a previous step to any modelling effort. The main objective of the experimental work was to determine the most common dispersion scenarios and to establish cause-effect relationships between the industrial emissions and their diffusion and transport. To this effect, the SO_2 is used as a tracer of itself (5) and of the polluted airmass as a whole.

In order to collect data in real time and document all conditions, the experimental program relied on two instrumented vehicles to complement the monitoring network. Each of these carried a COSPEC SO_2 remote sensor, a fast response SIGN/X SO_2 monitor and a dual pen chart recorder. A speed proportional counter to advance the recorder linearly with distance travelled, and a stabilized power inverter completed the installation. The vehicles used were small vans intended to keep up with prevailing traffic and take their measurements while moving along their preassigned routes in the shortest possible time.

The instruments were carefully calibrated at regular intervals, and in the case of the COSPEC, its optical alignment and electronic settings were carefully checked to minimize its diurnal baseline drift and to maintain null response to clouds (6). This procedure allowed the taking of measurements in a systematic manner twice a week under all weather conditions.

Field days started with the deployment of the instrumented vehicle(s) as soon as the light level allowed for COSPEC operation. This occurs at about sunrise for clear days and 15 to 20 minutes later for overcast conditions. At this time the air flow is still considered representative of night flow conditions. Operations continued until near sunset. It takes approximately one hour to complete a circuit over the right margin of the estuary and 1.5 to 2 h for the left side. Figure 1 shows the road network used. The going run used the roads parallel to the waterway, and the return was made in a square wave fashion in order to document the flow behaviour across the valley axis. Six to ten circuits per day (depending on the season) can be completed by one vehicle. With two vehicles, this can be increased to more than double.

Collected records were digitized, filed and plotted over the road network map of the area, as shown in Figures 1 to 6. Shown values are the SO_2 overhead burden (COSPEC,s in ppm-m) and ground concentrations (ug/m3) over the vertical of each point of the route. The folding directions chosen are only intended to minimize interference or overlap between profiles.

Finally, for each day of measurements the data were grouped with those of the fixed network and meteorological maps at 0600, 1200 and 1800 GMT. These were all co-analyzed and classified into similar patterns of dispersion behaviour.

3. RESULTS

With over 120 days of field measurements, taken during three years, six main dispersion patterns have emerged. In general terms these are:

#1. **Stagnation or re-entry cycles.** - Occur preferably with a High NE of Spain and weak gradient winds of S to SE component. This situation is favoured during the period of migrating warm core anticyclones from early fall to late winter. It occurs approximately 25% of the time and, if the situation persists for a few days, episodic conditions (4) may develop.

During these periods, a nightime ground-based inversion forms in the bottom of the valleys by a combination of radiational cooling and the accumulation of cold downslope drainage wind. A first blocking of the stabilized airmass occurs over Bilbao due to the roughness and heat island effects of the city (4). A second stable airmass accumulates NW of the city and drains toward the sea with most of the industrial emissions enbedded in it. This airmass can also become blocked, by convective overturning (7), if the sea temperature is higher than that of the draining air. This situation occurs more frequently from early fall to late winter.

Solar heating the following day initiates a fumigation and a sea breeze that re-enters the stagnated airmass (4). In mid to late winter, blocking of the drainage towards the sea still takes place. Solar heating, however, is too weak to re-enter the airmass but can produce a classical Hewson-type fumigation near midday. In late spring and midsummer, surface sea temperatures can be lower than at the bottom of the airmass. This is conductive to the formation of a shallow marine inversion over which the drainage flow slides out to the open sea.

Figures 1a to 1d show a re-entry cycle in late July. The drainage flow can be seen in 1a. The start of the re-entry with extensive fumigation is shown in 1b. Further advance of the reentering airmass, with clearing at the seashore, can be seen in 1c, and the end of the re-entry cycle, developing into a type 5 pattern, is shown in Figure 1d.

#2. **Direct Channeling.** - Occurs approximately 21% of the time with a High West of the Peninsula and/or a Mediterranean Low, and it is usually accompanied by rain and strong winds. It can occur throughout the year but it is more frequent in spring and summer.

Figure 2 shows one example of direct channeled flow over the left side of the estuary. The plume of the power plant, located in the port area, and other plumes can be observed travelling directly inland. Its most characteristic aspect with respect to the concentration field is that the observed ground concentrations start with zero values at the coastline and increase inland in a nearly monotonic fashion.

#3. **Southerly Winds.** - Occur approximately 20% of the time, with a Low W to N of Spain. They can be quite persistent during springtime, if a low zonal index exists, and "cut off Lows" stagnate over the British Isles. Strong adiabatic warming of the air takes place upon descending nearly 800 m from the central plateau. With these warm winds, clear skies prevail but nightime inversion formation tends to be inhibited as well as the stagnation of the polluted airmass which slides out over the (colder) sea.

Figure 3 shows one example of the pronounced topographical effects that can take place under southerly wind conditions. The port power plant plume (~200m stacks) can be seen trapped in the wake of the 455m hill behind it.

#4. **Westerly Winds.** - Occur with WSW to WNW gradient winds. It is associated with rain and the passage or occlusion of cold frontal systems (Low traveling via Northern France or the British channel). It happens approximately 12% of the time. It is transitional, lasting usually less than one day, and involves important topographical effects.

These effects are shown in Figure 4. The plume from the same power plant is downwashed over the port area and coast. Once down it is rechanneled inland along the waterway (and also along the Asua basin) by the surface winds which are forced to follow the terrain. This gives rise to simultaneous high concentrations all the way from the coast inland as opposed to a type #2 pattern where the observed concentrations show a gradual increase inland from (essentially) zero at the shoreline.

This is a most important difference which helps to distinguish it from a type #2 pattern in the signals from the fixed monitoring network, since at ground level (~<10 m) recorded wind directions are essentially the same in both cases, i.e. a NW flow component forced by the surface channeling.

#5. **Northerly Winds.**- Occur with gradient flow from the NNW to NNE with a High W of Spain or after the establishment of a strong sea breeze, reinforced by the formation of an Iberian Thermal Low (8), in late summer (see Type #1 pattern and below). It happens ~11% of the time and also gives rise to important topographical effects, as shown in Figures 5 and 6. The port power plant plume is lifted over and entrained in the wake of the hill behind it. In these conditions, ground level (~<10 m) winds are also channelled along the estuary in a general NW direction.

#6. **Other situations.** - Including (strong) Easterly winds and transitory periods amounting to less than 11% of the time.

4. EFFECTS OF THE IBERIAN THERMAL LOW

One of the most striking meso-meteorological effects observed occurs when the Iberian thermal low forms over the central plateau (8). An example of the formation and relaxation of this low pressure area is shown in Figure 7 for July 27th, 1982. It has been observed that the centre of this Low forms over North Africa in late Spring and migrates to the Spanish central plateau during the summer. This situation will last into early fall if the year is very dry. Its effects on the turning of the plume from a tall stack in Teruel were fist noticed by one of our colleagues in 1979 (9).

During the examination of the Bilbao data it also became apparent that a number of re-entry cycles in late summer (17 cases) had a much stronger forcing than just the local sea-breeze. As Figures 1d, 5 and 6 show, the plume from the port power plant ends its re-entry cycle in a direction nearly perpendicular to the coast and flows over a 455m ridge and a ~700 m mountain barrier on its way inland.

A search on the evolution of surface winds in nearby airports and the insolation for July 27th, 1982, are shown in Figures 8 and 9. The first of them shows that the flows of Bilbao and San Sebastian can be channelled along the mountain passes to Vitoria, accounting for the "Gorbea winds" which blow in this city at, or after, noon during the summer.

A preliminary study of this situation indicates that this inflow starts as soon as the pressure gradient between the Northern coast and the center of the Low becomes of the order of 8mb/~300 km (12mb/~400 Km have been observed). This study also suggests that the inward flow is rather shallow, limited to between ~700 to 800m in height, and is restricted to natural mountain passes. This would account for the fact that Santander, backed by a more extense and higher mountain range (up to ~2000m), does not notice the pull.

Through this process, the coastal emissions can enter the Castilian region, in late summer, during the afternoon, with typical fetches of 100 to 150 Km. Once over the mountain ranges and into the central plateau, these plumes would stabilize and participate in the local drainage flows during the night. This leaves them ready for the inward flow the following day. In dry summers, when the formation of the thermal low is most frequent, this pull- relapse mechanism could account for the inflow of an aged air mass towards the centre of the Peninsula and could explain the occurence of poor visibility episodes in Madrid in late summer. Further research in this area is now underway.

5. DISCUSSION OF LOCAL EFFECTS

The data can be used to explain some of the reasons for the "anomalous" situations reported by previous users of the monitoring network data.

In the first place it should be noted that, as a result of channelling, upper winds from W through N to NNE are observed as a general NW flow at the surface (including the airport in the Asua basin). The total percentage of NW surface winds is ~56%, as a result of: true NW upper flow (~21%), channeled upper Westerly (~12%) and northerly (~11%) flows, and ~12% for the sea breeze leg of the re-entry cycle (pattern #1).

Expected results from a Gaussian type formulation (10) with - D - stability, assuming all of these situations as true NW winds, indicate that the ground impingement of the port power plant emissions would start at ~14 to 16(+) km from the source and have little effect on the fixed network. It must also be indicated that: (a) unstable classified situations coincide with either Southerly winds -and flow out to the sea- or with the sea breeze period of the re-entry cycle, when those higher emissions stay most frequently in the return upper flow, and (b) this is a peak load plant with discontinuous operation which complicates matters further.

If the plant is not in operation, patterns 4 and 5 merge into a generalized # 2 pattern with minor variations as a function of the upper wind directions. As soon as the plant is operating, however, a splitting of this situation takes place. For the true NW flow (pattern type 2) the impact is as expected. However, things change significantly as soon as the upper winds are from the W or N, as patterns #4 and #5 show. It is obvious, therefore, that **any** modelling effort must include allowances for these topographically induced effects whenever the power is "on" and the upper winds have a cross-valley component.

This analysis is intended to show that the systematic use of moving laboratories equipped with remote sensors can be used to: (a) characterize the flow patterns in complex topographical settings, (b) help diagnose existing situations, and (c) guide the understanding of the dynamic processes involved for the development and/or application of dispersion models.

ACKNOWLEDGEMENTS

The experimental part of this work was financed by the Environmental Vicecouncelry of the Basque Goverment and the authors also thank Dr. Lucio Alonso for his helpful comments during the preparation of the manuscript.

A large number of people in addition to the authors have contributed to this work, particularly in the important task of taking the data and spending long days chasing plumes all about the place. We would therefore like to acknowledge the help of: Erasmo Otamendi, Pedro Luis Arias, Gotzon Gangoiti, Inés Apraiz, Fernando Fernandez, Fernando Gutierrez and last but not least, "Captain" Valentin Zabala.

REFERENCES

1. D.G. Fox, J.E. Fairobent, " NCAQ Panel Examines Uses and Limitations of Air Quality Models," Bull.Am.Meteorol.Soc.**62**:218 (1981).
2. R. Wilson, F. Shair, B.Reynolds, W. Greene," Characterization of Transport and Dispersion of Pollutants in a Narrow Mountain Valley Region by Means of an Atmospheric Tracer,"Atmos.Environ.**17**:1633 (1983).
3. B.W. Atkinson, "Meso-scale Atmospheric Circulations," Academic Press., New York.(1981).
4. M.M. Millán et al., "A fumigation episode in an industrialized estuary: Bilbao, November 1981," Atmos.Environ.**18**:563 (1984).
5. M.M. Millán, R.H.Hoff, "Remote Sensing of Air Pollutants by Correlation Spectroscopy: Instrumental Response Characteristics," Atmos. Eviron., **17**:852 (1978).
6. P.Camagni, S.Sandroni(Edts.),"Optical Remote Sensing of Air Pollution," Elsevier Science, New York.(1984).
7. G.J.McRae, F.H.Shair, J.H.Seinfeld, "Convective Downmixing of Plumes in a Coastal Environment," J.Appl.Meteorol.**20**:1312 (1981).
8. R. G. Barry, R. J. Chorley, "Atmosphere, Weather and Climate," Methuen, London,(1982).
9. L.A.Alonso. Personal communication on the results of the 1979 "Teruel" Power Plant plume survey.
10. D.B. Turner,"Workshop of Atmospheric Dispersion Estimates,"U.S.Environmental Protection Agency Publication AP-26,(1970).

FIGURE CAPTIONS

Figure 1.- A Re-entry cycle as documented by the data from the moving laboratories equipped with remote sensors. (a) General drainage towards the port.- (b) Massive re-entry of the polluted airmass as documented by the moving laboratory on the right side od the estuary. With the onset of the seabreeze the airmass is drawn inland and any plume individuality is lost. -(c) One hour later the sea breeze has pushed the pollution further inland and the coastline is now clear.- (d) With a well developed seabreeze and the initial airmass pushed further inland, some isolated plumes can now be observed in a general S or SE direction.

Figure 2.- Observed pattern with true NW winds. The port power plant plume travels directly inland along the estuary. Results from the moving laboratories show that ground concentrations are very low at the shoreline and grow progressively inland.

Figure 3.- With strong southerly winds, the plumes from the port power plant (~200 m stacks) are trapped in the turbulent cavity of the 455 m hill located behind it. The graphs show this effect as recorded by one of the moving laboratories.

Figure 4.- Downdrafting of the port power plant plume under (upper) West winds (NW at the surface). High ground concentrations prevail along the estuary due to the channelling of the surface flow.

Figures 5 and 6.- Observed pattern with (upper) N winds on July 27, and August 24, 1982. The power plant plume is lifted over and partially entrapped on the wake of the 455 m hill in its way inland.

Figure 7.- Surface pressure charts, showing the formation and relaxation of the Iberian Thermal Low for July 27, 1982.

Figure 8.- Hourly surface wind vectors in Bilbao, San Sebastian, Santander and Vitoria airports on July 27, 1982. Drainage flow prevails in Bilbao until the onset of the seabreeze at ~1100 h. This situation was confirmed by the data of the moving laboratories. San Sebastian shows onshore (N) winds from 0800 h onwards, veering to NE by 1400 h. In Vitoria, the wind blows towards the central plateau from 0800 h and shows a substantial increase in speed after the onset of the seabreezes in San Sebastian and Bilbao. Both of these airsheds are connected to the Vitoria area by mountain passes lower than 680 m. Santander, on the other hand, is backed by an extensive massif, ~100 Km wide and its connection to the central plateau is through three passes which reach heights above 1000 m. The air motions in this city do not appear to participate in the inland flow.

Figure 9.- Hours of insolation, maximum inland air temperatures and sea surface temperatures for July 27, 1982.

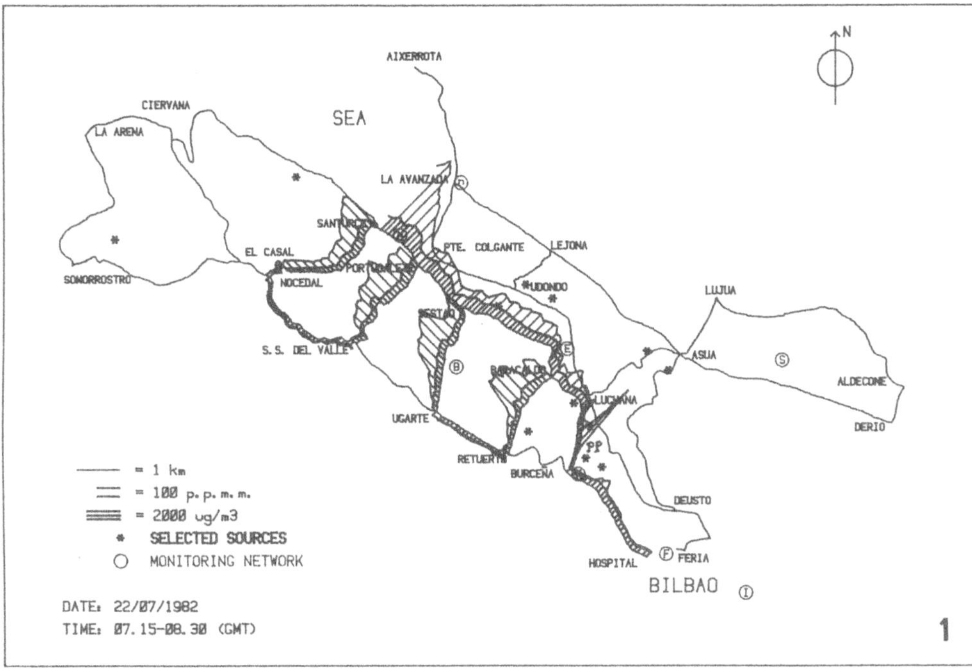
N
AIXERROTA
SEA
CIERVANA
LA ARENA
LA AVANZADA
SANTURCE
EL CASAL
PORTUGALETE
PTE. COLGANTE
LEJONA
SOMORROSTRO
NOCEDAL
UDONDO
LUJUA
SESTAO
S. S. DEL VALLE
ASUA
ALDECONE
BARACALDO
LUCHANA
DERIO
UGARTE
RETUERTO
BURCEÑA
DEUSTO
HOSPITAL
FERIA
BILBAO
= 1 km
= 100 p. p. m. m.
= 2000 ug/m3
SELECTED SOURCES
MONITORING NETWORK
DATE: 22/07/1982
TIME: 07.15-08.30 (GMT)
1

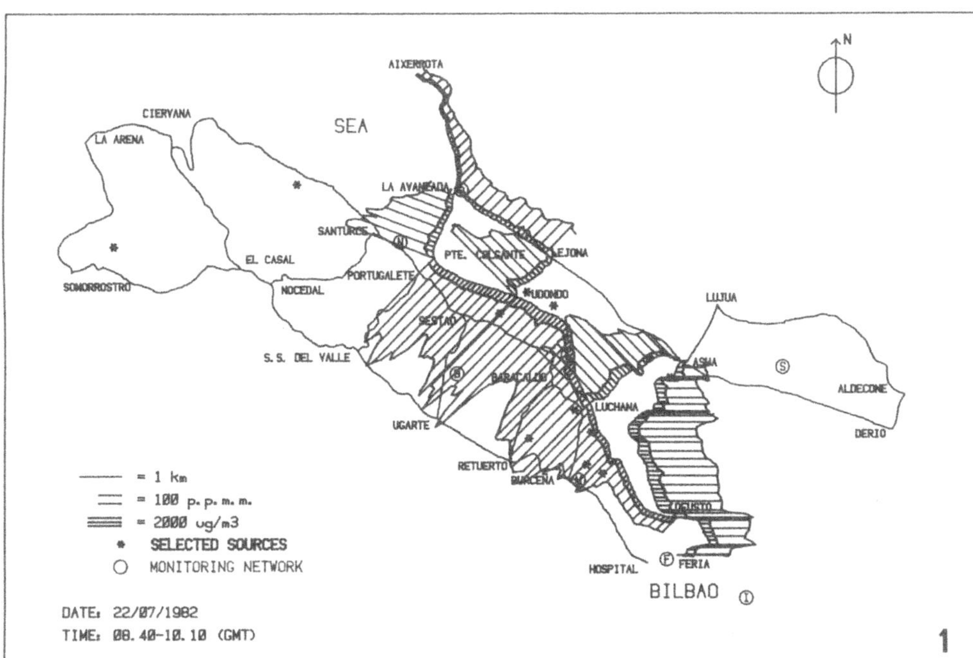
N
AIXERROTA
SEA
CIERVANA
LA ARENA
LA AVANZADA
SANTURCE
EL CASAL
PTE. COLGANTE
LEJONA
PORTUGALETE
SOMORROSTRO
NOCEDAL
UDONDO
LUJUA
SESTAO
S. S. DEL VALLE
ASUA
ALDECONE
BARACALDO
LUCHANA
DERIO
UGARTE
RETUERTO
BURCEÑA
DEUSTO
HOSPITAL
FERIA
BILBAO
= 1 km
= 100 p. p. m. m.
= 2000 ug/m3
SELECTED SOURCES
MONITORING NETWORK
DATE: 22/07/1982
TIME: 08.40-10.10 (GMT)
1

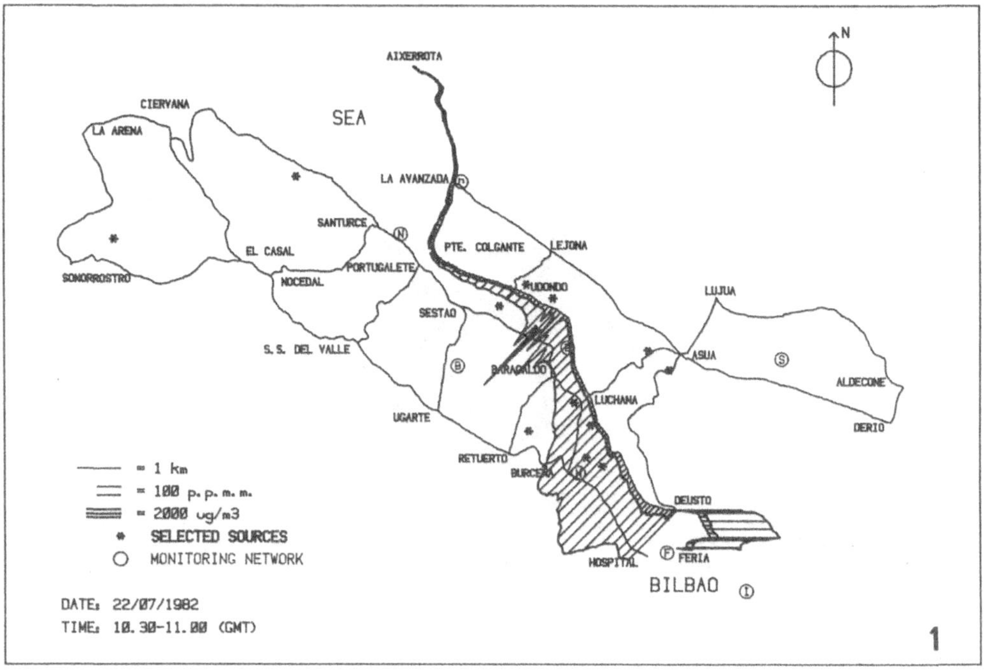
AIXERROTA
N
CIERVANA
LA ARENA
SEA
LA AVANZADA
SANTURCE
EL CASAL
SONORROSTRO
PORTUGALETE
NOCEDAL
PTE. COLGANTE
LEJONA
UDONDO
LUJUA
SESTAO
S. S. DEL VALLE
ASUA
BARACALDO
ALDECONE
LUCHANA
UGARTE
DERIO
RETUERTO
BURCENA
DEUSTO
FERIA
HOSPITAL
BILBAO
= 1 km
= 100 p. p. m. m.
= 2000 ug/m3
SELECTED SOURCES
MONITORING NETWORK
DATE: 22/07/1982
TIME: 10.30-11.00 (GMT)
1

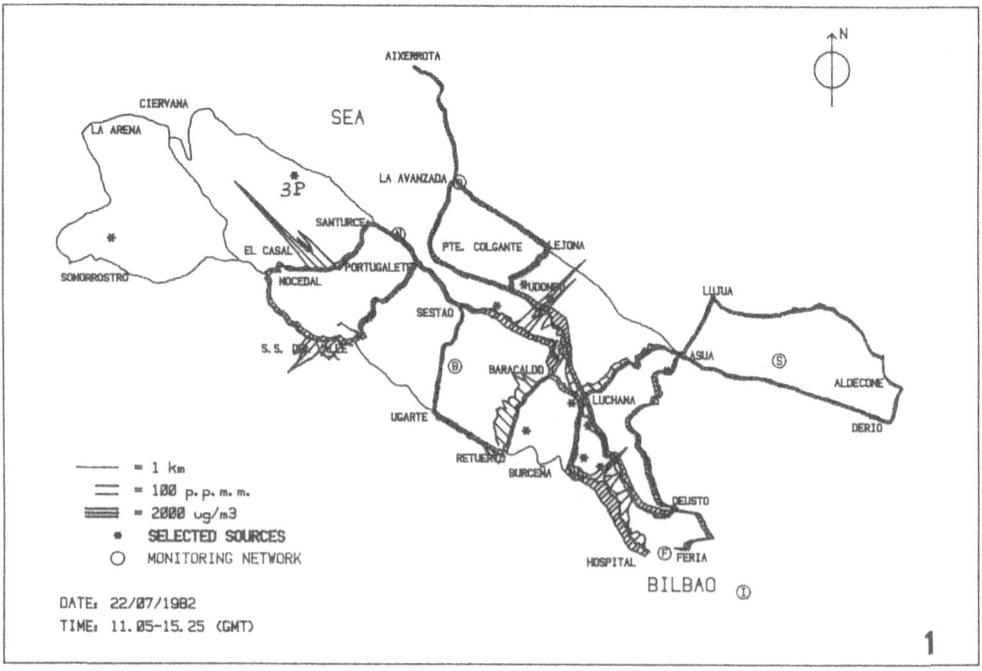
AIXERROTA
N
CIERVANA
LA ARENA
SEA
3P
LA AVANZADA
SANTURCE
EL CASAL
PTE. COLGANTE
LEJONA
SONORROSTRO
PORTUGALETE
NOCEDAL
UDONDO
LUJUA
SESTAO
S. S. DEL VALLE
ASUA
BARACALDO
ALDECONE
LUCHANA
UGARTE
DERIO
RETUERTO
BURCENA
DEUSTO
HOSPITAL
FERIA
BILBAO
= 1 km
= 100 p. p. m. m.
= 2000 ug/m3
SELECTED SOURCES
MONITORING NETWORK
DATE: 22/07/1982
TIME: 11.05-15.25 (GMT)
1

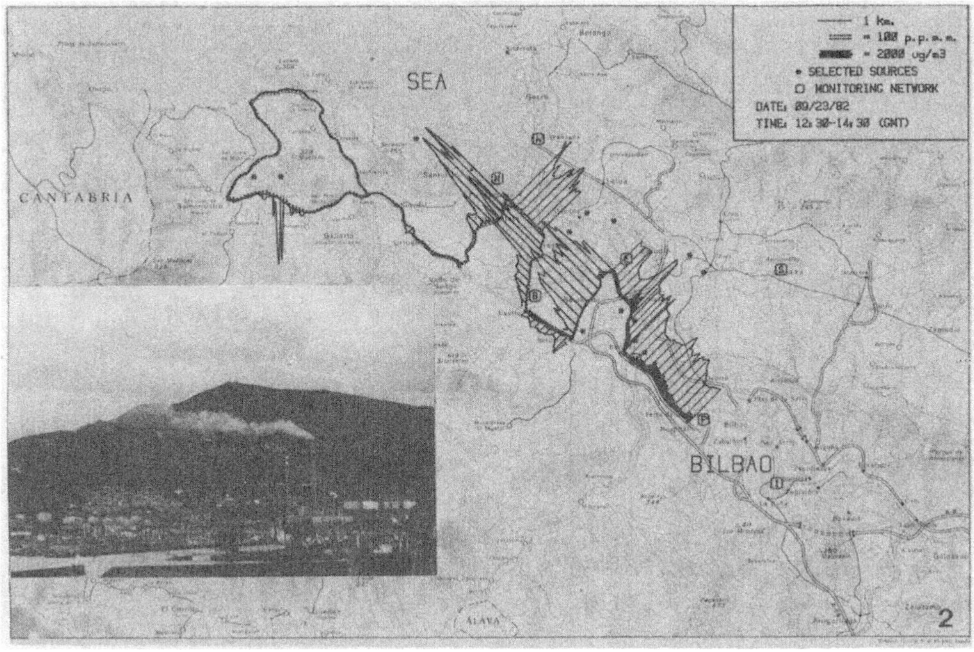

1 km.
= 100 p.p.m.m.
= 2000 ug/m3
SELECTED SOURCES
MONITORING NETWORK
DATE: 09/23/82
TIME: 12:30-14:30 (GMT)
SEA
CANTABRIA
BILBAO
ALAVA
2

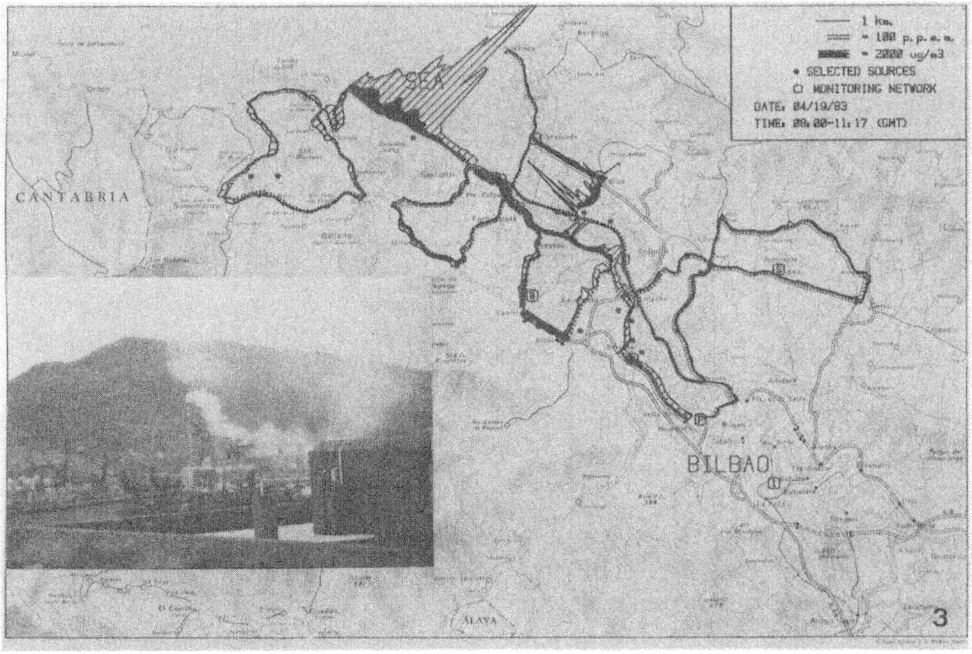

1 km.
= 100 p.p.m.m.
= 2000 ug/m3
SELECTED SOURCES
MONITORING NETWORK
DATE: 04/19/83
TIME: 08:00-11:17 (GMT)
SEA
CANTABRIA
BILBAO
ALAVA
3

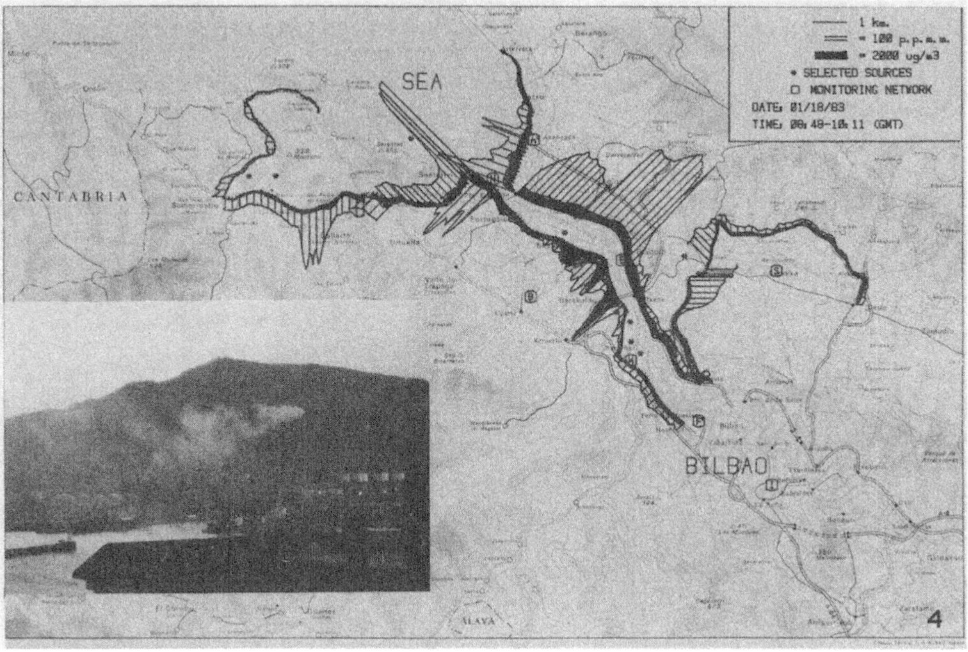
1 km.
= 100 p.p.m.m.
= 2000 ug/m3
* SELECTED SOURCES
□ MONITORING NETWORK
DATE: 01/18/83
TIME: 08:48-10:11 (GMT)
SEA
CANTABRIA
BILBAO
ALAVA
4

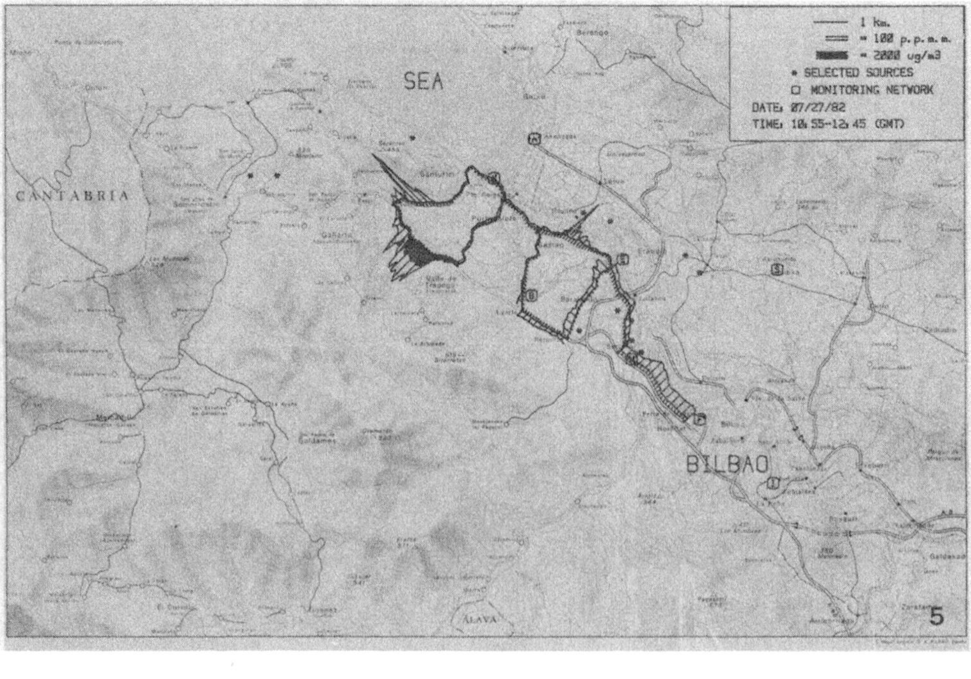
1 km.
= 100 p.p.m.m.
= 2000 ug/m3
* SELECTED SOURCES
□ MONITORING NETWORK
DATE: 07/27/82
TIME: 10:55-12:45 (GMT)
SEA
CANTABRIA
BILBAO
ALAVA
5

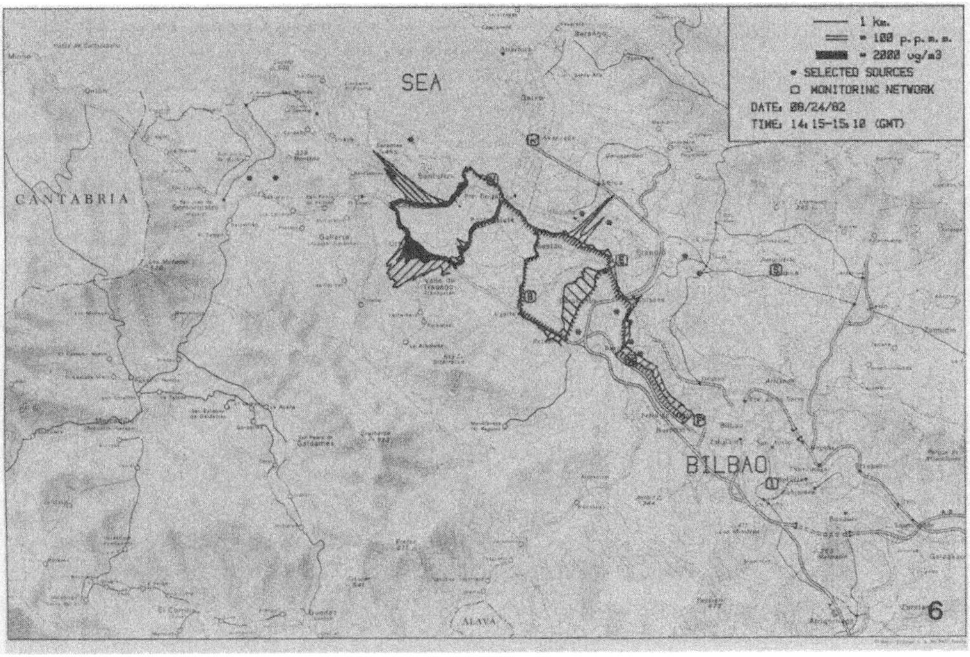

6

CHART SURFACE - 27-07-82

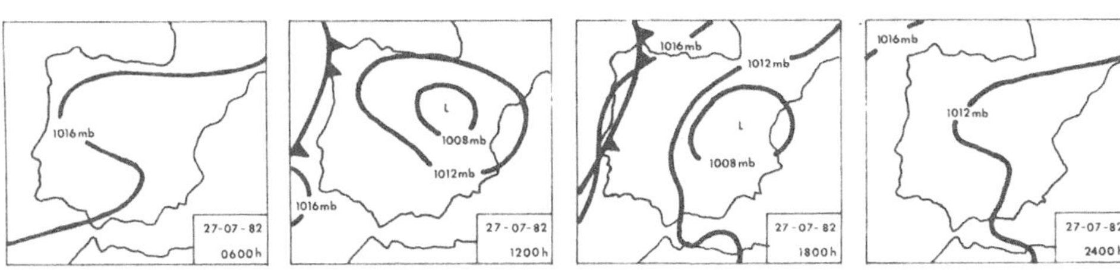

7

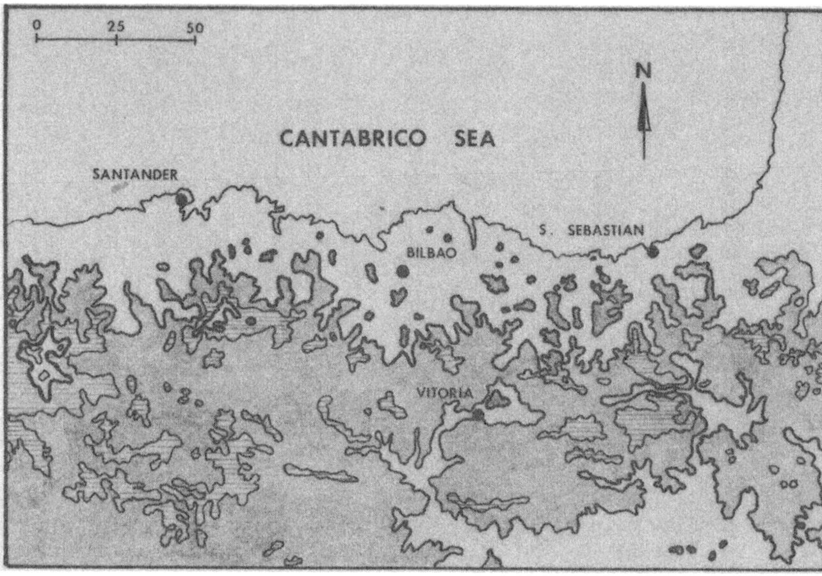

0 25 50
N
CANTABRICO SEA
SANTANDER
S. SEBASTIAN
BILBAO
VITORIA

2m/s
27-07-82
SANTANDER
S. SEBASTIAN
BILBAO
VITORIA

8

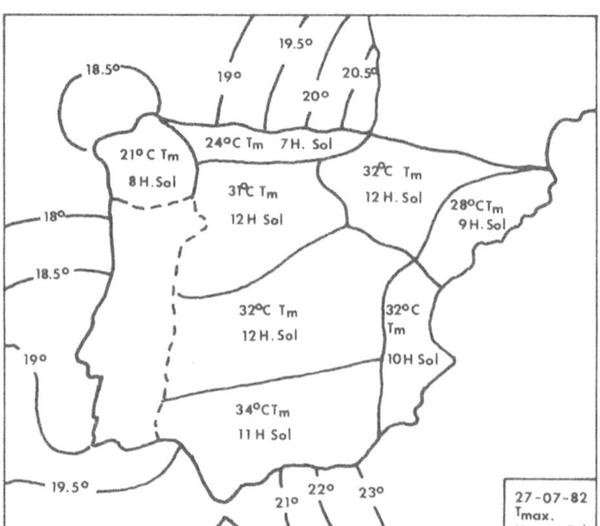

18.5°
19°
19.5°
20°
20.5°
21°C Tm
8H. Sol
24°C Tm 7H. Sol
31°C Tm
12H Sol
32°C Tm
12H. Sol
28°C Tm
9H. Sol
18°
18.5°
32°C Tm
12H. Sol
32°C
Tm
10H Sol
19°
34°C Tm
11H Sol
19.5°
21°
22°
23°
27-07-82
Tmax.
Horas Sol

9

WORKING PARTY 5
TRANSPORT AND MODELLING

Chairman: P. BESSEMOULIN

Model calculations of long-term changes in atmospheric hydroxyl, hydrogen peroxide and ozone from changes in the emissions of nitrogen oxides and methane

A mass balance of SO_2 over Baden-Würtemberg during the tulla experiment

Mesoscale modelling of nitric acid generation in night-time moist atmosphere

Long range modelling of the formation transport and deposition of photochemical oxidants and acidifying pollutants

Analysis of an acid deposition episode and preparation of meteorological input data for an eulerian longrange transport and deposition model

Physico-chemical properties of radiation fog in the Po Valley, Italy

Air sampling flights at low altitudes along the border between the Federal Republic of Germany and its neighbours

The oxidation of sulphur dioxide in the aqueous phase in plumes

Monte-Carlo methods for simulating diffusion in the atmospheric boundary layer

Modelling for langrangian turbulent dispersion of neutral and buoyant pollutants released in the atmospheric boundary layer

A three-dimensional mathematical model of dispersion in turbulent shear flow

The rapid rotation of the wind direction and sharp ozone peak features

Local scale plume model for nitrogen oxides - comparison with measurements

High PAN concentrations during non-summer periods : a study of two episodes in Creteil (Paris) - France

Verification of the pollutant transport model 'MODIS' using EPRI plains site data from a tall stack

Long-range transport of nitrogen and sulphur compounds

Modelisation de la chimie d'un panache

Echantillonnage et mesures sequentiels de pluies: conception d'un appareil équipé d'un automate programmable (SAPSEP)

SUMMARY BY THE CHAIRMAN

P. BESSEMOULIN
National Meteorological RC, F

Study of the transport and diffusion of atmospheric pollutants and the physical and chemical changes they undergo, falls into two complementary categories, i.e. field experiments and computer modelling. The work summarized here was presented orally and by means of posters at the Symposium and the Workshop held in Bilthoven from 23 to 25 September 1985, organized jointly with Cost Action 611 Working Party 4 (Pollutants cycles).

Two points should be noted as regards the general approach adopted by Working Party 5:

- the participation of several scientists involved in the large-scale international cooperation programmes (EMEP, PHOXA, TULLA etc.), which tends to enhance the integrating role of Cost 611 in the field of atmospheric pollutant transport modelling and empirical studies;

- the increasing attention devoted to the study of meso-meteorological geographical and/or time scales: almost half the talks and posters presented here concern these questions.

This concern should be seen in the light of the current interest shown by the national meteorological departments in the development of fine-mesh models which can be initialized on the basis of data of various origins and are aimed at providing better descriptions of the current weather conditions or improved short-term forecasts.

Work in recent years has generally fallen into the following two categories:
local scale;
meso-synoptic scale.

Local-scale studies have concentrated on the following three areas :

- modelling the transport and diffusion of passive pollutants, account being taken of the detailed structure of the atmospheric boundary layer. Methods have been proposed to make up for the inadequacies of the Gaussian models or to determine the relevant parameters objectively on the basis of the results of elaborate models;

- the study and modelling of reactive plumes, with particular reference to photooxidants and acid compounds, in dry or satured atmospheres;

- tracking and determination of the physical and chemical composition of the wet-deposit vectors, i.e. precipitation and fog.

The studies of phenomena involving the transport of pollutants over medium or great distances (transfrontier flows) can also be broadly divided into three groups:

- those concerned with determining the physical and chemical properties of the pollutants present at a given location in the light of more or less qualitative considerations based on a trajectory calculation;
- studies using models describing in relative detail the interactions between the discharge dynamics and the physico chemical and micro physical parameters involved. In the most recent work in this area account has increasingly been taken of the wide range of reactions which bring about the aqueous phase (which in turn ipso facto entails a detailed description of this phase).

The ultimate aim of these models is always the same, however complex they may be. Once they have been validated, they can help in deciding on a strategy for monitoring emissions on a European scale.

- empirical studies of transport, mainly by means of aircraft fitted out with appropriate instruments.

We can already draw a number of general conclusions regarding the various tools:

- we now have models which permit both concentrations and deposits (photooxidants and acid compounds) to be estimated on a climatological or episodic basis, to an often acceptable degree of accuracy;

- the importance of models for determing a strategy for emission monitoring is broadly recognized in view of the chemistry involved and its highly non-linear aspects. The need to consider photooxidants and acid compounds in conjunction has been clearly established;

- an aircraft fitted with the appropriate instruments is the most useful of the various experimental tools;

- the most sophisticated models require input data which are not all routinely available, since they demand, for example, a combination of traditional meteorological data and data relating to the fields analysed and/or scheduled for analysis, together with satellite information.

However, we schould not forget the less positive aspects such as the following:

- our understanding of even the climatology of certain compounds (e.g. O_3, NO_x etc.) is extremely limited or non-existent at the European scale;

- empirical measurement, in a network, of dry deposits still poses a challenge to the scientific community, although these account for the major proportion of the total deposits - and hence play an important role in the models;

- there are currently no high quality sets of meso-scale data available concerning all the relevant meteorological, physicochemical and microphysical parameters. Experiments such as TULLA can serve as a model for the major experimental programmes which will be called for in the next few years and which will probably require considerable involvement at the design stage by the people responsible for producing the models, since a model should, if possible, indicate what must be measured, where and how often.

In view of the spatial and temporal scales and the wide range of techniques involved (ground network, airborne equipment, remote sensing etc.), international cooperation will be essential if such projects are to be completed on a European scale.

Finally, we should stress the vital and complementary role of studies concerning the chemistry of the atmosphere, in view of the observed and/or modelled behaviour of certain substances which could have a major climatic impact, as in the case of radioactive gases (CO_2, O_3, CH_4 etc)., or chemical impact, such as the very probable increase of hydroxyl radicals resulting from the steady increase in NO_x emissions.

MODEL CALCULATIONS OF LONG-TERM CHANGES IN ATMOSPHERIC HYDROXYL, HYDROGEN PEROXIDE AND OZONE FROM CHANGES IN THE EMISSIONS OF NITROGEN OXIDES AND METHANE

Ø. HOV* and I.S.A. ISAKSEN**
*) Norwegian Institute for Air Research,
P.O. Box 130, N-2001 Lillestrøm, Norway
**) Institute of Geophysics, University of Oslo
P.O. Box 1022, Blindern, N-0315 Oslo 3, Norway

SUMMARY

Model calculations using a 2-dimensional, meridional model of the global troposphere show that an increase in the emissions of NOx or CH_4 or both, would enhance the conversion rate of NOx to nitrate more than the conversion rate of SO_2 to sulphate in the lower part of the troposphere at mid latitudes in the northern hemisphere where the majority of anthropogenic NOx- and SO_2-emissions take place.

1. INTRODUCTION

Sulphur and nitrogen oxides are converted into acids both through gas phase and liquid phase processes in the atmosphere. The reactions of the hydroxyl free radical (OH) with SO_2 and NO_2 are the major gas phase sources of sulphuric and nitric acids:

$$OH + SO_2 \rightarrow HOSO_2 \quad (1)$$

$$OH + NO_2 \rightarrow HONO_2 \quad (2)$$

where the $HOSO_2$ radical product goes through a chain-carrying step (Cox, 1974; Stockwell and Calvert, 1983)

$$HOSO_2 + O_2 \rightarrow HO_2 + SO_3 \quad (3)$$

and the SO_3-product reacts rapidly with water vapour to generate sulphuric acid:

$$SO_3 + H_2O \rightarrow H_2SO_4 \quad (4)$$

The rate constant for the OH radical reaction with NO_2 is about 10 times that for the SO_2 reaction.

The concentration of OH depends on the pollutant level and composition with solar radiation as the source for energy. The concentration is low at night and during the winter with little sunshine.

When species like H_2O_2 and O_3 are dissolved in cloud water, they readily oxidize dissolved SO_2. The H_2O_2-mechanism is rather insensitive to the acidity of the water, while the ozone-mechanism is very slow at low pH. Gaseous hydrogen peroxide is a major oxidant leading to sulphuric acid generation in cloud water (Penkett et al., 1979). Metal catalysis is probably not important for the formation of sulphate in regional air masses over Europe. This was concluded by Cox and Penkett (1982) from 20 years of records reported by Salmon et al. (1978) of manganese, iron, sulphate and nitrate in aerosol measured at Harwell, with a substantial decline in the concentration of Mn and Fe at the same time as sulphate showed no trend in the winter and an upward trend close to 4% per year in the summer.

The aerosol measurements reported by Salmon et al. (1978) from Harwell showed an increase in sulphate and nitrate for the time period 1957-1974. The annual mean concentration of SO_2 at Birkenes in South Norway has declined strongly since 1978, while the annual mean concentrations of sulphate in air and precipitation and nitrate in precipitation do not demonstrate a similar decline (SFT, 1984).

Measurements of sulphate and nitrate in snow from South Greenland 1895-1978 have shown an increasing trend in the deposition of both compounds with a large year-to-year fluctuation (Neftel et al., 1985).

Measurements of glacial acidity in the Canadian Arctic indicated little variation in the magnitude of air pollution in the first half of this century but a marked increase by about 75% between 1956 and 1977 (Barrie et al., 1985). This increase is probably linked to increasing long-range transport of products from the combustion of fossil fuels.

The measurements of ions in snow and ice in the Arctic referred to above, indicate that there was a substantial increase in acid deposition during a time period when the emissions of SO_2 and NOx in North America and Europe including the Soviet Union, increased significantly (1950-1975).

Over the last years observational evidence has been put forward of growing concentrations of O_3, CH_4 and perhaps CO. For the period 1978-1983 Blake and Rowland (1986) estimated the global CH_4-increase to be 1.1% per year. Recently, the concentration of CH_4 in the troposphere was deduced from infrared solar spectra recorded at Jungfraujoch in Switzerland in 1951, and it was found that the mean CH_4-mixing ratios had increased by 1.1 ± 0.2% per year for the time period 1951-1981 (Rinsland et al., 1985).

Increasing concentration of CO at some Northern Hemisphere locations has been reported. Khalil and Rasmussen (1984) believed that 3 1/2 years of data from Cape Meares, Oregon, showed about 6% per year increase in CO. Dianov-Klokov and Yurganow (1981) reported data indicating a 2% per year increase in winter time CO at northern temperate latitudes, while the concentration seems to have remained constant in the summer. According to the authors the increase could be connected to the anthropogenic emission of CO, which they claim goes up by 4% per year. Using similar methods as for CH_4 (infrared spectroscopy), Rinsland and Levine (1985) estimated an average increase of 2% per year in the free tropospheric concentration of CO over Europe for the period 1950-1977. Fraser et al. (1984) could not detect trends in the CO concentration in the Southern Hemisphere from measurements at Cape Grim, Tasmania.

In a comprehensive review of the seasonal behaviour of tropospheric ozone, as well as of trends and anthropogenic influence, Logan (1985) concluded that there is evidence for an increase in ozone in the middle troposphere over Europe over the past 15 years. The increase over North America and Japan is less obvious. The upward trend in ozone over Europe is mainly judged from the record of ozone sonde observations over Hohenpeißenberg in Bavaria (Attmannspacher et al., 1984) and on the results from the ozone network in the German Democratic Republic (Warmbt, 1979). The ozone sonde records indicate a 50-70% increase in the concentration of ozone above the boundary layer from 1967-1982.

Changes in the emissions of SO_2, NOx, CO and hydrocarbons both on a regional and global scale over the last decades have contributed to the long-term changes in the atmospheric concentration of trace gases. It has been shown through model calculations that the increase in atmospheric methane may be explained in parts through increasing direct emissions and in parts by the lowering of atmospheric OH as methane goes up (Isaksen and Hov, 1986). In this paper will be discussed how the oxidation mechanisms of SO_2 to sulphate and NOx to nitrate may have changed over the last decades and may change in the future, in the global atmosphere due to the effect of changes in the emissions of NOx and CH_4 on OH, H_2O_2, O_3 and the NO_3-radical.

2. MODEL DESCRIPTION

The model was a two-dimensional, meridional model with 19x20 grid points. It was developed by Isaksen and Rodhe (1978). In the vertical two different grid steps were used: 250 m up to 3.25 km, and 1 km from there and up to 17 km. In the horizontal the grid size was 10^0 latitude. The field of mean motion was taken from Newell et al. (1972). The coefficients of turbulent diffusion were adopted from Hidalgo and Crutzen (1977), with some modifications. In the boundary layer, the vertical diffusion coefficients reported by Isaksen and Rodhe (1978) were used. Across the equator in the middle and upper troposphere the horizontal diffusion coefficients were increased by a factor of 3 compared to the coefficients reported by Hidalgo and Crutzen (for heights above 4 km and between 20^0N and 20^0S latitude). This brought the K_y-coefficients closer to more recent values reported by Plumb and Mahlman (1985).

The boundary conditions were zero horizontal flux for all species across the vertical boundaries at 85^0S and 85^0N. At the lower boundary, fluxes of species were given either as emissions from natural or anthropogenic sources, or as ground removal. At the top boundary, fluxes were determined by stratospheric transport into the troposphere as a function of season and latitude. In Table 1, the most important upper and lower boundary conditions are summarized.

Table 1: Upper and lower boundary conditions in the meridional 2-D model of the global troposphere. +: indicates a source for the troposphere, -: loss, and 0: no influence (zero flux).

Boundary	Species	CO, NMHC[1]	O_3	NO	NO_2	PAN	HNO_3	CH_4	Other species
Top		0	+	+	+	0	+	-[2]	0
Surface	Emissions	+	0	+	+	0	0	+	0
	Deposition	≈0	-	0	-	-	-	0	0

[1] Nonmethane hydrocarbons.
[2] 5% of the surface source

2.1 Chemistry

Gas chemical processes, precipitation, interaction with particulate matter and the ground, and emissions of organic gases, nitrogen oxides, sulphur dioxide and other gases from natural and anthropogenic sources determine the distribution of atmospheric trace gases. In the chemistry calculation of ozone, hydroxyl, hydrogen peroxide, and the other secondary species, the anthropogenic emissions of nonmethane hydrocarbons were divided into groups of species which were represented by C_2H_6, n-C_4H_{10}, n-C_6H_{14}, C_2H_4, C_3H_6 and m-xylene. The volume fractions by compound were 2/9, 2/9, 1/9, 2/9, 1/9 and 1/9, respectively. The non-methane natural hydrocarbons are confined mainly to the atmospheric boundary layer at low latitudes. Only isoprene of the non-methane natural hydrocarbons was assumed to be of importance for the gas phase chemistry, and of the products of the initial reaction, 60% were assumed to go through further gas reactions. The rest was assumed to be made up by photochemically rather long-lived oxygenated intermediates like alcohols and peroxides, which may be removed by tropical rain (Zimmermann et al., 1978).

The photochemistry of the mixture of species resulting from emissions of HC, NOx and CO in tropospheric air followed to a large extent previously published schemes (see Isaksen et al., 1985 for further references). The calculation of photochemical processes, inclusion of an average cloud cover and numerical procedures were described in the same paper. Dry deposition of O_3, NO_2, PAN and HNO_3 was treated in terms of a deposition velocity, and a simple parameterization was used to derive the deposition velocity to be used at the lowest grid point (250 m) from the velocity at the reference height (1 m), see Isaksen et al., (1985) and Isaksen and Rodhe (1978). The deposition velocities were weighted by the fraction of land and sea in each latitudinal zone of the model, and there was furthermore a shift with season due to a change in surface with much lower deposition velocities during the winter.

Scavenging by cloud or precipitation was of interest mainly for HNO_3 in this model, and a hemispheric mean value of $5x10^{-6}$ s^{-1} was used as the first order decay rate coefficient during summer. During winter the value was reduced by a factor of 4. The scavenging rate coefficient had a maximum around 50^0N and 3 km height falling off to the north (Isaksen and Rodhe, 1978).

3. EMISSIONS

Direct anthropogenic release of CO through the burning of fossil fuel was taken to be 640 Mt/yr (Zimmermann et al., 1978). Indirectly, there is a significant production of CO through the degradation of CH_4, isoprene and anthropogenic, nonmethane hydrocarbons. The sources of methane were taken to be 450 Mt/yr, with 100 Mt/yr from fossil fuel, 100 Mt/yr from swamps and fresh water lakes, 100 Mt/yr from ruminants, 60 Mt/yr each from rice paddies and biomass burning, and the rest from oceans, tundra and termites.

The calculated concentration of CH_4 is discussed by Isaksen and Hov (1986). At 1 km height the average mixing ratio was 1.74 ppm in the northern and 1.665 ppm in the southern hemisphere, with an interhemispheric concentration ratio of 1.06. The calculated concentrations were approximately 5% higher than the measured average concentrations at the NOAA-GMCC stations (1.65 ppm for NH and 1.55 for the SH), but it should be kept in mind that the calculated CH_4-concentrations are steady state values obtained after 40 years of integration with present day's emissions and OH-concentration distribution (no feedback from CH_4 to OH in this case). Present atmospheric CH_4-levels are not in a steady state, as it will take more than 10 years to reach steady state if emissions and loss processes were kept constant. The calculated interhemispheric concentration ratio is consistent with the measurements made during Gametag in 1978 (Heidt et al., 1980) and the measurements reported by Singh and Salas (1982).

The global emission of NOx was taken as 29 Mt(N) per year, with 20 Mt(N) per year from anthropogenic sources, 4 Mt(N) per year from lightning and 5 Mt(N) per year from biomass burning. An additional source for the troposphere of NOx at the tropopause was assumed in the model calculations.

Anthropogenic, nonmethane hydrocarbon emissions were taken to be 90 Mt per year, assuming 40% coming from automobile exhaust, 30% from solvent usage and 30% from natural gas exploitation and distribution (Isaksen et al., 1985).

The NMHC/NOx ratio for the global, anthropogenic emissions is about 1.35 with 90 Mt and 20 Mt(N) as annual NMHC and NOx-emissions, respectively, when NOx is expressed as NO_2 in calculating the ratio.

The emissions of isoprene were taken to be 340 Mt/y, as estimated by Zimmermann et al. (1978). The isoprene emissions influence the photochemistry in the tropics considerably, while the impact at higher latitudes is slight. Terpenes are assumed to be of less importance for global tropospheric gas chemistry, and are not treated in the model calculations.

4. CALCULATIONS OF CONCENTRATION CHANGES OVER SEVERAL DECADES

The results with present day's emissions were discussed in some detail by Isaksen and Hov (1986). For methane, the calculated interhemispheric concentration ratio was 1.06 with an average mixing ratio of 1.74 ppm in the northern and 1.665 ppm in the southern hemisphere at 1 km height at steady state with present day's emissions. For ozone there was a maximum at 40^0 and 50^0N latitude in the calculated distribution for late spring with a minimum at low latitudes which was less pronounced than what can be seen in measured meridional ozone profiles.

For CO, for late spring the interhemispheric concentration ratio at mid latitude was about 5, and the maximum concentration in the northern hemisphere was a little more than 200 ppb. The calculated interhemispheric ratio was somewhat higher than the ratio estimated on the basis of measurements at Cape Grim (41^0S) and Cape Meares (45^0N) (Fraser et al., 1986; Rasmussen and Khalil, 1982). Observations in the free troposphere indicate a somewhat higher concentration level than what is calculated (Heidt et al., 1980), indicating that the sources of CO in the model are underestimated.

The lifetimes of OH, HCHO, PAN and NOx (=NO+NO_2) are short, and the concentration of these species reflects the local conditions in terms of emissions and meteorology. The chemistry is furthermore non-linear. This means that the concentrations calculated in a zonally averaged model is different from the more realistic concentrations calculated from a model covering specific longitude bands and then averaged over all longitudes. Isaksen et al. (1985) found that in oceanic air with very little NOx, OH in the lower atmosphere may be reduced by perhaps 50% compared to the calculated concentration in a zonally averaged model.

Long-term calculations were carried out with 1980 as the base year for two different cases: The anthropogenic NOx-emissions increasing by 3%/yr, that is the total NOx-emissions in the model increasing by 20/29 x 3%/yr ≈2.1%/yr, and for the methane concentrations increasing by 1.5%/yr. The calculations were speeded up by computing explicitly only the first year, the base year (1980) and the last year. If a model calculation was to be done with a 3% emission increase in NOx every year, the annual cycle was computed until convergence (negligible change in concentration distributions from one year to the next) with anthropogenic NOx emissions at 64% of the base year ($0.64 = 1.03^{-15}$), corresponding to 1965, and at 156%, corresponding to 1995. For methane, a 60 yr period was covered with

1950 as the first and 2010 as the last year. The response time to changes in the emissions was so fast that sufficient convergence was achieved already after one year of model simulation.

For methane it was chosen to change the concentration by approximately the amount observed to be the annual increase globally, rather than increasing the emission rate since it would require a calculation covering several decades for methane to reach steady state.

4.1 Increasing global, anthropogenic NOx-emissions

In Fig. 1 is shown the meridional distribution at the end of May of NOx (=NO + NO_2), O_3, OH and H_2O_2 with height in 1965 and 1995 in a 30 years' calculation with an annual increase in the anthropogenic NOx-emissions of 3%. In the southern hemisphere there was not much change in the concentrations, while in the lower part of the northern hemisphere up to 0.9%/yr increase in ozone was calculated, 1.4%/yr in OH, 2.3%/yr in NOx while there was a drop in H_2O_2 of 0.5%/yr or less.

The increase in NOx was largely as NO_2. This means that the formation of nitric acid through reaction (2) increased more than the increase in total NOx-emission in the lower troposphere at mid northerly latitudes. The rate of formation of nitric acid increased by nearly (2.3 + 1.4)%/yr, or 3.7%/yr, there.

The increase in OH that was calculated for increasing NOx-emissions would also enhance gas-phase sulphate production through reactions (1) followed by (3) and (4). The wet-phase conversion of dissolved SO_2 to sulphate would proceed at a slower rate, however, since it is mainly controlled by H_2O_2 which was calculated to decline as NOx goes up. This is due to the suppression of HO_2-radicals with increasing NO. At high Ph wet-phase sulphate formation could be enhanced since dissolved ozone than is active in the oxidation.

4.2 Increasing the atmospheric methane-concentration

In Fig. 2 is shown the meridional distribution at the end of May of O_3, OH and H_2O_2 with height in 1950 and 2010 in a calculation where the concentration of CH_4 increased by 1.5%/yr over 60 years. Changes in O_3, OH and H_2O_2 were calculated to occur throughout the troposphere and in both hemispheres.

In the lower part of the troposphere in the northern hemisphere O_3 was calculated to increase by about 0.4%/yr, OH to decrease by 0.5%/yr and H_2O_2 to increase by about 0.5%/yr.

The decrease in OH arose from the increase in its loss through reaction with CH_4. The concentration of hydroxyl decreased less than CH_4 went up, however, which means that the flux of material through the reaction

$$CH_4 + OH \rightarrow CH_3 + H_2O \tag{5}$$

increased and led to the formation of more O_3 and HO_2-radicals in the decomposition reactions of CH_4, and consequently also H_2O_2 went up since the production is determined by

$$HO_2 + HO_2 \rightarrow H_2O_2 + O_2 \tag{6}$$

The calculations showed that increasing CH_4 only, led to a decreasing conversion of NO_2 and SO_2 to nitrate and sulphate through the OH-reactions. The formation of nitrate in the dark would increase due to the O_3 increase, and the liquid phase conversion of SO_2 would likewise become more efficient.

4.3 Possible consequences for dark chemistry of nitrate formation

A review of the measured concentration of nitrate and sulphate in precipitation at some sites in the eastern U.S. has shown that there is a strong seasonal dependence in the sulphate deposition suggesting a dominating photochemical influence on the gas and liquid phase chemistry. For nitrate deposition on the other hand, there is no strong variation with season (Calvert et al., 1985). Measurements at Birkenes in Norway since 1974 in general supports the picture of less sulphate in precipitation during the winter compared to the summer, while for nitrate this is not the case (SFT, 1985). Calvert et al. (1985) suggest that the formation of nitrate is dominated by reaction (2) in summer, while in winter the following reaction sequence may dominate:

$$O_3 + NO_2 \rightarrow NO_3 + O_2 \qquad (7)$$

$$NO_3 + NO_2 \rightleftarrows N_2O_5 \qquad (8)$$

$$N_2O_5 + H_2O\ (\text{cloud water}) \rightarrow 2H^+ + 2NO_3^- \qquad (9)$$

The NO_3-radicals may also be captured by cloud water and react with soluble organic matter, e.g.

$$NO_3 + HOCH_2OH\ (aq) \rightarrow H^+\ (aq) + NO_3^-\ (aq) + HOCH_2O\ (aq) \qquad (10)$$

and where $HOCH_2O$ (aq) further reacts with dissolved O_2 to form formic acid and dissolved HO_2 which may produce dissolved hydrogen peroxide (Calvert et al., 1985).

Reaction (9) was not included in the model calculations, but the results from the model runs with changes in NOx-emissions and CH_4-concentrations can be used to get an indication about the change in the nitrate formation during the winter and during the night, provided that a process like reaction (9) is the main pathway for nitrate formation in the dark.

The rate of formation of the NO_3-radical at night is controlled by reaction (7). The formation rate of the NO_3-radical increases as the sum of the increases in NO_2 and O_3. The concentration of N_2O_5 is given by the equilibrium (8):

$$N_2O_5 = \frac{k_f \cdot NO_2 \cdot NO_3}{k_r}$$

and is shifted to the right which means that $N_2O_5 \gg NO_3$ at night at atmospheric temperatures. This means the formation rate of nitrate through reaction (9) at night in the lower mid latitude northern troposphere increases as the sum of the increases in NO_2 and O_3. It has to be remembered, however, that the inclusion of reaction (9) in the model with a reaction rate coefficient which on an annual basis makes reaction (9) comparable in significance to reaction (2), would have led to a smaller increase in NOx (= NO + NO_2) than calculated here.

It can be concluded that increasing the global NOx-emission enhances both the photochemical conversion of NO_2 to HNO_3 through reaction with OH, and even more enhances the efficiency of the dark reaction where N_2O_5 reacts with water droplets. The sulphate formation is increased since OH goes up, while the liquid phase conversion becomes less efficient due to a suppression of H_2O_2 as NOx increases.

4.4 Conclusions

A picture of the combined effect of increasing the anthropogenic NOx-emissions by 3%/yr and the CH_4-concentrations by 1.5%/yr can be gained by just adding the calculated annual changes in the concentrations together. At mid latitudes in the lower part of the troposphere in the northern hemisphere, O_3 would increase at a rate of 1.3%/yr, OH increase by 0.9%/yr, NOx would increase by 2.3%/yr while H_2O_2 would remain virtually unchanged.

The implication of these numbers is that at mid latitudes in the lower troposphere, the rate of formation of HNO_3 through the OH-route would increase somewhat more than the increase in NOx-emission.

The rate of formation of sulphate through the OH-route would increase by about 0.9%/yr plus the rate of increase in the SO_2 concentration, while the liquid phase conversion would increase approximately at the same rate as the increase in the SO_2 concentration.

An attempt has been made to estimate the magnitude of the increase in the nitrate formation at night and during the winter when it was assumed that reaction (9), where N_2O_5 reacts with water droplets, is the main process where nitrate is formed in the dark. It is suggested that with increasing NOx-emissions, the formation of nitrate through this route increases faster than the formation of nitric acid through the gas phase reaction between NO_2 and the hydroxyl radical.

These conclusions suggest that the winter mean concentration of the sum of nitrate and nitric acid in air and precipitation should have increased more than the summer values at mid latitudes in the northern hemisphere during years with increasing NOx-emissions. The findings also suggest that the fraction that nitric acid + nitrate makes up of lower tropospheric NOy (≈NOx + PAN + HNO_3 + nitrate), has increased more than the fraction that sulphate makes up of lower tropospheric SOx (=SO_2 + sulphate) during times with increasing SO_2 and NOx emissions.

ACKNOWLEDGEMENT

This work was partly funded by The Royal Norwegian Research Council for Science and Technology through a research contract with ØH (No. FK03.15772).

REFERENCES

Attmannspacher, W., R. Hartmannsgruber and P. Lang (1984) Langzeittendenzen des Ozons der Atmosphäre aufgrund der 1967 begonnenen Ozonmeßreihen am Meteorologischen Observatorium Hohenpeißenberg. Meteorol. Rdsch., 37, 193-199.

Barrie, L.A., D. Fisher and R.M. Koerner (1985) Twentieth century trends in Arctic air pollution revealed by conductivity and acidity observations in snow and ice in the Canadian high Arctic. Atmos. Environ., 19, 2055-2063.

Blake, D. and S. Rowland (1986) World-wide increase in tropospheric methane 1978-1983. J. Atm. Chem., 4, 43-62.

Cox, R.A. (1974) The photolysis of nitrous acid in the presence of carbon monoxide and sulphur dioxide. J. Photochem, 3, 291-304.

Cox, R.A. and S.A. Penkett (1982) Formation of atmospheric acidity. Proceedings, CEC Workshop on Acid Deposition, Ed. by S. Beilke and A.J. Elshout, Commission of the European Communities, Brussels, 58-83.

Dianov-Klokov, V.I. and L.N. Yurganow (1981) A spectroscopic study of the global space-time distribution of atmospheric CO. Tellus, 33, 262-273.

Fraser, P.J., P. Hyson and R.A. Rasmussen (1986). Methane, carbon monoxide and chloroform in the southern hemisphere. J. Atm. Chem., 4, 3-42.

Heidt, L.E., J.P. Krasnec, R.A. Lueb, W.H. Pollock, B.E. Henry and P.J. Crutzen (1980) Latitudinal distributions of CO and CH_4 over the Pacific. J. Geophys. Res., 85, 7329-7336.
Hidalgo, H. and P.J. Crutzen (1977) The tropospheric and stratospheric composition perturbed by NOx emissions of high altitude aircraft. J. Geophys. Res., 82, 5833-5866.
Hov, Ø. and I.S.A. Isaksen (1985) NOx- and HC-emission changes, episodic increase of ozone in the atmospheric boundary layer and long term change of ozone in the free troposphere. Manuscript in preparation.
Isaksen, I.S.A. and H. Rodhe (1978) A two-dimensional model for the global distribution of gases and aerosol particles in the troposphere. Report AC-47, Department of Meteorology, University of Stockholm.
Isaksen, I.S.A., Ø. Hov, S.A. Penkett and A. Semb (1985) Model analysis of the measured concentration of organic gases in the Norwegian Arctic. J. Atm. Chem., 3, 3-27.
Khalil, M.A.K. and R.A. Rasmussen (1984) Carbon monoxide in the earth's atmosphere: Increasing trend. Science, 224, 54-56.
Logan, J.A. (1985) Tropospheric ozone: Seasonal behaviour, trends and anthropogenic influence. J. Geophys. Res., 90, 10463-10482.
Neftel, A., J. Beer, H. Oeschger, F. Zürcher and R.C. Finkel (1985) Sulphate and nitrate concentrations in snow from South-Greenland 1895-1978. Nature, 314, 611-613.
Newell, R.E., J.W. Kidson, D.C. Vincent and G.J. Boer (1972) The general circulation of the tropical atmosphere and interactions with extratropical latitudes, vol. 1. MIT press, Cambridge, Mass.
Penkett, S.A., B.M.R. Jones, K.A. Brice and A.E.J. Eggleton (1979) The importance of atmospheric ozone and hydrogen peroxide in oxidising sulphur dioxide in cloud and rainwater. Atmos. Environ., 13, 123.
Plumb, R.A. and J.D. Mahlman (1985) The zonally-averaged transport characteristic of the GFDL general circulation/transport model. Manuscript, CSIRO, Aspendale, Vic., Australia.
Rasmussen, R.A. and M.A.K. Khalil (1982) Latitudinal distributions of trace gases in and above the boundary layer. Chemosphere, 11, 227-235.
Rinsland, C.P., J.S. Levine and T. Miles (1985) Concentration of methane in the troposphere deduced from 1951 infrared solar spectra. Nature, 318, 245-249.
Rinsland, C.P. and J.S. Levine (1985) Free tropospheric carbon monoxide concentrations in 1950 and 1951 deduced from infrared total coloumn amount measurements. Nature, 318, 250-254.
Salmon, L., D.H.F. Atkins, E.M.R. Fisher, C. Healy and D.V. Law (1978) Retrospective trend analysis of the content of U.K. air particulate material 1957-1974. The Science of the total Environm., 9, 161.
SFT (1985) Monitoring of the long-range transport of polluted air and precipitation. The Norwegian State Pollution Control Authority, Oslo (Report 201/85).
Singh, H.B. and L.J. Salas (1982) Measurement of selected light hydrocarbons over the Pacific Ocean: latitudinal and seasonal variation. Geophys. Res. Lett., 9, 842-845.
Stockwell, W.R. and J.G. Calvert (1983) The mechanism of the $HO-SO_2$-reaction. Atmos. Environ., 17, 2231-2235.
Warmbt, W. (1979) Ergebnisse langjähriger Messungen des bodennahen Ozons in der DDR. Z. Met., 29, 24-31.
Zimmermann, P.R., R.B. Chatfield, J. Fishman, P.J. Crutzen and P.L. Hanst (1978) Estimates of the production of CO and H_2 from the oxidation of hydrocarbon emissions from vegetation. Geophys. Res. Lett., 5, 679-682.

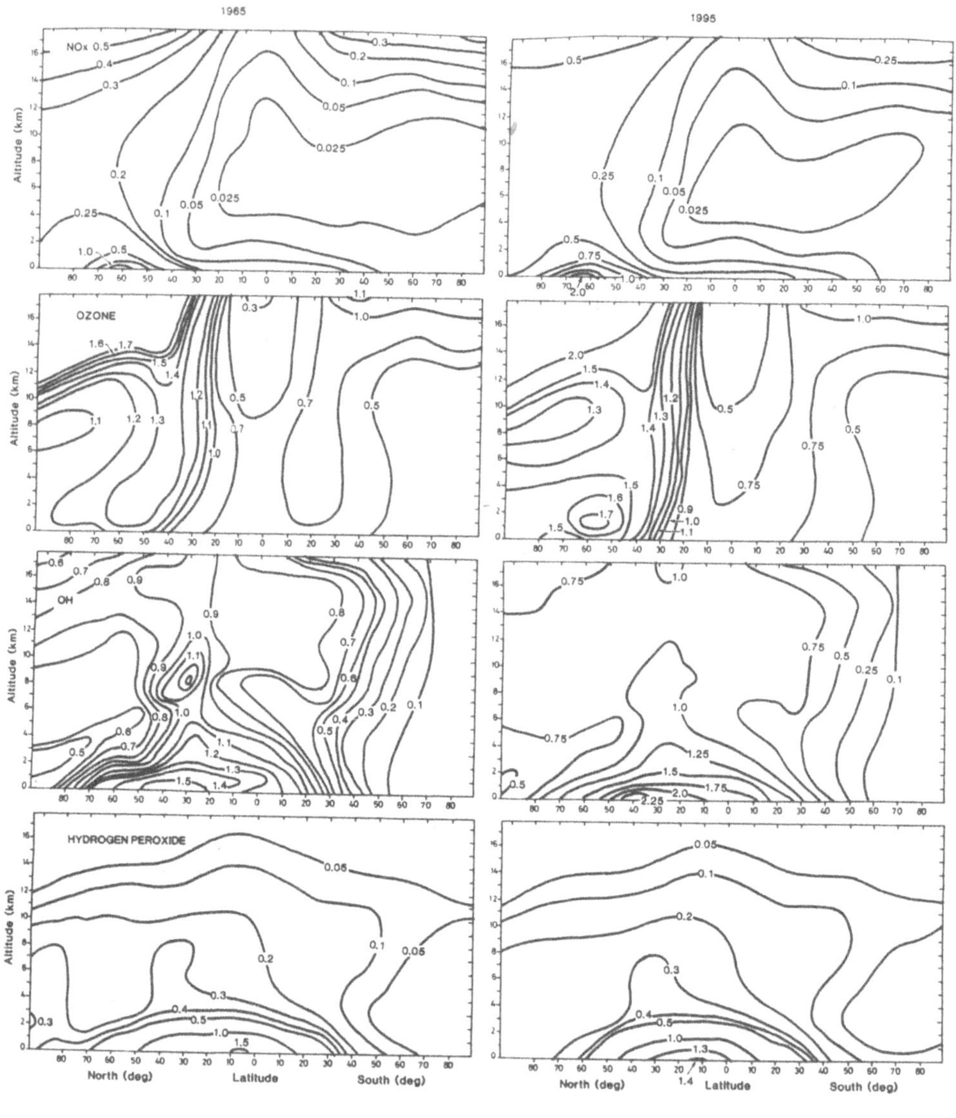

Fig. 1: The meridional concentration with height at the end of May for NOx (in ppb), O_3 (in 10^{12} molecules/cm^3), OH (in 10^6 molecules/cm^3) and H_2O_2 (in ppb) in 1965 (left column) and 1995 (right column) for a calculation where the global anthropogenic NOx-emissions increased by 3%/yr.

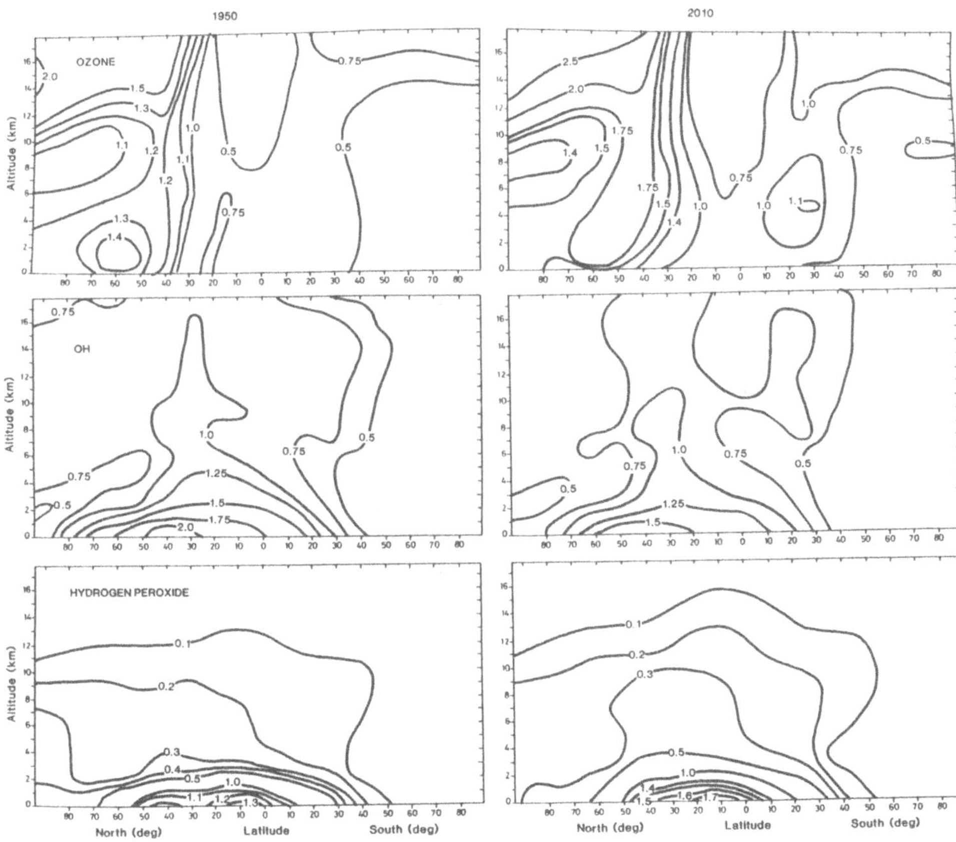

Fig. 2: The meridional concentration with height at the end of May for O_3 (in 10^{12} molecules/cm^3), OH (in 10^6 molecules/cm^3) and H_2O_2 (in ppb) in 1950 (left column) and 2010 (right column) for a calculation where the tropospheric concentration of CH_4 was increased by 1.5%/yr.

A MASS BALANCE OF SO_2 OVER BADEN-WÜRTTEMBERG DURING THE TULLA EXPERIMENT

K. Nester, M. Bär and F. Fiedler
Institut für Meteorologie und Klimaforschung
Kernforschungszentrum Karlsruhe/Universität Karlsruhe

Summary

The TULLA experiment was carried out during March 1985 over Baden-Württemberg. One of the aims of the experiment was the determination of a mass balance of SO_2 over Baden-Württemberg. The results of the SO_2 measurements in different levels along the boundaries of Baden-Württemberg on March 25, 1985 are used to determine the corresponding mass balance. To compare the measured concentrations along the flight patterns with model results, calculations with a three dimensional diffusion model have been performed. The flow field and the field of the dispersion coefficients are determined by a non-hydrostatic meso-scale model, which calculates these fields using the meteorological data measured in Stuttgart on that day. The emissions of SO_2, available as hourly averages at a 1 km x 1 km grid, and the SO_2-flux into Baden-Württemberg have been taken into account as sources. The concentration of SO_2 along the flight pattern derived from measurements and from the model are in good agreement. Although the influences of orography on the flow field are very strong the maximum and minimum concentrations at the outflow boundaries are located at the same positions.

1. INTRODUCTION

Phenomenological knowledge on flow conditions over complex terrain and the diffusion of atmospheric pollutants in mesoscale regions are still very crude. Most of the previous investigations are concentrated on the conditions of the transport and diffusion of pollutants in the lee of single sources up to a few decades of kilometers.

Only for long range transport problems some simplified models are available which present the possibility to estimate the relations over some thousands of kilometers by using homogenous vertical distributions. For those conditions, where the local distribution of the sources, the vertical distributions of the concentrations and the complex flow field are of importance only very few studies are available.

A large experiment called TULLA (Transport und Umwandlung von Luftschadstoffen im Lande Baden-Württemberg und aus Anrainerstaaten) was therefore carried out from March 18 to March 29, 1985. During intensive phases of the experiment pollutants such as SO_2, NO, NO_2, O_3 and some other species together with the meteorological measurements have been comprehensively measured in the lower troposphere. In addition carefully compiled emission data from single and area sources in Baden-Württemberg (B-W), on a time scale of one hour and within a grid of 1 x 1 km are available for the whole period of the TULLA experiment. All activities of the TULLA experiment aimed at improving our knowledge about the influence of changing emission, transport and dispersion conditions on the distribution of air pollutants and their dry deposition over irregular terrain like B-W.

In the frame of the general scientific aims of the Tulla experiment, (1) one aspect was the determination of the mass balance of SO_2 over B-W. As an example the data of March 25, 1985 were chosen.

2. THE METEOROLOGICAL SITUATION ON MARCH 25, 1985

During daytime of March 25, 1985, the atmosphere was almost neutrally stratified up to a height of 700 m. The wind in a level of 200 m above ground blew with a speed of about 6 m/s. The general direction was 220 degrees.

The wind field over B-W was calculated with the KAMM (Karlsruher Atmosphärisches Mesoskaliges Modell) model (2) using radiosonde measurement of Stuttgart at noon. The KAMM model is threedimensional, non-hydrostatic and instationary. A terrain following coordinate system is used. It calculates the flow, temperature and diffusion coefficient fields over an irregular terrain by solving

the Navier Stokes equations of motion,
the continuity equation and
the thermodynamic equation.

Figure 1 shows the wind field near ground level. The influence of the topography, which is shown in figure 2, on the flow field is clearly evident. Especially the channeling effect of the Upper Rhine Valley is simulated as expected (3).

3. THE SO_2 MEASUREMENTS ON MARCH 25, 1985

For the interpretation of the measured SO_2 concentration around B-W, it is necessary to know the distribution of the main sources, which is given in figure 2. Figure 2 contains also the topography together with some cities and rivers.

It can be seen that the sources are mainly concentrated in the Upper Rhine and Neckar Valley. The total SO_2 emission on that day was 30 t/h.

On March 25, 1985, four aircrafts were measuring SO_2 on a flight pattern around B-W. The lowest level was about 200 m above ground level, varying between 100 m and 300 m. The other aircrafts flew in levels of about 100 m distance above each other. The concentration distribution of SO_2 along the flight pattern measured with the lowest aircraft is shown in figure 3. The topography is also plotted. Due to the flow conditions at that day and the strength and locations of the sources, the highest concentration of 65 $\mu g/m^3$ is found in the northern part of B-W. The SO_2 measurements from the highest aircraft show that on that day the SO_2 is concentrated in the lowest part of the troposphere. Only some higher concentrations are observed, belonging to the plumes of elevated single sources.

4. THE DETERMINATION OF THE MASS BALANCE OF SO_2

The SO_2 measurements of the aircrafts are used to derive a SO_2 profile at each grid point of the model along the flight pattern. From this profile the SO_2 flux along the flight pattern - this means around B-W - is calculated with the simulated wind speeds.

	southern boundary	western boundary	northern boundary	eastern boundary	balance
SO_2 flux [t/h]	6.1	3.2	-20.3	-4.1	-15.1

The mass balance can be written as:

$$\frac{d}{dt} MSO_2 = F_I + F_O + D + T + E$$

with
F_I = input flux of SO_2 (9.3 t/h)
F_O = output flux of SO_2 (24.4 t/h)
E = emission of SO_2 (30 t/h)
D = dry deposition of SO_2
T = chemical transformation of SO_2

If we assume stationary conditions the terms D and T add up to a value of 14.9 t/h. The estimate of the dry deposition over B-W is based on the ground level concentration of SO_2 along the flight pattern, derived from the SO_2 measurements of the aircrafts. At each grid point inside the flight pattern the SO_2 concentration is interpolated. The weighting function was the inverse of the square distance to the grid points at the boundaries. Dry deposition velocities at every grid point were calculated by a dry deposition model. By multiplying the concentration at each grid point by the corresponding dry deposition velocity the dry deposition is determined. Figure 4 shows the result of this estimate. The main dry deposition can be observed in the northern part of B-W. The total dry deposition amounts to 10.8 t/h. This means that the chemical transformation delivers a loss of SO_2 of 3.1 t/h. On average, an air parcel containing SO_2-concentrations needs 7.5 hours to pass the experimental site. Under these conditions, a transformation rate of SO_2 of 1%/h is derived.

Although the figures are first estimates of the different contributions to the mass balance of SO_2, they indicate the right order of magnitude.

To compare the SO_2-measurements at the outflow boundaries, a calculation with the DRAIS (Dreidimensionales regionales Ausbreitungs- und Immissions-Simulationsmodell) model was performed. The input flux and the sources inside of B-W were taken into account. Figure 5 shows the SO_2 concentration distribution around B-W, which is comparable to that of figure 3. The agreement between both distributions is quite satisfying. The maximum and minimum concentrations are located at almost the same positions. At the outflow boundaries, the calculated SO_2 concentrations are lower than the measured ones. The mean relative error has a value of 0.6.

5. CONCLUSIONS

To emphasize the value of the data collected during the TULLA experiment, a first estimate of the SO_2 balance over Baden-Württemberg was carried out for March 25, 1985. It shows that the export of SO_2 is about 80 % of the total emission of SO_2. The difference between import and export of SO_2 is about half of the emissions. This means, that dry deposition and chemical transformation are important removal processes amounting to the other half. Dry deposition is much more important than chemical transformation. Although the results are plausible, it should be emphasized that they are preliminary till all data is analysed.

REFERENCES

(1) Fiedler, F. (1986). Ziele und Durchführung des TULLA-Experiments. Statuskolloquium 4.-7.3.1986 "Projekt Europäisches Forschungszentrum für Maßnahmen zur Luftreinhaltung". Kernforschungszentrum Karlsruhe.

(2) Dorwarth, G. (1986). Numerische Berechnung des Druckwiderstandes typischer Geländeformen. Wiss. Ber. Institut für Meteorologie und Klimaforschung, Universität Karlsruhe, Nr. 6.

(3) Fiedler, F. (1983). Einige Charakteristika der Strömung im Oberrheingraben. Wiss. Mitt. Meteor. Inst. Universität Karlsruhe, Nr. 4.

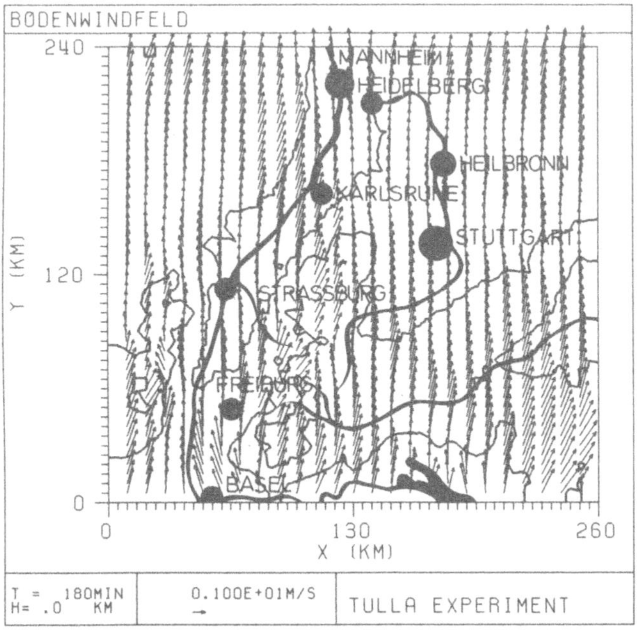

Fig. 1: Windfield at ground level (25.3.85)

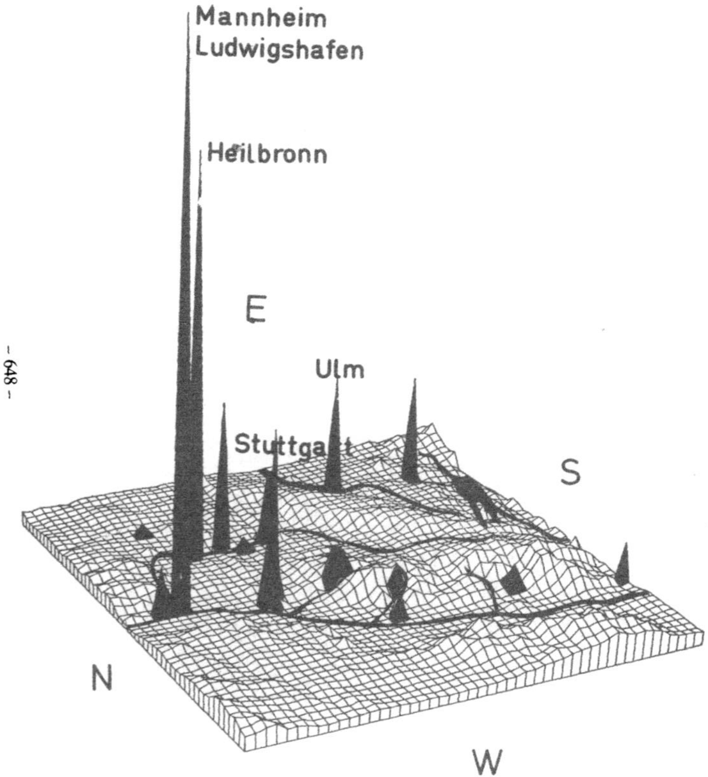

Fig. 2: SO_2-Emissions from single sources
(25.3.85)

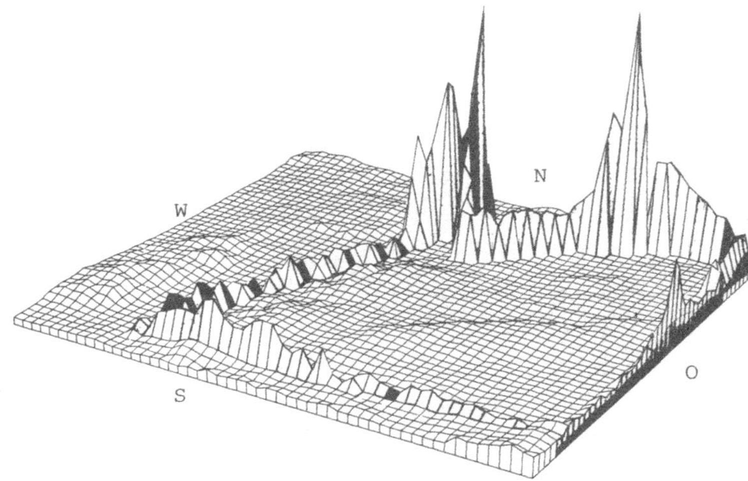

Fig. 3: Measured SO_2-concentration along the flight path
(lowest level)

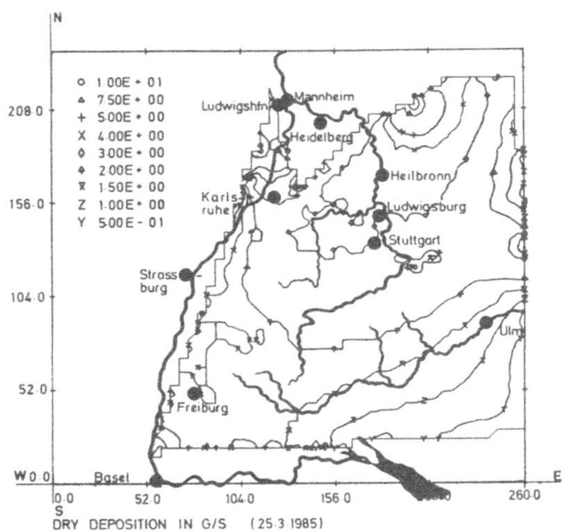

Fig. 4: Dry deposition in a grid area of 5 km x 5 km

Fig. 5: Calculated SO_2-concentration along the flight path as in figure 3

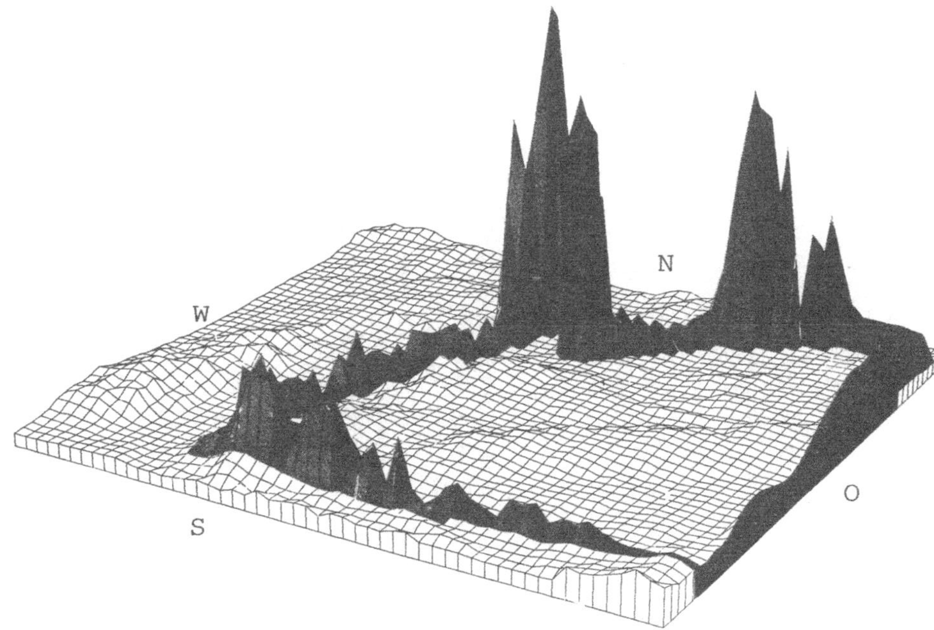

MESOSCALE MODELING OF NITRIC ACID GENERATION IN NIGHT-TIME MOIST ATMOSPHERE

N .CHAUMERLIAC and R. ROSSET
L.A.M.P.-O.P.G.C., LA/C.N.R.S. n° 267, Université de Clermont II,
B.P. 45, 63170 Aubière, France.

Summary

A three-dimensional mesoscale numerical model (5) with detailed semi-spectral microphysics has been applied to the study of night-time nitric acid production. Combining a chemical module and the mesoscale meteorological model, two-dimensional simulations have been performed for a typical case of moist advection. We have observed a threshold behaviour for NO_3 disappearance with relative humidity, around 60% already noticed experimentally.

1. INTRODUCTION

Nitrogen oxide removal and subsequent nitric acid generation are commonly separated into day-time and night-time processes (1), (2), (3). During the day, HNO_3 is generated mainly through $NO_2 + OH + M \longrightarrow HNO_3 + M$. During the night, HNO_3 can be generated either by homogeneous reaction, $N_2O_5 + H_2O$ (vapor) $\longrightarrow 2\ HNO_3$, or by heterogeneous scavenging reactions with aerosols and hydrometeors (4). Experimental estimates indicate that this latter reaction mechanism taking place within night-time fogs or clouds could remove NO_x and generate HNO_3 as efficiently as the day-time reaction with OH.

We use, as a framework, a meteorological mesoscale model (5) including a very detailed microphysical scheme for simulating orographic precipitation. This model has already been extended to the study of gas and aerosol removal by clouds and raindrops (6). Because of the semi-spectral parameterization for liquid water as well as for aerosols, this model seems especially appropriate for the study of night-time nitric acid production for both homogeneous and heterogeneous processes. As a first step, we proceed to a two-dimensional test in a typical case of moist advection by considering only homogeneous reaction mechanism for HNO_3 production.

In the first part of this paper, the complete mesoscale model allowing for non-linear interactions between dynamics, microphysics and aerosol physico-chemistry is presented. Then, a simplified version of this model is used to explore the influence of relative humidity on the NO_3 disappearance and HNO_3 generation in the gas phase. We conclude by outlining future developments of this study including heterogeneous processes which are now currently underway.

2. MICROPHYSICAL AND PHYSICOCHEMICAL FEATURES OF THE MESOSCALE MODEL

We shall focus on the microphysical and physicochemical aspects of the model, necessary for the understanding of this paper. For further details, the reader is referred to (5), (6), (7).

The model includes the following warm rain microphysical processes : nucleation, condensation, evaporation, accretion, autoconversion, selfcollection and sedimentation. We make use of the following activation spectrum to form cloud droplets :

$$N = CS^k,$$

where S is the supersaturation, C and k empirical constants related to the origin of the air mass (continental or maritime).

Liquid (both cloud and rain) water is assumed to be distributed lognormally with diameter. Thus,

$$dN = \frac{N}{\sqrt{2\pi}\,\sigma\,D} \quad \exp\left(-\frac{1}{2\sigma^2}\ln^2\frac{D}{D_0}\right) dD$$

is the number of droplets in the size range D to D + dD. Here and D_0 are distribution parameters. Only one of these two can be diagnosed given q and N. We close the microphysical system of equations by assuming a constant value of and by computing D_0. Assumed values for the cloud are $\sigma_c = 0.1575$ (continental case) or $\sigma_c = 0.2775$ (maritime case) and for the rain $\sigma_r = 0.547$.

The conversion rates of cloud droplets to raindrops are computed according to Berry and Reinhardt's parameterizations , which are appropriate to the log-normal distribution.

The predictive variables for microphysics as well as the microphysical processes are summarized in Figure 1.

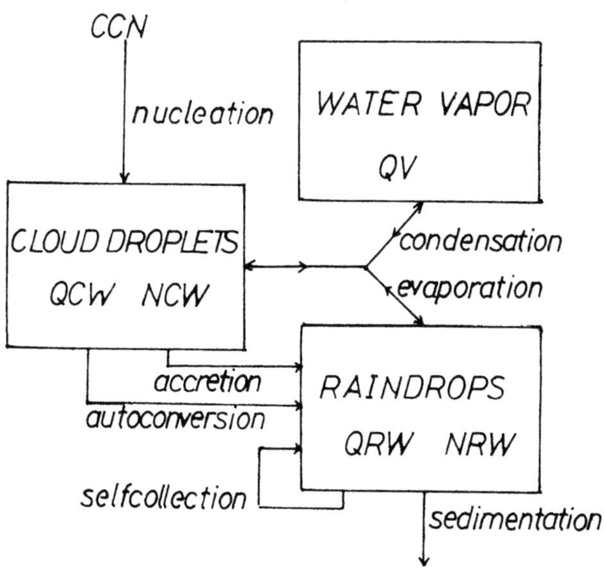

Figure 1 : Schematic diagram of microphysical processes.

Sulfate particles are also assumed to be distributed according to a log-normal function, setting the dispersion parameter σ to 0.69. Number concentration N and volume concentration q are the predictive variables for aerosol particles. We discriminate between three categories of particles : some are free in the air, others are attached to cloud droplets and the third category consists of aerosols removed by precipitation. Schematic diagram is provided in Figure 2 for describing the various processes at work between the three aerosol classes. Parameterizations developed for brownian, turbulent diffusion rates are taken from (8). The dynamical capture rate is calculated in a similar way following (9).

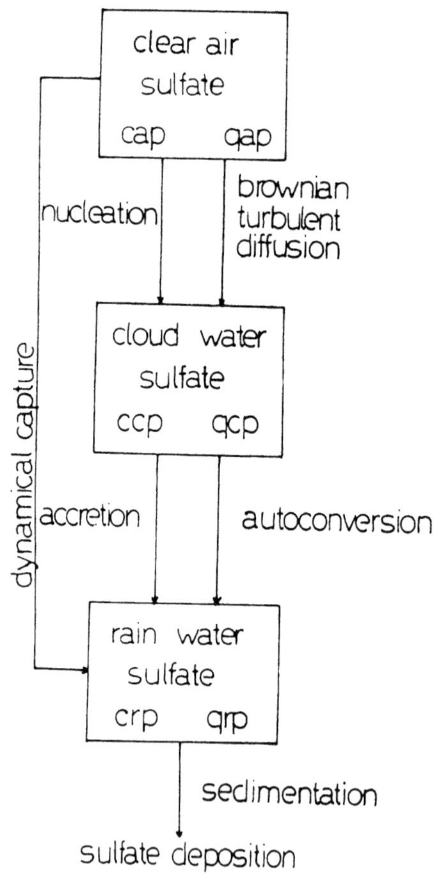

Figure 2 : Schematic diagram of aerosol physicochemical processes.

3. PRELIMANARY TEST OF NIGHT-TIME HNO_3 GENERATION IN THE PRESENCE OF HUMIDITY

During the night, there are no photolysis and reactions with OH become negligibly slow because of low OH concentration; furthermore, in the absence of local NO-sources, the gaseous reaction scheme for night-time HNO_3 production can be simply written as :

$$NO_2 + O_3 \longrightarrow NO_3 + O_2 \qquad (R1)$$

$$NO_3 + NO_2 \longrightarrow N_2O_5 \qquad (R2)$$

$$N_2O_5 \longrightarrow NO_3 + NO_2 \qquad (R3)$$

$$N_2O_5 + H_2O \longrightarrow 2HNO_{3(vapor)} \qquad (R4)$$

with the following reaction rates :

$$k_1 = 1.936\ 10^2 \exp(-2450/T)\ \text{ppm min}^{-1}$$

$$k_2 = 1.2906\ 10^3\ \text{ppm min}^{-1}$$

$$k_3 = 1.956\ 10^{16} \exp(-10840/T)\ \text{min}^{-1}$$

$$k_4 = 1.6133\ 10^{-6}\ \text{ppm min}^{-1}$$

A simulation with a two-dimensional version of the model, without orography, has been conducted for a period of six hours. We consider a two-layer atmosphere with an horizontal wind speed of 20 ms^{-1} except in the lower vertical levels (below 1 km height) where it decreases to zero at the ground level. To represent the advection of moist air, we define along the inflow boundary a temporal variation of the humidity such that the relative humidity is increasing from 50% to 80% over the six hour run.

Using initial concentrations of 0.075 ppm for NO_2 and 0.2335 ppm for O_3 (after (10), we first run a chemical module at each model grid point for the conditions (T,P, relative humidity) of the model. We then initialize the mesoscale model with the values obtained from the chemical module, for all the gaseous species after three hours integration using a Gear method.

In Figure 3, vertical cross-sections of NO_3 and relative humidity are provided. Differences between results obtained after 6 hours and 3 hours simulation time are shown. One can notice a slight correlation between the NO_3 and relative humidity fields with a good correspondence of the extrema of variation.

The time evolution of the same variables as shown in Figure 4 provides additional information. Here we display the evolution over the 6 h run of the gases and relative humidity. In the first two hours, NO_2 is converted to NO_3 and N_2O_5 is depleted to generate HNO_3 because of the relative humidity. But when relative humidity increases and is greater than 60% NO_3 is observed to disappear. This threshold effect has already been noticed by (10).

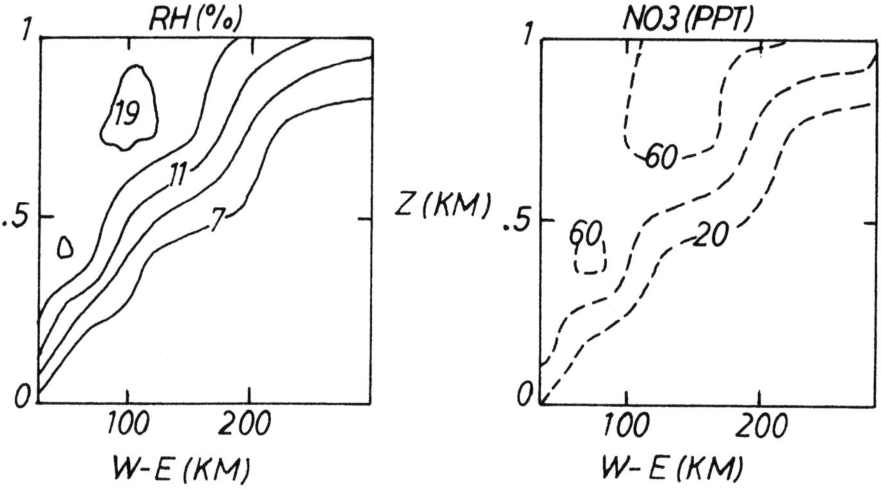

Figure 3 : Vertical cross-sections of
(1) NO_3 concentration (in ppt)
(2) relative humidity (in %)
expressed as the difference between results after 6 hours and after 3 hours. NO_3 concentration is decreasing (dashed lines) while relative humidity is increasing (solid lines)

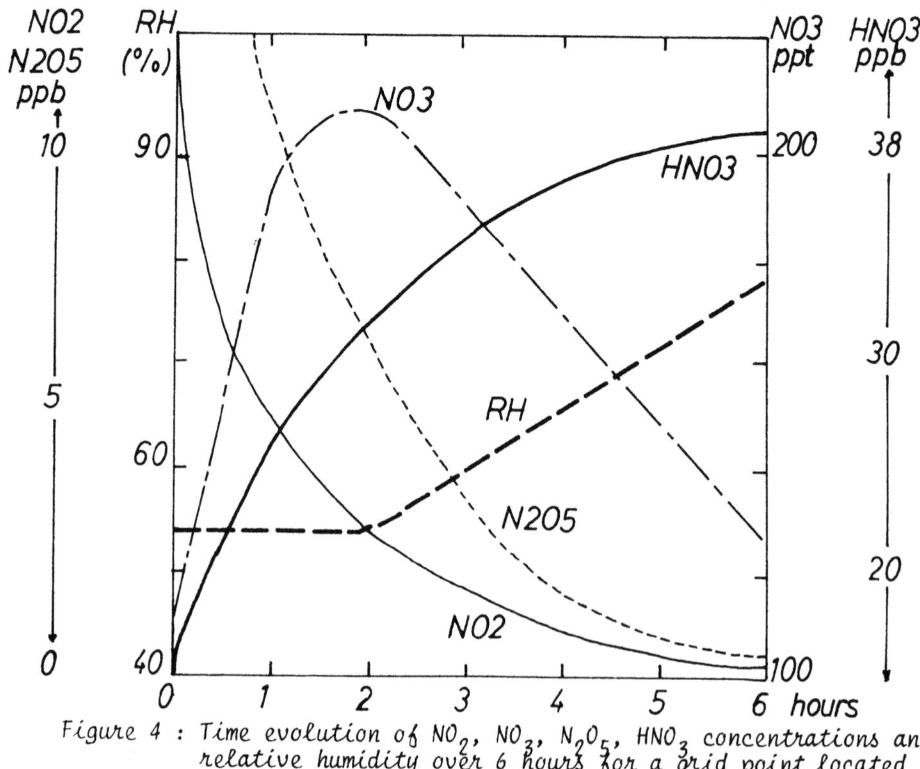

Figure 4 : Time evolution of NO_2, NO_3, N_2O_5, HNO_3 concentrations and relative humidity over 6 hours for a grid point located in the center of the model domain at a height of 700 m.

4. HETEROGENEOUS PROCESSES FOR NIGHT-TIME HNO_3 GENERATION

Actual levels of NO_3 observed at night are typically less than those that would be predicted on the basis of the previous reactions (R1) to (R4). Heterogeneous scavenging reactions with aerosols and hydrometeors should therefore be considered. (11), (4) including reactions such as :

$$N_2O_5 + H_2O_{(liq)} \longrightarrow 2HNO_3 \quad (R5)$$

$$NO_3 + H_2O_{(liq)} \longrightarrow 2HNO_3 \quad (R6)$$

which can take place rapidly on wet particle surfaces or on cloud droplets.

Parameterization for those heterogeneous processes need a detailed treatment of the diffusion rate of gases to particles (4). Those diffusion rates should then be integrated over the aerosol or cloud droplet spectra. This spectral information is included in the mesoscale meteorological model ((5), (6)) and seems appropriate for such parameterizations. Sensitivity tests could be performed upon the initial concentrations of NO_2, O_3, NO_3, on the particle and droplet sizes, and on the accommodation coefficient. For example, particle growth by vapor absorption should be included in the model as a function of the dry radius of the particles and relative humidity (12)

REFERENCES

(1) EHHALT, D.H. and J.W. DRUMMOND (1982) : The tropospheric cycle of NO_x. D. Reidel, Hingham, Mass., 219-251.

(2) PLATT, U. et al. (1981) : The diurnal variation of NO_3. J. Geophys. Res., 86, 11965-11970.

(3) CALVERT, J.G. et al. (1985) : Chemical mechanisms of acid generation. Nature, 317, 27-35.

(4) HEIKES, B. and A.M. THOMPSON (1983) : Effects of heterogeneous processes on NO_3 chemistry in the troposphere, J. Geophys. Res., 10, 10883-10895.

(5) NICKERSON, E.C. et al. (1986) : The numerical simulation of clouds, rain and airflow over the Vosges and Black Forest Mountains : a meso-model with parameterized microphysics, M.W.R., 114, n°2, 398-416.

(6) CHAUMERLIAC,N. et al. (1986) : Sulfur scavenging in a mesoscale meteorological model with semi-spectral microphysics. Submitted to J. Geophys. Res.

(7) CHAUMERLIAC, N. et al.(1986) : a 3D mesoscale model as a potential tool for the evaluation of sulfate particles scavenging. To appear in Annales Geophysical.

(8) DINGLE, A.N. and Y. LEE (1973). An analysis of in-cloud scavenging. J. Appl. Meteor., 12, 1295-1302.

(9) DANA, J. and J.M. HALES (1976). Statistical aspects of the washout of polydisperse aerosols. Atm. Env., 10, 45-50.

(10) PLATT, U. et al. (1980) : Detection of NO_3 in the polluted troposphere by differential optical absorption. Geophys. Res. Lett., 7, 89-92.

(11) CHAMEIDES, W.L. (1986) : Possible role of NO_3 in the night-time chemistry of a cloud; J. Geophys. Res., 19, 5331-5337.

(12) HANEL, G. (1976) : the properties of atmospheric aerosol particles as functions of the relative humidity at thermodynamic equilibrium with the surrounding moist air. Advances in Geophysics, 19, 73-188.

LONG RANGE MODELLING OF THE FORMATION TRANSPORT AND DEPOSITION OF PHOTOCHEMICAL OXIDANTS AND ACIDIFYING POLLUTANTS

R.M. STERN
Freie Universität Berlin
Institut für Geophysikalische Wissenschaften

P.J.H. BUILTJES
TNO Division of Technology for Society

Summary

Within the framework of the Dutch Air Quality Management System and the Dutch-German PHOXA-Project (Photochemical Oxidant And Acid Deposition Model Application) a Lagrangian and two Eulerian dispersion models have been and will be applied:

- The Measuring-Modeling Program Aerosols-Model (MPA) developed by TNO and RIVM, a rather simple two-layer trajectory model;
- the Regional Transport Model (RTM-III) developed by SAI, an Eulerian long range transport model of medium complexity especially designed for the simulation of photochemical oxidants;
- the Transport and Deposition of Acidifying Pollutants Model (TADAP) developed by ERT, an Eulerian long range transport model of high complexity with the main emphasis on the simulation of the dry and wet deposition of acidifying substances.

These three models of an increasing level of complexity are described in this paper. The models have been and will be used to investigate the effects of precursor emission controls in Central Europe on the concentrations of photochemical oxidants and on the deposition of acidifying pollutants. Some results of the application of the MPA-model to photochemical air pollution episodes in the Netherlands are given. Simulation results are also presented for the RTM-III-model, which has been applied to a photochemical smog episode in Central Europe. Furthermore, the planned application of the TADAP-model on an acid deposition episode is discussed.

1. INTRODUCTION

It has been known for some years now that the production, transport and deposition of secondary pollutants such as ozone, sulphate or nitrate is a regional problem. Observations made both in North America and in Europe have shown elevated levels of these pollutants spread over hundreds of kilometers. Measurents of the acidity of precipitation also clearly demonstrate the regional extent of the "acid rain" phenomena. Due to this wide distribution over large areas, it is important to derive a relationship between the concentrations of the secondary pollutants and the geographical distribution of the related precursor sources. An immediate implication of these findings is that control of secondary pollutants cannot be reached by limiting emissions on a city-by-city basis or even for densely populated and industrialized regions, such as Western Europe, even on a country-by-country basis i.e., regional and international control strategies are required.

This demand has been supported by recent model studies in Europe. The

results of calculations using the Airshed-model of Systems Applications Inc. in the Netherlands (1) and in the Federal Republic of Germany (2) have demonstrated the predominant effect of the regional transport of precursors and oxidants into the area investigated on the concentrations of photochemical oxidants within that area. The most important finding was that the implementation of emission controls only within the Netherlands or parts of the FRG, such as the Rhine-Ruhr area, will not be an effective measure for reducing oxidant levels within these areas. Therefore, significant improvement in photochemical air quality in Western Europe requires the cooperative efforts of several European countries. The same holds good for the control of sulphur compounds. This was shown by the calculations of Eliassen and Saltbones (3) carried out within the framework of the WMO/ECE-EMEP project. These calculations confirmed that, in most countries in Europe, the deposition of sulphur due to foreign sources represents an important contribution to the total deposition.
Based on these findings, the Netherlands and the FRG initiated programs to develop and apply mathematical models for the calculation of the transport, dispersion, chemical transformation and removal of air pollutants in the regional scale. The models are or will be integrated in respective Air Quality Management Systems and have been or will be used for the purposes of control strategy development. The main part of the model application will be performed within in the framework of the Dutch-German PHOXA programma (Photochemical Oxidant and Acid Deposition Model Application).
It is the object of the PHOXA programme to support political decisions on control strategies concerning the foregoingly quoted groups of substances by applying these models to various strategy options in order to identify their respective effectiveness on the concentration levels of ozone and nitrogen dioxide and on acid deposition.
The PHOXA programme consists of the following parts:

- Application and evaluation of regional dispersion models to simulate transport and formation of photochemical oxidants during episodes and deposition of acidifying substances under episodic as well as long term conditions.
- Establishment of the data bases necessary for the operation of the models.
- Application of the models using emission scenarios.

A detailed description of the PHOXA programme is given by Meinl and Builtjes (4).

In this paper, three models of increasing complexity used in the above context are described:

- The MPA-model (Measuring/Modelling Program Aerosols) developed by TNO and RIVM, a rather simple two-layer trajectory model.

This model is designed for the calculations of short term as well as long term averages for photochemical oxidants and acidifying pollutants for a certain number of receptor points.

- The Regional Transport Model (RTM-III) developed by SAI, an episodic Eulerian long range transport model of medium complexity especially designed for the simulation of photochemical oxidants.

An earlier version of this model has already been used to simulate high regional O_3 concentrations in the northeastern United States (5).
One of the reasons for choosing the model was that it represents a good compromise between model complexity and input needs, and another because it has shown itself to be the only one model, among the models suitable for control strategy development, which was available at the time of the

model selection (6).

- The Transport and Deposition of Acidifying Pollutants-Model (TADAP) developed by ERT, an episodic Eulerian long range transport model of high complexity with the main emphasis on the simulation of dry and wet deposition of acidifying pollutants.
  The development of this state-of-the-art long range transport model was jointly sponsored by the Ontario Ministry of the Environment, the Umweltbundesamt of the FRG and the Atmospheric Environment Service Canada. The design of this complex model is especially directed to regulatory applications. The problems to be addressed by the model are:

The source-receptor relationships for $SO_4^=$ and NO_3^- depositions for travel times of some days and travel distances in the range of 1000 km and the formation and transport of oxidants in the same space and time scales.

2. MODELLING AREA AND GENERAL DATA BASE

To be able to apply photochemical and acid deposition dispersion models, a general input data base is required. Considering the long range transport phenomena involved, a model area with the following geographical borderlines has been chosen:

- 10° longitude west to 24° longitude east
- 47°30' latitude north to 60° latitude north

The enclosed area extends over 3,129,000 km² and covers most of the heavily industrialized regions in Europe.

A general emission and land use data base has been created using grid cells with sizes as equal increments of 30' in longitude and 15' in latitude. This corresponds to grid cells of approximately 30 x 30 km.

The emission data base was compiled for the base year 1980 and is currently updated to the year 1982. It covers point sources and anthropogenic as well as natural area sources. The species considered are SO_2, $SO_4^=$, CO, NO, NO_2, NH_3 and VOC's. The following sources categories are identified: power plants, industrial point and area sources divided in 9 branches, traffic split up into local and highway road traffic, shipping, rail and air traffic, households, small-scale industry and natural sources, deciduous and coniferous forests as sources for hydrocarbon emissions, farming as a source for ammonia.

Special, time dependent factors are used to derive episodic specific emission data from the yearly emission figures. The total VOC's are split up in different hydrocarbon classes as needed by the respective chemical mechanism of the models.

A land use data base is needed to derive model inputs as roughness length scales, hydrocarbon emission from forests or for the calculation of deposition velocities. In the PHOXA-modelling area 10 land use categories are distinguished between. For each grid cell the percentage coverage for each category is given.

Additionally, both ambient air quality and deposition data together with meteorological data have been collected for the respective time periods to be modelled.

3. DESCRIPTION AND APPLICATION OF THE MPA-MODEL

Within the framework of the Measuring Program Aerosols, a receptor-oriented trajectory model has been developed by TNO and RIVM in cooperation with KNMI - The Royal Dutch Meteorological Institute and IMOU, University of Utrecht. The model has two layers in the vertical, a mixed layer transported along the ground level trajectory and a reservoir layer

transported above the mixing height. Both trajectories are calculated using the so-called Air Mass Transformation model (12). Both trajectories have a time duration of 72-hours, during the last 24-hours of which there is an exchange between the two layers. The area covered is the ECE/EMEP area. The emissions are taken from EMEP with the exception of the Netherlands and its surroundings. CO- and NH_3-emissions for the total area are also added. For the gasphase chemistry the CBM-III (see chapter 4) has been used with a more detailed sulphur and NO_x-chemistry. Aerosol-reactions and oxidationprocesses in cloudwater are also dealt with. The model has been used to calculate 5 different O_3-episodes between 1980-1984 for 3 Dutch receptor points. A comparison between calculated and measured maximum O_3-concentrations showed a reasonable agreement, with a correlation coefficient of 0.72.
Using the runs for the 5 episodes as a basis, several emission scenario calculations have been made. These calculations showed that hydro-camon emission reduction is much more benefical in bringing maximum O_3-concentrations down than NO_x-emission reduction. Down to about 50% HC-emission reduction a x% - HC-emission reduction leads to a 0,5 x% reduction in maximum O_3-concentrations. Down to about 50% NO_x-emission, reduction had no, or even an adverse effect on the O_3-levels.
Using estimates of historic emissions 'historic' runs have also been performed. They showed that between 1900 and 1955, hardly any increase in maximum calculated O_3-concentrations occured. However, between 1955 - 1975 a steadily increase in O_3-levels of about 2% a year was found, with a leveling of for later years. For more information about the MPA-model and its application see (7) and (8).

4. DESCRIPTION AND APLLICATION OF THE RTM-III-MODEL

4.1 Model structure

The SAI-Regional Transport Model is an Eulerian grid model based on the numerical solution of a 3½ layer multispecies diffusion equation. The vertical layers of the model are defined as follows (Fig. 1): A Mixed Layer (Layer M) beneath the base of the elevated temperature inversion and above the ground surface. The thickness of this layer can vary both temporally and spatially. A Lower Inversion Layer (Layer IL) immediately above the base of the inversion. The purpose of this layer is to keep track of emissions injected into the stable layer during the night and subsequently released to the mixed layer under daytime fumigation conditions. An Upper Inversion Layer (Layer IU) above the lower inversion layer and beneath a constant height, the top of the modeling region. This layer will serve as a reservoir for emissions penetrating the upper part of the inversion layer and aged pollutants entrained into this layer during multiple day episodes. A Surface Layer (Layer S) immediately above the ground representing the constant-flux surface layer. This layer, embedded in the mixed layer, will only accomodate modifications of vertical concentration profiles due to processes such as dry deposition and local NO/O_3 titration.
Both the mixed layer and the lower and upper inversion layers are prognostic layers whereas the surface layer is, as a part of the mixed layer, purely diagnostic. Vertical exchange of pollutants across the boundaries between the prognostic layers can occur as a result of the vertical changes of the heights of the layer boundaries and the vertical velocities induced by the large-scale divergences. The interface between the mixed layer and the lower-inversion layer is impermiable, i.e. no diffusive flux occurs.

Between lower and upper inversion layer a diffusive flux is parameterized with K-theory.
The advection terms of the diffusion equation are solved numerically using the Sharp and Smooth Transport Algorithm (SHASTA) developed by Boris and Book (9) or, as an user-specified option, the more accurate, but more costly to run Zalesak-scheme (10).

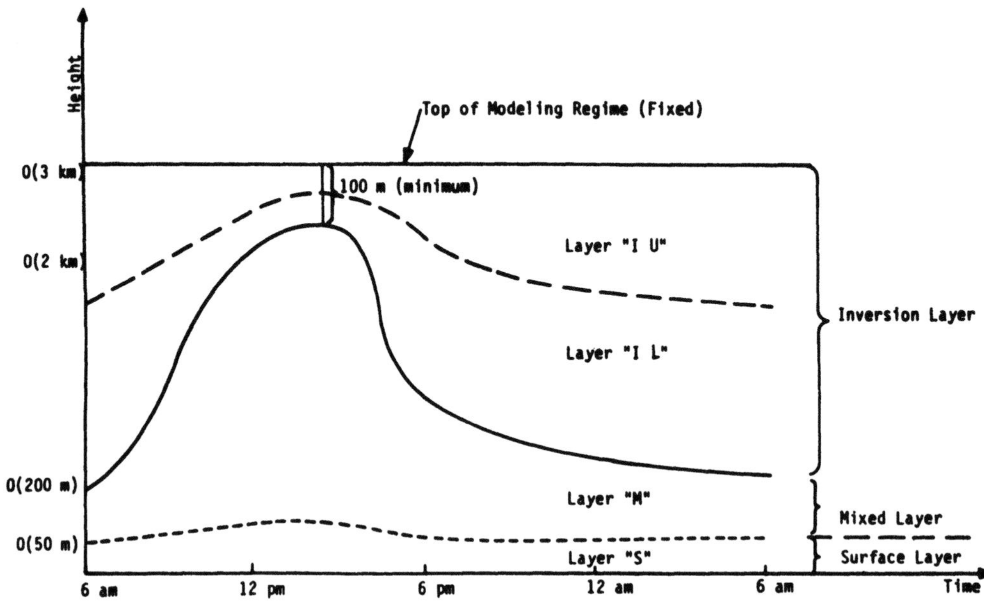

Fig. 1: Vertical structure of the RTM-III-model

The surface layer serves for the calculation of the surface depletion of pollutant masses. Transfer of gaseous pollutants from the atmosphere to the ground proceeds in three steps: The gases are transported by turbulent diffusion to a laminar sublayer just above the surface, by molecular diffusion through this laminar sublayer and interact finally by absorption or chemical reaction with the surface. By balancing the flux through the turbulent layer/laminar sublayer interface and the removal rate at the ground, the flux of species reaching the ground can be expressed in terms of the mixed layer concentrations (11).
The model contains a most recent version of the Carbon-Bond-Mechanism, the so called Carbon-Bond-IV, which has 70 reactions describing the chemistry of 17 transported species, using 9 different hydro-carbon classes. The CBM-IV is a condensed version of the CBM-X (expanded carbon-bond-mechanism), which is too large for use in Eulerian oxidant models. The photolysis rate constants of CBM-IV are diurnally and spatially varied as a function of the solar zenith angle and the cloud coverage.
To simulate the subgrid-scale effects of point source dispersion and chemestry processes, a point-source module embedded in a Gaussian puff framework is incorporated in the RTM-III-model. Puffs are released at regular intervals and tracked untill they are incorporated into the grid system. Chemical transformations within the plume are handled using simple photo-steady-state relationsships.

4.2 Model input data

There are four different types of input data: Emission, meterological, air quality and land use data.
Emission data are needed as point or area sources in hourly to 3-hourly time resolution.
Main meterological input data are layer averaged horizontal wind velocities for the three dimensional grid and mixing depths for each horizontal grid cell in hourly to 3-hourly time resolution. Surface data are needed in gridded form for temperature, windspeed, atmospheric pressure, relative humidity, rain fall and cloud cover in the same time resolution.
Air quality data are used to derive initial and boundary conditions and for eveluation purposes. The characterization of the surface in the form of land use classes is needed for the estimation of both the diffusion towards the surface and the absorption rate.

4.3 Application of the RTM-III-model

The first application of the RTM-III-model was carried out for the photochemical smog episode, 22-26 July 1980. The meteorological situation during this episode was characterized by a high pressure system moving from Central Europe in northern direction. This period was selected because measured maximum hourly ozone concentrations in many parts of Central Europe exceeded 100 ppb. Peak values were between 160-180 ppb.
Mixing height fields for the episode considered have been constructed by the Royal Dutch Meterological Institute (KNMI) using the Air Mass Transformation model (12) in combination with radio sonde data. The so determined maximum mixing height is 1900 m. Corresponding to the definition of the upper layers (Fig. 1) the top of the modelling region is set to 2200 m. The layer averaged windfields have been constructed by KNMI using the upper air data analyses of the ECMWF in combination with optimum interpolation schemes for ground level and 70 m height windfields.
All the meterological surface data have been derived from surface observations and the operational 4-dimensional analysis of the German Weather Service using a statistical interpolation scheme. The boundaries conditions have been calculated with the Eliassen/Hov-Lagrangian model (13) on the EMEPgrid with a grid size of 150 x 150 km employing the EMEP-emission inventory. For this purpose, the CBM-X-mechanism was incorporated in the model. To minimize the influence of the initialization of the RTM-III-model, the first day of the episode was run twice starting with estimated initial conditions.
The results of the RTM-III-model presented in the following are preliminary. The calculated concentrations are the result of the first run carried out by SAI using the input information as provided by the different working groups of PHOXA. To lower the computational loads, the gridsize was doubled for the first run, e.g. the horizontal grids have dimensions of approximately 60 km x 60 km. This first comparison with measurements shall help to identify those parameters which should be used in an input directed sensitivity analysis of the model. Fig. 2 until Fig. 4 show three examples for predicted and measured O_3 and NO_x-concentration time histories at one station in the FRG and two stations in the Netherlands. The station Langenbrugge (Fig. 2) is a background station in a rural environment. As can be seen, there is a good agreement between measured and calculated ozone concentrations. Most of the observed values are in the range between predicted mixed-layer average and predicted groundlevel (layer S, see Fig. 1) concentrations. Fig. 3 shows the comparison between observation and calculation for a grid cell in the nor-

thern, more rural part of the Netherlands. This grid cell contains three stations of the Dutch network. The variabality of the measurements in the grid cell is indicated by the bar around the mean value. For this grid cell the agreement for ozone concentrations is also quite good. An example for an ozone underprediction is given in Fig. 4, and shows the time history for a grid cell in the southern, more industrialized part of the Netherlands. This underprediction is most pronounced at July, 25th. Mixed layer averaged ozone deficits as shown in Fig. 4 are produced by the model mostly in the high industrialized regions in the Netherlands and in Belgium, where the measurements show a large spatial variability in the O_3 and NO_x fields (see Fig. 4) indicating the influence of nearby sources which cannot resolved by the grid system of the RTM-III-model. Sensitivity tests preformed with higher spatial resolution over these areas revealed that the unexplained variability in the ozone field can only be partly attributed to the coarse grid. Especially on July, 25th the meteorological input data may not represent the real atmospheric motion. Fig 5 shows an example for a hourly averaged predicted mixed layer O_3 field with inserted observations. The simulated ozone cloud exceeds 100 ppb over Belgium, the Netherlands and Western Germany. Comparison with the observations shows that position and gradients of the ozone cloud are predicted remarkably good. Some of the boxes with inserted observations display mean and range (i.g. minimum and maximum) in ozone measurements for a cluster of sites within and surrounding the box.
It is obvious, that the calculated concentrations, which are volume averages, agree reasonably well with the observed mean ozone concentration in a box, whereas the spatial variability in the observed ozone field cannot be reproduced by the model. Further sensitivity runs will be carried out to assess the results obtained in more detail. Two more episodes in 1982 will be alo calculated.
Some very preliminary conclusions which can be drawn from a number of runs of this first episode carried out and directed to the influence of HC - and NO_x-emissions on maximum O_3-concentrations indicate that for the total area the influence of HC - and NO_x-emission reduction on O_3-levels is similar to the influence found with the MPA-model. However these results also show marked differences in behaviour in different parts of the area.

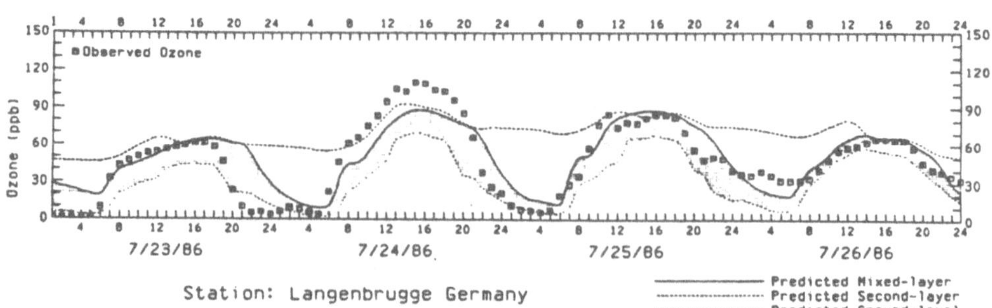

Fig. 2: Measured and calculated time history of O_3

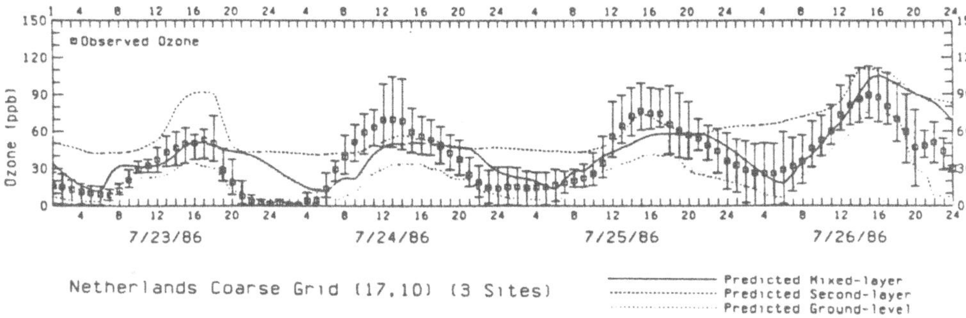

Fig. 3: Measured and calculated time history in a grid cell in the northern part of the Netherlands

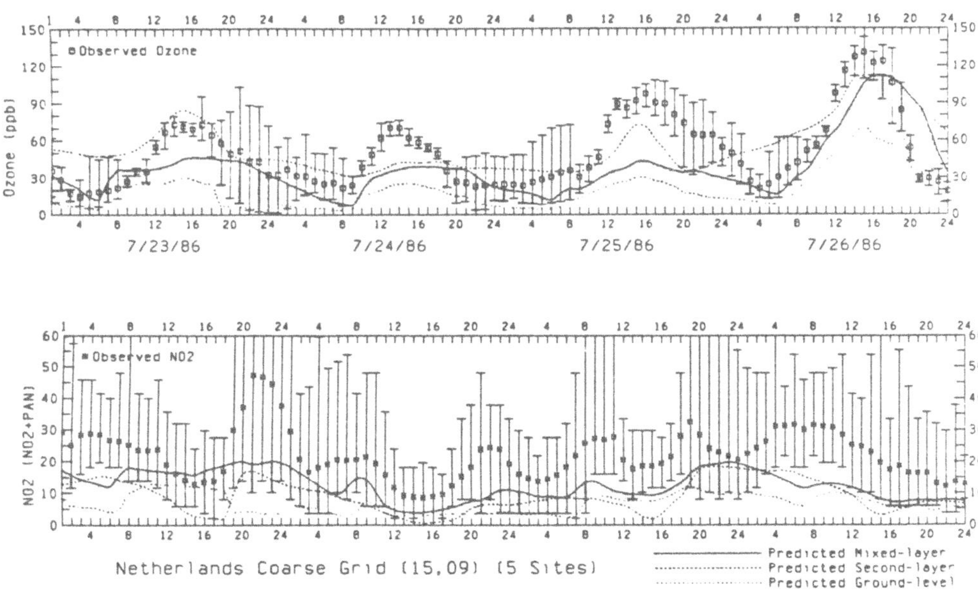

Fig. 4: Measured and calculated time history in a grid cell in the southern part of the Netherlands.

5. DESCRIPTION AND APPLICATION OF THE TADAP-MODEL

5.1. Model structure

The TADAP-model is a comprehensive, multi-layer, multi-species Eulerian grid model, which incorparates, as far as possible, the current understanding of the processes governing the transport, chemical transformation, and scavenging of acidifying species. The vertical grid system of the model is based on fixed layers extending up to approximately 10.000 m with a greater resolution near the surface. The transport of

the pollutants in the three-dimensional grid is handled by a sophisticated numerical solver for the solution of the advection-diffusion equation based on cubic splines.
The gas-phase chemistry uses a "lumped molecule" approach and is a condensed form of a very detailed mechanism (14). It consists of 90 reactions among 50 species, of which 30 are advected and includes the photochemistry that yields the radical pool which is responsible for the conversion of emitted SO_2 and NO_x to H_2SO_4 and HNO_3.
It also generates H_2O_2 and O_3, which are the primary oxidants in the aqueous phase. The aqueous-phase chemistry consists of 25 reactions among 13 species including the mass transfer between gas and aqueous phase and the aqueous phase oxidation of S (IV) by O_2, O_3 and peroxides. The characteristics of the aqueous phase, the concentrations of the different water phases in clouds and the conversion rates between these phases are derived from two cloud models. A one-dimensional steady-state model for stratiform clouds, which is driven by surface precipitation and other meteorological inputs, and a cumulus cloud model based on the parameterization proposed by Scott (15). The bulkwater technique developed by Kessler (16) is used for the microphysical formulation.
Dry deposition is modeled using a resistance approach in which the dry deposition velocity is expressed as the inverse of the sum of the atmospheric resistance, the deposition layer resistance and the surface layer resistance.

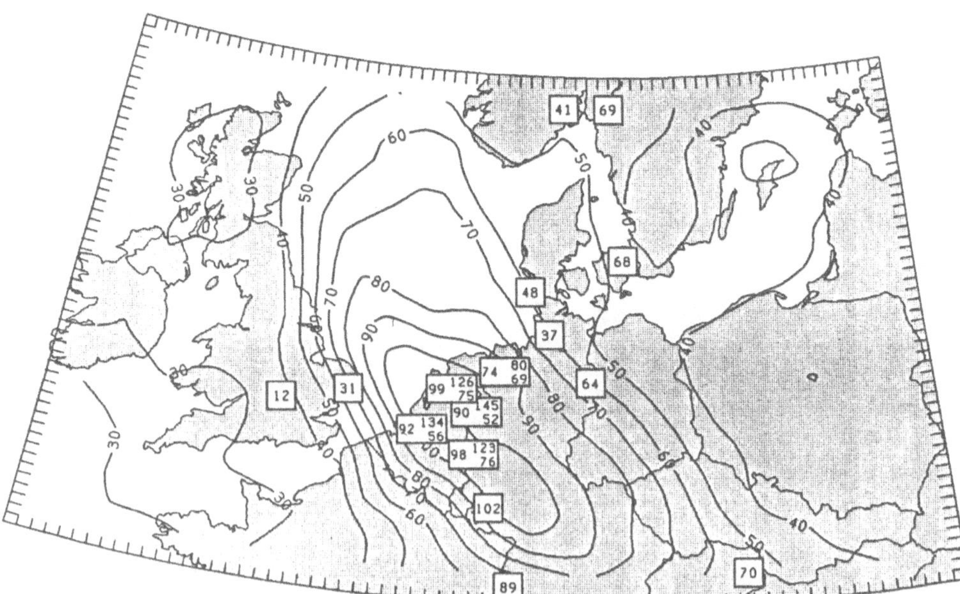

Hourly Averaged Predicted Mixed-Layer O3 Concentrations (ppb)
With Inserted Observations
For Hour Ending 1800 on 7 / 26 / 80

Fig. 5: Calculated ozone isopleths in the PHOXA-region (RTM-III-model)

Model output refers primary to variables which are directly related to the acid deposition problem: S (IV), S (VI), nitrates, H^+, NH_4^+, Cl^- in rain, and gas phase concentrations of SO2, NO_x/NO_y, O_3 and HC.
A more comprehensive model description can be found elsewhere (17).

5.2 Planned model application

TADAP will be applied in the PHOXA-area to an acid deposition episode, febr. 21 - March 10, 1982. Emission and land use data will be derived from the general data base of PHOXA. The very complex meteorological input data sets cannot be derived from observations because many of the needed variables such as cloud water content and the microphysical conversion rates are not measured routinely. Therefore, the regional weather prediction model of the German Weather Service, (the so called EUROPA-model), will be used as a driver for TADAP. A more comprehensive discussion of the meteorological input preparation and an analysis of the selected acid deposition episode can be found in an other paper of this symposium (18).

6. CONCLUSIONS AND RECOMMANDATIONS

- Calculations carried out using a two-layer trajectory model with receptor points in the Netherlands indicated that HC-emission control is much more effective than NO_x-control in abating maximum O3-concentrations.
- Evaluation of the first, preliminary results of the application of RTM-III to a photochemical episode showed that the model predicts the temporal and spatial pattern of ozone in the large scale reasonably well, especially in rural areas.
- Further sensitivity runs with RTM-III should be carried out, including application to other episodes to reveal in detail the O_3-behaviour in Central Europe, with emphasis on the possibly different behaviour in different parts of the area.
- The effect of possible emission abatement strategies should be evaluated taking into consideration both photochemistry and acid deposition simultaneously.

REFERENCES

1. Builtjes, P.J.H. et al. (1984). Evaluation of the performance of a photochemical dispersion model in practical applications. In: Air Pollution Modeling and its Application III. Plenum Press.
2. Stern, R. and Scherer, B. (1984). Anwendung eines complexen Ausbreitungsmodells zur Simulation eines photochemischen Smogepisode. Abschlussbericht FE-Vorhaben Nr. 104 02 551/01 Umweltbundesamt Berlin.
3. Eliassen, A. und Saltbones, J. (1983). Modeling of long range transport of sulphur over Europe: A two year model run and some model experiments. Atmos. Environ. 17,1457-1473.
4. Meinl, H. und Builtjes, P. (1984). Photochemical oxidant and acid deposition model application (PHOXA) within the framework of control strategy development. Umweltbundesamt, Berlin.
5. Liu, M, K et al. (1984). Development and application of a regional oxidant model to the north east United States. Atmos. Environ, 18 (6), 1145-1161.

6. EPA (1984). Proceedings of the EPA-OECD International conference on long range transport models for photochemical oxidants and their precursons, EPA-600/9-84-006.
7. V.d. Hout, K.D. e.a. (1985). Hydro-carbons in relation to air quality. Dutch Air Quality Management System no. 48, CMP-TNO 85/03 (In Dutch).
8. Builtjes, P.J.H. (1985). Summary-report hydro-carbons. Dutch Air Quality Management System no. 49. CMP-TNO 85/04 (In Dutch).
9. Boris, J.P. and Book, D.L. (1973). Flux corrected transport, I: SHASTA, an algorithum that works. J. Comp. Phys, 11, 38-69.
10. Zalesak, S.T. (1979). Fully multidimensional flux-corrected transport algorithus for fluids. J. Comp. Phys, 31, 335-362.
11. Durran, D.M.. A study of long-range air pollution problems related to coal development in the Northern Great Plains, Atmos. Environ. 13, 1021-1037.
12. Reiff, J. (1984). An air mass transformation model for short-range weather forecasting. Monthly weather review 112_r.
13. Eliassen, A. et al. (1982). A Lagrangian long range transport model with atmospheric boundary layer chemistry. J. Appl.Met., 21.
14. Atkinson, R. et al. (1982). An updated chemical mechanism for hydrocarbon/NO_x/SO_2 photooxidation suitable for inclusion in atmospheric simulation models. Atmos. Environ. 16,1341-1355.
15. Scott, B (1978). Parameterization of sulfate removal by precipitation. J. Appl. Met.,17.
16. Kessler, E (1969). On the distribution and concentration of water substances in atmospheric circulations. Meteor.Monogr. Vol.10,No.32.
17. Stern, R.M. (1986). Das "Transport and Deposition of Acidifying Pollutants" - Modell (TADAP) - Beschreibung und geplante Anwendung. Querschnittsseminar "Atmosphärische Prozesse." Reichstag Berlin published by Umweltbundesamt Berlin.
18. Scherer, B. and Scholl, M. (1986). Analysis of an acid deposition episode and preparation of meteorological input data for an Eulerian long range transport and deposition model.
4. European Symposium on Physico-Chemical Behaviour of Atmospheric Pollutants, Stresa 23-25 Sept., 1986.

ANALYSIS OF AN ACID DEPOSITION EPISODE AND PREPARATION OF METEOROLOGICAL INPUT DATA FOR AN EULERIAN LONGRANGE TRANSPORT AND DEPOSITION MODEL

Bernhard Scherer und Margit Scholl
Freie Universität Berlin
Institut für Geophysikalische Wissenschaften

Summary

A complex 3-dimensional acid deposition model will be applied to the episode 20.2. - 11.3.1982 over Central Europe. An overview is given of the present status of the meteorological input preparation. The episode is devided in two periods with similar meteorological features: During both Central Europe is under the influence of a high pressure system moving eastwards followed by rapidly moving frontal zones carrying precipitation. However, air parcel trajectories show different origins in both periods. Selected air quality measurements are presented to show characteristics of pollutant behaviour. The daily concentration of SO_2 in the first period is higher than in the second. The mean concentration of $SO_4^{=}$ in rain is increasing eastwards for the total of the episode. To obtain physical consistency, a complex weather forecast model from the DWD will be used as a preprocessor, allowing to bypass the internal cloud moduls of TADAP. A detailed meteorological analysis will be performed utilizing conventional network observations and satellite data to gain additional insight into dynamical processes and to compare measured and prognostic variables with special emphasis to cloud and precipitation parameters.

1. INTRODUCTION

As part of Dutch-German cooperation the Federal Republic of Germany (FRG) and the Netherlands support programmes to develop and apply mathematical models to calculate the transport, chemical transformation and deposition of air pollutants on a regional scale. The PHOXA programme (PHotochemical OXidant and Acid deposition model application) (1) as part of this cooperation utilizes several models of varying complexity and different physical and chemical structures (2; paper presented at this symposium), one being the Transport And Deposition of Acidifying Pollutants model (TADAP) developed by ERT (3). The preparation of its meteorological input is partially described in this paper.

The TADAP-model is especially designed to calculate the formation and deposition of acidifying substances. Evaluation runs are now being performed for eastern parts of the United States of America and Canada during the OSCAR-episode (Oxidant Scavenging Characteristics of April Rain, 1981). Within PHOXA the TADAP-model will be applied over Western Central Europe during a 20 day acid deposition episode from February 20 until March 11, 1982. A brief description of the modelling area is given in part 2 of the paper, main characteristics of the meteorological situation are described in part 3. Selected air quality data are presented in part 4 to show main features of pollutant behaviour and distribution during this episode.

To gain an optimum of physical consistency among the manifold meteorological input, a mesoscale dynamical weather forecast model - the EUROPA-MODELL (EUM) (4) - operated by the German Weather Service (DWD) is used to derive the the total meteorological input. A brief description of the meteorological input requirements and the procedures employed to create the data is given in part 5.

To support the interpretation of the model results a meteorological analysis will be performed utilizing network observations and METEOSAT data to derive cloud and precipitation parameters. Its methodology is also briefly discussed in part 5.

2. THE MODELLING AREA

The modelling area lies approx. between 47° and 60° N and between 10° W and 25° E (see surface charts in Fig.1). It covers Central Western Europe including Great Britain and Northern France in the West. The eastern boundary runs through western parts of the USSR and Rumania. The southern boundary runs through Central France, Switzerland, Austria, Yugoslavia and Hungary. In the North, Southern Sweden and Norway are included. Therefore, this region encloses most of the major industrial centers of Europe.

The grid used is on a polarstereographic projection (60° N, $\lambda_0 = 10°$ E), identical to the one used with the EUM. The size of a gridcell is 63.5 km^2. There are 42 gridcells from West to East and 26 gridcells from South to North and 10 layers in the vertical up to approx. 9000 m. To gain a high resolution in the boundary layer a semi-logarithmic coordinate is used in the vertical with 5 grid cells below 1000 m.

Air quality data are available from EMEP (5). There are 55 monitoring stations inside and several stations outside the region. However, not all stations report complete data records during the episode. For the Netherlands and FRG additional data are provided from local authorities. The routine meteorological observations from over 500 stations in the region were made available by the DWD.

3. SURVEY OF THE METEOROLOGICAL SITUATION

The episode can be devided in two periods with rather similar meteorological conditions: the first from February 20 until March 4, and the second from March 5 until March 11. During both periods Central Europe is more or less under the influence of a high pressure system which moves eastwards followed by rapidly moving fronts within a strong westerly flow carrying precipitation as rain and snow. A low pressure system over the northern Mediterranean leads in conjunction with the anticyclone over Central Eastern Europe to air mass transports over the south east boundary towards the center of the region.

Some of the meteorological features of the episode are depicted in Fig.1. At the start of the first period the blocking high is situated over Central Europe. It starts moving slowly to the East on February 21. While the anticyclone moves eastwards the low over the Mediterranean extends northwards on February 23. On its eastern side an area of snowfall is lead from Northern Yugoslavia over Austria into Germany, where it snows temporarily until February 26. On February 27, Central Europe is again under high pressure influence. The first warm front arrives over the central region on February 28 followed by cold fronts on March 1. More fronts cross over Central Europe until March 4 when the Atlantic low moves rapidly eastwards over Scandinavia. The dynamical development will be demonstrated by presenting an experimental film loop of three-hourly satellite images of the thermal infra-red METEOSAT channel.

A new high develops on March 6 and a cyclone over the Adriatic Sea leads to heavy precipitation over Yugoslavia and parts of Austria. The anticyclone over Central Europe moves eastwards and on March 10 again fronts cross over Europe in a strong westerly current. Windspeed over the region is higher during the second period.

At the start of the episode the daily mean temperature varies between 5°C and -5°C from western to eastern parts of the region. Approximately half the region from the FRG to the eastern boundary is under snow cover. At the end of the episode the daily mean temperature rises 3° to 5°C over the region and most of the snow disappears except at higher altitudes.

Precipitation varies during the course of the episode over the region (see Fig.2).

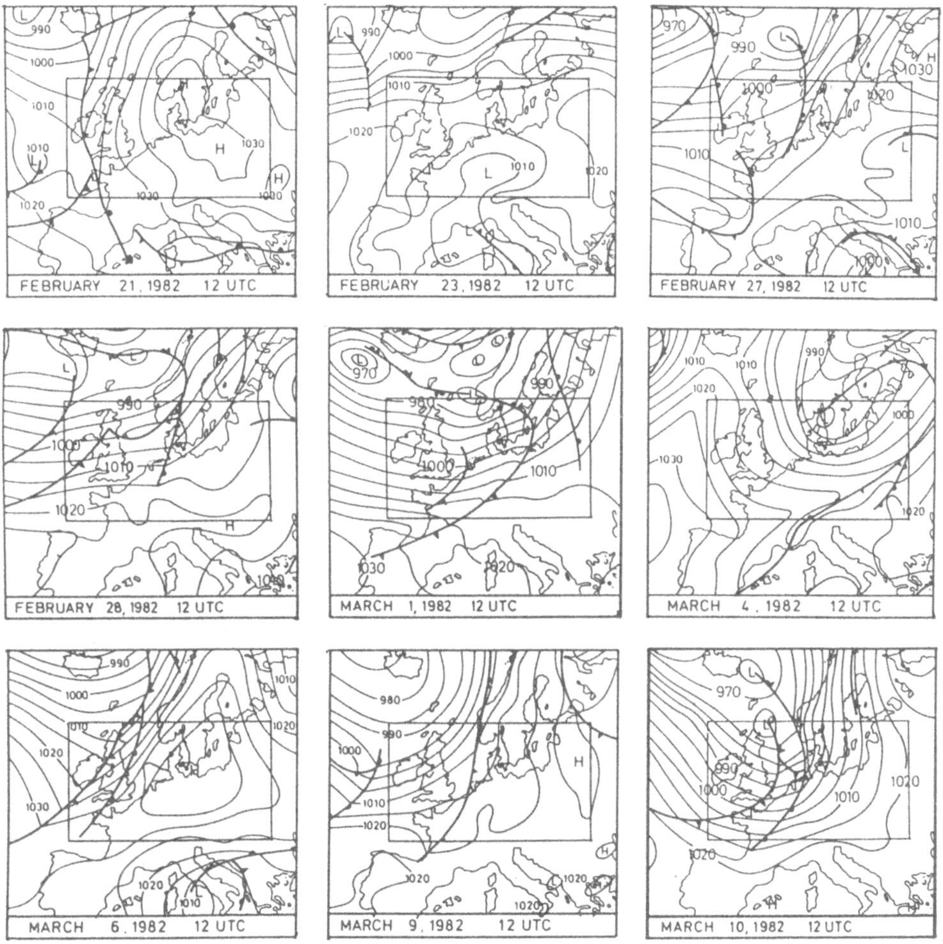

Fig. 1 Surface pressure charts with frontal analysis for 9 selected days of the episode. The modelling region is enclosed by the inner rectangular.

Ireland and the western parts of Britain and Norway receive over 100 mm up to 180 mm while Poland and parts of Russia and Czechoslavacia receive less than 20 mm. At the coastal regions and the mountainous regions at the southern and eastern boundaries precipitation varies ranges between 50 and 100 mm.

4. SELECTED MONITORING RESULTS AND ANALYSIS

The distinct division of the episode by meteorological conditions can be seen in the evolution of the daily mean SO_2-concentration in the FRG (Fig.3). The figure indicates that the concentration during the first part of the episode is higher than during the second period, which is more pronounced in the center and at southern stations. At the mountain sites, values are lower but also depict two distinct maxima. In the course of the episode the highest mean concentration of nearly 300 $\mu g/m^3$ was observed at the station Usingen on February 26. On that day the mean daily SO_2-concentration averaged over 17 stations in the center region e.g. Belgium, Netherlands and FRG, is approx. 130 $\mu g/m^3$. Starting values are high

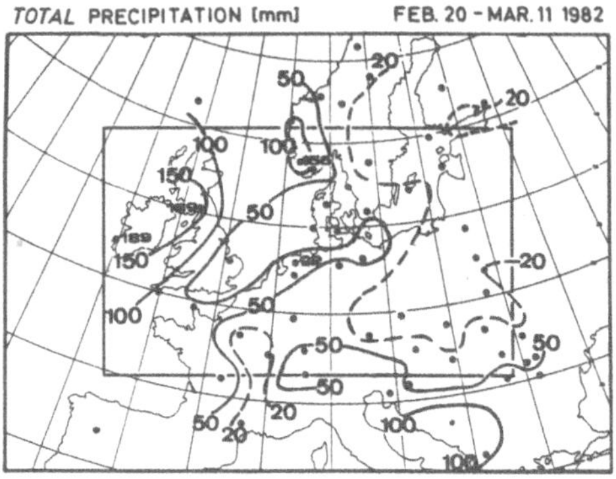

Fig. 2 Total amount of precipitation during the episode from February 20 until March 11

at the beginning which obviously is due to the meteorological situation before February 20 which allowed pollutants to accumulate over several days. After fronts and precipitation have passed over Europe concentrations rise from background values with the start of the second periode.

SO_2-concentration in the rest of the region is lower in the center as far as reported. In the northern and western parts the maximum values are 70 $\mu g/m^3$ in Sweden and 86 $\mu g/m^3$ in France. In the south eastern part Hungary reports the highest values of 117 $\mu g/m^3$ right at the start of the episode. The concentration behaves differently at those stations due to different meteorological conditions.

NO_2-concentrations in the central modelling area show overall a rather similar behaviour as SO_2 with the exception of Westerland where the maximum daily mean concentration of NO_2 is highest in the second part. Maximum values range between 12 $\mu g/m^3$ at mountain sites and 45 mg/m^3 at rural stations.

For three stations in the FRG backward trajectories at 850 hPa provided by the Norwegian Meteorological Institute (6) are shown for selected days in Fig.4. The date of arrival at the site is indicated at every trajectory. The time of arrival is 12 UTC (full lines) and 00 UTC (dashed lines) on the day indicated. Going backwards from the site, every mark (▶) indicates a 24 hours distance. The maximum lenght of the trajectories is 96 hours unless they leave the region earlier. During the first part, see left side of Fig.4, the trajectories have their origin in the eastern or south eastern parts of Europe. Trajectories from the Mediterranean reach only the southern sites of Deuselbach on February 24/25 and of Brotjacklriegel on February 25/26. Trajectories ending at these stations during the time of the observed SO_2-maximum are the shortest, indicating low windspeed. Thus the air had been moving slowly over areas with high emissions for 3 to 4 days, allowing pollutants to accumulate. When the strong westerly current developed over Central Europe the trajectories turn to the West. Air with origin over the Atlantic reached Westerland 24 hours before Brotjacklriegel in the South of the FRG.

In the second part of the episode (see the right side of Fig.4) windspeed is considerably higher than in the prior period indicated by the larger distance of the 24 hourly trajectories. The trajectories ending over Westerland and Deuselbach enter the region mainly from the North, cross over the central and south eastern parts and arrive at last from southern directions. During this part of the episode only most eastern sites of the FRG are reached by trajectories with their origin in the Mediteranean.

To give an impression of regional differences of $SO_4^=$ wet deposition, the total amount during the episode is shown in Fig. 5 a. The total deposition of SO_4 over Poland is approx. and low as over Ireland, although Poland receives only one tenth of the amount of precipitation than Ireland and the western parts of Britain (see Fig. 2). Consequently the mean concentration of $SO_4^=$ in rain during the episode increases eastwards (Fig. 5 b). Total deposition is high over industrialized areas in the Netherlands and Germany and in the mountainous region at the south eastern border but not neccessarily where high $SO_4^=$ concentration occurs. Only over Yugoslavia high concentrations combined with large amounts of precipitation lead to the highest deposition in Europe during this episode. The vertical distribution

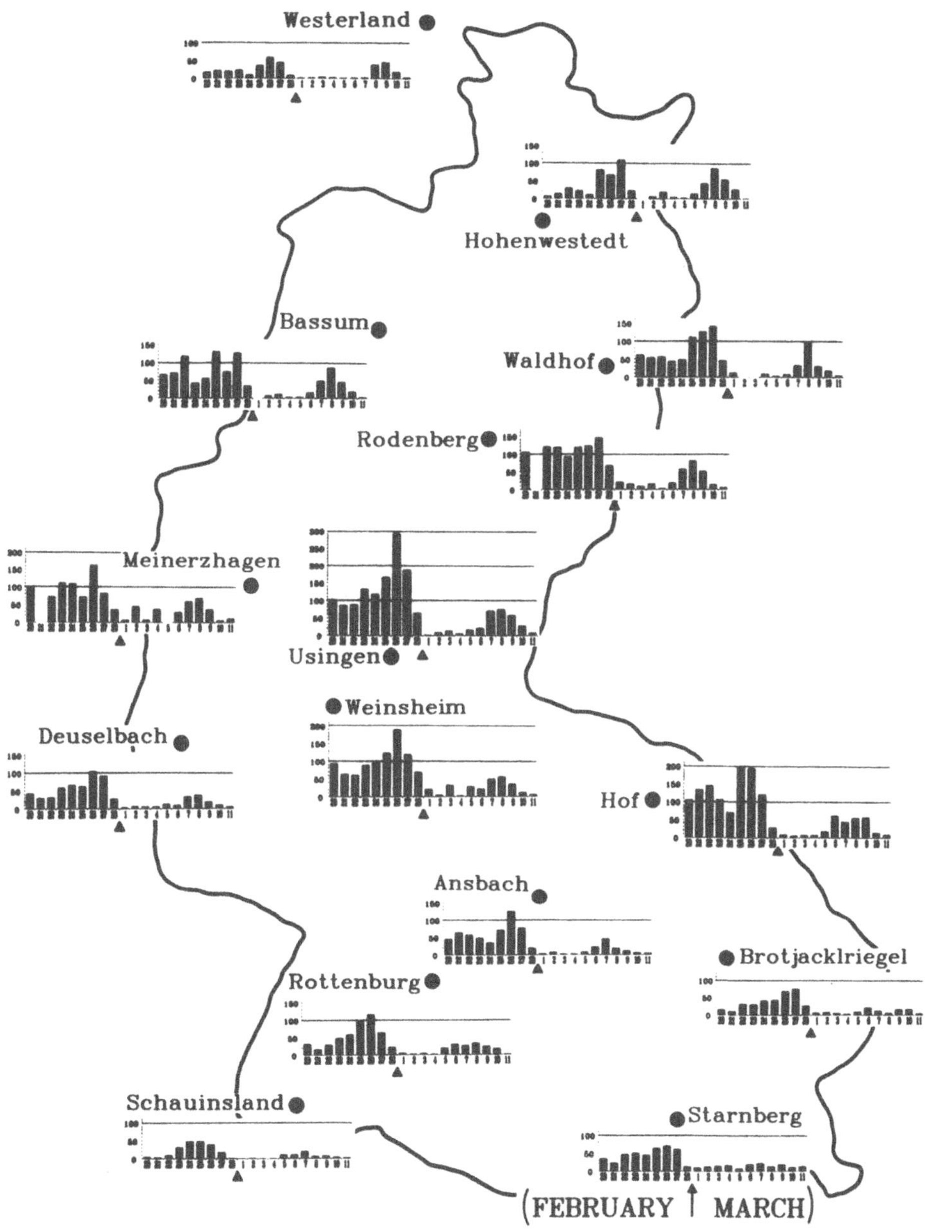

SOURCE: AIR QUALITY MONITORING SYSTEM, U B A (F.R.G.)

Fig. 3 Daily mean SO_2 concentration at 15 stations in the FRG. Stations characteristics: Rural sites are Westerland, Waldhof and Deuselbach. Mountain sites are Schauinsland (1200 m) and Brotjacklriegel (1000 m). All other stations are in the vicinity of urban areas (25 to 60 km) and may show rural characteristics as well as urban influence. Horinzontal lines in the station diagrams indicate concentration values at 100, 200 or 300 $\mu g/m^3$ respectively.

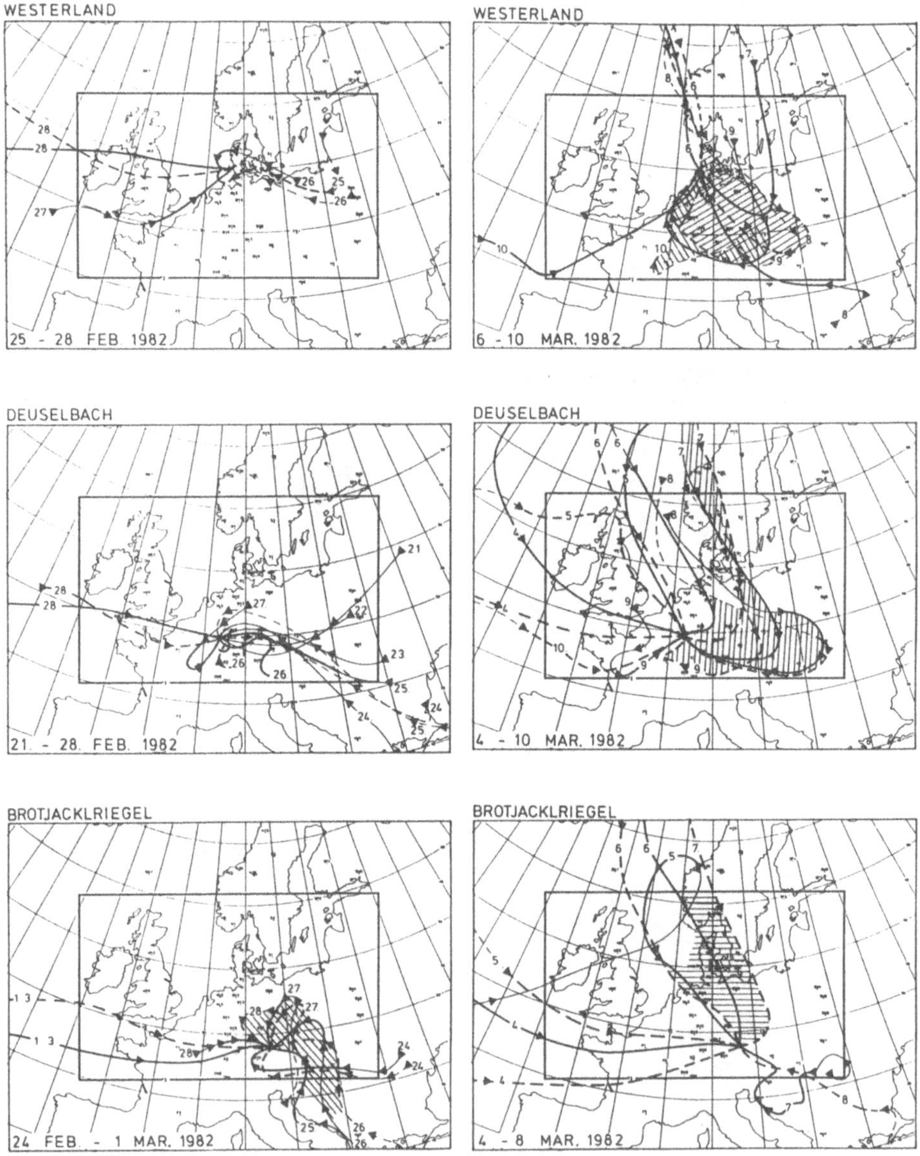

Fig. 4 Successive 96-hours backward trajectories at 850 hPa for three sites and selected days of the episode. Date of arrival is indicated at each trajectory. Time of arrival is 12 UTC (full lines) or 00 UTC (dashed lines). Trajectories arriving at station during days with higher SO_2-concentration lie within the hatched areas. See text for further explanations.

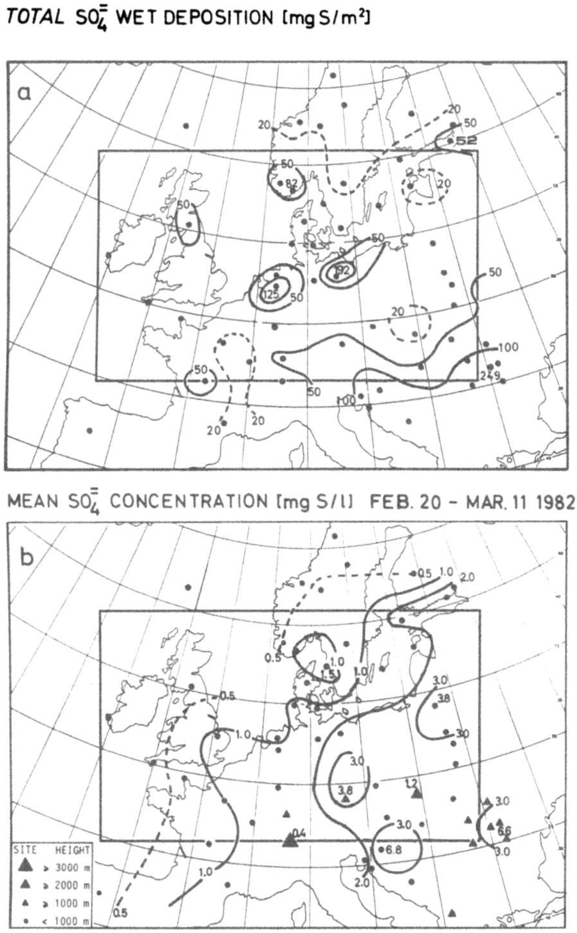

Fig. 5

a) Total amount of $SO_4^=$ deposition during the episode (February 20 - March 11)

b) Mean $SO_4^=$ concentration in rain during the episode

is indicated by the two mountain sites over 2000 m and 3000 m. Their measurements (1.2 $\mu g/m^3$ and 0.4 $\mu g/m^3$) show substantial lower concentrations than the surrounding stations including mountain sites up to 2000 m. Until now the available data have only been partially analysed. The analysis will be continued including other variables as pH, H^+, NO_3^- and NH_4^+.

5.1 PREPARATION OF THE METEOROLOGICAL INPUT FOR TADAP

The preparation of the meteorological input combines two major tasks as shown in the flow diagram in Fig.6.

- Utilization of a prognostic meteorological model output to derive the total meteorological input for the TADAP-model including micrometeorological cloud variables. This allows to bypass the two cloud modules of TADAP which are used to calculate the variables needed for the aqueous phase chemistry calculations. The German Weather Service performs the EUM runs and provides the required output. These data are further processed at the Freie Universität Berlin.
- Performance of a three dimensional objective analysis for the episode utilizing satellite data and conventional network observations. The analysis will improve the actual understanding of the meteorological situation during the episode, i.e. vital parts of the input as precipitation, cloud parameters, location of frontal zones etc. can be discussed qualitatively. As an alternative when the use of cloud modules is intended this analysis also provides the input of these modules. The combined information of both prognostic model results and analysis will be integrated into the evaluation of the TADAP model.

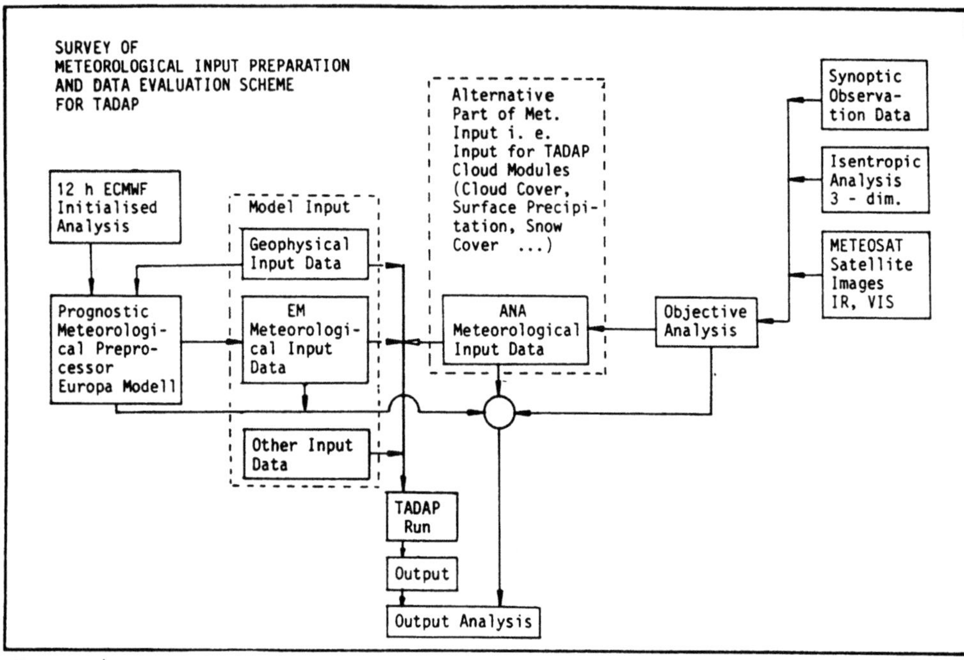

Fig. 6 (see text for further discussion)

5.2 Data requirements

The meteorological input of the TADAP-model consists of:

- surface meteorological data
- micrometeorological cloud variables
- wind field and vertical profiles of meteorological scalars.

Temporal resolution of all variables is one hour. Surface meteorological variables are wind, temperature and pressure, friction velocity, convective velocity scale and Monin-Obukhov length, the total sky cover, soil and water surface temperature, snow and ice cover, snow depth and albedo as well as the moisture content of soil and canopy layer.

Alternative inputs are the large and small scale precipitation, cloud base and top height, precipitation type and cloud type code. These variables are only needed if the micrometeorological cloud variables of the EUM are not used. The cloud variables consist of profiles of hydrometeor mixing rations e. g. cloud water, rain, snow/ice and the conversion rates between these fields. The scavenging rate for sub micron particles, the mass mean radii of the hydrometeors and their fall velocities are also needed.

The wind field is required in horizontal and vertical components at cell interfaces. Turbulent vertical diffusivity is also defined at cell interface. Other scalars as temperature, pressure and humidity are required in the cell center.

5.3 Methods

The EUM is presently prepared for operational use for short range weather forecast at the DWD. It is applied to an area enclosing Europe and the East Atlantic. Its special features are:

- a high horizontal (63.5 km) and vertical resolution (18 layers, where 6 layers are within the lowest 1200 m (7))
- detailed parameterization of hydrological and near surface processes
- detailed representation of the forcing functions at the surface.

Initial and boundary conditions are derived from large scale analysis (8). To limit prognostic errors, a special scheme is used: Every 12 hours a new forecast is started initialised with the latest analysis. After a spin up time of approx. 6 hours the old forecast is forced to approach the new one within successive steps to avoid discontinuities within the meteorological fields (9).

Besides the usual prognostic variables e.g. temperature, wind etc., the model predicts explicitly the liquid water content of clouds (10) and provides all microphysical variables required for the aqueous phase chemistry calculations in both stable and convective clouds. It provides also the water content of two soil layers extending down to the root zone and the surface conditions needed in the deposition calculations.

The TADAP-model input is derived from the one hourly EUM output. Vertically both models use different grid systems. Therefore the variables will be interpolated from the sigma-system of the EUM to the terrain following logarithmic coordinate system of TADAP with a spline interpolation scheme. The horizontal wind components have to be interpolated horizontally as well, since they are required at the cell interfaces. Some surface variables as temperature, wind, friction velocity etc. are derived externally after the model run, utilizing semi empirical relations (11). All other surface variables can be taken directly from the EUM-output.

5.4 Objective analysis and the use of satellite data

Fig. 7 shows the data flow and the desired outputs of the analysis in their approximate order of importance. The routine surface observations are used to analyse occurence of fog and fog type, the ground surface conditions and visibility. To gain three dimensional meteorological fields isentropic analysis are performed every 12 hours (12). The optimal vertical resolution of isentropic surface resolves many mesoscale features which are not usually depicted in large scale analysis. Isentropic analysis is utilized to derive fields of meteorological scalars to locate temperature inversions and frontal zones and to diagnose dynamical processes in detail.

A main disadvantage of the use of conventional network data is the fact that soundings are only available every 12 hours and cloud parameters cannot be derived from these observations completely. The gap can be filled up by using METEOSAT data of the visible channel in the 0.5 - 0.9 μm spectral band and the infra- red channel in the 10.5 - 12.5 μm 'window', which are processed every half hour.

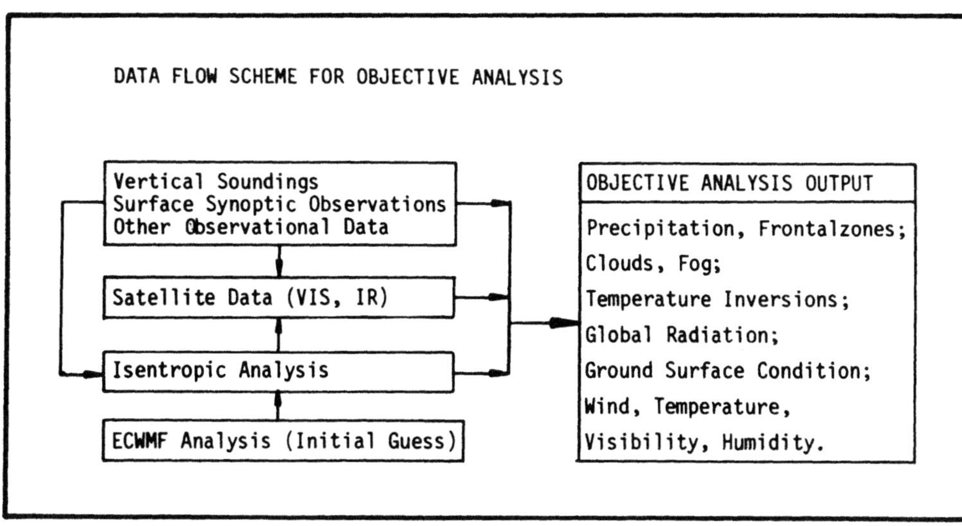

Fig. 7 (See text for further discussions)

We use an automatic pattern- analysis algorithmus to produce images with sharpedged, uniform radiance segments which are distributed into 10 classes of clouds (13). With the inclusion of these data and the aid of three-dimensional pressure and temperature fields, ground observations can be completed by informations about cloud top height, cloud analysis, frontal zones cloud systems and upper tropospheric winds etc. The knowledge of the actual vertical cloud extent and its development indicates the distribution of precipitation which itself is of major interest for the wet deposition. Moreover, criterias can be derived to interpolate observed values such as cloud amount and precipitation to the model grid, allowing to examine parts of the meteorological TADAP input.

Acknowledgements

The project is supported by the Umweltbundesamt. Valuable discussions with Mr. Stern are highly acknowledged. We thank Dr. Reimer who provides the isentropic analysis and Mr. Schubert for the development of the plot software. We would like to thank the Technical Group of the Institute of Geophysical Science for professional help. Mrs. Rothkegel typed the manuscript.

REFERENCES

1. Meinl, H. und Builtjes, P. (1984) Photochemical oxidant and acid deposition model application (PHOXA) within the framework of control strategy development. Umweltbundesamt, Berlin.
2. Stern, R.M. and Builtjes, P. (1986) Long range Modelling of the Formation, Transport and Deposition of Photochemical Oxidants and Acidifying Pollutants. European Symposium of Physico Chemical Behaviour of Atmospheric Pollutants, Stresa, Italy.
3. Stern, R.M. (1986) Das "Transport And Deposition of Acidifying Pollutants" Modell (TADAP). Beschreibung und geplante Anwendung. Querschnittsseminar "Atmosphärische Prozesse", Berlin, published by Umweltbundesamt, Berlin.
4. Schwirner, J.U., Müller, E., Linke, A., Majewski, D. (1983) Physikalisch-numerische Struktur des EUROPA-Modells. Ann. Meteor. NF 20, 227 - 228.
5. NILU Norwegian Institute for Air Research (1985) Preliminary Data Report EMEP/ECE, Oktober 1981 - March 1982.
6. Eliassen, A. Norwegian Meteorological Institute. Trajectories provided for PHOXA-modelling programme.
7. Müller, E. (1981) Turbulent flux parameterization in a regional-scale model. ECMWF Workshop on planetary boundary layer parameterization, 193-220.
8. Majewski, D. (1985) Balanced initial and boundary values for a limited area model. Beitr. Phys. Atm. 58, 147-159.
9. Jacobsen, I. (1986) Die Bereitstellung einer meteorologischen Datenbasis für das Ausbreitungsmodell TADAP mit dem EUROPA-Modell des DWD. Querschnittseminar "Atmosphärische Prozesse, Berlin, published by Umweltbundesamt, Berlin.
10. Müller, E. (1983) Parametrisierte Niederschlagsprozesse in einem regionalen Wettervorhersagemodell. Ann. Meteor. NF 20, 204 - 207.
11. Businger, J.A. et al. (1971) Flux profile relationships in the atmospheric surface layer. J. Atm. Sci. 28, 181 - 189.
12. Reimer, E. (1980) A test of objective analysis with optimum utilization of the radiosonde network in Central Europe. Contr. Atm. Physics, 53, 3.
13. Hilt, E.(1982) Überlegungen und Untersuchungen zur maschinellen Analyse meteorologischer Satellitenbilddaten, Dissertation, Berlin.

PHYSICO-CHEMICAL PROPERTIES OF RADIATION FOG IN THE PO VALLEY, ITALY

S. FUZZI, G. ORSI and G. NARDINI
Istituto FISBAT - C.N.R.
Via de'Castagnoli 1, 40126 Bologna (Italy)

M.C. FACCHINI
Istituto Chimico "G. Ciamician", Università di Bologna
Via Selmi 2, 40126 Bologna (Italy)

Summary

A field program is described, which is being carried out in the Po Valley by a group of European and American institutes in the attempt to describe the heterogeneous processes (processes involving gas, aqueous and particulate phases) occurring in fog. A complex network of instrumentation has been set up in the field during two consecutive experimental campaigns at a site in the Po Valley. The results obtained up to now are presented and the need for future experiments to complete the overall picture of radiation fog processes is underlined. An evaluation is also given regarding the role of radiation fog in chemical deposition.

1. INTRODUCTION

A great number of studies has been undertaken over the last few years in order to understand the complex phenomenon of acid deposition, namely the relationship between emission of pollutants into the atmosphere and deposition of acidifying chemical substances which cause environmental damage in ecologically sensitive areas (1). An important aspect of the complex cycle which links precursor emission and acid deposition is the chemical processing of pollutants, operated by the atmospheric dispersed liquid phase: fog and cloud droplets and precipitation elements.

In this paper, we present the initial results of a field program which attempts to describe the above mentioned processes in a particular, well defined atmospheric system: radiation fog. The program has been undertaken in collaboration with several European and American institutes: Atlanta University (AU), Atmospheric Sciences Research Center (ASRC), Boris Kidric Institute (KIBK), Centro Italiano Studi Esperienze (CISE), ENEL Centro Ricerche Termiche Nucleari (CRTN), Georgia Institute of Technology (GIT), Lawrence Berkeley Laboratory (LBL), National Center for Atmospheric Research (NCAR), Presidio Multizonale USL 28 (PMP), Technical University of Vienna (TUV), University of Frankfurt (UF), University of Padova (UP), University of Vienna

(UV). The project is being carried out at the FISBAT field station of S. Pietro Capofiume, in the eastern part of the Po Valley (SPC in Fig. 1). In

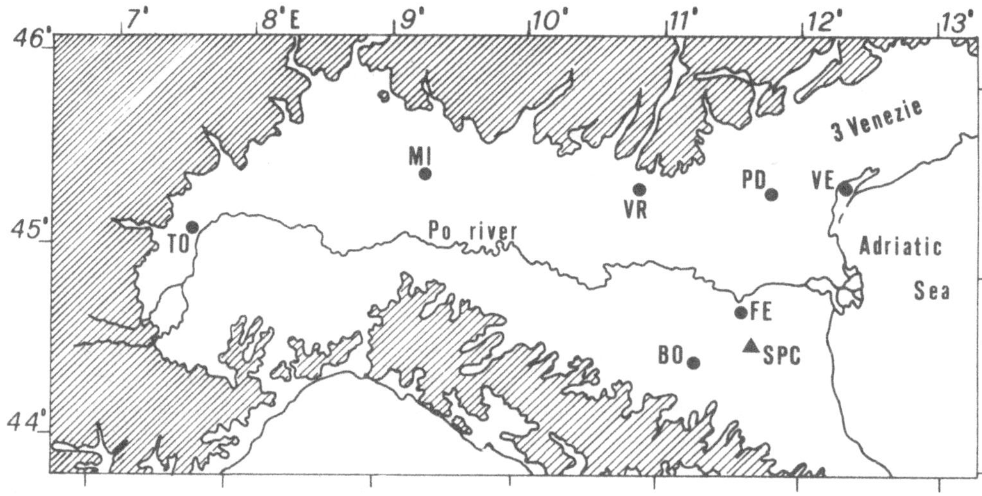

Fig. 1 - Map of the Po Valley, Italy.

this area, radiation fog frequently occurs from October through March as shown in Table 1.

month / time of the day (GMT)	0000	0600	1200	1800
October	6	6	1	1
November	8	8	4	6
December	12	12	10	12
January	12	14	10	12
February	8	10	5	5
March	2	4	1	0

Table 1 - Average number of days, during the fall-winter months, in which, at different times, the visibility in the area of S. Pietro Capofiume is less than 1 Km, due to fog occurrence (20 years statistics).

Radiation fog is formed by the steady cooling of air close to the ground, until the dew point is exceeded (2). Since radiation fog forms in conditions of high atmospheric stability, with minimal air mass exchange, it is possible to apply the mass balance criteria between the three phases involved (gas, liquid and particulate), in order to determine the physico-chemical transformations of the various components of the system. The study of a "cloud formed in contact with the ground" also allows the use of a widespread network of instrumentation which would not be possible on an aircraft.

2. EXPERIMENTAL PART

The network of instrumentation used during this program is continuously improved in function of the experimental results obtained during the measurements in progress. The present experimental set-up is shown in Fig. 2.

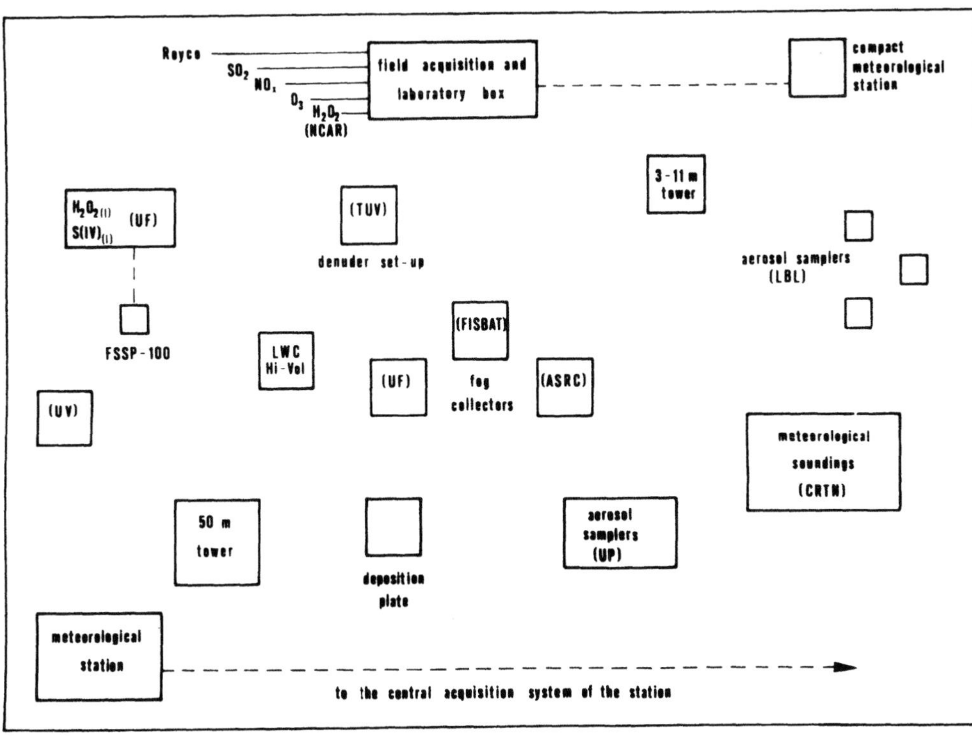

Fig. 2 - Schematic diagram of the instrumented field network. For the identification of the groups, see the text.

Only this summary will be proposed here; a detailed description of all experimental operations is reported elsewhere (3).

Two major campaign have been carried out up to now: November 84 and

Components		Range of concentration and mean concentration	
		November 84	November 85
H^+	(μeq l^{-1})	0.4 - 1148 (263)	0.2 - 3388 (320)
NH_4^+	"	279 - 4056 (1629)	519 - 5896 (1582)
K^+	"	6 - 103 (18)	14 - 302 (47)
Na^+	"	1 - 190 (34)	12 - 201 (40)
Ca^{2+}	"	11 - 454 (45)	12 - 235 (59)
Mg^{2+}	"	2 - 98 (14)	6 - 82 (20)
F^- [a]	"	9 - 135 (57)	23 - 467 (104)
Cl^-	"	48 - 331 (147)	53 - 1404 (201)
NO_2^-	"	0 - 104 (7)	0 - 26 (5)
NO_3^-	"	23 - 3340 (995)	172 - 2270 (555)
SO_4^{2-}	"	150 - 2066 (732)	326 - 3063 (908)
HCHO	(μM l^{-1})	23 - 105 (64)	1 - 53 (13)
S(IV)	"	n.m. ---	77 - 764 (310)
H_2O_2	"	< 0.1 ---	0.1 - 2 (1.2)
Fe	"	1 - 18 (6)	1 - 27 (8)
Mn	"	0 - 4 (1.1)	0.4 - 4 (1.6)

n.m. = not measured
[a] some weak organic acids co-elute with F^-

Table 2 - Mean concentration and range of the different components in fog.

November 85.

3. RESULTS AND DISCUSSION

In Table 2, a summary of fog water chemical composition during the two major field experiments is reported. The November 84 data set is almost completely processed and will be published soon. Some initial conclusions drawn from the analysis of these results are reported below.

a) Nucleation scavenging represents a major contribution to sulphate content in fog droplets, as already theoretically predicted (4).
b) Excess nitrate is frequently present in the liquid phase with respect to pre-fog aerosol. HNO_3 scavenged from the gas phase by the growing droplets could explain this finding; a second possibility is the occurrence of NO_x heterogeneous transformations (5). Some spot measurements of HNO_3 gas phase concentration (6) could not help discriminating between the two different processes.
c) Once fog is formed, the chemical composition of the system appears to be driven essentially by microphysical processes: condensation/evaporation/deposition of droplets. In fact, reducing the fog water concentrations to equivalent air concentrations, by taking into account the actual fog liquid water content (LWC), it is observed that oscillations occur in the trends of major ions concentration during the fog episodes. These oscil-

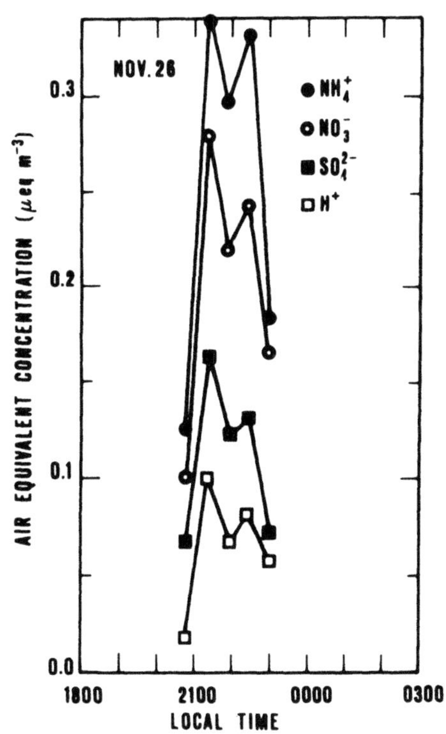

Fig. 3 - Temporal trends of major ion concentration during a typical fog episode. The concentrations were reduced to μeq per m^3 of air, by taking into account the actual LWC.

lations are in-phase with each other (see for example Fig. 3); this tends to rule out a significant contribution of aqueous phase reactions, which would lead to a different behaviour of nitrate with respect to sulphate.

d) Nevertheless, there is evidence of a potential for S(IV) to S(VI) liquid phase transformation in fog but this appears to take place at the very beginning of the fog episodes, or in the wet aerosol stage. The principal pathway for S(IV) transformation appears to be the transition metal catalyzed reaction. Only a minor contribution should be expected by O_3 and elemental carbon catalyzed reactions. H_2O_2 gas phase concentrations in the range of 100-200 pptv were measured in pre-fog conditions (7). H_2O_2 is, however, rapidly scavenged in the early stage of fog formation and its liquid phase concentration is always below detection limit (<0.1 μM).

These findings were considered only preliminary, since some important parameters were not continuously monitored (gas phase HNO_3 and NH_3, liquid phase S(IV)), some parameters would require better measuring sensitivity (instrumental SO_2 detection limit is 1 ppbv) and, finally, not all parameters have sufficient time resolution or frequency (fog LWC measurements by Hi-Vol filter technique (8), for example, have a frequency of 2-3 data per hour).

The processing of November 85 data is underway by the groups partecipating in the field operations. We are confident that with the bringing together of the new, more complete results, some of the preliminary conclusions will be assessed and some questions previously left open answered. Another major field experiment is now being planned for November 86.

During the project, some data were also taken on the contribution of radiation fog to chemical deposition in the Po Valley (9). Only data taken at the field station of S. Pietro Capofiume are available at the moment, but a program of measurements at other sites in the Po Valley will be initiated soon.

event	duration (hours)	TSP removal (%)
Nov. 23, 1984	12	78
Nov. 23-24, 1984	11	24
Nov. 24-26, 1984	27	68
Nov. 26, 1984	4	35
Nov. 27, 1984	12	50

Table 3 - TSP removal from the air during 5 fog episodes.

The first result obtained is an evaluation of the depletion of aerosol total mass (TSP) from the air due to fog. The results relative to five fog episodes in November 84 are reported in Table 3. The differences in removal efficiency are probably due to the different microphysical characteristics of the individual fog systems. Estimated wet deposition for major ions: ammonium, nitrate, sulphate, accounted for more than 10% of the input due to bulk precipitation over the fall-winter period (fog season in the Po Valley). On the other hand, the environmental effects of acid deposition (for example, injury to plant foliage) are more closely related to the H^+ concentration rather than to the total H^+ deposited (10). Thus, fog episodes characterized by low pH and much higher ionic loading of the droplets with respect to rain (one to two orders of magnitude), should be treated as intense pollution episodes, rather than long term budgets. Synergistic effects are also possible with pre-fog dry deposited substances, which could increase the potential effects of fog water deposition alone (11).

Acknowledgements - We wish to thank for their efforts all groups and individuals who partecipated in the field program described herein. This program is sponsored by "Progetto Finalizzato Energetica 2, Sottoprogetto Ambiente e Salute" and "Progetto Strategico Clima ed Ambiente dell'Area Mediterranea", both of the Italian National Research Council.

REFERENCES

(1) CALVERT J.G., LAZRUS A., KOK G.L., HEIKES B.G., WALEGA J.G., LIND J. and CANTRELL C.A. (1985) Chemical mechanisms of acid generation in the troposphere. Nature, 317, 27-35, and references reported therein.

(2) JIUSTO J.E. (1981) Fog structure. In: Clouds: Their Formation, Optical Properties and Effects (P.V. Hobbs and A. Deepak eds.), pp. 187-239, Academic Press, New York.

(3) FUZZI S. (Editor) Heterogeneous Atmospheric Chemistry Project. FISBAT C.N.R. Report, February 1986, Bologna, Italy.

(4) STELSON A.W., KIANG C.S. and FUZZI S. (1985) Relative times for sulfate accumulation in fog. Proceedings: Heterogeneous Processes in Source-Dominated Atmospheres, October 8-11, New York, New York.

(5) HEIKES B.G. and THOMPSON A.M. (1983) Effects of heterogeneous processes on NO_3, HONO and HNO_3 chemistry in the troposphere. J. Geophys. Res., 88, 10883-10895.

(6) PUXBAUM H., WEBER W. and PECH G. (1985) Occurrence of inorganic and organic acidic components at four source-dominated sites in Europe. Proceedings: Heterogeneous Processes in Source-Dominated Atmospheres, October 8-11, New York, New York.

(7) LIND J. and LAZRUS A. (1986) Measurements of hydrogen peroxide vapor at S. Pietro Capofiume. In: Heterogeneous Atmospheric Chemistry Project (S. Fuzzi ed.), FISBAT-C.N.R. Report, pp. 27-31, Bologna, Italy.

(8) PILIE' R.J., MACK E.J., KOKMOND W.C., ROGERS C.W. and EADIE W.J. (1975) The life cycle of valley fog. Part I: micrometeorological characteristics. J. Appl. Meteor., 14, 347-363.

(9) FUZZI S., ORSI G. and MARIOTTI M. (1985) Wet deposition due to fog in the Po Valley, Italy. J. Atmos. Chem., 3, 289-296.

(10) EVANS L.S., RAYNOR G.S. and JONES D.M.A. (1984) Frequency distributions for durations and volumes of rainfalls in the Eastern United States in relation to acidic precipitation. Water Air Soil Pollut., 23, 187-195.

(11) WISNIEWSKI J. (1982) The potential acidity associated with dews, frosts and fogs. Water Air Soil Pollut., 17, 361-377.

AIR SAMPLING FLIGHTS AT LOW ALTITUDES ALONG THE BORDER BETWEEN THE FEDERAL REPUBLIC OF GERMANY AND ITS NEIGHBOURS

S.BEILKE,R.BERG,W.GROSCH

Umweltbundesamt,Pilotstation Frankfurt

Federal Republic of Germany

and

A.H.BLOMMERS,F.W.JANSEN,J.LELIEVELD

Geosens B.V.,Rotterdam,The Netherlands

Abstract

This paper reports the first results on aircraft measurements carried out in order to study transboundary transport of SO_2,NO_x,O_3 and aerosol components across the border between the Federal Republic of Germany(FRG) and its neighbours.Based on the results of 25 flights covering a range of several meteorological conditions the following conclusions can be drawn:

High concentrations of SO_2 are found along the border between the FRG and its eastern neighbours for advection of air from east to the west,particularly in winter.During these cases the corresponding concentrations of NO_x are relatively small.On the other hand NO_x and SO_2 concentrations are about equal during eastbound transport across the eastern border.

Aircraft measurements of eastbound and westbound transboundary fluxes over the FRG western border during episodes, show similiar ratios between SO_2 and NO_x as observed along the eastern border.

The ozone levels along the eastern border show somewhat higher concentrations during airflows from east to west compared to situations during westerly flow conditions.This is most likely due to the prevailing meteorological conditions during the measurements,and the associated emission of hydrocarbons,both causing a higher photochemical activity in the east of the FRG.

In the plumes of large industrial area sources crossing the borderline both a generation and a removal of ozone was observed depending on emission characteristics and meteorological conditions.

Based on the results of 25 air sampling flights carried out on 16 days along the border between the FRG and its eastern and western neighbours the conclusion can be drawn that, during episodic conditions,the FRG is to a considerable extent a transit country for air pollutions,especially for SO_2 and O_3.

1. INTRODUCTION

In view of the problems of acid deposition and photochemical oxidants in the lower atmosphere over Europe,mean residence times of the relevant substances vary from one day to about one week.On these time scales it is reasonable to assume that such substances can be transported over hundreds of kilometers crossing national boundaries before being deposited,mainly somewhere in Europe.
In general,transboundary fluxes can be determined by modelling, by aircraft and ground based measurements or by a combined approach.
Numerous long-range transport model calculations have been carried out for different European emission areas trying to establish a source-receptor relationship in a quantitative way. Most models deal with an emission-deposition relationship as a long-term average by providing monthly or annual figures.With respect to the acidification problem,such long-term transport models describe the transport and deposition of sulfur compounds with an accuracy of a factor two or better(Van DOP,1983).
In contrast to long-term models,episodes with a time scale of a few hours to some days are much more difficult to simulate (KLUG et al.,1984).However,episodes with high concentrations and considerable deposition are common and environmentally significant(SMITH,1983).
An important aspect of the difficulties in modelling is the uncertainty of the emission inventories which are needed as an input to such models.Emission values on which models are based are not known precisely.The lack of precision varies with pollutant species and from region to region.
In contrast to the development of models,aircraft measurements in order to study transboundary fluxes are very scarce in Europe.
In 1985 an extensive measuring programme of several years has been started with the goal to determine the transboundary transport of air pollutants between the FRG and its neighbours using a combination of airborne and ground based measurements. During the first stage of this programme(1985-1986) series of orientation flights were made with the objective to establish an overall picture of the air pollution distribution,rather than an exact determination of transboundary fluxes along the entire border during different meteorological flow conditions.

In the years to come,the results,obtained in the first stage of the flight programme,will be used as a basis for more detailed studies of the transboundary fluxes.

This paper reports the first results of a study of aircraft measurements,carried out during the first stage of the flight programme,especially of the horizontal and vertical profiles of SO_2,NO_x,O_3 and aerosol components along the border between the FRG and its eastern and western neighbours during episodes. These concentration profiles are combined with the relevant meteorological data,so that transboundary fluxes of air pollutants can be assessed in a semi-quantitative way for the episodes investigated.

2. INSTRUMENTATION AND FLIGHT ROUTES

The airborne measurements were carried out with a piper Navajo Chieftain PA 31 equipped with instruments to measure the following chemical and meteorological parameters(table 1).

Table 1: Relevant aircraft instrumentation

Component	**Principle/Instrument**
Chemical components:	
SO_2 (Line integrals)	Correlation spectrometer (COSPEC IV), Barringer Research
SO_2 (in situ)	pulsed fluorescence monitor, Thermo Electron 43, modified
NO/NO_x (in situ)	chemoluminescence monitor, Thermo Electron 14D, modified
O_3 (in situ)	chemoluminescence monitor, Bendix
Aerosol constituents (SO_4^{2-}, NO_3^-, Cl^-, NH_4^+)	ECN designed filter packs, sampled with high volume pump. Ion chromatography.
Other parameters:	
pressure	transducer, Druck Ltd.
temperature	PT 100 sensor, Rosemount

Figure 1 presents the main flight tracks along the eastern and western border of the FRG with check points:

- flight route I along the FRG-German Democratic Republic(GDR) border between Heiligenhafen(Baltic Sea;checkpoint 1) and Bayreuth(Bavaria;checkpoint 21) representing a distance of ca. 620 km.
- flight route II along the border between the FRG and its western neighbours(NL,B,L,F) between Emden(North Sea; checkpoint 1) and Basel(Switzerland;checkpoint 29) over a distance of ca. 800 km.

Horizontal concentration profiles of the chemical compounds specified in table 1 were measured at the lowest possible flight levels(50-150 meters above the ground).

In order to get information about the vertical concentration profiles throughout the mixing layer,spirals were flown at certain horizontal distances.

3. CONCENTRATION PROFILES AND METEOROLOGICAL DATA

Until July 1986,25 flights were carried out on 16 days along flight routes I and II specified in figure 1.

3.1. Flights along the eastern border of the FRG

The figures 2a-d show the results of the measurements carried out on 4 July 1985 along the FRG-GDR border at easterly inflow. They represent a characteristic example of transport from east to west during the summer.Figure 2a summarizes the horizontal concentration profiles of the flight from Heiligenhafen(Baltic Sea) to Bayreuth(8.23 UTC to 10.56 UTC;checkpoints 1 to 21) whereas figure 2b shows the corresponding results of the return flight(12.36 UTC to 15.27 UTC;checkpoints 21 to 1).

The meteorological situation taken from the surface synoptic map of 4 July 1985,12.00 UTC shows an anticyclone centred over the Baltic Sea causing a steady southeasterly airflow near the earth surface across the eastern border.The air was rather warm(max.temp.25°C) and dry.There were practically no low clouds The hight of the mixing layer was between 900-1200 m for the first flight and between 1300-1500m during the return flight.

In figure 2c backward trajectories are presented arriving at Bad Hersfeld on 4 July 1985 at 10 a.m.The trajectories are calculated with the model of the Royal Netherlands Meteorological Institute(KNMI),for three different pressure levels(1000 hPa; 925 hPa;850 hPa).

Considering the meteorological situation and taking into account that the air masses have been transported some hundreds of kilometers over polluted areas in the southern part of the GDR (figure 2c),the conditions during the measurements were favourable for photochemical formation of ozone,other oxidants and of sulfate.
In figure 2d vertical concentration profiles are presented for SO_2,NO_x and O_3 measured at Othfresen(checkpoint 10) from 13.59 UTC to 14.10 UTC in the centre of a large SO_2 peak in a plume that is about 250 km wide.
As can be derived from figures 2a,2b and 2d,high concentrations of SO_2 are found along the southern part of the FRG-GDR border on this day,whereas the corresponding concentrations of NO_x are relatively small.
Figure 2b shows elevated concentrations of sulfate in the SO_2 plume over a flight path of more than 200 km.
The vertical profiles in figure 2d clearly show a correlation between the concentration lapse of O_3 and SO_2 maximum values occurring between 1000-1600 meters.This suggests that a considerable fraction of ozone is of anthropogenic origin.Similiar ozone concentration profiles were also reported by Van DUUREN et al.(1981) in the industrial plume of the Dutch Rijnmond district and by some authors in the plumes of some industrial areas in the FRG(FRICKE,1980;NEUBER et al.,1981).

An other example of transport from east to west during a winter episode is given in figures 3a-c.These figures show the results of aircraft measurements on 5 February 1986 along the FRG-GDR

border during which a flow of cold air masses(temperature max. 2°C) entered the FRG from the east.
There was a clear sky and the height of the mixing layer was between 500-600 meters.The backward trajectories arriving at Göttingen on 5 February 1986 at 10.00 UTC are presented in figure 3b.
Horizontal concentration profiles(figure 3a) measured along the 620 km flight track between Heiligenhafen(Baltic Sea) and Bayreuth(9.32 UTC to 11.52 UTC) revealed very high in situ and vertically integrated SO_2 concentrations together with elevated sulfate concentrations over a flight path of ca.350 km.

The vertical concentration profiles measured at Northeim(checkpoint 11) show inverse curves for SO_2 and NO_x compared to O_3 indicating an O_3 consumption due to oxidation of NO emitted together with SO_2 in the industrial areas east of the frontier (figure 3c).

As an example for transport from west to east,figures 4a,4b,4c and 4d summarize the results of two flights carried out on 2 August 1985 along the FRG-GDR border during westerly flow conditions.
The synoptic surface map for 2 August 1985,12.00 UTC,shows a low pressure system between Scotland and Iceland and an anticyclone with its center near the Azores causing a westerly to southwesterly flow of temperate polar air masses(17-22°C)across the FRG-GDR border.The sky was partly cloudy(4/8 to 7/8) with cumulus and stratocumulus.The height of the mixing layer was between 1000-1500 meters along the entire flight path.
Backward trajectories calculated for the pressure levels 1000, 925 and 850 hPa,arriving at Schlitz(checkpoint 16)on 2 August 1985 at10.00 UTC are given in figure 4c.
Horizontal concentration profiles are presented for the flight track from Heiligenhafen(Baltic Sea) to Bayreuth(figure 4a; 8.12 UTC to 10.36 UTC;checkpoints 1 to 21) and for the corresponding return flight(figure 4b;12.01 UTC to 14.46 UTC; checkpoint 21 to 1).
As is evident from these figures,all concentrations exept the O_3 concentration along the border were rather low during the monitoring flight.In the afternoon,concentrations of all species investigated were somewhat increased.In contrast to situations with easterly inflow,comparision of in situ SO_2 and NO_x concentrations show about the same low values.
The vertical concentration profiles for SO_2,NO_x and O_3 measured at Göttingen(checkpoint 12) from 13.22 UTC to 13.27 UTC(fig.4d) show rather low and nearby constant values throughout the mixing layer.

3.2. Flights along the western border of the FRG

In the second part of this section some results are presented of flights along the border between the FRG and its western neighbours(route II,figure 1).

An example for transport east to west is given in figures 5a-e, summarizing the results of a flight carried out on 3 March 1986 along the western border of the FRG(FRG - NL,B,L,F) between Emden(North Sea) and Saarbrücken(about 600 km;checkpoints 1 to

20) for steady easterly flow conditions.

The synoptic weather situation was characterized by an anticyclone over the western part of the USSR and a low pressure system over the Mediterranėan.As a result of this pressure distribution an easterly flow of relatively cold air masses(ca.0°C) passed the above mentioned frontier.The height of the mixing layer was 300-600 m over the northern part of the flighttrack and increased to 1200 m in the south.The sky was clear in the north and became cloudy in the south connected with light snowfall.

Backward trajectories for 3 pressure levels arriving at Venlo on 3 March 1986(12.00 UTC) had passed over some major industrial areas,such as the Ruhr area,the GDR Halle-Leipzig area and the southern part of Poland(figure 5c).

Horizontal concentration profiles along the 600 km flight track between Emden and Saarbrücken(9.57 UTC to 12.17 UTC;checkpoints 1 to 20) are given in figure 5a.As far as SO_2 is concerned a large fraction of this gas,transported across the whole flight path,can be attributed to long-range transport rather than to relatively nearby emissions.

Between checkpoints 11 and 15 the industrial plume of the Ruhr area is superimposed to the relatively high concentrations of SO_2 and NO_x(mainly NO_2) originating from more distant sources.

Another indication for the dominant role of the long-range transport of SO_2 is the concentration distribution of aerosol sulfate which differs considerably from the SO_2 profiles.The fact that the highest sulfate concentrations were measured outside of the plume of the industrial Ruhr area can be interpreted in terms of the relatively slow oxidation of SO_2 emitted at distant sources.

Another interesting feature is a large depletion of O_3 concentrations in the plume near the Ruhr area which is caused by the oxidation of NO to NO_2 by O_3.

In figure 5b the horizontal profiles of SO_2 and NO_x originating from the emissions of the Ruhr area are presented in more detail together with the corresponding O_3 profil.The profiles of SO_2, NO_x and NO_2 were obtained by subtracting the fraction attributable to long-range transport from the total concentrations.

A comparision of the NO_2 measured along the flighttrack with the corresponding O_3 deficit shows an 20-30 % excess of NO_2.A possible explanation for this phenomenon is a contribution of the reaction of NO with molecular oxygen in which NO_2 is formed. Such a contribution seems likely since NO_x emissions from power plants in the Ruhr area clearly dominate over those from traffic. The relative importance of the NO oxidation routes by O_2 and O_3 for NO_2 generation was first demonstrated in the plume of the Wilhelmshaven power plant(ELSHOUT and BEILKE,1984).

Vertical concentration profiles measured at Venlo(checkpoint 12) in the plume of the Ruhr area are given in figure 5d for SO_2, NO_x and O_3 and in figure 5e for NO,NO_2 and O_3.

These figures show very clearly the depletion of O_3 due to its reaction with NO.

An example for transport from west to east is given in figures 6a-c,summarizing the results of a flight carried out on 7 August 1985 along the western border between Emden(North Sea) and

Basel(flight track II in figure 1;checkpoints 1 to 29).

The synoptic surface map for 12.00 UTC on 7 August 1985 shows an anticyclone from the Azores towards northern Spain and a low pressure system near Iceland causing a westerly flow of polar maritime air masses.The sky was partly clouded(6/8) with cumulus and stratocumulus.The height of the mixing layer was between 600-1200 m along the entire flighttrack.
Backward trajectories arriving at Emmerich(checkpoint 10) on 7 August 1985,14.00 UTC,are shown in figure 6b.

As can be seen in figure 6a the horizontal concentrations of the chemical components are rather low and uniformly distributed for all components over the ca.800 km flight path between Emden and Basel(7.17 UTC to 10.19 UTC;checkpoints 1 to 29).

Similiar to the horizontal concentrations,the vertical concentration profiles for SO_2,NO_x and O_3 measured at Herongen (checkpoint 12) show rather low and constant concentrations throughout the mixing layer(figure 6c) resembling the profiles obtained on 2 August 1985 at the FRG-GDR border for westerly flow conditions.

4. CONCLUSIONS

Based on the results of a total of 25 air sampling flights carried out on 16 days along the border of the Federal Republic of Germany and its eastern and western neighbours some more general results can be summarized as follows:

High vertically integrated and in situ SO_2 concentrations are measured along the border between the Federal Republic of Germany and its eastern neighbours during airflow from east to west especially in wintertime when the depth of the mixing layer is shallow.The corresponding concentrations of aerosol sulfate were also high.
In contrast to SO_2 the concentrations of NO_x were relatively low at an easterly inflow of air masses being advected over the industrial regions in the east.
A comparision of the fluxes of SO_2 and NO_x across this border demonstrates a clear dominance of SO_2 over NO_x by a factor between 2 and 12(figure 7, I E) depending on the SO_2/NO_x emission ratios,photochemical activity of the lower atmosphere and wind velocity.The very high ratio of 12 determined on 30 August 1985 can most likely be explained by a high photochemical activity on this day(50-105 ppb O_3) in connection with low wind velocities resulting in high NO_x oxidation rates enhancing the SO_2/NO_x ratio with increasing distance from the main sources.Maximum NO_x oxidation rates in central Europe are of the order of 20%/h during summer days at full sunshine(ELSHOUT,1981) whereas the corresponding SO_2 oxidation rates are ca. 2 %/h(COX and PENKETT, 1983).Since the corresponding oxidation rates for SO_2 and NO_x are much lower in winter,such a shift of the SO_2/NO_x ratio towards higher values with increasing distance from the source should be much less pronounced in winter than in summer.
A clear dominance of SO_2 over NO_x for easterly incoming flow was also observed by PAFFRATH et al.(1985) during an extensive flight programme to investigate transboundary fluxes of air

pollutants in the northeastern region of Bavaria.

In contrast to the high concentrations of SO_2 and sulfate measured at easterly incoming flow,the concentrations of these components are lower for airflow from west to east along the border between the FRG and its eastern neighbours.Because air pollutant fluxes across the border can only by assessed in a semi empirical way during the first stage of this flight programme(1985-1986) they therefore are rather crude.The fluxes of SO_2 and sulfate across this border from west to east are considerably lower than the amounts of these components transported from east to west for the episodes investigated.
A comparision of the concentrations along and the fluxes across the border between FRG and GDR show similiar concentrations and fluxes for SO_2 and NO_x for transport from west to east.As seen in figure 7 (·I W),the results of seven episode aircraft measurements show flux ratios for SO_2/NO_x(by weight,NO_x as NO_2) between 0.7 and 1.3.

As far as transport across the border between the FRG and its western neighbours is concerned,four of the five episode aircraft measurements show a slight dominance of SO_2 fluxes over NO_x for easterly flow conditions(figure 7 , II E).

For transport from west to east three of the four measuring flights resulted in a dominance of NO_x over SO_2 transported across this border.
The concentrations of SO_2 and aerosol sulfate along and the fluxes across this border were considerably higher for airflow from east to west(5 episode flights) than for transport from west to east (4 episode flights).

As far as ozone is concerned somewhat higher concentrations are observed for airflow from east to west,probably caused by a higher photochemical activity,than for westerly flow conditions. In the plumes of large industrial source areas crossing the border both a generation and a removal of ozone is observed depending on emission characteristics and meteorological situation.

Another result of the episode aircraft measurements during episodes is that during such situations the FRG is to a considerable extent a transit country for air pollutants,particularly SO_2 and O_3.

5. REFERENCES

COX,R.A.and PENKETT,S.A.(1983)
Formation of atmospheric acidity.Proceedings:Acid Deposition-A Challenge for Europe,Karlsruhe,19-21 Sept.1983,pp.58-81.

ELSHOUT,A.(1981)
Transport and Modelling-Field Experiments.Proceedings:Second European Symposium,Varese,Italy,29 Sept.-1 Oct.1981.COST 61a bis,pp.657-659.

ELSHOUT,A.and BEILKE,S.(1984)
Oxidation of NO to NO_2 in flue gas plumes of power stations. Proceedings:Third European Symposium,Varese,Italy,10-12 April 1984.Project COST 61a bis,pp.535-543.

FRICKE,W.(1980)
Die Bildung und Verteilung von anthropogenem Ozon in der unteren Troposphäre.Dissertation,University of Frankfurt, 1980,133 pages.

KLUG,W. et al.(1984)
A comparision between four different interregional air pollutant models.Committee on the Challenges of Modern Society,NATO Report Nr.155.

NEUBER,E.;GEORGII,H.W.and MÜLLER,J.(1981)
Aircraft measurements of oxidants and precursors in the plumes of heavy industrialized areas.Proceedings:Second European Symposium,Varese,Italy,29 Sept.- 1 Oct.1981, pp.469-481, COST 61a bis.

PAFFRATH,D.;PETERS,W.and LIDL,J.(1985)
Untersuchungen der Luftschadstoffbelastung im Osten Bayerns-Meßergebnisse aus den Monaten August und September 1984. Research Report DFVLR,Nr.DFVLR-FB 85-03,136 pages.

SMITH,F.B.(1983)
Conditions pertaining to high acid concentrations in rain. Proceedings:VDI Colloquium on Acid Precipitation,Lindau, Germany,June 1983.VDI Berichte 500,pp.67-76.

Van DOP,H.(1983)
The residence and transport of pollutants in the atmosphere-a meteorological problem.Proceedings:Acid Deposition-A Challenge for Europe,Karlsruhe,19-21 Sept.1983,pp.48-57.

Van DUUREN,H.et al.(1981)
Measurements by aeroplane of the distribution of ozone and primary air pollutants.Proceedings:Second European Symposium, Varese,Italy,29 Sept.- 1 Oct.1981,pp.460-468.COST 61a bis.

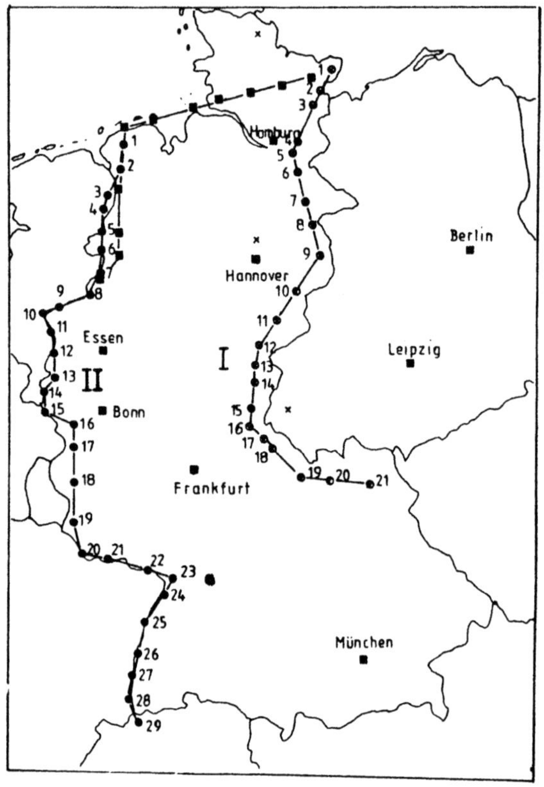

Figure 1 :

Flight routes along the border of the Federal Republic of Germany and its eastern and western neighbours.

⊗ : flight route I along the FRG-GDR border between Heiligenhafen(Baltic Sea;checkpoint 1) and Bayreuth (Bavaria;checkpoint 21) representing a distance of 620 km.

● : flight route II along the border between the FRG and its western neighbours(NL,B,L,F) between Emden(North Sea;checkpoint 1) and Basel(CH,checkpoint 29) representing a distance of ca.800 km.

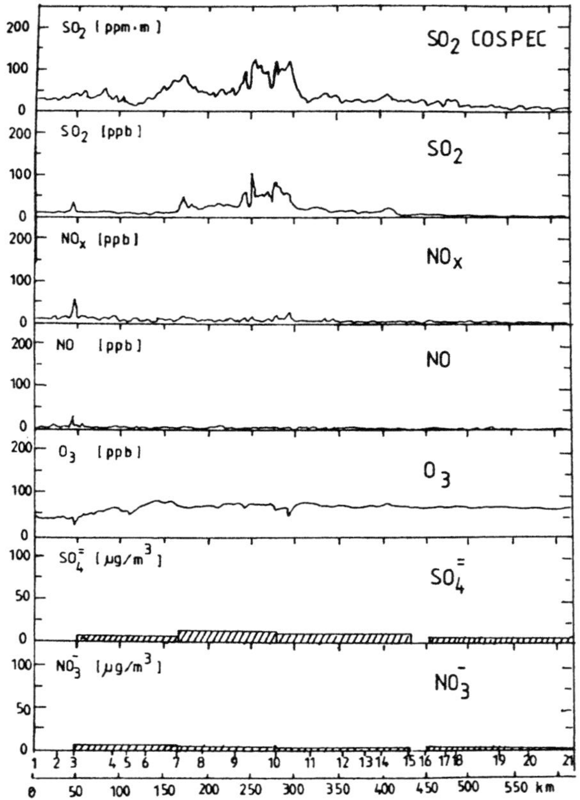

Figure 2a :

Horizontal concentration profiles of air pollutants measured on 4 July 1985 along the FRG-GDR border between Heiligenhafen and Bayreuth(flight route I,checkpoints 1 to 21;8.23 UTC to 10.56 UTC)for transport from east to west.Flight level between 50 and 150 m above ground level.

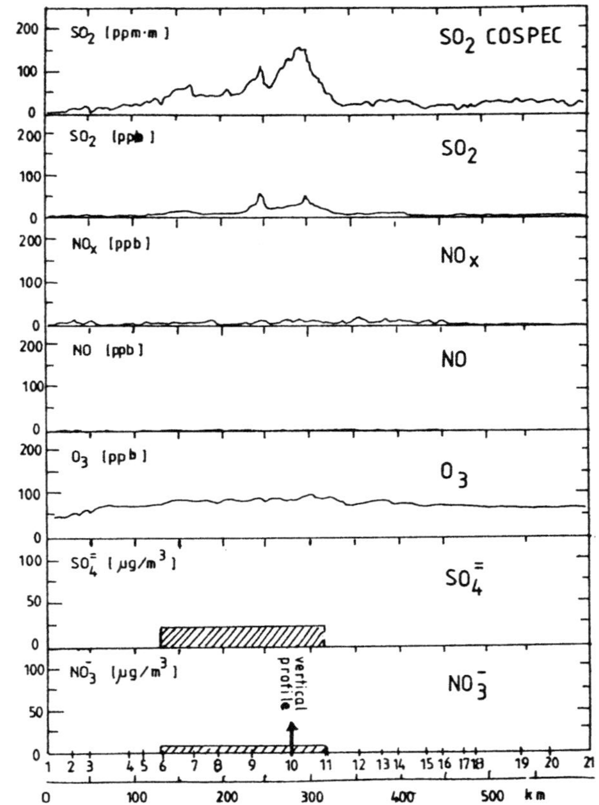

Figure 2b :

Horizontal concentration profiles of the return flight carried out on 4 July 1985 along the FRG-GDR border between Bayreuth and Heiligenhafen(checkpoints 21 to 1;12.36 UTC-15.27 UTC)for transport from east to west.
Flight level between 50-150 m above ground level.

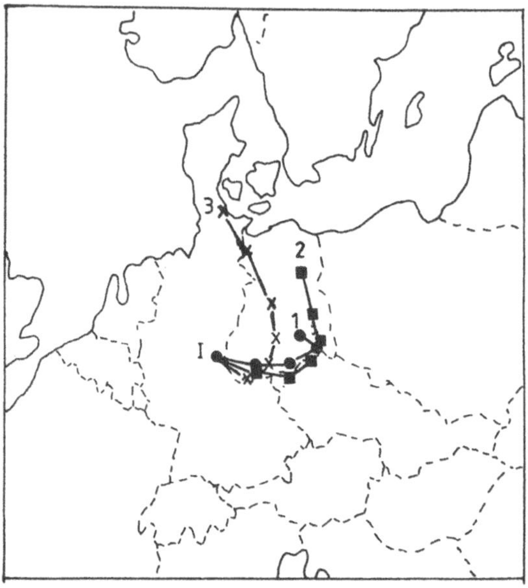

Figure 2c :

Backward trajectories arriving at Bad Hersfeld(flight route I; checkpoint 15) on 4 July 1985 at 10.00 UTC.
Each timestep represents 6 hours.
The trajectories were calculated for 3 different pressure levels (1 =1000 hPa; 2 =925 hPa;3 =850 hPa) using the model of the Royal Netherlands Meteorological Institute.

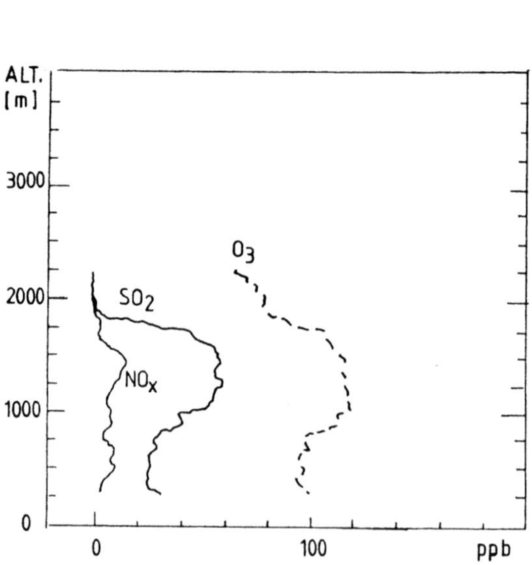

Figure 2d :

Vertical concentration profiles for SO_2,NO_x and O_3 measured at Othfresen(checkpoint 10;flight route I) on 4 July 1985,13.59 UTC to 14.19 UTC,in the center of the large scale SO_2 plume for easterly incoming flow.
Altitude above mean sea level.

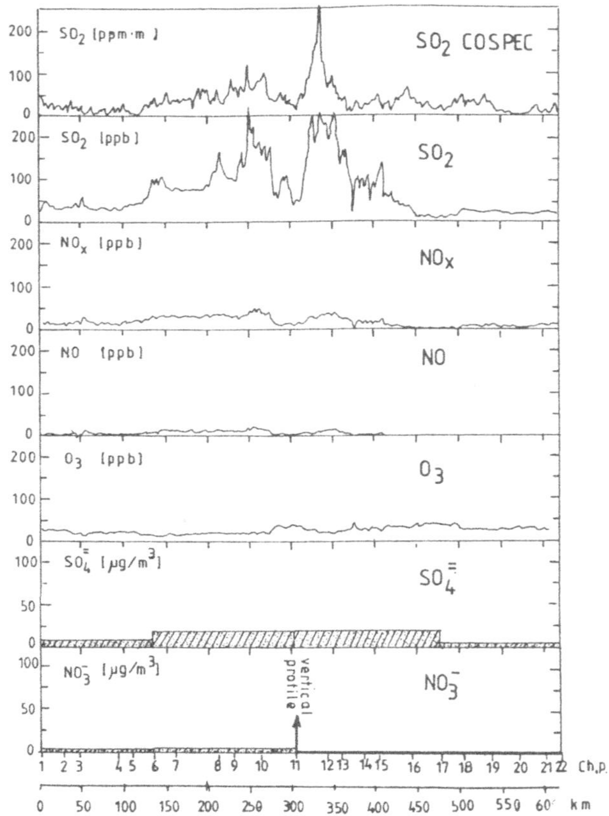

Figure 3a :

Horizontal concentration profiles of SO_2, NO_x, O_3, sulfate, and nitrate measured on 5 February 1986 along the FRG-GDR border between Heiligenhafen(Baltic Sea)and Bayreuth(flighttrack I; checkpoints 1 to 21;9.32 UTC to 11.52 UTC) for transport from east to west.
Flight level between 50 and 150 m above the ground.

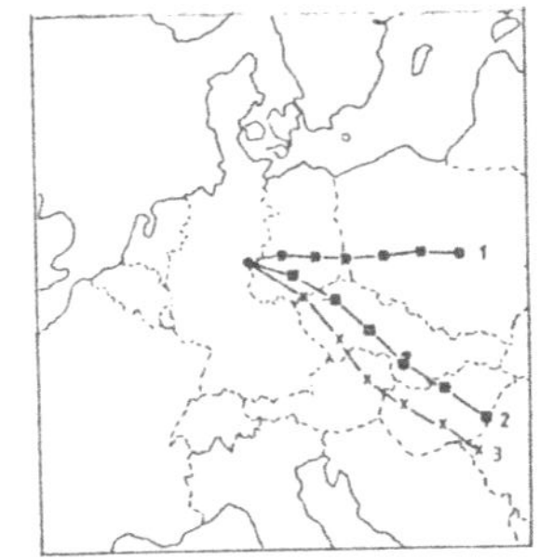

Figure 3b :

Backward trajectories arriving at Göttingen on 5 February 1986 at 10.00 UTC(1000,925 and 850 hPa).
Each timestep represents 6 hours.

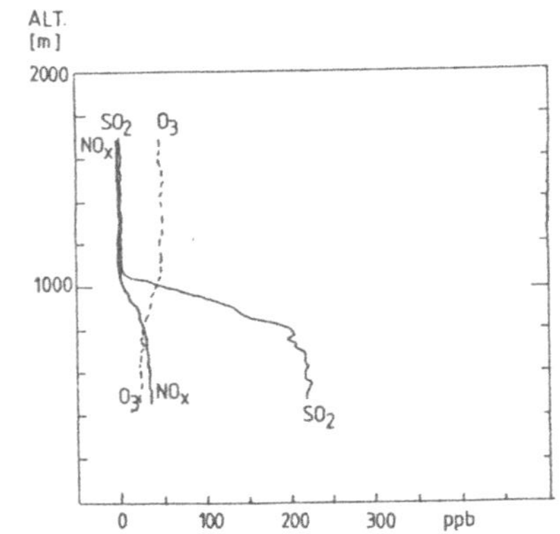

Figure 3c :

Vertical concentration profiles of SO_2, NO_x, and O_3 at Northeim (checkpoint 11;flight route I) on 5 February 1986,13.56 UTC to 14.13 UTC,for easterly incoming flow.Altitude above sea level.

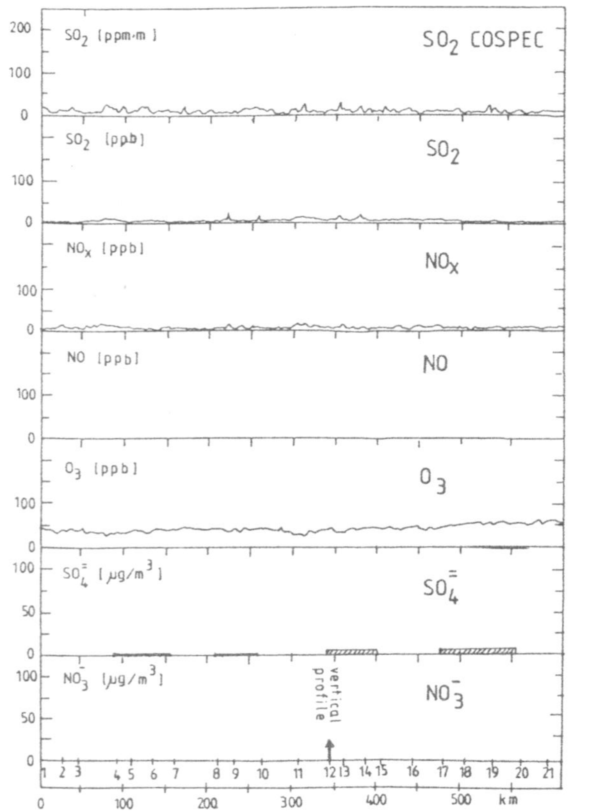

Figure 4a :

Horizontal concentration profiles of SO_2,NO_x,O_3,sulfate and nitrate measured on 2 August 1985 along the FRG-GDR border between Heiligenhafen(Baltic Sea) and Bayreuth(flight route I; checkpoints 1 to 21;8.12 UTC to 10.36 UTC) for transport from west to east.
Flight level between 50 and 150 m above ground level.

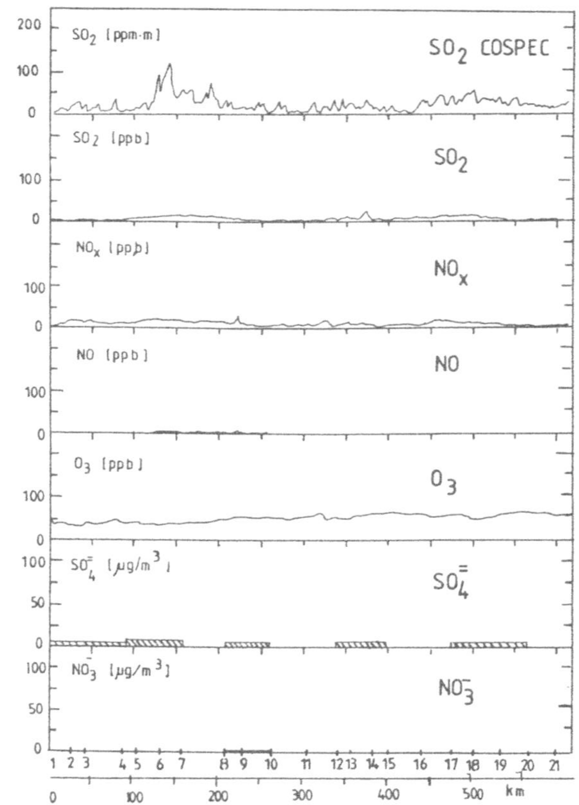

Figure 4b :

Horizontal concentration profiles of SO_2,NO_x,O_3,sulfate and nitrate measured on 2 August 1985 along the FRG-GDR border between Bayreuth and Heiligenhafen(return flight of 4a;flight route I;checkpoints 21 to 1;12.01 UTC to 14.46 UTC) for transport from west to east.
Flight level between 50 and 150 m above ground level.

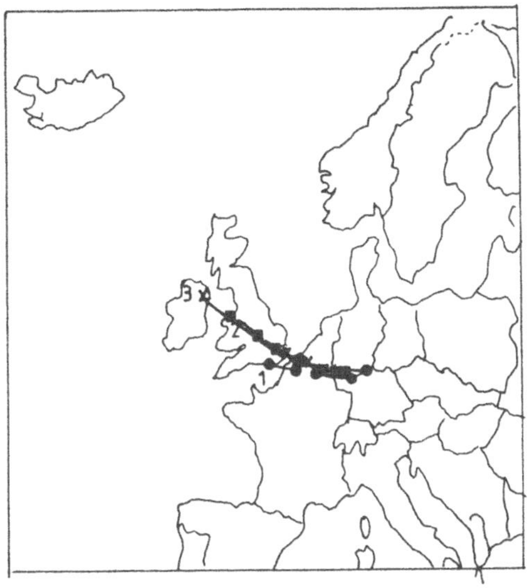

Figure 4c :

Backward trajectories arriving at Schlitz(flight route I; checkpoint 16) on 2 August 1985,10.00 UTC.
(1 = 1000 hPa; 2 =925 hPa; 3 =850 hPa)
Each timestep represents 6 hours.

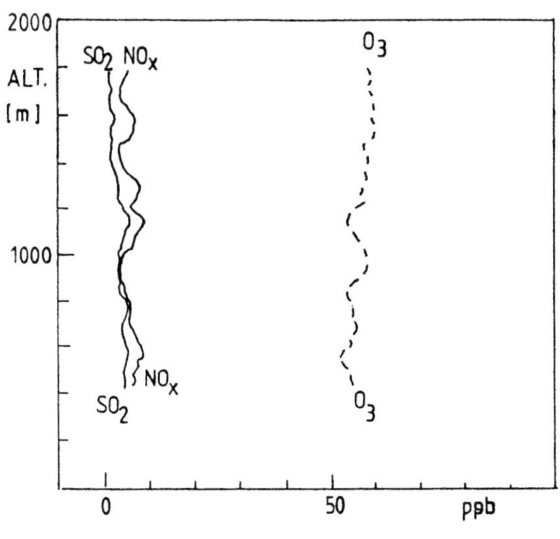

Figure 4d :

Vertical concentration profiles of SO_2,NO_x,and O_3 measured at Göttingen(flight route I;checkpoint 12) on 2 August 1985 from 13.22 UTC to 13.27 UTC for airflow from west to east.
Altitude above mean sea level.

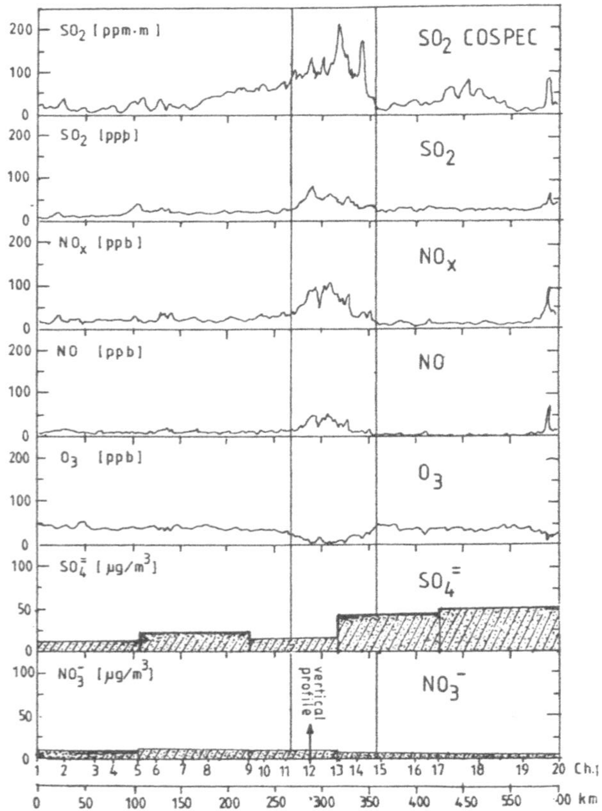

Figure 5a :

Horizontal concentration profiles of SO_2, NO_x, O_3, sulfate and nitrate measured on 3 March 1986 along the western border between Emden(North Sea) and Saarbrücken(flight route II; checkpoints 1 to 20;9.57 UTC to 12.17 UTC) for transport from east to west.
Flight level between 50 and 150 m above ground level.

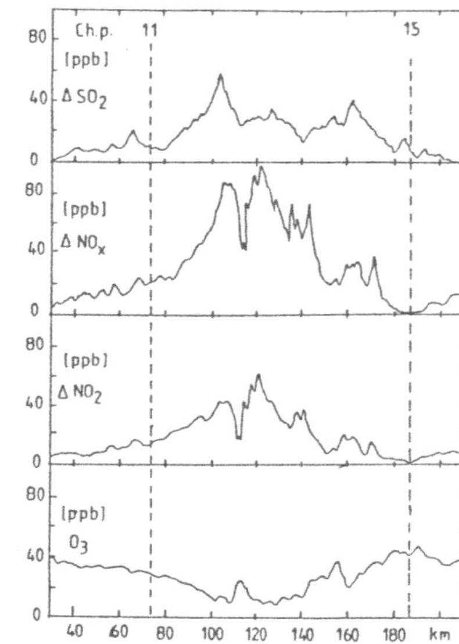

Figure 5b :

Horizontal profiles of SO_2, NO_x, and O_3 originating from or influenced by the emissions of the industrial Ruhr area (checkpoints 11 to 15 in figure 5a). The profiles of SO_2, NO_x, and NO_2 were determined by subtracting the concentrations attributable to long-range transport from the total concentrations. 3 March 1986.
Flight levels between 50 and 150 m above ground level.

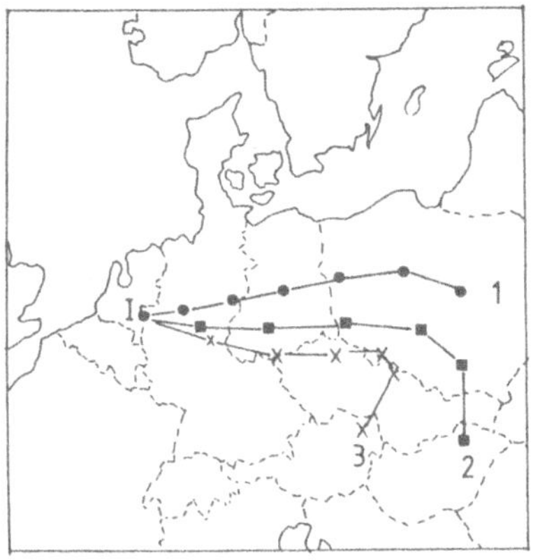

Figure 5c :

Backward trajectories arriving at Venlo(flight route II; checkpoint 12) on 3 March 1986,12.00 UTC
(1= 1000 hPa; 2= 925 hPa; 3= 850 hPa)
Each timestep represents 6 hours.

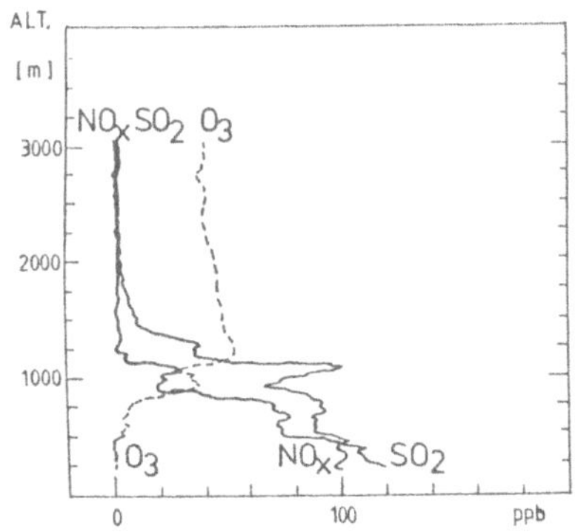

Figure 5d :

Vertical concentration profiles of SO_2, NO_x, and O_3 measured at Venlo(flight route II ; checkpoint 12) on 3 March 1986(16.22 UTC to 16.53 UTC) for transport from east to west.
Altitude above mean sea level.

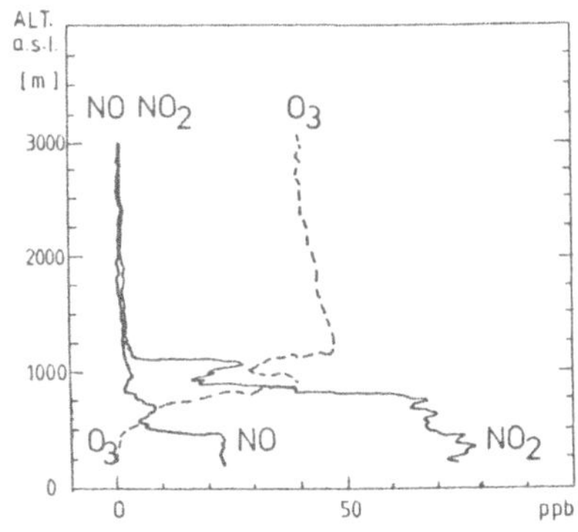

Figure 5e :

Vertical profiles of NO, NO_2 and O_3. All other facts as in figure 5d.

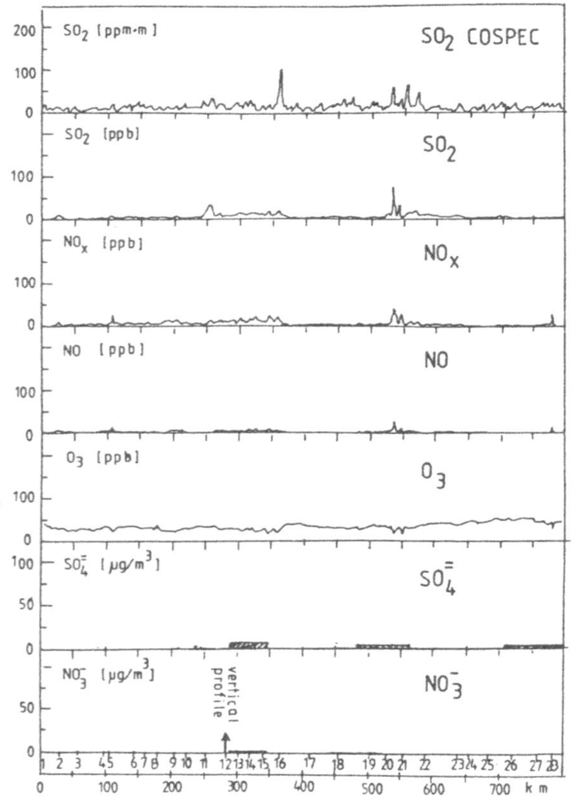

Figure 6a :

Horizontal concentration profiles of SO_2, NO_x, O_3, sulfate and nitrate measured on 7 August 1985 along the western border between Emden(North Sea) and Basel(flight route II; checkpoints 1 to 29;8.17 UTC to 11.19 UTC) for transport from west to east. Flight level between 50 and 100m above ground level.

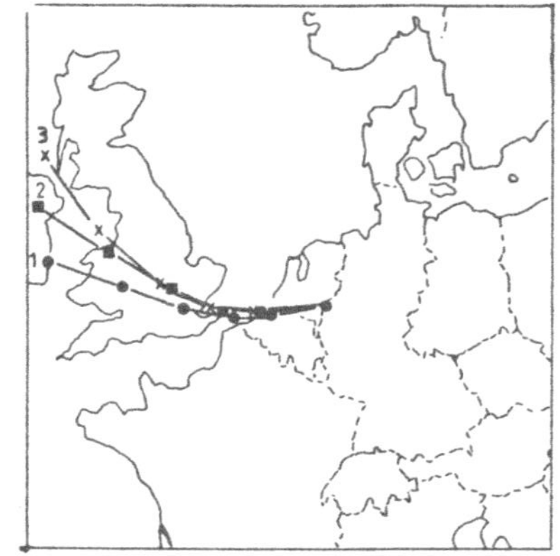

Figure 6b :

Backward trajectories arriving at Emmerich(flight route II; checkpoint 10) on 7 August 1985,14.00 UTC.
(1 = 1000 hPa;2 = 925 hPa; 3 = 850 hPa).

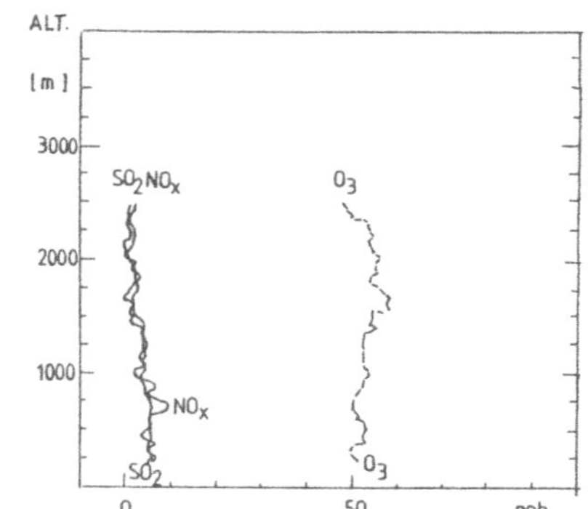

Figure 6c :

Vertical concentration profiles of SO_2, NO_x, and O_3 measured at Herongen(flight route II;checkpoint 12) on 7 August 1985(14.26 UTC to 14.48 UTC)for transport from west to east. Altitude above mean sea level.

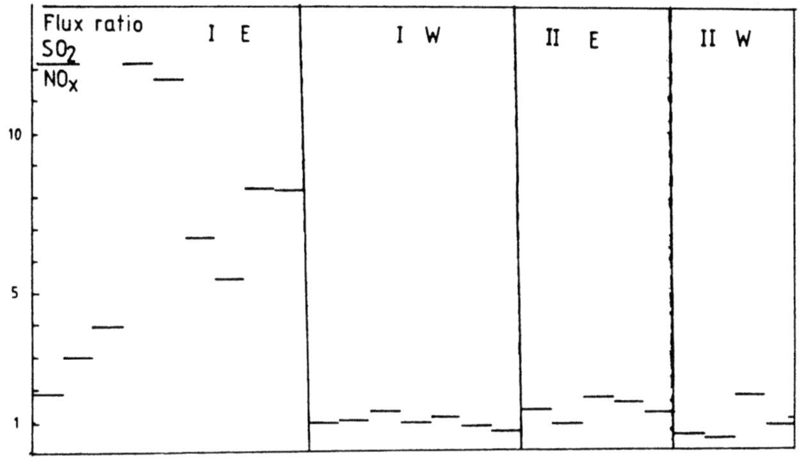

Figure 7 :

Comparision of the fluxes of SO_2 and NO_x(by weight,NO_x as NO_2) across the border of the Federal Republic of Germany and its eastern and western neighbours.

I E : Flight track I between the FRG and its eastern neighbours.Transport from east to west.

I W : Flight track I
Transport from west to east

II E : Flight route II between the FRG and its western neighbours.Transport from east to west.

II W : Flight route II
Transport from west to east.

THE OXIDATION OF SULPHUR DIOXIDE IN THE AQUEOUS PHASE IN PLUMES

F.G. RÖMER, A.A. VELDKAMP, H.J. SLANGEWAL
N.V. KEMA, Environmental Research Department
P.O. Box 9035, 6800 ET Arnhem, The Netherlands
W.A.J. VAN PUL
Agricultural University, Department of Physics and Meteorology
Duivendaal 1 and 2, 6701 AP Wageningen, The Netherlands
H.F.R. REIJNDERS
National Institute of Public Health and Environmental Hygiene,
Laboratory for Inorganic Chemistry
P.O. Box 1, 3720 BA BILTHOVEN, The Netherlands

Summary

The conversion rate of sulphur dioxide dissolved in cloud and rainwater is largely influenced by hydrogen peroxide. This conversion was studied using two approaches:

a. wash-out experiments under the plumes of a coal-fired power plant
b. aircraft measurements in the plumes originating from industrial areas.

At a distance of about ten kilometers from the power station distinct relationships have been found between sulphur-containing compounds and hydrogen peroxide in the aqueous phase. At larger distances from sources the conversion of sulphur dioxide into sulphate by hydrogen peroxide in the aqueous phase can be studied using an aircraft.

The results of both types of experiments show strong variation in sulphate formation due to several factors. Experience so far indicates that the oxidation of sulphur dioxide by hydrogen peroxide in the aqueous phase can be considerable. Relatively large amounts of sulphate (>50%) can be produced from the oxidation of sulphur dioxide by hydrogen peroxide in cloud and rainwater.

It was found that in rainwater collected under a plume the formaldehyde concentration decreased along with an increase in sulphite concentration. In this experiment an excess of sulphite was present besides formaldehyde. The decrease in formaldehyde suggests the formation of hydroxymethanesulphonate.

1. INTRODUCTION

The oxidation of sulphur dioxide in the liquid phase can be studied systematically in the laboratory. Extensive studies are reported in the literature (1). The advantage of such studies is that valuable kinetic data are made available for model calculations. The main oxidation pathways are supposed to be the oxidation through (hydrogen) peroxide, ozone and through metal ion catalysed reaction of oxygen.

The oxidation of sulphur dioxide in the real atmosphere is largely influenced by meteorology (season, time of the day) and location (pollutant/oxidant concentration). The concentration of sulphate in atmospheric water (cloud, rain) is a result of the scavenging of sulphate aerosols and sulphur dioxide, the latter being subsequently oxidized to sulphate. Therefore, it is virtually impossible to make reliable estimates of the contribution of the separate oxidants to the ultimate sulphate formation in relation to the distance from sources. On the other hand, if the concentrations of oxidant and sulphate are measured both inside and outside plumes, quantitative data can be ontained of the relative contribution of an oxidant to sulphate formation. This paper deals with the oxidation of sulphur dioxide by hydrogen peroxide in plumes in the aqueous phase. This effect is demonstrated in wash-out experiments under the plumes of a coal-fired power plant and through aircraft measurements.

2. EXPERIMENTAL

2.1 Wash-out experiments

In the period 1983 to 1986 five wash-out experiments were carried out under the plumes of Nijmegen coal-fired power plant (600 MW, stack height 150 m, and 2x240 MW, stack height 125 m). At a distance of about 11 km 60 to 80 samplers were placed under and outside the plume for periods of a few hours (figure 1). The distance between the samplers was 200 m. To minimize the contribution of dry deposition the experimental setup was chosen in such a way that sampling was started just before rainfall. A van equipped with monitors for NO, NO_2, SO_2, O_3 and a tracer (SF_6) was operated for plume detection at ground level.

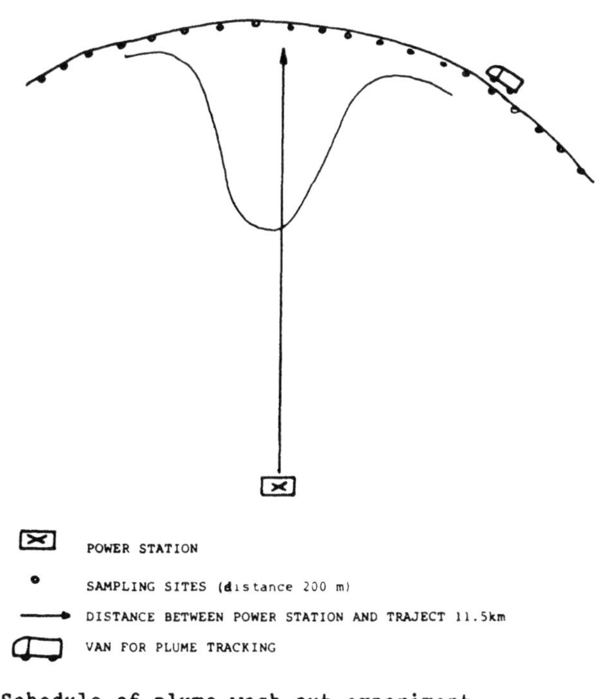

Figure 1 Schedule of plume wash-out experiment

The sample volumes and chemical composition of the rainwater samples were determined directly after the experiments. The analyses enclosed the measurement of main components (H_3O^+, NH_4^+, Na^+, K^+, Ca^{2+}, Mg^{2+}, F^-, Cl^-, NO_3^-, SO_4^{2-}) and hydrogen peroxide. On a few occasions formaldehyde and intermediate anions NO_2^- and SO_3^{2-} were also analysed. A description of the analytical methods is given elsewhere (2).

2.2 Aircraft measurements

From 1982 to 1985 a series of 12 aircraft measurements were carried out in the large scale plume of Rijnmond industrial area near Rotterdam.

Figure 2 Flight pattern and measuring location

Table I Meteorological conditions for flights 142, 235 and 239

flight	wind dd	wind ff	temperature (°C)	air mass type	cloud type	remarks
82-07-06 (142)	260	10	8	Pm	Sc, Cu	passage of cold front with rain
85-05-23 (235)	270	5	4	Pm	Sc, Cu	no frontal passage no rain
85-06-20 (239)	300	10	9	Pm	Sc and St, Cu fract.	passage of trough little rain

* mean values at flight level
dd direction (degrees)
ff speed ($m.s^{-1}$)

Tracks with lengths of about 50 km were flown upwind and downwind of this area (figure 2). During all measurements cloudy conditions were prevalent which constrained most of the time a flight altitude of at least 2000 ft. The meteorological conditions for three flights are summarized in table I. During the experiments the concentrations of NO, NO_2, SO_2 and O_3 were monitored. Cloud and rainwater were sampled separately and the samples were analysed mostly directly after each flight. The measurement of sulphate and nitrate aerosols and of nitric acid was carried out occasionally. These data are not reported here. Aircraft and instrumentation are described elsewhere (3, 4).

Cloud and rainwater samples were analysed for the main cationic and anionic components and hydrogen peroxide (2).

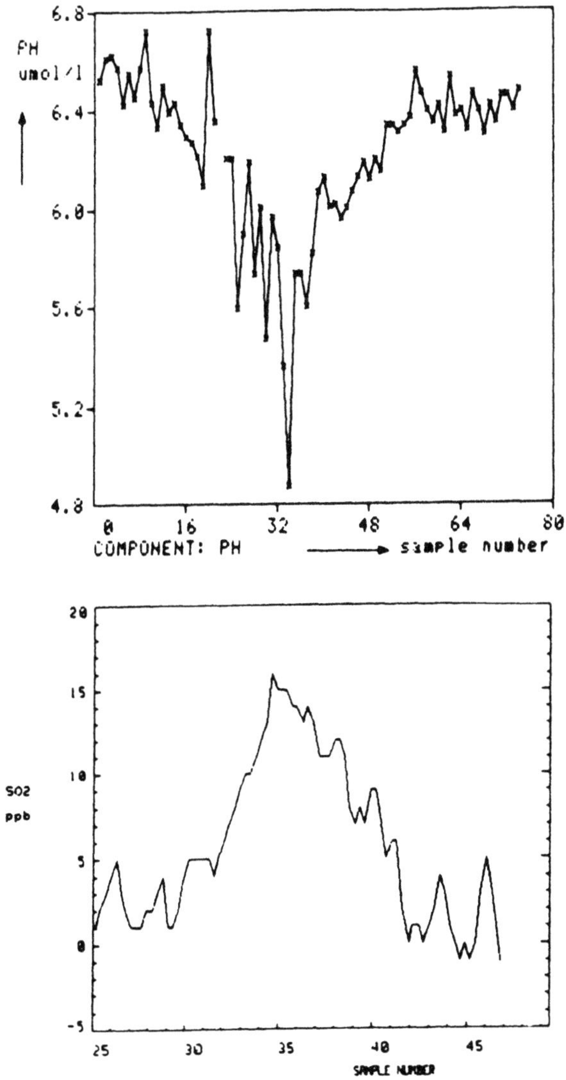

Figure 3 Profiles for pH and sulphur dioxide during plume wash-out (86-03-27)

3. RESULTS

3.1 Wash-out experiments

So far five experiments have been carried out, of which four generated usable information. Plume dimensions were derived from concentration profiles of plume components such as H_3O^+ and SO_4^{2-}. Because of the need to distinguish between background and plume samples, the most accurate results were obtained with the largest number of samplers. From successive plume crossings combined profiles for the different gases were composed. The dimensions derived from gas profiles sometimes agreed quite well with the dimensions from the water profiles. In general measurements of gases are only useful for plume tracking. In figure 3 an example is given of profiles of pH and the concentration of sulphur dioxide during an experiment on 20th September 1984.

The net contribution to the concentrations of H_3O^+, Cl^-, NO_3^- and SO_4^{2-} in rain under the plume is given in table II. Differences between the plume concentrations and that of the background were tested for statistical significance (test on the deviation in the mean, 95% confidence interval). The relative plume contribution by these components is expressed as the molar ratio of the net plume contribution/background concentration, which, together with the mean pH, are presented in table III. The percentage of wash-out of emitted sulphur dioxide together with some experimental conditions are summarized in table IV.

Table II Net plume contribution of some main components ($\mu mol.l^{-1}$)

date	hydronium	chloride	nitrate	sulphate
83-09-14	44.8	31.4	2.1	3.8
84-09-20	18.4	(-)	2.8	8.2
84-12-19	9.0 (5.7*)	(+)	(+)	16.9 ø
86-03-27	1.2*	3.2	1.8	15.6

(-) not measured
ø significance interval 75%
(+) not significant
* calculated from pH values

Table III Relative plume contributions of some main components (see text) and pH

date	pH		hydronium	chloride	nitrate	sulphate
	background	plume				
83-09-14	5.1	4.3	4.66	4.01	0.20	0.18
83-10-10*	5.2					
84-09-20	4.8	4.6	0.45	()	0.12	0.34
84-12-19	6.0	5.2	5.48	()	()	0.37
86-03-27	6.4	5.8	3.08	0.17	0.10	0.67

* no wash-out observed
() see table 1

Table IV Wash-out percentage ϕ of sulphur dioxide in relation to experimental conditions

date	wash-out SO_2 %	mean rain intensity $mm.h^{-1}$	rain period h	mean wind speed $m.s^{-1}$	mean temperature °C	pH background	H_2O_2 background $\mu mol.l^{-1}$
83-09-14	0.18	0.8	4.8	6.5	14.5	6.1	4.0
83-10-10	-	1.1	1.3	7.0	12.7	5.2	0.2
84-09-20	0.51	1.7	3.2	6.0	13.4	4.8	6.6
84-12-19	0.17	0.1	3.0	7.5	5.2	6.0	0.4
86-03-27*	0.52**	0.3	5.1	6.5	5.0	6.4	3.4

* 50% stack gas desulphurization
** 1.60% related to sum of S(IV) and S(VI)
ϕ (S(VI) in rain)/(S(IV) emitted) *%

In table V plume and background values are summarized for SO_4^{2-} and H_2O_2. The corresponding plume profiles are given in figure 4. In samples of the last experiment SO_3^{2-} and formaldehyde concentrations were measured. The plume profiles are presented in figure 5. Net plume and background weighted mean concentrations for these components are given in table VI.

3.2 Aircraft measurements

For flight numbers 142, 235 and 239 concentrations of sulphate ion and hydrogen peroxide in cloud water sampled at the same altitude are plotted in figure 6. The mean concentrations of sulphur dioxide and nitrogen oxides on the leeward tracks were always a little higher than on the upwind tracks. In the leeward tracks wide plumes were observed with, in some cases, high concentrations of sulphur dioxide and nitrogen oxides. The concentrations of ozone in these plumes were usually decreased. However, the mean values over the track did not vary much. Assuming a mean initial background (upwind) concentration of hydrogen peroxide in cloud water, the contribution of this oxidant to sulphate formation can be calculated. The results for two flights are given in table VII.

4. DISCUSSION

The main purpose of the wash-out experiments was the determination of the additional wet deposition of acidifying components in the surroundings of a coal-fired power plant. Valuable information of the conversion of sulphur dioxide can be obtained at the same time. In general the absolute values, as well as the relative plume contributions to wet deposition, vary strongly (tables II and III). This must be attributable to differences in meteorological conditions e.g. rain intensity and the chemical composition of background rain (table IV). The percentage of wash-out of sulphur dioxide at about 11 km from the stack amounted to less than 1%, comparable with values in the literature (5). There was no linear relationship between the emitted and deposited sulphur dioxide. It is notable that pH values are rather high because during the experiments the prevailing wind transported large amounts of ammonia from areas with

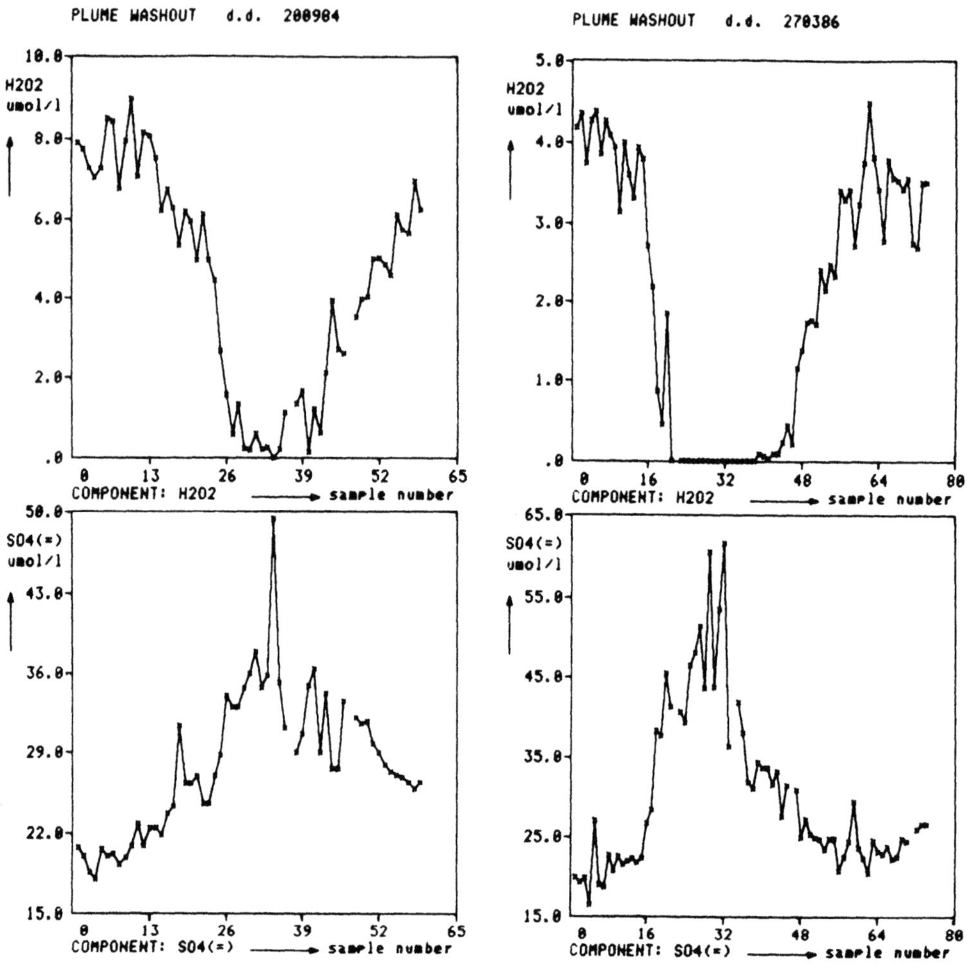

Figure 4 Plume profiles for sulphate ion and hydrogen peroxide concentrations (experiments on 84-09-20 and 86-03-27)

intensive cattle breeding. During one of the experiments (86-03-27) the relative contribution of ammonium under the plume amounted to 26% (net contribution 31.3 $\mu mol.l^{-1}$). This relative contribution could be discriminated with a confidence of 90%.

Looking at the hydrogen peroxide data the highest wash-out percentages (sulphur dioxide conversion) were measured for the highest oxidant concentrations (tables IV and V). However, some contribution of ozone to sulphate formation may also be expected for the pH range of interest (1). It has been observed that hydrogen peroxide contributes by up to 90% of the sulphate formation in the plume (table V). As the meteorological conditions largely vary many wash-out experiments have to be carried out in order to establish the relationship. However, in practice it is difficult to do this within a reasonable period. Therefore, model studies are necessary (6).

Table V Plume and background weighted mean concentrations ($\mu mol.l^{-1}$) for hydrogen peroxide and sulphate

date	background			plume			sulphate production by H_2O_2 oxidation (%)
	sulphate	H_2O_2	n*	sulphate	H_2O_2	n*	
83-09-14	20.6	4.1	14	24.4	0.6	21	92
83-10-10**	53.1	0.2	51				
84-09-20	24.2	6.5	30	32.4	2.0	27	55
84-12-19	45.7	0.4	23	62.6	0.1	30	2
86-03-27	23.2	3.3	42	38.8	0.4	31	19

* number of samples
** no wash-out observed

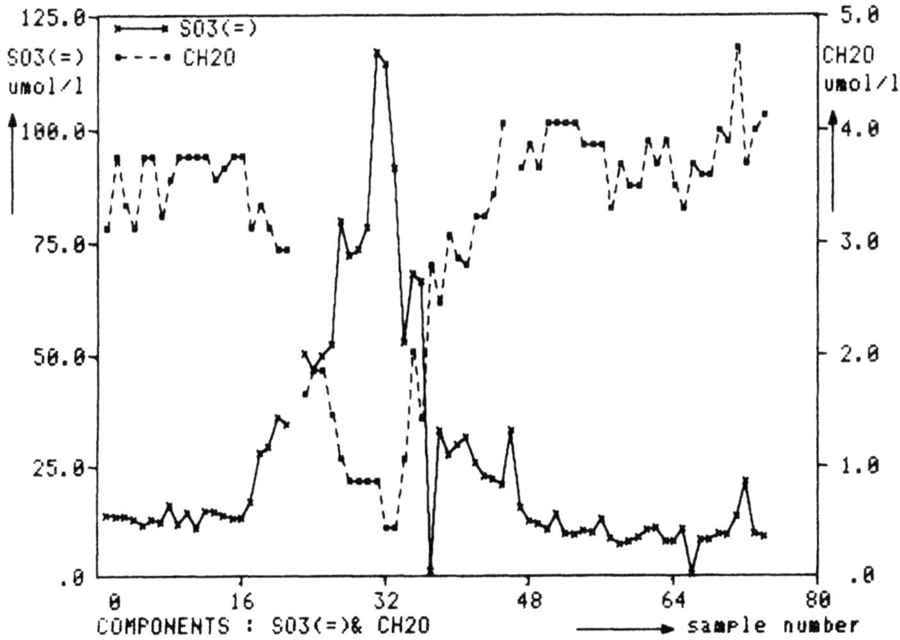

Figure 5 Plume profiles for sulphite ion and formaldehyde concentrations (86-03-27)

From figure 5 it follows that the concentration profiles in the plume for sulphite and formaldehyde are comparable with the profiles of other components e.g. sulphate, hydronium. Sulphite and formaldehyde concentrations in the plume and background correlated well ($r = 0.82$, $N = 74$). The decrease in formaldehyde in the plume amounts to 30% of the background concentration (table VI). This decrease must be attributable to the formation of hydroxymethanesulphonate (7). The supposed formation of hydroxymethanesulphonate is deduced from measurements in a plume because of the presence of high concentrations of sulphite relative to formaldehyde.

Table VI Plume and background weighted mean concentrations ($\mu mol.l^{-}$) for formaldehyde and sulphite (86-03-27)

background		plume	
sulphite	formaldehyde	sulphite	formaldehyde
11.6	3.8	44.9	2.4

From wash-out studies it is quite easily proved that hydrogen peroxide concentrations decreased in the plume. With aircraft measurements this effect is rather hard to illustrate and it is necessary to collect samples at least at one track upwind and two tracks downwind of the sources. A second demand is that during sampling, the flight altitude does not vary much. Moreover, the relation between the sulphate content and the hydrogen peroxide concentration can only be established over a large concentration range when a complete concentration spectrum is met in one flight, which is rare. From a data set of several flights the relation between hydrogen peroxide and sulphate can still be deduced. This was done for the three flights in figure 6.

Table VII Contribution of hydrogen peroxide to sulphate ion production in cloud water

flight	initial conmcentration*		final concentration		contribution %
	sulphate	H_2O_2	sulphate	h_2o_2	
142 (82-07-06)	20	56	150	< 1	43
239 (85-06-20)	76	22	175	4	18

* $\mu mol.l^{-1}$

5. CONCLUSION

The wet deposition of main components under plumes of a coal-fired power plant varied strongly in different wash-out experiments. During one of these experiments additional wash-out of ammonia in the plume was observed. Moreover, it was found that the formaldehyde concentration under the plume was lower than the background concentration. This was probably caused by the reaction of formaldehyde with an excess of sulphite present in the plume to form hydroxymethanesulphonate. The wet deposition of sulphate from a plume depends on several factors. This causes a non-linear relation between sulphur dioxide emissions and the wet deposition of sulphate. The oxidation of sulphite by hydrogen peroxide contributed from a

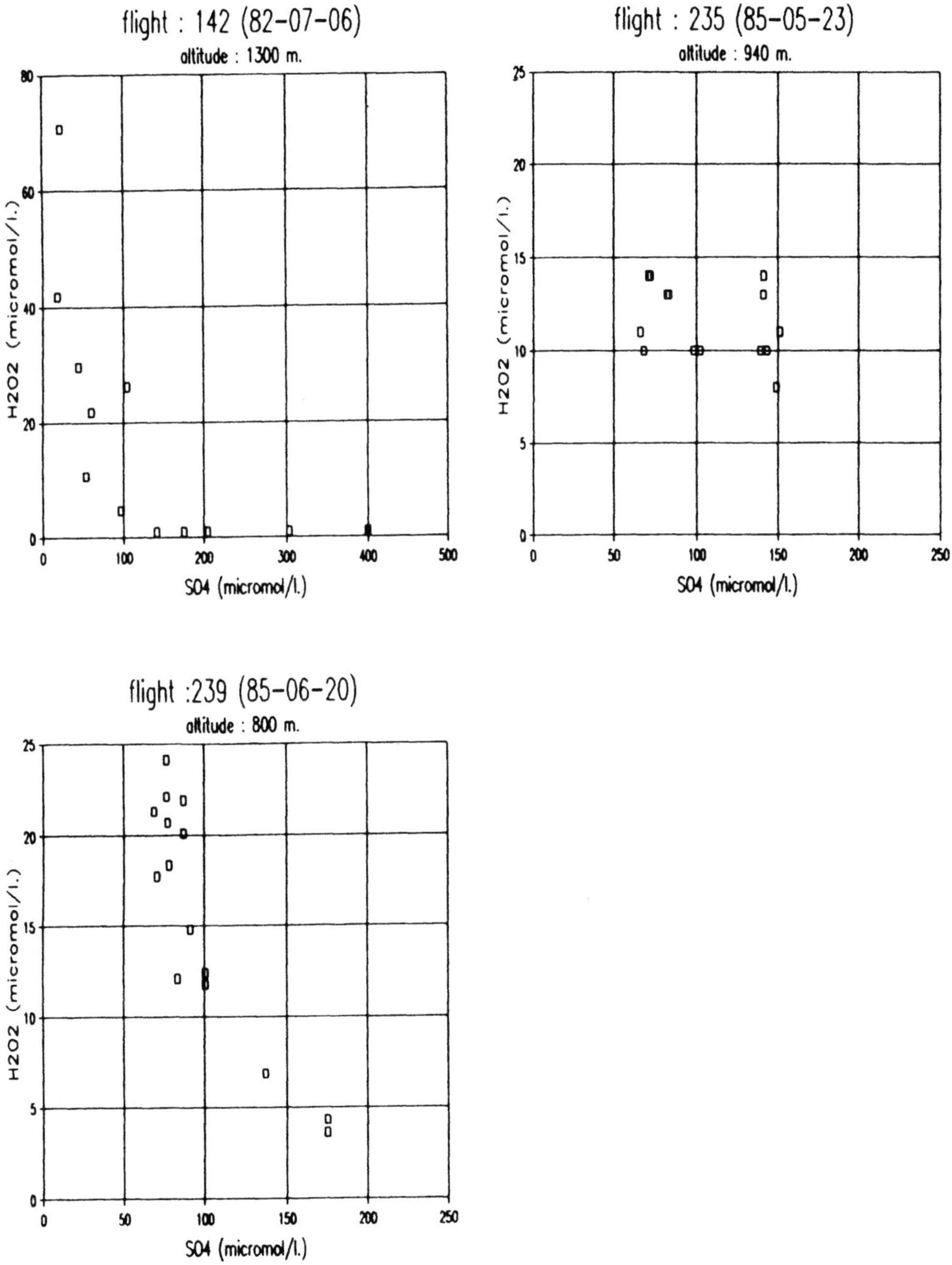

Figure 6 Relation between sulphate ion and hydrogen peroxide concentrations in cloud water of three flights

few to 90% to the sulphate formation in a plume (four wash-out experiments). This contribution of hydrogen peroxide in large plumes can also be established by aircraft measurements. However, to establish such a contribution from aircraft measurements many flights are needed.

ACKNOWLEDGEMENT

The personnel of Geosens B.V. (Rotterdam) is kindly acknowledged for their contribution to the aircraft experiments.

REFERENCES

(1) PENKETT, S.A., (1986). Laboratory studies of the multiphase S IV → S VI conversion rate. Chemistry of multiphase atmospheric systems, Jaeschke, W. ed., NATO-ASI Series, Springer Verlag, Berlin.

(2) REIJNDERS, H.F.R. et al. (1986). Characterization of cloud and rain water chemical composition. Proceedings of the Fourth European Symposium on Physico-Chemical Behaviour of Atmospheric Pollutants, Stresa (It.), September 1986, pp. .. - ..

(3) VILJEER, J.W. and DUUREN, H. VAN (1977). The measurement by aeroplane of atmospheric pollution, in particular in smoke plumes (in Dutch). Elektrotechniek, Vol. 55, pp. 540-547.

(4) RÖMER, F.G. et al. (1979). Messungen aus Flugzeugen. Verbreitung und Umwandlung primärer luftfremder Komponenten in Rauchfahnen. VGB-Konferenz Kraftwerk und Umwelt 1979, Essen (FRG), pp. 134-140.

(5) SLANINA, J., RÖMER, F.G. and ASMAN, W.A.H. (1982). Investigation of the source regions for acid deposition in The Netherlands. Proceedings of the CEC Workshop on Acid Deposition, Berlin, D. Reidel, Dordrecht, The Netherlands.

(6) JANSSEN,. A.J. and BRINK, H.M. TEN (1985). Plume wash-out: Modelling, calculation and validation. Netherlands Energy Research Foundation ECN, Petten NL, ECN-170.

(7) MUNGER, J.W.. TILLER. CHR. and HOFFMANN. M.R. (1986). Identification of hydroxymethanesulphonate in fog water. Science, Vol. 231, pp. 247-249.

MONTE-CARLO METHODS FOR SIMULATING DIFFUSION IN THE ATMOSPHERIC BOUNDARY LAYER

P. BESSEMOULIN
Centre National de Recherches Météorologiques
Toulouse, France

Summary

Gaussian diffusion models have problems in describing a given diffusion experiment because they cannot take into account the mean and turbulent structure of the atmospheric boundary layer (ABL). Despite the use of complex finite-difference schemes, it is known that finite-difference approximations of the A-D equation still introduce spurious numerical diffusion. On the other hand, the turbulent structure of the ABL may be such that the K theory is inadequate, particularly in the convection ABL where counter-gradient fluxes occur (André et al., 1978).
There is therefore a need for a method which insures conservation of mass, and has no computational diffusion. Here we propose to use a lagrangian statistical trajectory model based on Markov chain processes. The method consists in following the trajectory and velocity of a large number of particles (a few thousand) advected by the mean and turbulent flow field. Statistical averages allow for the computation of concentrations, structure of the plume, standard deviation of profiles, mean mass height...

1. STABLE AND NEUTRAL PLANETARY BOUNDARY LAYER : GAUSSIAN STATISTICS OF TURBULENCE

Following for instance Ley (1982), the vertical velocity fluctuation w' at time t+Δt is generated from the velocity at previous time t through a Markov chain equation :

$$w'(t+\Delta t) = R_w(\Delta t) \cdot w'(t) + (1-R_w^2)^{1/2} \sigma_w \cdot \Omega(t+\Delta t) + (1-R_w) \cdot T_{Lw} \frac{\partial \sigma_w^2}{\partial z}$$

where : - R_w is the lagrangian antocorrelation coefficient of the vertical velocity, which is assumed to follow an exponential law $\exp - \frac{\Delta t}{T_{Lw}}$

- T_{Lw} is the lagrangian integral scale
- σ_w is the standard deviation of w', this latter variable assumed to have a gaussian distribution (neutral or stable case)
- Ω(t+ Δt) is a random number taken from a gaussian distribution with unit variance and zero mean.

The third term on the right side of equation (1) has been used by several authors to avoid a non physical build-up of particles in regions with low values of σ_w.

The parameters appearing in the two first terms can easily be derived from the definitition of the autocorrelation function and from the expression of the standard deviation σ_w resulting from (1).

In a similar way, one can consider a relation for the longitudinal velocity fluctuation u', which is written slightly differently :

$$v(t+\Delta t) = \alpha \cdot v(t) + \beta \cdot \Omega(t+\Delta t) \qquad u'(t+\Delta t) = v(t+\Delta t) + \gamma w'(t+\Delta t)$$

where : - v is a dummy recursive variable

- α, β, γ are coefficients to be determined
- Ω has the same meaning as above.

Expressions for α, β, γ can be derived considering that :

$$R_u(\Delta t) = \frac{\overline{u'(t) \cdot u'(t+\Delta t)}}{\sigma_u^2} \quad \text{(autocorrelation for u')}$$

$$\frac{-U_*^2}{\sigma_u \sigma_w} = \frac{\overline{u'(t) \cdot w'(t)}}{\sigma_u \sigma_w} \quad (\overline{u'w'} \text{ correlation})$$

$$\sigma_u^2 = \overline{u'(t)^2} \quad \text{(turbulent horizontal kinetic energy)}$$

u_* being the friction velocity.

Trajectories of individual particles released from a point source at height H are computed by integrating the lagrangian equations :

$$\frac{dx_i}{dt} = \overline{U(z)} + u'(z,t) \tag{2}$$

$$\frac{dz_i}{dt} = \overline{W(z)} + w'(z,t) \tag{3}$$

$\overline{U}$ and $\overline{w}$ being the mean wind fields.

The time step Δt used to integrate (2) and (3) must be smaller than R_{Lw}. In the model we used $\Delta t \cdot T_{Lw}^{-1} = 0.1$, which has the advantage of leading to the same accuracy of the numerical scheme at all levels.

The atmospheric parameters U, σ_u, σ_w, T_{Lu}, T_{Lw}, U_* to be introduced in the model, have been chosen from numerous field experiments, in particular the Minnesota observation by Kaimal et al. (1976)

Outputs of the model are : concentration C(x,z), mean height of the plume, standard deviation of mass, and a parameter, named e, computed from a smoothing of the ground level concentrations from the model by the 2D formula :

$$\chi = \frac{C}{Q} = \sqrt{\frac{2}{\pi}} \cdot \frac{1}{\overline{U} \cdot e(x)} \cdot \exp -\frac{H^2}{2e^2(x)} \tag{4}$$

This formula gives the cross-wind integrated concentration as a function of downwind distance in the framework of the gaussian plume concept.

In other words, e is a parameter which gives with good accuracy the concentrations computed by the model, when introduced in the gaussian formulation (4). In this sense, e is the parameter most comparable with curves appearing in the literature.

Some of the model predictions have been first compared with field data :
- the Prairie Grass data re-examined by Nieuwstadt and Van Ulden (1978)
- Briggs curves
- Brookhaven curves

The agreement was considered satisfactory and consequently a systematic study was undertaken. This latter study was concerned with the influence of roughness length z_0, emission height H, and mean wind speed on vertical diffusion, for downwind distances up to 10,000 m.

General conclusion was that the equivalent parameter e was shown to :
- strongly increase with an increase of z_0 (Fig.1) ;
- moderately decrease with an increase of H (Fig. 2) ;
- strongly increase with an increase of U_* (Fig. 3).

2. UNSTABLE BOUNDARY LAYER

The turbulent unstable convective atmospheric boundary layer (CABL) has received considerable attention for the last ten years. From a pollution point of view, the interest lies mainly in the fact that peak concentrations can be found very near the emission zone, even if the stack is very high. As a consequence, gaussian models fail because the plume axis is far from being horizontal. On the other hand, the turbulent structure of the CABL is such that the K theory may be inadequate, particularly in the lower half of the layer where counter-gradient fluxes of energy occur (André et al. 1978).

At last, a Monte Carlo diffusion model, such as the one described above, implies that all turbulent fluctuations u' (resp. w') belong to a gaussian distribution with zero mean and σ_u^2 (resp. σ_w^2) variance. This is far from being the case in convective conditions, especially for the vertical velocity which is strongly positively skewed. Fortunately, several field experiments (see Taconet and Weill (1982), Lenschow (1980), Caughey et al(1983) have shown that motions in the CABL are well organized into "updrafts" and "downdrafts", and that the probability density function of w can be approximated by the sum of two gaussian distributions :

$$P(w) = \frac{RF_+}{\sqrt{2\pi}\,\sigma_+} \exp -\frac{(w-w_+)^2}{2\sigma_+^2} + \frac{RF_-}{\sqrt{2\pi}\,\sigma_-} \exp -\frac{(w-w_-)^2}{2\sigma_-^2} \qquad (5)$$

where : - RF_+ (resp. RF_-) is the relative frequency occurence of updrafts (resp. downdrafts) ;
- σ_+ (resp. σ_-) is the standard deviation of updrafts (resp. downdrafts) ;
- w_+ (resp. w_-) is the mean updraft (resp. downdraft) velocity.

The following relationships, associated with a closure relationship (for instance linking σ_+ and w_+ , σ_- and w_- or considering a given profile for w_+ and w_-) allows the computation of $RF_+(z)$, $RF_-(z)$, $\sigma_+(z)$, $\sigma_-(z)$, $w_+(z)$, $w_-(z)$:

$$\int_{-\infty}^{+\infty} P(w) \cdot dw = 1$$

$$\int_{-\infty}^{+\infty} w \cdot P(w)\, dw = 0 \qquad \text{(ensemble mean velocity is zero)}$$

$$\int_{-\infty}^{+\infty} w^2 \cdot P(w) \cdot dw = \sigma_w^2$$

$$\int_{-\infty}^{+\infty} w^3 \cdot P(w) \cdot dw = S_w^3 \qquad \text{(skewness)}$$

The Monte Carlo model used here is the following :

$$v(t+\Delta t) = v(t) \cdot \frac{\sigma_{wc}(z(t+\Delta t))}{\sigma_{wc}(z(t))} \cdot R_{wc}(t+\Delta t) + \Omega(t+\Delta t) \cdot \sigma_{wc}(z(t+\Delta t)) \left(1 - R_{wc}^2\right)^{1/2}$$

$$w(t+\Delta t) = w_c(z(t+\Delta t)) + v(t+\Delta t) \qquad (6)$$

$$z(t+\Delta t) = z(t) + w(t) \cdot \Delta t \qquad (7)$$

where - $w_c = w_+$ or w_-
- $\sigma_{wc} = \sigma_+$ or σ_-
- $R_{wc} = R_{w+}$ or R_{w-}

according to the mode of the particle. Other variables are the same as previously defined.

The simulation considers separately particles starting in updrafts (number proportional to RF_+) and in downdrafts (number proportional to RF_-) the total plume being simply the sum of the two contributions.

Two difficulties arise with this kind of modelling resulting in spurious accumulation of particles if no precautions are taken, due to :
- the vertical gradient of σ_w
- the vertical gradient of the convection velocity w_c (w+ or w-).

It can be shown than an additional term in (6) of the kind of the third one in the r.h.s. of (1) is needed to balance the effect of $\frac{\partial \sigma_w}{\partial z}$ and that accumulation can be avoided if particles are allowed to change their phase (ascending or descending) with a given probability depending on $\partial w_c / \partial z$.

This model is thus very similar to Baerentzen and Berkowicz's (1984); to validate the model, results from Willis and Deardoff's (1976, 1978) laboratory experiments have been used. In those measurements, particles were released in a convection water tank which has been shown to be a very good analogue of the atmospheric convective boundary layer (Willis and Deardorff (1974)). Cross wind integrated concentrations in the x - z plane can be compared to the results of the 2D dispersion model, provided the measured and computed quantities have been properly non dimensionalized :

$$X = \frac{w_*}{\overline{U}} \cdot \frac{x}{z_i} \qquad Z = \frac{z}{z_i} \qquad C(X, Z) = c(x, z) \cdot z_i^2 \cdot \frac{\overline{U}}{Q}$$

where - w_* is the convective velocity
- $\overline{U}$ is the mean wind speed
- Zi is the height of the CBL
- Q is the source strength
- c is the dimensional concentration

Two emissions heights have been simulated :
- H = 0.24 Zi (Fig. 4)
- H = 0.067 Zi (Fig. 5)

One can judge from Fig. 4 and 5 that in both cases the position and amplitude of isopleths are correctly predicted, especially the maximum concentration areas.

One main feature of the elevated release case is a rapid descent of the plume and a small vertical gradient of concentrations beyond X = 1.5. The discrepancies between model and experimental data can be ascribed to the top of the experimental convective layer not being impervious as in the numerical model.

The near ground release is characterized in both simulations by a concentration maximum just above the source release height and by an elevated maximum at X $\sim$ 1.5 - 1.7, just below the overlying stable layer. Again, the vertical concentration profile is very smooth beyond X $\sim$ 2.

3. CONCLUSIONS

A set of lagrangian diffusion models have been developped to be used in either a neutral or stable atmosphere, or in convective conditions.

A sensitivity study in the first case has shown the effects of release height, roughness and wind speed on the characteristics of the plume.

For a convective boundary layer, model results have compared favorably with laboratory results.

Those results will be improved in the future by applying very recent turbulent data related to the previous experiments (Deardorff and Willis, 1985).

REFERENCES

(1) ANDRE, J.C. et al. (1976) : Turbulence approximation for inhomogeneous flows : Part II. The numerical simulation of a penetrative convection experiment. J. Atmos. Sci. 33, pp 482-491.

(2) BAERENSTEN, J.H. and R. BERKOWICZ (1984) : Monte Carlo simulation of plume dispersion in the convective boundary layer. Atmospheric Environment 18, pp 701-712.

(3) CAUGHEY, S.J., H. KITCHEN and J.R. LEIGHTON (1983) : Turbulent structure in convective boundary layers and implication for diffusion. Boundary-Layer Met. 25, pp 345-352.
(4) DEARDORFF, J.W. and G.E. WILLIS (1985) : Further results from a laboratory model of the convective boundary layer. Boundary Layer Met.32, pp 205-236.
(5) KAIMAL, J.C., D.A. HAUGEN et al (1976) : Turbulent structure in the convective boundary layer. J. Atmos. Sci. 33, pp 2152-2169.
(6) LENSCHOW, D.H. and J.C. WYNGAARD, (1980) : Mean Field and second moment budgets in a baroclinic, convective boundary layer. J. Atmos. Sci. 37, pp 1313-1326.
(7) LEY, A.J. (1982) : A random walk simulation of two-dimensional turbulent diffusion in the neutral surface layer. Atmospheric Environment 16, pp 2799-2808.
(8) NIEUWSTADT, F.T.M. and A.P. VAN ULDEN (1978) : A numerical study of the vertical dispersion of passive contaminants from a continuous source in the atmospheric surface layer. Atmospheric Environment 12, pp 2119-2123.
(9) TACONET, O. and A. WEILL (1982) : Vertical velocity field in the convective boundary layer as observed with an acoustic doppler Sodar. Boundary-Layer Met. 23, pp 133-151.
(10) WILLIS, G.E. and J.W. DEARDORFF (1974) : A laboratory model of the unstable boundary layer. J. Atmos. Sci. 31, pp 1297-1307.
(11) WILLIS, G.E. and J.W. DEARDORFF (1976) : A laboratory model of diffusion into the convective planetary boundary layer. Quart. J.R. Met. Soc. 102, pp 427-445.
(12) WILLIS, G.E. and J.W. DEARDORFF (1978) : A laboratory study of dispersion from an elevated source within a modeled convective planetary boundary layer. Atmospheric Environment 12, pp 1305-1311.

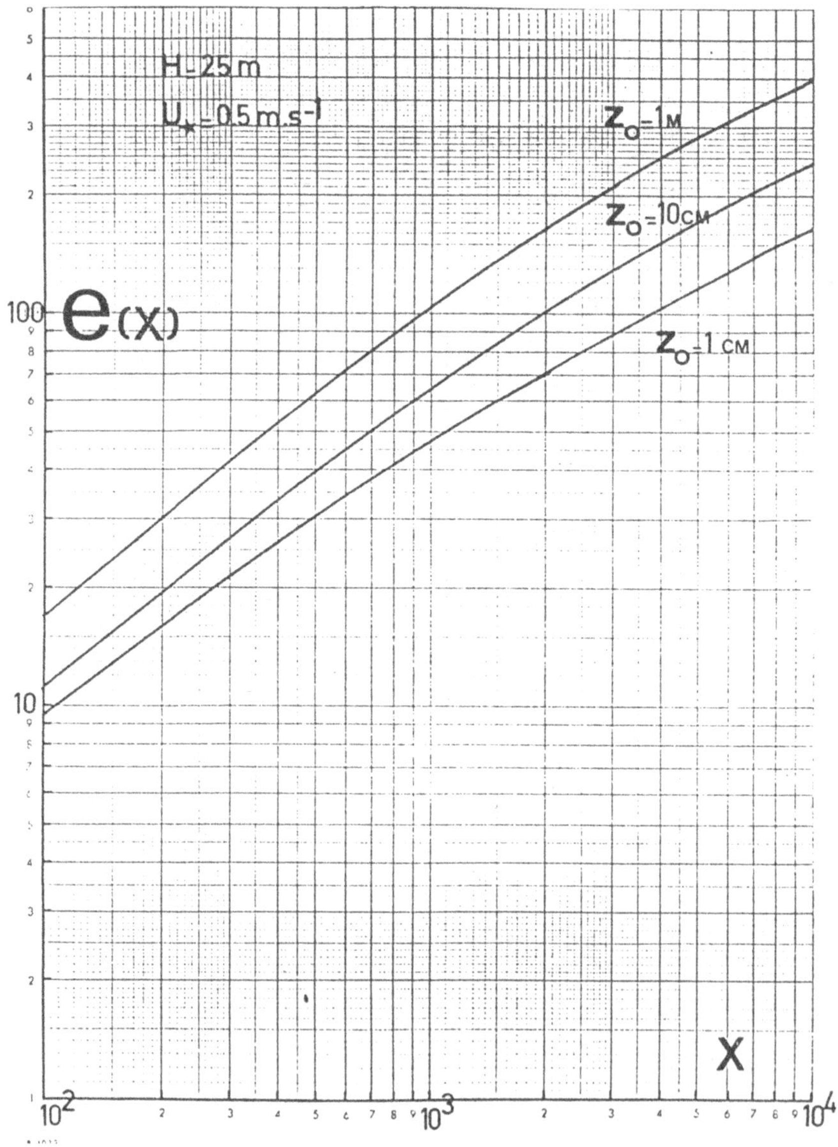

FIG. 1 : Influence of roughness

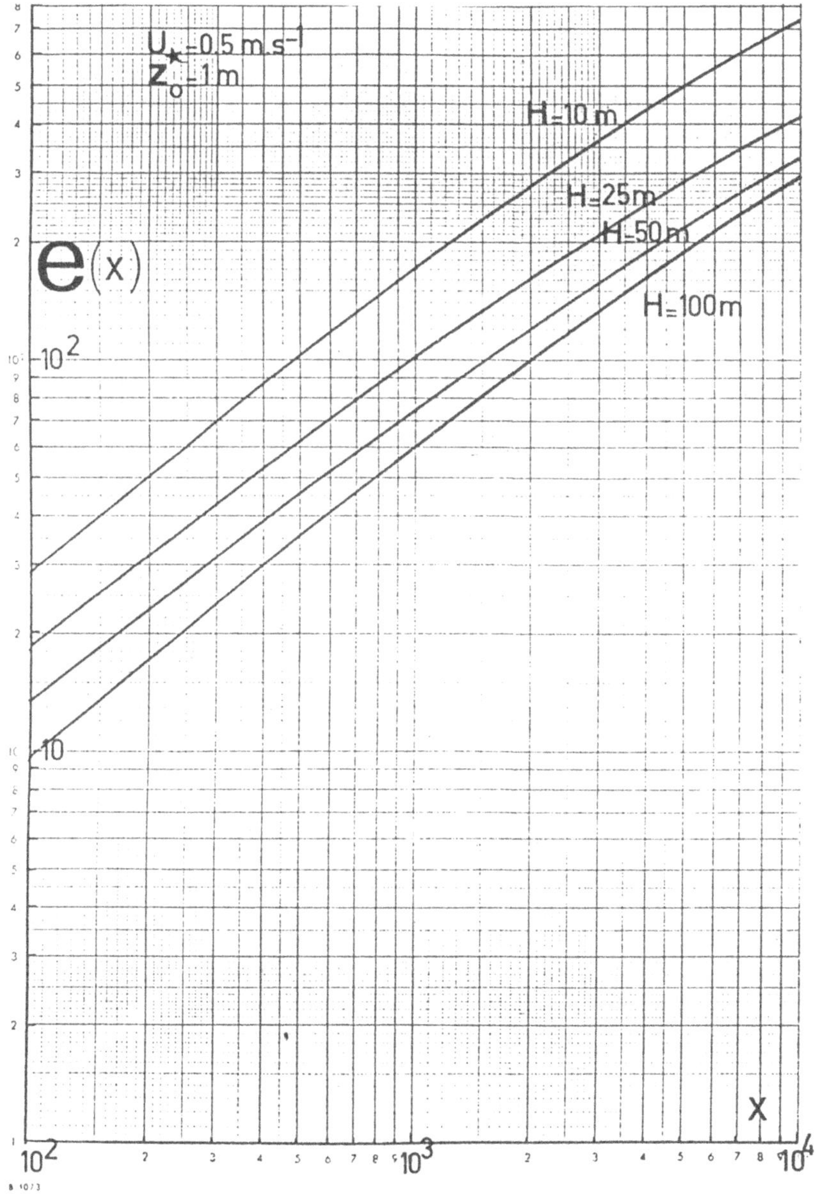

FIG. 2 : Influence of emission height

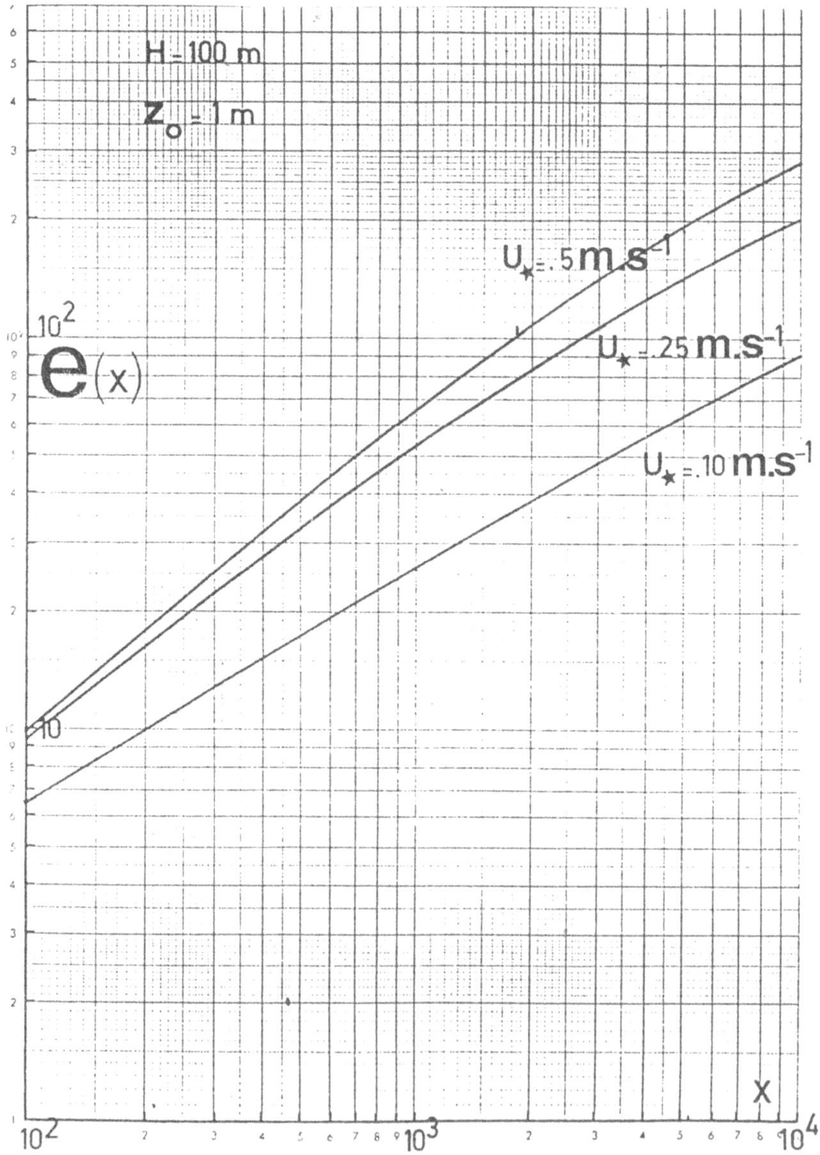

FIG. 3 : Influence of wind speed

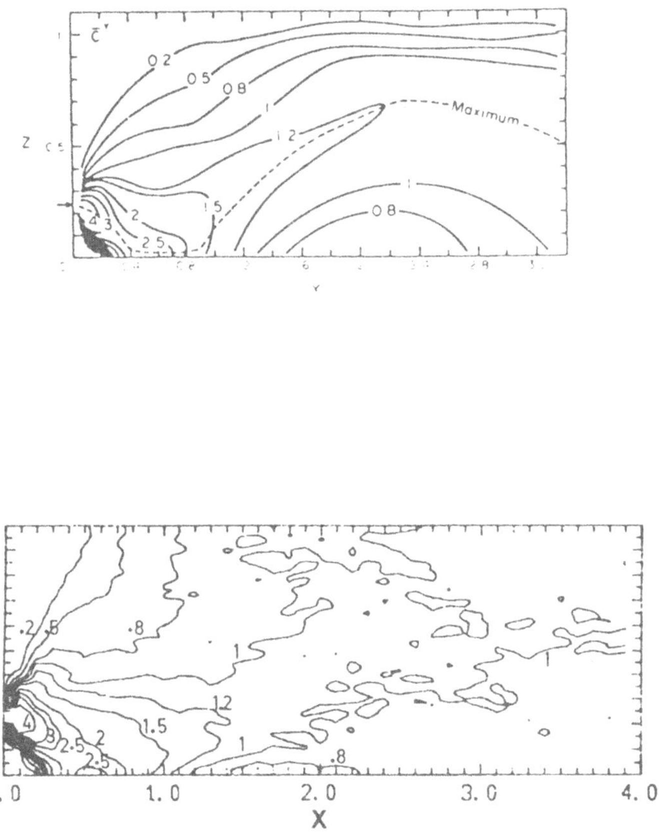

FIG. 4 : H = 0.24. Z_i

Smoothed contours of dimensionless cross-wind integrated concentration as a function of height and downwind distance.
Top : Laboratory results (Willis and Deardorff)
Bottom : Model results

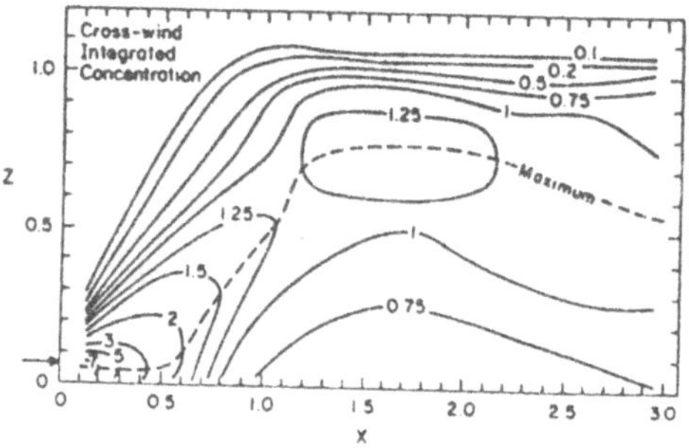

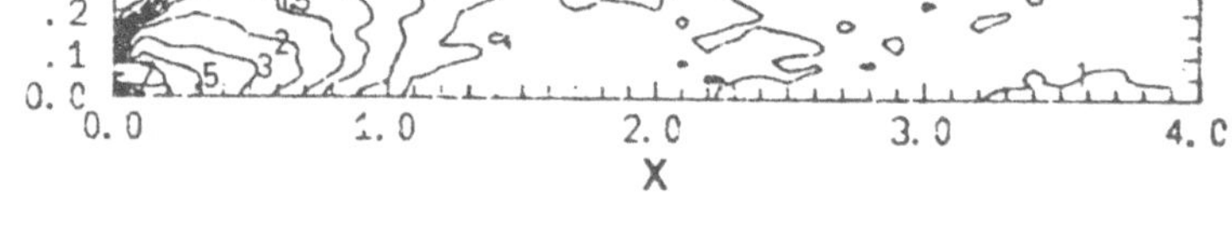

FIG. 5 : $H = 0.067\ Z_i$

Smoothed contours of dimensionless cross-wind integrated concentration as a function of height and downwind distance.
Top : Laboratory results (Willis and Deardorff)
Bottom : Model results

MODELLING OF LAGRANGIAN TURBULENT DISPERSION OF NEUTRAL AND BUOYANT POLLUTANTS RELEASED IN THE ATMOSPHERIC BOUNDARY LAYER

C. BORREGO
Departamento de Ambiente, Universidade de Aveiro
3800 AVEIRO - PORTUGAL

Summary

The present paper describes the non-isothermal atmospheric dispersion of neutral and buoyant pollutants, released from two-dimensional sources, in presence of obstacles or complex relief. The model calculations are divided in two parts. Firstly, the velocity field is solved using a first order model (eddy viscosity approach). Secondly, the concentration field is obtained with the use of a Lagrangian approach, in which the behaviour of a contaminant is simulated by "markers" whose displacement reproduces the statistics of the considered turbulent transport; the average history of a large number of marker releases then yields the ensemble solution of the diffusion equation. This model has the following advantages: (i) it take into account ground irregularities and the presence of buildings, giving in detail the velocity and temperature fields, this in comparison with the Gaussian model; (ii) it has a few member of equations with simple numerical scheme, which decrease the computational time and memory requirements, when compared with the Eulerian approach. The model is extensively validated by comparison with experimental data for neutral of buoyant gases, which provides a clear understanding of the potencial of such a model.

1. INTRODUCTION

Modelling of turbulent dispersion of a pollutant in the atmospheric boundary layer (as opposed to the simplest case of diffusion in a homogeneous flow) is a problem of great theoretical and pratical importance. First of all, the accuracy of the prediction is a proof of the correct understanding of the basic phenomena of turbulent transport and provides an ideal situation for the verification of the assumptions made. Second, such a tool is necessary to predict the possible consequences related to normal and/or accidental releases of gaseous pollutants in the atmosphere.

The oldest and simplest approach to the problem of turbulent dispersion is the Gaussian one, extensively described in Pasquill (1) and Borrego (2). Such a formulation is extremely simplified and is known not to give good results at a short and medium distances from the source (less than 1 km), where it is of special importance to be able to predict correctly the concentration of pollutant in the case of an accidental release of a dangerous (toxic or flammable) gas. Moreover, is also the situation where the eddy diffusivity model of the pollutant transport equation suffers from serious theoretical and pratical limitation (3).

Promising results, and in particular the elimination of some problems related to the discrete mesh in solving the Eulerian approach of transport equations, can be obtained with the use of a Lagrangian approach, firstly introduced by Taylor (4) for homogeneous flow fields, and more recently generalized to non-homogeneous ones by Durbin (5). In this approach, the behaviour of a contaminant is simulated by "markers" whose displacement reproduces the statistics of the considered turbulent transport; the average history of a large number of marker releases then yields the ensemble solution of the diffusion equation. Moreover, such an approach can be immediately and satisfactorily extended to take into account the gravity effect for buoyant pollutants (6).

2. THE DISPERSION PROBLEM IN TERMS OF A LAGRANGIAN MODEL

2.1: Basic equations of the stochastic approach

The basis of the stochastic approach to the dispersion of a pollutant lies in the following (5): suppose one knows the initial distribution of pollutant to be $C_o(y)$; also one knows an ensemble of random trajectories of fluid particles, each trajectory originating at some point y_o. If it is assumed that fluid particles retain their initial concentration $C_o(y_o)$, then the mean concentration at the point (y,t) is just the probability that a parcel observed at position y at time t originated at some arbitrary point y_o, times the concentration at that point $C_o(y_o)$, summed over all possible origins:

$$\bar{C}(y,t)=\int_{-\infty}^{\infty} P(y,t;y_o)\ C_o(y_o)dy_o \tag{1}$$

This equation uses the definition that the probability of a particle originating at y_o (actually between $y_o - dy_o/2$ and $y_o + dy_o/2$) is $P(y,t;y_o)dy_o$. Note that in equation (1) the final position of the random trajectories is fixed at y, while the initial position y_o is the random variable.

Therefore, the problem of determining $\bar{C}$ is the same of that following the trajectory of particles released in a turbulent velocity field (Lagrangian method):

$$Y(t) = \int_0^t \bar{V}(t')dt' \tag{2}$$

Which gives, in discretized form used in pratice,

$$Y(t) = \sum_{n\,o}^{t} V_n\ \Delta t. \tag{3}$$

In a stationary velocity field, the "markers" are advected by the mean flow and dispersed by the local turbulent motions. The first contribution is deterministic, while the second is random or contains, at least, a random contribution. As random motions along different axis are by definition uncorrelated, the velocity of the particles in the direction of the coordinate axis x_i can be written as

$$V_i = \bar{V}_i + V_{r,i} \tag{4}$$

where $\bar{V}_i$ is the mean field contribution and $V_{r,i}$ the "random like" one, to be modeled using the tools given by the stochastic theory.

2.2: Statistical approach to turbulent dispersion

A Markov process can be described as a time-dependent stochastic process in which the future is determined by the present, independently of the past. Thus when describing turbulent transport only turbulent diffusion is refered when a Markovian model of fluid trajectories (eddy diffusion) is considered, but in general is called turbulent dispersion. Furthermore, in the case of homogeneous turbulence the displacement of the particles under the action of turbulence are entirely random. Such a behaviour can be described by the Wiener process (or "drunkard's walk"):

$$dY(t) = a\left[Y(t)\right] dW_t \tag{5}$$

where Y(t) is the displacement of a particle in the direction Y, a(*) is some prescribed function of the turbulent field and dW_t is the infinitesimal step of the drunkward's walk, which is a Gaussian of mean zero and variance $\overline{W_t^2} = t$.

The same approach can be extended to the case of non-homogeneous turbulence, where the turbulent transport is no more entirely random but there is a mean tendency of the transported quantity to move in the direction of increasing turbulent length scales. Such an effect is given by (5):

$$dY(t) = b(Y)dt + a(Y)dW_t \tag{6}$$

where $\overline{b(Y)}$ is the mean Lagrangian velocity of the markers in the direction Y.

These coefficients can be found (5) using the Fokker-Planck equation for the transport of the probability P of equation (1):

$$a(Y) = \sqrt{2k(Y)} \qquad \overline{b(Y)} = \frac{\partial k(Y)}{\partial Y} \tag{7}$$

being k the eddy diffusivity. Then, the often-made distinction between "statistical" theory and "eddy diffusion" theory is fictitious: eddy diffusion theory is really just a Markovian statistical theory; the eddy diffusion equation is just the Fokker-Plank equation for the pdf of a Markovian random process.

When the time macroscale for the fluctuations of velocity, T_L, is different from zero, which is the case for an elevated source in a boundary layer, the process itself is no more Markovian. However, the Lagrangian approach can be extended to problems with time scale not iqual to zero, as indicated in (5), treating the turbulent velocity V_r as a Markovian process and obtain the Langevin equation:

$$dV_r(t) = \frac{-V_r(t)dt}{T_L} + \sigma_v\sqrt{\frac{2}{T_L}}\, dW_t\,. \tag{8}$$

Here σ_v is a turbulent velocity scale (constant); the initial condition for equation (8) is that $V_r(o)$ is a Gaussian random variable with mean zero and variance σ_v^2.

If the turbulent properties σ_v and T_L are assumed to be slowly varying in the region around the source, the reference (5) give an approximate solution for the displacement, which can be taken as the sum of a mean displacement with mean Lagrangian velocity:

$$\bar{V}_r = \frac{\partial(\sigma_v^2\, T_L)}{\partial Y}\left(\frac{t}{T_L} - 1 + e^{-t/T_L}\right) \tag{9}$$

and a random displacement along a Gaussian of mean zero and variance

$$\overline{Y^2} = \sigma_v^2\, T_L\, \frac{\partial T_L}{\partial Y}\left(e^{-t/T_L} - 1 + \frac{t}{T_L}\, e^{-t/T_L}\right) \tag{10}$$

By design, when $t \gg T_L$ this reproduces equations (7) with the eddy diffusivity $k = \sigma_v^2\, T_L$.

2.3: Random displacements and control volume

The random contribution in equation (4) requires the use of a random numbers generator respecting the conditions defined in §2.2. This mechanism is employed to generate at each time step random displacements following a Gaussian of mean zero and variance given by equation (3) for a ground sour ce, and a Gaussian mean zero and variance given by equation (10) for an elevated source.

Once known the trajectories, the mean concentration $\bar{C}(x_i)$ at a given location x_i in a fixed frame can be obtained by counting up the particles present in an appropriate control volume around the point x_i. In order to obtain consistently accurate results, it is necessary that the number of particles introduced in the field be sufficiently high (7) and that the control volumes are chosen correctly. Both parameters depend on the nature of the flow to be studied.

The number of particles in the field must be optimized by numerical tests and the concentration is calculated by counting the particles present in a control volume whose horizontal size is proportional to the distance from the source while the vertical one is proportional to the vertical scale of the plume.

2.4: Buoyant pollutants

One of the advantages in the modelling of dispersion through the displacement of markers is that the introduction of the buoyant effect is immediate: it is only necessary to add to the vertical displacement of each marker, as defined by equation (4), the contribution of the buoyancy force, to obtain (8):

$$Z_{i,g} = \frac{C_p}{\rho_a} \left(1 - \frac{\rho_a}{\rho_p}\right) g_i \Delta t^2 \tag{11}$$

where C_p is the local concentration, g_i the gravitational acceleration, ρ_a the surrounding ambient density and ρ_p the density of the pollutant.

3. VELOCITY FIELD

The most correct approach to the problem of estimating the velocity and turbulent field in presence of a buoyant pollutant would be to solve all together the equations for the transport of momentum, the dispersion and the turbulent field, to obtain directly the effect of the buoyancy on the flow. However, the code presently available is not geared to the task and requires the separate solution of the flow field: the velocity field is solved using a first order model (eddy viscosity approach), that is, following Nee and Kovasznay (9) and reference (8):

$$\frac{D\eta}{Dt} = \text{Diffusion} + \text{Production} - \text{Dissipation} \tag{12}$$

$$\frac{D}{Dt} = \frac{\partial}{\partial t} + V_i \frac{\partial}{\partial x_i} = \text{substantive derivative}$$

being $\eta = \nu + \nu_T$, the sum of the kinematic viscosity with the "apparent" or "eddy" viscosity.

The numerical method used to solve equation (12) lies in a grid of "marker-and-cells" type (10); since this technique is well known, it will not be explained here.

4. VALIDATION OF THE MODEL AND RESULTS

The model described above is validated by comparing its prediction to experimental data available in literature. The first such comparison is for the experiments by Raupach and Legg (11), which deals with an elevated line source in a simulated neutral atmospheric boundary layer. Heat is employed to simulate a passive pollutant. The experimental velocity profile, given in ref.(11), is used in the calculation, and the turbulent diffusivity obtained applying the Nee and Kovaszny transport equation for the turbulent viscosity adapted for an atmospheric boundary layer (8).

Comparison with present prediction is shown in figure 1 where concentrations are plotted as function of a non-dimensional source mass flow C*

$$C^* = \frac{Q}{\rho_a Z_s U_s} \tag{13}$$

where Q is the source intensity, U_s the flow velocity at the height of the source Z_s. The results presented show a good agreement between prediction and experiments: the initial spread of the plume is slightly underpredicted, due to the difficulty of selecting control volumes small enough in the immediate neighborhood of the source, but the agreement remains acceptable and improves quickly with distance from the source.

Reliable experimental data on buoyant gases dispersion are difficult to find in open literature. Therefore, the present two dimensional development is validated by comparison with experimental data obtained by the author in past few years. As the pratical interest of the model lies in its possible application to the prediction of heavy gas dispersal, a verification on experimental data for a very heavy gas ($C\,Br\,F_3$; molar mass 0.149 kg) is considered (12). The release is from an elevated line source (Z_s = = 0.05 δ, where δ is the boundary layer thickness) in an artificially thickned turbulent boundary layer.

For the numerical simulation the experimental velocity profile is taken as input and the turbulent diffusivity estimated with the already mentioned Nee and Kovasznay model. Due to the availability of an accurate knowledge of the flow field, the results of the comparison are quite good. Figure 2a) presents the calculated flow field for this particular case, where exists a wall (h = 0.1 δ) perturbing the flow; in figures 2 b), c) and d) are proposed comparisons between predicted and experimental concentration profiles at different locations downstream of the source. A consistently good agreement is found apart from a slight underestimation of the plume spread quite near to the source.

Results for dispersion of heavy and light gases from a point-like source in a non-adiabatic atmospheric boundary layer (under a temperature inversion at 60 m high) are presented in figure 3, as they are predicted by the model. There are no experimental data for comparison, but the results show the behaviour that are expected for such a case.

5. CONCLUSIONS

The full stochastic theory of turbulent dispersion has been implemented in a time marching Lagrangian model. This approach allows a representation of the phenomenon more consistent and complete than one yielded by the simple diffusivity models.

The predictions of the model have been compared to experimental data for pollutant releases from two-dimensional sources in simulated atmospheric boundary layers. Good results have been obtained for passive contaminants and heavy gas. The simulation of a heavy and light gas release from a point-like source in a non-adiabatic atmospheric boundary layer was also predicted.

It can then be concluded that this approach is a suitable method for the modelling and prediction of dispersion in the atmospheric boundary layer, with particular reference to pratical problems.

ACKNOWLEDGEMENTS

The author gratefully acknowledges the receipt of a fellowship from SEARN (Secretaria de Estado do Ambiente e Recursos Naturais).

REFERENCES

(1) PASQUILL, F. (1961). The estimation of windborne material. Meteorologi

cal Magazine, Vol. 90, 33-49.
(2) BORREGO, C. (1984). Modelo Gaussiano para a dispersão de poluentes na atmosfera. AMB-QA-(3)/84, Departamento de Ambiente, Universidade de Aveiro.
(3) BORREGO, C. et al. (1984). Comparação de alguns modelos de dispersão atmosférica: Aplicação a uma zona costeira - Relatório preliminar. AMB-QA(1)/84, Departamento de Ambiente, Universidade de Aveiro.
(4) TAYLOR, G. L. (1959). The present position of the theory of turbulent diffusion. Advances in Geophysics, Vol. 6, 101-111, Academic Press.
(5) DURBIN, P. A. (1983). Stochastic differential equations and turbulent dispersion. NASA RP 1103.
(6) BRADSHAW, P. and WOODS, J. D. (1976). Geophysical turbulence and buoyant flows in "Turbulence", ed. P. Bradshaw, Topics in Applied Physics, Vol. 12.
(7) BENOCCI, C. and OLIVARI, D. (1985). Lagrangian modeling of turbulent dispersion. IVK LS 1985-02.
(8) FOUSSAT, A. (1981). Modèle de dispersion atmosphérique non-isotherme d'un polluant gazeuse de densité quelconque en présence de non-uniformités orographiques. IVK TN 135.
(9) NEE, V. W. and KOVASZNAY, L.S.G. (1969). Simple phenomenological theory of turbulent shear flows. Phy. Fluids, Vol. 12, nº3, 473-492.
(10) WELCH, J. E. et al. (1965). The MAC method. Los Alamos Scientific Lab. Report LA-3425.
(11) RAUPACH, M. R. and LEGG, B. J. (1983). Turbulent dispersion from an elevated-line source: measurements of wind-concentration moments and budgets. J. Fluid Mech., Vol. 136, November, 111-137.
(12) BORREGO, C. and RIETHMULLER, M. L. (1980). Optical measurements of the diffusion of a heavy gas in turbulent boundary layers. IVK LS 1980-03.

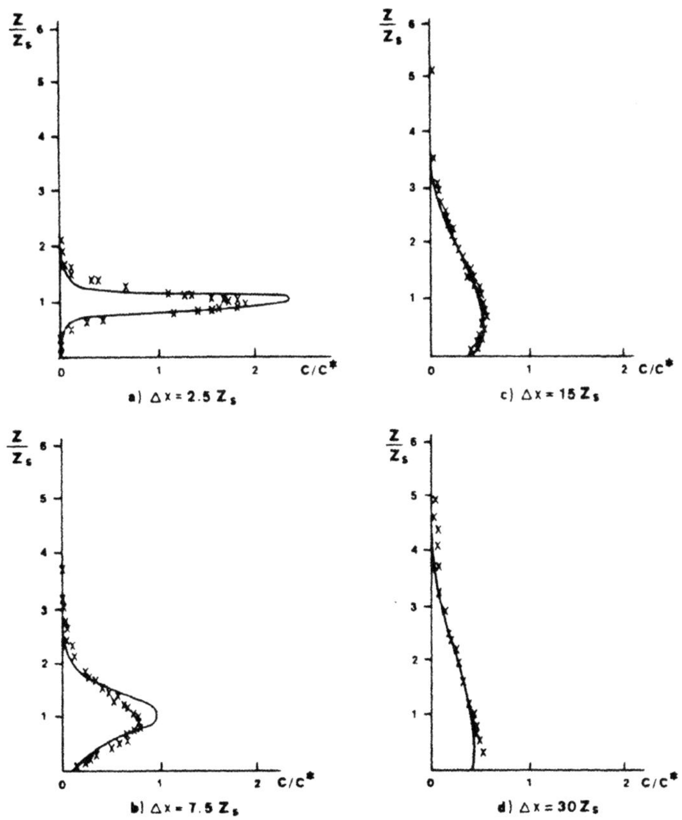

Fig. 1 - Non-dimensional concentration profiles: elevated line source in simulated atmospheric boundary layer.
—— prediction; x experiment (ref.11)

a) Velocity flow field

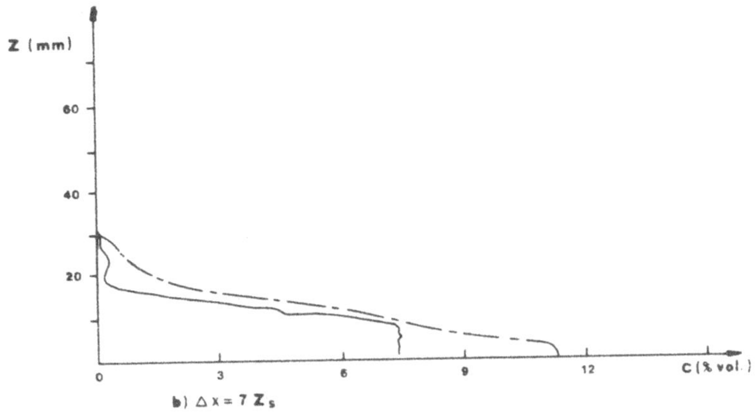

b) $\Delta x = 7\, Z_s$

Fig. 2 - Concentration profiles for heavy gas dispersion from a 2-D source in a simulated atmospheric boundary layer.
— - —prediction; —— experiment (ref. 12).

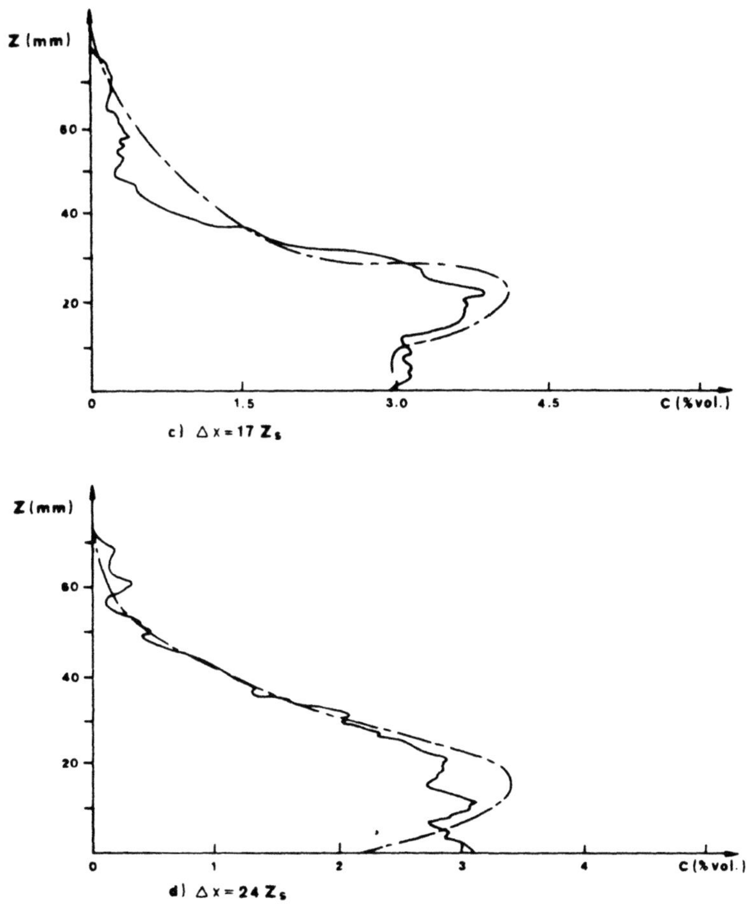

Fig. 2 - continuation (for legend, see preceding page).

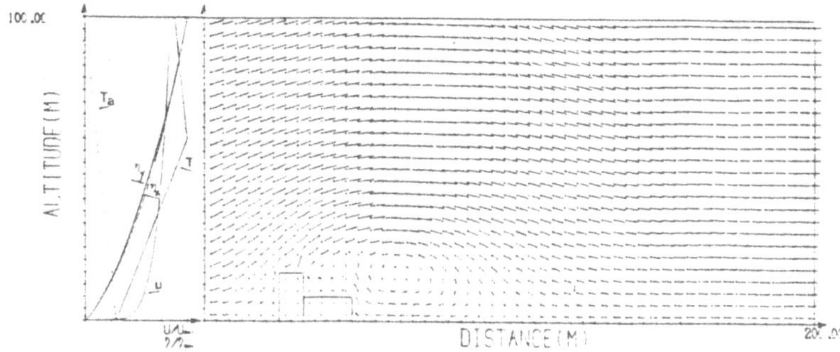

a) Velocity and temperature field and source configuration.

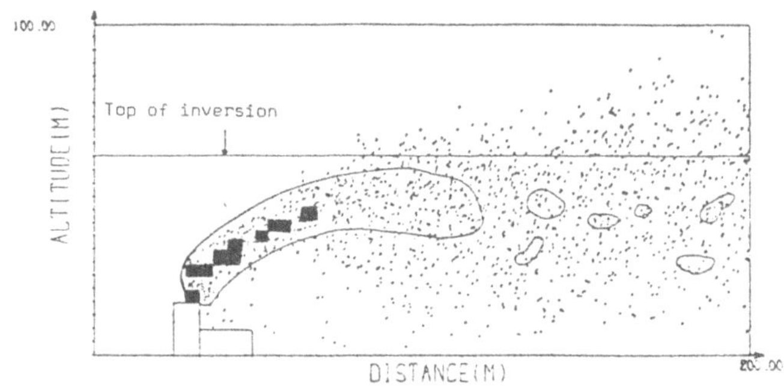

b) Heavy gas concentration, marker distribution.

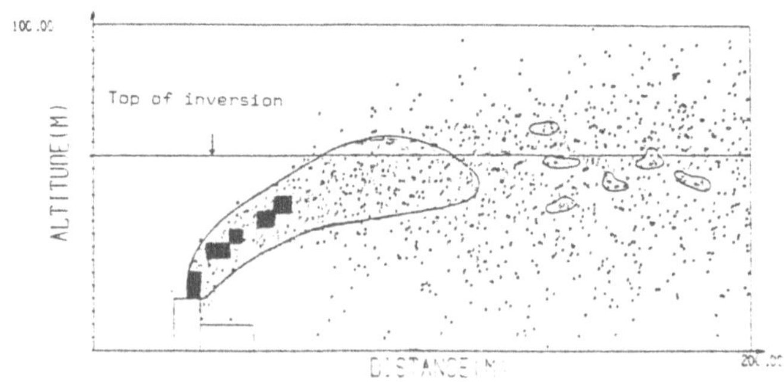

c) Light gas concentration, marker distribution.

Fig. 3 - Concentration calculations in a non-adiabatic atmospheric boundary layer (temperature inversion).

A THREE-DIMENSIONAL MATHEMATICAL MODEL OF DISPERSION IN TURBULENT SHEAR FLOW

T. TIRABASSI and M. TAGLIAZUCCA
(FISBAT-CNR, Bologna, Italy)

Summary

A three-dimensional model for continuous plume will be presented, where along-wind and vertical dispersion are based on recent analytical solutions of advection-diffusion equation, while the cross-wind diffusion is parametrized by a Gaussian multiplicative factor.
The proposed model appears as an intermediate approach between Gaussian Models and K Models. In fact it preserves the easiness of first and, at the same time, can be applied to an atmosphere with an inhomogeneous turbulence.
The model provides both pollutant concentration field and dry/wet deposition patterns, it uses linear chemistry and is applicable to any two-species reaction chain where this approximation is reasonable.
The model has been tested for sources in the surface layer using the Prairie-Grass experiment data. Moreover numerical tests comparing our results with an one-dimensional analytical solution that accepts an eddy exchange coefficient profile going to zero at the top of boundary layer have been performed.

1. INTRODUCTION

Transport and diffusion models of air pollution are based either on simple techniques, such as the Gaussian steady-state analytical equation, or on more complex algorithms, such as the K-theory differential equation. The Gaussian equation is an easy and fast method which, however, cannot properly simulate complex non-homogeneous conditions on a three-dimensional domain. The K-theory can virtually accept any complex three-dimensional meteorological input, but generally requires a finite-difference numerical integration which is computationally expensive and, sometimes, is affected by numerical advection errors.

However, the K-theory equation can be analytically solved in two dimensions under the following simplifying assumptions both for an unbounded (1), (2) and for a bounded atmosphere (3):

- dispersion in steady-state transport conditions;
- horizontal wind expressed as a power law function of height;
- vertical eddy diffusion coefficient expressed as a power law function of height.

We have adopted these solutions for the development of a model which allows the performance of a three-dimensional steady-state simulation using the Gaussian formula for the treatment of horizontal diffusion as proposed in (4).

The model provides both pollutant concentration field and dry/wet deposition patterns, it uses linear chemistry and is applicable to any two-species reaction chain where this approximation is reasonable.

2. THE ANALYTICAL SOLUTION

The steady-state, advection-diffusion equation that describes the dispersion of a passive material released by an elevated point source, advected by an x-directed mean wind profile u and diffused by the action of turbulent exchange coefficients K_x, K_y and K_z, is approximatively:

$$u \frac{\partial}{\partial_x} C = \frac{\partial}{\partial_z} (K_z \frac{\partial}{\partial_z}) C + \frac{\partial}{\partial_y} (K_y \frac{\partial}{\partial_y}) C \qquad (1)$$

if $\left| u \frac{\partial}{\partial_x} C \right| >> \left| \frac{\partial}{\partial_x} (K_x \frac{\partial}{\partial_x}) C \right|$

$$C(x,y,z) = \bar{C}(x,z) \frac{1}{\sqrt{2\pi}\,\sigma_y} \exp\left(-\frac{y^2}{2\sigma_y^2}\right) .$$

Unbounded atmosphere (H = + ∞)

$$\bar{C}(x,z) = \frac{Q(zH_e)^p h_0^\beta}{\lambda K_{z0}\, x} \exp\left(-\frac{u_0 h_0^r (z^\lambda + H_e^\lambda)}{\lambda^2 K_{z0}\, x}\right) I_{-\nu}\left(\frac{2u_0 h_0^r (zH_e)^q}{\lambda^2 K_{z0}\, x}\right)$$

Bounded atmosphere (H < + ∞)

$$\bar{C}(x,z) = \frac{2Qqh_0^\alpha}{H^{\alpha+1} u_o} \left\{ \lambda + \left(\frac{zR}{H}\right)^p \sum_{i=1}^{\infty} \left[\frac{J_{\gamma-1}(\sigma_{\gamma(i)} R^q)\, J_{\gamma-1}(\sigma_{\gamma(i)} (z/H)^q)}{J^2_{\gamma-1}(\sigma_{\gamma(i)})} \cdot \exp\left(-\frac{\sigma^2_{\gamma(i)} q^2 K_{z0}\, x}{H^\lambda h_0^r u_0}\right)\right]\right\} .$$

Conditions

$u(z) = u_0(z/h_0)^\alpha$ $\qquad$ $K_z = K_{z0}(z/h_0)^\beta$ $\qquad$ $K_y = u(z) \cdot f(x)$

Table I : Analytic solution equations. C is the pollution concentration, x is the along-wind direction, z is the height, Q the source emission, σ_y the lateral standard deviation, H_e the source effective height, u the wind velocity, K_z and K_y the vertical and lateral eddy exchange coefficients, h_0 the height where u_0 and K_{z0} are evaluated, H the mixing height, $\lambda=\alpha-\beta+2$, $\nu = (1-\beta)/\lambda$, $\gamma = (\alpha+1)/\lambda$, $r = \beta-\alpha$, $R = H_e/H$, $p = (1-\beta)/2$, $q = \lambda/2$, I_ν and J_ν represent the Bessel function and modified Bessel function of first kind and order ν, $\sigma_{\gamma(i)}$ the roots of J_ν and f(x) any function of x.

Unbounded atmosphere ($H = +\infty$)

$$\bar{C}(x,0) = \frac{Q}{\lambda^{\eta}} \frac{1}{\Gamma(\gamma)} \frac{h_0^{\eta}}{u_0^{\nu} (x K_{z0})^{\gamma}} \exp\left[- \frac{u_0 h_0^{r} H_e^{\lambda}}{\lambda^2 K_{z0} x} \right]$$

Bounded atmosphere ($H < +\infty$)

$$\bar{C}(x,0) = \frac{2 Q q h_0^{\alpha}}{H^{\alpha+1} u_0} \left\{ \gamma + R^p \sum_{i=1}^{\infty} \left[\frac{J_{\gamma-1}(\sigma_{\gamma(i)} R^q) \sigma_{\gamma(i)}^{\gamma-1}}{\Gamma(\gamma) J^2_{\gamma-1}(\sigma_{\gamma(i)}) 2^{\gamma-1}} \right.\right. .$$

$$\left.\left. . \exp\left(- \frac{\sigma^2_{\gamma(i)} q^2 K_{z0} x}{H^{\lambda} h_0^{r} u_0} \right) \right] \right\}$$

Table II: Analytic solution equations for ground level concentrations. The symbols are the same as in Table I but $\eta = (\alpha+\beta)/\lambda$ and Γ denotes the Gamma function.

The source condition is

$$C \longrightarrow \frac{Q}{u_e} \delta(z-H_e) \delta(y) \qquad \text{as} \quad x \longrightarrow 0$$

where $C=C(x,y,z)$ is the mean concentration field, Q is the emission intensity of the source placed at $(0,0, H_e)$, H_e is the effective height of the source ($H_e = H_s + DH$), u_e is the wind velocity at that height and DH is the plume rise, δ means the delta function.

The boundary conditions are:

$$K_z \frac{\partial C}{\partial z} = 0 \qquad \text{at } z = 0,H$$

where H is the height of the mixing layer;

$$u = u(z) = u_0 (z/h)^{\alpha}$$

$$K_z = K_z(z) = K_{x0}(z) (z/h)^{\beta}$$

$$K_y = u \cdot \text{(a function of x)}$$

where h is the height where u_0 and K_0 are evaluated .

The solutions of equation (1) are presented in Table I , while in Table II we show the solutions for ground level concentrations.

3. THE MODEL

The presented model utilizes the solution of Table I and II, and Briggs formulation (5) for computing the dynamic plume rise. The vertical atmospheric stability is evaluated either by Pasquill-Gifford stability classes or by the Monin-Obukhov length, while horizontal one either by Pasquill-Gifford stability or by the standard deviation of the horizontal

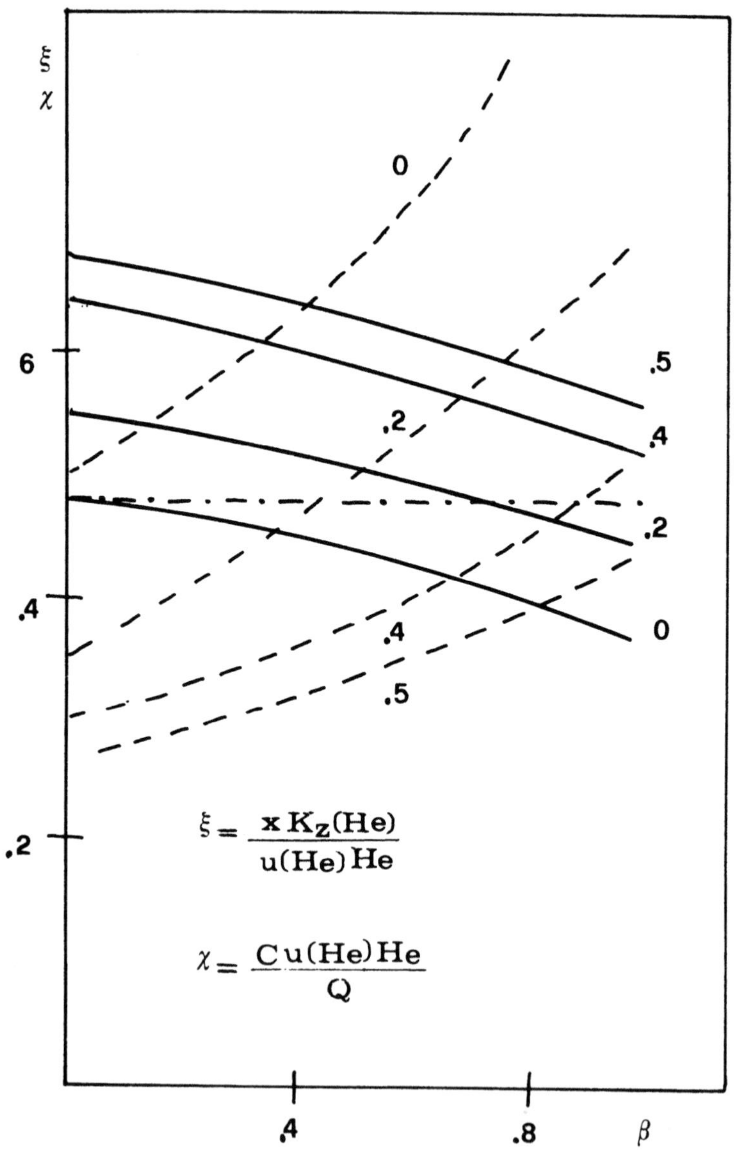

Figure 1.: Dimensionless position (dashed line) and value (continuous line) of maximum ground level concentration predicted by the two-dimension analytical solution as function of β for different α. Dash-dotted line is the maximum ground level concentration predicted by the Gaussian solution.

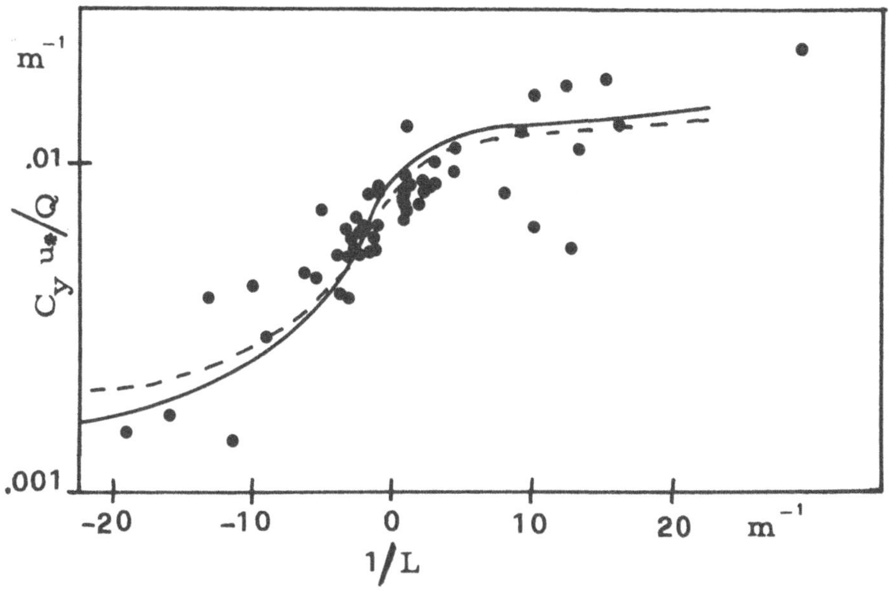

Figure 2: Observed (dots) and predicted by Dyer's parametrization (continuous line) and Businger's parametrization (dashed line) cross-wind integrated ground level concentration (C_y) normalised with u_x/Q as function of the inverse of Monin-Obukhov lenght (L) 200 meters downwind from the source.

wind direction. Lateral standard deviation σ_y is evaluated according to Briggs (6).

Both dry and wet deposition for the primary and secondary pollutants are simulated by first-order reaction schemes and are computed by an exponential reduction of the pollutant mass

$$M_i^{(new)} = M_i^{(old)} \exp\left[-P_{i,j}\ \Delta t/360{,}000\right]$$

where M_i indicates the primary (i=1) or the secondary (i=2) pollutant mass, j indicates dry (j=1) or wet (j=2) deposition, and $P_{i,j}$ is the corresponding percentage of reduction per hour (% h^{-1}). All mass differences $M_i^{(old)} - M_i^{(new)}$ are deposited and accumulated on the ground.

If the two $P_{i,1}$ are not directly specified as input values, they can be obtained from the deposition velocity values while $P_{i,2}$ can be obtained from precipitation data (7).

A first-order chemical reaction scheme is adopted, in which the chemical transformation term reduces the mass M_1 of primary pollutant and increases the mass M_2 of secondary pollutant for each time step according to

$$M_1^{(new)} = M_1^{(old)} \exp\left[-\ k\ \Delta t/360{,}000\right]$$

$$M_2^{(new)} = M_2^{(old)} + (w_2/w_1)\ M_1^{(old)} \left[1 - \exp\left(-k\ \Delta t/360{,}000\right)\right]$$

where k is the current chemical transformation factor expressed as a percentage of reduction per hour (% h^{-1}), and w_i are the pollutant molecular weights (i = 1,2).

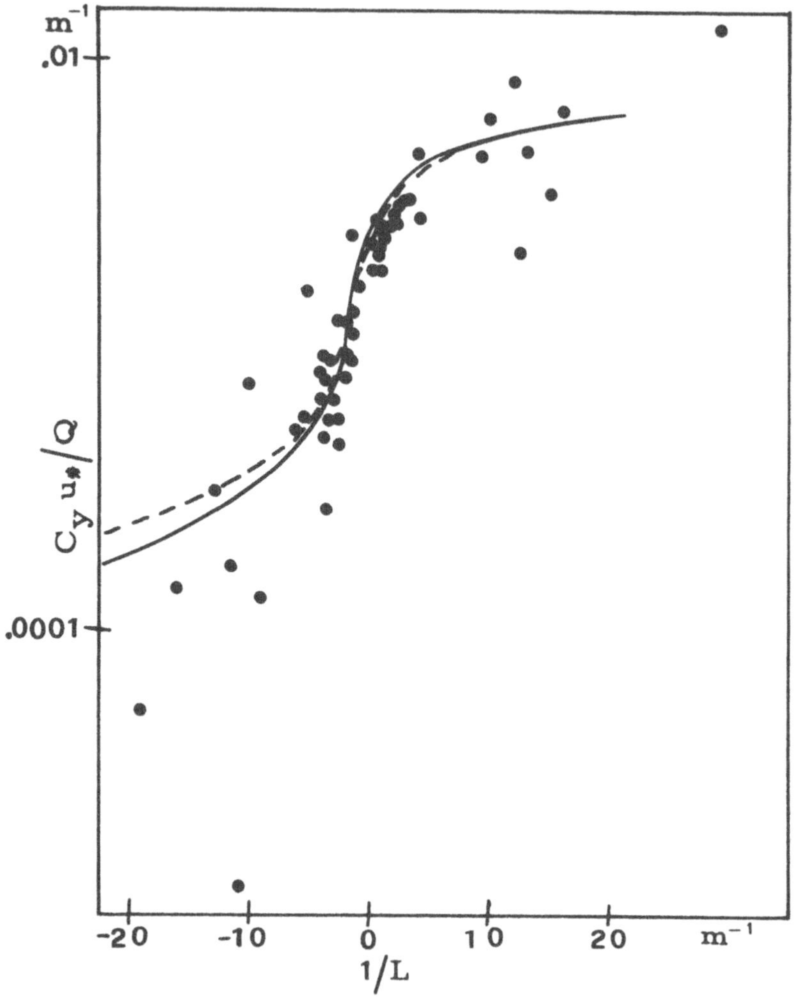

Figure 3: As figure 2 but 800 meters downwind from the source.

If the vertical wind profiles are available, they can be directly used to compute the exponent α fitting actual profiles with power law functions.

In particular, the exponent α of the power law profile of u is computed by fitting a power law to the measured wind profile. This process requires that the two profiles coincide at the effective source height H_e and to have the same average advective flow between the ground and H_e. That is,

$$u_p(H_e) = u_a(H_e)$$

$$\frac{1}{H_e} \int_0^{H_e} u_p(z)\, dz = \frac{1}{H_e} \int_0^{H_e} u_a(z)\, dz = \bar{u}$$

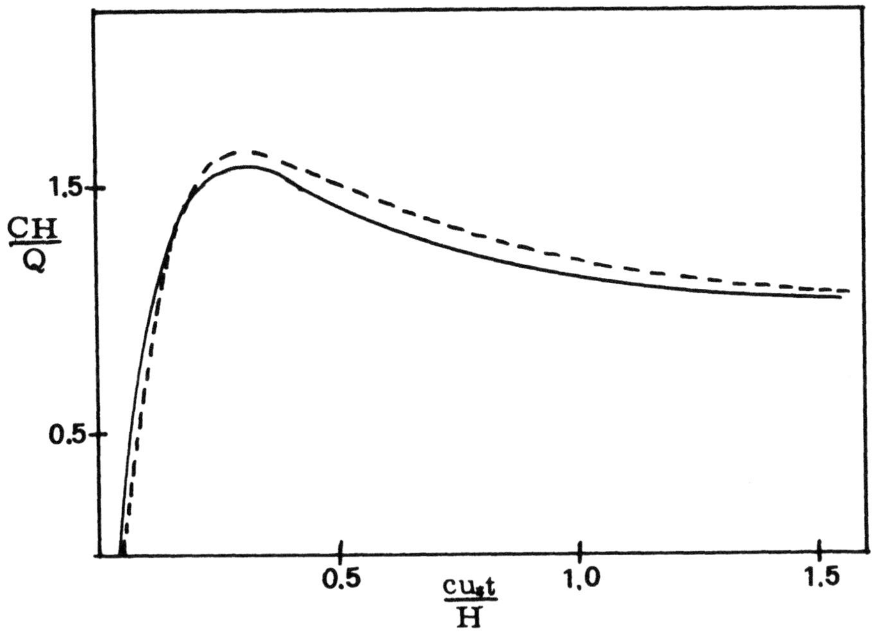

Figure 4: Comparison of ground level concentrations, for H_e = .25, between our model (continuous line) and the analytical solution assuming K_z approaching zero both at the surface and at the top of the boundary layer (14).

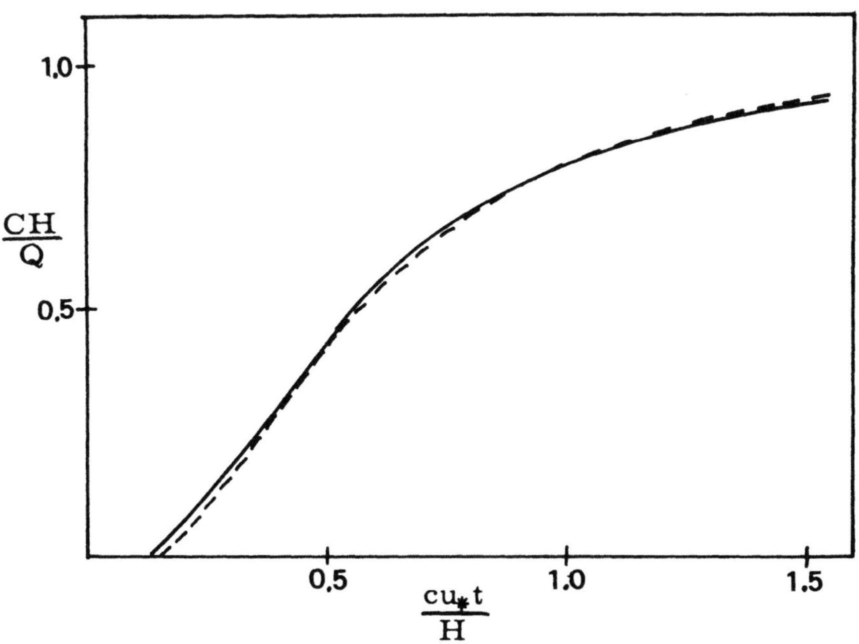

Figure 5: As figure 4 but H_e = .75.

where u_p is the power law fitting and u_a is the actual measured profile. It follows that

$$\alpha = (u(H_e) - \bar{u})/\bar{u}$$

On the contrary, if wind measurements are not available, theoretical profile are used.

The same approach is followed for the exponent β of the eddy exchange vertical profile. In particular we obtain:

$$\beta = \frac{K_z(H_e) - \bar{K}_z}{\bar{K}_z}$$

where $\bar{K}_z$ is the mean value of K_z from the ground to H_e.

We wish to outline that the above approach in evaluating the exponent of the eddy exchange coefficient vertical profile β is different from that applied in previous papers (8), (9). The present methodology seems to give better results for tall stack emissions.

4. MODEL EVALUATION AND CONCLUSIONS

In figure 1 are presented values and positions of maximum ground level concentrations predicted by the analytical solution (Table II) for different values of α and β. The effect of wind shear appears remarkable: the maximum ground level concentrations increase and approach to the source, according to the α increasing and β decreasing.

The proposed methodology has been tested for sources in the surface layer comparing the model results with the cross-wind integrated concentrations of the Prairie-Grass experiment (10), calculated by Van Ulden (11). In figure 2 and 3 the results of the model at two different distances from the source are presented. In the figures are reported values obtained applying the proposed methodology in evaluating β for similarity profiles of K_z obtained using flux-profile relationships proposed either by Businger (12) or by Dyer (13).

In order to evaluate the capability of the model to descrive advection-diffusion processes in a boundary layer characterized by a K_z profile going to 0 at H for any source height, we have compared the output of the model with that of an analytical solution of time-dependent, one-dimensional diffusion equation that accepts the following K_z profiles (14):

$$K_z = cu_* \, z(1 - z/H)$$

Here, c is an empirical constant and u_* is the friction velocity.

The above solution (14) can also describe the vertical dispersion from a stationary, continuous point source in a uniform wind field (α = 0) of magnitude u, when x = ut, where t is the time. In figures 4 and 5 the results for two different source heights are presented.

These preliminary evaluations have provided interesting results which confirm the potential utility of this approach as an alternative to the Gaussian equation. In particular we have shown, with numerical tests, that, althought power law profile of K_z cannot satisfy the condition $K_z \longrightarrow 0$ as $z \longrightarrow H$, nevertheless they provide an adequate approximation, at least for ground level concentrations.

REFERENCES

(1) YEH, G.T. and HUANG, C.H. (1975). Three-dimensional air pollutant modeling in the lower atmosphere. Boundary Layer Meteor., 9, 381-390.

(2) BERLYAND, M.Y. (1975). Modern problems of atmospheric diffusion and air pollution. Translated version by NEPC, USEPA, Raleigh, NC.

(3) DEMUTH, C. (1978). A contribution to the analytical steady solution of the diffusion equation for line sources. Atmos. Environ., 12, 1255-1258.

(4) HUANG, C.H. (1979). A theory of dispersion in turbulent shear flow, Atmos. Environ., 13, 453-463.

(5) BRIGGS, G.A. (1975). Plume rise prediction, Lectures on Air Pollution and Environmental Impact Analysis, American Meteorological Society, Boston, 59-111.

(6) BRIGGS, G.A. (1973). Diffusion Estimation for Small Emissions. USAEC Report ATDL-106.

(7) DRAXLER, R.R. and HEFFTER, J.L. (1981). Workbook for estimating the climatology of regional-continental scale atmospheric dispersion and deposition over the United States. NOAA Technical Memorandum ERL ARL-96.

(8) TIRABASSI, T., TAGLIAZUCCA M. and LUPINI R. (1982). A Non-Gaussian Model for Evaluating Ground Level Concentration by Steady Sources. Environmental System Analysis and Management, ed. by Rinaldi S., North-Holland Publishing Company, Amsterdam, 627-635.

(9) TIRABASSI, T., TAGLIAZUCCA, M. and ZANNETTI P. (1985). KAPPA-G, a non-Gaussian Plume Dispersion Model: Description and Evaluation against Tracer Measurements. JAPCA, 36, 592-596.

(10) BARAD, M.L. (1958). Project Prairie-Grass, a field program in diffusion. Geophys. Res. Pap. 59, AFCRC-TR-58-235.

(11) ULDEN, A.P. van (1978). Simple estimates for vertical diffusion from sources near the ground. Atmos. Environ., 12, 2125-2129.

(12) BUSINGER, J.A., WYNGAARD, J.C., IZUMI, Y. and BRADLEY, F.F. (1971). Flux profile relationships in the atmospheric surface layer. J. Atmos. Sci. 28, 181-189.

(13) DYER, A.J. (1974). A Review of flux-profile relationships. Boundary-Layer Meteorol. 7, 363-372.

(14) NIEUWSTADT F.T.M. (1980). An Analytic Solution of the Time-Dependent, One-Dimensional Diffusion Equation in the Atmospheric Boundary Layer, Atmos. Environ., 14, 1361-1364.

THE RAPID ROTATION OF THE WIND DIRECTION AND SHARP OZONE PEAK FEATURES.

G.TOUPANCE, P. PERROS, M. SOEDOMO
Laboratoire de Physicochimie de l'Environnement
Université PARIS VAL DE MARNE - F 94000 CRETEIL

Summary

The diurnal evolution of the ozone concentration follows generally the well known pseudosinusoidal curve. However, sharp accidents are often observed, that may have a variety of origins. Such accident are sometimes observed in natural areas. From field measurements obtained in different areas, it is shown that sharp increases in ozone concentration can be frequently attributed to the transit of a polluted air mass, during a short period of time, in a rapidly rotating wind system.

1 - INTRODUCTION

The diurnal profiles of ozone concentration follow generally a well known pseudosinusoidal variation. These profiles result from the competition of various mechanisms : photochemical formation, chemical and physical consumption, vertical atmospheric transfers.

In practice many deviations from this typical profile are observed. The most frequent is the presence of sharp peaks superimposed to the background sinusoidal profiles. Such accidents are observed both in rural and in urbanised areas, and can have various origins.

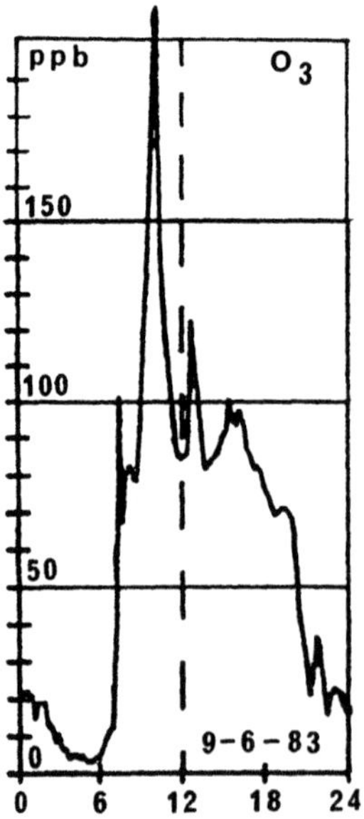

Fig. 1
Inversion Breakdown peak

In polluted areas the most frequent reason of a rapid variation of the ozone concentration is the breakdown of a morning temperature inversion.

During night ozone is consumed near the ground while an important ozone concentration may survive above the inversion layer. Early in the morning, additional quantities of ozone are produced at this level by photochemical processes; very few ozone is produced below the inversion layer, because of the large quantity of reducing agents, mainly NO, accumulated during the night. The sudden breakdown of the temperature inversion, during the morning, causes a sudden income of air rich in ozone. A sharp peak of ozone is then observed as shown (Fig. 1) on a profile obtained at Rognac (fig. 3) during the European Compaign at FOS-BERRE (France) in June 1983 (1). The further development of the atmospheric convection brings back the O3 concentration to the classical steady state value.

Such sharp peaks are observed in the areas of emission of precursors, during the middle of the morning.

2 - OZONE PEAKS DUE TO WIND ROTATION

Recently sharp peaks of ozone concentrations have been observed in rural areas. They occur later in the daytime, at noon and afternoon, than in the previously described situation.

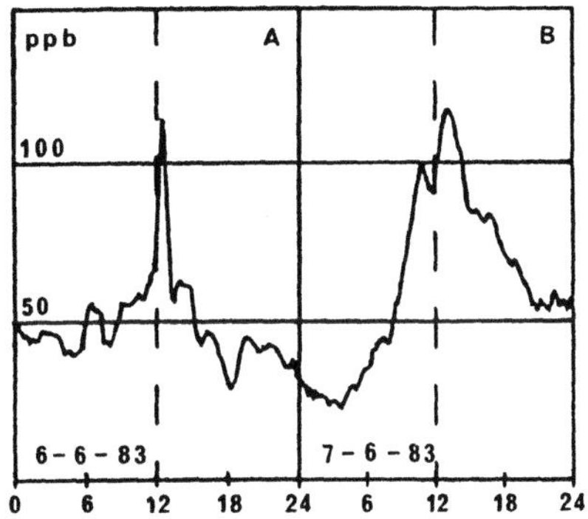

Fig. 2
O3 at BAUX de PROVENCE, 6-7 june 83

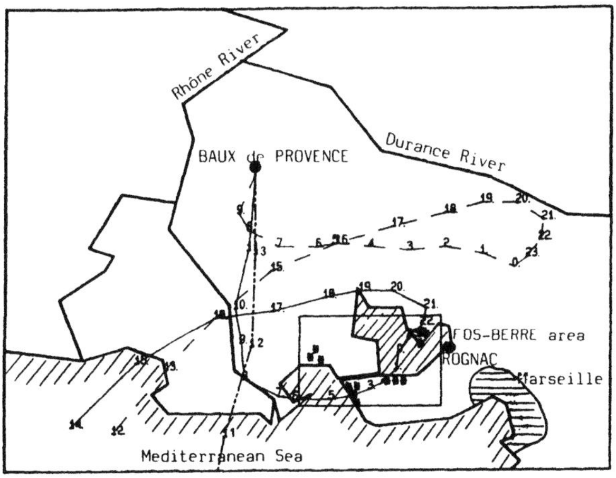

Fig. 3
FOS BERRE area: Trajectories of air masses reaching Baux de Provence at 10H, 12H, 14H the 6th of june 1983

Figure 2A represents an event observed at "Les Baux de Provence", in south of France, 40 km north west of the FOS BERRE area (fig. 3). The site is a traditional village situated on the top of a 350 m. high hill, in a very natural landscape. The background level observed (Fig. 2) is 40-50 ppb, both during night and daytime, and represents the background concentration of ozone in south of France during the considered period. A very sharp peak reaching 120 ppb is observed at noon on june 6th, and a similar dayly profile was recorded several times at Baux de Provence during the campaign (1). No local influence or local atmospheric convection was able to explain such accidents. A backtrajectory calculation of air masses reaching the site on the 6th of june at 10, 12, 14 h (local solar time), is given on fig. 3.

During the period of the study the area was characterised by a marked land and sea breeze system (2). The trajectories of the air masses were complex: before and after noon the origin of air is exclusively natural ; around noon, a puff of polluted air, enriched in precursors emitted during the previous night above the industrial area of FOS-BERRE and submitted to phochemical processes during the morning transfer to Baux de Provence, is blown over Baux de Provence and can thus explain the sharp peak observed at 12 h.

For comparison, fig. 2B represents the situation on the 7th of June when the wind direction was fixed, all day long, at south east (2). Baux de Provence was downwind the plume of the FOS-BERRE industrial area from the morning to the evening. The ozone curve is similar to the dayly classical pattern, typical of poluted areas, with a broad peak and O3 contrations higher than the background level from 8 to 20 h.

Similar situations have been observed recently in the Paris area. In the course of a study on O3 formation in the Paris region, ozone analysers have been installed in rural areas, upwind and downwind of Paris, about 50 km far from the center, in SW (Pinceloup), SE (Andrezel) and NE (Montagny) directions (fig 4 and 5).

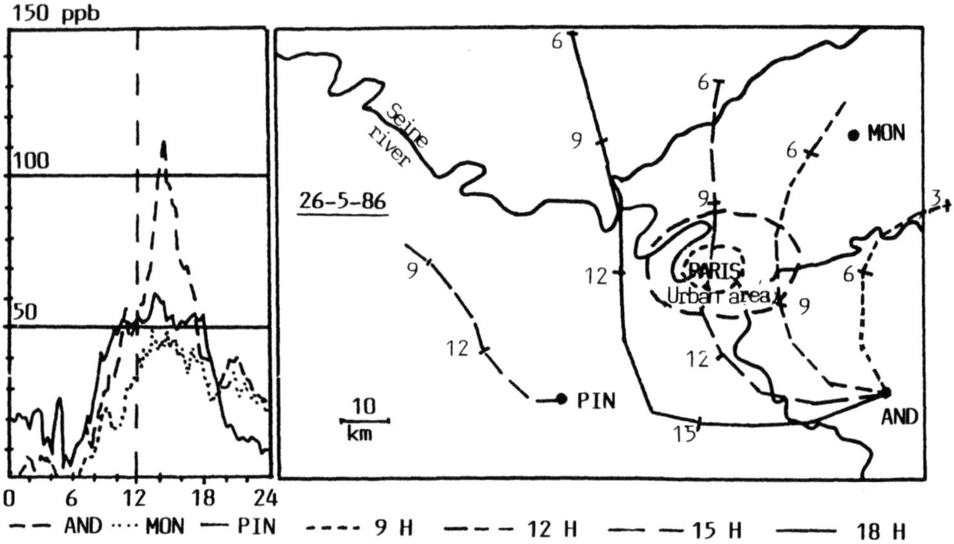

Fig. 4 : Ozone profiles in the Paris area the 26th of may 1986
A : O3 at Andrezel (AND), Montagny (MON), PINCELOUP (PIN)
B : Back trajectories for Andrezel at 9H, 12H, 15H, 18H.

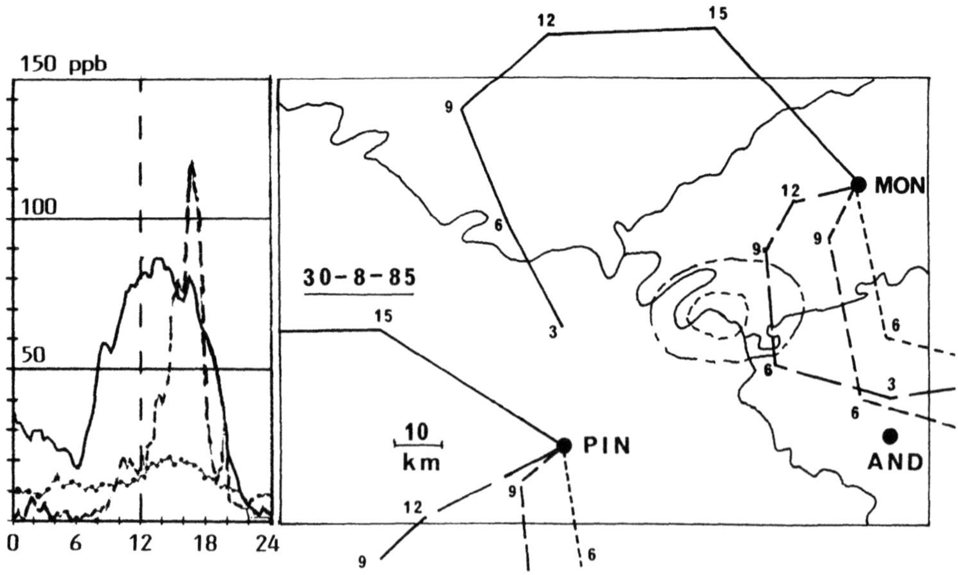

Fig. 5 : Ozone profiles in the Paris area the 30 of august 1985
A : O3 at Andrezel (AND), Montagny (MON), PINCELOUP (PIN)
B : Back trajectories for Montagny at 9H, 12H, 15H, 18H.

The diurnal profiles in these three sites generally follow a classical pattern, illustrated on figure 4 and 5 by the O3 profile at Pinceloup, both when the site is supplied by natural air as well as when it is inside the urban plume (3). However some sharp accidents have been sometimes recorded late in the afternoon; two typical events are given on figure 4 for Andrezel and on figure 5 for Montagny.

When the peak occur, the direction of wind is not from Paris and the maximum is stongly shifted to the middle or to the late afternoon. A reconstruction of the backtrajectories of air masses during these episodes give evidence of a situation similar to that discribed previously in the case of "Les Baux de Provence" : sharp peaks in ozone concentration are produced by a rapid rotation of the wind direction during the day ; the sites are exposed, during a short period of time, to air which had passed above the urbanized area some hours before. As the rate of photoproduction of ozone is maximum around noon, such peak episodes occur late in the afternoon contrary to episodes due to the breakdown of the temperature inversion which occur mainly in the morning. It should be noticed that the actual wind direction can be very different to that of the urbanized area.

3 - CONCLUSION

A large variety of diurnal patterns can be observed in field measurement. Sharp peaks of ozone are frequently recorded; some of thenm must be attributed to the brief passage of a puff of polluted air, which can be caused by the rapid rotation of a wind field. This observation may help in the interpretation of field measurements. It shows that natural areas may be submitted to high concentrations of pollutants through complex transfer mechanisms.

4 - REFERENCES

1. PERROS P., TOUPANCE G.: Proceedings of the 3rd European Symposium on Physico-Chemical Behaviour of Atmospheric Pollutants, Varese , Italy, 10-12 April 1984, B. VERSINO and G. ANGELETTI Edit., D. REIDEL Pub. Co. 1984, p 596-606.
2. AUGUSTIN H., BESSEMOULIN P., Proceedings of the 3rd European Symposium on Physico-Chemical Behaviour of Atmospheric Pollutants, Varese , Italy, 10-12 April 1984, B. VERSINO and G. ANGELETTI Edit., D. REIDEL Pub. Co., 1984, p 565-572.
3. G. TOUPANCE, P. PERROS, M. SOEDOMO, "Recherche sur la pollution oxydante dans le bassin Parisien", Ministry of the Environnement, Paris, France, Grant ENV 84070, May 1986.

The data used in this paper were collected with help of French Ministry of the Environment and of CNRS.

LOCAL SCALE PLUME MODEL FOR NITROGEN OXIDES - COMPARISON WITH MEASUREMENTS

CHRISTER PERSSON
Swedish Meteorological and Hydrological Institute

Summary

A plume dispersion model on local scale (0 - ~20 km) including atmospheric chemistry for nitrogen oxides is developed. The model is designed for and used in practical environmental planning for e.g. coalfired power plants. High concentrations of NO_2 are especially considered. The instantaneous plume dilution is described - assuming total mixing within the plume cross section - as a function of emission parameters and meteorological conditions. The interplay of emissions, instantaneous plume dilution, entrainment of polluted ambient air, meteorological conditions and atmospheric chemistry processes determines the production or loss of the different compounds. On a local scale only six chemical reactions need to be included. For the numerical solution we have used the method by Gear. A verification of the model by means of plume data from several European power plants has been performed. The verification shows that the model is capable of describing the NO_2-formation in power plant plumes in most weather situations in a fairly accurate way. No important systematic errors seem to be present and the correlation coefficient for NO_2-fractions of total NO_x is found to be about 0.7 out to at least 1 000 seconds after emission. On a local geographical scale the plume chemistry is controlled by the dispersion.

1. INTRODUCTION

The local environmental impact of emission of nitrogen oxides (NO_x) from a power plant depends both on the concentration and the chemical nature of the species produced by the emitted material. The NO_x-emission consists to about 90 - 95% of nitric oxide (NO) but it will more or less rapidly be oxidized in the atmosphere. In the first step it is oxidized mainly to nitrogen dioxide (NO_2). As the negative effects on health by NO_2 are much more serious than those of NO, the NO_2-formation has to be considered. NO_2 is also more directly related to the acidification of the ground since it is deposited - mainly through dry deposition to vegetation - much more efficiently than NO.

The present study is aiming at development of a practical air quality model for different nitrogen oxides, which can be used in environmental planning. The model is presented in detail by Persson and Funkquist (1), only an outline will be given below. The verification, which is shown, is a good basis for more extensive practical applications of the model in environmental planning.

2. SHORT PRESENTATION OF THE MODEL

The in-plume oxidation of NO is a complicated process depending on a number of chemical and physical parameters. In the model the instantaneous plume dilution and the simultaneous chemical reactions in the plume are described in a Lagrangian coordinate system following the wind, i.e. following a trajectory. We start with a segment of the plume, which is moved by the wind away from the source while, simultaneously the concentrations of different compounds are changed through dilution, entrainment of ambient polluted air, chemical reactions within the plume and deposition.

The following basic assumption are made:

- The distribution of the pollutants is homogeneous ("top hat" concentration distribution) within each cross section of the instantaneous plume.
- The effect of very short time turbulent concentration fluctuations of the chemical transformations can be neglected.
- Wind speed is constant with height within each cross section of the instantaneous plume.

The first two assumptions can in certain situations be critical. They will be further discussed in the verification section. With the assumption of a "top hat" concentration distribution, the problem to describe the instantaneous plume dispersion can be reduced to a decription of how the plume cross section area varies. The area is a function of the distance from the source and depends on the inherent turbulence within the plume (especially at the initial stage) and the atmospheric turbulence (at larger distances). At the initial stage the momentum and buoyancy fluxes from the source are of large importance for the instantaneous plume dispersion.

On a local scale (0 - ~20 km) only six reactions are of importance for the NO_x-chemistry. These are given in Table I together with their rate constants. The photo dissociation rate has been determined for 60°N at different times of the day and year according to a method developed by Isaksen et al (2). Other rate coefficients were obtained from Grennfelt and Sjödin (3).

TABLE I. Chemical reactions considered in the model. Reaction rates for three component reactions are given in cm^6 $molecules^{-2}s^{-1}$, for two component reactions in cm^3 $molecules^{-1}s^{-1}$ and dissociation rates in s^{-1}. T = temperature of the plume (°K).

	Reaction rates
R1 $NO_2 + h\nu \rightarrow NO + O(^3P)$	$j_1 = 0 - 6.2\ 10^{-3}$
R2 $O(^3P) + O_2 + M \rightarrow O_3 + M$	$k_2 = 1.1\ 10^{-34}\ e^{510/T}$
R3 $O_3 + NO \rightarrow NO_2 + O_2$	$k_3 = 2.1\ 10^{-12}\ e^{-1450/T}$
R4 $2NO + O_2 \rightarrow 2NO_2$	$k_4 = 1.5\ 10^{-40}\ e^{1780/T}$
R5 $NO + NO_2 + H_2O \rightarrow 2HNO_2$	$k_5 = 6.0\ 10^{-38}$
R6 $2HNO_2 \rightarrow NO + NO_2 + H_2O$	$k_6 = 1.9\ 10^{-11}\ e^{-5000/T}$

3. VERIFICATION

3.1 Selected cases of plume studies

The main difficulty with a verification of this kind is to obtain simultaneous data for all relevant parameters concerning power plants emissions, meteorological conditions, ambient ozone concentrations and different NO_x-compounds in the plume. The present verification will be based on studies from four different countries presented in the literature. These measurements were regarded as satisfactory for our purpose.

Data concerning plume measurements, emission and meteorological conditions for the 34 studied cases are given in Table II. Emission data and ozone concentrations were in most cases determined simultaneously with the plume studies, while meteorological data from the power plant site, with some exceptions, were not available.

3.2 Meteorological information

For most of the cases studied, data was obtained from a nearby routine meteorological observing station. Methods based on routine meteorological data were used to make a crude calculation of the sensible heat flux, H, the friction velocity, u_*, and the vertical wind profile. The net radiation at the ground, R_n, is calculated from solar elevation and cloudiness by a method developed by Nielsen et al (5). The sensible heat flux can then be determined from the empirical relation $H = 0.4 (R_n - 100)$, there H and R_n are given in Wm^{-2}. Using an empirical method from Omstedt (6), the sensible heat flux and the wind speed at 10 m height can be used to determine the stability, the convective velocity scale, w_*, and a typical value for the mixing height. In the present model, we use five stability classes and denote them with block letters. The classification is done in agreement with Weil and Brower (7). The classes are unstable - B, slightly unstable - C, neutral - D, slightly stable - E and stable - F.

A negative heat flux indicates stable stratification (classes E and F), where the well known Pasquill-Gifford-Turner classification scheme is used. The vertical wind profile is calculated in the way suggested by Businger et al (8).

Using meteorological data determined in this way from routine meteorological observations always creates uncertainty in the results. However, this method is relevant for the verification, since that is the only way of getting meteorological data in many practical applications. Also, a more sophisticated method of using routine meteorological data, where vertical temperature soundings of the atmosphere are included, has been given by Omstedt (9). That method can in practice easily be used for Swedish sites and in those cases improve the meteorological data used for the calculations.

Non-routine meteorological data were used for cases 24 - 32. These data were obtained from a 300 m high mast west of Helsinki, Tammelin (10).

3.3 Measurements from the Netherlands and West Germany

Field studies on the oxidation rates of NO in power plant plumes were carried out by N.V. KEMA, Arnhem, the Netherlands, with direct measurements of the concentrations of the gaseous components in the plume using an aircraft or a van equipped with instruments, and by UBA, Frankfurt, West Germany, with Barringer remote sensing spectrometers (COSPEC). Results from the measurements are presented by Elshout and Beilke (11) and (12). More detailed information about the measurements in the Netherlands and for Wilhelmshaven has also been supplied by Dr Elshout through private communications. Those studies therefore provide the most complete set of plume data, necessary emission data, ozone concentrations and meteorological data available for this verification.

For the power plants in West Germany no detailed emission data could be obtained but had to be estimated. Information from case no 3, which is a plant operating with roughly the same capacity, and similar plants was used.

Case no 23 should be treated separately since there is a very stable layer (strong inversion) aloft, which is of great importance for the results. The

TABLE II . Cases used in the comparison between model calculations and plume measurements. Name of power plant, time, available emission data, ozone concentration in the surrounding air and meteorological data are given. Wind speed at plume height (Uplume) and stability are calculated from routine meteorological data. For cases 1 - 12 also wind speed and stability given by KEMA are indicated (within parenthesis). Cases 1 - 11 refer to plants fired with oil or natural gas, where we assume an O_2-content of 2% in the exhaust gases. The other cases refer to coal fired power plants for which the O_2-content is assumed to be 4%. Emission velocity of the exhaust gases is put equal to $20\,ms^{-1}$ for all cases and the ratio NO_2/NO_x in the emission is assumed to be 5% in all cases except nos 33 and 34, where we assume 2%. h_0 = stack height.

V_0 = exhaust gas volume
T_0 = exhaust gas temperature
Mv = Maasvlakte on the North Sea 30 km W Rotterdam, with a total capacity of 1 050 MW h_0 = 175 m
Fl = Flevo on the IJsselmeer 45 km NE Amsterdam, total capacity 835 MW, h_0 = 150 m
Waalh and Gld = other powerstations in the Netherlands
Wilh = Wilhelmshaven on the North Sea, total capacity 700 MW h_0 = 280 m
Stau = Staudinger on the Main 20 km E Frankfurt, total capacity 1 500 MW operating at ≤ 900 MW, h_0 = 195, 250 m
Fin = Finnå 20 km W Helsinki on the south coast of Finland h_0 = 152 m
Asn = Asnäs 3 km S Kalundborg on the west coast of Sjælland h_0 = 120 m, (220 m)

Case no	Power plant	Day	Time	Emission data			O_3 (ppb)	Meteorological data		
				NO_x (gs^{-1}) as NO_2	V_0 (m^3s^{-1} at 0°C)	T_0 (°C)		Uplume (ms^{-1})	Temperature (°C)	Stability (Pasquill class)
Data from the Netherlands										
1	Mv	760608	11 - 14	111	193	130	125	6.0 (8.5)	28	B (C)
2	Fl	800423	12 - 15	445	301	130	55	6.1 (7.5)	8	C (C)
3	Mv	780815	9 - 12	197	360	130	40	14.0 (15)	21	C (C)
4	Mv	781129	13 - 16	217	377	130	20	5.4 (8)	2	D (C)
5	Waalh	810319	11 - 15	136	211	130	50	19.2 (15)	8	D (D)
6	Mv	780830	16 - 19	282	377	130	30	11.7 (10)	15	D (D)
7	Mv	751104	16 - 17	69.4	184	130	35	3.9 (5)	12	D (D)
8	Fl	811124	12 - 16	299	301	130	26	20.0 (17.5)	7	D (D)
9	Fl	800409	11 - 17	181	139	130	45	14.6 (15)	6	C (D)
10	Gld	800130	10 - 16	172	114	130	(5-) 25*	19.5 (15)	8	D (D)
11	Fl	810422	10 - 17	350	237	130	45	10.0 (8)	7	C (D)
Data from West Germany										
12	Wilh	820513	daytime	197^x	360^x	180^x	55	7.0 (7)	10	E^{xx}(D)
13	Stau	810730	"	"	"	"	35	6.4	21	C
14	Stau	810624	"	"	"	"	55	6.4	21	C
15	Stau	810728	"	"	"	"	25	2.0	20	C
16	Stau	800829	"	"	"	"	55	6.5	23	C
17	Stau	800821	"	"	"	"	45**	20.1	23	D

Case no	Power plant	Day	Time	Emission data			O_3 (ppb)	Meteorological data		
				NO_x (gs^{-1}) as NO_2	V_0 (m^3s^{-1} at 0^oC)	T_0 (^oC)		U_{plume} (ms^{-1})	Temperature (^oC)	Stability (Pasquill class)
18	Stau	800820	daytime	197^x	360^x	180^x	35	11.4	19	D
19	Stau	810721	"	"	"	"	5	3.1	17	D
20	Stau	810529	"	"	"	"	55	11.2	18	D
21	Stau	810722	"	"	"	"	10	6.5	19	C
22	Stau	810430	"	"	"	"	5	11.8	12	D
23	Stau	810804	"	"	"	"	70	3.0	28	B
Data from Finland										
24	Fin	811026	18 - 21	19.3	28	137	20***	4.5	1	E
25	Fin	811204	6 - 7	61.1	73	113	28***	10.8	2	E
26	Fin	811214	9 - 10	32.2	45	128	28***	8.5	-11	F
27	Fin	811214	10 - 12	32.2	45	128	20***	8.5	-9	F
28	Fin	820113	15 - 16	59.4	92	106	28***	3.3	-7	E
29	Fin	820113	17 - 18	59.4	92	106	28***	3.0	-6	E
30	Fin	820113	19 - 20	59.4	92	106	28***	5.8	-5	E
31	Fin	820128	20 - 21	78.7	95	105	32***	4.4	-13	F
32	Fin	820128	21 - 22	78.7	95	105	32***	6.0	-13	F
Data from Denmark										
33	Asn	840815	11 - 12	739	897	132	37	6.7	19	C
34	Asn	840815	12 - 14	591	717	126	37	5.9	20	C

* Probably 25 ppb is correct. The lower value seems to be influenced by a local NO-source.

** Ozone value obtained from a background station in Southern West Germany.

*** Estimated ozone values (see text).

^x Estimated emission data based on information from similar plants.

^xx The first few km slightly stable at plume level due to sea breeze. Further away slightly unstable.

inversion could not be obtained from the ground based routine meteorological data but was indicated in UBA's plume data collection.

In Table III statistical information of measured and calculated NO_2-fractions in plumes is compared. All cases from the Netherlands and West Germany, except for the above discussed case no 23, are included (22 cases). Since it is possible that the measured values in case no 7 could be incorrect, comparisons where case no 7 is excluded are also shown in Table III. The correlation coefficient between measured and calculated NO_2-fractions is around 0.75 for all plume travel times between 100 to 1 000 seconds, which must be regarded as a good correlation. Also, in the mean NO_2-fractions, there is a good agreement between measurements and model calculations, although the calculated NO_2-fractions are a little too low during the first 200 seconds. Also, the standard deviations are smaller for model calculated values than for measured during the first 200 seconds. The assumed constant value of 5% NO_2 in the emission can be one reason for that.

TABLE III. Comparisons between measured and calculated NO_2-fractions (%-units of total NO_x) in plumes expressed in terms of mean values, standard deviations and correlation coefficients. Data from the Netherlands and West Germany have been used.

	Time after emission									
	100 s		200 s		400 s		600 s		1 000 s	
All cases	*Meas.*	*Calc.*	*Meas.*	*Calc.*	*Meas.*	*Calc.*	*Meas.*	*Calc.*	*Meas.*	*Calc.*
Mean NO_2-fraction (%)	13.5	11.8	21.5	18.5	30.6	29.5	37.2	39.7	48.4	53.8
Stand.dev.(%)	10.3	5.6	15.2	11.9	17.5	16.6	16.2	18.9	22.0	21.2
Corr.coef.	0.69		0.77		0.69		0.52		0.54	
All cases except case no 7	*Meas.*	*Calc.*	*Meas.*	*Calc.*	*Meas.*	*Calc.*	*Meas.*	*Calc.*	*Meas.*	*Calc.*
Mean NO_2-fraction (%)	14.1	11.7	22.1	18.2	31.4	28.7	38.3	38.4	51.3	51.1
Stand.dev.(%)	10.2	5.9	15.2	12.1	17.7	16.7	16.2	18.8	22.0	21.4
Corr.coef.	0.74		0.81		0.77		0.64		0.77	
No of cases	21		22		19		15		8	

3.4 Measurements from Finland

During the winter 1981 - 82 the Swedish Water and Air Pollution Research Laboratory (IVL) had two ground-based stations for continuous measurements of NO and NO_2 near Finnå power plant in a western suburb of Helsinki. During several short periods these stations were influenced by the plume of the Finnå power plant.

Model calculations based on the observed stability in the rural environment always gave too high NO_x-concentrations in the plume compared to observed NO_x-concentrations. Is is plausible that the stratification in the more urban areas around the measuring stations is more unstable than that near the meteorological mast, located in a rural area. Assuming somewhat more unstable stratifications we get complete agreement for NO_x-concentrations. The NO_2-fractions given in Figure 1 are then obtained. There is a pronounced dependence on the NO_x-concentration in the plume. Low NO_x-concentration gives a high NO_2-fraction and vice versa. This clearly indicates that, in these cases, the entrainment of ambient ozone and its reaction with NO controls the NO_2-production.

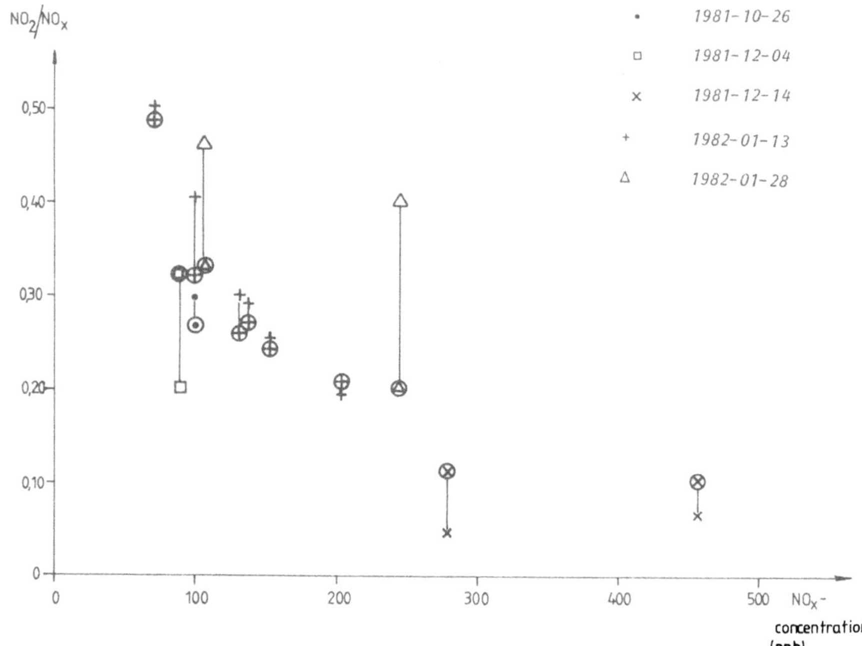

FIGURE 1. Observed and calculated NO_2/NO_x-ratio as a function of NO_x-concentration in the plume of Finnå. The stability in the calculation was adjusted to obtain complete agreement between observed and calculated NO_x-concentrations in the plume. Measured values are circled while calculated values are not.

4. DEVIATIONS FROM PHOTO-STATIONARY EQUILIBRIUM

The $NO_x - O_2$ system together with the incoming sunlight, described by reactions 1 - 3, create a photo-stationary equilibrium given by the equation

$$\psi = \frac{k_3 [NO][O_3]}{j_1 [NO_2]} = 1$$

when the mixing is complete. Observed values of ψ are generally >1 and it has been assumed to indicate inhomogeneous mixing of O_3-rich and NO-rich air.

Some examples of ψ-values obtained from the present model calculations are presented in Figure 2. The ψ-values clearly deviate from 1 and decreases with larger distances. Since the present model assumes homogeneous mixing within each cross-section of the plume, the calculated deviations from photo-stationary equilibrium are created only through different time rates for the entrainment and the chemical reactions. As long as the ψ-values are much larger than one, the chemical reactions in the plume must be slow compared to the entrainment.

Observed ψ-values by Elshout and Beilke (12) are higher than those calculated by the model for the same cases from the Netherlands, as indicated in Figure 2. It is plausible to assume that the model assumption of instantaneous homogeneous mixing could create discrepancies. However, our model calculations of NO- and NO_2-concentrations agree rather well with observed data from the Netherlands, as was shown above. Therefore, the disagreement in the ψ-values could perhaps also be due to that the observed ψ-values from the Netherlands are based on smaller dissociation values or too high ozone values are measured in the plume.

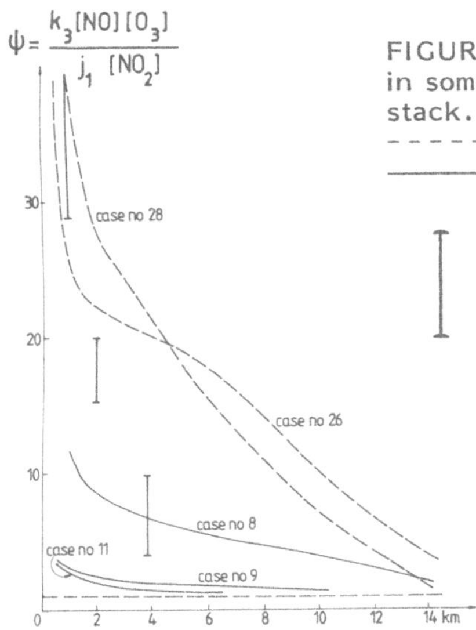

FIGURE 2. The photostationary state relationship in some plumes as a function of distance from the stack.

- - - - - - calculated for two cases from Finland

———— calculated for three cases from the Netherlands

I observed interval for the same three cases from the Netherlands, Elshout and Beilke (12)

5. IMPORTANCE OF THE O_2-REACTION

In Figure 3 a comparison between measured and calculated values of $[NO_2-(-O_3)]/NO_x$ is given. The excess of NO_2 with respect to the corresponding O_3 deficit can be assigned to the O_2-reaction. The other part of NO_2 is formed from the reactions of NO with O_3. The model calculation only gives a few extra %-units of NO_2 from the O_2-reaction, while measured values indicate more. The situations are similar for all cases 1 - 12. A reason for this discrepancy could be the model assumption of homogeneous mixing. However, an overestimate of measured O_3-concentration in the plume could also be of importance, which would be in agreement with the results for the photo-stationary ratio.

In order to test the importance of the O_2-reaction for the simulated NO_2-values in other situations than those included in the verification, we have performed some calculations for a rather large emission source. We assume flue gas emission of 200 Nm^3/s with 200 or 400 ppm NO_x and an ozone concentration of 30 ppb. It was shown that the O_2-reaction in those cases is of large importance for the calculated NO_2-production when the stratification is very stable. With an emitted NO_x-concentration of 400 ppm more than 70% of the produced NO_2 1 500 sec after emission comes from the O_2-reaction. For neutral and, especially, unstable stratification the O_2-reaction is of little importance.

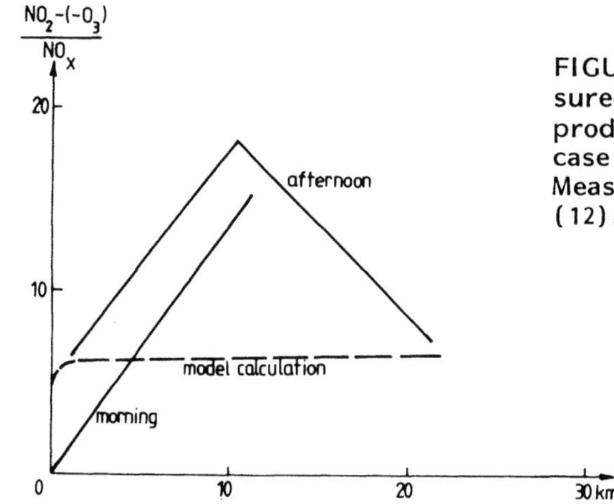

FIGURE 3. Comarison between measured and calculated amount of NO_2 produced by the O_2-reaction in case 12, Wilhelmshaven, cf text. Measured values from reference (12).

6. INHOMOGENEOUS MIXING

The model assumption of homogeneous mixing within each cross-section of the plume may be critical in some cases. In reality, in an instantaneous plume a more or less heterogeneous mixture of "plume-parcels" and "ambient air-parcels" will be formed by the turbulent eddies. It is molecular diffusion, which at the boundaries of the eddies leads to local mixing, that makes the reaction between e.g. NO and O_3 possible. It should be expected that the assumption of homogeneous mixing causes a somewhat faster (but not larger) formation of NO_2 through the O_3-reaction and a suppression of the formation through the O_2-reaction. The verification in this study of real power plant plumes shows, however, no pronounced systematic discrepancy, which can be attributed to the assumption of homogeneous mixing. The reason might be that in ordinary power plant plumes, under not too extreme weather conditions, these effects are too small to be revealed in a verification of this kind.

7. CONCLUSIONS

The present verification shows that the model is capable of describing the NO_2-formation in power plant plumes in most weather situations in a fairly accurate way. No important systematic errors seem to be present and the correlation coefficient for NO_2-fractions of total NO_x in large power plant plumes is found to be about 0.7 out to at least 1 000 seconds after emission. It is important to note that the calculations are based only on routine meteorological data, emission data and ozone concentrations in the ambient air. Thus, the calculations have been performed in the same way as in the case of practical environmental planning, which gives a realistic picture of the potential of the model for practical applications.

It is not possible from the verification to find any systematic discrepancies in the results due to the assumption of homogeneous mixing within the plume. However, it is obvious that the inhomogeneity, which can be found in real plumes, should be very important in certain extreme weather situations. For more normal weather situations and ordinary power plant plumes, the inhomogeneity seems to be of minor importance for the NO_2-production.

8. ACKOWLEDGEMENTS

Different measuring groups have provided me with data and made this verification possible. Dr Elshout and Mr Jansen at KEMA have supplied very valuable data from the Netherlands. Peringe Grennfelt and Bo Galle from the Swedish Water and Air Pollution Research Laboratory have been very helpful with data from Finland and Denmark and with ozone data.

REFERENCES

(1) PERSSON CH and FUNKQUIST L, 1984: Locale scale plume model for nitrogen oxides - Model description. Swedish Meteorological and Hydrological Institute. Report RMK 41.

(2) ISAKSEN I, MIDTBØ K, SUNDE J and CRUTZEN P, 1977: A simplified method to include molecular scattering and reflection in calculations of photon fluxes and photodissociation rates. Geoph. Norv. 31, 11 - 26.

(3) GRENNFELT P and SJÖDIN A, 1982: Icke-fotokemiska omvandlingsreaktioner för kväveoxider i atmosfären och i nederbörd. Swedish Water and Air Pollution Research Laboratory, B Publ 670.

(4) GIFFORD F.A, 1975: Atmospheric dispersion models for environmental pollution applications. In Lectures on Air Pollution and Environmental Impact Analyses. American Meteorological Society, Boston, Ma.

(5) NIELSEN L.B, PRAHM L.P, BERKOWICZ R and CONRADSEN K, 1981: Net incoming radiation estimated from hourly global radiation and/or cloud observations. J. Climat. 1, 255 - 272.

(6) OMSTEDT G, 1983: En undersökning av "The Maryland PPSP dispersion model for tall stacks". Swedish Meteorological and Hydrological Institute. FoU-Notis Nr 23.

(7) WEIL J.C and BROWER R.P, 1982: The Maryland PPSP dispersion model for tall stacks. PPSP-Mp-36, Martin Marietta Corp., Environmental Center, Balmore, Md 21227.

(8) BUSINGER J.A, WYNGAARD J.C, IZUMI Y, BRADLEY E.F, 1971: Flux profile relationships in the atmospheric surface layer. J. Atm. Sci. 28, 181 - 189.

(9) OMSTEDT G, 1984: An operational air pollution model using routine meteorological data. Swedish Meteorological and Hydrological Institute. Report RMK 40.

(10) TAMMELIN B, 1982: Personal communication. Finnish Meteorological Inst.

(11) ELSHOUT A.J and BEILKE S, 1983: Oxidation of NO to NO_2 in flue gas plumes of power stations. Paper presented at VGB Congress "Power Plants 1983", Hamburg.

(12) ELSHOUT A.J and BEILKE S, 1984: Die Oxidation von NO zu NO_2 in Abgasfahnen von Kraftwerken. VGB Kraftwerkstechnik 64, Heft 7.

HIGH PAN CONCENTRATIONS DURING NON-SUMMER PERIODS : A STUDY OF TWO EPISODES IN CRETEIL (Paris) - FRANCE.

N. TSALKANI, P. PERROS, G. TOUPANCE
Laboratoire de Physicochimie de l'Environnement
UNIVERSITE PARIS VAL DE MARNE - 94010 CRETEIL (France)

Summary

The effect of sunlight on air containing hydrocarbons and nitrogen oxides leads to a very complex series of reactions, which form, among other secondary pollutants, peroxyacetylnitrate (PAN). This compound is believed to be the most useful indicator of photochemical pollution, and has been monitored for several months at the University of Creteil (Paris, France). Although it is commonly accepted that high PAN levels occur more frequently in summer, our results show elevated PAN concentrations during non-summer periods, precisely during two episodes in October 1985 and February 1986. A close examination of these episodes indicates that, in addition to local photochemical production, long-range transport of polluted air masses can occur in winter, revealing the influence of emissions from central Europe.

1 - INTRODUCTION

Peroxyacetylnitrate (PAN) is an important secondary pollutant formed in ambient air photochemically by a series of reactions between hydrocarbons and oxides of nitrogen (1). It is generally considered as a better indicator of photochemical air pollution than ozone, which has also a significant natural source (2-4).

Its presence in urban atmospheres during photochemical episodes has been detected throughout the world, at concentrations reaching several ppb(v). World wide PAN measurements have also shown the ubiquity of this compound in the clean atmosphere, but at much lower levels, situated in the ppt(v) range (5,6). Long-range transport of air masses containing significant PAN levels has been reported (4,7) suggesting that PAN may play an important role as a reservoir for nitrogen oxides.

In general, attention has mainly been paid to summer observations, when PAN concentrations are expected to present a maximum, because of the enhanced overall photochemical activity. Nevertheless, high PAN levels during winter months have also been reported (7,8), suggesting that late-winter photochemical episodes may not be an uncommon phenomenon, but data are currently too sparse.

In this paper we examine two episodes of high PAN levels reaching 9 ppb, recorded in October 1985 and February 1986 in Creteil, a southeastern suburb of Paris. The results support the aforementioned possibility of significant photochemical activity during non-summer periods. The possibility of in situ photochemical formation or long-range transport is discussed, in relation to some meteorological parameters.

2 - EXPERIMENTAL

The experimental procedures of PAN analysis have already been discussed elsewhere (9), and will be reported in details in a forthcoming paper (10). Briefly, PAN was measured with an automated gas chromatograph equipped with an electron capture detector, which operated in the constant frequency mode. The PAN peak was separated on a 0.3 id x 60 cm glass column, filled with a mixture of 4.8% QF1 and 0.2% Diglycerol on 80-100 mesh Chromosorb G AW DMCS. Air samples (4 ml) were injected on a hourly basis into the column using a 6-port 2-position gas switching valve, actuated automatically by a timer. The column and injector were maintained at 30°C, while the detector operated at 60°C. A mixture of 90% Ar and 10% CH_4 at a flow rate of 25 ml/min. was used as carrier gas. Under these conditions, PAN eluted in ca. 4.5 min.

The gas chromatograph was calibrated with two different methods, in an attempt to minimize calibration uncertainties. In the first method, based on the IR spectrometry, we used published absorption coefficients (1) of the PAN IR bands to determine, from the IR spectrum, the concentration of PAN formed after a 15 min. photolysis of ethynitrite vapour in the presence of oxygen. A known volume of the irradiation mixture was then diluted adequately in order to obtain a PAN sample in the ppb range, suitable to calibrate the gas chromatograph. For the second method, we prepared pure gaseous PAN samples using the procedure reported by Gaffney et al (11). The PAN concentrations were determined by hydrolysis of the PAN samples in an alkaline solution and subsequent colorimetric analysis of the nitrite ions so produced, using the Saltzmann method.

The two calibration procedures showed a difference of 10%, which is considered very satisfactory, taking into accound their complete independency, as well as the difficulties in the preparation of pure PAN analytical standards, due to the instability of this compound. The detection limit of the GC-ECD system was about 30 ppt (twice the noise level).

3 - RESULTS AND DISCUSSION

Up to the present, most measurements of photochemical oxidants were reported for summer months, when formation of photochemical air pollution is expected to present a maximum, due to the elevated solar radiation. However, some limited observations of high oxidant concentrations during winter months have already been reported (7,8), indicating that photochemical episodes may also occur during non-summer periods.

PAN concentrations have been monitored automatically for several months in Creteil (Paris). The results so far obtained indicate a background level of about 0.3-0.5 ppb during periods of low photochemical activity (late automn and winter months). However, surprisingly high PAN levels, which attained a maximum of 9 ppb, were recorded during two episodes in October 1985 and February 1986. Figure 1 shows hourly PAN concentrations for the periods 18-31 October 1985 and 7-20 February 1986.

Ozone data were not available in Creteil at the time of the episodes. However, O_3 concentrations recorded in several stations in the Paris basin by the local air pollution monitoring network (AIRPARIF) were found to be very low during the two episodes, both in the centre of Paris as well as in rural stations. The differences in the location of these stations, compared to Creteil, do not allow a comparison between O_3 data and the observed high PAN concentrations to be made.

It can be clearly seen from Fig. 1 that the two episodes were not similar in nature. During the first episode, centred around 26 October,

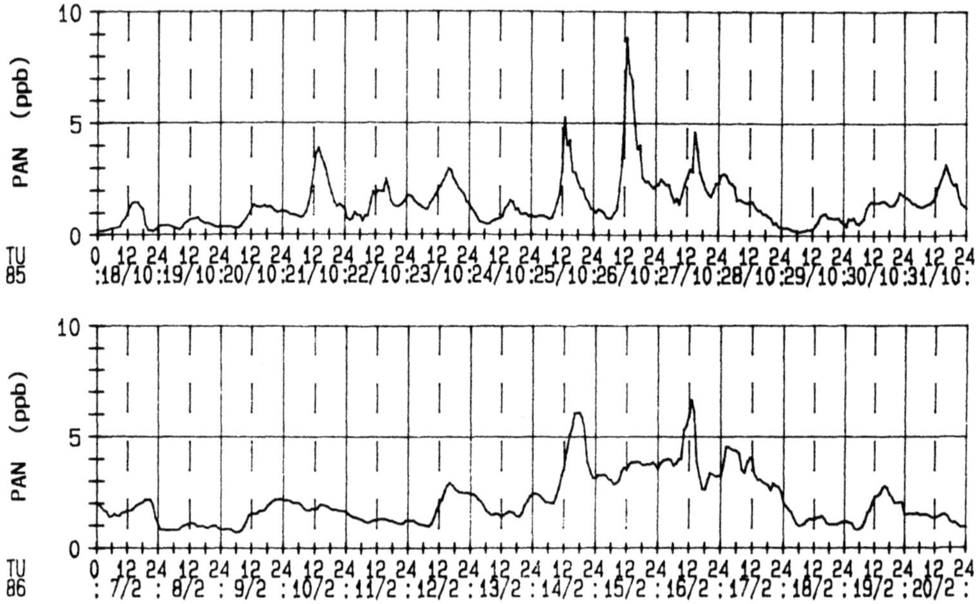

Fig. 1. Hourly PAN concentrations measured in Creteil (Paris) for the periods 18-31 October 1985 and 7-20 February 1986

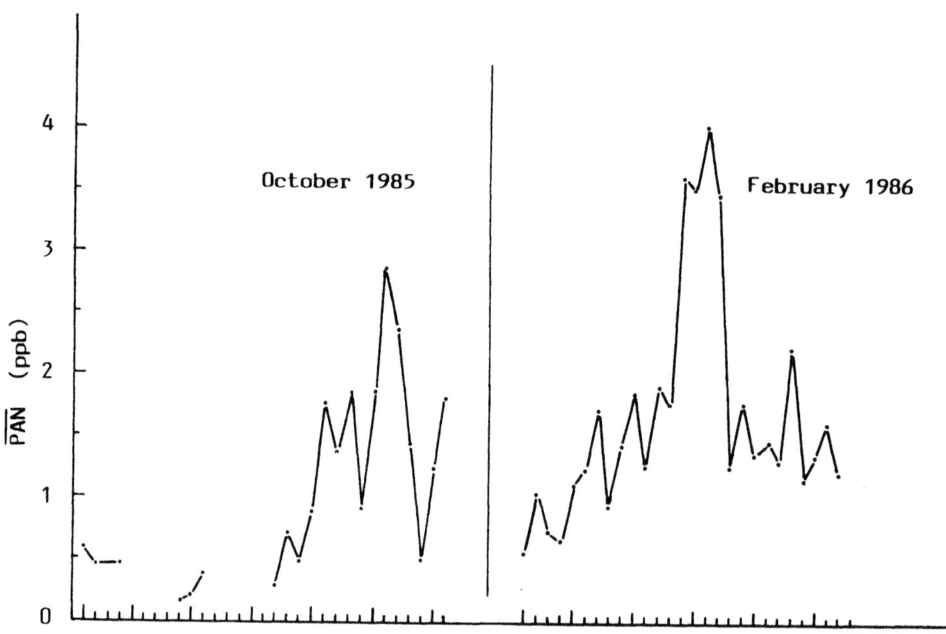

Fig. 2. Daily average PAN concentrations for October 1985 and February 1986 in Creteil

the PAN peaks were generally sharp, with the maximum situated around noon and a marked diurnal variation with relatively low night-time values. In February, on the contrary, there was no important diurnal variation with the exception of two days (14/2 and 16/2), and night-time values were maintained at high levels, exceeding 3 ppb during the episode. This resulted in higher daily average PAN concentrations in February than in October, as illustrated in Fig. 2.

(a) Examination of the October 1985 episode.

During the second half of October 1985, anticyclonic conditions persisted over central and northern Europe, interrupted by the passage of several fronts. In the Paris basin, the sky was generally clear and solar radiation attained an average maximum of 40 mW/cm^2 at noon. To these conditions, already favourable to the formation of photochemical oxidants, were added on 25 October a decrease of the wind velocity and a marked temperature inversion (fig.3). This inversion, situated between 350 and 750 m, isolated the boundary layer, thus facilitating the accumulation of pollutants. As one could expect under these conditions, PAN concentrations began to increase on October 25th, passing from a maximum of 9 ppb recorded at 1300h TU on the 26th (fig. 1). The early afternoon maximum on the 27th was somewhat lower (4.6 ppb), possibly due to a slight increase of the wind speed and therefore an increased dispersion of the pollutants within the mixing layer.

The night-time minima in PAN concentrations can be explained in terms of thermal decomposition of PAN, combined with dry deposition to the ground. During the episode, temperatures were in general warm, and varied between a minimum of 6°C, situated around 0600h TU and a maximum of 16°C around 1400h TU. At such temperatures, thermal decomposition of PAN and subsequent reaction with NO can be a main cause for the decay of PAN during the night. The contribution of dry deposition to the ground may also be of

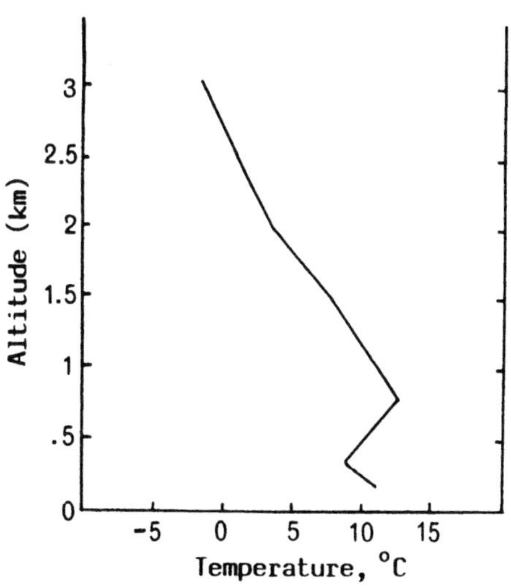

Fig. 3. Radiosonde ascent at Trappes (48.46N, 02.01E), 1200 TU, 25/10/85

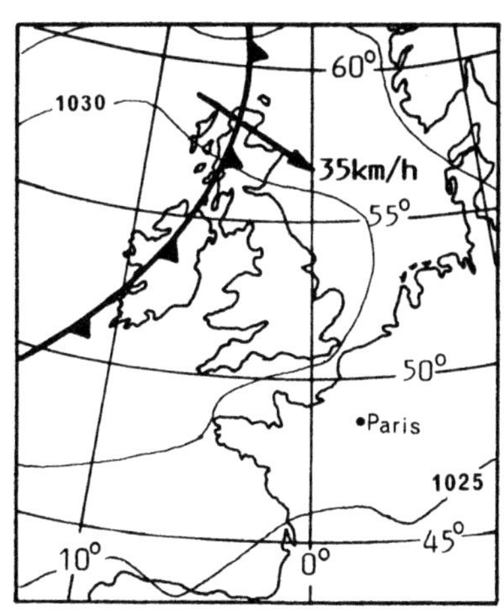

Fig. 4. Surface pressure chart for 1800 TU, 27/10/85. The cold front situated over Ireland reached northern France on the next day.

importance, although deposition velocities of PAN are relatively slow, 0.21 cm/s over grass and 0.26 cm/s over soil (2). It should be noted, however, that a decay in PAN mixing ratio may also be the result of an increase of the mixing height, leading to a better dilution of the polluted air mass. The low night-time values resulted in lower daily average PAN concentrations during the October 1985 episode than during the second episode (February 14-17, 1986), as shown in fig.2.

The marked diurnal variation of the PAN mixing ratio, combined with the low wind intensities and sufficient solar radiation, suggest that the observed high PAN levels were due to in situ photochemical production rather than long-range transport.

This first episode ended when a cold front, situated over England on October 27 (fig.4) moved south-eastwards and reached northern France on the 28th. A significant change of the meteorological conditions was indeed observed on October 28. The wind backed to the northern direction and increased in intensity, and solar radiation intensity and duration decreased significantly. This resulted in a slower PAN formation rate and marked the end of the episode.

(b) The February 1986 episode

During February 1986, temperatures were in general low, with daily averages varying from -6 to +4°C and night-time minima as low as -10°C. It is well known that temperature strongly influences the rate of PAN thermal decomposition, while its effect on PAN formation rate is not very important, since the activation energies of the reactions involved are generally small (7). Consequently, PAN stability increases with decreasing temperature, so for sufficient solar radiation PAN concentrations are expected to be higher at lower temperatures. For example, the PAN thermal decomposition rate $k_d = 10^{15.8}\exp(-13159.3/T)$ (12) leads to a PAN half-life time of 5.8h, 33h and 9.3 days at +10°C, 0°C and -10°C respectively.

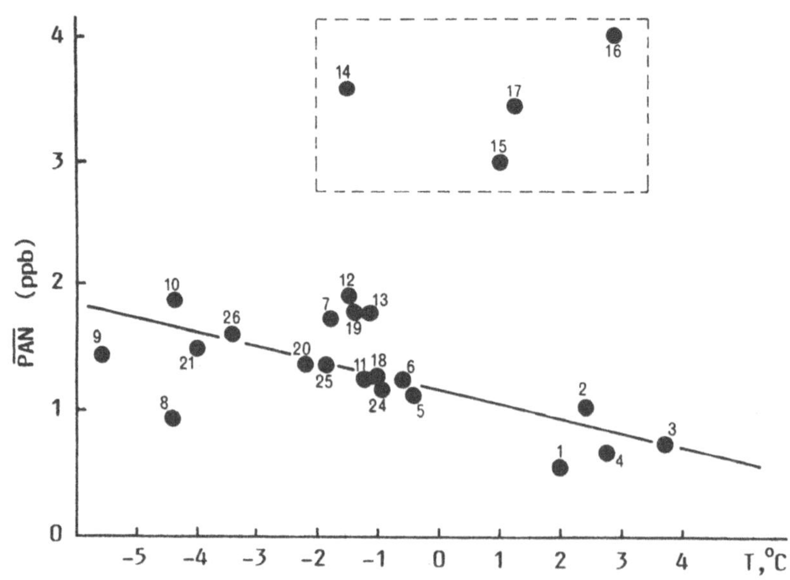

Fig. 5. Daily average PAN concentrations plotted against daily average temperatures for February 1986.

Figure 5 shows daily average PAN concentations plotted against daily average temperatures for February 1986. Indeed, PAN concentrations are higher at lower temperatures, and this correlation is clearly shown for most days of this month, except for the days of the episode, 14-17/2/1986. Apparently, temperature is not the only factor that accounts for those elevated PAN concentrations, and other parameters,such as the meteorological conditions, need to be examined.

As it was mentioned above, the February 1986 episode was quite different in character than the previous one. It was characterized by high night-time PAN concentrations, which resulted in higher average PAN levels than during the first episode (fig.2), and by the absence of a strong diurnal variation with maximum values around noon, except for the 16 February (the late afternoon peak on 14 February will be discussed in details later).

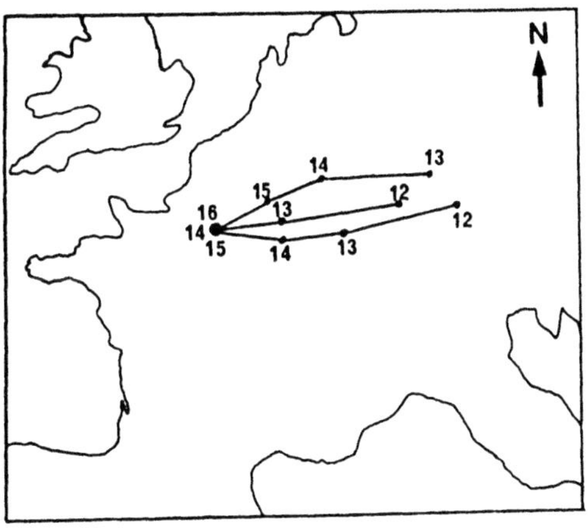

Fig. 6. Back-trajectory calculations for air masses arriving at Creteil (Paris) at 1200 TU on February 14, 15 and 16.

The meteorology during this episode was also different. An anticyclone, located over southern Scandinavia and influencing France during several days, moved north-eastwards on 13 February, giving way to a series of fronts that crossed France during the next days. At the same time, a depression developed over the Atlantic in the west of France, influencing most part of the country, and its combination with the high pressions maintained over north-eastern Europe created strong easterly winds. These winds brought to the Paris bassin air masses containing emissions from the industrial zones of West Germany, as indicated by back trajectory calculations (fig.6). On their way to France, the primary pollutants reacted under the influence of solar radiation to form PAN, and indeed PAN concentrations began to increase on 14 February, as the air mass reached the Paris basin, two days after rotation of the wind from north to east. A half-life time for PAN of more than one day at the prevailing temperatures (e.g. 2.7 days at -3°C) is well consistent with the hypothesis of long-range transport of PAN at a regional scale. Moreover, owing to the elevated wind speeds, the PAN formed on the way reached Paris before being completely deposited to the ground. It should be noted, however, that very strong winds would have enhanced the dispersion of the pollutants, and long-range transport of PAN could probably have not been observed.

The late afternoon peak in PAN concentration recorded on that day deserves a closer examination. The fact that it occured around 1800h TU exludes the possibility of local photochemical production, since solar radiation had dropped to zero at that time. A radiosonde ascent for Trappes (48.46N, 02.01E) at noon on 14 February shows a strong temperature inversion situated between 600 and 1100m (fig.7), which isolated the boundary layer. This inversion was presumably broken by the passage of one of the fronts that crossed France at those days, and air above the thermal inversion possibly containing high PAN concentrations could have been brought to the ground.

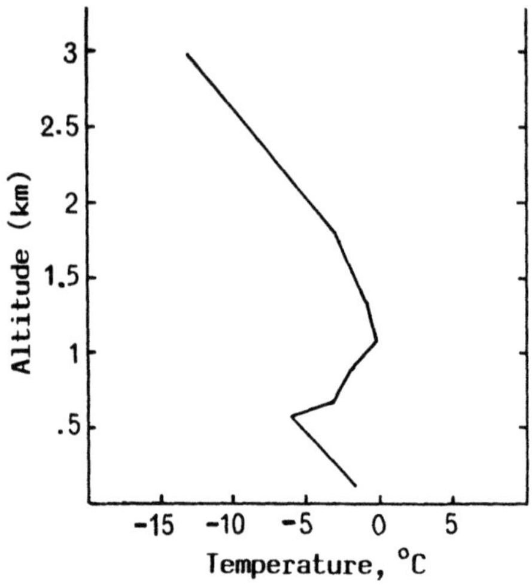

Fig. 7. Radiosonde ascent at Trappes (1200 TU, 14/2/86) showing an inversion between 600 and 1100 m.

These high PAN levels persisted on the next days, owing to the low temperatures. On 16 February, a second peak reaching 6.7ppb was recorded at 1300h TU (fig.1). Solar radiation was elevated on that day, so local photochemical production can account for that peak, which was superposed to long-range transport of PAN.

On 17 February the depression situated at the west of France during the previous days moved north-eastwards and reached northern France. In the afternoon of that day the wind diminished in intensity, backing progressively to the north-east, and a decay in PAN concentration was observed, marking the end of the episode.

4 - CONCLUSION

The results discussed in this paper, in agreement to those reported by Peake and Sandhu (7) and Brice et.al. (8), confirm the possibility that important photochemical episodes may occur during winter periods. Both in situ photochemical production or long-range transport in winter are possible, and the latter does not necessarily need anticyclonic conditions to occur. Simultaneous PAN measurements in several stations are needed, in order to elucidate the role of mainland European emissions in the regional-scale transport of PAN in winter.

REFERENCES

1. STEPHENS, E.R.: Adv. Environ. Sci. Techn., 1, 119-146 (1969)
2. PENKETT, S.A., SANDALLS, F.J., JONES, B.M.R.: VDI Berichte, 270, 47 (1977)
3. NIEBOER, H., VAN HAM, J.: Atmos. Environ., 10, 115-120 (1976)
4. NIELSEN, T., SAMUELSSON, U., GRENNFELT,P., THOMSEN, E.L.: Nature, 293(5833), 553-555 (1981)
5. SINGH, H.B., SALAS, L.J.: Atmos. Environ., 17, 1507-1516 (1983)
6. SINGh, H.B., SALAS, L.J.: Nature, 302(5906), 326-328 (1983)
7. PEAKE, E., SANDHU, H.S.: Can. J. Chem., 61, 927-935 (1983)
8. BRICE, K.A., PENKETT, S.A., ATKINS, D.H.F., SANDALLS, F.J., BAMBER, D.J., TUCK, A.F., VAUGHAN, G.: Atmos. Environ., 18, 2691-2702 (1984)
9. TSALKANI, N., TOUPANCE, G., PERROS, P.: Proceed. 2nd German-French Workshop on the study of chemical reactions with tropospheric interest, 11-13 March 1986, St-Hugues-de-Biviers (France)
10. TSALKANI, N., TOUPANCE, G.: PAN measurements in the atmosphere of Athens (under preparation)
11. GAFFNEY, J.S., FAJER, R., SENUM, G.I.:Atmos. Environ., 18, 215-218 (1984)
12. SCHURATH, U., KORTMAN, U., GLAVAS, S.: Proceed. 3rd European Symposium on Physico-chemical Behaviour of Atmospheric Pollutants, B. Versino & H. Ott (eds.), D. Reidel (publisher), p.27-37, 1984

Verification of the Pollutant Transport Model 'MODIS' using EPRI Plains Site Data from a tall Stack

G. Petersen, D. Eppel, H. Grassl
Forschungszentrum Geesthacht
Postfach 1160
D-2054 Geesthacht
Federal Republic of Germany

Summary

A comprehensive numerical model for the simulation of pollutant dispersion from a point source into the mixing layer of the atmosphere over flat terrain is described. A moment reduction technique is used (MODIS = MOment DIStribution) to combine the simplicity of the Gaussian plume description with the versatility of Eulerian grid formulations. Turbulent dispersion coefficients are parameterized in terms of mean square wind variances which in turn are obtained by a simplified second order closure model. The data base of the 'EPRI Plume Model Validation and Development Project' is used to validate the model for wind velocities above 0.5 m/s and for horizontal scales up to 10^2 km.

1. INTRODUCTION

In recent years various strategies for engineering the releases of pollutants into the atmosphere have been developed. The object of this paper is to present a numerical model which can be used as a tool to assess the environmental load caused by the pollutant plume from a single stack. The model MODIS is an extension of the elementary Gaussian model currently employed by administrations for regulatory work. While physical processes enter into the Gaussian model indirectly through the model parameters such as horizontal and vertical plume widths, average constant velocity, etc. the code MODIS contains parametrizations more closely related to the transport properties of the atmosphere, thus, allowing for the simulation of episodes. In section 2. a short resumee of the model formulation is given together with some remarks on the numerical technique used. Section 3. contains a presentation of the data treatment together with a demonstration of the model behaviour when compared to some of the EPRI data sets.

2. MODEL DESCRIPTION

The ensemble averaged transport diffusion equation for a scalar C is the common basis for most air quality models:

$$\frac{\partial}{\partial t} < C > + \nabla \cdot (< \vec{U} >< C >) = - \nabla < \vec{u}'c' > + \text{sinks} + \text{sources} \qquad (1)$$

To bring eq. (1) into a tractable form we assume that only a single point source is present inside the solution domain. Topographical variations of the lower boundary are not allowed. Removal processes must be describable as $- r \cdot < C >$. The turbulent fluxes, $< \overline{u}'c' >$, resulting from the separation of the quantities into averages and turbulent fluctuations,

$$c = \langle C \rangle + c', \quad \vec{U} = \langle \vec{U} \rangle + \vec{u}' \tag{2}$$

are parameterized by the familiar gradient ansatz:

$$\langle u_i'c' \rangle = - K_{ii} \frac{\partial}{\partial x_i} \langle C \rangle, \quad i = 1,2,3. \tag{3}$$

The formulation is completed be prescribing boundary conditions. On inflow boundaries the concentration is set to zero, and on outflow boundaries a simple radiation condition is applied. On the ground (lower boundary) we require the vertical flux to be given by a deposition velocity w_d:

$$K_{33} \frac{\partial}{\partial z} \langle C \rangle = w_d \langle C \rangle, \quad \text{at } z = 0, \quad (w_d > 0). \tag{4}$$

In order to reduce the dimensionality of the problem by one and still keep the essentially three-dimensional features of the plume a moment expansion on the unknown concentration is performed (1). Assuming the wind field to depend on height only an average wind direction α_o is defined by:

$$\alpha_o = \frac{\int dz\, \alpha(z) \left| \langle \vec{U}(z) \rangle \right|}{\int dz \left| \langle \vec{U}(z) \rangle \right|} \tag{5}$$

where the integration goes over the vertical extent of the solution domain Then, the direction orthogonal to the main wind direction, the y or cross-wind direction can be eliminated by the moment reduction. Defining the moments $c^{(n)}$ by

$$c^{(n)}(x,z,t) = \int_{-\infty}^{+\infty} y^n \langle c(x,y,z,t) \rangle \, dy, \quad n = 0,1,2,... \tag{6}$$

a set of coupled equations for these moments is obtained. The actual concentration field, assumed to have a Gaussian form in cross-wind direction,

$$\langle c(x,y,z,t) \rangle = \frac{\alpha_o}{\sqrt{2\pi\sigma_y^2(x,z,t)}} \exp \left[- \frac{(y - y_o(x,z,t))^2}{2\sigma_y^2(x,z,t)} \right], \tag{7}$$

can be reconstructed from the moments by:

$$\alpha_o = c^{(0)}, \quad y_o = \frac{c^{(1)}}{c^{(0)}}, \quad \sigma_y^2 = \frac{c^{(2)}c^{(0)} - c^{(1)2}}{c^{(0)2}}. \tag{8}$$

The quality of the approximation can be improved by taking more moments, however, at the expense of more storage and execution time requirements. In our opinion, the restriction to the lowest three moments is fully justified by the quality of the data available to us. Moreover, numerical tests with height constant wind fields have shown, that the error made by approximating the plume by only three moments is less than ten percent if the angular deviation of the wind from the main direction stays inside a cone of about sixty degrees (2).

The diffusion coefficients are determined via the turbulent velocity variances. Separating the turbulent energy into horizontal and vertical part,

$$e^h = \langle u'^2 + v'^2 \rangle, \quad e^v = \langle w'^2 \rangle, \quad e = e^h + e^v, \tag{9}$$

the diffusion coefficients K_{22} and K_{33} are assumed to be

$$K_{22} = \tau_y \, e^h, \quad K_{33} = \tau_z \, e^v \tag{10}$$

the factors of proportionality being the essential parameters to be fixed by comparison with experiments. After a variety of data fitting the best fit values for τ_y and τ_z,

$$\tau_y = 35 \text{ sec}, \quad \tau_z = 20 \text{ sec}, \tag{11}$$

were taken for all calculations, i.e. no further parameter adjustment was performed. The stability independence of the τ's indicates that much of the dispersive property of the turbulent atmosphere can be described by the dynamical equations used to determine the velocity variances:

$$\frac{\partial}{\partial t} e^h = \frac{\partial}{\partial z} \left[\sqrt{e}\lambda_1 \frac{\partial}{\partial z} e^h \right] - \frac{\sqrt{e}}{3l_1} \left[e^h - \frac{2}{3} e \right]$$
$$+ 2 \, C_o \sqrt{e} l_o \left\{ \left[\frac{\partial}{\partial z} \langle U \rangle \right]^2 + \left[\frac{\partial}{\partial z} \langle V \rangle \right]^2 \right\} - \frac{4}{3} \frac{e^{3/2}}{\Lambda^1} \tag{12}$$

$$\frac{\partial}{\partial t} e^v = \frac{\partial}{\partial z} \left[3 \sqrt{e}\lambda_1 \frac{\partial}{\partial z} e^v \right] - \frac{\sqrt{e}}{3l_1} \left[e^v - \frac{1}{3} e \right]$$
$$- 2 \, \alpha_h \, C_o \sqrt{e} l_o \frac{g}{\Theta} \frac{\partial}{\partial z} \Theta - \frac{2}{3} \frac{e^{3/2}}{\Lambda_1}. \tag{13}$$

The above equations contain the parameterized processes for turbulence production by shear and buoyancy together with turbulent dissipation and redistribution terms forcing the system towards homogeneous turbulence (3). Θ denotes the potential temperature. The length scales l_1, λ_1, and Λ_1 are connected to the universal length l_o and the constant C_o (4). The inverse Prandtl number α_h is determined from surface layer characteristics (5).

A main problem when treating buoyant plumes by far-field Eulerian grid models (to which MODIS belongs) is the proper location of the source in the numerical grid. As the point and value of maximum ground concentration are very sensitive to the effective source height which does not coincide with the stack height this ambiguity is eliminated by explicitly calculating the plume rise in the neighbourhood of the stack by a separate model (6). The code does not need extra data as those already used. The resulting plume center line H(x) is used to approximately determine a source distance dependent rise velocity by:

$$+ \, w_s = U_{av} \frac{\partial}{\partial x} H(x) \tag{14}$$

where U_{av} is the height averaged velocity, and x denotes the source distance.

Finite difference techniques are employed to discretize eqs. (6), (12), and (13). The advective terms are treated by an upwind scheme, all other derivatives being approximated by central differences. Time integration is done by a simple Euler procedure using a semi-implicit Crank-Nicolson algorithm for the diffusion terms. Depending on the grid Peclét number accuracy estimates can be given for simple cases (2).

3. APPLICATIONS

The field data were chosen from the intensive measurement program conducted downwind of the 187 m stack of the Kincaid power plant of the Commonwealth Edison Company located in the midwestern plains of the United States (7).

Available stack data are emission temperature and exit velocity together with emission rates for the SF_6 tracer. Measured ground concentrations stem from a 200 stations network of movable samplers. The hourly profiles of wind obtained from a 100 m meteorological tower and radio sonde ascents are smoothed by a least squares polynomial fit before the interpolation to the grid points is done. While the hourly averaged temperature profiles from the tower are directly taken as input for MODIS the sonde profiles of temperature are shifted by a constant in order to smoothly match the highest measured value of the tower to the lowest of the sonde because the data suggest a systematic error of the sonde values. The grid values are obtained by linear interpolation of the matched profiles . The time evolution of the data fields are calculated by linearly interpolating between the hourly profiles.

Two typical applications to data sets from the intensive measurement periods in spring and summer of 1980 and 1981 (7):

a) May 7, 1980 from 13:00 to 16:00
b) June 1, 1981 from 2:00 to 5:00

Case a) represents weakly stable mixing conditions capped by an inversion layer, and case b) shows the pronounced features of a night jet.

Case a) (May 7, 1980) is calculated with a grid of 80 points in x-direction and 26 points in the vertical. Horizontal grid spacing is Δx = 250 m. The vertical grid lines are unequally spaced between 2 m (the lowest tower value) and 3000 m (inside the inversion layer). The data input is given in Fig. 1. The location of the inversion layer above 2000 m allows the plume to rise up to that height over a distance of more than 5 km. Clearly, if such a dramatic change in the vertical location of the plume is not taken care of in a grid model no realistic ground concentrations can be expected. Fig. 2 shows the ground concentration pattern. Whereas the calculated pattern shows a good overall fit to the sampler values and the location of maximum ground concentration is less than 400 m away from the measured point the calculated maximum ground concentration falls short of the measured value by a factor of two. It may well be that the measured temperature profiles suggest a more unstable boundary layer than actually existed which would explain the lower maximum value.

Case b) (June 1, 1981) represents a well developed night jet situation as documented by the strong low level wind, strong horizontal wind shear and stable stratification depicted in Fig. 3. Under these conditions the plume is practically trapped and channelled. Fig. 4 indicates that the location and magnitude of the maximum ground concentration can be accepted as reliable because the outmost sampler arrays containing the maximum meas-

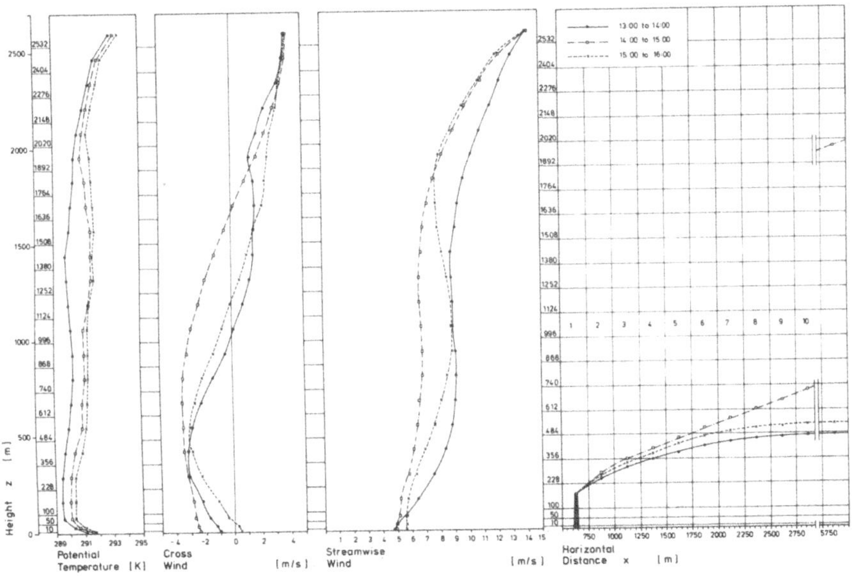

Fig. 1 Data input for May 7, 1980

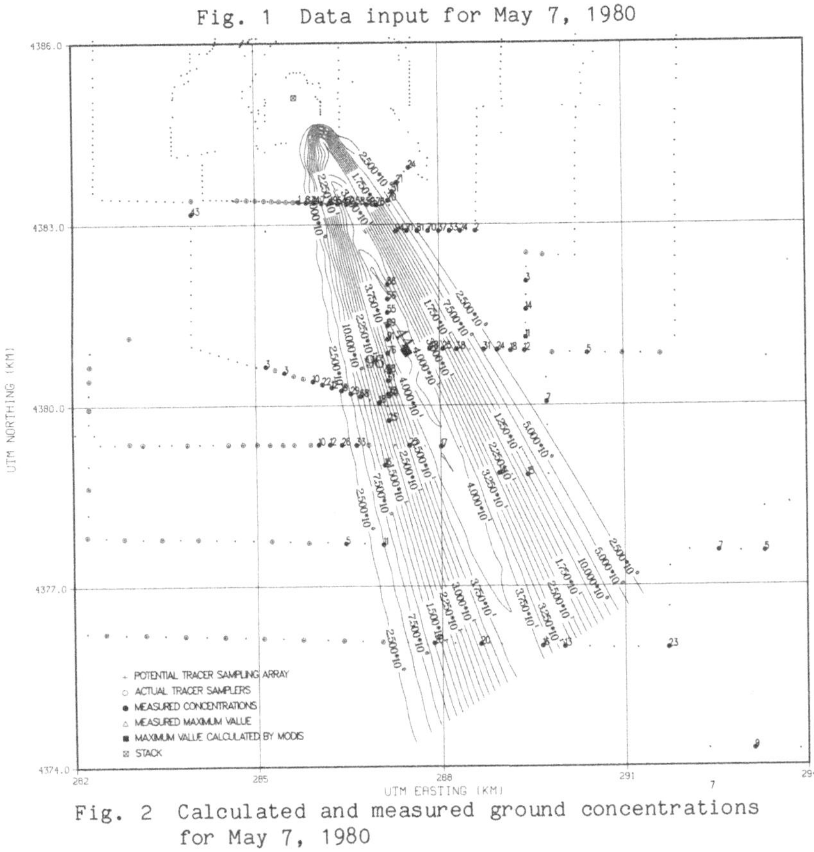

Fig. 2 Calculated and measured ground concentrations for May 7, 1980

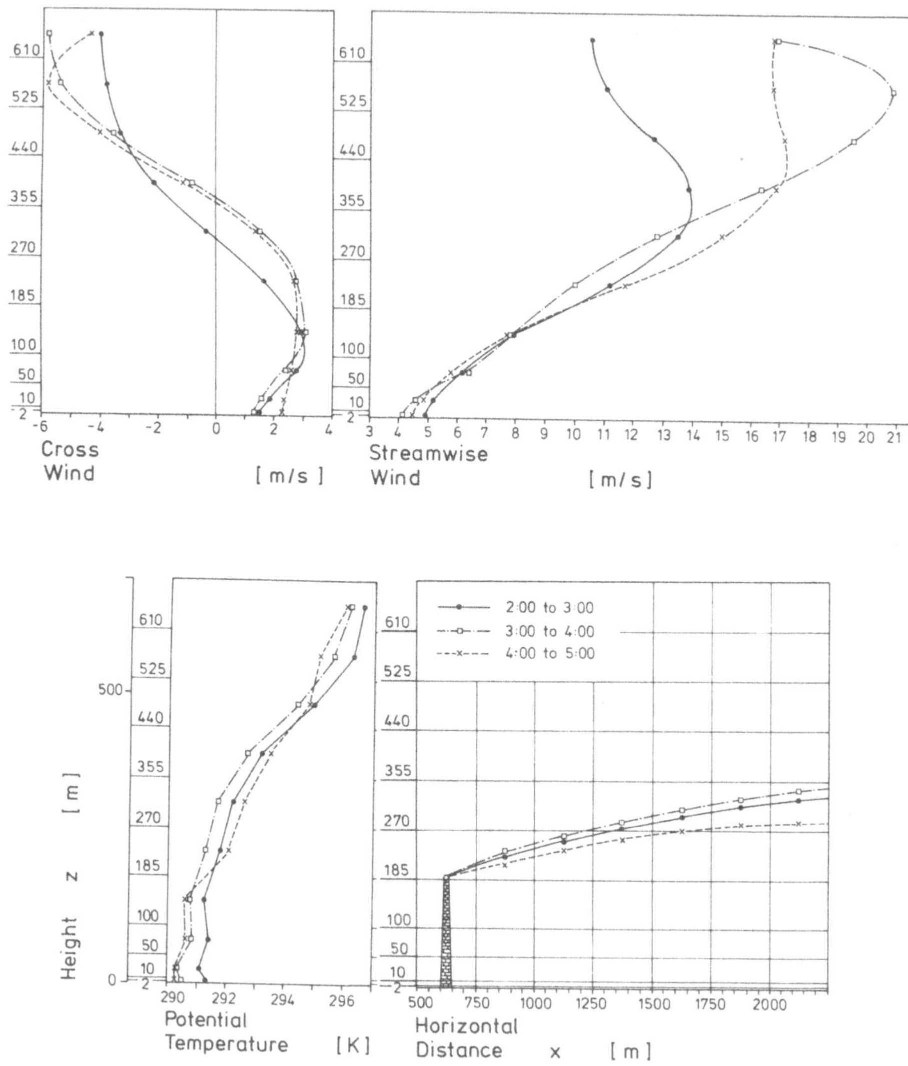

Fig. 3 Data input for June 1, 1981

ured value are nicely fitted despite the fact that the distance from the source to the maximum measured value is about 50 km, and the distance between source and calculated location of maximum ground concentration is more than 60 km.

4. CONCLUSION

An Eulerian plume model has been described together with examples of comparisons to EPRI data. The model can be used operationally requiring sequences of turbulence averaged wind and temperature profiles as input. The small lateral spread compares favourably with other grid models where either the lateral grid spacing is too crude close to the stack, or, if the stack region is well resolved too few grid points are available far away from the source. During the course of the validation calculations it became clear that a realistic plume rise must be incorporated into the grid model in order not to fall off orders of magnitude from the experimental data, a need which cannot be met by simple effective source height formulas. In our aproach an integral plume model can be successfully used to determine the plume rise despite the fact that close to the stack the assumptions underlying the gradient diffusion ansatz are violated.

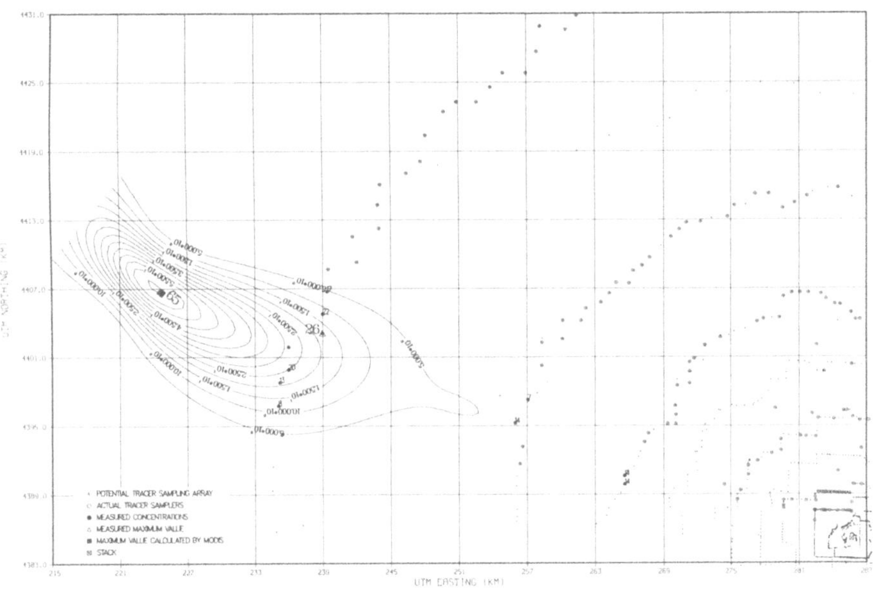

Fig. 4 Calculated and measured ground concentrations, June 1, 1981, 4:00 to 5:00

ACKNOWLEDGEMENTS

We are gratefully indebted to H. Graßl's valuable discussions and constructive criticism, and to H. Wilk for his expert drawings.

REFERENCES

(1) see e.g.: Saffman, P.G., The effect of wind shear on horizontal spread from an instantaneous ground source, Quart. J. Roy. Met. Soc 88 (1962) 382-393.
Gee, J.H., D.R. Davis, A note on horizontal dispersion from an instantaneous ground source, Quart. J. Roy., Met. Soc. 89 (1963) 542-545.
Kao, S.K., J.C. Doran, Turbulent diffusion in linear shear flow, Pageoph. 114 (1976) 357-363.
Manton, M.J., On the dispersion of particles in the atmosphere, Boundary Lay. Met. 17 (1979) 145-165
Mulholland, M., Simulation of tracer experiments using a numerical model for point sources in a sheared atmosphere, Atmos. Env. 14 (1980) 1347-1360.

(2) Eppel, D., J,. Häuser, A. Müller, Three-dimensional simulation model for the propagation of air pollutants, 5th Int. Clean Air Congress, Oct. 20 - 26, 1980, Buenos Aires.
Eppel, D., J. Häuser, H. Lohse, F. Tanzer, Pollutant dispersion in Ekman layer using moment-reduced transport equations in:
C. de Wispeleare, (Ed.), Air Pollution Modelling and its Applicat. II, Plenum Publ. Company 1983.

(3) Mellor, G.L., T. Yamada, Development of turbulence closure models for geophysical fluid problems, Revs. Geoph. and Space Phys. 20 (1982) 851-875.

(4) Zilitinkevitch, S.S., D.L. Laikhtman, A.S. Monin, Dynamics of the atmospheric boundary layer, Izv. Atmosph. and Oceanic Physics 3,3 (1967) 170-191.

(5) Businger, J.A. J.C. Wyngaard, Y. Izumi, E.F. Bradley, Flux-profile relationships in the atmospheric surface layer, J. Atmos. Sci. 33 (1971) 181-189.

(6) Schatzmann, M., An integral model of plume rise, Atmos. Environ. 13 (1979) 721-731.

(7) Second-order closure exercise for the Kincaid power plant plume, EPRI EA-3079, Project 1616-9, Final Report, July 1983.

LONG-RANGE TRANSPORT OF NITROGEN AND SULPHUR COMPOUNDS

M. FERM[1], K. FUGLSANG[2], H. ARESKOUG[3], J-E. HANSSEN[4] and H. LÄTTILÄ[5]

[1] Swedish Environmental Research Institute P.O. Box 5207, S-402 24 Göteborg, Sweden

[2] National Agency of Environmental Protection-Air Poll. Lab. Risø, DK-4000 Roskilde, Denmark. Present address: Technological Institute, P.O. Box 141, DK-2630, Tåstrup Denmark

[3] The National Swedish Environmental Protection Board -Air Res. Lab. Studsvik S-611 82 Nyköping, Sweden

[4] Norwegian Institute for Air Research, P.O. Box 130, N-2001 Lillestrøm, Norway

[5] Finnish Meteorological Institute, Sågaregatan 22 E, SF-00810 Helsinki, Finland

Summary

Total (particulate + gaseous) ammonium and total nitrate were measured on a 24-h basis continuously at six Nordic EMEP stations from August 15 to October 31, 1984. They were both measured using impregnated filters mounted in open face filter holders. High concentrations in air occurred episodically at several sites almost simultaneously. Calculated backtrajectories indicated that the air had passed over the western parts of Europe during the episodes. The concentrations of H^+, NH_4^+, NO_3^- and SO_4^{2-} in precipitation was also measured. From the data it is clear that NO_3^- and SO_4^{2-} are enriched in precipitation compared to NH_4^+. One possible explanation to this is the in-cloud oxidation of SO_2 and NO_2. Another contributing factor is a higher apparent scavenging rate for SO_4^{2-} and NO_3^- that may be caused by a lower concentration decrease with altitude.

1. INTRODUCTION

The deposition of nitrogen and sulphur species causes acidification and eutrofication at such a high rate that damages are suspected to occur and have been observed at several places in Europe.

There are several facts that complicates the actions against the deposition. The damages may occur far from the sources of the precursors to these pollutants. The damages are not immediate but come about after several decades of deposition. Other factors, e.g. climate and insect attacks, may be combined in the damage and it is then hard to estimate the contribution from the individual causes. Nevertheless, it is important to clearify the concentration distribution and deposition patterns of nitrogen and acid airborne species since decisions on action have to be based on knowledge of their transformation and transport paths. This project is a first experiment to show wether it is possible to use a simple filter technique to measure the total (gaseous + particulate) ammonium and nitrate concentrations in air on a routine basis at the EMEP stations. So far the nitrogen species have only been measured routinely in precipitation. The air concentrations are compared to air mass trajectories in order to estimate wether local sources or distant sources are responsible to the observed concentrations.

2. ANALYTICAL TECHNIQUES

Fine particles seldom collide with smoth surfaces or each other. They can therefore be of different chemical compositions, depending on their origin. They will rapidly (minutes) equilibrate themselves with the surrounding gas phase. When collected on a filter they are forced together and reactions changing their composition take place. This will also change the concentrations of different gases behind the filter. Either gases nor particles which have significant concentrations in the two phases can therefore be correctly separated by filtration. Techniques to separate the two phases have now been developed (the denuder techniques).

In connection with networks that works on a routine basis, it is important to have simple techniques. To be able to engage many stations it is also advantegeous if the technique is not expensive. Both these requirements are met in the impregnated filter techniques. Total ammonium is then collected on a oxalic acid impregnated filter and total nitrate on a sodium carbonate impregnated one. Using these filters reproducibility between the labs participating in the report have been investigated. The denuder technique was also used simultaneously (1).

The results were then found encouraging enough to make a pilot study at 6 stations run by these laboratories. In the pilot study, one laboratory, however, preferred a filter pack technique, which in a separate investigation (2) was found to be equivalent to the single filter techniques.

3. MEASURED DATA

Six EMEP stations were selected, namely: Keldsnor (54°44'N, 10°44'E) in Denmark, Birkenes (58°23'N, 8°15'E) and Jergul (69°24'N, 24°46'E) in Norway, Aspvreten (58°49'N, 17°24'E) and Rörvik (57°25'N, 11°56'E) in Sweden and Utö (59°47'N, 21°23'E) in Finland. The sampling started in August 15, and ended in October 31 (1985). Beside total concentrations of NH_4^+ and NO_3^- in air, gaseous SO_2, NO_2, particulate SO_4^{2-} and in precipitation pH, NH_4^+, NO_3^-, SO_4^{2-} and sea spray component like Cl^-, Na^+ or Mg^{2+} were measured.

4. CONCENTRATIONS IN AIR

The concentrations of total ammonium and total nitrate are shown for all stations during two months in figures 1 and 2, respectively. As can be seen, all stations except Jergul had an episode of high concentrations for several days at the end of September and in the beginning of October. There is also a small one-day episode, September 19. The concentration of particulate sulphate was also increased during the episodes (Fig. 3). Air mass back trajectories were calculated for the whole measuring periode by the Norwegian Meteorological Institute. The trajectories arriving at noon are shown for four episodic days in figure 4. As can be seen from the figure, the wind passed over the western parts of Europe at these occasions. When backtrajectories are used for the whole period it is better presented as a windrose showing the mean concentration in different sectors surrounding the stations. Eight sectors were used here. The results are shown in figures 5 and 6. The figures outside the sectors show the number of days which the mean is based on. Ammonium and nitrate seem to origin from the same directions (south and southwest) in these diagrams. The concentrations drop continuously to the north and Birkenes seem to be at the pereferi of the "plume".

5. COMPARISON BETWEEN CONCENTRATIONS IN AIR AND PRECIPITATION

NH_4^+, NO_3^- and SO_4^{2-} are incorporated in precipitation by several complicated processes. The most important are the so called in-cloud processes. The highly water soluble particles act as cloud condensation nuclei thus forming droplets which may grow to a precipitating size. Most of the water soluble particles will by this process be transferred to the aqueous phase. Highly water soluble gases will be almost quantitatively scavenged out in the clouds. Below-cloud both the rain out of particles and water soluble gases are rather inefficient.

The precursors NO_2 and SO_2 have a very small solubility in water but can be faster in-cloud oxidized to highly water soluble species. The oxidation of SO_2 then proceeds in the aqueous phase. H_2O_2 may then be the most important oxidizing agent. In-cloud formation of NO_3^- has been suggested as an important atmospheric process (3-5) but the mechanism is still not well understood.

The scavenging ratio is often used when the total scavenging efficiency for different species are compared. For particulate nonvolatile elements it is remarkably equal for different components at a certain site if the pollutants in question are mainly of distant origin (i.e. are well mixed within the mixing layer) (6). The scavenging ratio W is defined as the ratio between the concentration in precipitation divided by the concentration of its precursors in air. The time basis also affects the ratio. All concentrations were, however, measured on a 24 h basis (6^{00} GMT). The interpretation of individual ratios or its longtime average absolute value is complicated and perhaps not meaningful. It will here be used to see if any species is relatively more abundant in ground level air or precipitation. The average concentrations when data for all airborne and precipitation species were simultaneously available are given in table 1.

There are several precursors for NO_3^- and SO_4^{2-} in precipitation but the less watersoluble (those having a lower oxidation number) were not considered here. The scavenging ratios for NH_4^+ and NO_3^- were calculated as the concentration in precipitation divided by the total concentration in air. For SO_4^{2-} the excess concentration in precipitation (corrected for sea spray contribution using the Mg^{2+} or Cl^- concentrations) was divided by the particulate SO_4^{2-} concentration since there is no gaseous sulphur specie with oxidation number 6. The mean values for the whole period are shown in table 2. As can be seen $W(NH_4^+)$ is for all stations lower than $W(NO_3^-)$ and $W(SO_4^{2-})$. For four stations $W(NO_3^-)$ is higher than $W(SO_4^{2-})$ and for two stations the situation is reversed.

NH_3 is mainly emitted at ground level. After the emission NH_3 is dry deposited to the surrounding vegetations and soils or reacting with acid gases or particles. The particles are only slowly dry deposited and this is therefore the only process that can long-range transport NH_4^+. About 90% of the airborne NH_4^+ has been found in the particle phase at Rörvik and 10% in the gas phase in the few analyses (≈40) that have been carried out so far. The fraction in the gas phase decreases with temperature.

Airborne SO_4^{2-} (particulate) and NO_3^- (HNO_3 + particulate NO_3^-) are secondary pollutants which are formed even at high altitudes. If the reaction rates are determined by the concentration of oxidants they will get a different concentration profile compared to NH_4^+, with relatively higher concentration at higher altitude. This fact will contribute to higher scavenging ratios for NO_3^- and SO_4^{2-} compared to NH_4^+. In cloud oxidation of SO_2 and NO_2 will also lead to higher W for SO_4^{2-} and NO_3^-. The enrichment of NO_3^- and sulphate in precipitation compared to NH_4^+ was here calculated as

$$\Sigma = \frac{2[SO_4^{2-}]_{ex} + [NO_3^-]_p - (2[SO_4^{2-}]_a + [NO_3^-]_{tot})\,W(NH_4^+)}{2[SO_4^{2-}]_{ex} + [NO_3^-]_p}$$

where ex means sea spray corrected concentration, p precipitation, a airborne and tot the total airborne concentration. If $[NO_3^-]_{tot}$ and $[SO_4^{2-}]_a$ have a concentration profile proportional to that for NH_4^+ and are scavenged with the same efficiency as NH_4^+, Σ shows the average fractions of NO_3^- and SO_4^{2-} in precipitation that has been in-cloud oxidized and scavenged out. Σ = 0.60 means e.g. that 60% of the nitrate and sulphate concentrations (equivalents/l) are either caused by in-cloud oxidation or different concentration profiles.

The in-cloud oxidation and scavenging processes will lead to incorporation of hydrogen ions in precipitation. This is, however, not the only process leading to acid precipitation. Scavenging of acid particles and below cloud scavenging of HNO_3 will also form H^+. The fraction Ψ defined as

$$\Psi = \frac{[H^+]_p}{2[SO_4^{2-}]_{ex} + [NO_3^-]_p}$$

will thus be a maximum average fraction of SO_4^{2-} and NO_3^- that has been in-cloud oxidized and scavenged. There will be no important neutralization processes at remote places with low gaseous NH_3 concentrations. Σ and Ψ are shown as averages in table 2. The H^+ concentration was calculated from pH measurements. H^+ is the least stable ion in precipitation and may therefore contain a larger error than the others.

As can be seen from all stations except the two Norwegian ones, the enrichment of NO_3^- and SO_4^{2-} in precipitation relative to NH_4^+ (Σ) is higher than the maximum enrichment due to in-cloud oxidation (Ψ). This is also true for almost all individual values. Beside the in-cloud formation of NO_3^- and SO_4^{2-} there must consequently be a higher relative scavenging rate for SO_4^{2-} and NO_3^- than for NH_4^+. A possible reason for this fact is that they have different concentration profiles. The data from Norway neither contradict nor supports this idea because they have a high fraction of hydrogen ions which probably originated from below cloud scavenged HNO_3, acid particles or other species.

REFERENCES

(1) FERM, M., FUGLSANG, K., HANSSEN, J.E., LÄTTILÄ, H. and SÖDERLUND, R. (1985) Field intercomparisons of measurements of total ammonium, nitrate and sulphur concentrations. In Advancement in Air Pollution, Monitoring Equipment and Procedures. Proc. Freiburg, Federal Republic of Germany, June 3-6, 1985, pp. 52-66.

(2) FUGLSANG, K. (1986) A Filter Pack for Determination of total Ammonia, total Nitrate, Sulfur Dioxide and Sulfate in the Atmosphere. Report MSTL LUFT- A103, Risø National Laboratory, DK-4000 Roskilde, Danmark.

(3) HEGG, D.A:, HOBBS, P.V. and RADKE, L.F. (1984) Measurement of the scavenging of sulfate and nitrate in clouds. Atmospheric Environment 18, 1939-1946.

(4) LAZRUS, A.L., HAAGENSON, P.L., KOK, G.L., HUEBERT, B.J., KREITZBERG, C.W., LIKENS, G.E., MOHNEN, V.A., WILSON, W.E. and WINCHESTER, J.W. (1983) Acidity in air and water in a case of warm frontal precipitation. Atmospheric Environment 17, 581-591.

(5) BARRIE, L. (1985) Scavenging Ratios, Wet Deposition, and In-Cloud Oxidation. An Application to the oxides of Sulphur and Nitrogen. Journal of Geophysical Research 90, 5789-5799.

(6) PEIRSON, D.H., CAWSE, P.A., SALMON, L. and CAMBRAY, R.S. (1973) Trace Elements in the Atmospheric Environment. Nature 241, 252-256.

Table 1. Some average concentrations in air and precipitation when simultaneous data have been available.

	Concentration in air nmole m^{-3}				Precipitation, conc. in µM				
Station	No. of observations	$[NH_4^+]_{tot}$	$[NO_3^-]_{tot}$	$[SO_4^{2-}]_a$	mm	$\{H^+\}$	$[NH_4^+]$	$[NO_3^-]$	$[SO_4^{2-}]_{ex}$
Aspvreten	25	84	24	25	3.0	62	62	60	44
Birkenes	18	72	41	29	14.9	53	38	33	23
Jergul	9	15	3	10	4.0	27	7	6	13
Keldsnor	18	269	110	57	2.5	34	59	54	32
Rörvik	33	102	68	30	5.8	73	64	88	53
Utö	19	95	52	47	3.6	51	78	91	81

Table 2. **Scavenging ratios, fractions of the concentrations of SO_4^{2-} and NO_3^- in precipitation that have been enriched compared to NH_4^+ (Σ) and the maximum in-cloud formation of SO_4^{2-} and NO_3^- (Ψ) at the six EMEP stations. The average concentrations are given in table 1.**

Station	$W(NH_4^+)\times10^{-6}$	$W(NO_3^-)\times10^{-6}$	$W(SO_4^{2-})\times10^{-6}$	Σ	Ψ
Aspvreten	1.10	5.61	4.21	0.66	0.48
Birkenes	0.53	1.34	0.94	0.43	0.73
Jergul	0.42	1.78	1.31	0.73	0.92
Keldsnor	0.31	1.41	0.88	0.59	0.31
Rörvik	0.84	2.07	2.94	0.58	0.40
Utö	1.48	2.30	3.42	0.53	0.20

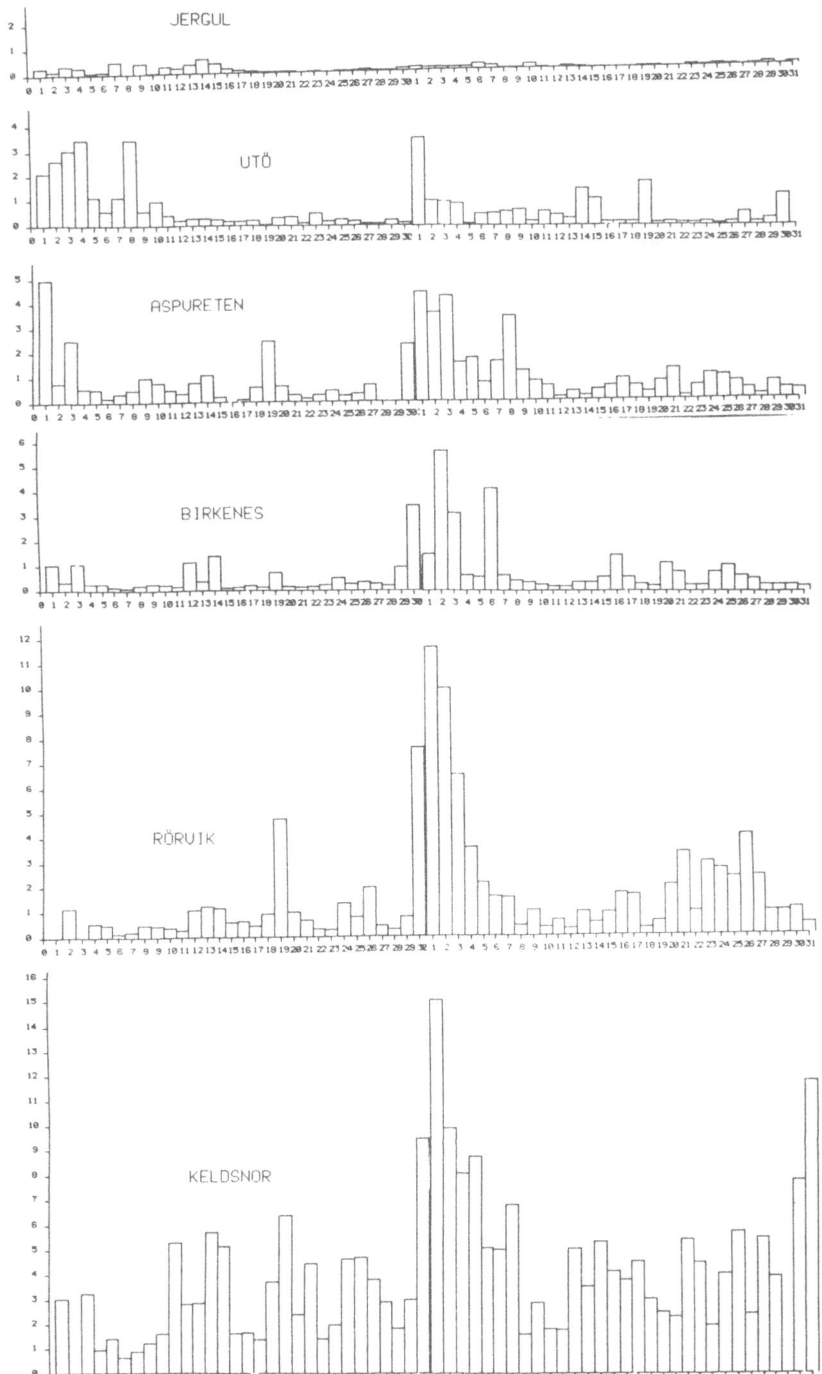

Figure 1.
TOTAL AIRBORNE AMMONIUM µG N/M^3 SEPTEMBER-OCTOBER 1985

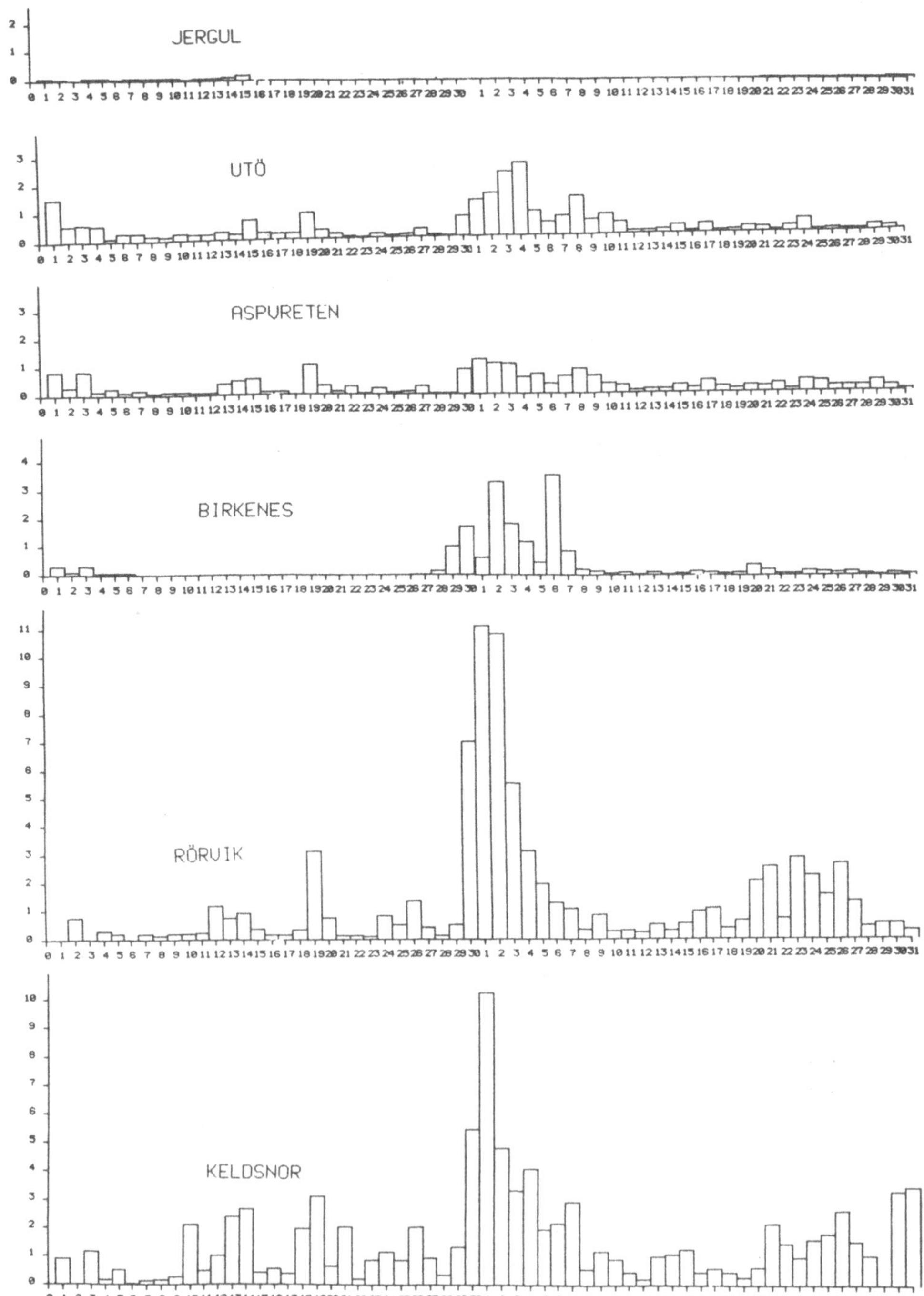

Figure 2.

TOTAL AIRBORNE NITRATE µg N/m³ SEPTEMBER-OCTOBER 1985

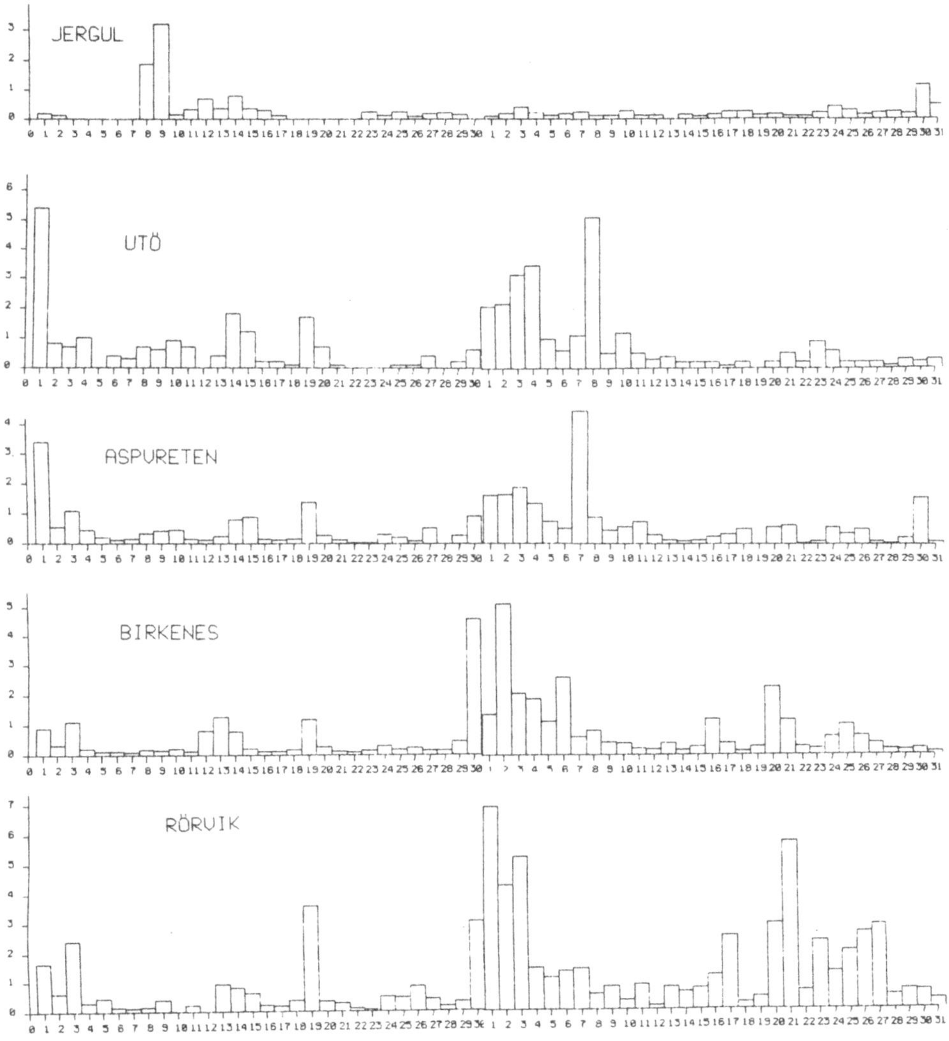

Figure 3. PARTICULATE SO_4 µG S/M^3 SEPTEMBER-OCTOBER 1985

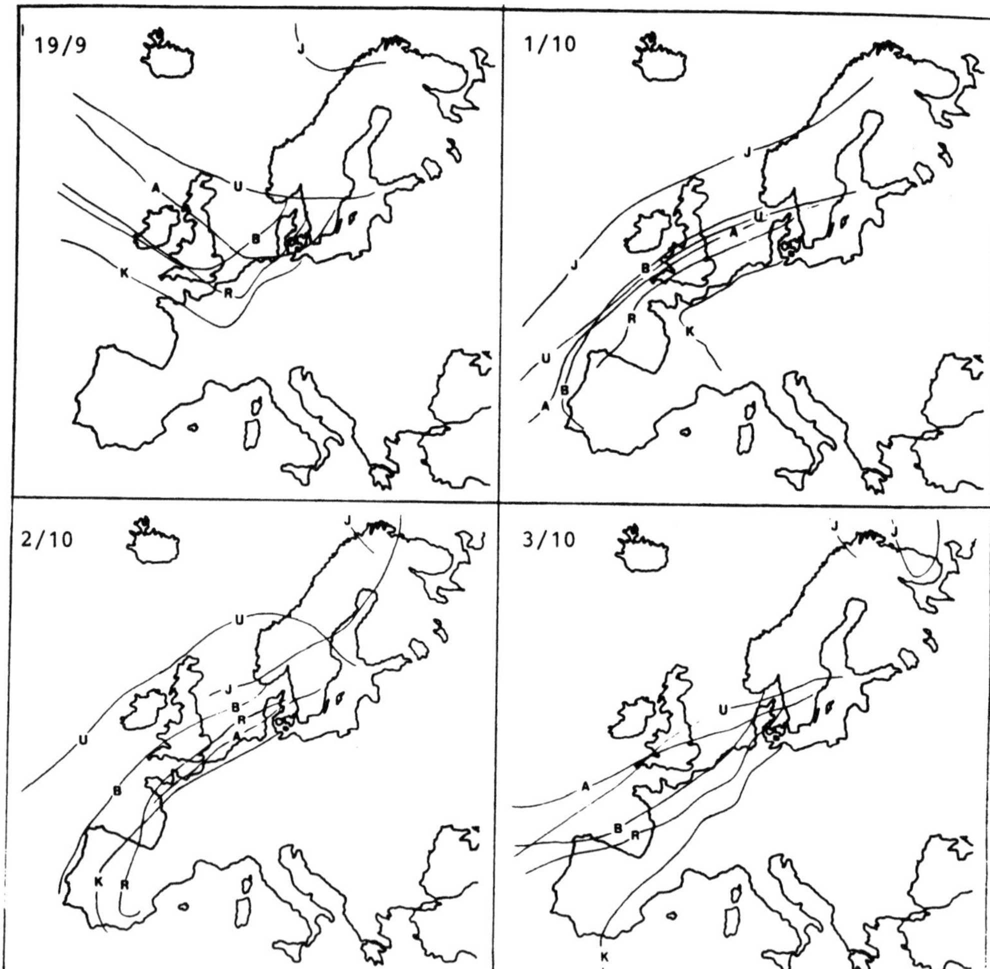

Figure 4. Air mass trajectories arriving at noon to the different stations (which are denoted by the first letter of their names).

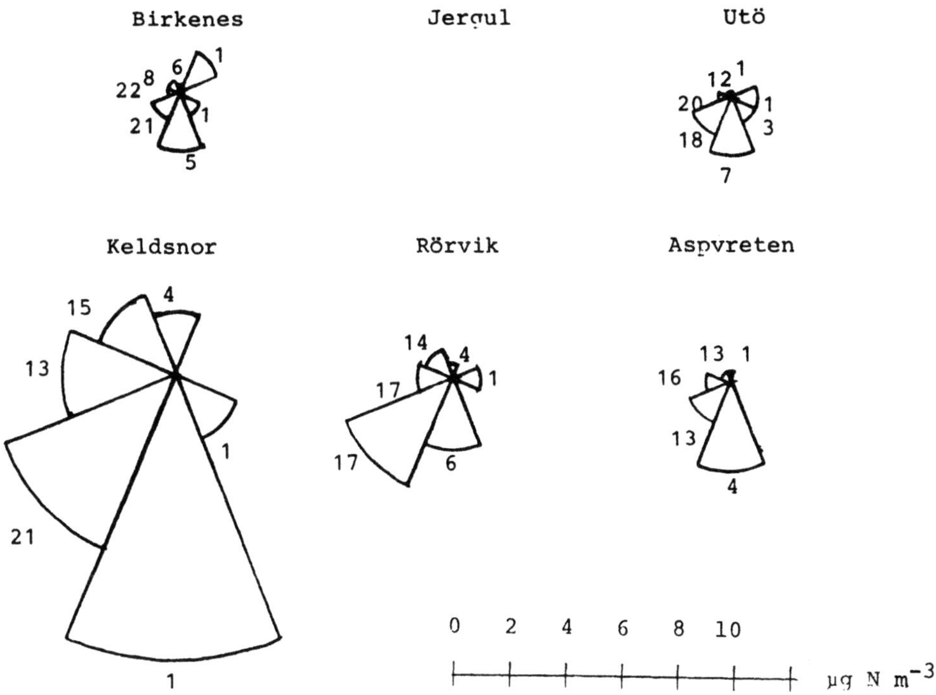

Figure 5. Average total NH_4^+ concentration in air at each sector. (North is upwards.) The number of measurements are indicated outside the sectors.

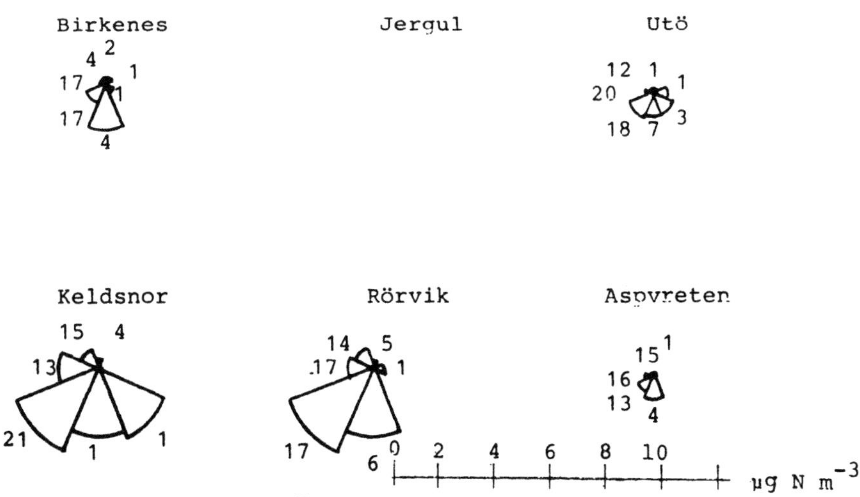

Figure 6. Avarage total NO_3^- concentration in air at each sector. (North is upwards.) The number of measurements are indicated outside the sectors.

MODELISATION DE LA CHIMIE D'UN PANACHE

E. JOOS
Electricité de France
Direction des Etudes et Recherches
6, quai Watier - 78400 CHATOU

Summary

The main tool of the EDF program concerning the study of physicochemical processes which occur when a power plant plume interacts with the atmosphere, is the code PARADE which has been developped to simulate acid products formation in the plume area during its transport and its dispersion in the ambient air.
This paper pertains to the evaluation of this model, for unsaturated atmospheres, using experimental data collected during a field study. Mono and multi-parameter sensitivity analyses of acid species concentrations are also performed.
Now, a cloud chemistry module, incorporated into PARADE, has to be tested.

1. INTRODUCTION

La complexité et l'interdépendance des processus mis en jeu au cours des interactions physicochimiques atmosphère-panache de centrale à charbon, lors du transport et de la dispersion de ce dernier, sont telles que leur étude, afin de déterminer quantitativement la contribution du panache aux retombées acides, nécessite d'utiliser une approche numérique.

A cet effet, EDF s'est engagé dans le développement du code PARADE (PAnache Réactif en Atmosphère avec DEpôts) qui comporte plusieurs étapes :

1. Incorporation de la chimie en phase gazeuse et prise en compte de la formation d'aérosols à partir de cette phase.
2. Prise en compte des interactions physicochimiques panache-nuages non précipitants.
3. Prise en compte des interactions panache-pluie (nuages précipitants).

Cette communication se rapporte plus particulièrement au premier point et concerne la formation des aérosols de sulfate et de nitrate à partir de la phase gazeuse, pour des atmosphères non saturées.

Après une brève description du code PARADE, nous présentons un exemple de simulation d'évaluation des capacités du modèle, à l'aide d'une configuration étudiée expérimentalement, puis nous donnons quelques résultats d'analyses de sensibilité mono et multiparamètres.

2. PRESENTATION DU CODE PARADE

Nous avons, comme modèle de base, retenu le code RPM (Reactive Plume Model) qui utilise une approche lagrangienne basée sur une rangée de cellules contigües, perpendiculaires à la direction du vent et qui se développent en largeur et en hauteur selon une distribution gaussienne pour les composés inertes (1). Les espèces chimiques sont supposées mélangées de façon homogène dans chaque cellule ce qui implique que le modèle ne fournit pas de résolution verticale. De plus, ces espèces sont soumises à des réactions chimiques dans chacune des cellules, et à la diffusion se produisant au niveau des parois inter-cellules et entre la cellule externe et l'atmosphère. Les dépôts secs sont traités au moyen de vitesses de dépôt dépendant du temps. Le modèle est initialisé par des profils de concentration gaussiens lorsque la surhauteur est atteinte.

En ce qui concerne le traitement des réactions chimiques en phase gazeuse, nous avons choisi la version la plus récente du Carbon Bond Mechanism, le CBM IV (2).

La formation d'aérosols à partir des espèces présentes dans la phase gazeuse n'étant pas traitée initialement dans le RPM, nous lui avons incorporé le module MARS (Model for an Aerosol Reacting System) développé par SAXENA et al. (3).

MARS détermine la composition en sulfate/nitrate/ammonium/eau d'un aérosol en équilibre avec la phase gazeuse environnante. Il calcule, en utilisant les équilibres thermodynamiques, la distribution des espèces chimiques entre la phase gazeuse et la phase "aérosol" à partir des concentrations totales (gaz + aérosols) d'acide sulfurique, d'acide nitrique et d'ammoniac, de l'humidité relative et de la température.

A partir de certaines hypothèses et en fonction du cas considéré, MARS détermine d'abord les principales espèces susceptibles de se former avant d'effectuer le calcul de la composition chimique détaillée. Cette procédure en 2 étapes permet d'économiser du temps, au niveau de l'exécution du programme, par rapport à des modèles plus détaillés en comparaison desquels MARS fournit des résultats très satisfaisants.

De plus, MARS prend en compte la dissolution du SO_2 gazeux dans une éventuelle phase liquide de l'aérosol, suivie de l'oxydation par H_2O_2, O_3 et O_2 (catalysée par Mn^{2+} et Fe^{3+}).

Dans la suite de cette communication, nous appellerons PARADE le modèle global RPM+MARS.

3. EVALUATION DU CODE PARADE

Afin d'évaluer les capacités du code et de le valider en partie, il est nécessaire de disposer de données expérimentales cohérentes. Dans ce but, EDF a réalisé une série de 4 campagnes sur site avec la mise en oeuvre d'un avion muni notamment d'analyseurs à court temps de réponse, environ 4 secondes, pour SO_2, NO, NO_x, et O_3 (4).

Pour ce travail, nous avons retenu un vol effectué au cours de la première campagne et susceptible d'être modélisé par PARADE dans sa forme actuelle (absence de nuages, pas de cisaillement de vent...).

Le tableau 1 présente les caractéristiques de l'atmosphère pour ce vol durant lequel la présence d'une inversion basée au sol et prenant de l'altitude au cours de l'après-midi a eu pour conséquence des dépôts négligeables.

L'ozone n'ayant pu être mesuré correctement durant la première campagne, nous avons supposé une concentration atmosphérique vraisemblable compte tenu du moment de l'année et de l'ensoleillement lors du vol. D'autre part, aucune donnée concernant les concentrations d'aérosols ou d'acide nitrique n'est disponible pour ce vol, les prélèvements n'ayant débuté que lors de la deuxième campagne.

Dans le but de minimiser les écarts entre mesures et calculs en ce qui concerne la dispersion proprement dite, et donc de valider surtout la partie chimique du code, en particulier pour la phase gazeuse, l'évolution de la largeur et de l'épaisseur du panache ont été fournies au modèle.

En ce qui concerne l'initialisation, la valeur moyenne expérimentale pour la première coupe, réalisée généralement à quelques kilomètres de la cheminée, est donnée, pour les différents gaz analysés, au modèle qui en déduit des profils gaussiens sur les 10 cellules composant la section du panache.

Dans le tableau 2 sont rassemblées les valeurs moyennes des concentrations mesurées de chaque composé lors des différentes coupes ainsi que les valeurs moyennes calculées correspondantes. Dans cette comparaison, la très faible réactivité de SO_2 permet de l'utiliser comme référence pour le traitement de la dispersion par le modèle ; ce n'est que lorsque l'accord entre valeurs calculées et valeurs mesurées est correct pour SO_2 que l'on peut comparer les gaz "réactifs", NO, NO_2, O_3. Les valeurs entre parenthèses ont été obtenues en ne faisant intervenir que la dispersion et ne prennent donc pas en compte les effets de la chimie. Toutes les heures sont exprimées en temps universel.

Pour ce vol, l'accord entre les moyennes expérimentales et calculées est excellent et l'impact des transformations chimiques est important ainsi que le montre la comparaison entre les concentrations calculées avec et sans chimie.

En ce qui concerne les espèces acides formées dans la zone du panache, les valeurs moyennes au bout de 2 heures sont les suivantes : aérosols de sulfate (AERS) : 0,6 $\mu g/m^3$, aérosols de nitrate (AERN) : 1,1 $\mu g/m^3$; l'acide nitrique (HNO_3), qui a réagi en quasi-totalité avec l'ammoniac, présente une concentration de 0,15 ppb. Pour cette simulation, la composition de l'atmosphère en composés hydrocarbonés non méthaniques a été supposée égale à celle d'une atmosphère moyennement polluée et la concentration en ammoniac a été fixée à 2 ppb.

La figure 1 présente l'évolution des aérosols de sulfate, de nitrate, et de l'acide nitrique dans le panache dont, pour des raisons de symétrie, seule une moitié est considérée. Il faut noter que la concentration d'AERS, nulle initialement, passe par un maximum au bout de 65 minutes tandis que celle d'AERN, qui provient de HNO_3, lui-même directement dépendant de NO_2, culmine à 105 minutes. Ceci est dû au fait que la formation d'AERS est régie par la concentration de SO_2 disponible qui, relativement importante au départ, décroît par la suite sous l'effet de la dispersion, le rôle de la chimie étant négligeable. Pour NO_2, il faut prendre en compte les processus de formation à partir de NO qui expliquent le retard du maximum d'AERN sur celui d'AERS, la dispersion l'emportant plus tard. Dans cette simulation, ainsi que dans toutes les autres, les concentrations ambiantes pour NH_3, H_2O_2, Fe^{3+} et Mn^{2+} sont, respectivement, 2 ppb, 1 ppb, 0,5 $\mu g/m^3$ et 0,5 $\mu g/m^3$. La concentration ambiante choisie en NH_3 est suffisante pour neutraliser la quasi- totalité de HNO_3, ainsi que le révèle la figure 1 où le maximum d'acide nitrique dans le panache n'atteint que 0,16 ppb.

Pour déterminer la contribution nette maximale du panache, pour une simulation donnée, une seconde simulation est effectuée sans le panache mais dans les mêmes conditions atmosphériques. Les résultats des 2 simulations sont alors comparés.

Dans le cas de ce vol, la contribution nette maximale du panache, déterminée après avoir effectué une simulation atmosphérique, est de 0,45 $\mu g/m^3$ pour AERS. La quasi-totalité des nitrates et de l'acide nitrique proviennent de l'atmosphère. Ceci s'explique, d'une part par l'importante concentration atmosphérique de NO_x (40 ppb), d'autre part par un taux de réaction de $NO_2 + OH \longrightarrow HNO_3$ identique dans les 2 cas ; en effet, dans le cas de la simulation avec panache, la concentration plus forte en NO_2 est compensée par une concentration en radicaux OH plus faible, dans le même rapport, à cause de la présence importante de SO_2. Ainsi, le taux de formation de HNO_3 reste pratiquement constant, avec ou sans panache.

4. ETUDES DE SENSIBILITE

Les résultats de l'exemple de validation de la chimie en phase gazeuse s'étant révélés satisfaisants, il nous a paru intéressant d'effectuer des études de sensibilité mono et multiparamètres. Dans le premier type d'étude, la valeur d'un paramètre donné change par rapport à celle d'une simulation de référence. Dans le second cas, il n'y a plus de référence puisque plusieurs paramètres peuvent, simultanément, prendre chacun des valeurs comprises entre deux limites extrêmes. Ce type d'étude a plutôt pour but de comparer l'influence des fluctuations de plusieurs paramètres d'entrée sur les solutions du modèle.

Pour ces tests de sensibilité, nous avons utilisé, pour la dispersion du panache, la configuration rencontrée lors du vol.

L'étude de sensibilité monoparamètre concerne l'humidité relative (HR), 5 valeurs ayant été injectées dans PARADE : 50, 80, 90, 95 et 97 %.

Les courbes pour AERS (figure 2), passant par un maximum au bout de 70 minutes, révèlent l'importance des réactions d'oxydation de SO_2 en phase aqueuse qui ne peuvent jouer un rôle notable que pour des humidités supérieures à 90 %, le contenu en eau liquide étant alors, à la température considérée, suffisamment important. Après 70 minutes, lorsque le maximum en AERS est atteint, le rapport de la concentration d'eau liquide sur celle d'aérosols de sulfate est, pour les 5 HR, et en suivant un ordre croissant de 50 à 97 %, 0, 4,7, 11,4, 23,8, 27,0 respectivement. La concentration moyenne maximale en AERS est de l'ordre de 0,7 $\mu g/m^3$ pour une humidité de 90 % et de 1,7 $\mu g/m^3$ pour 97 %.

La dépendance de la formation des nitrates particulaires vis-à-vis de l'humidité (figure 2) est plus faible et n'est due en fait qu'à un changement des conditions de l'équilibre thermodynamique régissant l'apparition de ces composés.

La forme des courbes de concentration d'AERN, étroitement corrélée à la formation d'acide nitrique, révèle très nettement 2 parties distinctes : la première où domine la réaction de formation NO_2+OH, et la seconde, après la tombée de la nuit - au bout de 175 minutes, à partir de 16 h 45 - où la chimie des radicaux NO_3, photolysés le jour, devient prépondérante, avec la réaction : $NO_3+NO_2 \longrightarrow N_2O_5$. Néanmoins, il existe encore une incertitude sur la nature exacte de ce processus de formation nocturne de l'acide nitrique, ainsi que sur sa constante cinétique. Pour ce travail nous avons pris la valeur de MORRIS et NIKI (5).

Pour l'analyse de sensibilité multiparamètre, nous avons retenu la technique FAST qui associe à chaque paramètre incertain une fréquence spécifique dans l'espace de la transformée de Fourier du système (6).

Les sensibilités des solutions sont alors déterminées en résolvant le système d'équations pour des valeurs discrètes de la variable de la transformée de Fourier, et en calculant ensuite les coefficients de Fourier associés à chaque fréquence correspondant à un paramètre.

Cette approche permet une analyse de sensibilité globale non linéaire de systèmes dont certains paramètres sont sujets à une forte incertitude, par exemple certaines constantes cinétiques du schéma réactionnel, ou susceptibles de varier d'une configuration à l'autre. L'influence respective des paramètres testés, sur les concentrations calculées des aérosols de sulfate et de nitrate, est traduite par des variances partielles, la somme de celles-ci, à un temps donné, devant être égale à 1. Plus la variance partielle correspondant à un paramètre donné est grande plus l'influence de ses variations est importante.

Pour effectuer ce travail, nous avons utilisé le code développé par Mc RAE et al. (7).

L'analyse concerne l'impact, sur la formation d'espèces acides dans la zone du panache, de la variation possible des 3 paramètres suivants :

O_3 : 30-60 ppb

T : 5-15° C

Humidité relative : 50-95 %.

La figure 3 montre que la concentration d'AERS est plus sensible aux variations de la température qu'à celles de l'humidité, cette situation s'inversant après 70 minutes. Quant aux variations d'ozone, leur impact est toujours faible.

Pour AERN (figure 3), les variations de température et d'humidité relative ont pratiquement la même importance après 40 minutes de simulation. Il est intéressant de noter que l'influence des variations de la concentration d'ozone est négligeable devant l'impact des fluctuations de la température et de l'humidité relative. Ceci implique que la concentration en radical OH n'est pas très dépendante de la photolyse de O_3 mais résulte surtout de la chimie des NO_x/composés hydrocarbonés ambiants. Il est donc impératif d'inclure un traitement satisfaisant de cette chimie dans les modèles traitant la formation des aérosols de sulfate et de nitrate.

Notons que l'étude de 3 paramètres implique pour FAST d'effectuer 27 fois le code PARADE avec, à chaque fois, un triplet différent (O_3, T, HR), le but étant d'échantillonner les 3 intervalles de variation.

Au cours de ces exécutions, pour un temps de 70 minutes, les valeurs extrêmes ont été : 3,7-4,7 $\mu g/m^3$ pour AERS, 0,0-0,9 $\mu g/m^3$ pour AERN.

5. CONCLUSION.

La simulation d'évaluation du code PARADE présentée dans cette communication, ainsi que les analyses de sensibilité, font partie d'un travail destiné à définir, dans le cas d'atmosphères non saturées, les configurations les plus favorables à la formation de produits acides dans le panache (8). La présence d'eau liquide, dans la zone du panache, permettant notamment une oxydation plus rapide du SO_2 en sulfate, il est également important de prendre en compte dans PARADE l'action des nuages et de la pluie. Un module de chimie des nuages vient d'être incorporé au code et sera testé prochainement, en particulier à l'aide de données expérimentales.

REFERENCES BIBLIOGRAPHIQUES.

1) D.A. STEWART, M.K. LIU.
Development and application of a reactive plume model.
Atmospheric Environment, 15, 11, 2377.

2) G.Z. WHITTEN, G.M. GERY.
Development of CBM-X mechanism for urban and regional AQSM.
Rapport EPA, 1985.

3) P. SAXENA, A.B. HUDISCHEWWSKYJ, C. SEIGNEUR, J.H. SEINFELD.
A comparative study of equilibrium approaches to the chemical characterization of secondary aerosols.
Publié dans "Atmospheric Environment".

4) G. MAFFIOLO, E. JOOS, R. QUESNEL, P.F. THOMAS.
Development and testing of short response time SO2, NOX and O3 analysers. Soumis au "Journal of Air Pollution Control Association" pour publication.

5) E.D. MORRIS Jr, J. NIKI.
Reaction of dinitrogen pentoxide with water.
Journal of Physical Chemistry 77, 1929, 1973.

6) R.I. CUKIER, H.B. LEVINE, K.E. SCHULER.
Nonlinear sensitivity analysis of multi-parameter model systems.
J. Comp. Phys. 26, 1, 1978.

7) G.J. Mc RAE, J.W. TILDEN, J.H. SEINFELD.
Global sensitivity analysis - A computational implementation of the Fourier Amplitude Sensitivity Test (FAST).
Computers and Chemical Engineering, 6, 1, 15, 1982.

8) E. JOOS, A. MENDONCA, C. SEIGNEUR.
Evaluation of a reactive plume model with power plant plume data. Application to the sensitivity analysis of sulfate and nitrate formation.
Soumis pour publication à "Atmospheric Environment".

Mois	Octobre
Puissance de la centrale	800 MW
Vitesse du vent	5 m/s
O_3 atm.	40* ppb
NO atm.	12 ppb
NO_2 atm.	29 ppb
SO_2 atm.	25 ppb
Humidité relative	50 %
Température	7° C
Stabilité de l'atmosphère	neutre

* concentration supposée.

Tableau 1 : Caractéristiques de l'atmosphère lors du vol.

Temps	Concentration en ppb					
	SO_2		NO		NO_2	
	M*	C*	M	C	M	C
0 (7 km)	165	165	76	76	42	42
1 h 10 (28 km)	100	100 (100)	34,5	33 (46,5)	43,5	49 (36)
1 h 50 (40 km)	57,5	65,5 (65,5)	14	15 (30)	43	48 (33)

* M : mesurée ; C : calculée.

Tableau 2 : comparaisons calculs-mesures pour le vol.
Début de simulation : 14 h.

a)

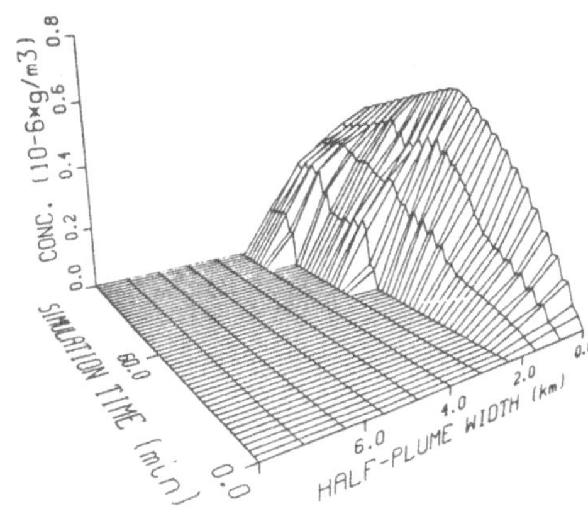

b)

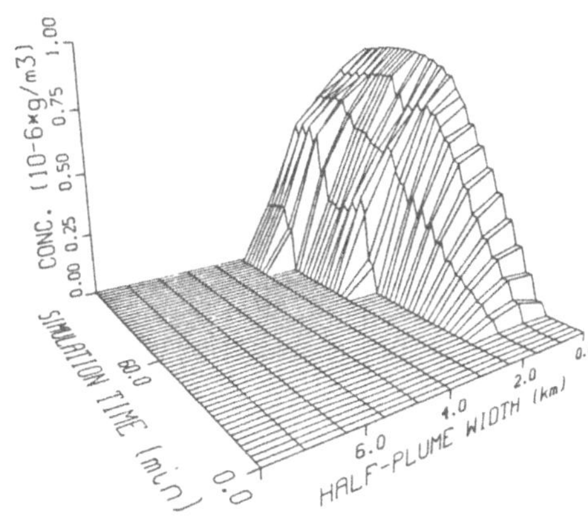

c)

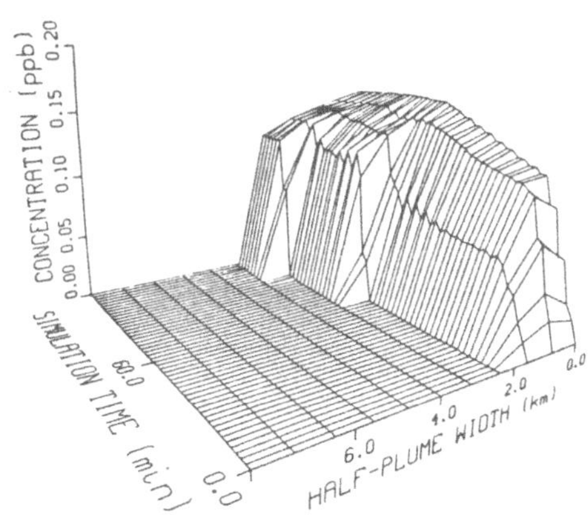

Figure 1 : Evolution dans le panache des concentrations calculées, d'aérosols de sulfate (a), d'aérosols de nitrate (b), et d'acide nitrique (c).

a)

b)

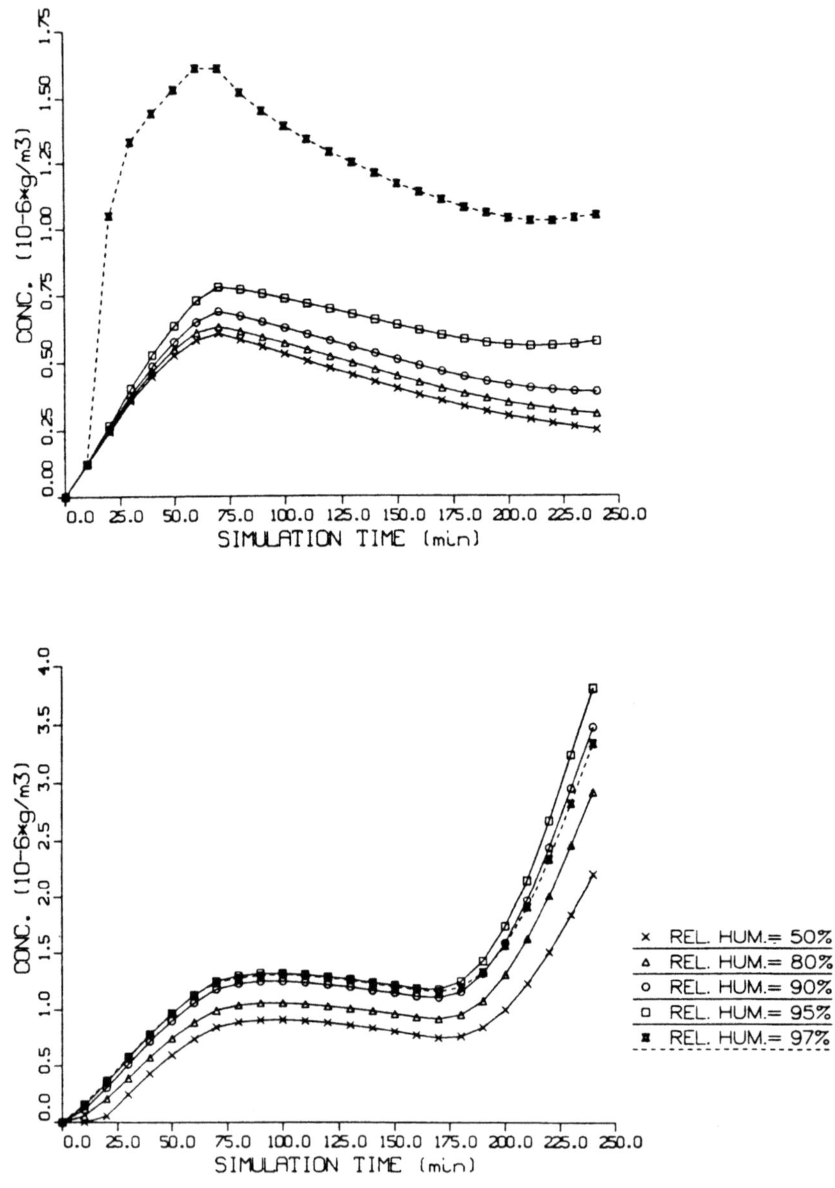

Figure 2 : Evolution dans le panache de la concentration moyenne d'aérosols de sulfate (a) et d'aérosols de nitrate (b) en fonction de l'humidité relative.

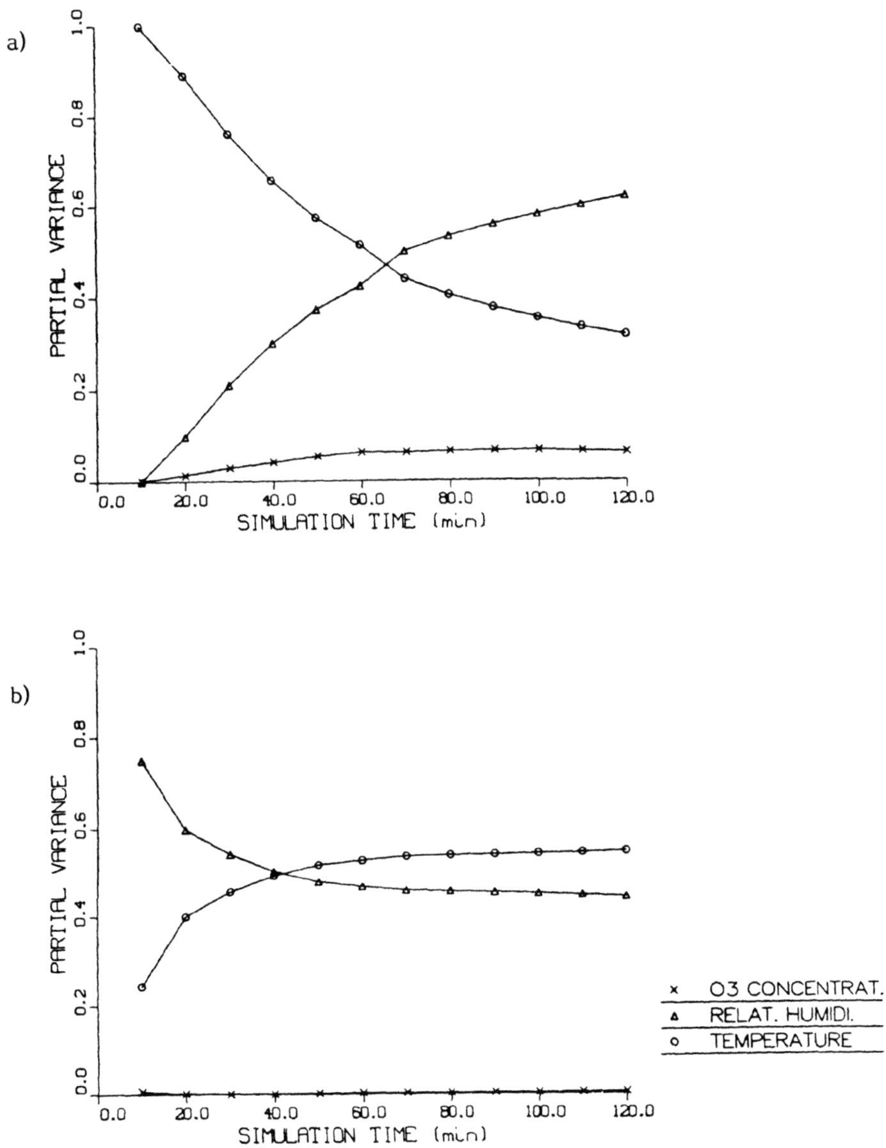

Figure 3 : Analyse de sensibilité multiparamètre concernant l'influence de l'humidité, de la température et de la concentration ambiante d'ozone pour les aérosols de sulfate (a) et les aérosols de nitrate (b).

ECHANTILLONNAGE ET MESURES SEQUENTIELS DE PLUIES : CONCEPTION D'UN APPAREIL EQUIPE D'UN AUTOMATE PROGRAMMABLE (SAPSEP)

P. MASNIERE
Electricité de France
Direction des Etudes et Recherches
6, Quai Watier - 78400 CHATOU

Summary

In order to undertake a detailed investigation of rains within the framework of the french program "DEFORPA" about symptoms and mechanisms involved in the forests decline, "Electricité de France" has built a sequential wet deposition collector. Its design takes into consideration the site chosen for experiments (in Vosges forests), far from laboratories. A programmable automaton operates the different sequences of the sample line ; precipitation events are broken into increments of 0,1 mm (smaller than 0,1 mm for the last one) which are analysed (pH and conductivity), and, afterwards, automatically preserved in a deep freeze unit for later chemical analysis. A rinsing sequence is programmed at the end of each precipitation event. The device has an operating autonomy of several weeks : more than 200 samples may be collected.

1 - INTRODUCTION

Depuis le début de 1983, le dépérissement du sapin et de l'épicéa se manifeste clairement dans les VOSGES et présente des symptômes identiques à ceux relevés antérieurement en FORET NOIRE et en BAVIERE. Actuellement, ce dépérissement affecte d'autres régions.

Un certain nombre d'hypothèses sont avancées pour tenter d'expliquer ce déclin. Elles font intervenir l'histoire climatologique récente, l'influence de la pollution atmosphérique (SO_2, NO_x, photooxydants...) durant les dernières décennies, une perturbation du cycle nutritif liée à l'acidification des sols et/ou à la dégradation des aiguilles et des feuilles, l'intervention de parasites ou de virus... La possibilité d'une tendance sur quelques décennies à l'acidification des dépôts atmosphériques humides (les "pluies acides"), et qui résulterait d'apports polluants d'origine anthropogénique, est aussi mise en cause dans le cadre d'une explication complexe des dégâts observés.

Afin de tester cette éventualité, il a semblé judicieux d'analyser systématiquement les pluies observées au cours des quelques années durant lesquelles se dérouleront les recherches prévues dans le cadre du programme français "DEFORPA" (programme de recherches concernant le Dépérissement des Forêts attribué à la Pollution Atmosphérique, mis en place par le Ministère de l'Environnement).

Trois modes d'échantillonnage de pluies sont envisagés : prélèvements cumulés (recueillis soit dans des collecteurs constamment ouverts, soit dans des collecteurs équipés d'obturateurs et se découvrant en cas de précipitations) et prélèvements séquentiels à l'aide d'un pluviomètre automatisé.

Cet appareil dénommé SAPSEP (Station Automatique de Prélèvements Séquentiels et d'Etude de Pluies) a été mis au point par EDF. Il est piloté par un automate programmable. Sa conception tient compte :

- de la possibilité d'échantillonner les précipitations avec des pas de 1/10 mm de hauteur de pluie.
- de la nécessité d'une large automatisation des processus et des mesures de pH et de conductivité (datage des événements et enregistrement des données).
- de l'éloignement des laboratoires d'analyses et du suivi épisodique de la marche de l'appareil (possibilité de réfrigération de plus de 200 échantillons séquentiels à fins d'analyses ultérieures ; rinçage automatique des circuits en fin de pluie).
- de la possibilité de coupure d'alimentation du réseau électrique (installation de batteries de secours permettant de terminer normalement le cycle de mesures et fermer le couvercle).

L'utilisation de ce pluviomètre devrait permettre une recherche de relations entre les paramètres physico-chimiques de pluies d'une part, et les paramètres météorologiques, l'évolution des teneurs en divers ions dans l'aérosol atmosphérique et celle des polluants gazeux d'autres part.

L'appareil est par ailleurs équipé d'un collecteur destiné à recueillir les retombées atmosphériques sèches ; ce dernier est obturé lorsque le collecteur de précipitations est découvert. Les retombées sèches sont rassemblées après rinçage du récepteur toutes les 2 à 3 semaines et sont analysées ultérieurement.

2 - CARACTERISTIQUES GENERALES DU PLUVIOMETRE SEQUENTIEL AUTOMATIQUE

L'appareil est schématisé dans la figure 1.

2.1 - Alimentation électrique

L'ensemble est alimenté en 220 V. La plupart des organes (hormis l'automate programmable) est alimenté en courant continu (12 et 24 V). Des batteries rechargeables de sauvegarde permettent une certaine autonomie (choisie de 15 mm) de fonctionnement en cas de rupture d'alimentation du réseau (le module de sorties de l'automate fonctionnant en 220 V continuant d'être alimenté grâce à un onduleur) ; un cycle de mesures peut alors être bouclé et le couvercle refermé.

2.2 - Alimentation en air comprimé

Vannes et vérins sont reliés à des bouteilles d'air comprimé. La pression dans les circuits est d'environ 10 bars (suffisante pour actionner les verrins). Les vannes permettent d'alimenter et de vidanger les cellules d'échantillonnage et de mesures. Les verrins assurent d'une part le mouvement (mise en mouvement vertical et rotation) du couvercle fermant alternativement 2 collecteurs de retombées humides et sèches (cf 2.4) et d'autre part le mouvement vertical du distributeur d'échantillons dans les tubes portés par le dispositif de congélation. Le bras distributeur est mis en rotation à l'aide d'un moteur pas à pas qui assure également le passage de l'une des couronnes de tubes d'échantillonnage à l'autre.

2.3 - Automate programmable - séquences d'échantillonnage

L'ensemble des commandes est relié à un automate programmable (Renault-Automatismes A.P. SMC 50) : détecteurs des pluies, ouverture/fermeture du couvercle, détecteurs de proximité, vannes du circuit, répartition des échantillons, pompe de rinçage, datage des événements et des séquences de mesure, acquisition des résultats de mesures du pH et conductivité (réalisée actuellement sur imprimante).

Les différentes phases du processus sont visualisées à l'aide de voyants sur la façade de l'automate. La mise en équilibre du liquide dans les cellules avant les cycles de mesures, les temporisations de plongée du bec verseur dans les tubes de fermeture du couvercle avec l'arrêt de la pluie, ainsi que celles d'ouverture de vannes, sont variables et accessibles par l'intermédiaire de l'imprimante. Cette dernière permet également de situer un incident de fonctionnement, de ramener le bec verseur au niveau du 1er tube lors du chargement périodique du congélateur avec une nouvelle série de tubes.

2.4 - Détecteurs de précipitations

Les 2 détecteurs de précipitations équipant l'appareil et qui commandent l'ouverture/fermeture, sont constitués de 2 peignes métalliques emboîtés, délimitant respectivement des surfaces de 14 et 100 cm^2. Ces peignes sont constitués de barrettes métalliques de 2 mm de diamètre, espacées de 0,5 mm. Des résistances chauffantes (respectivement 80 et 20 W) assurent l'évaporation des gouttelettes éventuelles, elles préviennent par ailleurs la prise en compte de la rosée.

2.5 - Collecteur de retombées atmosphériques

Les collecteurs de retombées sèches et humides ont une forme conique. Les sections de collecte sont respectivement de 0,039 m^2 et 0,38 m^2 (un cône de section 0,039 m^2 peut également être adapté pour les retombées humides) ; on peut effectuer ainsi des prélèvements séquentiels de durées très différentes, à volume échantillonné identique. La collecte est effectuée à environ 0,20 m au-dessus du toit de l'appareil.

2.6 - Surface des circuits d'échantillonnage

L'ensemble du circuit emprunté par les pluies collectées est téflonné (collecteur, tuyauteries, vannes) ; seules les cellules de mesures et les tubes d'échantillonnage sont en pyrex.

2.7 - Détermination des séquences de pluies

Un récipient d'échantillonnage, situé entre le collecteur et les cellules de mesures, est équipé de 2 détecteurs de proximité permettant de signaler la position d'un flotteur. Le premier est situé à une hauteur telle que le volume de pluie échantillonné soit d'environ 40 ml, le second correspond à environ 80 ml. Le premier détermine le pas séquentiel choisi pour l'étude fine des pluies (et correspond à une hauteur de précipitation d'environ 0,1 mm). Le second détermine le volume d'eau de rinçage (désionisée) délivré par une pompe en fin de pluie ; il peut également intervenir comme sauvegarde, en cas de fortes précipitations, si le volume de 40 ml n'est pas vidangé du fait du temps nécessaire à l'analyse de l'échantillon précédent dans les cellules de mesures situées en aval. Le volume de la dernière séquence sera variable (0 à 40 ml). La vidange est par conséquent programmée en fin de détection de pluie.

2.8 - Mesures de pH et conductivité

Chaque échantillon séquentiel (40 ml en général, et variable en fin de pluie) est divisé en 2 volumes à l'aide d'un trop plein : environ 30 ml sont rassemblés dans une cellule où plongent des sondes de température et de conductivité, le reste passe dans une cellule dotée d'une électrode pour la mesure du pH. Après quelques dizaines de secondes, (stabilisation de la solution), les mesures sont effectuées parallèlement sur les 2 fractions :

- les mesures de conductivité sont effectuées avec correction de température (25° C) à l'aide d'un conductimètre TACUSSEL (CDIT-NUM) relié à une électrode du type COM1/G ; les résultats sont visualisés par lecture directe (affichage numérique) et restitués sur l'imprimante. La gamme 0 - 200 $\mu s.cm^{-1}$ est utilisé du fait des conductivités généralement rencontrées dans les eaux de pluie.
- les mesures de pH sont effectuées à l'aide d'une électrode combinée ROSEMOUNT, modèle 650 2 045, reliée à l'automate à travers un transmetteur. La gamme sélectionnée permet des mesures de pH 2 à pH 8. Le système de vidange des 2 cellules de mesures est tel que l'électrode pH ne peut rester plus de quelques minutes hors d'une fraction de pluie ou d'eau de rinçage.

Lorsque 5 séries de mesures de pH et conductivité sont accomplies et enregistrées durant les 30 secondes imparties, les 30 ml contenus dans les cellules de mesure de conductivité sont préservées pour y effectuer ultérieurement une analyse chimique (détermination des teneurs ioniques par absorption atomique et chromatographie d'échanges d'ions notamment).Ils sont échantillonnés dans l'une des 252 éprouvettes en pyrex plongeant dans un système de refroidissement (< 0° C) et répartis sur 4 cercles concentriques. Un bec verseur plonge alors dans les éprouvettes lors de chaque vidange de la cellule de mesure (l'ouverture de chaque tube est protégée par une feuille de caoutchouc téflonné, fendue en croix). L'eau de rinçage (volume limité à 30 ml, du fait du système de trop-plein reliant les cellules de mesures) est également accumulée dans les tubes.

2.9 - Echantillonnage des retombées sèches

Les retombées sèches, accumulés durant une période déterminée (2 à 3 semaines) dans le collecteur, sont entraînées dans un flacon en polyéthylène lors d'un rinçage manuel à l'eau désionisée. La solution recueillie, dont on mesure le volume, est analysée après conservation éventuelle à basse température. Une masse de retombées en diverses espèces chimiques solubles, par unité de surface et unité de temps, est ainsi déterminée.

3 - CONCLUSION

E.D.F. participant au programme français DEFORPA dans le cadre de la Convention Environnement signée entre le Ministère de l'Environnement et le Ministère de l'Energie, a développé un pluviomètre à ouverture automatique original.

Celui-ci est destiné d'une part à effectuer des prélèvements séquentiels de pluies (par pas de 0,1 mm) et d'autre part à recueillir alternativement les retombées atmosphériques sèches (rassemblées après rinçage du collecteur toutes les 2 à 3 semaines).

Un automate programmable gère la commande des divers actionneurs (déplacement du couvercle, électrovannes, déplacement du bras distributeur d'échantillons au-dessus des tubes destinés à les conserver dans une unité de congélation, pompe d'eau de rinçage), et la mesure de la conductivité et du pH sur les prélèvements successifs.

Les résultats des mesures sont actuellement restitués sur imprimante (mais pourraient être stockés sur d'autres supports). Compte-tenu des séquences de rinçage, environ 200 prélèvements séquentiels de pluies peuvent être recueillis et conservés plusieurs semaines dans une unité de congélation afin de limiter les possibilités de transformations chimiques et de développements d'algues ; ils sont analysés ultérieurement en laboratoire (chromatographie ionique, absorption atomique). L'approvisionnement prévu en air comprimé et le nombre de tubes d'échantillonnage, permettent d'envisager un autonomie de fonctionnement supérieure en général, à 3 semaines. Les batteries de secours permettent d'autre part un fonctionnement de l'appareil (y compris la fin du cycle de prélèvements), durant 15 minutes en cas de coupure d'alimentation du réseau électrique.

Cette station dénommée SAPSEP (Station Automatique de Prélèvements Séquentiels et d'Etudes de Pluies) est installée dans un secteur forestier dépérissant des VOSGES (DONON) afin de permettre l'étude de pluies dans le cadre du programme français DEFORPA.

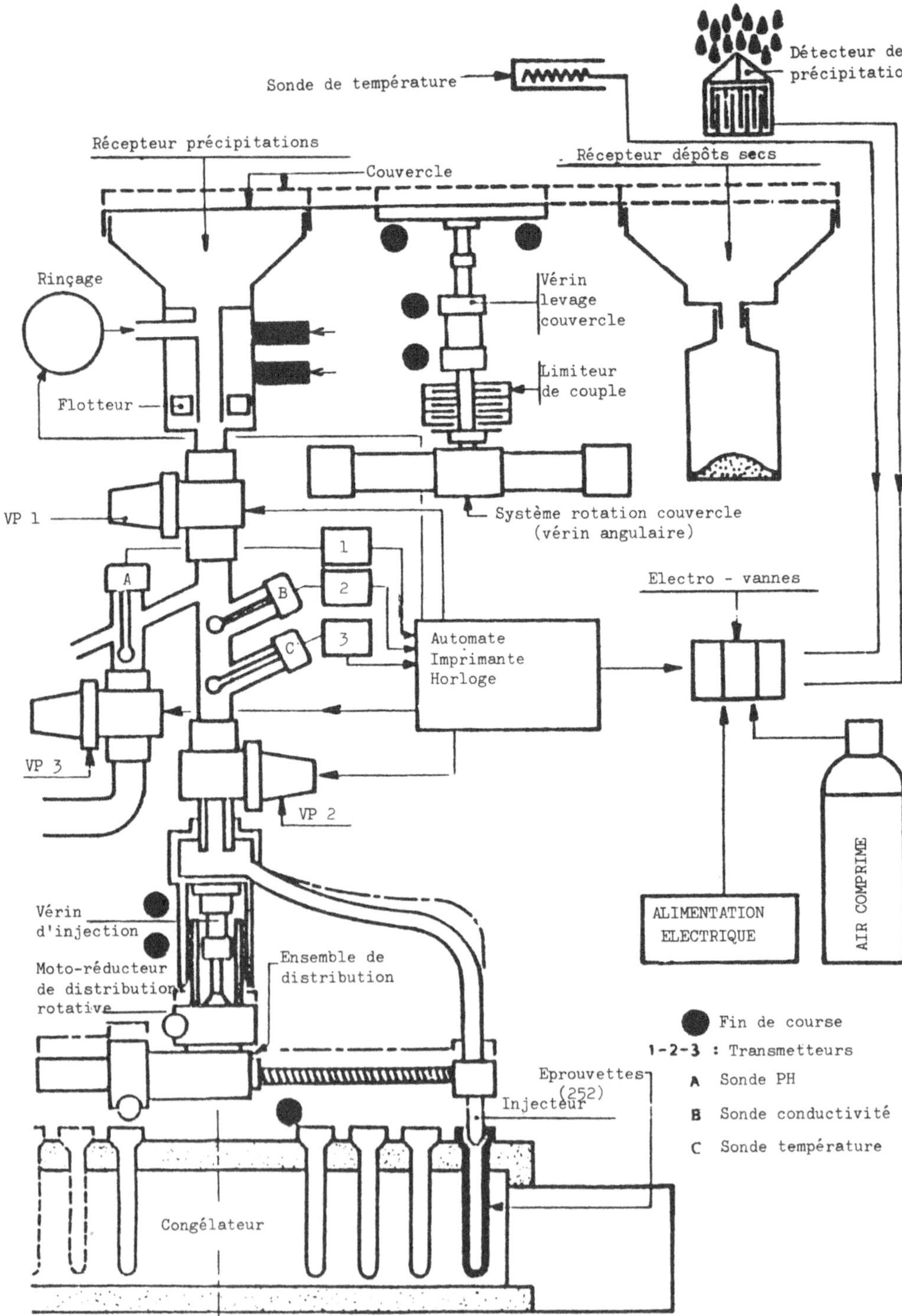

FIGURE 1: Schéma du pluviomètre à ouverture automatique E.D.F.

LIST OF PARTICIPANTS

ALLEGRINI, I.
Istituto Inquinamento Atmosferico
del CNR
C.P. 10
Monterotondo Staz.
I - 00016 ROMA

ANDERSEN, H.V.
National Agency of Environmental
Protection
Air Pollution Laboratory
Risö National Laboratory
DK - 4000 ROSKILDE

ANGELETTI, G.
Commission of the European
Communities
Directorate General Science,
Research and Development
200, rue de la Loi
B - 1049 BRUXELLES

BACHMANN, K.
Technische Hochschule Darmstadt, FBB
Hochschule Str., 4
D - 6100 DARMSTADT

BARNES, I.
Bergische Universität/GH Wuppertal
Physikalische-Chemie/FB9
Gauss-Str., 20
D - 5600 WUPPERTAL 1

BARTAIRE, J.G.
Commission of the European
Communities
Directorate General Environment,
Consumer Protection and Nuclear
Safety
200, rue de la Loi
B - 1049 BRUXELLES

BAUMANN, U.
Gesellschaft für Strahlen- und
Umweltforschung
Schulstrasse, 10
D - 8050 FREIBURG

BECKER, K.H.
Bergische Universität/GH Wuppertal
Physikalische-Chemie/FB9
Gauss-Str., 20
D - 5600 WUPPERTAL 1

BEILKE, S.
Umweltbundesamt-Pilotstation
Frankfurt
Frankfurter Str., 135
D - 6050 OFFENBACH

BESSEMOULIN, P.
CNRM
42, av. Coriolis
F - 31057 TOULOUSE Cedex

BISHOP, G.
Commission of the European
Communities
Joint Research Centre
Ispra Establishment
I - 21020 ISPRA (VA)

BORREGO, C.
University of Aveiro
Department of Environment
P - 3800 AVEIRO

BOURDEAU, P.
Commission of the European Communities
Directorate General Science, Research and Development
200, rue de la Loi
B - 1049 BRUXELLES

BRIMBLECOMBE, P.
University of East Anglia
School of Environmental Sciences
UK - NORWICH NR4 7TJ

BUERGERMEISTER, S.
Universität Frankfurt
Inst. für Meteorologie
Feldberg-Str., 47
D - 6000 FRANKFURT

CARLIER, P.
Université Paris VII
Laboratoire de Physico-Chimie de l'Atmosphère
2, Place Jussieu
F - 75251 PARIS Cedex 05

CECINATO, A.
Istituto Inquinamento Atmosferico del CNR
C.P. 10
Monterotondo Staz.
I - 00016 ROMA

CHAUMERLIAC, N.
Université de Clermont II
LAMP - OPGC
Complexe Scientifique de Cézeaux
B.P. 45
F - 63171 AUBIÈRE

CICCIOLI, P.
Istituto Inquinamento Atmosferico del CNR
C.P. 10
Monterotondo Staz.
I - 00016 ROMA

CIESLIK, S.
Commission of the European Communities
Joint Research Centre
Ispra Establishment
I - 21020 ISPRA (VA)

COLIN, J.L.
Université Paris VII
Laboratoire de Physico-Chimie de l'Atmosphère
2, Place Hussieu
F - 75251 PARIS Cedex 05

COX, R.A.
AERE-Harwell
EMS Division B. 551
UK - DIDCOT OX11 ORA

DE LA SERNA, J.
Escuela Nacional de Sanidad
Ciudad Universitaria
E - 28040 MADRID

DE SANTIS, F.
Istituto Inquinamento Atmosferico del CNR
C.P. 10
Monterotondo Staz.
I - 00016 ROMA

DESMETH, M.
Science Policy Office
Wetenshapstraat, 8
B -1040 BRUSSELS

DEUMLING, D.
Hatzfeldt'sche Verwaltung
Schloß Schönstein
D - 5248 WISSEN

DI PALO, C.
Istituto Inquinamento Atmosferico del CNR
C.P. 10
Monterotondo Staz.
I - 00016 ROMA

DOLLARD, G.
AERE-Harwell
UK - DIDCOT OX11 ORA

DUTOT, A.L.
Université Paris Val-de-Marne
Laboratoire de Physico-Chimie de l'Environnement
ave Général de Gaulle
F - 94010 CRETEIL Cedex

FACCHINI, M.C.
Università di Bologna
Istituto Chimico
Via Selmi, 2
I - 40126 BOLOGNA

FERM, M.
Swedish Environmental Research Institute
P.O. Box 5207
S - 40224 GOTHENBURG

FLYGER, H.
National Agency of Environmental Protection
Air Pollution Laboratory
Risø National Laboratory
DK - 4000 ROSKILDE

FUCHS, G.
Technische Hochschule Darmstadt
FB8
Hochschule Str., 4
D - 6100 DARMSTADT

FUZZI, S.
Istituto Fisbat-CNR
Via Castagnoli, 1
I - 40126 BOLOGNA

GEISS, F.
Commission of the European Communities
Joint Research Centre
Ispra Establishment
I - 21020 ISPRA (VA)

GEORGII, H.W.
Universität Frankfurt
Inst. für Meteorologie
Feldberg-Str., 47
D - 6000 FRANKFURT/MAIN

GISLER, B.
EAWAG
Ueberlandstr., 133
CH - 8600 DUEBENDORF

GLAVAS, S.
University of Patras
Dept. of Chemistry
GR - 26110 PATRAS

GRANAT, L.
University of Stockholm
Dept. of Meteorology
S - 10691 STOCKHOLM

GUILLOT, P.
Commission of the European Communities
Directorate General Science, Research and Development
200, rue de la loi
B - 1049 BRUXELLES

HARTMANN, D.
Universität Göttingen
Institut für Phys.-Chemie
D - 3400 GÖTTINGEN

HJORTH, J.
Commission of the European Communities
Joint Research Centre
Ispra Establishment
I - 21020 ISPRA (VA)

HOV, Ø
Norwegian Institute for Air Research
P.O. Box 130
N - 2001 LILLESTRØM

IOANNILLI, E.
ENEL - Laboratorio Centrale DCO
Via Nino Bixio, 39
I - 29100 PIACENZA

JACOB, P.
Universität Dortmund
Postfach 500 500
D - 4600 DORTMUND 50

JANSSEN, L.H.J.M.
NV - KEMA
Postbus 9035
NL - 6800 ET ARNHEM

JENKINS, G.J.
Dept. of Environment
B354 Romney House
Marsham Street
UK - LONDON SW1P 3PY

JOOS, E.
Electricité de France
Dept. EAA
6, Quai Watier
F - 78400 CHATOU

KALLEND, A.S.
C.E.G.B.
C.E.R.L., Kelvin Avenue
UK - LEATHERHEAD, SURREY KT22 7SE

KARLSSON, V.
Finnish Meteorological Institute
Shaajankatu 22 E
SF - 00810 HELSINKI

KIRSCHMER, P.
Landesanstalt für Immissionsschutz
des Landes Nordrhein - Westfalen
Wallnever Str., 6
D - 4300 ESSEN 1

KLEIN, Th.
Bergische Universität/GH Wuppertal
Gauss-Str., 20
D - 5600 WUPPERTAL 1

KLUG, W.
Technische Hochschule
Institut für Meteorology
Hochschule Str., 4
D - 6100 DARMSTADT

KOTZIAS, D.
Commission of the European
Communities
Joint Research Centre
Ispra Establishment
I - 21020 ISPRA (VA)

KULMALA, A.
Finnish Meteorological Institute
Shaajankatu 22 E
SF - 00810 HELSINKI

LATELLA, A.
Via Monte d'oro, 16
I - 00186 ROMA

LE BRAS, G.
CNRS-CRCCHT
av. de la Recherche Scientifique
F - 45071 ORLÉANS Cedex 02

LEFEVRE, R.
Université Paris XII - LMAA
av. Général de Gaulle
F - 94010 CRETEIL

LIBERTI, A.
Istituto Inquinamento Atmosferico
del CNR
C.P. 10
Monterotondo Staz.
I - 00016 ROMA

LJUNGSTROM, E.
Chalmers University of Technology
Dept. of Inorganic Chemistry
S - 41296 GÖTEBORG

MASNIERE, P.
Electricité de France
Dept. EAA
6, Quai Watier
F - 7800 CHATOU

McELROY, W.J.
C.E.G.B. - CERL
Kelvin Avenue
UK - LEATHERHEAD, SURREY KT22 7SE

MILLAN M. MILLAN
School of Engineering
Alda de Urquijo S/N
E - BILBAO (Vizcaya)

MOORTGAT, G.
Max-Planck-Inst. für Chemie
Air Chemistry Division
Saarstrasse, 23
D - 6500 MAINZ

MORELLI, J.
Ecole normale Supérieure
Institut de Biogéochimie Marine
46, rue d'Ulm
F - 75230 PARIS Cedex 05

MOSELLO, R.
CNR - Ist. Idrobiologia
Largo Tonolli, 50/52
I - 28048 PALLANZA

MULLER, M.
Ministère de l'Environnement
Service de la Recherche (SRETIE)
14, bvd du Général Leclerc
F - 92524 NEUILLY/SEINE Cedex

NEFTEL, A.
University of Bern
Physics Institute
Sidlerstrasse, 5
CH - 3012 BERN

NESTER, K.
Kernforschungszentrum Karlsruhe
Universität Karlsruhe
Institut für Meteorologie und
Kernforschung
Postfach 3640
D - 7500 KARLSRUHE 1

NGUYEN, B.C.
Centre des Faibles Radioactivités
Domaine du CNRS
B.P. 1
F - 91190 GIF-SUR-YVETTE

NIELSEN, O.J.
Risø National Laboratory
Chemistry Dept.
DK - 4000 ROSKILDE

NIEMITZ, K.J.
Hoechst Angewandte Physik
Postfach 800320
D - 6230 FRANKFURT/MAIN 80

NODOP, K.
Universität Frankfurt
Inst. für Meteorologie und Geophysik
c/o NILU
P.O. Box 130
N - 2001 LILLESTRØM

OCKELMANN, G.
Universität Frankfurt
Institut für Meteorologie
Feldbergstrasse, 47
D - 6000 FRANKFURT 1

OTT, H.
Commission of the European
Communities
Directorate General Science,
Research and Development
200, rue de la Loi
B - 1049 BRUXELLES

PAFFRATH, M.
Universität München
Institut für Meteorologie
Theresienstrasse, 37
D - 8000 MÜNCHEN 40

PAYRISSAT, M.
Commision of the European
Communities
Joint Research Centre
Ispra Establishment
I - 21020 ISPRA (VA)

PALOMINO, I.
Junta de Energia Nuclear
Proteccion Radiologica y
Medioambiente
Avenida Complutense
E - 28040 MADRID

PAPENBROCK, Th.
Ruhr Universität Bochum
Physikalische Chemie I
D - 46630 BOCHUM

PERRIN, M.L.
Commissariat à l'Energie Atomique
IPSN/DPT/SPIN, CEN/FAR
B.P. 6
F - 92265 FONTENAY-aux-ROSES Cedex

PERROS, P.
Université Paris XII - LPCE
av. Général de Gaulle
F - 94010 CRETEIL Cedex

PETERSEN, G.
GKSS Forschungszentrum
Postfach 1160
D - 2054 GEESTHACHT

POHJOLA, V.
Finnish Meteorological Institute
Sahaajankatu 22E
SF - 00840 HELSINKI

POREMSKI, J.
Umweltbundesamt
Bismark Platz, 1
D - 1000 BERLIN 33

POSSANZINI, M.
Istituto Inquinamento Atmosferico
del CNR
C.P. 10
Monterotondo Staz.
I - 00016 ROMA

REIJNDERS, H.F.R.
National Institute of Public Health
and Environmental Hygiene
Postbus 1
NL - 3720 BA BILTHOVEN

REITER, R.
Consulting Bureau Reiter
Fritz Müllerstrasse, 54
D - 8100 GARMISCH-PARTENKIRCHEN

RESTELLI, G.
Commission of the European
Communities
Joint Research Centre -
Ispra Establishment
I - 21020 ISPRA (Va)

ROKKJAER, J.
International Agency of
Environmental Protection - Air
Pollution Laboratory
Risø National Laboratory
DK - 4000 ROSKILDE

ROEMER, F.G.
NV-KEMA
Postbus 9035
NL - 6800 ET ARNHEM

ROSSET, R.
Université de Clermont II
LAMP-OPGC
Complexe Scientifique des Cézeaux
B.P. 45
F - 63170 AUBIÈRE

ROTONDO', P.P.
Commission of the European
Communities
Directorate General
Telecommunications, Information
Industries and Innovation
L - 2920 LUXEMBOURG

SCHERER, B.
Freie Universität Berlin
Institut für Geophysikalische
Wissenschaften
Fachrichtung Meteorologie
Thiel-Allee, 50
D - 1000 BERLIN 33

SCHMIDT, V.
Bergische Universität/GH Wuppertal
Physikalische-Chemie/FB 9
Gauss-Strasse, 20
D - 5600 WUPPERTAL 1

SCHOLL, M.
Freie Universität Berlin
Institut für Geophysikalische
Wissenschaften
Fachrichtung Meteorologie
Thiel-Allee, 50
D - 1000 BERLIN 33

SCHURATH, U.
Universität Bonn
Institut für Physikalische Chemie
Wegeler Strasse, 12
D - 5300 BONN

SCOTT, J.A.
University College Dublin
Physics Department
Belfield
IRL - DUBLIN 4

SIDEBOTTOM, H.
University College Dublin
Chemistry Department
Belfield
IRL - DUBLIN 4

SIGG, A.
University of Bern
Physics Institute
Sidlerstrasse, 5
CH - 3012 BERN

SJODIN, A.
Swedish Environmental Research
Institute
P.O. Box 5207
S - 40224 GOETEBORG

SLANINA, J.
Netherlands Energy Research
Foundation
Postbus 1
NL - 1755 ZG PETTEN

SOUCHU, P.
Université Bretagne Occidentale
Institut d'Etudes Marines
Faculté des Sciences
F - 29283 BREST

STANGL, H.
Commission of the European
Communities
Joint Research Centre
Ispra Establishment
I - 21020 ISPRA (VA)

STAUBES, R.
Universität Frankfurt
Institut für Meteorologie
Feldbergstrasse, 47
D - 6000 FRANKFURT / MAIN

STERN, R.
Freie Universität Berlin
Thiel-Allee, 50
D - 1000 BERLIN 33

STIEF-TAUCH, P.
Commission of the European
Communities
Directorate General Environmental,
Consumer Protection and Nuclear
Safety
200, rue de la Loi
B - 1049 BRUXELLES

STUHL, F.
Ruhr-Universität Bochum
Physikalische Chemie I
D - 4630 BOCHUM

TARTARI, G.
CNR - Ist. Ricerca sulle Acque
I - 20047 BRUGHERIO (MI)

TEN BRINK, H.
Netherlands Energy Research
Foundation
Postbus 1
NL - 1755 ZG PETTEN

TILLE, J.W.
Technische Hochschule Darmstadt-FBB
Hochschule Strasse, 4
D - 6100 DARMSTADT

TIRABASSI, T.
CNR - Istituto FISBAT
Via dei Castagnoli, 1
I - 40126 BOLOGNA

TSALKANI, N.
Université Paris Val-de-Marne
Laboratoire de Physico-Chimie
de l'Environnement
av. Général de Gaulle
F - 94010 CRÉTEIL

VAN CLEUVENBERGEN, R.
University of Antwerp
Dept. of Chemistry
Universiteitsplein, 1
B - 2610 WILRIJK

VAN DINGENEN, R.
State University Ghent (RUG)
Proeftuinstr. 86
B - 9000 GHENT

VERSINO, B.
Commission of the European
Communities
Joint Research Centre
Ispra Establishment
I - 21020 ISPRA (VA)

VINCKIER, C.
Katholisch Universiteit van Leuven
Dept. of Chemistry
Celestijnenlaan, 200F
B - 3030 HEVERLEE

WALSH, J.J.
University College Dublin
Dept. of Chemical Engineering
Upper Merrion St.
IRL - DUBLIN 2

WAYNE, R.P.
University of Oxford
Physical Chemistry Laboratory
South Parks Road
UK - OXFORD OX1 3QZ

WEBER, E.
Bundesministerium des Innern
Graurheindorfer Strasse, 198
D - 5300 BONN 1

WINKLER, P.
DWD - Meteorologisches Observatorium
Frahmredder, 95
D - 2000 HAMBURG 65

ZURCHER, F.
EAWAG
Uberlandstrasse, 133
CH - 8600 DÜBENDORF

INDEX OF AUTHORS

GPSR Compliance

The European Union's (EU) General Product Safety Regulation (GPSR) is a set of rules that requires consumer products to be safe and our obligations to ensure this.

If you have any concerns about our products, you can contact us on ProductSafety@springernature.com

In case Publisher is established outside the EU, the EU authorized representative is:

Springer Nature Customer Service Center GmbH
Europaplatz 3
69115 Heidelberg, Germany

Batch number: 10607216

Printed by Printforce, the Netherlands